Introduction to Probability Models

OPERATIONS RESEARCH:
VOLUME TWO

DUXBURY

Introduction to Probability Models

OPERATIONS RESEARCH:

VOLUME TWO

FOURTH EDITION

Wayne L. Winston

INDIANA UNIVERSITY

THOMSON

—★—™

BROOKS/COLE

Australia ■ Canada ■ Mexico ■ Singapore

Spain ■ United Kingdom ■ United States

THOMSON

BROOKS/COLE

Publisher: Curt Hinrichs

Assistant Editor: Ann Day

Editorial Assistant: Katherine Brayton

Technology Project Manager: Burke Taft

Marketing Manager: Joseph Rogove

Marketing Assistant: Jessica Perry

Advertising Project Manager: Tami Strang

Print/Media Buyer: Jessica Reed

Permissions Editor: Bob Kauser

Production Project Manager: Hal Humphrey

Production Service: Hoyt Publishing Services

Text Designer: Kaelin Chappell

Copy Editors: David Hoyt and Erica Lee

Illustrator: Electronic Illustrators Group

Cover Designer: Lisa Langhoff

Cover Image: PhotoDisc

Cover Printer: Transcontinental Printing, Louiseville

Compositor: ATLIS Graphics

Printer: Transcontinental Printing, Louiseville

Printed in Canada

1 2 3 4 5 6 7   07 06 05 04 03

For more information about our products, contact us at:
Thomson Learning Academic Resource Center
1-800-423-0563
For permission to use material from this text, contact us by:
Phone: 1-800-730-2214   **Fax:** 1-800-730-2215
Web: http://www.thomsonrights.com

Library of Congress Control Number 2003104157
Student Edition with InfoTrac College Edition:
ISBN 0-534-40572-X
Student Edition without InfoTrac College Edition:
0-534-42339-0

Brooks/Cole—Thomson Learning
10 Davis Drive
Belmont, CA 94002
USA

Asia
Thomson Learning
5 Shenton Way #01-01
UIC Building
Singapore 068808

Australia/New Zealand
Thomson Learning
102 Dodds Street
Southbank, Victoria 3006
Australia

Canada
Nelson
1120 Birchmount Road
Toronto, Ontario M1K 5G4
Canada

Europe/Middle East/Africa
Thomson Learning
High Holborn House
50/51 Bedford Row
London WC1R 4LR
United Kingdom

Latin America
Thomson Learning
Seneca, 53
Colonia Polanco
11560 Mexico D.F.
Mexico

Spain/Portugal
Paraninfo
Calle Magallanes, 25
28015 Madrid
Spain

Contents

v

Preface

In today's world, knowledge of how to model uncertainty is very useful in making good decisions. This book presents a comprehensive introduction to the variety of problems and techniques required to model and analyze such decisions. In recent years, software to model and solve problems involving uncertainty has become widely available. Its use is illustrated throughout the book. Like most tools, however, it is useless unless the user understands its application and purpose. Users must ensure that the mathematical input accurately reflects the real-life problems to be solved and that the numerical results are correctly applied to solve them. This text takes great care to develop the necessary theory to understand how to properly apply the methods. However, the focus of the book is ultimately on how to apply probability models to real problems. With this in mind, this book emphasizes modeling and the interpretation of software output.

Intended Audience

This book is intended for advanced undergraduates or beginning graduate students who have had courses in elementary calculus and an introductory course in probability and statistics. A formal course in probability theory is not required. Chapter 1 provides a review of the probability and calculus required for the book.

The emphasis is on the development of concepts and their illustration through applications from the areas of manufacturing, finance, and operations. This book extends the coverage of probabilistic models that appear in *Operations Research: Applications and Algorithms,* 4th Edition. It provides a more comprehensive and contemporary introduction to the concepts. This text should be suitable for the following courses:

- Courses in probability models or stochastic processes, offered in departments of industrial engineering, operations research, mathematics, schools of business, and master's programs in finance.

- A second course in an operations research sequence.

- Courses in financial engineering, which should find the material sufficient for an introductory course.

Making Teaching and Learning Easier

The following features help to make this book reader friendly:

- To provide immediate feedback to students, problems are placed at the end of the sections, and most chapters conclude with review problems. There are some 800 problems, grouped according to difficulty: Group A for practice of basic techniques, Group B for underlying concepts, and Group C for mastering the theory independently.

- The book avoids excessive theoretical formulas in favor of word problems and interesting problem applications. Many problems are based on published applications of probability models. Each chapter includes several examples to guide the student step by step through even the most complex topics.

- Each section is as self-contained as possible, which allows the instructor to be extremely flexible in designing a course. The Instructor's Notes identify which portions of the book must be covered as prerequisites to each section.

- To help students review for exams, each chapter has a summary of concepts and formulas. A *Student Solutions Manual* is available, providing worked-out solutions to selected problems. The *Student Solutions Manual* may be purchased separately or packaged with the text at a nominal additional price.

- Instructors who adopt this text in their courses may receive the *Instructor's Suite CD-ROM.* This CD contains complete solutions to every problem in the text, Microsoft PowerPoint slides, and Instructor's Notes.

- The book contains instruction for using the software contained on the CD that accompanies the book. All of the files needed for examples and exercises are also included.

Coverage and Organization

This book is completely self-contained; all the necessary mathematical background is reviewed in Chapter 1. Each chapter is designed to be modular, allowing a course to tailor the book to its needs. For the reader familiar with *Operations Research,* we have denoted where new material has been added since the 3rd Edition. This text contains 800 problems, including 200 new problems not appearing in the OR book.

The book is written to accommodate a variety of courses. Only Chapters 1, 2, 5, 8, and 9 are required material. Beyond these chapters, the instructor may emphasize a variety of subjects: inventory and dynamic programming (Chapters 3, 4, 6, and 7), simulation (Chapters 10–12), or financial applications (Chapters 13, 14, and 16).

Chapter 1 reviews the topics in calculus and basic probability that are needed throughout the rest of the text. This chapter includes coverage of moment-generating functions and the use of the Excel normal distribution functions.

Chapter 2 presents the classical approach to decision making under uncertainty, with topics such as decision trees and utility theory. It includes a discussion of Kahneman's Nobel Prize-winning work on decision making under uncertainty. Also covered is the analytic hierarchy process for decision making with multiple objectives.

Chapters 3 and 4 cover deterministic and probabilistic inventory models in OR. New material in Chapter 3 includes coverage of Roundy's power-of-two inventory policies and multiple-product EOQ models.

Chapter 5 covers discrete-time Markov chains.

Chapters 6 and 7 update the coverage of deterministic and probabilistic dynamic programming in OR.

Chapter 8 is an extensive revision of the previous material on queuing theory. Additions include the use of Excel functions to compute Poisson and exponential probabilities, Buzen's algorithm for closed queuing networks, approximations for $G/G/s$ queuing systems, and calculating transient probabilities for nonstationary queuing systems.

Chapter 9 provides an introduction to simulation concepts.

In the last decade, user-friendly simulation software has become widely available. With these software packages, students can solve complex, realistic problems. Chapters 10–12 cover both spreadsheet-based and process-modeling simulation packages.

Chapter 10 explains the use of the Process Model simulation package for queuing simulations, including how the queuing models of Chapter 8 can be simulated with Process Model.

Chapter 11 covers spreadsheet simulation using Palisade Corporation's Excel add-in @Risk, to model capital budgeting, project management, reliability problems, and other topics.

Chapter 12 presents Palisade Corporation's Excel add-in Riskoptimizer, used to optimize spreadsheet simulations. Riskoptimizer is essentially a Solver for Monte Carlo simulation.

In the last 30 years, financial engineering concepts have revolutionized corporate finance and investments. Chapters 13, 14, and 16 are devoted to financial engineering topics.

Chapter 13 covers important topics in option pricing theory, including the use of real options in capital budgeting. Chapter 14 develops the scenario approach to portfolio optimization. Chapter 16 contains an introduction to stochastic calculus and stochastic control. It covers Ito's Lemma and traces the original derivation of the Black–Scholes option pricing formula. Chapter 15 provides thorough coverage of forecasting.

Use of the Computer

In deference to the virtually universal usage of Excel, this software is featured throughout the book when appropriate. When Excel's native capabilities are limited, the text discusses add-in software that builds on the capabilities of Excel, or uses stand-alone software.

The CD accompanying the book contains several valuable software packages.

- **@Risk.** A professional Monte Carlo simulation add-in for Excel by Palisade Corporation.
- **Riskoptimizer.** Software that employs optimization in simulation models. It is featured in Chapter 12.
- **Process Model.** This discrete-event simulation software is easy to learn and use. It is illustrated in Chapter 11. Process Model was generously provided by Process Model Inc.

Software illustrations appear at the ends of sections, to give maximum flexibility to instructors who wish to employ different software packages in their courses.

Acknowledgments

Many people have played significant roles in the development of this text. My views on teaching this material were greatly influenced by the many excellent teachers I have had, including Gordon Bradley, Eric Denardo, John Little, Robert Mifflin, Martin Shubik, and Harvey Wagner. In particular, I would like to acknowledge Professor Wagner's *Principles of Operations Research,* which taught me more about this than any other single book.

Thanks go to all the people at Duxbury Press who worked on the book, especially our outstanding editor, Curt Hinrichs.

I greatly appreciate the editing and production skills of David Hoyt, and the fine typesetting of ATLIS Graphics.

I owe a great debt to the reviewers of the second edition, whose comments greatly improved the quality of the manuscript: Esther Arkin, State University of New York at Stony Brook; James W. Chrissis, Air Force Institute of Technology; Rebecca E. Hill, Rochester Institute of Technology; and James G. Morris, University of Wisconsin, Madison.

Thanks to the 146 survey respondents who provided valuable feedback on the course and its emerging needs. They include Nikolaos Adamou, University of Athens & Sage Graduate School; Jeffrey Adler, Rensselaer Polytechnic Institute; Victor K. Akatsa, Chicago State University; Steven Andelin, Kutztown University of Pennsylvania; Badiollah R. Asrabadi, Nicholls State University; Rhonda Aull-Hyde, University of Delaware; Jonathan Bard, Mechanical Engineering; John Barnes, Virginia Commonwealth University; Harold P. Benson, University of Florida; Elinor Berger, Columbus College; Richard H. Bernhard, North Carolina State University; R. L. Bulfin, Auburn University; Laura Burke, Lehigh University; Jonathan Caulkins, Carnegie Mellon University; Beth Chance, University of the Pacific; Alan Chesen, Wright State University; Young Chun, Louisiana State University; Chia-Shin Chung, Cleveland State University; Ken Currie, Tennessee Technological University; Ani Dasgupta, Pennsylvania State University; Nirmil Devi, Embry-Riddle Aeronautical University; James Falk, George Washington University; Kambiz Farahmand, Texas A & M University–Kingsville; Yahya Fathi, North Carolina State University; Steve Fisk, Bowdoin College; William P. Fox, United States Military Academy; Michael C. Fu, University of Maryland; Saul I. Gass, University of Maryland; Ronald Gathro, Western New England College; Perakis Georgia, Massachusetts Institute of Technology; Alan Goldberg, California State University–Hayward; Jerold Griggs, University of South Carolina; David Grimmett, Austin Peay State University; Melike Baykal Gursoy, Rutgers University; Jorge Haddock, Rensselaer Polytechnic Institute; Jane Hagstrom, University of Illinois; Carl Harris, George Mason University; Miriam Heller, University of Houston; Sundresh S. Heragu, Rensselaer Polytechnic Insitute; Rebecca E. Hill, Rochester Institute of Technology; David Holdsworth, Alaska Pacific University; Elained Hubbard, Kennesaw State College; Robert Hull, Western Illinois University; Jeffrey Jarrett, University of Rhode Island; David Kaufman, University of Massachusetts; Davook Khalili, San Jose University; Morton Klein, Columbia University; S. Kumar, Rochester Institute of Technology; David Larsen, University of New Orleans; Mark Lawley, University of Alabama; Kenneth D. Lawrence, New Jersey Institute of Technology; Andreas Lazari, Valdosta State University; Jon Lee, University of Kentucky; Luanne Lohr, University of Georgia; Joseph Malkovitch, York College; Masud Mansuri, California State University–Fresno; Steven C. McKelvey, Saint Olaf College; Ojjat Mehri, Youngstown State University; Robert Mifflin, Washington State University; Katya Mints, Columbia University; Rafael Moras, St. Mary's University; James G. Morris, University of Wisconsin, Madison; Frederic Murphy, Temple University; David Olson, Texas A & M University; Mufit Ozden, Miami University; R. Gary Parker, Georgia Institute of Technology; Barry Pasternack, California State University, Fullerton; Walter M. Patterson, Lander University; James E Pratt, Cornell University; B. Madhu Rao, Bowling Green State University; T. E. S. Raghavan, University of Illinois–Chicago; Gary Reeves, University of South Carolina; Gaspard Rizzuto, University of Southwestern; David Ronen, University of Missouri–St. Louis; Paul Savory, University of Nebraska, Lincoln; Jon Schlosser, New Mexico Highlands University; Delray Schultz, Millersville University; Richard Serfozo, Georgia Technological Institute; Morteza Shafi-Mousavi, Indiana University–South Bend; Dooyoung Shin, Mankato State University; Ronald L. Shubert, Elizabethtown College; Joel Sobel, University of California, San Diego; Manbir S. Sodhi, University of Rhode Island; Ariela Sofer, George Mason University; Toni M. Somers, Wayne State University; Robert

Stark, University of Delaware; Joseph A. Svestka, Cleveland State University; Alexander Sze, Concordia College; Roman Sznajder, University of Maryland–Baltimore County; Bijan Vasigh, Embry-Riddle Aeronautical University; John H. Vande Vate, Georgia Technological University; Richard G. Vinson, University of South Alabama; Jin Wang, Valdodsta State University; Zhongxian Wang, Montclair State University; Robert C. Williams, Alfred University; Arthur Neal Willoughby, Morgan State; Shmuel Yahalom, SUNY–Maritime College; James Yates, University of Central Oklahoma; Bill Yurcik, University of Pittsburgh.

Thanks also go to reviewers of the previous editions: Sant Arora, Harold Benson, Warren J. Boe, Bruce Bowerman, Jerald Dauer, S. Selcuk Erenguc, Yahya Fathi, Robert Freund, Irwin Greenberg, John Hooker, Sidney Lawrence, Patrick Lee, Edward Minieka, Joel A. Nachlas, David L. Olson, Sudhakar Pandit, David W. Pentico, Bruce Pollack-Johnson, Michael Richey, Gary D. Schudder, Lawrence Seiford, Michael Sinchcomb, and Paul Stiebitz.

I retain responsibility for all errors and would love to hear from users of the book. I can be reached at

Indiana University
Department of Operations and Decision Technology
Kelley School of Business
Room 570
Bloomington, IN 47405

Wayne Winston (Winston@indiana.edu)

About the Author

Wayne Winston

Wayne L. Winston is Professor of Operations & Decision Technologies in the Kelley School of Business at Indiana University, where he has taught since 1975. Wayne received his B.S. degree in mathematics from MIT and his Ph.D. degree in operations research from Yale. He has written the successful textbooks *Operations Research: Applications and Algorithms; Introduction to Mathematical Programming; Simulation Modeling with @RISK; Practical Management Science;* and *Financial Models Using Simulation and Optimization.* Wayne has published over 20 articles in leading journals and has won many teaching awards, including the school-wide MBA award four times. His current interest is in showing how spreadsheet models can be used to solve business problems in all disciplines, particularly in finance and marketing.

Wayne enjoys swimming and basketball, and his passion for trivia won him an appearance several years ago on the television game show *Jeopardy,* where he won two games. He is married to the lovely and talented Vivian. They have two children, Gregory and Jennifer.

1

Review of Calculus and Probability

We review in this chapter some basic topics in calculus and probability, which will be useful in later chapters.

1.1 Review of Differential Calculus

Limits

The idea of a **limit** is one of the most basic ideas in calculus.

DEFINITION ■ The equation

$$\lim_{x \to a} f(x) = c$$

means that as x gets closer to a (but not equal to a), the value of $f(x)$ gets arbitrarily close to c. ■

It is also possible that $\lim_{x \to a} f(x)$ may not exist.

EXAMPLE 1   Limits

1 Show that $\lim_{x \to 2} x^2 - 2x = 2^2 - 2(2) = 0$.

2 Show that $\lim_{x \to 0} \frac{1}{x}$ does not exist.

Solution **1** To verify this result, evaluate $x^2 - 2x$ for values of x that are close to, but not equal to, 2.

2 To verify this result, observe that as x gets near 0, $\frac{1}{x}$ becomes either a very large positive number or a very large negative number. Thus, as x approaches 0, $\frac{1}{x}$ will not approach any single number.

Continuity

DEFINITION ■ A function $f(x)$ is **continuous** at a point a if

$$\lim_{x \to a} f(x) = f(a)$$

If $f(x)$ is not continuous at $x = a$, we say that $f(x)$ is **discontinuous** (or has a discontinuity) at a. ■

EXAMPLE 2   **Bakeco**

Bakeco orders sugar from Sugarco. The per-pound purchase price of the sugar depends on the size of the order (see Table 1). Let

$$x = \text{number of pounds of sugar purchased by Bakeco}$$
$$f(x) = \text{cost of ordering } x \text{ pounds of sugar}$$

Then

$$f(x) = 25x \text{ for } 0 \leq x < 100$$
$$f(x) = 20x \text{ for } 100 \leq x \leq 200$$
$$f(x) = 15x \text{ for } x > 200$$

For all values of x, determine if x is continuous or discontinuous.

Solution From Figure 1, it is clear that

$$\lim_{x \to 100} f(x) \quad \text{and} \quad \lim_{x \to 200} f(x)$$

do not exist. Thus, $f(x)$ is discontinuous at $x = 100$ and $x = 200$ and is continuous for all other values of x satisfying $x \geq 0$.

TABLE 1
Price of Sugar Paid by Bakeco

Size of Order	Price per Pound
$0 \leq x < 100$	25¢
$100 \leq x \leq 200$	20¢
$x > 200$	15¢

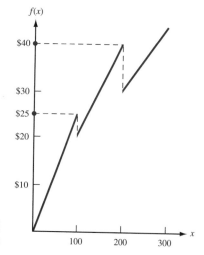

FIGURE 1
Cost of Purchasing Sugar for Bakeco

TABLE 2
Rules for Finding the Derivative of a Function

Function	Derivative of Function
a	0
x	1
$af(x)$	$af'(x)$
$f(x) + g(x)$	$f'(x) + g'(x)$
x^n	nx^{n-1}
e^x	e^x
a^x	$a^x \ln a$
$\ln x$	$\frac{1}{x}$
$[f(x)]^n$	$n[f(x)]^{n-1} f'(x)$
$e^{f(x)}$	$e^{f(x)} f'(x)$
$a^{f(x)}$	$a^{f(x)} f'(x) \ln a$
$\ln f(x)$	$\dfrac{f'(x)}{f(x)}$
$f(x)g(x)$	$f(x)g'(x) + f'(x)g(x)$
$\dfrac{f(x)}{g(x)}$	$\dfrac{g(x)f'(x) - f(x)g'(x)}{g(x)^2}$

Differentiation

DEFINITION ■   The **derivative** of a function $f(x)$ at $x = a$ (written $f'(a)$) is defined to be

$$\lim_{\Delta x \to 0} \frac{f(a + \Delta x) - f(a)}{\Delta x} \quad ■$$

If this limit does not exist, $f(x)$ has no derivative at $x = a$.

We may think of $f'(a)$ as the slope of $f(x)$ at $x = a$. Thus, if we begin at $x = a$ and increase x by a small amount Δ (Δ may be positive or negative), then $f(x)$ will increase by an amount approximately equal to $\Delta f'(a)$. If $f'(a) > 0$, $f(x)$ is increasing at $x = a$, whereas if $f'(a) < 0$, $f(x)$ is decreasing at $x = a$. The derivatives of many functions can be found via application of the rules in Table 2 (a represents an arbitrary constant). Example 3 illustrates the use and interpretation of the derivative.

EXAMPLE 3   Derivatives

If a company charges a price p for a product, it can sell $3e^{-p}$ thousand units of the product. Then $f(p) = 3,000pe^{-p}$ is the company's revenue if it charges a price p.

1   For what values of p is $f(p)$ decreasing? For what values of p is $f(p)$ increasing?

2   Suppose the current price is \$4 and the company increases the price by 5¢. By approximately how much would the company's revenue change?

Solution   We have

$$f'(p) = -3,000pe^{-p} + 3,000e^{-p} = 3,000e^{-p}(1 - p)$$

1 For $p < 1, f'(p) > 0$ and $f(p)$ is increasing, whereas for $p > 1, f'(p) < 0$ and $f(p)$ is decreasing.

2 Using the interpretation of $f'(4)$ as the slope of $f(p)$ at $p = 4$ (with $\Delta p = 0.05$), we see that the company's revenue would increase by approximately

$$0.05(3,000e^{-4})(1 - 4) = -8.24$$

In actuality, of course, the company's revenue would increase by

$$f(4.05) - f(4) = 3,000(4.05)e^{-4.05} - 3,000(4)e^{-4}$$
$$= 211.68 - 219.79 = -8.11$$

Higher Derivatives

We define $f^{(2)}(a) = f''(a)$ to be the derivative of the function $f'(x)$ at $x = a$. Similarly, we can define $f^{(n)}(a)$ (if it exists) to be the derivative of $f^{(n-1)}(x)$ at $x = a$. Thus, for Example 3,

$$f''(p) = 3,000e^{-p}(-1) - 3,000e^{-p}(1 - p)$$

Taylor Series Expansion

In our study of queuing theory (Chapter 8) and stochastic calculus (Chapter 16), we will need the Taylor series expansion of a function $f(x)$. Given that $f^{(n+1)}(x)$ exists for every point on the interval $[a, b]$, Taylor's theorem allows us to write for any h satisfying $0 \leq h \leq b - a$,

$$f(a + h) = f(a) + \sum_{i=1}^{i=n} \frac{f^{(i)}(a)}{i!} h_i + \frac{f^{(n+1)}(p)}{(n + 1)!} h^{n+1} \tag{1}$$

where (1) will hold for some number p between a and $a + h$. Equation (1) is the **nth-order Taylor series expansion** of $f(x)$ about a.

EXAMPLE 4 | **Taylor Series Expansion**

Find the first-order Taylor series expansion of e^{-x} about $x = 0$.

Solution Since $f'(x) = -e^{-x}$ and $f''(x) = e^{-x}$, we know that (1) will hold on any interval $[0, b]$. Also, $f(0) = 1, f'(0) = -1$, and $f''(x) = e^{-x}$. Then (1) yields the following first-order Taylor series expansion for e^{-x} about $x = 0$:

$$e^{-h} = f(h) = 1 - h + \frac{h^2 e^{-p}}{2}$$

This equation holds for some p between 0 and h.

Partial Derivatives

We now consider a function f of $n > 1$ variables $(x_1, x_2, \ldots, x_n)$, using the notation $f(x_1, x_2, \ldots, x_n)$ to denote such a function.

The **partial derivative** of $f(x_1, x_2, \ldots, x_n)$ with respect to the variable x_i is written $\dfrac{\partial f}{\partial x_i}$, where

$$\frac{\partial f}{\partial x_i} = \lim_{\Delta x_i \to 0} \frac{f(x_1, \ldots, x_i + \Delta x_i, \ldots, x_n) - f(x_1, \ldots, x_i, \ldots, x_n)}{\Delta x_i}$$ ■

Intuitively, if x_i is increased by Δ (and all other variables are held constant), then for small values of Δ, the value of $f(x_1, x_2, \ldots, x_n)$ will increase by approximately $\Delta \dfrac{\partial f}{\partial x_i}$. We find $\dfrac{\partial f}{\partial x_i}$ by treating all variables other than x_i as constants and finding the derivative of $f(x_1, x_2, \ldots, x_n)$. More generally, suppose that for each i, we increase x_i by a small amount Δx_i. Then the value of f will increase by approximately

$$\sum_{i=1}^{i=n} \frac{\partial f}{\partial x_i} \Delta x_i$$

EXAMPLE 5   Partial Derivative

The demand $f(p, a) = 30{,}000 p^{-2} a^{1/6}$ for a product depends on $p =$ product price (in dollars) and $a =$ dollars spent advertising the product. Is demand an increasing or a decreasing function of price? Is demand an increasing or a decreasing function of advertising expenditure? If $p = 10$ and $a = 1{,}000{,}000$, by how much (approximately) will a \$1 cut in price increase demand?

Solution

$$\frac{\partial f}{\partial p} = 30{,}000(-2p^{-3})a^{1/6} = -60{,}000 p^{-3} a^{1/6} < 0$$

$$\frac{\partial f}{\partial a} = 30{,}000 p^{-2}\left(\frac{a^{-5/6}}{6}\right) = 5{,}000 p^{-2} a^{-5/6} > 0$$

Thus, an increase in price (with advertising held constant) will decrease demand, while an increase in advertising (with price held constant) will increase demand. Since

$$\frac{\partial f}{\partial p}(10, 1{,}000{,}000) = -60{,}000\left(\frac{1}{1{,}000}\right)(1{,}000{,}000)^{1/6} = -600$$

a \$1 price cut will increase demand by approximately $(-1)(-600)$, or 600 units.

We will also use *second-order partial derivatives* extensively. We use the notation $\dfrac{\partial^2}{\partial x_i \partial x_j}$ to denote a second-order partial derivative. To find $\dfrac{\partial^2}{\partial x_i \partial x_j}$, we first find $\dfrac{\partial f}{\partial x_i}$ and then take its partial derivative with respect to x_j. If the second-order partials exist and are everywhere continuous, then

$$\frac{\partial^2 f}{\partial x_i \partial x_j} = \frac{\partial^2 f}{\partial x_j \partial x_i}$$

EXAMPLE 6   Second-Order Partial Derivatives

For $f(p, a) = 30{,}000 p^{-2} a^{1/6}$, find all second-order partial derivatives.

Solution

$$\frac{\partial^2 f}{\partial p^2} = -60,000(-3p^{-4})a^{1/6} = \frac{180,000a^{1/6}}{p^4}$$

$$\frac{\partial^2 f}{\partial a^2} = 5,000p^{-2}\left(\frac{-5a^{-11/6}}{6}\right) = -\frac{25,000p^{-2}a^{-11/6}}{6}$$

$$\frac{\partial^2 f}{\partial a \partial p} = 5,000(-2p^{-3})a^{-5/6} = -10,000p^{-3}a^{-5/6}$$

$$\frac{\partial^2 f}{\partial p \partial a} = -60,000p^{-3}\left(\frac{a^{-5/6}}{6}\right) = -10,000p^{-3}a^{-5/6}$$

Observe that for $p \neq 0$ and $a \neq 0$,

$$\frac{\partial^2 f}{\partial a \partial p} = \frac{\partial^2 f}{\partial p \partial a}$$

PROBLEMS

Group A

1 Find $\lim_{h \to 0} \dfrac{3h + h^2}{h}$.

2 It costs Sugarco 25¢/lb to purchase the first 100 lb of sugar, 20¢/lb to purchase the next 100 lb, and 15¢ to buy each additional pound. Let $f(x)$ be the cost of purchasing x pounds of sugar. Is $f(x)$ continuous at all points? Are there any points where $f(x)$ has no derivative?

3 Find $f'(x)$ for each of the following functions:

a xe^{-x}

b $\dfrac{x^2}{x^2 + 1}$

c e^{3x}

d $(3x + 2)^{-2}$

e $\ln x^3$

4 Find all first- and second-order partial derivatives for $f(x_1, x_2) = x_1^2 e^{x_2}$.

5 Find the second-order Taylor series expansion of $\ln x$ about $x = 1$.

Group B

6 Let $q = f(p)$ be the demand for a product when the price is p. For a given price p, the price elasticity E of the product is defined by

$$E = \frac{\text{percentage change in demand}}{\text{percentage change in price}}$$

If the change in price (Δp) is small, this formula reduces to

$$E = \frac{\frac{\Delta q}{q}}{\frac{\Delta p}{p}} = \left(\frac{p}{q}\right)\left(\frac{dq}{dp}\right)$$

a Would you expect $f(p)$ to be positive or negative?

b Show that if $E < -1$, a small decrease in price will increase the firm's total revenue (in this case, we say that demand is **elastic**).

c Show that if $-1 < E < 0$, a small price decrease will decrease total revenue (in this case, we say that demand is **inelastic**).

7 Suppose that if x dollars are spent on advertising during a given year, $k(1 - e^{-cx})$ customers will purchase a product ($c > 0$).

a As x grows large, the number of customers purchasing the product approaches a limit. Find this limit.

b Can you give an interpretation for k?

c Show that the sales response from a dollar of advertising is proportional to the number of potential customers who are not purchasing the product at present.

8 Let the total cost of producing x units, $c(x)$, be given by $c(x) = kx^{1-b}$ ($0 < b < 1$). This cost curve is called the **learning**, or **experience cost curve.**

a Show that the cost of producing a unit is a decreasing function of the number of units that have been produced.

b Suppose that each time the number of units produced is doubled, the per-unit product cost drops to r% of its previous value (because workers learn how to perform their jobs better). Show that $r = 100(2^{-b})$.

9 If a company has m hours of machine time and w hours of labor, it can produce $3m^{1/3}w^{2/3}$ units of a product. At present, the company has 216 hours of machine time and 1,000 hours of labor. An extra hour of machine time costs $100, and an extra hour of labor costs $50. If the company has $100 to invest in purchasing additional labor and machine time, would it be better off buying 1 hour of machine time or 2 hours of labor?

1.2 Review of Integral Calculus

In our study of random variables, we often require a knowledge of the basics of integral calculus, which will be briefly reviewed in this section.

Consider two functions: $f(x)$ and $F(x)$. If $F'(x) = f(x)$, we say that $F(x)$ is the **indefinite integral** of $f(x)$. The fact that $F(x)$ is the indefinite integral of $f(x)$ is written

$$F(x) = \int f(x)\, dx$$

The following rules may be used to find the indefinite integrals of many functions (C is an arbitrary constant):

$$\int (1)\, dx = x + C$$

$$\int af(x)\, dx = a \int f(x)\, dx \qquad\qquad (a \text{ is any constant})$$

$$\int [f(x) + g(x)]\, dx = \int f(x)\, dx + \int g(x)\, dx$$

$$\int x^n\, dx = \frac{x^{n+1}}{n+1} + C \qquad\qquad (n \neq -1)$$

$$\int x^{-1}\, dx = \ln x + C$$

$$\int e^x\, dx = e^x + C$$

$$\int a^x\, dx = \frac{a^x}{\ln a} + C \qquad\qquad (a > 0, a \neq 1)$$

$$\int [f(x)]^n f'(x)\, dx = \frac{[f(x)]^{n+1}}{n+1} + C \qquad\qquad (n \neq -1)$$

$$\int f(x)^{-1} f'(x)\, dx = \ln f(x) + C$$

For two functions $u(x)$ and $v(x)$,

$$\int u(x)v'(x)\, dx = u(x)v(x) - \int v(x)u'(x)\, dx \qquad \text{(Integration by parts)}$$

$$\int e^{f(x)} f'(x)\, dx = e^{f(x)} + C$$

$$\int a^{f(x)} f'(x)\, dx = \frac{a^{f(x)}}{\ln a} + C \qquad\qquad (a > 0, a \neq 1)$$

The concept of an integral is important for the following reasons. Consider a function $f(x)$ that is continuous for all points satisfying $a \leq x \leq b$. Let $x_0 = a$, $x_1 = x_0 + \Delta$, $x_2 = x_1 + \Delta, \ldots, x_i = x_{i-1} + \Delta, x_n = x_{n-1} + \Delta = b$, where $\Delta = \frac{b-a}{n}$. From Figure 2, we see that as Δ approaches zero (or equivalently, as n grows large),

$$\sum_{i=1}^{i=n} f(x_i)\, \Delta$$

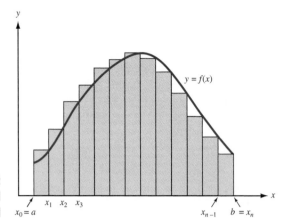

FIGURE 2
Relation of Area and
Definite Integral

$x_0 = a$ x_1 x_2 x_3 x_{n-1} $b = x_n$

will closely approximate the area under the curve $y = f(x)$ between $x = a$ and $x = b$. If $f(x)$ is continuous for all x satisfying $a \leq x \leq b$, it can be shown that the area under the curve $y = f(x)$ between $x = a$ and $x = b$ is given by

$$\lim_{\Delta \to 0} \sum_{i=1}^{i=n} f(x_i)\Delta$$

which is written as

$$\int_a^b f(x) \, dx$$

or the **definite integral** of $f(x)$ from $x = a$ to $x = b$. The **Fundamental Theorem of Calculus** states that if $f(x)$ is continuous for all x satisfying $a \leq x \leq b$, then

$$\int_a^b f(x) \, dx = F(b) - F(a)$$

where $F(x)$ is any indefinite integral of $f(x)$. $F(b) - F(a)$ is often written as $[F(x)]_a^b$. Example 7 illustrates the use of the definite integral.

EXAMPLE 7   **Customer Arrivals at a Bank**

Suppose that at time t (measured in hours, and the present $t = 0$), the rate $a(t)$ at which customers enter a bank is $a(t) = 100t$. During the next 2 hours, how many customers will enter the bank?

Solution   Let $t_0 = 0$, $t_1 = t_0 + \Delta$, $t_2 = t_1 + \Delta$, ..., $t_n = t_{n-1} + \Delta = 2$ (of course, $\Delta = \frac{2}{n}$). Between time t_{i-1} and time t_i, approximately $100t_i\Delta$ customers will arrive. Therefore, the total number of customers to arrive during the next 2 hours will equal

$$\lim_{\Delta \to 0} \sum_{i=1}^{i=n} 100t_i\Delta$$

(see Figure 3). From the Fundamental Theorem of Calculus,

$$\lim_{\Delta \to 0} \sum_{i=1}^{i=n} 100t_i\Delta = \int_0^2 (100t) \, dt = [50t^2]_0^2 = 200 - 0 = 200$$

Thus, 200 customers will arrive during the next 2 hours.

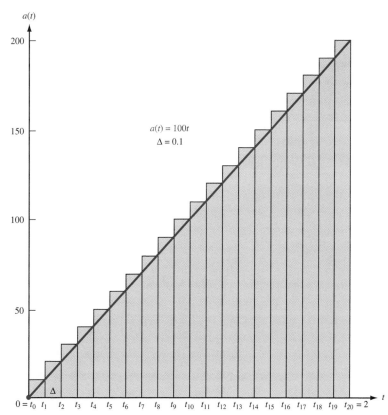

FIGURE 3
Relation of Total
Arrivals in Next 2 Hours
to Area under $a(t)$
Curve

PROBLEMS

Group A

1 The present is $t = 0$. At a time t years from now, I earn income at a rate e^{2t}. How much money do I earn during the next 5 years?

2 If money is continuously discounted at a rate of $r\%$ per year, then \$1 earned t years in the future is equivalent to e^{-rt} dollars earned at the present time. Use this fact to determine the present value of the income earned in Problem 1.

3 At time 0, a company has I units of inventory in stock. Customers demand the product at a constant rate of d units per year (assume that $I \geq d$). The cost of holding 1 unit of stock in inventory for a time Δ is \$$h\Delta$. Determine the total holding cost incurred during the next year.

1.3 Differentiation of Integrals

In our study of inventory theory in Chapter 4, we will have to differentiate a function whose value depends on an integral. Let $f(x, y)$ be a function of variables x and y, and let $g(y)$ and $h(y)$ be functions of y. Then

$$F(y) = \int_{g(y)}^{h(y)} f(x, y) \, dx$$

is a function only of y. **Leibniz's rule for differentiating an integral** states that

$$\text{If} \quad F(y) = \int_{g(y)}^{h(y)} f(x, y) \, dx, \quad \text{then}$$

$$F'(y) = h'(y)f(h(y), y) - g'(y)f(g(y), y) + \int_{g(y)}^{h(y)} \frac{\partial f(x, y)}{\partial y} \, dx$$

Example 8 illustrates Leibniz's rule.

EXAMPLE 8   **Leibniz's Rule**

For

$$F(y) = \int_{1}^{y^2} \frac{y \, dx}{x}$$

find $F'(y)$.

Solution   We have that $f(x, y) = \dfrac{y}{x}$, $h(y) = y^2$, $h'(y) = 2y$, $\dfrac{\partial f}{\partial y} = \dfrac{1}{x}$, $g(y) = 1$, $g'(y) = 0$. Then

$$F'(y) = 2y \left(\frac{y}{y^2} \right) - 0 \left(\frac{y}{1} \right) + \int_{1}^{y^2} \frac{dx}{x}$$

$$= 2 + [\ln x]_{1}^{y^2} = 2 + \ln y^2 - 0 = 2 + 2 \ln y$$

PROBLEMS

Group A

For each of the following functions, use Leibniz's rule to find $F'(y)$:

1 $F(y) = \int_{y}^{y^2} (2y + x) \, dx$

2 $F(y) = \int_{0}^{y} yx^2 \, dx$

3 $F(y) = \int_{0}^{y} 6(5 - x)f(x) \, dx + \int_{y}^{\infty} 4(x - 5)f(x) \, dx$

1.4   Basic Rules of Probability

In this section, we review some basic rules and definitions that you may have encountered during your previous study of probability.

DEFINITION ■   Any situation where the outcome is uncertain is called an **experiment.** ■

For example, drawing a card from a deck of cards would be an experiment.

DEFINITION ■   For any experiment, the **sample space** S of the experiment consists of all possible outcomes for the experiment. ■

For example, if we toss a die and are interested in the number of dots showing, then $S = \{1, 2, 3, 4, 5, 6\}$.

DEFINITION ■

> An **event** E consists of any collection of points (set of outcomes) in the sample space. ■
>
> A collection of events $E_1, E_2, \ldots, E_n$ is said to be a **mutually exclusive** collection of events if for $i \neq j$ ($i = 1, 2, \ldots, n$ and $j = 1, 2, \ldots, n$), E_i and E_j have no points in common. ■

With each event E, we associate an event $\overline{E}$. $\overline{E}$ consists of the points in the sample space that are not in E. With each event E, we also associate a number $P(E)$, which is the probability that event E will occur when we perform the experiment. The probabilities of events must satisfy the following rules of probability:

Rule 1 For any event E, $P(E) \geq 0$.

Rule 2 If $E = S$ (that is, if E contains all points in the sample space), then $P(E) = 1$.

Rule 3 If $E_1, E_2, \ldots, E_n$ is a mutually exclusive collection of events, then

$$P(E_1 \cup E_2 \cup \cdots \cup E_n) = \sum_{k=1}^{k=n} P(E_k)$$

Rule 4 $P(\overline{E}) = 1 - P(E)$.

DEFINITION ■

> For two events E_1 and E_2, $P(E_2|E_1)$ (the **conditional probability** of E_2 given E_1) is the probability that the event E_2 will occur given that event E_1 has occurred. Then
>
> $$P(E_2|E_1) = \frac{P(E_1 \cap E_2)}{P(E_1)} \qquad (2)$$
>
> Suppose events E_1 and E_2 both occur with positive probability. Events E_1 and E_2 are **independent** if and only if $P(E_2|E_1) = P(E_2)$ (or equivalently, $P(E_1|E_2) = P(E_1)$). ■

Thus, events E_1 and E_2 are independent if and only if knowledge that E_1 has occurred does not change the probability that E_2 has occurred, and vice versa. From (2), E_1 and E_2 are independent if and only if

$$\frac{P(E_1 \cap E_2)}{P(E_1)} = P(E_2) \qquad \text{or} \qquad P(E_1 \cap E_2) = P(E_1)\,P(E_2) \qquad (3)$$

EXAMPLE 9   **Drawing a Card**

Suppose we draw a single card from a deck of 52 cards.

1   What is the probability that a heart or spade is drawn?

2   What is the probability that the drawn card is not a 2?

3   Given that a red card has been drawn, what is the probability that it is a diamond? Are the events

$$E_1 = \text{red card is drawn}$$
$$E_2 = \text{diamond is drawn}$$

independent events?

4 Show that the events

$$E_1 = \text{spade is drawn}$$
$$E_2 = 2 \text{ is drawn}$$

are independent events.

Solution   **1**   Define the events

$$E_1 = \text{heart is drawn}$$
$$E_2 = \text{spade is drawn}$$

E_1 and E_2 are mutually exclusive events with $P(E_1) = P(E_2) = \frac{1}{4}$. We seek $P(E_1 \cup E_2)$. From probability rule 3,

$$P(E_1 \cup E_2) = P(E_1) + P(E_2) = (\tfrac{1}{4}) + (\tfrac{1}{4}) = \tfrac{1}{2}$$

2   Define event $E = $ a 2 is drawn. Then $P(E) = \frac{4}{52} = \frac{1}{13}$. We seek $P(\bar{E})$. From probability rule 4, $P(\bar{E}) = 1 - \frac{1}{13} = \frac{12}{13}$.

3   From (2),

$$P(E_2|E_1) = \frac{P(E_1 \cap E_2)}{P(E_1)}$$

$$P(E_1 \cap E_2) = P(E_2) = \frac{13}{52} = \frac{1}{4}$$

$$P(E_1) = \frac{26}{52} = \frac{1}{2}$$

Thus,

$$P(E_2|E_1) = \frac{\frac{1}{4}}{\frac{1}{2}} = \frac{1}{2}$$

Since $P(E_2) = \frac{1}{4}$, we see that $P(E_2|E_1) \neq P(E_2)$. Thus, E_1 and E_2 are not independent events. (This is because knowing that a red card was drawn increases the probability that a diamond was drawn.)

4 $P(E_1) = \frac{13}{52} = \frac{1}{4}$, $P(E_2) = \frac{4}{52} = \frac{1}{13}$, and $P(E_1 \cap E_2) = \frac{1}{52}$. Since $P(E_1) P(E_2) = P(E_1 \cap E_2)$, E_1 and E_2 are independent events. Intuitively, since $\frac{1}{4}$ of all cards in the deck are spades and $\frac{1}{4}$ of all 2's in the deck are spades, knowing that a 2 has been drawn does not change the probability that the card drawn was a spade.

PROBLEMS

Group A

1 Suppose two dice are tossed (for each die, it is equally likely that 1, 2, 3, 4, 5, or 6 dots will show).

   a What is the probability that the total of the two dice will add up to 7 or 11?

   b What is the probability that the total of the two dice will add up to a number other than 2 or 12?

   c Are the events

$$E_1 = \text{first die shows a 3}$$
$$E_2 = \text{total of the two dice is 6}$$

independent events?

   d Are the events

$$E_1 = \text{first die shows a 3}$$
$$E_2 = \text{total of the two dice is 7}$$

independent events?

   e Given that the total of the two dice is 5, what is the probability that the first die showed 2 dots?

   f Given that the first die shows 5, what is the probability that the total of the two dice is even?

1.5 Bayes' Rule

An important decision often depends on the "state of the world." For example, we may want to know whether a person has tuberculosis. Then we would be concerned with the probability of the following states of the world:

$$S_1 = \text{person has tuberculosis}$$

$$S_2 = \text{person does not have tuberculosis}$$

More generally, n mutually exclusive states of the world $(S_1, S_2, \ldots, S_n)$ may occur. The states of the world are **collectively exhaustive:** $S_1, S_2, \ldots, S_n$ include all possibilities. Suppose a decision maker assigns a probability $P(S_i)$ to S_i. $P(S_i)$ is the **prior probability** of S_i. To obtain more information about the state of the world, the decision maker may observe the outcome of an experiment. Suppose that for each possible outcome O_j and each possible state of the world S_i, the decision maker knows $P(O_j|S_i)$, the **likelihood** of the outcome O_j given state of the world S_i. Bayes' rule combines prior probabilities and likelihoods with the experimental outcomes to determine a post-experimental probability, or **posterior probability,** for each state of the world. To derive Bayes' rule, observe that (2) implies that

$$P(S_i|O_j) = \frac{P(S_i \cap O_j)}{P(O_j)} \tag{4}$$

From (2), it also follows that

$$P(S_i \cap O_j) = P(O_j|S_i)P(S_i) \tag{5}$$

The states of the world $S_1, S_2, \ldots, S_n$ are collectively exhaustive, so the experimental outcome O_j (if it occurs) must occur with one of the S_i (see Figure 4). Since $S_1 \cap O_j$, $S_2 \cap O_j, \ldots, S_n \cap O_j$ are mutually exclusive events, probability rule 3 implies that

$$P(O_j) = P(S_1 \cap O_j) + P(S_2 \cap O_j) + \cdots + P(S_n \cap O_j) \tag{6}$$

The probabilities of the form $P(S_i \cap O_j)$ are often referred to as **joint probabilities,** and the probabilities $P(O_j)$ are called **marginal probabilities.** Substituting (5) into (6), we obtain

$$P(O_j) = \sum_{k=1}^{k=n} P(O_j|S_k)P(S_k) \tag{7}$$

Substituting (5) and (7) into (4) yields **Bayes' rule:**

$$P(S_i|O_j) = \frac{P(O_j|S_i)P(S_i)}{\sum_{k=1}^{k=n} P(O_j|S_k)P(S_k)} \tag{8}$$

The following example illustrates the use of Bayes' rule.

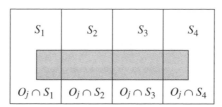

FIGURE 4
Illustration of Equation (6)

$$P(O_j) = P(O_j \cap S_1) + P(O_j \cap S_2)$$
$$+ P(O_j \cap S_3) + P(O_j \cap S_4)$$

Shaded area = outcome O_j

EXAMPLE 10   Bayes' Rule

Suppose that 1% of all children have tuberculosis (TB). When a child who has TB is given the Mantoux test, a positive test result occurs 95% of the time. When a child who does not have TB is given the Mantoux test, a positive test result occurs 1% of the time. Given that a child is tested and a positive test result occurs, what is the probability that the child has TB?

Solution The states of the world are

$$S_1 = \text{child has TB}$$
$$S_2 = \text{child does not have TB}$$

The possible experimental outcomes are

$$O_1 = \text{positive test result}$$
$$O_2 = \text{nonpositive test result}$$

We are given the prior probabilities $P(S_1) = .01$ and $P(S_2) = .99$ and the likelihoods $P(O_1|S_1) = .95$, $P(O_1|S_2) = .01$, $P(O_2|S_1) = .05$, and $P(O_2|S_2) = .99$. We seek $P(S_1|O_1)$. From (8),

$$P(S_1|O_1) = \frac{P(O_1|S_1)P(S_1)}{P(O_1|S_1)P(S_1) + P(O_1|S_2)P(S_2)}$$

$$= \frac{.95(.01)}{.95(.01) + .01(.99)} = \frac{95}{194} = .49$$

The reason a positive test result implies only a 49% chance that the child has TB is that many of the 99% of all children who do not have TB will test positive. For example, in a typical group of 10,000 children, 9,900 will not have TB and $.01(9,900) = 99$ children will yield a positive test result. In the same group of 10,000 children, $.01(10,000) = 100$ children will have TB and $.95(100) = 95$ children will yield a positive test result. Thus, the probability that a positive test result indicates TB is $\frac{95}{95+99} = \frac{95}{194}$.

PROBLEMS

Group A

1 A desk contains three drawers. Drawer 1 contains two gold coins. Drawer 2 contains one gold coin and one silver coin. Drawer 3 contains two silver coins. I randomly choose a drawer and then randomly choose a coin. If a silver coin is chosen, what is the probability that I chose drawer 3?

2 Cliff Colby wants to determine whether his South Japan oil field will yield oil. He has hired geologist Digger Barnes to run tests on the field. If there is oil in the field, there is a 95% chance that Digger's tests will indicate oil. If the field contains no oil, there is a 5% chance that Digger's tests will indicate oil. If Digger's tests indicate that there is no oil in the field, what is the probability that the field contains oil?

Before Digger conducts the test, Cliff believes that there is a 10% chance that the field will yield oil.

3 A customer has approached a bank for a loan. Without further information, the bank believes there is a 4% chance that the customer will default on the loan. The bank can run a credit check on the customer. The check will yield either a favorable or an unfavorable report. From past experience, the bank believes that $P(\text{favorable report being received})|$ customer will default$) = \frac{1}{40}$, and $P(\text{favorable report}|$ customer will not default$) = \frac{99}{100}$. If a favorable report is received, what is the probability that the customer will default on the loan?

4 Of all 40-year-old women, 1% have breast cancer. If a woman has breast cancer, a mammogram will give a positive indication for cancer 90% of the time. If a woman does not have breast cancer, a mammogram will give a positive indication for cancer 9% of the time. If a 40-year-old woman's mammogram gives a positive indication for cancer, what is the probability that she has cancer?

5 Three out of every 1,000 low-risk 50-year-old males have colon cancer. If a man has colon cancer, a test for hidden blood in the stool will indicate hidden blood half the time. If he does not have colon cancer, a test for hidden blood in the stool will indicate hidden blood 3% of the time. If the hidden-blood test turns out positive for a low-risk 50-year-old male, what is the chance that he has colon cancer?

Group B

6 You have made it to the final round of "Let's Make a Deal." You know there is $1 million behind either door 1, door 2, or door 3. It is equally likely that the prize is behind any of the three. The two doors without a prize have nothing behind them. You randomly choose door 2, but before door 2 is opened Monte reveals that there is no prize behind door 3. You now have the opportunity to switch and choose door 1. Should you switch? Assume that Monte plays as follows: Monte knows where the prize is and will open an empty door, but he cannot open door 2. If the prize is really behind door 2, Monte is equally likely to open door 1 or door 3. If the prize is really behind door 1, Monte must open door 3. If the prize is really behind door 3, Monte must open door 1. What is your decision?

1.6 Random Variables, Mean, Variance, and Covariance

The concepts of random variables, mean, variance, and covariance are employed in several later chapters.

DEFINITION ■ A **random variable** is a function that associates a number with each point in an experiment's sample space. We denote random variables by boldface capital letters (usually **X**, **Y**, or **Z**). ■

Discrete Random Variables

DEFINITION ■ A random variable is **discrete** if it can assume only discrete values $x_1, x_2, \ldots$. A discrete random variable **X** is characterized by the fact that we know the probability that $\mathbf{X} = x_i$ (written $P(\mathbf{X} = x_1)$). ■

$P(\mathbf{X} = x_i)$ is the **probability mass function** (pmf) for the random variable **X**.

DEFINITION ■ The **cumulative distribution function** $F(x)$ for any random variable **X** is defined by $F(x) = P(\mathbf{X} \leq x)$. For a discrete random variable **X**,

$$F(x) = \sum_{\substack{\text{all } x \\ \text{having } x_k \leq x}} P(\mathbf{X} = x_k) \quad ■$$

An example of a discrete random variable follows.

EXAMPLE 11 **Tossing a Die**

Let **X** be the number of dots that show when a die is tossed. Then for $i = 1, 2, 3, 4, 5, 6$, $P(\mathbf{X} = i) = \frac{1}{6}$. The cumulative distribution function (cdf) for **X** is shown in Figure 5.

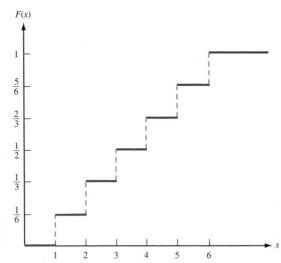

FIGURE 5
Cumulative Distribution Function for Example 11

Continuous Random Variables

If, for some interval, the random variable **X** can assume all values on the interval, then **X** is a **continuous** random variable. Probability statements about a continuous random variable **X** require knowing **X**'s **probability density function** (pdf). The probability density function $f(x)$ for a random variable **X** may be interpreted as follows: For Δ small,

$$P(x \leq \mathbf{X} \leq x + \Delta) \cong \Delta f(x)$$

From Figure 6, we see that for a random variable **X** having density function $f(x)$,

$$\text{Area 1} = P(a \leq \mathbf{X} \leq a + \Delta) \cong \Delta f(a)$$

and

$$\text{Area 2} = P(b \leq \mathbf{X} \leq b + \Delta) \cong \Delta f(b)$$

Thus, for a random variable **X** with density function $f(x)$ as given in Figure 6, values of **X** near a are much more likely to occur than values of **X** near b.

From our previous discussion of the Fundamental Theorem of Calculus, it follows that

$$P(a \leq \mathbf{X} \leq b) = \int_a^b f(x) \, dx$$

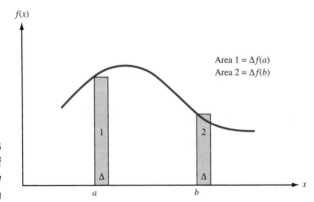

FIGURE 6
Illustration of Probability Density Function

Thus, for a continuous random variable, any area under the random variable's pdf corresponds to a probability. Using the concept of area as probability, we see that the cdf for a continuous random variable $\mathbf{X}$ with density $f(x)$ is given by

$$F(a) = P(\mathbf{X} \le a) = \int_{-\infty}^{a} f(x)\,dx$$

EXAMPLE 12 **Cumulative Distribution Function**

Consider a continuous random variable $\mathbf{X}$ having a density function $f(x)$ given by

$$f(x) = \begin{cases} 2x & \text{if } 0 \le x \le 1 \\ 0 & \text{otherwise} \end{cases}$$

Find the cdf for $\mathbf{X}$. Also find $P(\frac{1}{4} \le \mathbf{X} \le \frac{3}{4})$.

Solution For $a \le 0$, $F(a) = 0$. For $0 \le a \le 1$,

$$F(a) = \int_{0}^{a} 2x\,dx = a^2$$

For $a \ge 1$, $F(a) = 1$. $F(a)$ is graphed in Figure 7.

$$P(\tfrac{1}{4} \le \mathbf{X} \le \tfrac{3}{4}) = \int_{1/4}^{3/4} 2x\,dx = [x^2]_{1/4}^{3/4} = (\tfrac{9}{16}) - (\tfrac{1}{16}) = \tfrac{1}{2}$$

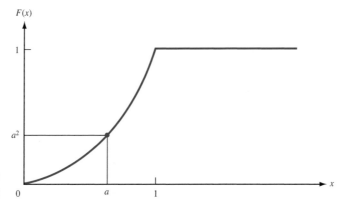

FIGURE 7
Cumulative Distribution Function for Example 12

Mean and Variance of a Random Variable

The **mean** (or expected value) and **variance** are two important measures that are often used to summarize information contained in a random variable's probability distribution. The mean of a random variable $\mathbf{X}$ (written $E(\mathbf{X})$) is a measure of central location for the random variable.

Mean of a Discrete Random Variable

For a discrete random variable $\mathbf{X}$,

$$E(\mathbf{X}) = \sum_{\text{all } k} x_k P(\mathbf{X} = x_k) \tag{9}$$

Mean of a Continuous Random Variable

For a continuous random variable,

$$E(\mathbf{X}) = \int_{-\infty}^{\infty} x f(x) \, dx \tag{10}$$

Observe that in computing $E(\mathbf{X})$, each possible value of a random variable is weighted by its probability of occurring. Thus, the mean of a random variable is essentially the random variable's center of mass.

For a function $h(\mathbf{X})$ of a random variable $\mathbf{X}$ (such as $\mathbf{X}^2$ and $e^{\mathbf{X}}$), $E[h(\mathbf{X})]$ may be computed as follows. If $\mathbf{X}$ is a discrete random variable,

$$E[h(\mathbf{X})] = \sum_{\text{all } k} h(x_k) P(\mathbf{X} = x_k) \tag{9'}$$

If $\mathbf{X}$ is a continuous random variable,

$$E[h(\mathbf{X})] = \int_{-\infty}^{\infty} h(x) f(x) \, dx \tag{10'}$$

The variance of a random variable $\mathbf{X}$ (written as var $\mathbf{X}$) measures the dispersion or spread of $\mathbf{X}$ about $E(\mathbf{X})$. Then var $\mathbf{X}$ is defined to be $E[\mathbf{X} - E(\mathbf{X})]^2$.

Variance of a Discrete Random Variable

For a discrete random variable $\mathbf{X}$, (9') yields

$$\text{var } \mathbf{X} = \sum_{\text{all } k} [x_k - E(\mathbf{X})]^2 P(\mathbf{X} = x_k) \tag{11}$$

Variance of a Continuous Random Variable

For a continuous random variable $\mathbf{X}$, (10') yields

$$\text{var } \mathbf{X} = \int_{-\infty}^{\infty} [x - E(\mathbf{X})]^2 f(x) \, dx \tag{12}$$

Also, var $\mathbf{X}$ may be found from the relation

$$\text{var } \mathbf{X} = E(\mathbf{X}^2) - E(\mathbf{X})^2 \tag{13}$$

For any random variable $\mathbf{X}$, $(\text{var } \mathbf{X})^{1/2}$ is the **standard deviation** of $\mathbf{X}$ (written σ_x).

Examples 13 and 14 illustrate the computation of mean and variance for a discrete and a continuous random variable.

EXAMPLE 13   Discrete Random Variable

Consider the discrete random variable $\mathbf{X}$ having $P(\mathbf{X} = i) = \frac{1}{6}$ for $i = 1, 2, 3, 4, 5, 6$. Find $E(\mathbf{X})$ and var $\mathbf{X}$.

Solution

$$E(\mathbf{X}) = (\tfrac{1}{6})(1 + 2 + 3 + 4 + 5 + 6) = \tfrac{21}{6} = \tfrac{7}{2}$$
$$\text{var } \mathbf{X} = (\tfrac{1}{6})[(1 - 3.5)^2 + (2 - 3.5)^2 + (3 - 3.5)^2$$
$$+ (4 - 3.5)^2 + (5 - 3.5)^2 + (6 - 3.5)^2] = \tfrac{35}{12}$$

EXAMPLE 14 | Continuous Random Variable

Find the mean and variance for the continuous random variable $\mathbf{X}$ having the following density function:

$$f(x) = \begin{cases} 2x & \text{if } 0 \le x \le 1 \\ 0 & \text{otherwise} \end{cases}$$

Solution

$$E(\mathbf{X}) = \int_0^1 x(2x)\, dx = \left[\frac{2x^3}{3}\right]_0^1 = \frac{2}{3}$$

$$\text{var } \mathbf{X} = \int_0^1 \left(x - \frac{2}{3}\right)^2 2x\, dx = \int_0^1 \left(x^2 - \frac{4x}{3} + \frac{4}{9}\right) 2x\, dx$$

$$= \left[\frac{2x^4}{4} - \frac{8x^3}{9} + \frac{8x^2}{18}\right]_0^1 = \frac{1}{18}$$

Independent Random Variables

DEFINITION ■ Two random variables $\mathbf{X}$ and $\mathbf{Y}$ are **independent** if and only if for any two sets A and B,

$$P(\mathbf{X} \in A \text{ and } \mathbf{Y} \in B) = P(\mathbf{X} \in A)P(\mathbf{Y} \in B) \quad ■$$

From this definition, it can be shown that $\mathbf{X}$ and $\mathbf{Y}$ are independent random variables if and only if knowledge about the value of $\mathbf{Y}$ does not change the probability of any event involving $\mathbf{X}$. For example, suppose $\mathbf{X}$ and $\mathbf{Y}$ are independent random variables. This implies that where $\mathbf{Y} = 8$, $\mathbf{Y} = 10$, $\mathbf{Y} = 0$, or $\mathbf{Y} =$ anything else, $P(\mathbf{X} \ge 10)$ will be the same. If $\mathbf{X}$ and $\mathbf{Y}$ are independent, then $E(\mathbf{XY}) = E(\mathbf{X})E(\mathbf{Y})$. (The random variable $\mathbf{XY}$ has an expected value equal to the product of the expected value of $\mathbf{X}$ and the expected value of $\mathbf{Y}$.)

The definition of independence generalizes to situations where more than two random variables are of interest. Loosely speaking, a group of n random variables is independent if knowledge of the values of any subset of the random variables does not change our view of the distribution of any of the other random variables. (See Problem 5 at the end of this section.)

Covariance of Two Random Variables

An important concept in the study of financial models is covariance. For two random variables $\mathbf{X}$ and $\mathbf{Y}$, the **covariance** of $\mathbf{X}$ and $\mathbf{Y}$ (written cov($\mathbf{X}$, $\mathbf{Y}$)) is defined by

$$\text{cov}(\mathbf{X}, \mathbf{Y}) = E\{[\mathbf{X} - E(\mathbf{X})][\mathbf{Y} - E(\mathbf{Y})]\} \tag{14}$$

If $\mathbf{X} > E(\mathbf{X})$ tends to occur when $\mathbf{Y} > E(\mathbf{Y})$, and $\mathbf{X} < E(\mathbf{X})$ tends to occur when $\mathbf{Y} < E(\mathbf{Y})$, then cov($\mathbf{X}$, $\mathbf{Y}$) will be positive. On the other hand, if $\mathbf{X} > E(\mathbf{X})$ tends to occur when $\mathbf{Y} < E(\mathbf{Y})$, and $\mathbf{X} < E(\mathbf{X})$ tends to occur when $\mathbf{Y} > E(\mathbf{Y})$, then cov($\mathbf{X}$, $\mathbf{Y}$) will be negative. The value of cov($\mathbf{X}$, $\mathbf{Y}$) measures the association (actually, linear association) between random variables $\mathbf{X}$ and $\mathbf{Y}$. It can be shown that if $\mathbf{X}$ and $\mathbf{Y}$ are independent random variables, then cov($\mathbf{X}$, $\mathbf{Y}$) = 0. (However, cov($\mathbf{X}$, $\mathbf{Y}$) = 0 can hold even if $\mathbf{X}$ and $\mathbf{Y}$ are not independent random variables. See Problem 6 at the end of this section for an example.)

EXAMPLE 15   Gotham City Summers

Each summer in Gotham City is classified as being either a rainy summer or a sunny summer. The profits earned by Gotham City's two leading industries (the Gotham City Hotel and the Gotham City Umbrella Store) depend on the summer's weather, as shown in Table 3. Of all summers, 20% are rainy, and 80% are sunny. Let **H** and **U** be the following random variables:

$\mathbf{H}$ = profit earned by Gotham City Hotel during a summer

$\mathbf{U}$ = profit earned by Gotham City Umbrella Store during a summer

Find cov($\mathbf{H}$,$\mathbf{U}$).

Solution   We find that

$$E(\mathbf{H}) = .2(-1,000) + .8(2,000) = \$1,400$$
$$E(\mathbf{U}) = .2(4,500) + .8(-500) = \$500$$

With probability .20, Gotham City has a rainy summer. Then

$$[\mathbf{H} - E(\mathbf{H})][\mathbf{U} - E(\mathbf{U})] = (-1,000 - 1,400)(4,500 - 500) = -9,600,000(\text{dollars})^2$$

With probability .80, Gotham City has a sunny summer. Then

$$[\mathbf{H} - E(\mathbf{H})][\mathbf{U} - E(\mathbf{U})] = (2,000 - 1,400)(-500 - 500) = -600,000(\text{dollars})^2$$

Thus,

$$\text{cov}(\mathbf{H},\mathbf{U}) = E\{[\mathbf{H} - E(\mathbf{H})][\mathbf{U} - E(\mathbf{U})]\} = .20(-9,600,000) + .80(-600,000)$$
$$= -2,400,000(\text{dollars})^2$$

The fact that cov($\mathbf{H}$,$\mathbf{U}$) is negative indicates that when one industry does well, the other industry tends to do poorly.

TABLE 3
Profits for Gotham City Covariance

Type of Summer	Hotel Profit	Umbrella Profit
Rainy	−$1,000	$4,500
Sunny	$2,000	−$500

Mean, Variance, and Covariance for Sums of Random Variables

From given random variables $\mathbf{X}_1$ and $\mathbf{X}_2$, we often create new random variables (c is a constant): $c\mathbf{X}_1$, $\mathbf{X}_1 + c$, $\mathbf{X}_1 + \mathbf{X}_2$. The following rules can be used to express the mean, variance, and covariance of these random variables in terms of $E(\mathbf{X}_1)$, $E(\mathbf{X}_2)$, var $\mathbf{X}_1$, var $\mathbf{X}_2$, and cov($\mathbf{X}_1, \mathbf{X}_2$). Examples 16 and 17 illustrate the use of these rules.

$$E(c\mathbf{X}_1) = cE(\mathbf{X}_1) \tag{15}$$
$$E(\mathbf{X}_1 + c) = E(\mathbf{X}_1) + c \tag{16}$$
$$E(\mathbf{X}_1 + \mathbf{X}_2) = E(\mathbf{X}_1) + E(\mathbf{X}_2) \tag{17}$$
$$\text{var } c\mathbf{X}_1 = c^2\text{var } \mathbf{X}_1 \tag{18}$$
$$\text{var}(\mathbf{X}_1 + c) = \text{var } \mathbf{X}_1 \tag{19}$$

If X_1 and X_2 are independent random variables,

$$\text{var}(X_1 + X_2) = \text{var } X_1 + \text{var } X_2 \qquad (20)$$

In general,

$$\text{var}(X_1 + X_2) = \text{var } X_1 + \text{var } X_2 + 2\text{cov}(X_1, X_2) \qquad (21)$$

For random variables $X_1, X_2, \ldots, X_n$,

$$\text{var}(X_1 + X_2 + \cdots + X_n) = \text{var } X_1 + \text{var } X_2 + \cdots + \text{var } X_n + \sum_{i \neq j} \text{cov}(X_i, X_j) \qquad (22)$$

Finally, for constants a and b,

$$\text{cov}(aX_1, bX_2) = ab\,\text{cov}(X_1, X_2) \qquad (23)$$

EXAMPLE 16   **Tossing a Die: Mean and Variance**

I pay \$1 to play the following game: I toss a die and receive \$3 for each dot that shows. Determine the mean and variance of my profit.

Solution   Let X be the random variable representing the number of dots that show when the die is tossed. Then my profit is given by the value of the random variable $3X - 1$. From Example 13, we know that $E(X) = \frac{7}{2}$ and var $X = \frac{35}{12}$. In turn, Equations (16) and (15) yield

$$E(3X - 1) = E(3X) - 1 = 3E(X) - 1 = 3\left(\tfrac{7}{2}\right) - 1 = \tfrac{19}{2}$$

From Equations (19) and (18), respectively,

$$\text{var}(3X - 1) = \text{var}(3X) = 9(\text{var } X) = 9\left(\tfrac{35}{12}\right) = \tfrac{315}{12}$$

EXAMPLE 17   **Gotham City Profit: Mean and Variance**

In Example 15, suppose I owned both the hotel and the umbrella store. Find the mean and the variance of the total profit I would earn during a summer.

Solution   My total profits are given by the random variable $H + U$. From Equation (17) and Example 15,

$$E(H + U) = E(H) + E(U) = 1{,}400 + 500 = \$1{,}900$$

Now

$$\text{var } H = .2(-1{,}000 - 1{,}400)^2 + .8(2{,}000 - 1{,}400)^2 = 1{,}440{,}000(\text{dollars})^2$$
$$\text{var } U = .2(4{,}500 - 500)^2 + .8(-500 - 500)^2 = 4{,}000{,}000(\text{dollars})^2$$

From Example 15, $\text{cov}(H, U) = -2{,}400{,}000$ (dollars)^2. Then Equation (21) yields

$$\begin{aligned}
\text{var}(H + U) &= \text{var } H + \text{var } U + 2\text{cov}(H, U)\\
&= 1{,}440{,}000(\text{dollars})^2 + 4{,}000{,}000(\text{dollars})^2 - 2(2{,}400{,}000)(\text{dollars})^2\\
&= 640{,}000(\text{dollars})^2
\end{aligned}$$

Thus, $H + U$ has a smaller variance than either H or U. This is because by owning both the hotel and umbrella store, we will always have, regardless of the weather, one industry that does well and one that does poorly. This reduces the spread, or variability, of our profits.

PROBLEMS

Group A

1 I have 100 items of a product in stock. The probability mass function for the product's demand D is $P(D = 90) = P(D = 100) = P(D = 110) = \frac{1}{3}$.

   a Find the mass function, mean, and variance of the number of items sold.

   b Find the mass function, mean, and variance of the amount of demand that will be unfilled because of lack of stock.

2 I draw 5 cards from a deck (replacing each card immediately after it is drawn). I receive \$4 for each heart that is drawn. Find the mean and variance of my total payoff.

3 Consider a continuous random variable X with the density function (called the *exponential density*)

$$f(x) = \begin{cases} e^{-x} & \text{if } x \geq 0 \\ 0 & \text{otherwise} \end{cases}$$

   a Find and sketch the cdf for X.

   b Find the mean and variance of X. (*Hint:* Use integration by parts.)

   c Find $P(1 \leq X \leq 2)$.

4 I have 100 units of a product in stock. The demand D for the item is a continuous random variable with the following density function:

$$f(d) = \begin{cases} \frac{1}{40} & \text{if } 80 \leq d \leq 120 \\ 0 & \text{otherwise} \end{cases}$$

   a Find the probability that supply is insufficient to meet demand.

   b What is the expected number of items sold? What is the variance of the number of items sold?

5 An urn contains 10 red balls and 30 blue balls.

   a Suppose you draw 4 balls from the urn. Let X_i be the number of red balls drawn on the ith ball ($X_i = 0$ or 1). After each ball is drawn, it is put back into the urn. Are the random variables X_1, X_2, X_3, and X_4 independent random variables?

   b Repeat part (a) for the case in which the balls are not put back in the urn after being drawn.

Group B

6 Let X be the following discrete random variable: $P(X = -1) = P(X = 0) = P(X = 1) = \frac{1}{3}$. Let $Y = X^2$. Show that $\text{cov}(X, Y) = 0$, but X and Y are not independent random variables.

1.7 The Normal Distribution

The most commonly used probability distribution in this book is the normal distribution. In this section, we discuss some useful properties of the normal distribution.

DEFINITION ■   A continuous random variable X has a normal distribution if for some μ and $\sigma > 0$, the random variable has the following density function:

$$f(x) = \frac{1}{\sigma(2\pi)^{1/2}} \exp\left[-\frac{(x - \mu)^2}{2\sigma^2}\right] \quad ■$$

If a random variable X is normally distributed with a mean μ and variance σ^2, we write that X is $N(\mu, \sigma^2)$. It can be shown that for a normal random variable, $E(X) = \mu$ and var $X = \sigma^2$ (the standard deviation of X is σ). The normal density functions for several values of σ and a single value of μ are shown in Figure 8.

For any normal distribution, the normal density is symmetric about μ (that is, $f(\mu + a) = f(\mu - a)$). Also, as σ increases, the probability that the random variable assumes a value within c of μ (for any $c > 0$) decreases. Thus, as σ increases, the normal distribution becomes more spread out. The properties are illustrated in Figure 8.

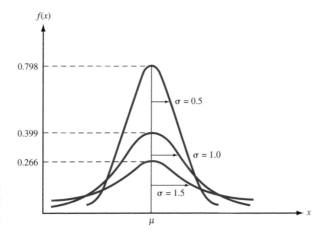

FIGURE 8
Some Examples of
Normal Distributions

Useful Properties of Normal Distributions

Property 1   If $\mathbf{X}$ is $N(\mu, \sigma^2)$, then $c\mathbf{X}$ is $N(c\mu, c^2\sigma^2)$.

Property 2   If $\mathbf{X}$ is $N(\mu, \sigma^2)$, then $\mathbf{X} + c$ (for any constant c) is $N(\mu + c, \sigma^2)$.

Property 3   If $\mathbf{X}_1$ is $N(\mu_1, \sigma_1^2)$, $\mathbf{X}_2$ is $N(\mu_2, \sigma_2^2)$, and $\mathbf{X}_1$ and $\mathbf{X}_2$ are independent, then $\mathbf{X}_1 + \mathbf{X}_2$ is $N(\mu_1 + \mu_2, \sigma_1^2 + \sigma_2^2)$.

Finding Normal Probabilities via Standardization

If $\mathbf{Z}$ is a random variable that is $N(0, 1)$, then $\mathbf{Z}$ is said to be a standardized normal random variable. In Table 4, $F(z) = P(\mathbf{Z} \leq z)$ is tabulated. For example,

$$P(\mathbf{Z} \leq -1) = F(-1) = .1587$$

and

$$P(\mathbf{Z} \geq 2) = 1 - P(\mathbf{Z} \leq 2) = 1 - F(2) = 1 - .9772 = .0228.$$

If $\mathbf{X}$ is $N(\mu, \sigma^2)$, then $(\mathbf{X} - \mu)/\sigma$ is $N(0, 1)$. This follows, because by property 2 of the normal distribution, $\mathbf{X} - \mu$ is $N(\mu - \mu, \sigma^2) = N(0, \sigma^2)$. Then by property 1, $\frac{x-\mu}{\sigma}$ is $N(\frac{0}{\sigma}, \frac{\sigma^2}{\sigma^2}) = N(0, 1)$. The last equality enables us to use Table 4 to find probabilities for any normal random variable, not just an $N(0, 1)$ random variable. Suppose $\mathbf{X}$ is $N(\mu, \sigma^2)$ and we want to find $P(a \leq \mathbf{X} \leq b)$. To find this probability from Table 4, we use the following relations (this procedure is called **standardization**):

$$P(a \leq \mathbf{X} \leq b) = P\left(\frac{a - \mu}{\sigma} \leq \frac{\mathbf{X} - \mu}{\sigma} \leq \frac{b - \mu}{\sigma}\right)$$

$$= P\left(\frac{a - \mu}{\sigma} \leq \mathbf{Z} \leq \frac{b - \mu}{\sigma}\right)$$

$$= F\left(\frac{b - \mu}{\sigma}\right) - F\left(\frac{a - \mu}{\sigma}\right)$$

The Central Limit Theorem

If $\mathbf{X}_1, \mathbf{X}_2, \ldots, \mathbf{X}_n$ are independent random variables, then for n sufficiently large (usually $n \geq 30$ will do, but the actual size of n depends on the distributions of $\mathbf{X}_1, \mathbf{X}_2, \ldots, \mathbf{X}_n$),

TABLE **4**
Standard Normal Cumulative Probabilities[†]

z	0.00	0.01	0.02	0.03	0.04	0.05	0.06	0.07	0.08	0.09
−3.8	0.0001	0.0001	0.0001	0.0001	0.0001	0.0001	0.0001	0.0001	0.0001	0.0001
−3.7	0.0001	0.0001	0.0001	0.0001	0.0001	0.0001	0.0001	0.0001	0.0001	0.0001
−3.6	0.0002	0.0002	0.0001	0.0001	0.0001	0.0001	0.0001	0.0001	0.0001	0.0001
−3.5	0.0002	0.0002	0.0002	0.0002	0.0002	0.0002	0.0002	0.0002	0.0002	0.0002
−3.4	0.0003	0.0003	0.0003	0.0003	0.0003	0.0003	0.0003	0.0003	0.0003	0.0002
−3.3	0.0005	0.0005	0.0005	0.0004	0.0004	0.0004	0.0004	0.0004	0.0004	0.0003
−3.2	0.0007	0.0007	0.0006	0.0006	0.0006	0.0006	0.0006	0.0005	0.0005	0.0005
−3.1	0.0010	0.0009	0.0009	0.0009	0.0008	0.0008	0.0008	0.0008	0.0007	0.0007
−3.0	0.0014	0.0013	0.0013	0.0012	0.0012	0.0011	0.0011	0.0011	0.0010	0.0010
−2.9	0.0019	0.0018	0.0018	0.0017	0.0016	0.0016	0.0015	0.0015	0.0014	0.0014
−2.8	0.0026	0.0025	0.0024	0.0023	0.0023	0.0022	0.0021	0.0021	0.0020	0.0019
−2.7	0.0035	0.0034	0.0033	0.0032	0.0031	0.0030	0.0029	0.0028	0.0027	0.0026
−2.6	0.0047	0.0045	0.0044	0.0043	0.0041	0.0040	0.0039	0.0038	0.0037	0.0036
−2.5	0.0062	0.0060	0.0059	0.0057	0.0055	0.0054	0.0052	0.0051	0.0049	0.0048
−2.4	0.0082	0.0080	0.0078	0.0076	0.0073	0.0071	0.0069	0.0068	0.0066	0.0064
−2.3	0.0107	0.0104	0.0102	0.0099	0.0096	0.0094	0.0091	0.0089	0.0087	0.0084
−2.2	0.0139	0.0136	0.0132	0.0129	0.0125	0.0122	0.0119	0.0116	0.0113	0.0110
−2.1	0.0179	0.0174	0.0170	0.0166	0.0162	0.0158	0.0154	0.0150	0.0146	0.0143
−2.0	0.0228	0.0222	0.0217	0.0212	0.0207	0.0202	0.0197	0.0192	0.0188	0.0183
−1.9	0.0287	0.0281	0.0274	0.0268	0.0262	0.0256	0.0250	0.0244	0.0239	0.0233
−1.8	0.0359	0.0351	0.0344	0.0336	0.0329	0.0322	0.0314	0.0307	0.0301	0.0294
−1.7	0.0446	0.0436	0.0427	0.0418	0.0409	0.0401	0.0392	0.0384	0.0375	0.0367
−1.6	0.0548	0.0537	0.0526	0.0516	0.0505	0.0495	0.0485	0.0475	0.0465	0.0455
−1.5	0.0668	0.0655	0.0643	0.0630	0.0618	0.0606	0.0594	0.0582	0.0571	0.0559
−1.4	0.0808	0.0793	0.0778	0.0764	0.0749	0.0735	0.0721	0.0708	0.0694	0.0681
−1.3	0.0968	0.0951	0.0934	0.0918	0.0901	0.0885	0.0869	0.0853	0.0838	0.0823
−1.2	0.1151	0.1131	0.1112	0.1093	0.1075	0.1057	0.1038	0.1020	0.1003	0.0985
−1.1	0.1357	0.1335	0.1314	0.1292	0.1271	0.1251	0.1230	0.1210	0.1190	0.1170
−1.0	0.1587	0.1562	0.1539	0.1515	0.1492	0.1469	0.1446	0.1423	0.1401	0.1379
−0.9	0.1841	0.1814	0.1788	0.1762	0.1736	0.1711	0.1685	0.1660	0.1635	0.1611
−0.8	0.2119	0.2090	0.2061	0.2033	0.2005	0.1977	0.1949	0.1922	0.1894	0.1867
−0.7	0.2420	0.2389	0.2358	0.2327	0.2297	0.2266	0.2236	0.2206	0.2177	0.2148
−0.6	0.2743	0.2709	0.2676	0.2643	0.2611	0.2578	0.2546	0.2514	0.2483	0.2451
−0.5	0.3085	0.3050	0.3015	0.2981	0.2946	0.2912	0.2877	0.2843	0.2810	0.2776
−0.4	0.3446	0.3409	0.3372	0.3336	0.3300	0.3264	0.3228	0.3192	0.3156	0.3121
−0.3	0.3821	0.3783	0.3745	0.3707	0.3669	0.3632	0.3594	0.3557	0.3520	0.3483
−0.2	0.4207	0.4168	0.4129	0.4090	0.4052	0.4013	0.3974	0.3936	0.3897	0.3859
−0.1	0.4602	0.4562	0.4522	0.4483	0.4443	0.4404	0.4364	0.4325	0.4286	0.4247
−0.0	0.5000	0.4960	0.4920	0.4880	0.4840	0.4801	0.4761	0.4721	0.4681	0.4641

Source: Reprinted by permission from David E. Kleinbaum, Lawrence L. Kupper, and Keith E. Muller, *Applied Regression Analysis and Other Multivariable Methods*, 2nd edition. Copyright © 1988 PWS-KENT Publishing Company.

[†]*Note:* Table entry is the area under the standard normal curve to the left of the indicated z-value, thus giving $P(Z \leq z)$.

TABLE **4**
Standard Normal Cumulative Probabilities (Continued)

z	0.00	0.01	0.02	0.03	0.04	0.05	0.06	0.07	0.08	0.09
0.0	0.5000	0.5040	0.5080	0.5120	0.5160	0.5199	0.5239	0.5279	0.5319	0.5359
0.1	0.5398	0.5438	0.5478	0.5517	0.5557	0.5596	0.5636	0.5675	0.5714	0.5753
0.2	0.5793	0.5832	0.5871	0.5910	0.5948	0.5987	0.6026	0.6064	0.6103	0.6141
0.3	0.6179	0.6217	0.6255	0.6293	0.6331	0.6368	0.6406	0.6443	0.6480	0.6517
0.4	0.6554	0.6591	0.6628	0.6664	0.6700	0.6736	0.6772	0.6808	0.6844	0.6879
0.5	0.6915	0.6950	0.6985	0.7019	0.7054	0.7088	0.7123	0.7157	0.7190	0.7224
0.6	0.7257	0.7291	0.7324	0.7357	0.7389	0.7422	0.7454	0.7486	0.7517	0.7549
0.7	0.7580	0.7611	0.7642	0.7673	0.7703	0.7734	0.7764	0.7794	0.7823	0.7852
0.8	0.7881	0.7910	0.7939	0.7967	0.7995	0.8023	0.8051	0.8078	0.8106	0.8133
0.9	0.8159	0.8186	0.8212	0.8238	0.8264	0.8289	0.8315	0.8340	0.8365	0.8389
1.0	0.8413	0.8438	0.8461	0.8485	0.8508	0.8531	0.8554	0.8577	0.8599	0.8621
1.1	0.8643	0.8665	0.8686	0.8708	0.8729	0.8749	0.8770	0.8790	0.8810	0.8830
1.2	0.8849	0.8869	0.8888	0.8907	0.8925	0.8943	0.8962	0.8980	0.8997	0.9015
1.3	0.9032	0.9049	0.9066	0.9082	0.9099	0.9115	0.9131	0.9147	0.9162	0.9177
1.4	0.9192	0.9207	0.9222	0.9236	0.9251	0.9265	0.9279	0.9292	0.9306	0.9319
1.5	0.9332	0.9345	0.9357	0.9370	0.9382	0.9394	0.9406	0.9418	0.9429	0.9441
1.6	0.9452	0.9463	0.9474	0.9484	0.9495	0.9505	0.9515	0.9525	0.9535	0.9545
1.7	0.9554	0.9564	0.9673	0.9582	0.9591	0.9599	0.9608	0.9616	0.9625	0.9633
1.8	0.9641	0.9649	0.9656	0.9664	0.9671	0.9678	0.9686	0.9683	0.9699	0.9706
1.9	0.9713	0.9719	0.9726	0.9732	0.9738	0.9744	0.9750	0.9756	0.9762	0.9767
2.0	0.9772	0.9778	0.9783	0.9788	0.9793	0.9798	0.9803	0.9808	0.9812	0.9817
2.1	0.9821	0.9826	0.9830	0.9834	0.9838	0.9842	0.9846	0.9850	0.9854	0.9857
2.2	0.9861	0.9864	0.9868	0.9871	0.9875	0.9878	0.9881	0.9884	0.9887	0.9890
2.3	0.9893	0.9896	0.9898	0.9901	0.9904	0.9906	0.9909	0.9911	0.9913	0.9916
2.4	0.9918	0.9920	0.9922	0.9924	0.9927	0.9929	0.9931	0.9932	0.9934	0.9936
2.5	0.9938	0.9940	0.9941	0.9943	0.9945	0.9946	0.9948	0.9949	0.9951	0.9952
2.6	0.9953	0.9955	0.9956	0.9957	0.9959	0.9960	0.9961	0.9962	0.9963	0.9964
2.7	0.9965	0.9966	0.9967	0.9968	0.9969	0.9970	0.9971	0.9972	0.9973	0.9974
2.8	0.9974	0.9975	0.9976	0.9977	0.9977	0.9978	0.9979	0.9979	0.9980	0.9981
2.9	0.9981	0.9982	0.9982	0.9983	0.9984	0.9984	0.9985	0.9985	0.9986	0.9986
3.0	0.9986	0.9987	0.9987	0.9988	0.9988	0.9989	0.9989	0.9989	0.9990	0.9990
3.1	0.9990	0.9991	0.9991	0.9991	0.9992	0.9992	0.9992	0.9992	0.9993	0.9993
3.2	0.9993	0.9993	0.9994	0.9994	0.9994	0.9994	0.9994	0.9995	0.9995	0.9995
3.3	0.9995	0.9995	0.9995	0.9996	0.9996	0.9996	0.9996	0.9996	0.9996	0.9997
3.4	0.9997	0.9997	0.9997	0.9997	0.9997	0.9997	0.9997	0.9997	0.9997	0.9998
3.5	0.9998	0.9998	0.9998	0.9998	0.9998	0.9998	0.9998	0.9998	0.9998	0.9998
3.6	0.9998	0.9998	0.9999	0.9999	0.9999	0.9999	0.9999	0.9999	0.9999	0.9999
3.7	0.9999	0.9999	0.9999	0.9999	0.9999	0.9999	0.9999	0.9999	0.9999	0.9999
3.8	0.9999	0.9999	0.9999	0.9999	0.9999	0.9999	0.9999	0.9999	0.9999	0.9999
3.9	1.0000									

the random variable $\mathbf{X} = \mathbf{X}_1 + \mathbf{X}_2 + \cdots + \mathbf{X}_n$ may be closely approximated by a normal random variable $\mathbf{X}'$ that has $E(\mathbf{X}') = E(\mathbf{X}_1) + E(\mathbf{X}_2) + \cdots + E(\mathbf{X}_n)$ and var $\mathbf{X}' =$ var $\mathbf{X}_1 +$ var $\mathbf{X}_2 + \cdots +$ var $\mathbf{X}_n$. This result is known as the Central Limit Theorem. When we say that $\mathbf{X}'$ closely approximates $\mathbf{X}$, we mean that $P(a \le \mathbf{X} \le b)$ is close to $P(a \le \mathbf{X}' \le b)$.

Finding Normal Probabilities with Excel

Probabilities involving a standard normal variable can be determined with Excel, using the =NORMSDIST function. The S in NORMSDIST stands for *standardized normal.* For example, $P(\mathbf{Z} \le -1)$ can be found by entering the formula

$$=\text{NORMSDIST}(-1)$$

Normal.xls

Excel returns the value .1587. See Figure 9 and file Normal.xls.

The =NORMDIST function can be used to determine a normal probability for any normal (not just a standard normal) random variable. If $\mathbf{X}$ is $N(\mu, \sigma^2)$, then entering the formula

$$=\text{NORMSDIST}(a,\mu,\sigma,1)$$

will return $P(\mathbf{X} \le a)$. The "1" ensures that Excel returns the cumulative normal probability. Changing the last argument to "0" causes Excel to return the height of the normal density function for $\mathbf{X} = a$. As an example, we know that IQs follow $N(100, 225)$. The fraction of people with IQ's of 90 or less is computed with the formula

$$=\text{NORMDIST}(90,100,15,1)$$

Excel yields .2525. See Figure 9 and file Normal.xls.

The height of the density for $N(100, 225)$ for $\mathbf{X} = 100$ is computed with the formula

$$=\text{NORMDIST}(100,100,15,0)$$

Excel yields .026596.

By varying the first argument in the =NORMDIST function, we may graph a normal density. See Figure 10 and sheet density of file Normal.xls.

Consider a given normal random variable $\mathbf{X}$, with mean μ and standard deviation σ. In many situations, we want to answer questions such as the following. (1) Eli Lilly believes that the year's demand for Prozac will be normally distributed, with $\mu = 60$ million d.o.t. (days of therapy) and $\sigma = 5$ million d.o.t. How many units should be produced this year if Lilly wants to have only a 1% chance of running out of Prozac? (2) Family income in Bloomington is normally distributed, with $\mu = \$30,000$ and $\sigma = \$8,000$. The poorest 10% of all families in Bloomington are eligible for federal aid. What should the aid cutoff be?

In the first example, we want the 99th percentile of Prozac demand. That is, we seek the number $\mathbf{X}$ such that there is only a 1% chance that demand will exceed $\mathbf{X}$ and a 99% chance

	E	F	G	H
7				
8				
9	P(Z<=-1)	0.158655	normsdist(-1)	
10	P(IQ<90)	0.252492	normdist(90,100,15,1)	
11	density for IQ=100	0.026596	normdist(100,100,15,0)	

FIGURE 9

FIGURE **10**

	D	E	F	G	H	I	J	K	L	M	N	O	P
1													
2		IQ	Density										
3		45	3.2018E-05										
4		50	0.00010282										
5		55	0.00029546										
6		60	0.00075973										
7		65	0.00174813										
8		70	0.0035994										
9		75	0.00663181										
10		80	0.010934										
11		85	0.01613138										
12		90	0.02129653										
13		95	0.02515888										
14		100	0.02659615										
15		105	0.02515888										
16		110	0.02129653										
17		115	0.01613138										
18		120	0.010934										
19		125	0.00663181										
20		130	0.0035994										
21		135	0.00174813										
22		140	0.00075973										
23		145	0.00029546										
24		150	0.00010282										
25		155	3.2018E-05										

Normal Density for IQ's

that it will be less than **X**. In the second example, we want the 10th percentile of family income in Bloomington. That is, we seek the number **X** such that there is only a 10% chance that family income will be less than **X** and a 90% chance that it will exceed **X**.

Suppose we want to find the pth percentile (expressed as a decimal) of a normal random variable **X** with mean μ and standard deviation σ. Simply enter the following formula into Excel:

$$=\text{NORMINV}(p,\mu,\sigma)$$

This will return the number x having the property that $P(\mathbf{X} \le x) = p$, as desired. We now can solve the two examples described above.

EXAMPLE 18   **Prozac Demand**

Eli Lilly believes that the year's demand for Prozac will be normally distributed, with $\mu = 60$ million d.o.t. (days of therapy) and $\sigma = 5$ million d.o.t. How many units should be produced this year if Lilly wants to have only a 1% chance of running out of Prozac?

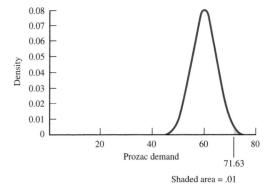

FIGURE 11
99th Percentile of
Prozac Demand

Solution   Letting $\mathbf{X}$ = annual demand for Prozac, we seek a value x such that $P(\mathbf{X} \geq x) = .01$ or $P(\mathbf{X} \leq x) = .99$. Thus, we seek the 99th percentile of Prozac demand, which we find (in millions) with the formula

$$=\text{NORMINV}(.99,60,5)$$

Excel returns 71.63, so Lilly must produce 71,630,000 d.o.t. This assumes, of course, that Lilly begins the year with no Prozac on hand. If the company had a beginning inventory of 10 million d.o.t., it would need to produce 61,630,000 d.o.t. during the current year. Figure 11 displays the 99th percentile of Prozac demand.

EXAMPLE 19   **Family Income**

Family income in Bloomington is normally distributed, with $\mu = \$30{,}000$ and $\sigma = \$8{,}000$. The poorest 10% of all families in Bloomington are eligible for federal aid. What should the aid cutoff be?

Solution   If $\mathbf{X}$ = income of a Bloomington family, we seek an x such that $P(\mathbf{X} \leq x) = .10$. Thus, we seek the 10th percentile of Bloomington family income, which we find with the statement

$$=\text{NORMINV}(.10,30000,8000)$$

Excel returns \$19,747.59. Thus, aid should be given to all families with incomes smaller than \$19,749.59. Figure 12 displays the 10th percentile of family income.

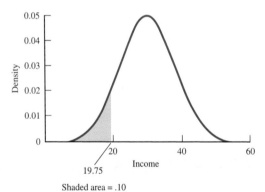

FIGURE 12
10th Percentile of
Family Income

EXAMPLE 20   **Stocking Chocolate Bars**

Daily demand for chocolate bars at the Gillis Grocery has a mean of 100 and a variance of 3,000 (chocolate bars)^2. At present, the store has 3,500 chocolate bars in stock. What is the probability that the store will run out of chocolate bars during the next 30 days? Also, how many should Gillis have on hand at the beginning of a 30-day period if the store wants to have only a 1% chance of running out during the 30-day period? Assume that the demands on different days are independent random variables.

Solution   Let

$$\mathbf{X}_i = \text{demand for chocolate bars on day } i \quad (i = 1, 2, \ldots, 30)$$
$$\mathbf{X} = \text{number of chocolate bars demanded in next 30 days}$$

Gillis will run out of stock during the next 30 days if $\mathbf{X} \geq 3{,}500$. The Central Limit Theorem implies that $\mathbf{X} = \mathbf{X}_1 + \mathbf{X}_2 + \cdots + \mathbf{X}_{30}$ can be closely approximated by a normal distribution $\mathbf{X}'$ with $E(\mathbf{X}') = 30(100) = 3{,}000$ and var $\mathbf{X}' = 30(3{,}000) = 90{,}000$ and

$\sigma_{\mathbf{X}'} = (90{,}000)^{1/2} = 300$. Then we approximate the probability that Gillis will run out of stock during the next 30 days by

$$P(\mathbf{X}' \geq 3{,}500) = P\left(\frac{\mathbf{X}' - 3000}{300} \geq \frac{3{,}500 - 3{,}000}{300}\right)$$
$$= P(\mathbf{Z} \geq 1.67) = 1 - P(\mathbf{Z} \leq 1.67)$$
$$= 1 - F(1.67) = 1 - .9525 = .0475$$

Let c = number of chocolate bars that should be stocked to have only a 1% chance of running out of chocolate bars within the next 30 days. We seek c satisfying $P(\mathbf{X}' \geq c) = .01$, or

$$P\left(\frac{\mathbf{X}' - 3{,}000}{300} \geq \frac{c - 3{,}000}{300}\right) = .01$$

This is equivalent to

$$P\left(\mathbf{Z} \geq \frac{c - 3{,}000}{300}\right) = .01$$

Since $F(2.33) = P(\mathbf{Z} \leq 2.33) = .99$,

$$\frac{c - 3{,}000}{300} = 2.33 \quad \text{or} \quad c = 3{,}699$$

Thus, if Gillis has 3,699 chocolate bars in stock, there is a 1% probability that the store will run out during the next 30 days. (We have defined running out of chocolate bars as having no chocolate bars left at the end of 30 days.)

Alternatively, we could find the probability that the demand is at least 3,500 with the Excel formula

$$=1 - \text{NORMDIST}(3500{,}3000{,}300{,}1)$$

This formula returns .0475.

We could also have used Excel to determine the level that must be stocked to have a 1% chance of running out as the 99th percentile of the demand distribution. Simply use the formula

$$= \text{NORMINV}(.99{,}3000{,}300)$$

This formula returns the value 3,699.

PROBLEMS

Group A

1 The daily demand for milk (in gallons) at Gillis Grocery is $N(1{,}000, 100)$. How many gallons must be in stock at the beginning of the day if Gillis is to have only a 5% chance of running out of milk by the end of the day?

2 Before burning out, a light bulb gives $\mathbf{X}$ hours of light, where $\mathbf{X}$ is $N(500, 400)$. If we have 3 bulbs, what is the probability that they will give a total of at least 1,460 hours of light?

Group B

3 The number of traffic accidents occurring in Bloomington in a single day has a mean and a variance of 3. What is the probability that during a given year (365-day period), there will be at least 1,000 traffic accidents in Bloomington?

4 Suppose that the number of ounces of soda put into a Pepsi can is normally distributed, with $\mu = 12.05$ oz and $\sigma = .03$ oz.

    a Legally, a can must contain at least 12 oz of soda. What fraction of cans will contain at least 12 oz of soda?

b What fraction of cans will contain under 11.9 oz of soda?

c What fraction of cans will contain between 12 and 12.08 oz of soda?

d 1% of all cans will contain more than _____ oz.

e 10% of all cans will contain less than _____ oz.

f Pepsi controls the mean content in a can by setting a timer. For what mean should the timer be set so that only 1 in 1,000 cans will be underfilled?

g Every day, Pepsi produces 10,000 cans. The government inspects 10 randomly chosen cans per day. If at least two are underfilled, Pepsi is fined $10,000. Given that $\mu = 12.05$ oz and $\sigma = .03$ oz, what is the chance that Pepsi will be fined on a given day?

5 Suppose the annual return on Disney stock follows a normal distribution, with mean .12 and standard deviation .30.

a What is the probability that Disney's value will decrease during a year?

b What is the probability that the return on Disney during a year will be at least 20%?

c What is the probability that the return on Disney during a year will be between -6% and 9%?

d There is a 5% chance that the return on Disney during a year will be greater than or equal to _____.

e There is a 1% chance that the return on Disney during a year will be less than _____.

f There is a 95% chance that the return on Disney during a year will be between _____ and _____.

6 The daily demand for six-packs of Coke at Mr. D's follows a normal distribution, with a mean of 120 and a standard deviation of 30. Every Monday, the delivery driver delivers Coke to Mr. D's. If the store wants to have only a 1% chance of running out of Coke by the end of the week, how many six-packs should be ordered for the week? (Assume that orders can be placed Sunday at midnight.)

7 The Coke factory fills bottles of soda by setting a timer on a filling machine. It has been observed that the number of ounces the machine puts in a bottle has a standard deviation of .05 oz. If 99.9% of all bottles are to have at least 16 oz of soda, to what amount should the average amount be set? (*Hint:* Use the Excel Goal Seek feature.)

8 We assemble a large part by joining two smaller parts together. In the past, the smaller parts we have produced have had a mean length of 1″ and a standard deviation of .01″. Assume that the lengths of the smaller parts are normally distributed and are independent.

a What fraction of the larger parts are more than 2.05″ in diameter?

b What fraction of the larger parts are between 1.96″ and 2.02″ in diameter?

9 Weekly Ford sales follow a normal distribution, with a mean of 50,000 cars and a standard deviation of 14,000 cars.

a There is a 1% chance that Ford will sell more than _____ cars during the next year.

b The chance that Ford will sell between 2.4 and 2.7 million cars during the next year is _____.

10 Warren Dinner has invested in nine different investments. The profits earned on the different investments are independent. The return on each investment follows a normal distribution, with a mean of $500 and a standard deviation of $100.

a There is a 1% chance that the total return on the nine investment is less than _____.

b The probability that Warren's total return is between $4,000 and $5,200 is _____.

1.8 *z*-Transforms

Consider a discrete random variable **X** whose only possible values are nonnegative integers. For $n = 0, 1, 2, \ldots$, let $P(\mathbf{X} = n) = a_n$. We define (for $|z| \le 1$) the *z*-**transform of X** (call it $p_{\mathbf{X}}^T(z)$) to be

$$E(z^{\mathbf{X}}) = \sum_{n=0}^{n=\infty} a_n z^n$$

To see why *z*-transforms are useful, note that

$$\left[\frac{dp_{\mathbf{X}}^T(z)}{dz}\right]_{z=1} = \left[\sum_{n=1}^{n=\infty} n z^{n-1} a_n\right]_{z=1} = E(\mathbf{X})$$

Also note that

$$\left[\frac{d^2 p_{\mathbf{X}}^T(z)}{dz^2}\right]_{z=1} = \left[\sum_{n=1}^{n=\infty} n(n-1) z^{n-2} a_n\right]_{z=1} = E(\mathbf{X}^2) - E(\mathbf{X})$$

This implies that we can find the mean, the second moment ($E(X^2)$), and variance of X from the following relationships:

$$E(X) = \left[\frac{dp_X^T(z)}{dz}\right]_{z=1} \tag{24}$$

$$E(X^2) = \left[\frac{d^2 p_X^T(z)}{dz^2}\right]_{z=1} + \left[\frac{dp_X^T(z)}{dz}\right]_{z=1} \tag{25}$$

$$\text{var } X = \left[\frac{d^2 p_X^T(z)}{dz^2}\right]_{z=1} + \left[\frac{dp_X^T(z)}{dz}\right]_{z=1} - \left(\left[\frac{dp_X^T(z)}{dz}\right]_{z=1}\right)^2 \tag{26}$$

The following examples illustrate the power of z-transforms.

EXAMPLE 21 z-Transform for the Binomial Random Variable

Suppose we toss a coin n times, and the probability of obtaining heads each time is p. Let $q = 1 - p$. If successive coin tosses are independent events, then the mass function describing the random variable X = number of heads is the well-known **binomial** random variable defined by

$$P(X = j) = \frac{n!}{j!(n-j)!} p^j (q)^{n-j}, j = 0, 1, 2, \ldots, n$$

The z-transform for the random variable X is given by

$$p_X^T(z) = \sum_{j=0}^{j=n} \frac{n!}{j!(n-j)!} p^j (q)^{n-j} z^j = \sum_{j=0}^{j=n} \frac{n!}{j!(n-j)!} (pz)^j (q)^{n-j} = (pz + q)^n$$

We can now use the z-transform to determine the mean and variance of the binomial random variable. Note that

$$\frac{dp_X^T(z)}{dz} = np(pz + q)^{n-1} \quad \text{and} \quad \frac{d^2 p_X^T(z)}{dz^2} = n(n-1)p^2(pz + q)^{n-2}$$

For $z = 1$, we find

$$\frac{dp_X^T(z)}{dz} = np \quad \text{and} \quad \frac{d^2 p_X^T(z)}{dz^2} = n(n-1)p^2$$

Then from (24), we find $E(X) = np$, and from (26), we find that var $X = n(n-1)p^2 + np - (np)^2 = npq$.

EXAMPLE 22 z-Transform for a Geometric Random Variable

Let the random variable X be defined as the number of coin tosses needed to obtain the first heads, given that successive tosses are independent, the probability that each toss is heads is given by p, and the probability that each coin is tails is given by $q = 1 - p$. Then X follows a **geometric random variable**, where $P(X = j) = pq^{j-1}$ ($j = 1, 2, \ldots, n$). Then $p_X^T(z) = \sum_{j=1}^{j=\infty} pq^{j-1} z^j$. For $x < 1$, we know that $a + ax + ax^2 + \cdots = \frac{a}{1-x}$. Therefore, $p_X^T(z) = \frac{pz}{1-qz}$. We find that

$$\frac{dp_X^T(z)}{dz} = \frac{p}{(1-qz)^2} \quad \text{and} \quad \frac{d^2 p_X^T(z)}{dz^2} = \frac{2\,pq}{(1-qz)^3}$$

Letting $z = 1$ (Equation (24)) tells us that

$$E(\mathbf{X}) = \frac{p}{p^2} = \frac{1}{p} \quad \text{and} \quad \text{var } \mathbf{X} = \frac{2\,pq}{p^3} + \frac{1}{p} - \frac{1}{p^2} = \frac{q}{p^2}$$

Suppose $\mathbf{X}_1, \mathbf{X}_2, \ldots, \mathbf{X}_n$ are independent random variables. Let $\mathbf{S} = \mathbf{X}_1 + \mathbf{X}_2 + \cdots + \mathbf{X}_n$. Then it is easy to prove (see Problem 2) that

$$p_{\mathbf{S}}^T(z) = p_{\mathbf{X}_1}^T(z) \cdots p_{\mathbf{X}_n}^T(z) \tag{27}$$

To see the usefulness of this result, reconsider Example 21. Let $\mathbf{X}_i =$ number of heads on the ith toss of a coin. Then number of heads on n tosses of a coin is given by $\mathbf{X}_1 + \mathbf{X}_2 + \cdots + \mathbf{X}_n$. For each $\mathbf{X}_i$, we have that $p_{\mathbf{X}_i}^T(z) = pz + q$. Then from (27), we find that $p_{\mathbf{X}}^T(z) = (pz + q)^n$. Of course, this agrees with the z-transform we obtained in Example 21.

PROBLEMS

Group A

1 For a given μ, the **Poisson random variable** has the mass function $P(\mathbf{X} = k) = e^{-\mu} \dfrac{\mu^n}{n!}$ ($k = 0, 1, 2, \ldots$). Find the mean and variance of a Poisson random variable.

Group B

2 Prove Equation (27).

3 Suppose we toss a coin. Successive coin tosses are independent and yield heads with probability p. The negative binomial random variable with parameter k assumes a value n if it takes n failures until the kth success occurs. Use z-transforms to determine the probability mass function for the negative binomial random variable.

Hint: The number of ways of making k choices from the numbers $0, 1, \ldots, n$ add up to n is given by $\dfrac{(n + k - 1)!}{n!(k - 1)!}$.

SUMMARY   Rules for Finding the Derivative of a Function

Function	Derivative of Function
a	0
x	1
$af(x)$	$af'(x)$
$f(x) + g(x)$	$f'(x) + g'(x)$
x^n	nx^{n-1}
e^x	e^x
a^x	$a^x \ln a$
$\ln x$	$\dfrac{1}{x}$
$[f(x)]^n$	$n[f(x)]^{n-1}f'(x)$
$e^{f(x)}$	$e^{f(x)}f'(x)$
$a^{f(x)}$	$a^{f(x)}f'(x) \ln a$
$\ln f(x)$	$\dfrac{f'(x)}{f(x)}$
$f(x)g(x)$	$f(x)g'(x) + f'(x)g(x)$
$\dfrac{f(x)}{g(x)}$	$\dfrac{g(x)f'(x) - f(x)g'(x)}{g(x)^2}$

Formulas for Determining Indefinite Integrals

$$\int (1)\, dx = x + C$$

$$\int af(x)\, dx = a \int f(x)\, dx \qquad \text{(a is any constant)}$$

$$\int [f(x) + g(x)]\, dx = \int f(x)\, dx + \int g(x)\, dx$$

$$\int x^n\, dx = \frac{x^{n+1}}{n+1} + C \qquad (n \neq -1)$$

$$\int x^{-1}\, dx = \ln x + C$$

$$\int e^x\, dx = e^x + C$$

$$\int a^x\, dx = \frac{a^x}{\ln a} + C \qquad (a > 0,\, a \neq 1)$$

$$\int [f(x)]^n f'(x)\, dx = \frac{[f(x)]^{n+1}}{n+1} + C \qquad (n \neq -1)$$

$$\int f(x)^{-1} f'(x)\, dx = \ln f(x) + C$$

For two functions $u(x)$ and $v(x)$,

$$\int u(x)v'(x)\, dx = u(x)v(x) - \int v(x)u'(x)\, dx \qquad \text{(Integration by parts)}$$

$$\int e^{f(x)} f'(x)\, dx = e^{f(x)} + C$$

$$\int a^{f(x)} f'(x)\, dx = \frac{a^{f(x)}}{\ln a} + C \qquad (a > 0,\, a \neq 1)$$

Leibniz's Rule for Differentiating an Integral

$$\text{If} \qquad F(y) = \int_{g(y)}^{h(y)} f(x, y)\, dx, \qquad \text{then}$$

$$F'(y) = h'(y)f(h(y),\, y) - g'(y)f(g(y),\, y) + \int_{g(y)}^{h(y)} \frac{\partial f(x, y)}{\partial y}\, dx$$

Probability

Basic Rules

Rule 1 For any event E, $P(E) \geq 0$.

Rule 2 If $E = S$ (that is, if E contains all points in the sample space), then $P(E) = 1$.

Rule 3 If $E_1, E_2, \ldots, E_n$ is a mutually exclusive collection of events, then

$$P(E_1 \cup E_2 \cup \cdots \cup E_n) = \sum_{k=1}^{k=n} P(E_k)$$

Rule 4 $P(\bar{E}) = 1 - P(E)$.

Formula for Conditional Probability

$$P(E_2|E_1) = \frac{P(E_1 \cap E_2)}{P(E_1)} \tag{2}$$

Bayes' Rule

$$P(S_i|O_j) = \frac{P(O_j|S_i)P(S_i)}{\sum\limits_{k=1}^{k=n} P(O_j|S_k)P(S_k)} \tag{8}$$

Random Variables, Mean, Variance, and Covariance

Mean of a Discrete Random Variable

$$E(\mathbf{X}) = \sum_{\text{all } k} x_k P(\mathbf{X} = x_k) \tag{9}$$

Mean of a Continuous Random Variable

$$E(\mathbf{X}) = \int_{-\infty}^{\infty} xf(x)\, dx \tag{10}$$

Variance of a Discrete Random Variable

$$\text{var } \mathbf{X} = \sum_{\text{all } k} [x_k - E(\mathbf{X})]^2\, P(\mathbf{X} = x_k) \tag{11}$$

Variance of a Continuous Random Variable

$$\text{var } \mathbf{X} = \int_{-\infty}^{\infty} [x - E(\mathbf{X})]^2 f(x)\, dx \tag{12}$$

Covariance of Two Random Variables

$$\text{cov}(\mathbf{X}, \mathbf{Y}) = E\{[\mathbf{X} - E(\mathbf{X})][\mathbf{Y} - E(\mathbf{Y})]\} \tag{14}$$

Mean, Variance, and Covariance for Sums of Random Variables

$$E(c\mathbf{X}_1) = cE(\mathbf{X}_1) \tag{15}$$
$$E(\mathbf{X}_1 + c) = E(\mathbf{X}_1) + c \tag{16}$$
$$E(\mathbf{X}_1 + \mathbf{X}_2) = E(\mathbf{X}_1) + E(\mathbf{X}_2) \tag{17}$$
$$\text{var } c\mathbf{X}_1 = c^2 \text{var } \mathbf{X}_1 \tag{18}$$
$$\text{var}(\mathbf{X}_1 + c) = \text{var } \mathbf{X}_1 \tag{19}$$

If $\mathbf{X}_1$ and $\mathbf{X}_2$ are independent random variables,

$$\text{var}(\mathbf{X}_1 + \mathbf{X}_2) = \text{var } \mathbf{X}_1 + \text{var } \mathbf{X}_2 \tag{20}$$

In general,

$$\mathrm{var}(\mathbf{X}_1 + \mathbf{X}_2) = \mathrm{var}\ \mathbf{X}_1 + \mathrm{var}\ \mathbf{X}_2 + 2\mathrm{cov}(\mathbf{X}_1, \mathbf{X}_2) \qquad (21)$$

For random variables $\mathbf{X}_1, \mathbf{X}_2, \ldots, \mathbf{X}_n,$

$$\mathrm{var}(\mathbf{X}_1 + \mathbf{X}_2 + \cdots + \mathbf{X}_n) = \mathrm{var}\ \mathbf{X}_1 + \mathrm{var}\ \mathbf{X}_2 + \cdots \qquad (22)$$

$$+ \mathrm{var}\ \mathbf{X}_n + \sum_{i \neq j} \mathrm{cov}(\mathbf{X}_i, \mathbf{X}_j) \qquad (23)$$

$$\mathrm{cov}(a\mathbf{X}_1, b\mathbf{X}_2) = ab\ \mathrm{cov}(\mathbf{X}_1, \mathbf{X}_2) \qquad (24)$$

Useful Properties of the Normal Distribution

Property 1 If $\mathbf{X}$ is $N(\mu, \sigma^2)$, then $c\mathbf{X}$ is $N(c\mu, c^2\sigma^2)$.

Property 2 If $\mathbf{X}$ is $N(\mu, \sigma^2)$, then $\mathbf{X} + c$ (for any constant c) is $N(\mu + c, \sigma^2)$.

Property 3 If $\mathbf{X}_1$ is $N(\mu_1, \sigma_1^2)$, $\mathbf{X}_2$ is $N(\mu_2, \sigma_2^2)$, and $\mathbf{X}_1$ and $\mathbf{X}_2$ are independent, then $\mathbf{X}_1 + \mathbf{X}_2$ is $N(\mu_1 + \mu_2, \sigma_1^2 + \sigma_2^2)$.

If $\mathbf{X}$ is $N(\mu, \sigma^2)$, then

$$P(a \leq \mathbf{X} \leq b) = F\left(\frac{b - \mu}{\sigma}\right) - F\left(\frac{a - \mu}{\sigma}\right)$$

where $F(x) = P(\mathbf{Z} \leq x)$ and $\mathbf{Z}$ is $N(0, 1)$.

z-Transforms

We define (for $|z| \leq 1$) the **z-transform of X** (call it $p_\mathbf{X}^T(z)$) to be

$$E(z^\mathbf{X}) = \sum_{n=0}^{n=\infty} a_n z^n$$

We can find the mean, the second moment ($E(\mathbf{X}^2)$), and variance of $\mathbf{X}$ from the following relationships:

$$E(\mathbf{X}) = \left[\frac{dp_\mathbf{X}^T(z)}{dz}\right]_{z=1} \qquad (24)$$

$$E(\mathbf{X}^2) = \left[\frac{d^2 p_\mathbf{X}^T(z)}{dz^2}\right]_{z=1} + \left[\frac{dp_\mathbf{X}^T(z)}{dz}\right]_{z=1} \qquad (25)$$

$$\mathrm{var}\ \mathbf{X} = \left[\frac{d^2 p_\mathbf{X}^T(z)}{dz^2}\right]_{z=1} + \left[\frac{dp_\mathbf{X}^T(z)}{dz}\right]_{z=1} - \left(\left[\frac{dp_\mathbf{X}^T(z)}{dz}\right]_{z=1}\right)^2 \qquad (26)$$

REVIEW PROBLEMS

Group A

1 Let $f(x) = xe^{-x}$.
  a Find $f'(x)$ and $f''(x)$.
  b For what values of x is $f(x)$ increasing? Decreasing?
  c Find the first-order Taylor series expansion for $f(x)$ about $x = 1$.

2 Let $f(x_1, x_2) = x_1 \ln(x_2 - x_1)$. Determine all first-order and second-order partial derivatives.

3 Some t years from now, air conditioners are sold at a rate of t per year. How many air conditioners will be sold during the next five years?

4 Let **X** be a continuous random variable with density function

$$f(x) = \begin{cases} \frac{4-x}{k} & \text{if } 0 \le x \le 4 \\ 0 & \text{otherwise} \end{cases}$$

a What is k?

b Find the cdf for **X**.

c Find $E(\mathbf{X})$ and var **X**.

d Find $P(2 \le \mathbf{X} \le 5)$.

5 Let $\mathbf{X}_i$ be the price (in dollars) of stock i one year from now. $\mathbf{X}_1$ is $N(15, 100)$ and $\mathbf{X}_2$ is $N(20, 2025)$. Today I buy three shares of stock 1 for $12/share and two shares of stock 2 for $17/share. Assume that $\mathbf{X}_1$ and $\mathbf{X}_2$ are independent random variables.

a Find the mean and variance of the value of my stocks one year from now.

b What is the probability that one year from now I will have earned at least a 30% return on my investment?

c If $\mathbf{X}_1$ and $\mathbf{X}_2$ were not independent, why would it be difficult to answer parts (a) and (b)?

Group B

6 An airplane has four engines. On a flight from New York to Paris, each engine has a 0.001 chance of failing. The plane will crash if at any time two or fewer engines are working properly. Assume that the failures of different engines are independent.

a What is the probability that the plane will crash?

b Given that engine 1 will not fail during the flight, what is the probability that the plane will crash?

c Given that engine 1 will fail during the flight, what is the probability that the plane will not crash?

7 Suppose that each engine can be inspected before the flight. After inspection, each engine is labeled as being in either good or bad condition. You are given that

P(inspection says engine is in good condition | engine will fail) = .001

P(inspection says engine is in bad condition | engine will fail) = .999

P(inspection says engine is in good condition | engine will not fail) = .995

P(inspection says engine is in bad condition | engine will not fail) = .005

a If the inspection indicates the engine is in bad condition, what is the probability that the engine will fail on the flight?

b If an inspector randomly inspects an engine (that is, with probability .001 she chooses an engine that is about to fail, and with probability .999 she chooses an engine that is not about to fail), what is the probability that she will make an error in her evaluation of the engine?

REFERENCES

Allen, R. *Mathematical Analysis for Economists.* New York: St. Martin's Press, 1938. An advanced calculus review.

Byrkit, D., and S. Shamma. *Calculus for Business and Economics.* New York: Van Nostrand, 1981. A review of calculus at a beginning level.

Chiang, A. *Fundamental Methods of Mathematical Economics.* New York: McGraw-Hill, 1978. An advanced calculus review.

Harnett, D. *Statistical Methods.* Reading, Mass.: Addison-Wesley, 1982. A review of probability.

Winkler, R., and W. Hays. *Statistics: Probability, Inference, and Decision.* Chicago: Holt, Rinehart &Winston, 1971. A review of probability.

2

Decision Making under Uncertainty

We have all had to make important decisions where we were uncertain about factors that were relevant to the decisions. In this chapter, we study situations in which decisions are made in an uncertain environment.

The following model encompasses several aspects of making a decision in the absence of certainty. The decision maker first chooses an action a_i from a set $A = \{a_1, a_2, \ldots, a_k\}$ of available actions. Then the state of the world is observed; with probability p_j, the state of the world is observed to be $s_j \in S = \{s_1, s_2, \ldots, s_n\}$. If action a_i is chosen and the state of the world is s_j, the decision maker receives a reward r_{ij}. We refer to this model as the *state-of-the-world decision-making model*.

This chapter presents the basic theory of decision making under uncertainty: the widely used Von Neumann–Morgenstern utility model, and the use of decision trees for making decisions at different points in time. We close by looking at decision making with multiple objectives.

2.1 Decision Criteria

In this section, we consider four decision criteria that can be used to make decisions under uncertainty.

EXAMPLE 1   Newspaper Vendor

News vendor Phyllis Pauley sells newspapers at the corner of Kirkwood Avenue and Indiana Street, and each day she must determine how many newspapers to order. Phyllis pays the company 20¢ for each paper and sells the papers for 25¢ each. Newspapers that are unsold at the end of the day are worthless. Phyllis knows that each day she can sell between 6 and 10 papers, with each possibility being equally likely. Show how this problem fits into the state-of-the-world model.

Solution   In this example, the members of $S = \{6, 7, 8, 9, 10\}$ are the possible values of the daily demand for newspapers. We are given that $p_6 = p_7 = p_8 = p_9 = p_{10} = \frac{1}{5}$. Phyllis must choose an action (the number of papers to order each day) from $A = \{6, 7, 8, 9, 10\}$.

If Phyllis purchases i papers and j papers are demanded, then i papers are purchased at a cost of $20i¢$, and $\min(i, j)$ papers are sold for 25¢ each.[†] Thus, if Phyllis purchases i papers and j papers are demanded, she earns a net profit of r_{ij}, where

$$r_{ij} = 25i - 20i = 5i \qquad (i \leq j)$$
$$r_{ij} = 25j - 20i \qquad (i \geq j)$$

The values of r_{ij} are tabulated in Table 1.

[†]$\min(i, j)$ is the smaller of i and j.

TABLE **1**
Rewards for News Vendor

Papers Ordered	Papers Demanded				
	6	7	8	9	10
6	30¢	30¢	30¢	30¢	30¢
7	10¢	35¢	35¢	35¢	35¢
8	−10¢	15¢	40¢	40¢	40¢
9	−30¢	−5¢	20¢	45¢	45¢
10	−50¢	−25¢	0¢	25¢	50¢

Dominated Actions

Why did we not consider the possibility that Phyllis would order 1, 2, 3, 4, 5, or more than 10 papers? Answering this question involves the idea of a dominated action.

DEFINITION ■ An action a_i is **dominated** by an action $a_{i'}$ if for all $s_j \in S$, $r_{ij} \leq r_{i'j}$, and for some state $s_{j'}$, $r_{ij'} < r_{i'j'}$. ■

If action a_i is dominated, then in no state of the world is a_i better than $a_{i'}$, and in at least one state of the world a_i is inferior to $a_{i'}$. Thus, if action a_i is dominated, there is no reason to choose a_i ($a_{i'}$ would be a better choice).

If Phyllis orders i papers ($i = 1, 2, 3, 4, 5$), she will earn (for all states of the world) a profit of $5i¢$. From the table of rewards, we see that, for $i = 1, 2, 3, 4, 5$, ordering 6 papers dominates ordering i papers ($j' = 6, 7, 8, 9,$ or 10 will do). Similarly, the reader should check that ordering i papers ($i > 11$) is dominated by ordering 10 papers (see Problem 3 at the end of this section). A quick check shows that none of the actions in $A = \{6, 7, 8, 9, 10\}$ are dominated. Thus, Phyllis should indeed choose her action from $A = \{6, 7, 8, 9, 10\}$.

We now discuss four criteria that can be used to choose an action.

The Maximin Criterion

For each action, determine the worst outcome (smallest reward). The maximin criterion chooses the action with the "best" worst outcome.

DEFINITION ■ The **maximin criterion** chooses the action a_i with the largest value of $\min_{j \in S} r_{ij}$. ■

For Example 1, we obtain the results in Table 2. Thus, the maximin criterion recommends ordering 6 papers. This ensures that Phyllis will, no matter what the state of the world, earn a profit of at least 30¢. The maximin criterion is concerned with making the worst possible outcome as pleasant as possible. Unfortunately, choosing a decision to mitigate the worst case may prevent the decision maker from taking advantage of good fortune. For example, if Phyllis follows the maximin criterion, she will never make less than 30¢, but she will never make more than 30¢.

TABLE 2
Computation of Maximin Decision for News Vendor

Papers Ordered	Worst State of the World	Reward in Worst State of the World
6	6, 7, 8, 9, 10	30¢
7	6	10¢
8	6	−10¢
9	6	−30¢
10	6	−50¢

The Maximax Criterion

For each action, determine the best outcome (largest reward). The maximax criterion chooses the action with the "best" best outcome.

DEFINITION ■ The **maximax criterion** chooses the action a_i with the largest value of $\max_{j \in S} r_{ij}$. ■

For Example 1, we obtain the results in Table 3. Thus, the maximax criterion would recommend ordering 10 papers. In the best state (when 10 papers are demanded), this yields a profit of 50¢. Of course, making a decision according to the maximax criterion leaves Phyllis open to the disastrous possibility that only 6 papers will be demanded, in which case she loses 50¢.

Minimax Regret

The minimax regret criterion (developed by L. J. Savage) uses the concept of opportunity cost to arrive at a decision. For each possible state of the world s_j, find an action $i^*(j)$ that maximizes r_{ij}. That is, $i^*(j)$ is the best possible action to choose if the state of the world is actually s_j. Then for any action a_i and state s_j, the opportunity loss or regret for a_i in s_j is $r_{i^*(j),j} - r_{ij}$. For example, if $j = 7$ papers are demanded, the best decision is to order $i^*(7) = 7$ papers, yielding a profit of $r_{77} = 7(25) - 7(20) = 35$¢. Suppose we chose to order $i = 6$ papers. Since $r_{67} = 6(25) - 6(20) = 30$¢, the opportunity loss or regret for $i = 6$ and $j = 7$ is $35 - 30 = 5$¢. Thus, if we order 6 papers and 7 papers are demanded, in hindsight we realize that by making the optimal choice (ordering 7 papers) for the actual state of the world (7 papers demanded), we would have done 5¢ better than we did by ordering 6 papers. Table 4 shows the opportunity cost or regret matrix for Example 1.

TABLE 3
Computation of Maximax Decision for News Vendor

Papers Ordered	State Yielding Best Outcome	Best Outcome
6	6, 7, 8, 9, 10	30¢
7	7, 8, 9, 10	35¢
8	8, 9, 10	40¢
9	9, 10	45¢
10	10	50¢

The minimax regret criterion chooses an action by applying the minimax criterion to the regret matrix. In other words, the minimax regret criterion attempts to avoid disappointment over what might have been. From the regret matrix in Table 4, we obtain the minimax regret decision in Table 5. Thus, the minimax regret criterion recommends ordering 6 or 7 papers.

The Expected Value Criterion

The expected value criterion chooses the action that yields the largest expected reward. For Example 1, the expected value criterion would recommend ordering 6 or 7 papers (see Table 6).

The decision-making criteria discussed in this section may seem reasonable, but many people make decisions without using any of them. A more comprehensive model of individual decision making, the Von Neumann–Morgenstern utility model, is discussed in Section 2.2.

TABLE 4
Regret Matrix for News Vendor

| Papers Ordered | Papers Demanded | | | | |
	6	7	8	9	10
6	$30 - 30 = 0¢$	$35 - 30 = 5¢$	$40 - 30 = 10¢$	$45 - 30 = 15¢$	$50 - 30 = 20¢$
7	$30 - 10 = 20¢$	$35 - 35 = 0¢$	$40 - 35 = 5¢$	$45 - 35 = 10¢$	$50 - 35 = 15¢$
8	$30 + 10 = 40¢$	$35 - 15 = 20¢$	$40 - 40 = 0¢$	$45 - 40 = 5¢$	$50 - 40 = 10¢$
9	$30 + 30 = 60¢$	$35 + 5 = 40¢$	$40 - 20 = 20¢$	$45 - 45 = 0¢$	$50 - 45 = 5¢$
10	$30 + 50 = 80¢$	$35 + 25 = 60¢$	$40 - 0 = 40¢$	$45 - 25 = 20¢$	$50 - 50 = 0¢$

TABLE 5
Computation of Minimax Regret Decision for News Vendor

Papers Ordered	Maximum Regret
6	20¢
7	20¢
8	40¢
9	60¢
10	80¢

TABLE 6
Computation of Expected Value Decision for News Vendor

Papers Ordered	Expected Reward
6	$\frac{1}{5}(30 + 30 + 30 + 30 + 30) = 30¢$
7	$\frac{1}{5}(10 + 35 + 35 + 35 + 35) = 30¢$
8	$\frac{1}{5}(-10 + 15 + 40 + 40 + 40) = 25¢$
9	$\frac{1}{5}(-30 - 5 + 20 + 45 + 45) = 15¢$
10	$\frac{1}{5}(-50 - 25 + 0 + 25 + 50) = 0¢$

PROBLEMS

Group A

1 Pizza King and Noble Greek are two competing restaurants. Each must determine simultaneously whether to undertake small, medium, or large advertising campaigns. Pizza King believes that it is equally likely that Noble Greek will undertake a small, a medium, or a large advertising campaign. Given the actions chosen by each restaurant, Pizza King's profits are as shown in Table 7. For the maximin, maximax, and minimax regret criteria, determine Pizza King's choice of advertising campaign.

TABLE 7

Pizza King Chooses	Noble Greek Chooses		
	Small	Medium	Large
Small	$6,000	$5,000	$2,000
Medium	$5,000	$6,000	$1,000
Large	$9,000	$6,000	$0

2 Sodaco is considering producing a new product: Chocovan soda. Sodaco estimates that the annual demand for Chocovan, **D** (in thousands of cases), has the following mass function: $P(\mathbf{D} = 30) = .30$, $P(\mathbf{D} = 50) = .40$, $P(\mathbf{D} = 80) = .30$. Each case of Chocovan sells for $5 and incurs a variable cost of $3. It costs $800,000 to build a plant to produce Chocovan. Assume that if $1 is received every year (forever), this is equivalent to receiving $10 at the present time. Considering the reward for each action and state of the world to be in terms of net present value, use each decision criterion of this section to determine whether Sodaco should build the plant.

3 For Example 1, show that ordering 11 or more papers is dominated by ordering 10 papers.

Group B

4 Suppose that Pizza King and Noble Greek stop advertising but must determine the price they will charge for each pizza sold. Pizza King believes that Noble Greek's price is a random variable **D** having the following mass function: $P(\mathbf{D} = \$6) = .25$, $P(\mathbf{D} = \$8) = .50$, $P(\mathbf{D} = \$10) = .25$. If Pizza King charges a price p_1 and Noble Greek charges a price p_2, Pizza King will sell $100 + 25(p_2 - p_1)$ pizzas. It costs Pizza King $4 to make a pizza. Pizza King is considering charging $5, $6, $7, $8, or $9 for a pizza. Use each decision criterion of this section to determine the price that Pizza King should charge.

5 Alden Construction is bidding against Forbes Construction for a project. Alden believes that Forbes's bid is a random variable **B** with the following mass function: $P(\mathbf{B} = \$6,000) = .40$, $P(\mathbf{B} = \$8,000) = .30$, $P(\mathbf{B} = \$11,000) = .30$. It will cost Alden $6,000 to complete the project. Use each of the decision criteria of this section to determine Alden's bid. Assume that in case of a tie, Alden wins the bidding. (*Hint:* Let p = Alden's bid. For $p \le 6,000$, $6,000 < p \le 8,000$, $8,000 < p \le 11,000$, and $p > 11,000$, determine Alden's profit in terms of Alden's bid and Forbes's bid.)

2.2 Utility Theory

We now show how the Von Neumann–Morgenstern concept of a utility function can be used as an aid to decision making under uncertainty.

Consider a situation in which a person will receive, for $i = 1, 2, \ldots, n$, a reward r_i with probability p_i. This is denoted as the **lottery** $(p_1, r_1; p_2, r_2; \ldots; p_n, r_n)$. A lottery is often represented by a tree in which each branch stands for a possible outcome of the lottery, and the number on each branch represents the probability that the outcome will occur. Thus, the lottery $(\frac{1}{4}, \$500; \frac{3}{4}, \$0)$ could be denoted by

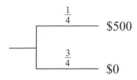

Suppose we are asked to choose between two lotteries (L_1 and L_2). With certainty, lottery L_1 yields $10,000:

$$L_1 \underline{\quad\quad 1 \quad\quad} \$10,000$$

Lottery L_2 consists of tossing a coin. If heads comes up, we receive \$30,000, and if tails comes up, we receive \$0:

$$L_2 \begin{cases} \xrightarrow{\frac{1}{2}} \$30{,}000 \\ \xrightarrow{\frac{1}{2}} \$0 \end{cases}$$

L_1 yields an expected reward of \$10,000, and L_2 yields an expected reward of $(\frac{1}{2})(30{,}000) + (\frac{1}{2})(0) = \$15{,}000$. Although L_2 has a larger expected value than L_1, most people prefer L_1 to L_2 because L_1 offers the certainty of a relatively large payoff, whereas L_2 yields a substantial $(\frac{1}{2})$ chance of earning a reward of \$0. In short, most people prefer L_1 to L_2 because L_1 involves less risk (or uncertainty) than L_2.

Our goal is to determine a method that a person can use to choose between lotteries. Suppose he or she must choose to play L_1 or L_2 but not both. We write $L_1 p L_2$ if the person prefers L_1. We write $L_1 i L_2$ if he or she is indifferent between choosing L_1 and L_2. If $L_1 i L_2$, we say that L_1 and L_2 are **equivalent lotteries**. Finally, we write $L_2 p L_1$ if the decision maker prefers L_2.

Suppose we ask a decision maker to rank the following lotteries:

$$L_1 \xrightarrow{\quad 1 \quad} \$10{,}000 \qquad L_2 \begin{cases} \xrightarrow{\ .50\ } \$30{,}000 \\ \xrightarrow{\ .50\ } \$0 \end{cases}$$

$$L_3 \xrightarrow{\quad 1 \quad} \$0 \qquad L_4 \begin{cases} \xrightarrow{\ .02\ } -\$10{,}000 \\ \xrightarrow{\ .98\ } \$500 \end{cases}$$

The Von Neumann–Morgenstern approach to ranking these lotteries is as follows. Begin by identifying the most favorable (\$30,000) and the least favorable ($-$\$10,000) outcomes that can occur. For all other possible outcomes ($r_1 = \$10{,}000$, $r_2 = \$500$, and $r_3 = \$0$), the decision maker is asked to determine a probability p_i such that he or she is indifferent between two lotteries:

$$\xrightarrow{\quad 1 \quad} r_i \qquad \text{and} \qquad \begin{cases} \xrightarrow{\ p_i\ } \$30{,}000 \\ \xrightarrow{\ 1 - p_i\ } -\$10{,}000 \end{cases}$$

Suppose that for $r_1 = \$10{,}000$, the decision maker is indifferent between

$$\xrightarrow{\quad 1 \quad} \$10{,}000 \qquad \text{and} \qquad \begin{cases} \xrightarrow{\ .90\ } \$30{,}000 \\ \xrightarrow{\ .10\ } -\$10{,}000 \end{cases} \tag{1}$$

and for $r_2 = \$500$, indifferent between

$$\xrightarrow{\quad 1 \quad} \$500 \qquad \text{and} \qquad \begin{cases} \xrightarrow{\ .62\ } \$30{,}000 \\ \xrightarrow{\ .38\ } -\$10{,}000 \end{cases} \tag{2}$$

and for $r_3 = \$0$, indifferent between

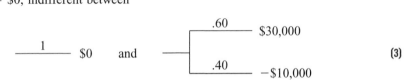

$$\text{(3)}$$

Using (1)–(3), the decision maker can construct lotteries L_1', L_2', L_3', and L_4' such that $L_i'iL_i$ and each L_i' involves only the best (\$30,000) and the worst (−\$10,000) possible outcomes. Thus, from (1), we find that L_1iL_1', where

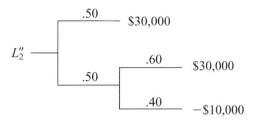

From (3), we find that L_2iL_2'', where

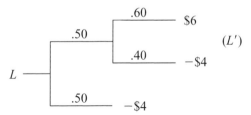

L_2'' is a compound lottery in which with probability .50 we receive \$30,000 and with probability .50 we play a lottery yielding a .60 chance at \$30,000 and a .40 chance at −\$10,000. More formally, a lottery L is a **compound lottery** if for some i, there is a probability p_i that the decision maker's reward is to play another lottery L'. The following is an example of a compound lottery:

Thus, with probability .50, L yields a reward of −\$4, and with probability .50, L causes us to play L'. If a lottery is not a compound lottery, it is a **simple lottery.**

Returning to our discussion of L_2'', we observe that L_2'' is a lottery that yields a .50 + .50(.60) = .80 chance at \$30,000 and a .40(.50) = .20 chance at −\$10,000. Thus, $L_2iL_2''iL_2'$, where

Similarly, using (3), we find that L_3iL_3', where

Using (2), we find that the decision maker is indifferent between L_4 and L_4'', where

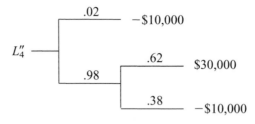

In actuality, however, L_4'' yields a $.98(.62) = .6076$ chance at \$30,000 and a $.02 + .38(.98) = .3924$ chance at $-\$10,000$. Thus, $L_4 i L_4'' i L_4'$, where

```
           .6076
              ──── $30,000
    L′₄ ──┤
           .3924
              ──── −$10,000
```

Since $L_i i L_i'$, we may rank L_1, L_2, L_3, and L_4 by ranking L_1', L_2', L_3', and L_4'. Consider two lotteries whose only possible outcomes are \$30,000 (the most favorable outcome) and $-\$10,000$ (the least favorable outcome). If he or she is given a choice between two lotteries of this type, the decision maker simply chooses the lottery with the larger chance of receiving the most favorable outcome. Applying this idea to L_1' through L_4' yields $L_1' p L_2' p L_4' p L_3'$. Since $L_i i L_i'$, we may conclude that $L_1 p L_2 p L_4 p L_3$.

We now give a more formal description of the process that we have used to rank L_1, L_2, L_3, and L_4. The **utility** of the reward r_i, written $u(r_i)$, is the number q_i such that the decision maker is indifferent between the following two lotteries:

```
   1                       qᵢ
    ──── rᵢ     and    ──────── Most favorable outcome
                    ──┤
                        1 − qᵢ
                      ──────── Least favorable outcome
```

This definition forces $u(\text{least favorable outcome}) = 0$ and $u(\text{most favorable outcome}) = 1$. For our possible payoffs of \$30,000, $-\$10,000$, \$0, \$500, and \$10,000, we first find that $u(\$30,000) = 1$ and $u(-\$10,000) = 0$. Then (1)–(3) yield $u(\$10,000) = .90$, $u(\$500) = .62$, and $u(\$0) = .60$. The specification of $u(r_i)$ for all rewards r_i is called the decision maker's **utility function.**

For a given lottery $L = (p_1, r_1; p_2, r_2; \ldots; p_n, r_n)$, define the expected utility of the lottery L, written $E(U$ for $L)$, by

$$E(U \text{ for } L) = \sum_{i=1}^{i=n} p_i u(r_i)$$

Thus, in our example

$$E(U \text{ for } L_1) = 1(.90) = .90$$
$$E(U \text{ for } L_2) = .50(1) + .50(.60) = .80$$
$$E(U \text{ for } L_3) = 1(.60) = .60$$
$$E(U \text{ for } L_4) = .02(0) + .98(.62) = .6076$$

Recall that we found that $L_i i L_i'$, where L_i' yielded an $E(U$ for $L_i)$ chance at \$30,000 and a $1 - E(U$ for $L_i)$ chance at $-\$10,000$. Thus, in choosing between lotteries L_1', L_2', L_3', and L_4' (or equivalently, L_1, L_2, L_3, and L_4), we simply chose the lottery with the largest expected utility. Given two lotteries L_1 and L_2, we may choose between them via the expected utility criteria:

$$L_1 p L_2 \quad \text{if and only if} \quad E(U \text{ for } L_1) > E(U \text{ for } L_2)$$
$$L_2 p L_1 \quad \text{if and only if} \quad E(U \text{ for } L_2) > E(U \text{ for } L_1)$$
$$L_1 i L_2 \quad \text{if and only if} \quad E(U \text{ for } L_2) = E(U \text{ for } L_1)$$

Von Neumann–Morgenstern Axioms

Von Neumann and Morgenstern proved that if a person's preferences satisfy the following axioms, then he or she should choose between lotteries by using the expected utility criterion.

Axiom 1: Complete Ordering Axiom

For any two rewards r_1 and r_2, one of the following must be true: The decision maker (1) prefers r_1 to r_2, (2) prefers r_2 to r_1, or (3) is indifferent between r_1 and r_2. Also, if the person prefers r_1 to r_2 and r_2 to r_3, then he or she must prefer r_1 to r_3 (transitivity of preferences).

In our discussion, we used the Complete Ordering Axiom to determine the most and least favorable outcomes.

Axiom 2: Continuity Axiom

If the decision maker prefers r_1 to r_2 and r_2 to r_3, then for some $c(0 < c < 1)$, $L_1 i L_2$, where

In our informal discussion, we used the Continuity Axiom when we found, for example, that $L_3 i L_3'$, where

Axiom 3: Independence Axiom

Suppose the decision maker is indifferent between rewards r_1 and r_2. Let r_3 be any other reward. Then for any c $(0 < c < 1)$, $L_1 i L_2$, where

L_1 and L_2 differ only in that L_1 has a probability c of yielding a reward r_1, whereas L_2 has a probability c of yielding a reward r_2. Thus, the Independence Axiom implies that the decision maker views a chance c at r_1 and a chance c at r_2 to be of identical value, and this view holds for all values of c and r_3. We applied the Independence Axiom when we used (3) to claim that $L_2 i L_2''$, where

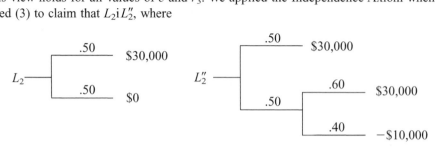

Axiom 4: Unequal Probability Axiom

Suppose the decision maker prefers reward r_1 to reward r_2. If two lotteries have only r_1 and r_2 as their possible outcomes, he or she will prefer the lottery with the higher probability of obtaining r_1.

We used the Unequal Probability Axiom when we concluded, for example, that L_1' was preferred to L_2' (because L_1' had a .90 chance at $30,000$ and L_2' had only a .80 chance at $30,000$).

Axiom 5: Compound Lottery Axiom

Suppose that when all possible outcomes are considered, a compound lottery L yields (for $i = 1, 2, \ldots, n$) a probability p_i of receiving a reward r_i. Then LiL', where L' is the simple lottery $(p_1, r_1; p_2, r_2; \ldots; p_n, r_n)$.

For example, consider the following compound lottery:

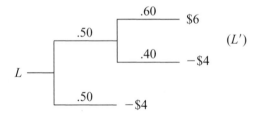

L yields a $.50 + .50(.40) = .70$ chance at $-\$4$ and a $.50(.60) = .30$ chance at $\$6$. Thus, LiL'', where

$$
L'' \quad
\begin{cases}
.70 & -\$4 \\
.30 & \$6
\end{cases}
$$

In our informal discussion, we used the Compound Lottery Axiom when, for example, we stated that the compound equivalent of L_2 (L_2'')

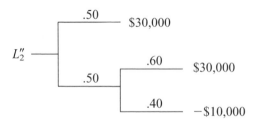

was equivalent to the following simple lottery:

$$.50 + .50(.60) = .80 \quad \$30,000$$

$$.50(.40) = .20 \quad -\$10,000$$

Why We May Assume u(Worst Outcome) = 0 and u(Best Outcome) = 1

Up to now, we have assumed that u(least favorable outcome) = 0 and u(most favorable outcome) = 1. Even if a decision maker's utility function does not have these values, we can transform his or her utility function (without changing the preferences among lotteries) into a utility function having u(least favorable outcome) = 0 and u(most favorable outcome) = 1.

LEMMA 1

Given a utility function $u(x)$, define for any $a > 0$ and any b the function $v(x) = au(x) + b$. Given any two lotteries L_1 and L_2, it will be the case that

1 A decision maker using $u(x)$ as his or her utility function will have $L_1 p L_2$ if and only if a decision maker using $v(x)$ as his or her utility function will have $L_1 p L_2$.

2 A decision maker using $u(x)$ as his or her utility function will have $L_1 i L_2$ if and only if a decision maker using $v(x)$ as his or her utility function will have $L_1 i L_2$.

Proof Let

$$L_1 = (p_1, r_1; p_2, r_2; \dots; p_n, r_n)$$
$$L_2 = (p'_1, r'_1; p'_2, r'_2; \dots; p'_m, r'_m)$$

Suppose the decision maker using $u(x)$ prefers L_1 to L_2. Then by the expected utility criterion, we know that

$$\sum_{i=1}^{i=n} p_i u(r_i) > \sum_{i=1}^{i=m} p'_i u(r'_i) \qquad (4)$$

Now the $v(x)$ decision maker will have $L_1 p L_2$ if

$$\sum_{i=1}^{i=n} p_i [au(r_i) + b] > \sum_{i=1}^{i=m} p'_i [au(r'_i) + b] \qquad (5)$$

Since

$$\sum_{i=1}^{i=n} p_i = \sum_{i=1}^{i=m} p'_i = 1$$

(5) simplifies to

$$a \sum_{i=1}^{i=n} p_i u(r_i) + b > a \sum_{i=1}^{i=m} p'_i u(r'_i) + b \qquad (6)$$

Since $a > 0$, (6) follows from (4). Thus, if the $u(x)$ decision maker has $L_1 p L_2$, the $v(x)$ decision maker has $L_1 p L_2$. Similarly, if (6) holds, then (4) will hold. Thus, if the $v(x)$ decision maker has $L_1 p L_2$, the $u(x)$ decision maker will also have $L_1 p L_2$. A similar argument can be used to prove part (2) of Lemma 1.

Using Lemma 1, we can show that without changing how an individual ranks lotteries, we can transform the decision maker's utility function into one having u(least favorable outcome) $= 0$ and u(most favorable outcome) $= 1$. To illustrate, let's reconsider ranking lotteries L_1–L_4. Suppose our decision maker's utility function had $u(-\$10,000) = -5$ and $u(\$30,000) = 10$. Define $v(x) = au(x) + b$. Choose a and b so that $v(\$30,000) = 10a + b = 1$ and $v(-\$10,000) = -5a + b = 0$. Then $a = \frac{1}{15}$ and $b = \frac{1}{3}$. Then by Lemma 1, the utility function $v(x) = \frac{u(x)}{15} + \frac{1}{3}$ will yield the same ranking of lotteries as does $u(x)$, and we will have constructed $v(x)$ so that $v(\$30,000) = 1$ and $v(-\$10,000) = 0$. Thus, we see that without loss of generality, we may assume that u(least favorable outcome) $= 0$ and u(most favorable outcome) $= 1$.

Estimating an Individual's Utility Function

How might we estimate an individual's (call her Jill) utility function? We begin by assuming that the least favorable outcome (say, $-\$10,000$) has a utility of 0 and that the most favorable outcome (say, \$30,000) has a utility of 1. Next we define a number $x_{1/2}$ having $u(x_{1/2}) = \frac{1}{2}$. To determine $x_{1/2}$, ask Jill for the number (call it $x_{1/2}$) that makes her indifferent between

$$\underline{\hspace{3cm}}\; x_{1/2} \quad \text{and}$$

$$\begin{array}{l} \frac{1}{2} \quad \$30,000 \quad \text{(Most favorable outcome)} \\[1mm] \frac{1}{2} \quad -\$10,000 \quad \text{(Least favorable outcome)} \end{array}$$

Since Jill is indifferent between the two lotteries, they must have the same expected utility. Thus, $u(x_{1/2}) = (\frac{1}{2})(1) + (\frac{1}{2})(0) = \frac{1}{2}$.

This procedure yields a point $x_{1/2}$ having $u(x_{1/2}) = \frac{1}{2}$. Suppose Jill states that $x_{1/2} = -\$3,400$. Using $x_{1/2}$ and the least favorable outcome ($-\$10,000$) as possible outcomes, we can construct a lottery that can be used to determine the point $x_{1/4}$ having a utility of $\frac{1}{4}$ (that is, $u(x_{1/4}) = \frac{1}{4}$). Point $x_{1/4}$ must be such that Jill is indifferent between

$$\underline{\hspace{3cm}}\; x_{1/4} \quad \text{and}$$

$$\begin{array}{l} \frac{1}{2} \quad x_{1/2} = -\$3,400 \\[1mm] \frac{1}{2} \quad -\$10,000 \quad \text{(Least favorable outcome)} \end{array}$$

Then $u(x_{1/4}) = (\frac{1}{2})(\frac{1}{2}) + (\frac{1}{2})(0) = \frac{1}{4}$. Thus, $x_{1/4}$ will satisfy $u(x_{1/4}) = \frac{1}{4}$. Suppose Jill states that $x_{1/4} = -\$8,000$. This gives us another point on Jill's utility function.

Jill can now use the $x_{1/2}$ and \$30,000 outcomes to construct a lottery that will yield a value $x_{3/4}$ satisfying $u(x_{3/4}) = \frac{3}{4}$. (How?) Suppose that $x_{3/4} = \$8,000$. Similarly, outcomes of $x_{1/4}$ and $-\$10,000$ can be used to construct a lottery that will yield a value $x_{1/8}$ satisfying $u(x_{1/8}) = \frac{1}{8}$. Now Jill's utility function can be approximated by drawing a curve (smooth, we hope) joining the points

$$(-\$10,000, 0), (x_{1/8}, 1/8), (x_{1/4}, 1/4), \ldots, (\$30,000, 1)$$

The result is shown in Figure 1. Unfortunately, if a decision maker's preferences violate any of the preceding axioms (such as transitivity), this procedure may not yield a smooth curve. If it does not yield a relatively smooth curve, more sophisticated procedures for assessing utility functions must be used (see Keeney and Raiffa (1976)).

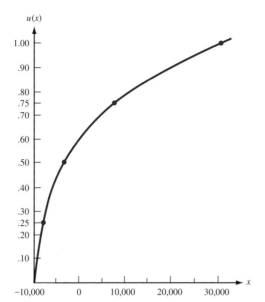

FIGURE 1
Jill's Utility Function

Relation between an Individual's Utility Function and His or Her Attitude toward Risk

A decision maker's utility function contains information about his or her attitude toward risk. To discuss this information, we need to define the concepts of a lottery's certainty equivalent and risk premium.

DEFINITION ■ The **certainty equivalent** of a lottery L, written $CE(L)$, is the number $CE(L)$ such that the decision maker is indifferent between the lottery L and receiving a certain payoff of $CE(L)$. ■

For example, we saw earlier that Jill was indifferent between

$$\underline{\hspace{2cm}1\hspace{2cm}} -\$3,400 \quad \text{and} \quad L \begin{cases} \frac{1}{2} & \$30,000 \\ \frac{1}{2} & -\$10,000 \end{cases}$$

Thus, $CE(L) = -\$3,400$.

DEFINITION ■ The **risk premium** of a lottery L, written $RP(L)$, is given by $RP(L) = EV(L) - CE(L)$, where $EV(L)$ is the expected value of the lottery's outcomes. ■

For example, if

$$L \begin{cases} \frac{1}{2} & \$30,000 \\ \frac{1}{2} & -\$10,000 \end{cases}$$

then $EV(L) = (\frac{1}{2})(\$30,000) + (\frac{1}{2})(-\$10,000) = \$10,000$. We have already seen that $CE(L) = -\$3,400$. Thus, $RP(L) = 10,000 - (-3,400) = \$13,400$; Jill values L at \$13,400 less than its expected value, because she does not like the large degree of uncertainty that is associated with the reward yielded by L.

Let a **nondegenerate lottery** be any lottery in which more than one outcome can occur. With respect to attitude toward risk, a decision maker is

1　**Risk-averse** if and only if for any nondegenerate lottery L, $RP(L) > 0$

2　**Risk-neutral** if and only if for any nondegenerate lottery L, $RP(L) = 0$

3　**Risk-seeking** if and only if for any nondegenerate lottery L, $RP(L) < 0$

An individual's attitude toward risk depends on the concavity (or convexity) of his or her utility function.

DEFINITION ■　A function $u(x)$ is said to be **strictly concave** (or **strictly convex**) if for any two points on the curve $y = u(x)$, the line segment joining those two points lies entirely (with the exception of its endpoints) below (or above) the curve $y = u(x)$. ■

If $u(x)$ is differentiable, then $u(x)$ will be strictly concave if and only if $u''(x) < 0$ for all x and $u(x)$ will be strictly convex if and only if $u''(x) > 0$ for all x. It can easily be shown that a decision maker with a utility function $u(x)$ is

1　Risk-averse if and only if $u(x)$ is strictly concave

2　Risk-neutral if and only if $u(x)$ is a linear function (if $u(x)$ is both convex and concave)

3　Risk-seeking if and only if $u(x)$ is strictly convex

To illustrate these definitions, we show that a decision maker with a concave utility function $u(x)$ exhibits risk-averse behavior (has $RP(L) > 0$). Consider a binary lottery L (a lottery with only two possible outcomes):

$$L \quad \begin{array}{c} \underline{\quad p \quad} \ x_1 \\ \underline{\quad 1-p \quad} \ x_2 \end{array} \qquad (\text{Assume } x_1 < x_2)$$

Suppose $u(x)$ is strictly concave. Then, from Figure 2, we see that

$$E(U \text{ for } L) = p\,u(x_1) + (1 - p)u(x_2) = y\text{-coordinate of point 1}$$

Since $CE(L)$ is the value x^* having $u(x^*) = E(U$ for $L)$, Figure 2 shows that $CE(L) < EV(L)$, so $RP(L) > 0$. This follows because the strict concavity of $u(x)$ implies that the line segment joining the points $(x_1, u(x_1))$ and $(x_2, u(x_2))$ lies below the curve $u(x)$.

We can also give an algebraic proof that $u(x)$ strictly concave implies that $RP(L) = EV(L) - CE(L) > 0$. Recall that for

$$L \quad \begin{array}{c} \underline{\quad p \quad} \ x_1 \\ \underline{\quad 1-p \quad} \ x_2 \end{array}$$

$EV(L) = px_1 + (1 - p)x_2$. Now the strict concavity of $u(x)$ implies that $u[px_1 + (1 - p)x_2] > pu(x_1) + (1 - p)u(x_2) = E(U$ for $L)$. Thus, the decision maker prefers $px_1 + (1 - p)x_2 = EV(L)$ with certainty to the prospect of playing L. The certainty equivalent

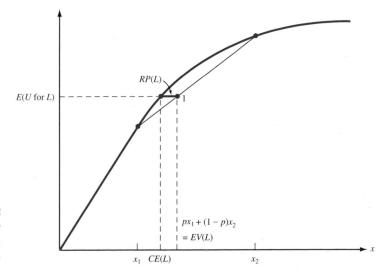

FIGURE 2
Why a Concave Utility Function Implies Risk-Averse Behavior

$RP(L)$

$E(U \text{ for } L)$

1

$px_1 + (1 - p)x_2$
$= EV(L)$

$x_1 \quad CE(L)$

x_2

x

of L must be less than $px_1 + (1 - p)x_2 = EV(L)$. This implies that $RP(L) = EV(L) - CE(L) > 0$, and the decision maker exhibits risk-averse behavior. In Problem 4 at the end of this section, the reader will be asked to show that if $u(x)$ is strictly convex, the decision maker exhibits risk-seeking behavior.

If the decision maker is risk-neutral (that is, $u(x) = ax + b$), he or she chooses among lotteries via the expected reward criterion of Section 2.1 (see Problem 5 at the end of this section). Thus, when ranking lotteries, a risk-neutral decision maker considers only the expected value (and not the risk) of the lotteries.

Example 2 illustrates the concepts of risk premium, certainty equivalent, and risk aversion.

EXAMPLE 2 **Joan's Assets**

Joan's utility function for her asset position x is given by $u(x) = x^{1/2}$. Currently, Joan's assets consist of \$10,000 in cash and a \$90,000 home. During a given year, there is a .001 chance that Joan's home will be destroyed by fire or other causes. How much would Joan be willing to pay for an insurance policy that would replace her home if it were destroyed?

Solution Let $x =$ annual insurance premium. Then Joan must choose between the following lotteries:

Asset Position

L_1: Buy insurance $\quad \dfrac{1}{} \quad (\$100,000 - x)$

$\qquad\qquad\qquad\qquad .001 \quad \$100,000 - \$90,000 = \$10,000$

L_2: Don't buy insurance

$\qquad\qquad\qquad\qquad .999 \quad \$100,000$

Joan will prefer L_1 to L_2 if L_1's expected utility exceeds L_2's expected utility. Thus, $L_1 p L_2$ if and only if

$$(100,000 - x)^{1/2} > .001(10,000)^{1/2} + .999(100,000)^{1/2}$$
$$> .10 + 315.91154$$
$$> 316.01154$$

Squaring both sides of the last inequality we find that L_1pL_2 if and only if

$$100,000 - x > (316.01154)^2$$
$$x < \$136.71$$

Thus, Joan would pay up to $136.71 for insurance. Of course, if $p = \$136.71$, L_1iL_2.
Let's compute the risk premium for L_2:

$$EV(L_2) = .001(10,000) + .999(100,000) = \$99,910$$

(an expected loss of $100,000 - 99,910 = \$90$). Since $E(U$ for $L_2) = 316.01154$, we can find $CE(L_2)$ from the relation $u(CE(L_2)) = 316.01154$, or $[CE(L_2)]^{1/2} = 316.01154$. Thus, $CE(L_2) = (316.01154)^2 = \$99,863.29$, and

$$RP(L_2) = EV(L_2) - CE(L_2) = 99,910 - 99,863.29 = \$46.71$$

Therefore, Joan is willing to pay for annual home insurance $46.71 more than the expected loss of $90. (Recall that Joan was willing to pay up to $90 + 46.71 = \$136.71$ to avoid the risk involved in her home being destroyed.) Joan exhibits risk-averse behavior $(RP(L_2) > 0)$. Since

$$u''(x) = \frac{-x^{-3/2}}{4} < 0$$

$u(x)$ is strictly concave, and $RP(L) > 0$ would hold for any nondegenerate lottery.

In reality, many people exhibit both risk-seeking behavior (they purchase lottery tickets, go to Las Vegas) and risk-averse behavior (they buy home insurance). A person whose utility function contains both convex and concave segments may exhibit both risk-averse and risk-seeking behavior. Consider a decision maker whose utility function $u(x)$ for change in current asset position is given in Figure 3. If forced to choose between

$$L_1 \xrightarrow{\quad 1 \quad} 0 \qquad \text{and} \qquad L_2 \begin{cases} \xrightarrow{.10} \$2,500 \\ \xrightarrow{.90} -\$300 \end{cases}$$

what would this person do?

From Figure 3, we find that $u(0) = .20$, $u(2,500) = .50$, and $u(-300) = .18$. Thus, $E(U$ for $L_1) = .20$ and $E(U$ for $L_2) = .10(.50) + .90(.18) = .212$. Thus, L_2pL_1. This means that L_2 has a certainty equivalent of at least $0. Since $EV(L_2) = -\$20$, this implies that $RP(L_2) = EV(L_2) - CE(L_2) < 0$. The decision maker exhibits risk-seeking behavior in this situation, because for changes in asset position between $0 and $2,500, $u(x)$ is a convex function.

Now suppose the decision maker can, for $200, insure himself against a loss of $2,000, which occurs with probability .08. Then he must choose between

$$L_3 \xrightarrow{\quad 1 \quad} -\$200 \qquad \text{and} \qquad L_4 \begin{cases} \xrightarrow{.08} -\$2,000 \\ \xrightarrow{.92} \$0 \end{cases}$$

From Figure 3, $u(-200) = .19$, $u(0) = .20$, and $u(-2,000) = 0$. Thus, $E(U$ for $L_3) = .19$ and $E(U$ for $L_4) = .80(0) + .92(.20) = .184$, and L_3pL_4. This shows that $CE(L_4) < -\$200$. Since $EV(L_4) = .08(-2,000) + .92(0) = -\$160$, $RP(L_4) = EV(L_4) - CE(L_4) > 0$, and the decision maker is exhibiting risk-averse behavior, because $u(x)$ is concave for $-2,000 < x < 0$. Thus, if his utility function has both convex and concave segments, a person can exhibit both risk-seeking and risk-averse behavior.

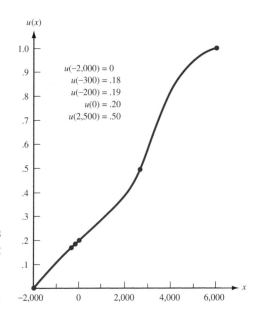

FIGURE 3
A Utility Function That Exhibits Both Risk-Seeking and Risk-Averse Behavior

$u(-2,000) = 0$
$u(-300) = .18$
$u(-200) = .19$
$u(0) = .20$
$u(2,500) = .50$

Exponential Utility

Classes of "ready-made" utility functions have been developed. One important class is called **exponential utility** and has been used in many financial investment analyses. An exponential utility function has only one adjustable numerical parameter, and there are straightforward ways to discover the most appropriate value of this parameter for a particular individual or company. So the advantage of using an exponential utility function is that it is relatively easy to assess. The drawback is that exponential utility functions do not capture all types of attitudes toward risk. Nevertheless, their ease of use has made them popular.

An exponential utility function has the following form:

$$U(x) = 1 - e^{-x/R}$$

Here, x is a monetary value (a payoff if positive, a cost if negative), $U(x)$ is the utility of this value, and $R > 0$ is an adjustable parameter called the **risk tolerance.** Basically, the risk tolerance measures how much risk the decision maker will tolerate. The larger the value of R, the less risk averse the decision maker is. That is, a person with a large value of R is more willing to take risks than a person with a small value of R.

To assess a person's (or company's) exponential utility function, we need only assess the value of R. There are a couple of tips for doing this. First, it has been shown that the risk tolerance is approximately equal to that dollar amount R such that the decision maker is indifferent between the following two options:

- Option 1: Obtain no payoff at all
- Option 2: Obtain a payoff of R dollars or a loss of $R/2$ dollars, depending on the flip of a fair coin

For example, if I am indifferent between a bet where I win $1,000 or lose $500, with probability 0.5 each, and not betting at all, then my R is approximately $1,000. From this criterion it certainly makes intuitive sense that a wealthier person (or company) ought to have a larger value of R. This has been found in practice.

A second tip for finding R is based on empirical evidence found by Ronald Howard, a prominent decision analyst. Through his consulting experience with several large companies, he discovered tentative relationships between risk tolerance and several financial variables—net sales, net income, and equity. (See Howard (1992).) Specifically, he found that R was approximately 6.4% of net sales, 124% of net income, and 15.7% of equity for the companies he studied. For example, according to this prescription, a company with net sales of $30 million should have a risk tolerance of approximately $1.92 million. Howard admits that these percentages are only guidelines. However, they do indicate that larger and more profitable companies tend to have larger values of R, which means that they are more willing to take risks involving given dollar amounts.

PROBLEMS

Group A

1 Suppose my utility function for asset position x is given by $u(x) = \ln x$.

a Am I risk-averse, risk-neutral, or risk-seeking?

b I now have $20,000 and am considering the following two lotteries:

L_1: With probability 1, I lose $1,000.

L_2: With probability .9, I gain $0.

With probability .1, I lose $10,000.

Determine which lottery I prefer and the risk premium of L_2.

2 Answer Problem 1 for a utility function $u(x) = x^2$.

3 Answer Problem 1 for a utility function $u(x) = 2x + 1$.

4 Show that a decision maker who has a strictly convex utility function will exhibit risk-seeking behavior.

5 Show that a decision maker who has a linear utility function will rank two lotteries according to their expected value.

6 A decision maker has a utility function for monetary gains x given by $u(x) = (x + 10,000)^{1/2}$.

a Show that the person is indifferent between the status quo and

L: With probability $\frac{1}{3}$, he or she gains $80,000

With probability $\frac{2}{3}$, he or she loses $10,000

b If there is a 10% chance that a painting valued at $10,000 will be stolen during the next year, what is the most (per year) that the decision maker would be willing to pay for insurance covering the loss of the painting?

7 Patty is trying to determine which of two courses to take. If she takes the operations research course, she believes that she has a 10% chance of receiving an A, a 40% chance for a B, and a 50% chance for a C. If Patty takes a statistics course, she has a 70% chance for a B, a 25% chance for a C, and a 5% chance for a D. Patty is indifferent between

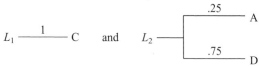

She is also indifferent between

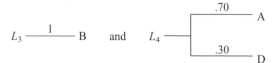

If Patty wants to take the course that maximizes the expected utility of her final grade, which course should she take?

8 We are going to invest $1,000 for a period of 6 months. Two potential investments are available: T-bills and gold. If the $1,000 is invested in T-bills, we are certain to end the 6-month period with $1,296. If we invest in gold, there is a $\frac{3}{4}$ chance that we will end the 6-month period with $400 and a $\frac{1}{4}$ chance that we will end the 6-month period with $10,000. If we end up with x dollars, our utility function is given by $u(x) = x^{1/2}$. Should we invest in gold or T-bills?

9 We now have $5,000 in assets and are given a choice between investment 1 and investment 2. With investment 1, 80% of the time we increase our asset position by $295,000, and 20% of the time we increase our asset position by $95,000. With investment 2, 50% of the time we increase our asset position by $595,000, and 50% of the time we increase our asset position by $5,000. Our utility function for final asset position x is $u(x)$. We are given the following values for $u(x)$: $u(0) = 0$, $u(640,000) = .80$, $u(810,000) = .90$, $u(0) = 0$, $u(90,000) = .30$, $u(1,000,000) = 1$, $u(490,000) = .7$.

a Are we risk-averse, risk-seeking, or risk-neutral? Explain.

b Will we prefer investment 1 or investment 2?

10 My current income is $40,000. I believe that I owe $8,000 in taxes. For $500, I can hire a CPA to review my tax return; there is a 20% chance that she will save me $4,000 in taxes. My utility function for (disposable income) = (current income) − (taxes) − (payment to accountant) is given by $\sqrt{x}$ where x is disposable income. Should I hire the CPA?

Group B

11[†] (The Allais Paradox) Suppose we are offered a choice between the following two lotteries:

L_1: With probability 1, we receive $1 million.

L_2: With probability .10, we receive $5 million.

With probability .89, we receive $1 million.

With probability .01, we receive $0.

Which lottery do we prefer? Now consider the following two lotteries:

[†]Based on Allais (1953).

L_3: With probability .11, we receive \$1 million.

With probability .89, we receive \$0.

L_4: With probability .10, we receive \$5 million.

With probability .90, we receive \$0.

Which lottery do we prefer? Suppose (like most people), we prefer L_1 to L_2. Show that L_3 must have a larger expected utility than L_4.

12 (The St. Petersburg Paradox) Let L represent the following lottery. I toss a coin until it comes up heads. If the first heads is obtained on the nth toss of the coin, I receive a payoff of $\$2^n$.

a If I were a risk-neutral decision maker, what would be the certainty equivalent of L? Is this reasonable?

b If a decision maker's utility function for increasing wealth by x dollars is given by $u(x) = \log_2(x)$, what would be the certainty equivalent of L?

13 Joe is a risk-averse decision maker. Which of the following lotteries will he prefer?

L_1: With probability .10, Joe loses \$100.

With probability .90, Joe receives \$0.

L_2: With probability .10, Joe loses \$190.

With probability .90, Joe receives \$10.

14[†] (The Ellsberg Paradox) An urn contains 90 balls. It is known that 30 are red and that each of the other 60 is either yellow or black. One ball will be drawn at random from the urn. Consider the following four options:

Option 1 We receive \$1,000 if a red ball is drawn.
Option 2 We receive \$1,000 if a yellow ball is drawn.
Option 3 We receive \$1,000 if a yellow or black ball is drawn.
Option 4 We receive \$1,000 if a red or black ball is drawn.

a Explain why most people prefer option 1 over option 2 and also prefer option 3 over option 4.

b If we prefer option 1 to option 2, explain why we should also prefer option 4 over option 3.

15 Although the Von Neumann–Morgenstern axioms seem plausible, there are many reasonable situations in which people appear to violate these axioms. For example, suppose

[†]Based on Ellsberg (1961).

TABLE 8

	Starting Salary	Location	Opportunity for Advancement
Job 1	E	S	G
Job 2	G	E	S
Job 3	S	G	E

a recent college graduate must choose between three job offers on the basis of starting salary, location of job, and opportunity for advancement. Given two job offers that are satisfactory with regard to all three attributes, the graduate will decide between two job offers by choosing the one that is superior on at least two of the three attributes. Suppose he or she has three job offers and has rated each one as shown in Table 8 (E = excellent, G = good, and S = satisfactory). Show that the graduate's preferences among these jobs violate the Complete Ordering Axiom.

Group C

16 Suppose my utility function for my asset position is $u(x) = x^{1/2}$. I have \$10,000 at present. Consider the following lottery:

L: With probability $\frac{1}{2}$, L yields a payoff of \$1,025.

With probability $\frac{1}{2}$, L yields a payoff of −\$199.

a If I don't have the right to play L, find an equation that when solved would yield the amount I would be willing to pay for the right to play L. This is called the **buying price** of lottery L.

b If I have the right to play L, what is the least I would accept from somebody who wanted to buy the right to play L? (After someone else buys L, I can't play L.) This is called the **selling price** of lottery L.

c Answer part (b) for the case that I have \$1,000.

d Suppose that my utility function for my asset position is $u(x) = 1 - e^{-x}$. Show that for all possible asset positions, the buying price of L and the selling price of L will remain the same. Show that for all asset positions, the buying price of L will equal the selling price of L.

2.3 Flaws in Expected Maximization of Utility: Prospect Theory and Framing Effects

The axioms underlying expected maximization of utility (EMU) seem reasonable, but in practice people's decisions often deviate from the predictions of EMU. Psychologists Tversky and Kahneman[‡] (1981) developed **prospect theory** and **framing effects for values** to try and explain why people deviate from the predictions of EMU.

[‡]In 2002, Kahneman received the Nobel Prize for Economics, in large part honoring his work with Tversky. Tversky was not awarded the prize because he died in 1996 (Nobel Prizes are not given posthumously).

Prospect Theory

Here is one example of a decision that cannot be explained by EMU. Ask a person to choose between lottery 1 and lottery 2:

Lottery 1: $30 for certain
Lottery 2:   80% chance at $45 and 20% chance at $0

Most people prefer lottery 1 to lottery 2. Next ask the same person to choose between lottery 3 and lottery 4:

Lottery 3:   20% chance at $45 and 80% chance at $0
Lottery 4:   25% at $30 and 75% chance at $0

Most people choose lottery 3 over lottery 4. Now let $u(0) = 0$ and $u(45) = 1$. A decision maker following EMU will choose lottery 1 over lottery 2 if and only if $u(30) > .8$. A decision maker following EMU will choose lottery 3 over lottery 4 if and only if $.2 > .25u(30)$ or $u(30) < .8$. This implies that a believer in EMU cannot choose lottery 1 over lottery 2 and lottery 3 over lottery 4. Thus, for this situation, the choices of most people contradict EMU. Tversky and Kahneman developed prospect theory to explain the decision-making paradox we have just described. Prospect theory assumes that we do not treat probabilities as they are given in a decision-making problem. Instead, the decision maker treats a probability p for an event as a "distorted" probability $\Pi(p)$. A $\Pi(p)$ function that seems to explain many paradoxes is shown in Figure 4.

The shape of the $\Pi(p)$ function in the figure implies that individuals are more sensitive to changes in probability when the probability of an event is small (near 0) or large (near 1). The equation we used to construct our $\Pi(p)$ curve is $\Pi(p) = 1.89799p - 3.55995p^2 + 2.662549p^3$. How does prospect theory explain our paradox? From the values of $\Pi(p)$ given in Figure 5, we can compare the expected "prospects" of lottery 1 versus lottery 2 and lottery 3 versus lottery 4.

Prospect for lottery 1: $u(30)$
Prospect for lottery 2:   .602
Prospect for lottery 3:   .258
Prospect for lottery 4:   .293$u(30)$.

Thus, lottery 1 is preferred to lottery 2 if $u(30) > .602$, while lottery 3 is preferred to lottery 4 if $.258 > .293u(30)$ or $u(30) < .258/.293 = .88$. Our paradox evaporates, because for many people, $u(30)$ will be between .602 and .88!

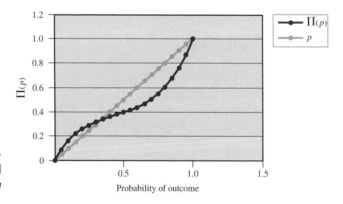

FIGURE 4
Weight Applied to Utility

	C	D
15	0.1	0.156803
16	0.15	0.213497
17	0.2	0.258382
18	0.25	0.293455
19	0.3	0.320713
20	0.35	0.342153
21	0.4	0.359771
22	0.45	0.375565
23	0.5	0.391531
24	0.55	0.409667
25	0.6	0.431969
26	0.65	0.460434
27	0.7	0.497059
28	0.75	0.543841
29	0.8	0.602778
30	0.85	0.675865
31	0.9	0.7651
32	0.95	0.872479
33	1	1

FIGURE 5
Lottery Prospects

Framing

The idea of framing is based on the fact that people often set their utility function from the standpoint of a frame or status quo from which they view the current situation. Most people's utility functions treat a loss of a given value as being more serious than a gain of an identical value. This is reflected in the utility function shown in Figure 6, which is convex for losses and concave for gains.

To see how framing can explain the failure of EMU, consider the following problem that Tversky and Kahneman gave to a group of students. The US is preparing for the outbreak of a disease that is expected to kill 600 people. Two alternative programs have been proposed:

Program I: 200 people are saved.

Program II: With probability $\frac{1}{3}$, 600 people are saved.

Most students preferred program I, probably because with program II there is a large risk of saving nobody. Since the programs are phrased in terms of lives saved, most people take the frame or reference point for this problem to be no lives saved or 600 people dead. Since the effect of each program is expressed in gains, and the utility function is concave for gains, we find that $u(200) = u((\frac{2}{3}) 0 + (\frac{1}{3})600)) > (\frac{1}{3})u(600) + (\frac{2}{3})u(0) = (\frac{1}{3})u(600)$. This implies, of course, that the person chooses program I over program II.

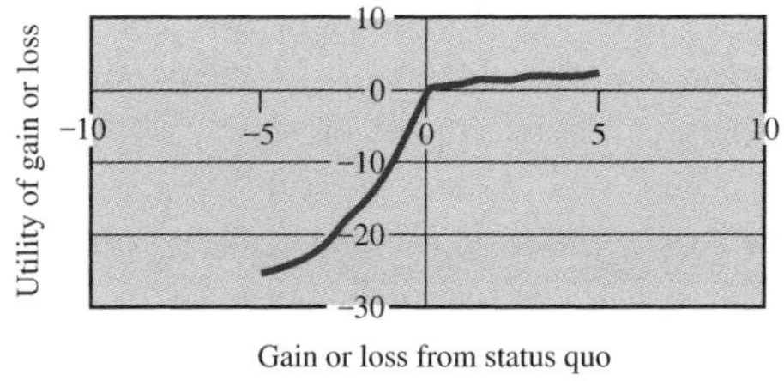

FIGURE 6
Utility Function for Framing

Utility of gain or loss

Gain or loss from status quo

Next, Tversky and Kahneman rephrased the problem as follows:

Program I: 400 people die.

Program II: With probability $\frac{2}{3}$, 600 people die.

Now most people choose program II. Note that both program I's are identical, as are both program II's. Why do most people choose program II for the second phrasing of the alternatives? The second phrasing shifts most people's reference points from "No lives saved" (in first phrasing) to "Nobody dies." The outcomes are expressed as losses (deaths), so the convexity of the utility curve for losses implies that

$$(\tfrac{2}{3})u(-600) = (\tfrac{2}{3})u(-600) + (\tfrac{1}{3})u(0) > u((\tfrac{2}{3})(-600) + \tfrac{1}{3}(0)) = u(-400)$$

This implies, of course, that the person chooses program II over program I.

PROBLEMS

Group A

1 Explain how prospect theory and/or framing explains the Allais Paradox. (See Problem 11 of Section 2.2.)

2 Suppose a decision maker has a utility function $u(x) = x^{1/3}$. We flip a fair coin and receive $10 for heads and $0 for tails.

a Using expected utility theory, determine the certainty equivalent of this lottery.

b Using $\Pi(p) = 1.89799p - 3.55995p^2 + 2.662549p^3$, use prospect theory to determine the certainty equivalent of the lottery.

c Intuively explain why your answer in part (b) is smaller than your answer in part (a).

d What implications does this problem have for the method used in Section 2.2 to estimate a person's utility function?

3 You are given a choice between lottery 1 and lottery 2. You are also given a choice between lottery 3 and lottery 4.

Lottery 1:   A sure gain of $240

Lottery 2:   25% chance to gain $1,000 and 75% chance to gain nothing

Lottery 3:   A sure loss of $750

Lottery 4:   A 75% chance to lose $1,000 and a 25% chance of losing nothing

84% of all people prefer lottery 1 over lottery 2, and 87% choose lottery 4 over lottery 3.

a Explain why the choice of lottery 1 over lottery 2 and lottery 4 over lottery 3 contradicts expected utility maximization. (*Hint:* Compare lottery 1 + lottery 4 to lottery 2 + lottery 3.)

b Can you explain this anomalous behavior?

4 Tversky and Kahneman asked 72 respondents to choose between lottery 1 and lottery 2 and lottery 3 and lottery 4.

Lottery 1:   A .001 chance at winning $5,000 and a .999 chance of winning $0

Lottery 2:   A sure gain of $5

Lottery 3:   A .001 chance of losing $5,000 and a .999 chance of losing $0

Lottery 4:   A sure loss of $5

More than 75% of all participants preferred lottery 1 to lottery 2 and lottery 4 to lottery 3.

a Which choices would be made by a risk-averse decision maker?

b Which choices would be made by a risk-seeking decision maker?

c How does the observed behavior of the participants contradict expected utility maximization?

d How does prospect theory resolve the contradiction?

2.4 Decision Trees

Often, people must make a series of decisions at different points in time. Then decision trees can be used to determine optimal decisions. A decision tree enables a decision maker to decompose a large complex decision problem into several smaller problems.

EXAMPLE 3   Colaco Marketing

Colaco currently has assets of $150,000 and wants to decide whether to market a new chocolate-flavored soda, Chocola. Colaco has three alternatives:

Alternative 1   Test market Chocola locally, then utilize the results of the market study to determine whether or not to market Chocola nationally.

Alternative 2   Immediately (without test marketing) market Chocola nationally.

Alternative 3   Immediately (without test marketing) decide not to market Chocola nationally.

In the absence of a market study, Colaco believes that Chocola has a 55% chance of being a national success and a 45% chance of being a national failure. If Chocola is a national success, Colaco's asset position will increase by $300,000, and if Chocola is a national failure, Colaco's asset position will decrease by $100,000.

If Colaco performs a market study (at a cost of $30,000), there is a 60% chance that the study will yield favorable results (referred to as a *local success*) and a 40% chance that the study will yield unfavorable results (referred to as a *local failure*). If a local success is observed, there is an 85% chance that Chocola will be a national success. If a local failure is observed, there is only a 10% chance that Chocola will be a national success. If Colaco is risk-neutral (wants to maximize its expected final asset position), what strategy should the company follow?

Solution   To draw a decision tree that represents Colaco's problem, we begin at the present and proceed toward future events and decisions. The decision tree in Figure 7 is constructed with two kinds of forks: decision forks (denoted by □) and event forks (denoted by ○).

A **decision fork** represents a point in time when Colaco has to make a decision. Each branch emanating from a decision fork represents a possible decision. An example of a decision fork occurs when Colaco must determine whether or not to test market Chocola.

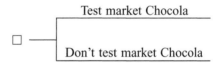

An **event fork** is drawn when outside forces determine which of several random events will occur. Each branch of an event fork represents a possible outcome, and the number on each branch represents the probability that the event will occur. For example, if Colaco decides to test market Chocola, the company faces the following event fork when observing the results of the test market study:

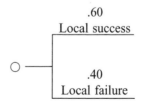

A branch of a decision tree is a **terminal branch** if no forks emanate from the branch. Thus, the branches indicating National success and National failure are terminal branches of Colaco's decision tree. Since we are maximizing expected final asset position at each terminal branch, we must enter the final asset position that will result if the path leading to the given terminal branch occurs. For example, the terminal branch National failure that follows Local failure leads to a final asset position of $150,000 - 30,000 - 100,000 = \$20,000$. If we were maximizing expected revenues, we would enter revenues on each terminal branch.

To determine the decisions that will maximize Colaco's expected final asset position, we work backward (sometimes called "folding back the tree") from right to left.[†] At each

---

[†]See Chapters 5 and 6 for an explanation of working backward (often called *dynamic programming*).

FIGURE 7
Colaco's Decision Tree (Risk-Neutral)

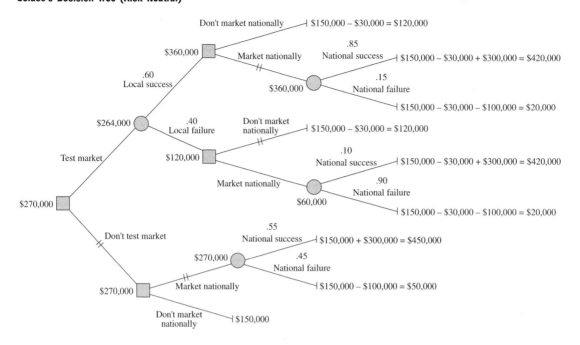

event fork, we calculate the expected final asset position and enter it in ○. At each decision fork, we denote by ‖ the decision that maximizes the expected final asset position and enter the expected final asset position associated with that decision in □. We continue working backward in this fashion until we reach the beginning of the tree. Then the optimal sequence of decisions can be obtained by following the ‖.

We begin by determining the expected final asset positions for the following three event forks:

1 Market nationally after Local success. Here we have an expected final asset position of .85(420,000) + .15(20,000) = $360,000.

2 Market nationally after Local failure. Here we have an expected final asset position of .10(420,000) + .90(20,000) = $60,000.

3 Market nationally after Don't test market. Here we have an expected final asset position of .55(450,000) + .45(50,000) = $270,000.

We may now evaluate three decision forks:

1 Decision after Local success. Market nationally yields a larger expected final asset position than Don't market nationally, so we ‖ Market nationally and enter an expected final asset position of $360,000.

2 Decision after Local failure. Don't market nationally yields a larger expected final asset position than Market nationally, so we ‖ Don't market nationally and enter an expected final asset position of $120,000.

3 Decision for Don't test market. Market nationally yields a larger expected final asset position than Don't market nationally, so we ‖ Market nationally and enter an expected final asset position of $270,000.

We now must evaluate the event fork emanating from the Test market decision. This event fork yields an expected final asset position of .60(360,000) + .40(120,000) = $264,000, which is entered in ○.

All that remains is to determine the correct decision at the decision fork Test market versus Don't test market. We have found that Test market yields an expected final asset position of $264,000, and Don't test market yields an expected final asset position of $270,000. Thus, we ‖ Don't test market and enter $270,000 in □.

We have now reached the beginning of the tree and have found that Colaco's optimal decision is Don't test market and then Market nationally. This strategy will yield an expected final asset position of $270,000. Observe that the decision tree also tells us that if we had test marketed and then acted optimally (Market nationally after Local success and Don't market nationally after Local failure), we would have obtained an expected final asset position of $264,000.

Incorporating Risk Aversion into Decision Tree Analysis

Note that Colaco's optimal strategy yields a .45 chance that the company will end up with a relatively small final asset position of $50,000. On the other hand, the strategy of test marketing and acting optimally on the results of the test market study yields only a (.60)(.15) = .09 chance that Colaco's asset position will be below $100,000. (Why?) Thus, if Colaco is a risk-averse decision maker, the strategy of immediately marketing nationally may not reflect the company's preference.

To illustrate how risk aversion may be incorporated into decision tree analysis, suppose that Colaco has the risk-averse utility function $u(x)$ in Figure 8 (x = final asset position). (How do we know that this utility function exhibits risk aversion?) *To determine Colaco's optimal decisions (that is, the decisions that maximize expected utility), simply replace each final asset position x_0 with its utility $u(x_0)$. Then at each event fork, compute the expected utility of Colaco's final asset position, and at each decision fork, choose the branch having the largest expected utility.*

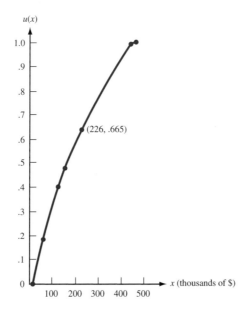

FIGURE 8
Colaco's Utility Function

We find from Figure 8 that $u(\$450,000) = 1$, $u(\$420,000) = .99$, $u(\$150,000) = .48$, $u(\$120,000) = .40$, $u(\$50,000) = .19$, and $u(\$20,000) = 0$. Substituting these values into the decision tree of Figure 7 yields the decision tree in Figure 9. We compute the expected utility at the following three event forks:

1 Market nationally after Local success. Here we have an expected utility of $.85(.99) + .15(0) = .8415$.

2 Market nationally after Local failure. Here we have an expected utility of $.10(.99) + .90(0) = .099$.

3 Market nationally after Don't test market. Here we have an expected utility of $.55(1) + .45(.19) = .6355$.

We may now evaluate three decision forks:

1 Decision after Local success. Market nationally yields a larger expected utility than Don't market nationally, so for this fork we ‖ Market nationally and enter an expected utility of .8415.

2 Decision after Local failure. Don't market nationally yields a larger expected utility than Market nationally, so for this fork we ‖ Don't market nationally and enter an expected utility of .40.

3 Decision for Don't test market. Market nationally yields a larger expected utility than Don't market nationally, so for this fork we ‖ Market nationally and enter an expected utility of .6355.

We now must evaluate the event fork emanating from the Test market decision. This event fork yields an expected utility of $.60(.8415) + .40(.40) = .6649$, which is entered in ○. All that remains is to determine the correct decision at the decision fork Test market versus Don't test market. We know that Test market yields an expected utility of .6649,

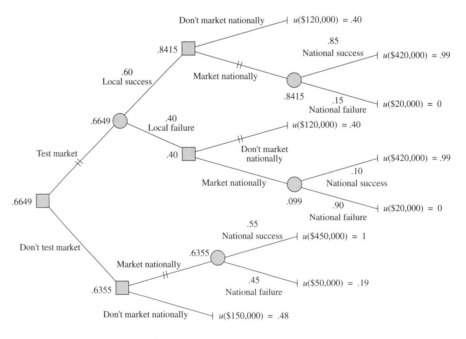

FIGURE 9
Colaco's Decision Tree
(Risk-Averse)

$u(\$226,000) = .6649$, so this situation is equivalent to a certain asset position of $226,000.

and Don't test market yields an expected utility of .6355, so we ‖ Test market and enter an expected utility of .6649 in □.

We have now reached the beginning of the tree and have found that Colaco's optimal decision is to begin by test marketing. If a local success is observed, then Colaco should market Chocola nationally; if a local failure is observed, then Colaco should not market Chocola nationally. This optimal strategy yields only a .60(.15) = .09 chance that Colaco will have a final asset position of less than $100,000. This reflects the risk-averse nature of the utility function in Figure 8. Also, we see from Figure 8 that $u(\$226,000) = .665$. Since Colaco views the current situation as having an expected utility of .6649, this means that the company considers the current situation equivalent to a certain asset position of $226,000. Thus, if somebody offered to pay more than $226,000 - 150,000 = \$76,000$ to buy the rights to Chocola, Colaco should take the offer. This is because receiving more than $76,000 for the rights to Chocola would bring Colaco's asset position to more than $150,000 + 76,000 = \$226,000$, and this situation has a higher expected utility than .665.

Expected Value of Sample Information

Decision trees can be used to measure the value of sample or test market information. To illustrate how this is done, we again assume that Colaco is risk-neutral. What is the value of the information that would be obtained by test marketing Chocola?

We begin by determining Colaco's expected final asset position if the company acts optimally and the test market study is costless. We call this expected final asset position Colaco's **expected value with sample information** (EVWSI). From Figure 7, we see that if we Test market and then act optimally, we will now have an expected final asset position of $264,000 + 30,000 = \$294,000$. Since $294,000 is larger than the expected asset position of the Don't test market branch ($270,000), we find that EVWSI = $294,000.

We next determine the largest expected final asset position that Colaco would obtain if the test market study were not available. We call this the **expected value with original information** (EVWOI). From the Don't test market branch of Figure 7, we find EVWOI = $270,000. Now the expected value of the test market information, referred to as **expected value of sample information** (EVSI), is defined to be EVSI = EVWSI − EVWOI.

In the Colaco example, EVSI is the most that Colaco can pay for the test market information and still be at least as well off as without the test market information. Thus, for the Colaco example, EVSI = $294,000 - 270,000 = \$24,000$. Since the cost of the test market study ($30,000) exceeds EVSI, Colaco should not (as we already know) conduct the test market study.

Expected Value of Perfect Information

We can modify the analysis used to determine EVSI to find the value of perfect information. By **perfect information** we mean that all uncertain events that can affect Colaco's final asset position still occur with the given probabilities (so there is still a .55 chance of Chocola being a national success and a .45 chance that Chocola will be a national failure), but Colaco finds out whether Chocola is a national success or a national failure *before* making the decision to market Chocola nationally or not. This information can then be used to determine Colaco's optimal marketing strategy. Thus, **expected value with perfect information** (EVWPI) is found by drawing a decision tree in which the decision maker has perfect information about which state has occurred before making a decision. Then the **expected value of perfect information** (EVPI) is given by EVPI = EVWPI − EVWOI.

FIGURE 10
**Expected Value with
Perfect Information
(EVWPI) for Colaco**

For the Colaco example, we find from Figure 10 that EVWPI = $315,000. Then EVPI = 315,000 − 270,000 = $45,000. Thus, a perfect (one that was always correct) test marketing study would be worth $45,000. EVPI is a useful upper bound on the value of sample or test market information; that is, no sample or test market information (no matter how good) can be worth more than $45,000.

EXAMPLE 4   **Art Dealer**

An art dealer's client is willing to buy the painting *Sunplant* at $50,000. The dealer can buy the painting today for $40,000 or can wait a day and buy the painting tomorrow (if it has not been sold) for $30,000. The dealer may also wait another day and buy the painting (if it is still available) for $26,000. At the end of the third day, the painting will no longer be available for sale. Each day, there is a .60 probability that the painting will be sold. What strategy maximizes the dealer's expected profit?

Solution   The decision tree for this example is given in Figure 11. The key to drawing this decision tree is that each day, the dealer must choose between buying the painting and waiting another day. Of course, waiting might mean that the dealer may never be able to buy the painting. As we see from the decision tree, the dealer should buy the painting on the first day.

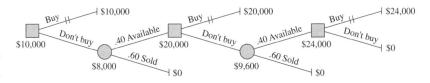

FIGURE 11
**Decision Tree for
Example 4**

PROBLEMS

Group A

1 Oilco must determine whether or not to drill for oil in the South China Sea. It costs $100,000, and if oil is found, the value is estimated to be $600,000. At present, Oilco believes there is a 45% chance that the field contains oil. Before drilling, Oilco can hire (for $10,000) a geologist to obtain more information about the likelihood that the field will contain oil. There is a 50% chance that the geologist will issue a favorable report and a 50% chance of an unfavorable report. Given a favorable report, there is an

80% chance that the field contains oil. Given an unfavorable report, there is a 10% chance that the field contains oil. Determine Oilco's optimal course of action. Also determine EVSI and EVPI.

2 The decision sciences department is trying to determine which of two copying machines to purchase. Both machines will satisfy the department's needs for the next ten years. Machine 1 costs $2,000 and has a maintenance agreement, which, for an annual fee of $150, covers all repairs. Machine 2 costs $3,000, and its annual maintenance cost is a random variable. At present, the decision sciences department believes there is a 40% chance that the annual maintenance cost for machine 2 will be $0, a 40% chance it will be $100, and a 20% chance it will be $200.

Before the purchase decision is made, the department can have a trained repairer evaluate the quality of machine 2. If the repairer believes that machine 2 is satisfactory, there is a 60% chance that its annual maintenance cost will be $0 and a 40% chance it will be $100. If the repairer believes that machine 2 is unsatisfactory, there is a 20% chance that the annual maintenance cost will be $0, a 40% chance it will be $100, and a 40% chance it will be $200. If there is a 50% chance that the repairer will give a satisfactory report, what is EVSI? If the repairer charges $40, what should the decision sciences department do? What is EVPI?

3 I am managing the Chicago Cubs. Suppose there is a runner on first base with nobody out and we want to determine whether we should bunt. Assume that a bunt will yield one of two results: (1) With probability .80, the bunt will be successful, in which case the batter is out and the runner on first base advances to second base. (2) With probability .20 the bunt is unsuccessful and the runner on first base is out trying to advance to second base and the batter is safe at first base.

The expected number of runs that the Cubs will score in an inning in various situations is given in Table 9.

a If our goal is to maximize the expected number of runs scored in an inning, should we bunt? Despite this answer, why do you think teams bunt?

b If we are considering stealing second base with nobody out, what chance of success is needed for stealing second to be an optimal decision?

4 The Nitro Fertilizer Company is developing a new fertilizer. If Nitro markets the product and it is successful, the company will earn a $50,000 profit; if it is unsuccessful, the company will lose $35,000. In the past, similar products have been successful 60% of the time. At a cost of $5,000, the effectiveness of the new fertilizer can be tested. If the test result is favorable, there is an 80% chance that the fertilizer will be successful. If the test result is unfavorable, there is only a 30% chance that the fertilizer will be successful. There is a 60% chance of a favorable test result and a 40% chance of an unfavorable test result. Determine Nitro's optimal strategy. Also find EVSI and EVPI.

TABLE 9

On-Base Situation	Number of Outs	Expected Number of Runs
Runner on first	0	0.813
Runner on first	1	0.498
Runner on second	1	0.671
Runner on second	0	1.194
No base runners	1	0.243

5 During the summer, Olympic swimmer Adam Johnson swims every day. On sunny summer days, he goes to an outdoor pool, where he may swim for no charge. On rainy days, he must go to a domed pool. At the beginning of the summer, he has the option of purchasing a $15 season pass to the domed pool, which allows him use for the entire summer. If he doesn't buy the season pass, he must pay $1 each time he goes there. Past meteorological records indicate that there is a 60% chance that the summer will be sunny (in which case there is an average of 6 rainy days during the summer) and a 40% chance the summer will be rainy (an average of 30 rainy days during the summer).

Before the summer begins, Adam has the option of purchasing a long-range weather forecast for $1. The forecast predicts a sunny summer 80% of the time and a rainy summer 20% of the time. If the forecast predicts a sunny summer, there is a 70% chance that the summer will actually be sunny. If the forecast predicts a rainy summer, there is an 80% chance that the summer will actually be rainy. Assuming that Adam's goal is to minimize his total expected cost for the summer, what should he do? Also find EVSI and EVPI.

6 Pete is considering placing a bet on the NCAA playoff game between Indiana and Purdue. Without any further information, he believes that each team has an equal chance to win. If he wins the bet, he will win $10,000; if he loses, he will lose $11,000. Before betting, he may pay Bobby $1,000 for his inside prediction on the game; 60% of the time, Bobby will predict that Indiana will win and 40% of the time, Bobby will predict that Purdue will win. When Bobby says that IU will win, IU has a 70% chance of winning, and when Bobby says that Purdue will win, IU has only a 20% chance of winning. Determine how Pete can maximize his total expected profit. What is EVSI? What is EVPI?

7 Erica is going to fly to London on August 5 and return home on August 20. It is now July 1. On July 1, she may buy a one-way ticket (for $350) or a round-trip ticket (for $660). She may also wait until August 1 to buy a ticket. On August 1, a one-way ticket will cost $370, and a round-trip ticket will cost $730. It is possible that between July 1 and August 1, her sister (who works for the airline) will be able to obtain a free one-way ticket for Erica. The probability that her sister will obtain the free ticket is .30. If Erica has bought a round-trip ticket on July 1 and her sister has obtained a free ticket, she may return "half" of her round-trip to the airline. In this case, her total cost will be $330 plus a $50 penalty. Use a decision tree approach to determine how to minimize Erica's expected cost of obtaining round-trip transportation to London.

8 I am a contestant on the TV show *Remote Jeopardy*, which works as follows. I am first asked a question about Stupid Videos. If I answer correctly, I earn $100. I believe that I have an 80% chance of answering such a question correctly. If I answer incorrectly, the game is over, and I win nothing. If I answer correctly, I may leave with $100 or go on and answer a question about Stupid TV Shows. If I answer this question correctly, I earn another $300, but if I answer incorrectly, I lose all previous earnings and am sent home. My chance of answering this question correctly is .60. If I answer the Stupid TV Shows question correctly, I may leave with my "earnings" or go on and answer a question about Statistics. If I answer this question correctly, I earn another $500, but if I answer it incorrectly, I lose all previous earnings and am sent home. My chance of answering this question correctly is .40. Draw a decision tree that can be used to maximize my expected earnings. What are my expected earnings?

Group B

In many decision tree problems, the decision maker's goal is to maximize the probability of a favorable event occurring. To incorporate this goal into a decision tree, simply give a reward of 1 to any terminal branch that results in the favorable event occurring and a reward of 0 to any terminal branch that results in the favorable event not occurring. Then maximizing expected reward is the same as maximizing the probability that the favorable event will occur. Use this idea to solve the next two problems.

9 The American chess master Jonathan Meller is playing the Soviet expert Yuri Gasparov in a two-game exhibition match. Each win earns a player one point, and each draw earns a half point. The player who has the most points after two games wins the match. If the players are tied after two games, they play until one wins a game; then the first player to win a game wins the match. During each game, Meller has two possible approaches: to play a daring strategy or to play a conservative strategy. His probabilities of winning, losing, and drawing when he follows each strategy are shown in Table 10. To maximize his probability of winning the match, what should the American do?

10 Yvonne Delaney is playing Chris Becker a single point for the women's world tennis championship. She has won the coin toss and elected to serve. If she tries a hard serve, her probability of getting the serve into play is .60. Given that the hard serve is in play, she has a .60 chance of winning the point. If she tries a soft serve, her probability of getting the serve in play is .90, but if the soft serve is in play, her probability of winning the point is only .50. To maximize her probability of winning the point, what should Yvonne do?

11[†] The Indiana Hoosiers trail the Purdue Boilermakers by a 14–0 score late in the fourth quarter of a football game. Indiana's guardian angel has informed Indiana that before the game ends they will have the ball two more times, and they will score a touchdown each time. The Indiana coach is indifferent between a tie and the following lottery: a 40% chance at beating Purdue and a 60% chance at losing to Purdue. Indiana's kicker has never missed an extra point, and Indiana has been successful on 35% of all two-point conversion attempts. After each touchdown (worth six points), Indiana must decide whether or not to attempt a one-point or a two-point conversion. Help the Indiana coach maximize his expected utility.

12 Edwina, a commodities broker, has acquired an option to buy 1,000 oz of gold at $50/oz. If she takes the option and if Congress relaxes import quotas, she can sell the gold for $80/oz. If she takes the option and Congress does not relax the import quotas, however, the company will lose $10/oz.

TABLE 10

Strategy	Win	Loss	Draw
Daring	.45	.55	0
Conservative	0	.10	.90

Edwina believes that there is a 50% chance that the government will relax the quota. She also has the option of waiting until Congress decides whether to relax the import quota. If she adopts this strategy, however, there is a 70% chance that some other broker will have already taken the option.

a If Edwina is risk-neutral, what should she do?

b If Edwina's utility function for a change x in her asset position is given by $u(x) = (10,000 + x)^{1/2}$, what should she do?

13 We are going to see the movie *Fatal Repulsion*. There are three parking lots we may park in. One is one block east of the theater (call this lot -1); one lot is directly behind the theater (lot 0); and one lot is one block west of the theater (lot 1). We are approaching the theater from the east. There is an 80% chance that lot -1 will have a vacant space, a 60% chance that lot 0 will, and an 80% chance that lot 1 will. Once we pass a lot, we can't go back to it. Assume that when we are at a given parking lot, we can determine whether it has any vacant spaces, but we can't see any of the other lots. Our dates for the evening will assess us a penalty equal to the distance (in blocks) that we park from the theater. If we find no space, they will assess a penalty of 10 (and never go out with us again). What strategy minimizes our expected penalty? Answer the same question if there is a 70% chance that lot 0 has a vacant space.

14[‡] A patient enters the hospital with severe abdominal pains. Based on past experience, Doctor Craig believes there is a 28% chance that the patient has appendicitis and a 72% chance that the patient has nonspecific abdominal pains. Dr. Craig may operate on the patient now or wait 12 hours to gain a more accurate diagnosis. In 12 hours, Dr. Craig will surely know whether the patient has appendicitis. The problem is that in the meantime, the patient's appendix may perforate (if he has appendicitis), thereby making the operation much more dangerous. Again based on past experience, Dr. Craig believes that if he waits 12 hours, there is a 6% chance that the patient will end up with a perforated appendix, a 22% chance the patient will end up with "normal" appendicitis, and a 72% chance that the patient will end up with nonspecific abdominal pain. From past experience, Dr. Craig assesses the probabilities shown in Table 11 of the patient dying. Assume that Dr. Craig's goal is to maximize the probability that the patient will survive. Use a decision tree to help Dr. Craig make the right decision.

15 a Suppose you are given a choice between the following options:

A_1: Win $30 for sure

A_2: 80% chance of winning $45 and 20% chance of winning nothing

[†]Based on Porter (1967).

[‡]Based on Clarke (1981).

TABLE 11

Situation	Probability That Patient Will Die
Operation on patient with appendicitis	.0009
Operation on patient with nonspecific abdominal pain	.0004
Operation on perforated appendix	.0064
No operation on patient with nonspecific abdominal pain	0

B_1: 25% chance of winning \$30

B_2: 20% chance of winning \$45

Most people prefer A_1 to A_2 and B_2 to B_1. Explain why this behavior violates the assumption that decision makers maximize expected utility.

b Now suppose you play the following game: You have a 75% chance of winning nothing and a 25% chance of playing the second stage of the game. If you reach the second stage, you have a choice of two options (C_1 and C_2), but your choice must be made now, before you reach the second stage.

C_1: Win \$30 for sure

C_2: 80% chance of winning \$45

Most people choose C_1 over C_2 and B_2 to B_1 (from part (a)). Explain why this again violates the assumption of expected utility maximization. Tversky and Kahneman (1981) speculate that most people are attracted to the sure \$30 in the second stage, even though the second stage may never be reached! Note that B_1 and C_1 both give \$30 with the same probability, and B_2 and C_2 both yield \$45 with the same probability. It appears that people do not act very rationally![†]

16 You have just been chosen to appear on Hoosier Millionaire! The rules are as follows: There are four hidden cards. One says "STOP" and the other three have dollar amounts of \$150,000, \$200,000, and \$1,000,000. You get to choose a card. If the card says "STOP," you win no money. At any time you may quit and keep the largest amount of money that has appeared on any card you have chosen, or continue. If you continue and choose the stop card, however, you win no money. As an example, you may first choose the \$150,000 card, then the \$200,000 card, and then you may choose to quit and receive \$200,000!

a If you goal is to maximize your expected payoff, what strategy should you follow?

b My utility function for an increase in cash satisfies $u(0) = 0$, $u(\$40,000) = .25$, $u(\$120,000) = .50$, $u(\$400,000) = .75$, and $u(\$1,000,000) = 1$. After drawing a curve through these points, determine a strategy that maximizes my expected utility. You might want to use your own utility function.

[†]Based on Tversky and Kahneman (1981).

2.5 Bayes' Rule and Decision Trees

The Colaco example and many other decision tree problems share several common features.

There are several states of the world. Different states of the world result in different payoffs to the decision maker. In the Colaco example, the two states of the world were that Chocola is a national success (*NS*) or a national failure (*NF*). We are also given (before the test marketing, if any, is done) estimates of the probabilities of each state of the world. These are called **prior probabilities.** In the Colaco example, the prior probabilities are $p(NS) = .55$ and $p(NF) = .45$.

In different states of the world, different decisions may be optimal. In the Colaco example, the company should market nationally if the state of the world is *NS* and not market nationally if the state of the world is *NF*.

It may be desirable to purchase information that gives the decision maker more foreknowledge about the state of the world. This may enable the decision maker to make better decisions. For instance, in the Colaco example, the information obtained from test marketing might help Colaco decide whether or not Chocola should be marketed nationally.

The decision maker receives information by observing the outcomes of an experiment. Let $s_1, s_2, \ldots, s_n$ denote the possible states of the world, and let $o_1, o_2, \ldots, o_m$ be the possible outcomes of the experiment. Often, the decision maker is given the conditional probabilities $p(s_i|o_j)(i = 1, 2, \ldots, n; j = 1, 2, \ldots, m)$. Given knowledge of the outcome of the experiment, these probabilities give new values for the probability of each state of the world. The probabilities $p(s_i|o_j)$ are called **posterior probabilities.**

In the Colaco example, the experiment was the test-marketing procedure, and the two possible outcomes were LF = local failure and LS = local success. The posterior probabilities were given to be

$$p(NS|LS) = .85, \qquad p(NS|LF) = .10,$$
$$p(NF|LS) = .15, \qquad p(NF|LF) = .90$$

Thus, the knowledge of a local test market success would greatly increase Colaco's estimate of the probability of national success, and the knowledge of a local test market failure would greatly decrease Colaco's estimate of the probability of a national success. The posterior probabilities just listed were used to define the event forks in the decision tree that followed the action Test market.

In many situations, however, we may be given the prior probabilities $p(s_i)$ for each state of the world, and instead of being given the posterior probabilities $p(s_i|o_j)$, we might be given the **likelihoods** $p(o_j|s_i)$. For each state of the world, the likelihoods give the probability of observing each experimental outcome. Thus, in the Colaco example, we might be given the prior probabilities $p(NS) = .55$ and $p(NF) = .45$ and the likelihoods $p(LS|NS) = \frac{51}{55}, p(LF|NS) = \frac{4}{55}, p(LS|NF) = \frac{9}{45}$, and $p(LF|NF) = \frac{36}{45}$.

To clarify the meaning of likelihoods, suppose that 55 products that have been national successes had previously been test marketed; of these 55 products, 51 were local successes and 4 were local failures. This would have led us to estimate $p(LS|NS)$ as $\frac{51}{55}$ and $p(LF|NS)$ as $\frac{4}{55}$.

To complete the decision tree in Figure 7, we still need to know the posterior probabilities $p(NS|LS)$, $p(NF|LS)$, $p(NS|LF)$, and $p(NF|LF)$. With the help of Bayes' rule (see Section 1.5), we can use the prior probabilities and likelihoods to determine the needed posterior probabilities. To begin the computation of the posterior probabilities, we need to determine the joint probabilities of each state of the world and experimental outcome (that is, we must determine $p(NS \cap LS)$, $p(NS \cap LF)$, $p(NF \cap LS)$, and $p(NF \cap LF)$). We obtain these joint probabilities by using the definition of conditional probability:

$$p(NS \cap LS) = p(NS)p(LS|NS) = .55(\tfrac{51}{55}) = .51$$
$$p(NS \cap LF) = p(NS)p(LF|NS) = .55(\tfrac{4}{55}) = .04$$
$$p(NF \cap LS) = p(NF)p(LS|NF) = .45(\tfrac{9}{45}) = .09$$
$$p(NF \cap LF) = p(NF)p(LF|NF) = .45(\tfrac{36}{45}) = .36$$

Next we compute the probability of each possible experimental outcome (often called a *marginal probability*) $p(LS)$ and $p(LF)$:

$$p(LS) = p(NS \cap LS) + p(NF \cap LS) = .51 + .09 = .60$$
$$p(LF) = p(NS \cap LF) + p(NF \cap LF) = .04 + .36 = .40$$

Now Bayes' rule can be applied to obtain the desired posterior probabilities:

$$p(NS|LS) = \frac{p(NS \cap LS)}{p(LS)} = \frac{.51}{.60} = .85$$

$$p(NF|LS) = \frac{p(NF \cap LS)}{p(LS)} = \frac{.09}{.60} = .15$$

$$p(NS|LF) = \frac{p(NS \cap LF)}{p(LF)} = \frac{.04}{.40} = .10$$

$$p(NF|LF) = \frac{p(NF \cap LF)}{p(LF)} = \frac{.36}{.40} = .90$$

These posterior probabilities can be used to complete the decision tree in Figure 7.

In summary, to find posterior probabilities, we go through the following three-step process:

Step 1 Determine the joint probabilities of the form $p(s_i \cap o_j)$ by multiplying the prior probability ($p(s_i)$) times the likelihood ($p(o_j|s_i)$).

Step 2 Determine the probabilities of each experimental outcome $p(o_j)$ by summing up all joint probabilities of the form $p(s_k \cap o_j)$.

Step 3 Determine each posterior probability ($p(s_i|o_j)$) by dividing the joint probability ($p(s_i \cap o_j)$) by the probability of the experimental outcome o_j ($p(o_j)$).

We now give a complete example of a decision tree analysis that requires use of Bayes' rule.

EXAMPLE 5   **Fruit Computer Company**

Fruit Computer Company manufactures memory chips in lots of ten chips. From past experience, Fruit knows that 80% of all lots contain 10% (1 out of 10) defective chips, and 20% of all lots contain 50% (5 out of 10) defective chips. If a good (that is, 10% defective) batch of chips is sent on to the next stage of production, processing costs of $1,000 are incurred, and if a bad batch (50% defective) is sent on to the next stage of production, processing costs of $4,000 are incurred. Fruit also has the alternative of reworking a batch at a cost of $1,000. A reworked batch is sure to be a good batch. Alternatively, for a cost of $100, Fruit can test one chip from each batch in an attempt to determine whether the batch is defective. Determine how Fruit can minimize the expected total cost per batch. Also compute EVSI and EVPI.

Solution   We will multiply costs by -1 and work with maximizing $-$(total cost). This enables us to use the EVSI and EVPI formulas of Section 2.4. There are two states of the world:

$$G = \text{batch is good}$$
$$B = \text{batch is bad}$$

We are given the following prior probabilities:

$$p(G) = .80 \quad \text{and} \quad p(B) = .20$$

Fruit has the option of performing an experiment: inspecting one chip per batch. The possible outcomes of the experiment are

$$D = \text{defective chip is observed}$$
$$ND = \text{nondefective chip is observed}$$

We are given the following likelihoods:

$$p(D|G) = .10, \quad p(ND|G) = .90, \quad p(D|B) = .50, \quad P(ND|B) = .50$$

To complete the decision tree in Figure 12, we need to determine the posterior probabilities $p(B|D)$, $p(G|D)$, $p(B|ND)$, and $p(G|ND)$. We begin by computing joint probabilities:

$$p(D \cap G) = p(G)p(D|G) = .80(.10) = .08$$
$$p(D \cap B) = p(B)p(D|B) = .20(.50) = .10$$
$$p(ND \cap G) = p(G)p(ND|G) = .80(.90) = .72$$
$$p(ND \cap B) = p(B)p(ND|B) = .20(.50) = .10$$

We then compute the probability of each experimental outcome:

$$p(D) = p(D \cap G) + p(D \cap B) = .08 + .10 = .18$$
$$p(ND) = p(ND \cap G) + p(ND \cap B) = .72 + .10 = .82$$

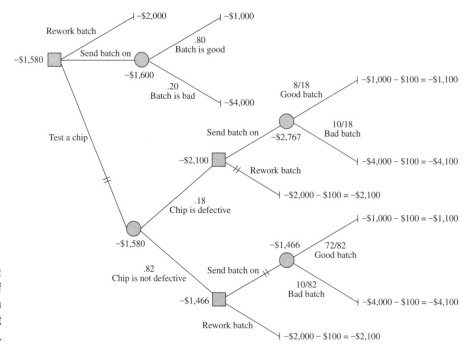

FIGURE 12
Illustration of Use of Bayes' Rule in Decision Tree for Fruit Computer Co.

Then we use Bayes' rule to determine the required posterior probabilities:

$$p(B|D) = \frac{p(D \cap B)}{p(D)} = \frac{.10}{.18} = \frac{5}{9}$$

$$p(G|D) = \frac{p(D \cap G)}{p(D)} = \frac{.08}{.18} = \frac{4}{9}$$

$$p(B|ND) = \frac{p(ND \cap B)}{p(ND)} = \frac{.10}{.82} = \frac{10}{82}$$

$$p(G|ND) = \frac{p(ND \cap G)}{p(ND)} = \frac{.72}{.82} = \frac{72}{82}$$

These posterior probabilities are used to complete the tree in Figure 12. Straightforward computations show that the optimal strategy is to test a chip. If the chip is defective, rework the batch. If the chip is not defective, send the batch on. An expected cost of $1,580 is incurred.

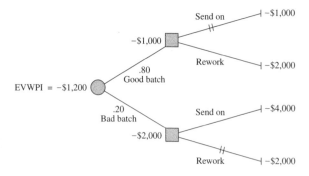

FIGURE 13
Expected Value with Perfect Information (EVWPI) for Fruit Computer

To find EVSI, suppose that testing one chip in a batch were costless. Then the Test chip branch of the tree would have its expected value increased by \$100 (to −\$1,480). Then we would have EVWSI = −\$1,480 and EVWOI = −\$1,600. Then EVSI = EVWSI − EVWOI = −\$1,480 − (−\$1,600) = \$120.

To find EVPI, we use the tree in Figure 13. We find EVWPI = −\$1,200. Then EVPI = EVWPI − EVWOI = −\$1,200 − (−\$1,600) = \$400.

Using LINGO to Compute Posterior Probabilities

The following LINGO program can be used to compute the posterior probabilities for Example 5 (or any other situation).

```
MODEL:
  1]SETS:
  2]ST/G,B/:PR;
  3]OUT/D,ND/:MARG;
  4]SXO(ST,OUT):POST,JOINT,LIKE;
  5]ENDSETS
  6]DATA:
  7]PR=.8,.2;
  8]LIKE=.1,.9,.5,.5;
  9]ENDDATA
 10]@FOR(SXO(I,J):JOINT(I,J)=PR(I)*LIKE(I,J););
 11]@FOR(OUT(J):MARG(J)=@SUM(ST(I):JOINT(I,J)););
 12]@FOR(SXO(I,J):POST(I,J)=JOINT(I,J)/MARG(J););
 13]END
```

Line 2 defines the states (G and B) and associates a prior probability with each state. The prior probabilities are input by the user in the DATA section of the program. Line 3 defines the set of possible experimental outcomes and associates a marginal probability (MARG) with each outcome. The values of MARG are computed in line 11. In line 4, we create the set SXO, consisting of (G, D), (G, ND), (B, D), and (B, ND), and associate with each member of this set the following:

1 A posterior probability (POST); for example, POST(G, D) = $p(G|D)$. The values of POST are computed in line 12.

2 A joint probability (JOINT); for example, JOINT(G, ND) = $p(G \cap ND)$. The values of JOINT are computed in line 10.

3 A likelihood (LIKE); for example, LIKE(B, D) = $p(D|B)$. The likelihoods are input in the DATA section.

Lines 6–9 input the relevant data. Recall that attributes with multiple subscripts (such as SXO) are stored so that the rightmost subscripts advance most rapidly. This helps us determine the order in which to input the values of LIKE.

In line 10, we compute all joint probabilities JOINT(I, J) by multiplying the prior probability PR(I) by the likelihood LIKE(I, J). In line 11 we compute each marginal probability MARG(J) by summing over I all joint probabilities involving J. In line 12, we compute each posterior probability POST(I, J) (really this is $p(I|J)$) by dividing $p(I \cap J)$ by the marginal probability of outcome J ($p(J)$).

PROBLEMS

Group A

1 A customer has approached a bank for a $50,000 one-year loan at 12% interest. If the bank does not approve the loan, the $50,000 will be invested in bonds that earn a 6% annual return. Without further information, the bank feels that there is a 4% chance that the customer will totally default on the loan. If the customer totally defaults, the bank loses $50,000. At a cost of $500, the bank can thoroughly investigate the customer's credit record and supply a favorable or unfavorable recommendation. Past experience indicates that

$$p(\text{favorable recommendation}|$$
$$\text{customer does not default}) = \tfrac{77}{96}$$
$$p(\text{favorable recommendation}|$$
$$\text{customer defaults}) = \tfrac{1}{4}$$

How can the bank maximize its expected profits? Also find EVSI and EVPI.

2 A nuclear power company is deciding whether or not to build a nuclear power plant at Diablo Canyon or at Roy Rogers City. The cost of building the power plant is $10 million at Diablo and $20 million at Roy Rogers City. If the company builds at Diablo, however, and an earthquake occurs at Diablo during the next five years, construction will be terminated and the company will lose $10 million (and will still have to build a power plant at Roy Rogers City). A priori, the company believes there is a 20% chance that an earthquake will occur at Diablo during the next five years. For $1 million, a geologist can be hired to analyze the fault structure at Diablo Canyon. He will either predict that an earthquake will occur or that an earthquake will not occur. The geologist's past record indicates that he will predict an earthquake on 95% of the occasions for which an earthquake will occur and no earthquake on 90% of the occasions for which an earthquake will not occur. Should the power company hire the geologist? Also find EVSI and EVPI.

3 Farmer Jones must determine whether to plant corn or wheat. If he plants corn and the weather is warm, he earns $8,000; if he plants corn and the weather is cold, he earns $5,000. If he plants wheat and the weather is warm, he earns $7,000; if he plants wheat and the weather is cold, he earns $6,500. In the past, 40% of all years have been cold and 60% have been warm. Before planting, Jones can pay $600 for an expert weather forecast. If the year is actually cold, there is a 90% chance that the forecaster will predict a cold year. If the year is actually warm, there is an 80% chance that the forecaster will predict a warm year. How can Jones maximize his expected profits? Also find EVSI and EVPI.

4 The NBS television network earns an average of $400,000 from a hit show and loses an average of $100,000 on a flop. Of all shows reviewed by the network, 25% turn out to be hits and 75% turn out to be flops. For $40,000, a market research firm will have an audience view a pilot of a prospective show and give its view about whether the show will be a hit or a flop. If a show is actually going to be a hit, there is a 90% chance that the market research firm will predict the show to be a hit. If the show is actually going to be a flop, there is an 80% chance that the market research firm will predict the show to be a flop. Determine how the network can maximize its expected profits. Also find EVSI and EVPI.

5 We are thinking of filming the Don Harnett story. We know that if the film is a flop, we will lose $4 million, and if the film is a success, we will earn $15 million. Beforehand, we believe that there is a 10% chance that the Don Harnett story will be a hit. Before filming, we have the option of paying the noted movie critic Roger Alert $1 million for his view of the film. In the past, Alert has predicted 60% of all actual hits to be hits and 90% of all actual flops to be flops. We want to maximize our expected profits. Use a decision tree to determine our best strategy. What is EVSI? What is EVPI?

Group B

6 Abdul has one die in his left hand and one in his right hand. One die has six dots painted on each face, and the other has one dot painted on two of the faces and six dots painted on each of the other four faces. Greta is to pick one die (either "left" or "right") and will receive $10 for each dot painted on the die that is picked. Before choosing, Greta may pay Abdul $15, and he will toss the die in his left hand and tell her how many dots are painted on the face that comes up. Use a decision tree to determine how to maximize Greta's profit. Also determine EVSI and EVPI.

7 Pat Sajork has two drawers. One drawer contains three gold coins, and the other contains one gold coin and two silver coins. We are allowed to choose one drawer, and we will be paid $500 for each gold coin and $100 for each silver coin in that drawer. Before choosing, we may pay Pat $200, and he will draw a randomly selected coin (each of the six coins has an equal chance of being chosen) and tell us whether it is gold or silver. For instance, Pat may say that he drew a gold coin from drawer 1. Should we pay Pat $200? What is EVSI? What is EVPI?

8 Joe owns a coin that is either a fair coin or a two-headed coin. Imelda believes that there is a $\tfrac{1}{2}$ chance that the coin is two-headed. She must guess what kind of coin Joe has. If she guesses correctly, she pays Joe nothing, but if she guesses incorrectly, she must pay Joe $2. Before guessing, she may pay 30¢ to see the result of a single coin toss (heads or tails). Determine how Imelda should minimize her expected loss. Also determine EVSI and EVPI.

9 The government is attempting to determine whether immigrants should be tested for a contagious disease. Let's assume that the decision will be made on a financial basis. Assume that each immigrant who is allowed into the country and has the disease costs the United States $100,000, and each immigrant who enters and does not have the disease will contribute $10,000 to the national economy. Assume that 10% of all potential immigrants have the disease. The government may admit all immigrants, admit no immigrants, or test immigrants for the disease before determining whether they should be admitted. It costs $100 to test a person for the disease; the test result is either positive or negative. If the test result is positive, the person *definitely* has the disease. However, 20% of all people who *do* have the disease test negative. A person who does not have the disease always tests negative. The government's goal is to maximize (per potential immigrant) expected benefits minus expected costs. Use a decision tree to aid in this undertaking. Also determine EVSI and EVPI.

10[†] Many colleges face the problem of whether athletes should be tested for drug use. Define

†Based on Feinstein (1990).

c_1 = Cost if athlete is falsely accused of drug use

c_2 = Cost if a drug user is not identified

c_3 = Cost due to invasion of privacy if a nonuser is tested

Suppose that 5% of all athletes are drug users, and that the test used is 90% reliable. This means that if an athlete uses drugs, there is a 90% chance that the test will detect it, and if the athlete does not use drugs, there is a 90% chance that the test will show no drug use.

a If $c_1 = 10$, $c_2 = 5$, and $c_3 = 1$, should the college test athletes for drugs?

b Prove that if $c_1 > c_2 > c_3$, then the college should not test for drugs.

2.6 Decision Making with Multiple Objectives

In previously considered decision problems, the decision maker made a choice based on how each possible action affected a single variable (or attribute). For example, in the news vendor problem, the number of papers ordered was determined by how this affected Phyllis's profits. Similarly, in the Colaco example, Colaco's decision depended on how each of its strategies affected its final asset position.

In many situations, however, the action chosen depends on how each possible action affects more than one attribute or variable. Four examples follow. (1) Suppose that Joe Bunker wants to buy a new car. In choosing which car to buy, Joe may consider the following attributes of each car:

Attribute 1 Size of car

Attribute 2 Fuel economy of car (miles per gallon)

Attribute 3 Style of car

Attribute 4 Price of car

(2) Suppose Joe Bunker has just graduated from the nation's top business school, Business School (B.S.) University, and has received five job offers. In choosing which to accept, Joe will consider the following attributes of each job:

Attribute 1 Starting salary of job

Attribute 2 Location of job

Attribute 3 Degree of interest Joe has in doing the work involved in the particular job

Attribute 4 Long-term opportunities associated with job

(3) Gotham City must determine where to locate a new jetport. In determining the site three factors (or attributes) must be considered:

Attribute 1 Accessibility of jetport for residents of Gotham City

Attribute 2 Degree of noise pollution caused by the jetport (if the jetport is placed in a densely populated area, noise pollution will be more serious than if the jetport is placed in a sparsely populated area)

Attribute 3 Size of the jetport (determined in part by the amount of land available at the jetport site)

(4) Wivco Toy Corporation is introducing a new product (a globot). Wivco must determine the price to charge for each globot. Two factors (market share and profits) will affect the pricing decision.

In these four examples, the decision maker chooses an action by determining how each possible action affects the relevant attributes. Such problems are called **multiattribute decision problems.**

Multiattribute Decision Making in the Absence of Uncertainty: Goal Programming

Suppose a woman believes that there are n attributes that will determine her decision. Let $x_i(a)$ be the value of the ith attribute associated with an alternative a. She associates a value $v(x_1(a), x_2(a), \ldots, x_n(a))$ with the alternative a. The function $v(x_1, x_2, \ldots, x_n)$ is the decision maker's **value function.** If A represents the decision maker's set of possible decisions, then she should choose the alternative a^* (with level x_i^* of attribute i) satisfying

$$\max_{a \in A} v(x_1(a), x_2(a), \ldots, x_n(a)) = v(x_1^*, x_2^*, \ldots, x_n^*)$$

Alternatively, the decision maker can associate a cost $c(x_1(a), x_2(a), \ldots, x_n(a))$ with the alternative a. The function $c(x_1, x_2, \ldots, x_n)$ is her **cost function.** If A represents the decision maker's set of possible decisions, then she should choose the alternative a^* (with level x_i^* of attribute i) satisfying

$$\min_{a \in A} c(x_1(a), x_2(a), \ldots, x_n(a)) = c(x_1^*, x_2^*, \ldots, x_n^*)$$

A particular form of the value or cost function is of special interest.

DEFINITION ■ A value function $v(x_1, x_2, \ldots, x_n)$ is an **additive value function** if there exist n functions $v_1(x_1), v_2(x_2), \ldots, v_n(x_n)$ satisfying

$$v(x_1, x_2, \ldots, x_n) = \sum_{i=1}^{i=n} v_i(x_i) \quad ■ \tag{7}$$

A cost function $c(x_1, x_2, \ldots, x_n)$ is an **additive cost function** if there exist n functions $c_1(x_1), c_2(x_2), \ldots, c_n(x_n)$ satisfying

$$c(x_1, x_2, \ldots, x_n) = \sum_{i=1}^{i=n} c_i(x_i) \quad ■ \tag{8}$$

Under what conditions will a decision maker have an additive value (or cost) function? Before answering this question, we need some more definitions.

DEFINITION ■ An attribute (call it attribute 1) is **preferentially independent** (pi) of another attribute (attribute 2) if preferences for values of attribute 1 do not depend on the value of attribute 2. ■

To illustrate the concept of preferential independence, we consider Joe's search for a job following graduation. In this situation, attribute 1 would be preferentially independent of attribute 2 if, for any possible job location, a higher starting salary is preferred to a lower salary.

As another illustration of preferential independence, suppose that the Griswold family is trying to determine how to spend Sunday afternoon. Let the two relevant attributes be

Attribute 1 Choice of activity (either picnic or go to see movie *Antarctic Vacation*)

Attribute 2 Sunday afternoon's weather (either sunny or rainy)

Suppose that on a sunny day, the picnic is preferred to the movie, but on a rainy day, the movie is preferred to the picnic. Then attribute 1 is not preferentially independent of attribute 2.

DEFINITION ■ If attribute 1 is pi of attribute 2, and attribute 2 is pi of attribute 1, then attribute 1 is **mutually preferentially independent** (mpi) of attribute 2. ■

Again refer to Joe's search for a job. Suppose Joe's five job offers are located in Los Angeles, Chicago, Dallas, New York, and Indianapolis. If, for any given salary level, Joe prefers to work in Los Angeles, then attribute 2 is pi of attribute 1. If attribute 1 were also pi of attribute 2, then attributes 1 and 2 would be mpi.

The concept of mutual preferential independence can be generalized to sets of attributes.

DEFINITION ■ A set of attributes S is **mutually preferentially independent** (mpi) of a set of attributes S' if (1) the values of the attributes in S' do not affect preferences for the values of attributes in S, and (2) the values of the attributes in S do not affect preferences for the values of attributes in S'. ■

In the example of Joe's purchase of a new car, let S = attributes 1 and 2, and S' = attributes 3 and 4. Then for S to be mpi of S', it must be the case that (1) Joe's preferences for size and fuel economy are unaffected by a car's style and price, and (2) Joe's preferences for car style and price are unaffected by the car's size and fuel economy. Thus, if S and S' were mpi, we could conclude that if for a given style and price level, Joe preferred A_1 (a large car getting 15 mpg) to A_2 (a small car getting 25 mpg), then for any style and price level, Joe would prefer A_1 to A_2.

DEFINITION ■ A set of attributes $1, 2, \ldots, n$ is **mutually preferentially independent** (mpi) if for all subsets S of $\{1, 2, \ldots, n\}$, S is mpi of $\bar{S}$. ($\bar{S}$ is all members of $\{1, 2, \ldots, n\}$ that are not included in S.) ■

It is easy to see that if there are only two attributes (1 and 2), the attributes are mpi if and only if attribute 1 is mpi of attribute 2.

The following result gives a condition ensuring that the decision maker will have an additive value (or cost) function.

THEOREM 1

If the set of attributes $1, 2, \ldots, n$ is mpi, the decision maker's preferences can be represented by an additive value (or cost) function.

This is not an obvious result. (For a proof, see Keeney and Raiffa (1976, Chapter 3).) To illustrate the result, suppose that the decision maker's value function for two attributes is given by

$$v(x_1, x_2) = x_1 + x_1 x_2 + x_2 \tag{9}$$

A decision maker with value function (2) would, for example, prefer (6, 6) to (4, 8) (because $v(6, 6) = 48$ and $v(4, 8) = 44$). The reader should verify that for (2), attribute 1 is pi of attribute 2, and attribute 2 is pi of attribute 1 (see Problem 3 at the end of this section). Thus, attributes 1 and 2 are mpi, and Theorem 1 implies that the decision maker's preferences can be represented by an additive function. To demonstrate this, define new attributes $1'$ and $2'$ as

$$\text{Value of attribute } 1' = x_1' = x_1 + x_2$$
$$\text{Value of attribute } 2' = x_2' = x_1 - x_2$$

Consider the additive value function

$$v'(x_1', x_2') = x_1' + \frac{(x_1')^2}{4} - \frac{(x_2')^2}{4}$$

It is easy to show that $v'(x_1', x_2') = v(x_1, x_2)$. Thus, $v'(x_1', x_2')$ represents the decision maker's preferences. For example, of the following two alternatives

$$\text{Alternative 1: } x_1 = 6, x_2 = 6$$
$$\text{Alternative 2: } x_1 = 4, x_2 = 8$$

we already know that the decision maker prefers alternative 1. In terms of the new attributes $1'$ and $2'$, we have

$$\text{Alternative 1: } x_1' = 12, x_2' = 0$$
$$\text{Alternative 2: } x_1' = 12, x_2' = -4$$

Then

$$v'(12, 0) = \text{value of alternative 1} = 12 + \frac{12^2}{4} = 48$$

$$v'(12, -4) = \text{value of alternative 2} = 12 + \frac{12^2}{4} - \frac{(-4)^2}{4} = 44$$

Therefore, in this example, the additive value function $v'(x_1', x_2')$ replicates the decision maker's preferences.

Multiattribute Utility Functions

In Section 2.2, we described how the Von Neumann–Morgenstern utility theory could be used to make decisions under uncertainty when only one attribute affected the decision maker's preference and attitude toward risk. In this section, we discuss the extension of utility theory to situations in which more than one attribute affects the decision maker's preferences and attitude toward risk. Even in this case, a decision maker who subscribes to the Von Neumann–Morgenstern axioms will still choose the lottery or the alternative that maximizes his or her expected utility. When more than one attribute affects a decision maker's preferences, the person's utility function is called a **multiattribute utility function.** We restrict ourselves here to explaining how to assess and use multiattribute utility functions when only two attributes are operative. The reader seeking a more detailed discussion of multiattribute utility functions is referred to Bunn (1984) and (at a more advanced level) the classic work by Keeney and Raiffa (1976).

Suppose a decision maker's preferences and attitude toward risk depend on two attributes. Let

$$x_i = \text{level of attribute } i$$
$$u(x_1, x_2) = \text{utility associated with level } x_1 \text{ of attribute 1 and level } x_2 \text{ of attribute 2}$$

How can we find a utility function $u(x_1, x_2)$ such that choosing a lottery or alternative that maximizes the expected value of $u(x_1, x_2)$ will yield a decision consistent with the decision maker's preferences and attitude toward risk?

In general, determination of $u(x_1, x_2)$ (or, in the case of n attributes, determination of $u(x_1, x_2, \ldots, x_n)$) is a difficult matter. Under certain conditions, however, the assessment of a utility function $u(x_1, x_2)$ is greatly simplified.

Properties of Multiattribute Utility Functions

DEFINITION ■ Attribute 1 is **utility independent** (ui) of attribute 2 if preferences for lotteries involving different levels of attribute 1 do not depend on the level of attribute 2. ■

Let's reconsider the problem of Wivco Toy Corporation. Wivco is introducing a new product (a globot) and must determine what price to charge for each globot. Two factors (market share and profits) will affect Wivco's pricing decision. Let

$$x_1 = \text{Wivco's market share}$$
$$x_2 = \text{Wivco's profits (millions of dollars)}$$

Suppose that Wivco is indifferent between

$$L_1 \quad \overset{\frac{1}{2}}{\underset{\frac{1}{2}}{\text{——}}} \quad \begin{array}{l} 10\%, \$5 \\ 30\%, \$5 \end{array} \qquad \text{and} \qquad L_1' \overset{1}{\text{——}} 16\%, \$5$$

If attribute 1 (market share) is ui of attribute 2 (profit), Wivco would also be indifferent between

$$L_2 \quad \overset{\frac{1}{2}}{\underset{\frac{1}{2}}{\text{——}}} \quad \begin{array}{l} 10\%, \$20 \\ 30\%, \$20 \end{array} \qquad \text{and} \qquad L_2' \overset{1}{\text{——}} 16\%, \$20$$

In short, if market share is ui of profit, then for any level of profits, a $\frac{1}{2}$ chance at a 10% market share and a $\frac{1}{2}$ chance at a 30% market share has a certainty equivalent of a 16% market share.

DEFINITION ■ If attribute 1 is ui of attribute 2, and attribute 2 is ui of attribute 1, then attributes 1 and 2 are **mutually utility independent** (mui). ■

If attributes 1 and 2 are mui, it can be shown that the decision maker's utility function $u(x_1, x_2)$ must be of the following form:

$$u(x_1, x_2) = k_1 u_1(x_1) + k_2 u_2(x_2) + k_3 u_1(x_1)u_2(x_2) \qquad (10)$$

In (10), k_1, k_2, and k_3 are constants, and $u_1(x_1)$ and $u_2(x_2)$ are functions of x_1 and x_2, respectively. Equation (10) is often called a **multilinear utility function.**

To show that a multilinear utility function exhibits mui, we assume that Wivco's utility function is of the form (10) and that Wivco is indifferent between L_1 and L_1'. If Wivco exhibits mui, then Wivco should also be indifferent between L_2 and L_2'. Using (10), we can now show that $L_1 i L_1'$ implies $L_2 i L_2'$. First, (10) and $L_1 i L_1'$ imply

$$\frac{1}{2}[k_1 u_1(10) + k_2 u_2(5) + k_3 u_1(10)u_2(5)]$$
$$+ \frac{1}{2}[\, k_1 u_1(30) + k_2 u_2(5) + k_3 u_1(30)u_2(5)]$$
$$= k_1 u_1(16) + k_2 u_2(5) + k_3 u_1(16)u_2(5)$$

Simplifying this equation yields (if $k_1 \neq 0$)

$$\tfrac{1}{2}[u_1(10) + u_1(30)] = u_1(16) \tag{11}$$

Using (11), we find

$$
\begin{aligned}
E(U \text{ for } L_2) &= \tfrac{1}{2}[k_1 u_1(10) + k_2 u_2(20) + k_3 u_1(10)u_2(20)] \\
&\quad + \tfrac{1}{2}[k_1 u_1(30) + k_2 u_2(20) + k_3 u_1(30)u_2(20)] \\
&= k_1 u_1(16) + k_2 u_2(20) + k_3 u_1(16)u_2(20) \\
&= E(U \text{ for } L_2')
\end{aligned}
$$

Thus, we see that a multilinear utility function of the form (10) implies that attribute 1 is ui of attribute 2. Similarly, it can be shown that (10) implies that attribute 2 is ui of attribute 1. Thus, (10) implies that attributes 1 and 2 are mui. It can also be shown that if x_1 and x_2 are mui, then $u(x_1, x_2)$ must be of the form (10) (see Keeney and Raiffa (1976)).

THEOREM 2

Attributes 1 and 2 are mui if and only if the decision maker's utility function $u(x_1, x_2)$ is a multilinear function of the form

$$u(x_1, x_2) = k_1 u_1(x_1) + k_2 u_2(x_2) + k_3 u_1(x_1)u_2(x_2) \tag{10}$$

The determination of a decision maker's utility function $u(x_1, x_2)$ can be further simplified if it exhibits **additive independence.** Before defining additive independence, we must define x_1 (best) or x_2 (best) to be the most favorable level of attribute 1 or 2 that can occur; also, x_1 (worst) or x_2 (worst) is the least favorable level of attribute 1 or 2 that can occur.

DEFINITION ■ A decision maker's utility function exhibits **additive independence** if the decision maker is indifferent between

Essentially, additive independence of attributes 1 and 2 implies that preferences over lotteries involving only attribute 1 (or only attribute 2) depend only on the marginal distribution for possible values of attribute 1 (or of attribute 2) and do not depend on the joint distribution of the possible values of attributes 1 and 2.

If attributes 1 and 2 are mui and the decision maker's utility function exhibits additive independence, it is easy to show that in (10), $k_3 = 0$ must hold. As in Section 2.2, simply scale $u_1(x_1)$ and $u_2(x_2)$ such that

$$
\begin{array}{ll}
u_1(x_1(\text{best})) = 1 & u_1(x_1(\text{worst})) = 0 \\
u_2(x_2(\text{best})) = 1 & u_2(x_2(\text{worst})) = 0
\end{array}
$$

Now (10) implies

$$
\begin{array}{ll}
u(x_1(\text{best}), x_2(\text{best})) = k_1 + k_2 + k_3 & u(x_1(\text{worst}), x_2(\text{worst})) = 0 \\
u(x_1(\text{best}), x_2(\text{worst})) = k_1 & u(x_1(\text{worst}), x_2(\text{best})) = k_2
\end{array}
$$

Then additive independence implies that

$$\tfrac{1}{2}(k_1 + k_2 + k_3) + \tfrac{1}{2}(0) = \tfrac{1}{2}(k_1) + \tfrac{1}{2}(k_2)$$
$$k_3 = 0$$

Thus, if attributes 1 and 2 are mui and the decision maker's utility function exhibits additive independence, his or her utility function is of the following additive form:

$$u(x_1, x_2) = k_1 u_1(x_1) + k_2 u_2(x_2) \tag{12}$$

Assessment of Multiattribute Utility Functions

If attributes 1 and 2 are mui, how can we determine $u_1(x_1)$, $u_2(x_2)$, k_1, k_2, and k_3? To determine $u_1(x_1)$ and $u_2(x_2)$, we apply the technique that was used to assess utility functions in Section 2.2. We illustrate by determining $u_1(x_1)$. Let $u_1(x_1(\text{best})) = 1$ and $u_1(x_1(\text{worst})) = 0$. Next determine a value of attribute 1 (call it $x_1(\tfrac{1}{2})$) having $u_1(x_1(\tfrac{1}{2})) = \tfrac{1}{2}$. By the definition of $u_1(x_1(\tfrac{1}{2}))$ and mui, the decision maker is (for any value of x_2) indifferent between

Thus, $x_1(\tfrac{1}{2})$ may be determined from the fact that the certainty equivalent of L is $(x_1(\tfrac{1}{2}), x_2)$. In a similar fashion, we can determine values $x_1(\tfrac{1}{4})$ and $x_1(\tfrac{3}{4})$ of the first attribute satisfying $u_1(x_1(\tfrac{1}{4})) = \tfrac{1}{4}$ and $u_1(x_1(\tfrac{3}{4})) = \tfrac{3}{4}$. Continuing in this fashion, we may approximate $u_1(x_1)$ and $u_2(x_2)$.

To find k_1, k_2, and k_3, we begin by rescaling $u_1(x_1)$, $u_2(x_2)$, and $u(x_1, x_2)$ such that

$$u(x_1(\text{best}), x_2(\text{best})) = 1, \qquad u(x_1(\text{worst}), x_2(\text{worst})) = 0,$$
$$u_1(x_1(\text{best})) = 1, \qquad u_1(x_1(\text{worst})) = 0, \qquad u_2(x_2(\text{best})) = 1, \qquad u_2(x_2(\text{worst})) = 0$$

Now (10) yields

$$u(x_1(\text{best}), x_2(\text{worst})) = k_1(1) + k_2(0) + k_3(0) = k_1$$

Thus, k_1 can be determined from the fact that the decision maker is indifferent between

Similarly (see Problem 3 at the end of this section), $u(x_1(\text{worst}), x_2(\text{best})) = k_2$ and k_2 can be determined from the fact that the decision maker is indifferent between

To determine k_3, observe that from (10) and

$$u(x_1(\text{best}), x_2(\text{best})) = u_1(x_1(\text{best})) = u_2(x_2(\text{best})) = 1$$

we find that

$$1 = u(x_1(\text{best}), x_2(\text{best})) = k_1(1) + k_2(1) + k_3(1) = k_1 + k_2 + k_3$$

Thus, $k_1 + k_2 + k_3 = 1$, or $k_3 = 1 - k_1 - k_2$. Of course, if the decision maker's utility function exhibits additive independence, then $k_3 = 0$.

The procedure to be used in assessing a multiattribute utility function (when there are two attributes) may be summarized as follows:

Step 1 Check whether attributes 1 and 2 are mui. If they are, go on to step 2. If the attributes are not mui, the assessment of the multiattribute utility function is beyond the scope of our discussion. (See Keeney and Raiffa (1976, Section 5.7).)

Step 2 Check for additive independence.

Step 3 Assess $u_1(x_1)$ and $u_2(x_2)$.

Step 4 Determine k_1, k_2, and (if there is no additive independence) k_3.

Step 5 Check to see whether the assessed utility function is really consistent with the decision maker's preferences. To do this, set up several lotteries and use the expected utility of each to rank the lotteries from most to least favorable. Then ask the decision maker to rank the lotteries from most to least favorable. If the assessed utility function is consistent with the decision maker's preferences, the ranking of the lotteries obtained from the assessed utility function should closely resemble the decision maker's ranking of the lotteries.

Example 6 illustrates the assessment and use of a multiattribute utility function.

EXAMPLE 6　　**Fruit Computer Co.**

Fruit Computer Company is certain that its market share during 2005 will be between 10% and 50% of the microcomputer market. Fruit is also sure that its profits during 2005 will be between $5 million and $30 million. Assess Fruit's multiattribute utility function where $u(x_1, x_2)$, where

$$x_1 = \text{Fruit's market share during 2005}$$
$$x_2 = \text{Fruit's profit during 2005 (in millions of dollars)}$$

Solution　We begin by checking for mui. It is helpful to draw a diagram (see Figure 14) that displays various levels of each attribute. First, we check whether attribute 1 (market share) is ui of attribute 2 (profit). We ask Fruit for the certainty equivalent of a $\frac{1}{2}$ chance at the

FIGURE 14
Possible Levels of Each Attribute for Fruit Computer Company

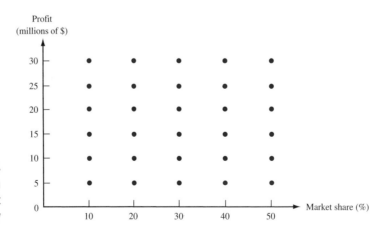

worst market share (10%) and a $\frac{1}{2}$ chance at the best market share (50%), with x_2 fixed at some level (say $x_2 = \$15$ million). Suppose the certainty equivalent of

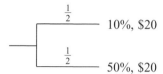

is (30%, \$15). To determine whether attribute 1 is ui of attribute 2, we fix attribute 2 at some other level (say, $x_2 = \$20$ million) and find Fruit's certainty equivalent for the following lottery:

```
         1
         2 ──── 10%, $20
    ┌────
    │    1
    └────2 ──── 50%, $20
```

If attribute 1 is ui of attribute 2, the certainty equivalent of this lottery should be close to (30%, \$20). For other values of x_2 (say, $x_2 = \$5, \$10, \$25,$ and \$30), we check if the certainty equivalent of the lottery

```
         1
         2 ──── 10%, x₂
    ┌────
    │    1
    └────2 ──── 50%, x₂
```

is close to (30%, x_2). Suppose this is the case. Then we repeat this procedure with other values of market share replacing 10% and 50%. If similar results ensue, then attribute 1 is ui of attribute 2. In analogous fashion, we can determine whether attribute 2 is ui of attribute 1. If attribute 1 is ui of attribute 2, and attribute 2 is ui of attribute 1, the two attributes are mui. Let's assume that attributes 1 and 2 are (at least approximately) mui and proceed to step 2 (checking for additive independence).

To check for additive independence, we must determine whether Fruit is indifferent between

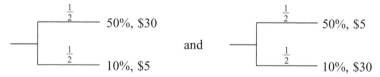

Suppose that Fruit is not indifferent between these lotteries. Then Fruit's utility function will not exhibit additive independence. We now know that $u(x_1, x_2)$ may be written as

$$u(x_1, x_2) = k_1 u_1(x_1) + k_2 u_2(x_2) + k_3 u_1(x_1) u_2(x_2)$$

We now proceed to step 3 (assessing $u_1(x_1)$ and $u_2(x_2)$). Suppose we obtain the results shown in Figure 15. To complete the assessment of Fruit's multiattribute utility function, we must determine k_1, k_2, and k_3 (step 4). To find k_1, we ask Fruit to determine the number k_1 that makes Fruit indifferent between

```
                            k₁
                        ┌──────── 50%, $30
    ──1── 50%, $5  and  ─┤
                        └──────── 10%, $5
                         1 − k₁
```

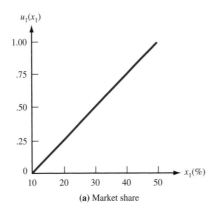

(a) Market share

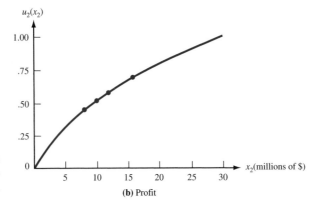

(b) Profit

FIGURE 15
$u_1(x_1)$ and $u_2(x_2)$ for
Fruit Computer

Suppose that for $k_1 = 0.6$, Fruit is indifferent between these two lotteries. Similarly, k_2 is the number that makes Fruit indifferent between

$$\frac{1}{\quad} \quad 10\%, \$30 \qquad \text{and} \qquad \begin{array}{l} k_2 \quad 50\%, \$30 \\ 1 - k_2 \quad 10\%, \$5 \end{array}$$

Suppose that for $k_2 = 0.5$, Fruit is indifferent between these two lotteries. Now $k_3 = 1 - k_1 - k_2 = -0.1$, and Fruit's multiattribute utility function is

$$u(x_1, x_2) = 0.6u_1(x_1) + 0.5u_2(x_2) - 0.1u_1(x_1)u_2(x_2) \tag{13}$$

where $u_1(x_1)$ and $u_2(x_2)$ are sketched in Figure 15. Note that

$$\frac{\partial u(x_1, x_2)}{\partial x_1} = 0.6u_1'(x_1) - 0.1u_1'(x_1)u_2(x_2)$$

Thus, as Fruit's profit increases, we see that the utility gained from an additional point of market share decreases. Similarly, if $k_3 > 0$, then as profit increases, the benefit gained from an additional point of market share would increase. As outlined in step 5, we should now check whether this multiattribute utility function is consistent with Fruit's preferences.

Use of Multiattribute Utility Functions

To illustrate how a multiattribute utility function might be used, suppose that Fruit must determine whether to mount a small or a large advertising campaign during the coming year. Fruit believes there is a $\frac{1}{2}$ probability that its main rival, CSL Computers, will mount a small TV ad campaign and a $\frac{1}{2}$ probability that CSL will mount a large TV ad campaign. At the end of the current year, Fruit's market share and profits (in millions of dollars) will be as shown in Table 12. Fruit must determine which of the following lotteries has a larger expected utility: From Figure 15, we find that $u_1(15) = 0.125$, $u_1(25) = 0.375$, $u_1(35) = 0.625$, $u_2(8) = 0.45$, $u_2(10) = 0.53$, $u_2(12) = 0.58$, and $u_2(16) = 0.70$. Then

$$u(25\%, \$16) = 0.6(.375) + 0.5(.7) - 0.1(.375)(.7) = .549$$
$$u(15\%, \$12) = 0.6(.125) + 0.5(.58) - 0.1(.125)(.58) = .358$$
$$u(35\%, \$8) = 0.6(.625) + 0.5(.45) - 0.1(.625)(.45) = .572$$
$$u(25\%, \$10) = 0.6(.375) + 0.5(.53) - 0.1(.375)(.53) = .470$$

Then

$$E(U \text{ for small ad campaign}) = (\tfrac{1}{2})(.549) + (\tfrac{1}{2})(.358) = .454$$
$$E(U \text{ for large ad campaign}) = (\tfrac{1}{2})(.572) + (\tfrac{1}{2})(.470) = .521$$

Thus, during the current year, Fruit should mount a large ad campaign.

TABLE 12
Effect of Advertising on Market Share and Profit

	CSL Chooses	
Fruit Chooses	Small Ad Campaign	Large Ad Campaign
Small ad campaign	25%, $16	15%, $12
Large ad campaign	35%, $8	25%, $10

PROBLEMS

Group A

1 National Express Carriers is interested in two attributes:

Attribute 1 The average cost of delivering a letter (known to be between $1 and $5)

Attribute 2 Percentage of all letters reaching their destination on time (known to be between 70% and 100%)

a Would National's multiattribute utility function exhibit mui?

b Would National's utility function be additive?

c Assume that attributes 1 and 2 exhibit mui. Suppose National is indifferent between ($1, 70%) for certain and the following lottery:

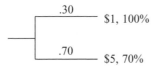

Also assume that National is indifferent between ($5, 100%) for certain and

Find National's multiattribute utility function. Express National's multiattribute utility function in terms of $u_1(x_1)$ and $u_2(x_2)$.

2 Keeney and Raiffa (1976) discuss the assessment of a blood bank's multiattribute utility function. For simplicity, we assume that the blood bank must determine at the beginning of each week how many pints of blood should be ordered. Any blood left over at the end of the week spoils (it is outdated). For the blood bank, two attributes of interest are as follows:

Attribute 1 Number of pints of blood by which ordered blood falls short of the week's demand (the weekly shortage). The weekly shortage is known to be always between 0 and 10 pints.

Attribute 2 Number of pints of blood that are outdated (known to be always between 0 and 10 pints)

Assume that attributes 1 and 2 exhibit mui.

a Suppose the blood bank is indifferent between

and between

Let x_1 = value of attribute 1, and x_2 = value of attribute 2. Also suppose that

$$u_1(x_1) = .58 \exp\left(1 - \frac{x_1}{10}\right) - .58$$

and

$$u_2(x_2) = 1 - \frac{x_2^2}{100}$$

Determine the blood bank's multiattribute utility function.

b Suppose that each week there is a $\frac{1}{2}$ chance that the demand for blood will be 25 pints and a $\frac{1}{2}$ chance it will be 35 pints. Would the blood bank be better off ordering 28 pints, 30 pints, or 32 pints?

3 Show that the method for determining k_2 described in the text is valid.

4 Gotham City is trying to determine how many ambulances it should have and how to staff them. Each ambulance may be staffed with paramedics or emergency medical technicians. Paramedics are considered to provide better service and are paid higher salaries. Budgetary limitations have forced the city to choose between the following two alternatives:

Alternative 1 Four ambulances, two staffed with emergency medical technicians and two staffed with paramedics

Alternative 2 Three ambulances, all staffed with paramedics

The city authorities believe that the following two attributes determine the city's satisfaction with ambulance service:

Attribute 1 Time until an ambulance reaches a patient

Attribute 2 Percentage of ambulance calls handled by paramedics. Assume that Gotham City's multiattribute utility function $u(x_1, x_2)$ exhibits mui and that

$$u_1(x_1) = 1 - \frac{x_1^2}{900} \quad \text{and} \quad u_2(x_2) = \frac{x_2^2}{10,000}$$

The time for an ambulance to reach a patient is always between 0 and 30 minutes. The city authorities are indifferent between

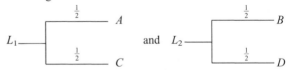

The city authorities are also indifferent between

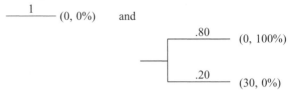

Assume that if an ambulance is available when a call comes in, then the ambulance will arrive in 5 minutes; if an ambulance is not available when a call comes in, it will arrive in 20 minutes. With three ambulances, one will be immediately available 60% of the time, and with four ambulances, one will be immediately available 80% of the time.

a Determine the city authorities' multiattribute utility function.

b Which alternative should they choose?

5 Public service Indiana (PSI) is considering two sites for a nuclear power plant. The following two attributes will influence its determination about where to build the plant:

Attribute 1 Cost of the plant (in millions of dollars)

Attribute 2 Acres of land damaged by building the plant

Assume that PSI's multiattribute utility function is given by $u(x_1, x_2) = .70u_1(x_1) + .20u_2(x_2) + .10u_1(x_1)u_2(x_2)$, where $u_1(x_1) = .1 + \exp(-.1x_1)$ and $u_2(x_2) = 2.5 - 2.5 \exp(.0006x_2 - .48)$.

Two locations for the power plant are under consideration. Location 1 is equivalent to the following lottery:

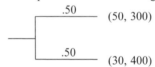

and location 2 is equivalent to the following lottery:

Which location should be chosen?

Group B

6 Consider the four points A, B, C, and D in Figure 16. Assume that more of each attribute is desirable and that a decision maker's utility function exhibits mui. Consider the following two lotteries:

L_1 ⎰ $\frac{1}{2}$ — A ⎱ $\frac{1}{2}$ — C and L_2 ⎰ $\frac{1}{2}$ — B ⎱ $\frac{1}{2}$ — D

a Show that if $k_3 > 0$, then $L_1 p L_2$.

b Show that if $k_3 < 0$, then $L_2 p L_1$.

c Show that if the decision maker exhibits additive independence ($k_3 = 0$), then $L_1 i L_2$.

d Let attribute 1 = performance of Germany on the eastern front near the end of World War II, and attribute 2 = performance of Germany on the western front. A high level of an attribute means that Germany did well,

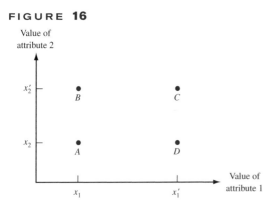

FIGURE 16

Value of
attribute 2

Value of
attribute 1

and a low level of an attribute means that Germany did poorly. Suppose that Germany will suffer defeat if it performs poorly on either front. If these attributes exhibit mui, what would be the sign of k_3?

e General Motors has domestic and international divisions. Let attribute 1 = profits in the domestic division and attribute 2 = profits in the international division. Suppose General Motors is reasonably happy if at least one division has a good year but is very unhappy if both divisions have a bad year. If these attributes exhibit mui, what would be the sign of k_3?

2.7 The Analytic Hierarchy Process

In Section 2.6, we discussed situations in which a decision maker chooses between alternatives on the basis of how well the alternatives meet various objectives. For example, in determining which job offer to accept, a job seeker (call her Jane) might choose between the offers by determining how well each one meets the following four objectives:

Objective 1 High starting salary (SAL)

Objective 2 Quality of life in city where job is located (QL)

Objective 3 Interest in work (IW)

Objective 4 Job location near family and relatives (NF)

When multiple objectives are important to a decision maker, it may be difficult to choose between alternatives. For example, one job offer may offer the highest starting salary, but it may score poorly on the other three objectives. Another job offer may meet objectives 2–4 but have a low starting salary. In such a case, it may be difficult for Jane to choose between job offers. Thomas Saaty's **analytic hierarchy process** (AHP) provides a powerful tool that can be used to make decisions in situations involving multiple objectives.

To illustrate how the AHP works, let's suppose that Jane has three job offers and must determine which offer to accept. For the ith objective (in this example, $i = 1, 2, 3, 4$), the AHP generates (by a method to be described shortly) a weight w_i ($i = 1, 2, 3, 4$) for the ith objective. For convenience, the chosen weights always sum to 1. Suppose that for this example, we have found Jane's weights to be

$$w_1 = .5115, \qquad w_2 = .0986, \qquad w_3 = .2433, \qquad w_4 = .1466$$

(These weights fail to add up to 1 due to rounding.) The weights indicate that a high starting salary is the most important objective, followed by interest in work, nearness to family, and quality of life in the city where the job is located.

Next suppose (again by a method that is soon to be described) that Jane can determine how well each job "scores" on each objective. For example, suppose Jane determines that each job scores on each objective as shown in Table 13. For example, job 1 best meets the objective of a high starting salary but "scores" worst on all other objectives.

Given Jane's weights and the score of each job on each objective, how can she determine which job offer to accept? For the jth job offer ($j = 1, 2, 3$), compute job offer j's overall score as follows:

$$\sum_{i=1}^{i=4} w_i \, (\text{job offer } j\text{'s score on objective } i)$$

TABLE **13**
Jane's "Score" for Each Job and Objective

Objective	Job 1	Job 2	Job 3
Salary	.571	.286	.143
Quality of life	.159	.252	.589
Interest in work	.088	.669	.243
Proximity to family	.069	.426	.506

Now choose the job offer with the highest overall score. Note that the overall score gives more weight to a job offer's score on the more important objectives. Computing each job's overall score, we obtain

$$\text{Job 1 overall score} = .5115(.571) + .0986(.159) + .2433(.088)$$
$$+ .1466(.069) = .339$$
$$\text{Job 2 overall score} = .5115(.286) + .0986(.252) + .2433(.669)$$
$$+ .1466(.426) = .396$$
$$\text{Job 3 overall score} = .5115(.143) + .0986(.589) + .2433(.243)$$
$$+ .1466(.506) = .265$$

Thus, the AHP would indicate that Jane should accept job 2.

Obtaining Weights for Each Objective

Suppose there are n objectives. We begin by writing down an $n \times n$ matrix (known as the **pairwise comparison matrix**) A. The entry in row i and column j of A (call it a_{ij}) indicates how much more important objective i is than objective j. "Importance" is to be measured on an integer-valued 1–9 scale, with each number having the interpretation shown in Table 14. For all i, it is necessary that $a_{ii} = 1$. If, for example, $a_{13} = 3$, objective 1 is weakly more important than objective 3. If $a_{ij} = k$, then for consistency, it is necessary that $a_{ji} = \frac{1}{k}$. Thus, if $a_{13} = 3$, then $a_{31} = \frac{1}{3}$ must hold.

Suppose that Jane has identified the following pairwise comparison matrix for her four objectives (SAL = high salary; QL = high quality of life; IW = interest in work; NF = nearness to family):

$$
\begin{array}{c}
\\ SAL \\ QL \\ IW \\ NF
\end{array}
\begin{array}{cccc}
SAL & QL & IW & NF \\
\left[\begin{array}{cccc}
1 & 5 & 2 & 4 \\
\frac{1}{5} & 1 & \frac{1}{2} & \frac{1}{2} \\
\frac{1}{2} & 2 & 1 & 2 \\
\frac{1}{4} & 2 & \frac{1}{2} & 1
\end{array}\right]
\end{array}
$$

Unfortunately, some of Jane's pairwise comparisons are inconsistent. To illustrate the meaning of consistency, note that since $a_{13} = 2$, she feels SAL is twice as important as IW. Since $a_{32} = 2$, she also believes that IW is twice as important as QL. Consistency of preferences would imply that Jane should feel that SAL is $2(2) = 4$ times as important as QL. Since $a_{12} = 5$, however, Jane believes that SAL is 5 times as important as QL. This shows that Jane's pairwise comparisons exhibit a slight inconsistency. Slight inconsistencies are common and do not cause serious difficulties. An index that can be used to measure the consistency of Jane's preferences will be discussed later in this section.

TABLE **14**
Interpretation of Entries in a Pairwise Comparison Matrix

Value of a_{ij}	Interpretation
1	Objective i and j are of equal importance.
3	Objective i is weakly more important than objective j.
5	Experience and judgment indicate that objective i is strongly more important than objective j.
7	Objective i is very strongly or demonstrably more important than objective j.
9	Objective i is absolutely more important than objective j.
2, 4, 6, 8	Intermediate values—for example, a value of 8 means that objective i is midway between strongly and absolutely more important than objective j.

Suppose there are n objectives. Let w_i = the weight given to objective i. To describe how the AHP determines the w_i's, let's suppose the decision maker is perfectly consistent. Then her pairwise comparison matrix should be of the following form:

$$A = \begin{bmatrix} \dfrac{w_1}{w_1} & \dfrac{w_1}{w_2} & \cdots & \dfrac{w_1}{w_n} \\[2ex] \dfrac{w_2}{w_1} & \dfrac{w_2}{w_2} & \cdots & \dfrac{w_2}{w_n} \\[2ex] \vdots & \vdots & & \vdots \\[2ex] \dfrac{w_n}{w_1} & \dfrac{w_n}{w_2} & \cdots & \dfrac{w_n}{w_n} \end{bmatrix} \tag{13}$$

For example, suppose that $w_1 = \frac{1}{2}$ and $w_2 = \frac{1}{6}$. Then objective 1 is three times as important as objective 2, so

$$a_{12} = \frac{w_1}{w_2} = 3$$

Now suppose that a consistent decision maker has a pairwise comparison matrix A of the form (13). How can we recover the vector $\mathbf{w} = [w_1 \quad w_2 \quad \cdots \quad w_n]$ from A? Consider the system of n equations

$$A\mathbf{w}^T = \Delta\mathbf{w}^T \tag{14}$$

where Δ is an unknown number and $\mathbf{w}^T$ is an unknown n-dimensional column vector. For any number Δ, (14) always has the trivial solution $\mathbf{w} = [0 \quad 0 \quad \cdots \quad 0]$. It can be shown that if A is the pairwise comparison matrix of a perfectly consistent decision maker (that is, if A is of the form (13)) and we do not allow $\Delta = 0$, then the only nontrivial solution to (14) is $\Delta = n$ and $\mathbf{w} = [w_1 \quad w_2 \quad \cdots \quad w_n]$. This shows that for a consistent decision maker, the weights w_i can be obtained from the only nontrivial solution to (14). Now suppose that the decision maker is not perfectly consistent. Let $\Delta_{\max}$ be the largest number for which (14) has a nontrivial solution (call this solution $\mathbf{w}_{\max}$). If the decision maker's comparisons do not deviate very much from perfect consistency, we would expect $\Delta_{\max}$ to be close to n and $\mathbf{w}_{\max}$ to be close to $\mathbf{w}$. Saaty verified that this intuition is indeed correct and suggested approximating $\mathbf{w}$ by $\mathbf{w}_{\max}$. Saaty also proposed measuring the decision maker's consistency by looking how close $\Delta_{\max}$ is to n. The software package Expert Choice gives (among other outputs) exact values of $\Delta_{\max}$ and $\mathbf{w}_{\max}$ and a measure of the decision maker's consistency.

In what follows, we outline a simple method (easily implemented on any spreadsheet) that can be used to approximate Δ_{max} and $\mathbf{w}_{max}$ and an index of consistency.

To approximate $\mathbf{w}_{max}$, we use the following two-step procedure:

Step 1 For each of A's columns, do the following. Divide each entry in column i of A by the sum of the entries in column i. This yields a new matrix (call it A_{norm}, for normalized) in which the sum of the entries in each column is 1. For Jane's pairwise comparison matrix, step 1 yields

$$A_{norm} = \begin{bmatrix} .5128 & .5000 & .5000 & .5333 \\ .1026 & .1000 & .1250 & .0667 \\ .2564 & .2000 & .2500 & .2667 \\ .1282 & .2000 & .1250 & .1333 \end{bmatrix}$$

Step 2 To find an approximation to $\mathbf{w}_{max}$ (to be used as our estimate of $\mathbf{w}$), proceed as follows. Estimate w_i as the average of the entries in row i of A_{norm}. This yields (as previously stated)

$$w_1 = \frac{.5128 + .5000 + .5000 + .5333}{4} = .5115$$

$$w_2 = \frac{.1026 + .1000 + .1250 + .0667}{4} = .0986$$

$$w_3 = \frac{.2564 + .2000 + .2500 + .2667}{4} = .2433$$

$$w_4 = \frac{.1282 + .2000 + .1250 + .1333}{4} = .1466$$

Intuitively, why does w_1 approximate the weight that objective 1 (salary) should be given? The percentage of the weight that SAL is given in pairwise comparisons of each objective to SAL is .5128. Similarly, .50 represents the percentage of total weight that SAL is given in pairwise comparisons of each objective to QL. Thus, we see that the four numbers averaged to obtain w_1 each represents in some way a measure of the total weight attached to SAL. Thus, averaging these numbers should give a good estimate of the percentage of the total weight that should be given to SAL.

Checking for Consistency

We can now use the following four-step procedure to check for the consistency of the decision maker's comparisons. (From now on, $\mathbf{w}$ denotes our estimate of the decision maker's weights.)

Step 1 Compute $A\mathbf{w}^T$. For our example, we obtain

$$A\mathbf{w}^T = \begin{bmatrix} 1 & 5 & 2 & 4 \\ \frac{1}{5} & 1 & \frac{1}{2} & \frac{1}{2} \\ \frac{1}{2} & 2 & 1 & 2 \\ \frac{1}{4} & 2 & \frac{1}{2} & 1 \end{bmatrix} \begin{bmatrix} .5115 \\ .0986 \\ .2433 \\ .1466 \end{bmatrix} = \begin{bmatrix} 2.0775 \\ 0.3959 \\ 0.9894 \\ 0.5933 \end{bmatrix}$$

Step 2 Compute

$$\frac{1}{n} \sum_{i=1}^{i=n} \frac{i\text{th entry in } A\mathbf{w}^T}{i\text{th entry in } \mathbf{w}^T}$$

$$= \left(\frac{1}{4}\right) \left\{ \frac{2.0775}{.5115} + \frac{.3959}{.0986} + \frac{.9894}{.2433} + \frac{.5933}{.1466} \right\}$$

$$= 4.05$$

TABLE 15

Values of the
Random Index (RI)

n	RI
2	0
3	.58
4	.90
5	1.12
6	1.24
7	1.32
8	1.41
9	1.45
10	1.51

Step 3 Compute the **consistency index** (CI) as follows:

$$CI = \frac{(\text{Step 2 result}) - n}{n-1} = \frac{4.05 - 4}{3} = .017$$

Step 4 Compare CI to the random index (RI) for the appropriate value of n, shown in Table 15.

For a perfectly consistent decision maker (see Problem 5), the ith entry in $A\mathbf{w}^T = n$ (ith entry of $\mathbf{w}^T$). This implies that a perfectly consistent decision maker has CI = 0. The values of RI in Table 15 give the average value of CI if the entries in A were chosen at random, subject to the constraint that all diagonal entries must equal 1 and

$$a_{ij} = \frac{1}{a_{ji}}$$

If CI is sufficiently small, the decision maker's comparisons are probably consistent enough to give useful estimates of the weights for his or her objective function. If $\frac{CI}{RI} < .10$, the degree of consistency is satisfactory, but if $\frac{CI}{RI} > .10$, serious inconsistencies may exist, and the AHP may not yield meaningful results. In our example, $\frac{CI}{RI} = \frac{.017}{.90} = .019 < .10$; thus, Jane's pairwise comparison matrix does not exhibit any serious inconsistencies.

Finding the Score of an Alternative for an Objective

We have now described how to determine the objective function weights that we earlier used to help Jane determine which job offer to accept. We now determine how well each job "satisfies" or "scores" on each objective. To determine these scores, we construct for each objective a pairwise comparison matrix in which the rows and columns are Jane's possible decisions (in this case, job offers). For SAL, suppose we obtain the following pairwise comparison matrix:

$$
\begin{array}{c c c c}
 & \text{Job 1} & \text{Job 2} & \text{Job 3} \\
\begin{array}{c} \text{Job 1} \\ \text{Job 2} \\ \text{Job 3} \end{array} &
\left[\begin{array}{ccc}
1 & 2 & 4 \\
\frac{1}{2} & 1 & 2 \\
\frac{1}{4} & \frac{1}{2} & 1
\end{array} \right]
\end{array}
$$

Thus, for example, with respect to salary, job 1 is better (between weakly and strongly) than job 3. We can now apply our procedure for generating weights to the SAL pairwise comparison matrix. We obtain

$$A_{\text{norm}} = \begin{bmatrix} .571 & .571 & .571 \\ .286 & .286 & .286 \\ .143 & .143 & .143 \end{bmatrix}$$

This yields $\mathbf{w} = [.571 \ .286 \ .143]$. These weights indicate how well each job "scores" with respect to the SAL objective. As previously stated in Table 13, we obtain

$$\text{Job 1 salary score} = .571$$
$$\text{Job 2 salary score} = .286$$
$$\text{Job 3 salary score} = .143$$

Since all three columns of the pairwise comparison matrix for salary are identical, Jane's pairwise comparisons for salary exhibit perfect consistency.

Suppose Jane's pairwise comparison matrix for quality of life (QL) is as follows:

$$\begin{array}{c} \\ \text{Job 1} \\ \text{Job 2} \\ \text{Job 3} \end{array} \begin{array}{ccc} \text{Job 1} & \text{Job 2} & \text{Job 3} \end{array} \\ \begin{bmatrix} 1 & \frac{1}{2} & \frac{1}{3} \\ 2 & 1 & \frac{1}{3} \\ 3 & 3 & 1 \end{bmatrix}$$

Then

$$A_{\text{norm}} = \begin{bmatrix} \frac{1}{6} & \frac{1}{9} & \frac{1}{5} \\ \frac{1}{3} & \frac{2}{9} & \frac{1}{5} \\ \frac{1}{2} & \frac{6}{9} & \frac{3}{5} \end{bmatrix}$$

and we obtain

$$\text{Job 1 quality of life score} = \frac{\frac{1}{6} + \frac{1}{9} + \frac{1}{5}}{3} = .159$$

$$\text{Job 2 quality of life score} = \frac{\frac{1}{3} + \frac{2}{9} + \frac{1}{5}}{3} = .252$$

$$\text{Job 3 quality of life score} = \frac{\frac{1}{2} + \frac{6}{9} + \frac{3}{5}}{3} = .589$$

For interest in work, suppose the pairwise comparison matrix is as follows:

$$\begin{array}{c} \\ \text{Job 1} \\ \text{Job 2} \\ \text{Job 3} \end{array} \begin{array}{ccc} \text{Job 1} & \text{Job 2} & \text{Job 3} \end{array} \\ \begin{bmatrix} 1 & \frac{1}{7} & \frac{1}{3} \\ 7 & 1 & 3 \\ 3 & \frac{1}{3} & 1 \end{bmatrix}$$

It can easily be shown that

$$\text{Job 1 interest in work score} = .088$$
$$\text{Job 2 interest in work score} = .669$$
$$\text{Job 3 interest in work score} = .243$$

Finally, for nearness to family, suppose the pairwise comparison matrix is as follows:

$$\begin{array}{c} \\ \text{Job 1} \\ \text{Job 2} \\ \text{Job 3} \end{array} \begin{array}{ccc} \text{Job 1} & \text{Job 2} & \text{Job 3} \\ \left[\begin{array}{ccc} 1 & \frac{1}{4} & \frac{1}{7} \\ 4 & 1 & 2 \\ 7 & 2 & 1 \end{array}\right] \end{array}$$

Routine calculations yield

Job 1 score for nearness to family = .069

Job 2 score for nearness to family = .426

Job 3 score for nearness to family = .506

As described earlier, we can now "synthesize" the objective weights with the scores of each job on each objective to obtain an overall score for each alternative (in this case, each job offer). As before, we find that job offer 2 is most preferred, followed by job offer 1, with job offer 3 the least preferred.

We close by noting that AHP has been applied by decision makers in countless areas, including accounting, finance, marketing, energy resource planning, microcomputer selection, sociology, architecture, and political science. See Zahedi (1986) and Saaty (1988) for a discussion of applications of AHP.

Implementing AHP on a Spreadsheet

AHP.xls

Figure 17 illustrates how easy it is to implement AHP on a spreadsheet (file AHP.xls). Enter in the pairwise comparison matrix for objectives in B7:E10. In B12 enter the formula =B7/**SUM**(B$7:B$10) and copy this to the range B12:E15, yielding A_{norm} for objectives. Compute the weight for salary in F12 with the command **AVERAGE**(B12:E12). Copy this to F12:F15 to compute the weights of the remaining objectives. In a similar fashion, the normalized matrices and weights for each objective are obtained.

To determine the score for job 1, enter into F17 the formula

=F$12 * F21 + F$13 * F29 + F$14 * F37 + F$15 * F45

Copying this formula to F17:F19 computes the score for jobs 2 and 3. Again, we see that job 2 receives the highest score (indicated by ****).

To compute the consistency index for the pairwise comparison matrix for objectives, the Excel matrix multiplication function MMULT is used, computing $A\mathbf{w}^T$ in the range C2:C5. In the range D2:D5 compute (ith entry in $A\mathbf{w}^T$)/(ith entry in $\mathbf{w}^T$). Finally, in E2 compute the CI, using the formula (**AVERAGE**(D2:D5) − 4)/3.

Mmult.xls

Using the Excel MMULT function, it is easy to multiply matrices. To illustrate, we will use Excel to find the matrix product AB (see Figure 18 and file Mmult.xls). We proceed as follows:

Step 1 Enter A and B in D2:F3 and D5:E7, respectively.

Step 2 Select the range (D9:E10) in which the product AB will be computed.

Step 3 In the upper left-hand corner (D9) of the selected range, type the formula

= MMULT(D2:F3,D5:E7)

Then hit **CONTROL SHIFT ENTER** (not just ENTER), and the desired matrix product will be computed. Note that MMULT is an array function, not an ordinary spreadsheet function. This explains why we must preselect the range for AB and use CONTROL SHIFT ENTER.

FIGURE 17
AHP Spreadsheet

	A	B	C	D	E	F	G
1		CONSISTENCY	INDEX	AwT/wT	CI		
2	IMPLEMENTING		2.0774038	4.0610902	0.0158569		
3	AHP	AwT=	0.3958173	4.0160976			
4	ON		0.9894231	4.0671937			
5	A SPREADSHEET		0.5932692	4.0459016			
6	OBJECTIVES MATRIX	SAL	QL	IW	NF		
7	SAL	1	5	2	4		
8	QL	0.2	1	0.5	0.5		
9	IW	0.5	2	1	2		
10	NF	0.25	2	0.5	1		
11	ANORM(OBJECTIVES)	SAL	QL	NF	IW	WEIGHTS	
12	SAL	0.512820513	0.5	0.5	0.5333333	0.5115385	SAL
13	QL	0.102564103	0.1	0.125	0.0666667	0.0985577	QL
14	NF	0.256410256	0.2	0.25	0.2666667	0.2432692	IW
15	IW	0.128205128	0.2	0.125	0.1333333	0.1466346	NF
16	SALARY MATRIX	JOB1	JOB2	JOB3			
17	JOB1	1	2	4	JOB1SC=	0.3395156	
18	JOB2	0.5	1	2	JOB2SC=	0.3960857	****
19	JOB3	0.25	0.5	1	JOB3SC=	0.2643988	
20	ANORM(SALARY)	JOB1	JOB2	JOB3		WEIGHTS	
21	JOB1	0.571428571	0.5714286	0.5714286		0.5714286	JOB1
22	JOB2	0.285714286	0.2857143	0.2857143		0.2857143	JOB2
23	JOB3	0.142857143	0.1428571	0.1428571		0.1428571	JOB3
24	QL MATRIX	JOB1	JOB2	JOB3			
25	JOB1	1	0.5	0.3333333			
26	JOB2	2	1	0.3333333			
27	JOB3	3	3	1			
28	ANORM(QL)	JOB1	JOB2	JOB3		WEIGHTS	
29	JOB1	0.166666667	0.1111111	0.2		0.1592593	JOB1
30	JOB2	0.333333333	0.2222222	0.2		0.2518519	JOB2
31	JOB3	0.5	0.6666667	0.6		0.5888889	JOB3
32	IW MATRIX	JOB1	JOB2	JOB3			
33	JOB1	1	0.1428571	0.3333333			
34	JOB2	7	1	3			
35	JOB3	3	0.3333333	1			
36	ANORM(IW)	JOB1	JOB2	JOB3		WEIGHTS	
37	JOB1	0.090909091	0.0967742	0.0769231		0.0882021	JOB1
38	JOB2	0.636363636	0.6774194	0.6923077		0.6686969	JOB2
39	JOB3	0.272727273	0.2258065	0.2307692		0.243101	JOB3
40	NF MATRIX	JOB1	JOB2	JOB3			
41	JOB1	1	0.25	0.1428571			
42	JOB2	4	1	2			
43	JOB3	7	2	1			
44	ANORM(NF)	JOB1	JOB2	JOB3		WEIGHTS	
45	JOB1	0.083333333	0.0769231	0.0454545		0.0685703	JOB1
46	JOB2	0.333333333	0.3076923	0.6363636		0.4257964	JOB2
47	JOB3	0.583333333	0.6153846	0.3181818		0.5056333	JOB3

	A	B	C	D	E	F
1	MatrixMultiplication					
2				1	1	2
3			A	2	1	3
4						
5			B	1	1	
6				2	3	
7				1	2	
8						
9				5	8	
10			C	7	11	
11						

FIGURE 18

PROBLEMS

Group A

1 Each professor's annual salary increase is determined by performance in three areas: teaching, research, and service to the university. The administration has come up with the following pairwise comparison matrix for these objectives:

$$\begin{array}{c c c c} & \text{Teaching} & \text{Research} & \text{Service} \\ \begin{array}{c} \text{Teaching} \\ \text{Research} \\ \text{Service} \end{array} & \left[\begin{array}{c c c} 1 & \frac{1}{3} & 5 \\ 3 & 1 & 7 \\ \frac{1}{5} & \frac{1}{7} & 1 \end{array} \right] \end{array}$$

The administration has compared two professors with regard to their teaching, research, and service over the past year. The pairwise comparison matrices are as follows. For teaching:

$$\begin{array}{c c c} & \text{Professor 1} & \text{Professor 2} \\ \begin{array}{c} \text{Professor 1} \\ \text{Professor 2} \end{array} & \left[\begin{array}{c c} 1 & 4 \\ \frac{1}{4} & 1 \end{array} \right] \end{array}$$

For research:

$$\begin{array}{c c c} & \text{Professor 1} & \text{Professor 2} \\ \begin{array}{c} \text{Professor 1} \\ \text{Professor 2} \end{array} & \left[\begin{array}{c c} 1 & \frac{1}{3} \\ 3 & 1 \end{array} \right] \end{array}$$

For service:

$$\begin{array}{c c c} & \text{Professor 1} & \text{Professor 2} \\ \begin{array}{c} \text{Professor 1} \\ \text{Professor 2} \end{array} & \left[\begin{array}{c c} 1 & 6 \\ \frac{1}{6} & 1 \end{array} \right] \end{array}$$

a Which professor should receive a bigger raise?

b Does the AHP indicate how large a raise each professor should be given?

c Check the pairwise comparison matrix for consistency.

2 A business is about to purchase a new personal computer. Three objectives are important in determining which computer should be purchased: cost, user-friendliness, and software availability. The pairwise comparison matrix for these objectives is as follows:

$$\begin{array}{c c c c} & \text{Cost} & \begin{array}{c}\text{User-}\\\text{friendliness}\end{array} & \begin{array}{c}\text{Software}\\\text{availability}\end{array} \\ \begin{array}{c} \text{Cost} \\ \text{User-friendliness} \\ \text{Software availability} \end{array} & \left[\begin{array}{c c c} 1 & \frac{1}{4} & \frac{1}{5} \\ 4 & 1 & \frac{1}{2} \\ 5 & 2 & 1 \end{array} \right] \end{array}$$

Three computers are being considered for purchase. The performance of each computer with regard to each objective is indicated by the following pairwise comparison matrices. For cost (low cost is good, high cost is bad!):

$$\begin{array}{c c c c} & \text{Computer 1} & \text{Computer 2} & \text{Computer 3} \\ \begin{array}{c} \text{Computer 1} \\ \text{Computer 2} \\ \text{Computer 3} \end{array} & \left[\begin{array}{c c c} 1 & 3 & 5 \\ \frac{1}{3} & 1 & 2 \\ \frac{1}{5} & \frac{1}{2} & 1 \end{array} \right] \end{array}$$

For user-friendliness:

$$\begin{array}{c c c c} & \text{Computer 1} & \text{Computer 2} & \text{Computer 3} \\ \begin{array}{c} \text{Computer 1} \\ \text{Computer 2} \\ \text{Computer 3} \end{array} & \left[\begin{array}{c c c} 1 & \frac{1}{3} & \frac{1}{2} \\ 3 & 1 & 5 \\ 2 & \frac{1}{5} & 1 \end{array} \right] \end{array}$$

For software availability:

$$\begin{array}{c c c c} & \text{Computer 1} & \text{Computer 2} & \text{Computer 3} \\ \begin{array}{c} \text{Computer 1} \\ \text{Computer 2} \\ \text{Computer 3} \end{array} & \left[\begin{array}{c c c} 1 & \frac{1}{3} & \frac{1}{7} \\ 3 & 1 & \frac{1}{5} \\ 7 & 5 & 1 \end{array} \right] \end{array}$$

a Which computer should be purchased?

b Check the pairwise comparison matrices for consistency.

3 Woody is ready to select his mate for life and has determined that beauty, intelligence, and personality are the key factors in selecting a satisfactory mate. His pairwise comparison matrix for these objectives is as follows:

$$\begin{array}{c c c c} & \text{Beauty} & \text{Intelligence} & \text{Personality} \\ \begin{array}{c} \text{Beauty} \\ \text{Intelligence} \\ \text{Personality} \end{array} & \left[\begin{array}{c c c} 1 & 3 & 5 \\ \frac{1}{3} & 1 & 3 \\ \frac{1}{5} & \frac{1}{3} & 1 \end{array} \right] \end{array}$$

Three women (Jennifer Lopez, Britney Spears, and Mandy Moore) are begging to be Woody's mate. His views of these women's beauty, intelligence and personality are given in the following pairwise comparison matrices.

Beauty:

	Jennifer	Britney	Mandy
Jennifer	1	5	3
Britney	$\frac{1}{5}$	1	$\frac{1}{2}$
Mandy	$\frac{1}{3}$	2	1

Intelligence:

	Jennifer	Britney	Mandy
Jennifer	1	$\frac{1}{6}$	$\frac{1}{4}$
Britney	6	1	2
Mandy	4	$\frac{1}{2}$	1

Personality:

	Jennifer	Britney	Mandy
Jennifer	1	4	$\frac{1}{4}$
Britney	$\frac{1}{4}$	1	$\frac{1}{9}$
Mandy	4	9	1

a Whom should Woody choose as his lifetime mate?

b Evaluate all pairwise comparison matrices for consistency.

4 In determining where to invest my money, two objectives—expected rate of return and degree of risk—are considered equally important. Two investments (1 and 2) have the following pairwise comparison matrices: Expected return:

	Investment 1	Investment 2
Investment 1	1	$\frac{1}{2}$
Investment 2	2	1

Degree of risk:

	Investment 1	Investment 2
Investment 1	1	3
Investment 2	$\frac{1}{3}$	1

a How should I rank these investments?

b Now suppose another investment (investment 3) is available. Suppose the pairwise comparison matrices for these investments are as follows. Expected return:

	Invest-ment 1	Invest-ment 2	Invest-ment 3
Investment 1	1	$\frac{1}{2}$	4
Investment 2	2	1	8
Investment 3	$\frac{1}{4}$	$\frac{1}{8}$	1

Degree of risk:

	Invest-ment 1	Invest-ment 2	Invest-ment 3
Investment 1	1	3	$\frac{1}{2}$
Investment 2	$\frac{1}{3}$	1	$\frac{1}{6}$
Investment 3	2	6	1

c Observe that the entries in the comparison matrices for investments 1 and 2 have not changed. How should I now rank the investments? Contrast my ranking of investments 1 and 2 with the answer from part (a).

5 Show that for a perfectly consistent decision maker, the ith entry in $A\mathbf{w}^T = n$ (ith entry of $\mathbf{w}^T$).

6 A consumer is trying to determine which type of frozen dinner to eat. He considers three attributes to be important: taste, nutritional value, and price. Nutritional value is considered to be determined by cholesterol and sodium levels. Three types of dinners are under consideration. The pairwise comparison matrix for the three attributes is as follows:

	Taste	Nutrition	Price
Taste	1	3	$\frac{1}{2}$
Nutrition	$\frac{1}{3}$	1	$\frac{1}{5}$
Price	2	5	1

Between the three frozen dinners the pairwise comparison matrix for each attribute is as follows. For taste:

	Dinner 1	Dinner 2	Dinner 3
Taste 1	1	5	3
Taste 2	$\frac{1}{5}$	1	$\frac{1}{2}$
Taste 3	$\frac{1}{3}$	2	1

For sodium:

	Dinner 1	Dinner 2	Dinner 3
Sodium 1	1	$\frac{1}{7}$	$\frac{1}{3}$
Sodium 2	7	1	2
Sodium 3	3	$\frac{1}{2}$	1

For cholesterol:

	Dinner 1	Dinner 2	Dinner 3
Cholesterol 1	1	$\frac{1}{8}$	$\frac{1}{4}$
Cholesterol 2	8	1	2
Cholesterol 3	4	$\frac{1}{2}$	1

For price:

	Dinner 1	Dinner 2	Dinner 3
Price 1	1	4	$\frac{1}{2}$
Price 2	$\frac{1}{4}$	1	$\frac{1}{6}$
Price 3	2	6	1

To determine how each dinner rates on nutrition you will need the following pairwise comparison matrix for cholesterol and sodium:

	Cholesterol	Sodium
Cholesterol	1	5
Sodium	$\frac{1}{5}$	1

Which frozen dinner would he prefer? (*Hint:* Nutrition score for a dinner = (score of dinner on sodium) * (weight for sodium) + (score for dinner on cholesterol) * (weight for cholesterol).)

7 You are trying to determine which MBA program to attend. You have been accepted at two programs: Indiana and Northwestern. You have chosen three attributes to use in helping you make your decision:

Attribute 1 Cost
Attribute 1 Starting salary
Attribute 1 Ambience of school (can we party there?!!)

Your pairwise comparison matrix for these attributes is as follows:

	Cost	Starting salary	Ambience
Cost	1	$\frac{1}{4}$	2
Starting salary	4	1	7
Ambience	$\frac{1}{2}$	$\frac{1}{7}$	1

For each attribute the pairwise comparison matrix for Indiana and Northwestern is as follows. For cost:

	Indiana	Northwestern
Indiana	1	6
Northwestern	$\frac{1}{6}$	1

For starting salary:

	Indiana	Northwestern
Indiana	1	$\frac{1}{3}$
Northwestern	3	1

For ambience:

	Indiana	Northwestern
Indiana	1	4
Northwestern	$\frac{1}{4}$	1

Which MBA program should you attend?

8 You have been hired by Arthur Ross to determine which of the following accounts receivable procedures should be used in an audit of the Keating Five and Dime Store:

a Analytic review

b Confirmations

c Test of subsequent collections (receipts)

The three criteria used to distinguish between the procedures are as follows:

a Reliability

b Cost

c Validity

The pairwise comparison matrix for the three criteria is as follows:

Reliability	1	5	7
Cost	$\frac{1}{5}$	1	2
Validity	$\frac{1}{7}$	$\frac{1}{2}$	1

For the reliability criterion the pairwise comparison matrix of the three procedures is as follows:

Analytical review	1	$\frac{1}{6}$	$\frac{1}{2}$
Confirmations	6	1	4
Test of subsequent collections	2	$\frac{1}{4}$	1

For the cost criterion the pairwise comparison matrix of the three procedures is as follows:

Analytical review	1	5	3
Confirmations	$\frac{1}{5}$	1	$\frac{1}{2}$
Test of subsequent collections	$\frac{1}{3}$	2	1

For the validity criterion the pairwise comparison matrix of the three procedures is as follows:

Analytical review	1	3	2
Confirmations	$\frac{1}{3}$	1	$\frac{1}{2}$
Test of subsequent collections	$\frac{1}{2}$	2	1

Use the AHP to determine which auditing procedure should be used. Also check the first pairwise comparison matrix for consistency.[†]

9 You are trying to determine which of two secretarial candidates (Jack and Jill) to hire. The three objectives that are important to your decision are personality, typing ability, and intelligence. You have assessed the following pairwise comparison matrix:

	Personality	Typing ability	Intelligence
Personality	1	$\frac{1}{4}$	$\frac{1}{3}$
Typing ability	4	1	$\frac{1}{2}$
Intelligence	3	2	1

The "score" of each employee on each objective is as follows:

	Personality	Typing ability	Intelligence
Jack	.4	.6	.2
Jill	.6	.4	.8

If you follow the AHP method which employee should be hired?

[†]Based on Lin, Mock, and Wright (1984).

S U M M A R Y Decision Criteria

In the **state-of-the-world model,** the decision maker first chooses an action a_i from a set $A = \{a_1, a_2, \ldots, a_k\}$ of available actions. With probability p_j the state of the world is observed to be $s_j \in S = \{s_1, s_2, \ldots, s_n\}$. If action a_i is chosen and the state of the world is s_j, the decision maker receives a reward r_{ij}.

The **maximin criterion** chooses the action a_i with the largest value of $\min_{j \in S} r_{ij}$. The **maximax criterion** chooses the action a_i with the largest value of $\max_{j \in S} r_{ij}$. In each state, the **minimax regret criterion** chooses an action by applying the minimax criterion to the regret matrix. The **expected value criterion** chooses the decision that yields the largest expected reward.

Utility Theory

A decision maker who subscribes to the Von Neumann–Morgenstern axioms, when facing a choice between several lotteries, should choose the lottery with the largest expected utility.

The **certainty equivalent** of a lottery L, written $CE(L)$, is the number $CE(L)$ such that the decision maker is indifferent between the lottery L and receiving a certain payoff of $CE(L)$. For a given lottery L, the **risk premium,** written $RP(L)$, is given by $RP(L) = EV(L) - CE(L)$.

A decision maker is **risk-averse** if and only if for any nondegenerate lottery L, $RP(L) > 0$. A risk-averse decision maker has a strictly concave utility function. A decision maker is **risk-neutral** if and only if for any nondegenerate lottery L, $RP(L) = 0$. A risk-neutral decision maker has a linear utility function. A decision maker is **risk-seeking** if and only if for any nondegenerate lottery L, $RP(L) < 0$. A risk-seeking decision maker has a strictly convex utility function.

Prospect Theory and Framing

Tversky and Kahneman resolved several flaws in EMU by developing prospect theory and framing. Prospect theory assumes that we do not treat probabilities as they are given in a decision-making problem. Instead, the decision maker treats a probability p for an event as a "distorted" probability $\Pi(p)$. The idea of framing is based on the fact that people often set their utility function from the standpoint of a frame or status quo from which they view the current situation.

Decision Trees

To determine the optimal decisions in a decision tree, we work backward (folding back the tree) from right to left. First assume that the decision maker is risk-neutral and wants to maximize final asset position. At each event fork, we calculate the expected final asset position and enter it in $\bigcirc$. At each decision fork, we denote by $\|$ the decision that maximizes the expected final asset position and enter the expected final asset position associated with that decision in $\square$. We continue working backward in this fashion until we reach the beginning of the tree. Then the optimal sequence of decisions can be obtained by following the $\|$.

To incorporate a decision maker's utility function into a decision tree analysis, simply replace each final asset position x_0 by its utility $u(x_0)$. Then at each event fork, compute expected utility, and at each decision fork, choose the branch having the largest expected utility.

The **expected value of sample information** (EVSI) measures the value associated with test or sample information: EVSI = EVWSI − EVWOI. **Expected value with perfect information** (EVWPI) is found by drawing a decision tree in which the decision maker has perfect information about which state has occurred before the decision must be made. Then the **expected value of perfect information** (EVPI) is given by EVPI = EVWPI − EVWOI.

Bayes' Rule and Decision Trees

We use Bayes' rule in decision tree analysis when we are given prior probabilities and (for each state of the world) the likelihood that an experimental outcome will occur. Bayes' rule is then used to compute the probability that each experimental outcome will occur and (for each experimental outcome) the posterior probability of each state of the world. Then the decision tree analysis proceeds as already described.

Decision Making with Multiple Objectives

Attribute 1 is **preferentially independent** (pi) of attribute 2 if preferences for values of attribute 1 do not depend on the value of attribute 2.

A set of attributes S is **mutually preferentially independent** (mpi) of a set of attributes S' if (1) the values of the attributes in S' do not affect preferences for the values of attributes in S; (2) the values of the attributes in S do not affect preferences for the values of attributes in S'.

A set of attributes $1, 2, \ldots, n$ is mutually preferentially independent (mpi) if for all subsets S of $\{1, 2, \ldots, n\}$, S is mpi of $\bar{S}$. ($\bar{S}$ is all members of $\{1, 2, \ldots, n\}$ that are not included in S.)

THEOREM 1

If the set of attributes $1, 2, \ldots, n$ is mpi, the decision maker's preferences can be represented by an additive value (or cost) function.

Multiattribute Utility Functions

Attribute 1 is **utility independent** (ui) of attribute 2 if preferences for lotteries involving different levels of attribute 1 do not depend on the level of attribute 2.

If attribute 1 is ui of attribute 2, and attribute 2 is ui of attribute 1, then attributes 1 and 2 are **mutually utility independent** (mui).

THEOREM 2

Attributes 1 and 2 are mui if and only if the decision maker's utility function $u(x_1, x_2)$ is a multilinear function of the form

$$u(x_1, x_2) = k_1 u_1(x_1) + k_2 u_2(x_2) + k_3 u_1(x_1) u_2(x_2) \qquad (10)$$

A decision maker's utility function exhibits **additive independence** if the decision maker is indifferent between

and

If attributes 1 and 2 are mpi and the decision maker's utility function exhibits additive independence, the decision maker's utility function is of the following additive form:

$$u(x_1, x_2) = k_1 u_1(x_1) + k_2 u_2(x_2)$$

The following procedure is used to assess multiattribute utility functions:

Step 1 Check whether attributes 1 and 2 are mui. If they are, go on to step 2. If the attributes are not mui, the assessment of the multiattribute utility function is beyond the scope of our discussion.

Step 2 Check for additive independence.

Step 3 Assess $u_1(x_1)$ and $u_2(x_2)$.

Step 4 Determine k_1, k_2, and (if there is no additive independence) k_3.

Step 5 Check to see whether the assessed utility function is really consistent with the decision maker's preferences. To do this, set up several lotteries and use the expected utility of each lottery to rank the lotteries from most to least favorable. Then ask the decision maker to rank the lotteries from most to least favorable. If the assessed utility function is consistent with the decision maker's preferences, the ranking of lotteries obtained from the assessed utility function should closely resemble the decision maker's ranking of the lotteries.

Analytic Hierarchy Process (AHP)

The AHP is often used to make decisions in situations when there are multiple objectives. Given a pairwise comparison matrix A, we can approximate the weights for each attribute as follows:

Step 1 For each of A's columns, do the following. Divide each entry in column i of A by the sum of the entries in column i. This yields a new matrix(A_{norm}), in which the sum of the entries in each column is 1.

Step 2 To find an approximation to $\mathbf{w}_{max}$, which will be used as our estimate of $\mathbf{w}$, proceed as follows. Estimate w_i as the average of the entries in row i of A_{norm}. To find the best decision, determine an overall score for a decision as follows:

$$\text{Decision score} = \sum_i w_i(\text{decision score on objective } i)$$

Now choose the decision with the largest score.

To check for consistency in pairwise comparision matrices, we use the following four-step process. ($\mathbf{w}$ denotes our estimate of the decision maker's weights.)

Step 1 Compute $A\mathbf{w}^T$.

Step 2 Compute

$$\frac{1}{n}\sum_{i=1}^{i=n}\frac{i\text{th entry in } A\mathbf{w}^T}{i\text{th entry in } \mathbf{w}^T}$$

Step 3 Compute the **consistency index** (CI) as follows:

$$CI = \frac{(\text{Step 2 result}) - n}{n - 1}$$

Step 4 Compare CI to the random index (RI) for the appropriate value of n. If $\frac{CI}{RI} < .10$, the degree of consistency is satisfactory, but if $\frac{CI}{RI} > .10$, serious inconsistencies may exist, and the AHP may not yield meaningful results.

REVIEW PROBLEMS

Group A

1 We have $1,000 to invest. All the money must be placed in one of three investments: gold, stock, or money market certificates. If $1,000 is placed in an investment, the value of the investment one year from now depends on the state of the economy (see Table 16). Assume that each state of the economy is equally likely. For each of the following decision criteria, determine the optimal decision:

a maximin

b maximax

TABLE 16

Value of $1,000	State 1	State 2	State 3
Money market certificate	$1,100	$1,100	$1,100
Stock	$1,000	$1,100	$1,200
Gold	$1,600	$300	$1,400

TABLE 17

	Economy Has Good Year	Economy Has Bad Year
Yield on stocks	22%	10%
Yield on bonds	16%	14%

c minimax regret

d expected value

2 In Problem 1, suppose that the utility function for the value of the investment (x) one year from now is given by $u(x) = \ln x$. Determine which investment we should choose. Could we have predicted this answer without a table of logarithms?

3 Consider the following four lotteries:

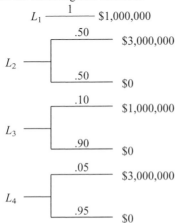

$$L_1 \xrightarrow{1} \$1,000,000$$

L_2
- .50 — $3,000,000
- .50 — $0

L_3
- .10 — $1,000,000
- .90 — $0

L_4
- .05 — $3,000,000
- .95 — $0

a Most people prefer L_1 to L_2 and L_4 to L_3. Explain why.

b Suppose a decision maker subscribes to the Von Neumann–Morgenstern axioms and prefers L_1 to L_2. Show that he or she must also prefer L_3 to L_4.

4 Jay Boyville Corporation is being sued by Lark Dent. Lark can settle out of court and win $40,000, or go to court. If Lark goes to court, there is a 30% chance that she will win the case. If she wins, a small and a large settlement are equally likely (a small settlement nets $50,000, and a large settlement nets $300,000).

a If Lark is risk-neutral, what should she do? What should Lark do if her utility function for an increase x in her cash position is given by $u(x) = x^{1/2}$?

b For $10,000, Lark can hire a consultant who will predict who will win the trial. The consultant is correct 90% of the time. Should she hire the consultant? (Assume Lark is risk-neutral.) What is EVSI?

c If Lark is risk-neutral, what is EVPI?

5 Rollo Megabux has $1 million to invest in stocks or bonds. The percentage yield on each investment during the coming year depends on whether the economy has a good or a bad year (see Table 17). It is equally likely that the economy will have a good or a bad year.

a If Rollo is risk-neutral, how should he invest his money?

b For $10,000, Rollo can hire a consulting firm to forecast the state of the economy. The consulting firm's forecasts have the following properties:

P(good forecast|economy good) = .80

P(good forecast|economy bad) = .20

Should Rollo hire the consulting firm? What are EVSI and EVPI?

6 Willy Mutton has three potential bank robberies lined up. His chance of success and the size of the take are given in Table 18: These robberies must be attempted in order; if you "pass" on a robbery you may not go on to the next robbery. If Willie is caught, he loses all his money. What strategy maximizes his expected "take"?

7 Let

x_1 = undergraduate grade point average (GPA) of a student applying to State U's MBA program

x_2 = GMAT score of the same student

Suppose that preference between applicants is based on the following value function:

$$v(x_1, x_2) = 200x_1 + x_2 - 0.1x_2(x_1)^2$$

a Would the MBA program prefer a student with a 3.8 GPA and a 500 GMAT score to a student with a 3.0 GPA and a 710 GMAT score?

b Does this value function exhibit mutual preferential independence?

8 The Pine Valley Board of Education is trying to determine its multiattribute utility function with respect to the following attributes:

Attribute 1 Average score of students on an English achievement test

Attribute 2 Average score of students on a mathematics achievement test

TABLE 18

Robbery	Chance of Success	Size of Take (in millions of dollars)
1	.60	7
2	.80	6
3	.70	5

The board believes that both attributes range between 70% and 90% correct answers. The board is indifferent between

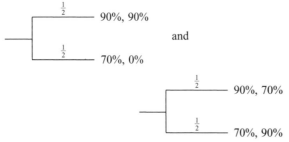

$\frac{1}{2}$ — 90%, 90%

and

$\frac{1}{2}$ — 70%, 0%

$\frac{1}{2}$ — 90%, 70%

$\frac{1}{2}$ — 70%, 90%

For any level x_2 of attribute 2, the board is also indifferent between

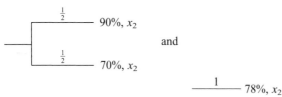

$\frac{1}{2}$ — 90%, x_2

and

$\frac{1}{2}$ — 70%, x_2

1 — 78%, x_2

For any level x_1 of attribute 1, the board is indifferent between

.65 — x_1, 90%

and

.35 — x_1, 70%

1 — x_1, 76%

The board is also indifferent between

1 — 90%, 70%   and

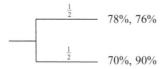

.40 — 90%, 90%

.60 — 70%, 70%

Finally, the board is indifferent between

1 — 70%, 90%   and

.60 — 90%, 90%

.40 — 70%, 70%

The board must decide which of two instructional techniques should be utilized in the Pine Valley schools. Technique 1 is equivalent to the following lottery:

$\frac{1}{2}$ — 78%, 76%

$\frac{1}{2}$ — 70%, 90%

Technique 2 is equivalent to the following lottery:

.60 — 90%, 90%

.40 — 70%, 70%

Would the board prefer technique 1 or technique 2?

9 BeatTrop Foods is trying to choose one of three companies to merge with. In making this decision seven factors are important:

Factor 1 Contribution to profitability
Factor 2 Growth potential
Factor 3 Labor environment
Factor 4 R&D ability of company
Factor 5 Organizational fit
Factor 6 Relative size
Factor 7 Industry commonality

The pairwise comparison for these factors is as follows:

$$
\begin{array}{c|ccccccc}
 & 1 & 2 & 3 & 4 & 5 & 6 & 7 \\
\hline
1 & 1 & 3 & 7 & 5 & 1 & 7 & 1 \\
2 & \frac{1}{3} & 1 & 9 & 1 & 1 & 5 & 1 \\
3 & \frac{1}{7} & \frac{1}{9} & 1 & \frac{1}{7} & \frac{1}{5} & \frac{1}{2} & \frac{1}{4} \\
4 & \frac{1}{5} & 1 & 7 & 1 & \frac{1}{4} & 7 & \frac{1}{3} \\
5 & 1 & 1 & 5 & 4 & 1 & 5 & 3 \\
6 & \frac{1}{7} & \frac{1}{5} & 2 & \frac{1}{7} & \frac{1}{5} & 1 & \frac{1}{6} \\
7 & 1 & 1 & 4 & 3 & \frac{1}{3} & 6 & 1 \\
\end{array}
$$

The three contenders for merger have the following pairwise comparison matrices for each factor:

Factor 1
$$
\begin{array}{c|ccc}
 & 1 & 2 & 3 \\
\hline
1 & 1 & 9 & 3 \\
2 & \frac{1}{9} & 1 & \frac{1}{5} \\
3 & \frac{1}{3} & 5 & 1 \\
\end{array}
$$

Factor 2
$$
\begin{array}{c|ccc}
 & 1 & 2 & 3 \\
\hline
1 & 1 & 7 & 4 \\
2 & \frac{1}{7} & 1 & \frac{1}{3} \\
3 & \frac{1}{4} & 3 & 1 \\
\end{array}
$$

Factor 3
$$
\begin{array}{c|ccc}
 & 1 & 2 & 3 \\
\hline
1 & 1 & \frac{1}{5} & \frac{1}{3} \\
2 & 5 & 6 & 2 \\
3 & 3 & \frac{1}{2} & 1 \\
\end{array}
$$

Factor 4
$$
\begin{array}{c|ccc}
 & 1 & 2 & 3 \\
\hline
1 & 1 & 6 & 3 \\
2 & \frac{1}{6} & 1 & \frac{1}{2} \\
3 & \frac{1}{3} & 2 & 1 \\
\end{array}
$$

Factor 5
$$
\begin{array}{c|ccc}
 & 1 & 2 & 3 \\
\hline
1 & 1 & \frac{1}{9} & \frac{1}{5} \\
2 & 9 & \frac{1}{2} & 4 \\
3 & 5 & \frac{1}{4} & 1 \\
\end{array}
$$

Factor 6
$$
\begin{array}{c|ccc}
 & 1 & 2 & 3 \\
\hline
1 & 1 & \frac{1}{7} & \frac{1}{4} \\
2 & 7 & 1 & 3 \\
3 & 3 & \frac{1}{3} & 1 \\
\end{array}
$$

Factor 7
$$
\begin{array}{c|ccc}
 & 1 & 2 & 3 \\
\hline
1 & 1 & \frac{1}{7} & \frac{1}{3} \\
2 & 7 & 1 & 3 \\
3 & 3 & \frac{1}{3} & 1 \\
\end{array}
$$

Use the AHP to determine the company with which BeatTrop should prefer to merge.

10 You are trying to determine which city to live in. New York and Chicago are under consideration. Four objectives will determine your decision: affordability of housing,

TABLE 19

Affordability of housing	.50
Cultural opportunities	.10
Quality of schools and universities	.20
Crime level	.20

TABLE **20**

	New York	Chicago
Affordability of housing	.30	.70
Cultural opportunities	.70	.30
Crime level	.40	.60

cultural opportunities, quality of schools and universities, and crime level. The weight for each objective is in Table 19. For each objective (except for quality of schools and universities) New York and Chicago scores are as given in Table 20. Suppose that the score for each city on the quality of schools and universities depends on two things: a score on public school quality and a score on university quality. The pairwise comparison matrix for public school and university quality is as follows:

$$\begin{array}{c} \\ \text{Public school quality} \\ \text{University quality} \end{array} \begin{array}{cc} \text{Public school} & \text{University} \\ \text{quality} & \text{quality} \\ \left[\begin{array}{cc} 1 & 4 \\ \frac{1}{4} & 1 \end{array} \right] \end{array}$$

To see how each city scores on public school quality and university quality use the following pairwise comparison matrices. For public school quality:

$$\begin{array}{c} \\ \text{New York} \\ \text{Chicago} \end{array} \begin{array}{cc} \text{New York} & \text{Chicago} \\ \left[\begin{array}{cc} 1 & 4 \\ \frac{1}{4} & 1 \end{array} \right] \end{array}$$

For university quality:

$$\begin{array}{c} \\ \text{New York} \\ \text{Chicago} \end{array} \begin{array}{cc} \text{New York} & \text{Chicago} \\ \left[\begin{array}{cc} 1 & \frac{1}{3} \\ 3 & 1 \end{array} \right] \end{array}$$

You should now be able to come up with a score for each city on the quality of schools and universities objective. Now determine where you should live.

Group B

11 In Problem 5, suppose Rollo cannot hire the consulting firm, and his utility function for ending cash position is $u(x) = \ln x$. How much money should he invest in stocks and bonds?

12 At present, littering is punished by a $50 fine, and there is a 10% chance that a litterer will be brought to justice. To cut down on littering, Gotham City is considering two alternatives:

Alternative 1 Raise the littering fine by 20% (to $60).
Alternative 2 Hire more police and increase by 20% the probability that a litterer will be brought to justice (to a 12% probability that a litterer will be caught).

Assuming that all Gotham City residents are risk-averse, which alternative will lead to a larger reduction in littering?

Group C

13[†] In Section 2.2, we discussed the concept of the risk premium of a lottery and a risk-averse decision maker. In many situations, we would like to measure the degree of risk aversion associated with a utility function, and how a decision maker's risk aversion depends on his or her wealth. In this problem, we develop **Pratt's measure of absolute risk aversion.** Consider Ivana, who has initial wealth W and utility function $u(w)$ for final wealth position w. She has placed money in a small investment. The investment will increase her wealth by a random amount $\mathbf{X}$, with $E(\mathbf{X}) = 0$. We want to investigate how the risk premium of $\mathbf{X}$ depends on W. Let $RP(W, \mathbf{X})$ be the risk premium associated with investment $\mathbf{X}$ if the decision maker's wealth is W.

a Explain why $RP(W, \mathbf{X})$ satisfies the following equation:
$E(\text{Utility for wealth level of } W + \mathbf{X})$
$= \text{utility of wealth level } [W - RP(W, \mathbf{X})]$

b Perform a second-order Taylor series expansion on $E(\text{utility for wealth level of } W + \mathbf{X})$ about W.

c Perform a first-order Taylor series expansion on utility of wealth level $[W - RP(W, \mathbf{X})]$ about W.

d Equating the answers in (b) and (c) (disregard the remainder terms), show that
$$RP(W, \mathbf{X}) = \frac{-\text{var}(\mathbf{X})u''(W)}{2u'(W)}$$

e Pratt's measure of absolute risk aversion at wealth level W, called $ARA(W)$, is defined to be twice the amount of risk premium per unit of variance when a decision maker is faced with a small lottery that has a zero expected value. Use your answer in part (d) to explain why
$$ARA(W) = \frac{-u''(W)}{u'(W)}$$

f If $ARA(W)$ is an increasing function of W, then $u(w)$ is said to exhibit increasing risk aversion, and if $ARA(W)$ is a decreasing function of W, then $u(w)$ exhibits decreasing risk aversion. Is increasing or decreasing risk aversion more consistent with most people's behavior? Determine whether the following utility functions exhibit increasing or decreasing risk aversion:

g $u(w) = \ln w$
h $u(w) = w^{1/2}$
i $u(w) = aw - bw^2$, where $w < \frac{a}{2b}$. Explain how the answer indicates that a quadratic utility function probably is not an accurate representation of most people's preferences.

[†]Based on Pratt (1964).

REFERENCES

The following books discuss decision making under uncertainty at an intermediate level:

Bunn, D. *Applied Decision Analysis.* New York: McGraw-Hill, 1984.

Vatter, P., et al. *Quantitative Methods in Management: Text and Cases.* Homewood, Ill.: Irwin, 1978.

Winkler, R. *Introduction to Bayesian Inference and Decision.* New York: Holt, Rinehart & Winston, 1972.

For discussion of decision making under uncertainty at a more advanced level, readers should consult the next five books:

French, S. *Decision Theory.* New York: Wiley, 1986.

Keeney, R., and H. Raiffa. *Decision Making with Multiple Objectives.* New York: Wiley, 1976.

Raiffa, H. *Decision Analysis.* Reading, Mass.: Addison-Wesley, 1968.

Watson, S., and D. Buede. *Decision Synthesis.* Cambridge, England: Cambridge Press, 1987.

Winterfeldt, D., and W. Edwards. *Decision Analysis and Behavioral Research.* Cambridge, England: Cambridge Press, 1986.

Allais, M. "Le comportement de l'homme rationnel devant le risque: Critique des postulats et axioms de l'école Americaine," *Econometrica* 21(1953):503–546.

Clarke, J. "Applications of Decision Analysis to Clinical Medicine," *Interfaces* 17(no. 2, 1981):27–34.

Ellsberg, D. "Risk, Ambiguity, and the Savage Axioms," *Quarterly Journal of Economics* 75(1961):643–669.

Feinstein, C. "Deciding Whether to Test Student Athletes for Drug Use," *Interfaces* 20(no. 3, 1990):80–87.

Howard, R. "Heathens, Heretics, and Cults: The Religious Spectrum of Decision Aiding," *Interfaces* 22(no. 6, 1992):15–27.

Porter, R. "Extra-Point Strategy in Football," *American Statistician* 21(1967):14–15.

Pratt, J. "Risk Aversion in the Small and the Large," *Econometrica* 32(1964):122–136.

Tversky, A., and D. Kahneman. "The Framing of Decisions and the Psychology of Choice," *Science* 211(1981): 453–458.

Alternative approaches to multiattribute decision making under certainty are discussed in the following:

Steuer, R. *Multiple Criteria Optimization.* New York: Wiley, 1985.

Zeleny, M. *Multiple Criteria Decision Making.* New York: McGraw-Hill, 1982.

Zionts, S., and J. Wallenius. "An Interactive Programming Method for Solving the Multiple Criteria Problem," *Management Science* 22(1976): 652–663.

The following are recommended for elementary discussions of multiattribute utility theory:

Bunn, D. *Applied Decision Analysis.* New York: McGraw-Hill, 1984.

Keeney, R. "An Illustrated Procedure for Accessing Multi-attributed Utility Functions," *Sloan Management Review* 14(1972):37–50.

This classic gives a comprehensive discussion of multi-attribute utility theory:

Keeney, R., and H. Raiffa. *Decision Making with Multiple Objectives.* New York: Wiley, 1976.

The following references discuss the AHP:

Golden, B., E. Wasil, and P. Harkey. *The Analytic Hierarchy Process.* Heidelberg, Germany: Springer-Verlag, 1989.

Lin, W., T. Mock, and A. Wright. "The Use of AHP as an Aid in Planning the Nature and Extent of Audit Procedures," *Auditing: A Journal of Practice and Theory* 4(no. 1, 1984):89–99.

Saaty, T. *The Analytic Hierarchy Process.* Pittsburgh, Pa.: 1988.

Zahedi, F. "The Analytic Hierarchy Process—a Survey of the Method and Its Applications," *Interfaces* 16(no. 4, 1986):96–108.

3

Deterministic EOQ Inventory Models

In this chapter, we begin our formal study of inventory modeling. In Volume 1, we described how linear programming can be used to solve certain inventory problems. Our study of inventory will continue in Chapters 4, 6, and 7.

We begin by discussing some important concepts of inventory models. Then we develop versions of the famous economic order quantity (EOQ) model that can be used to make optimal inventory decisions when demand is deterministic (known in advance). In Chapters 4 and 7, we discuss models in which demand is allowed to be random.

3.1 Introduction to Basic Inventory Models

To meet demand on time, companies often keep on hand stock that is awaiting sale. The purpose of inventory theory is to determine rules that management can use to minimize the costs associated with maintaining inventory and meeting customer demand. Inventory models answer the following questions. (1) When should an order be placed for a product? (2) How large should each order be?

Costs Involved in Inventory Models

The inventory models considered in this book involve some or all of the following costs.

Ordering and Setup Cost

Many costs associated with placing an order or producing a good internally do not depend on the size of the order or on the production run. Costs of this type are referred to as the *ordering and setup cost.* For example, ordering cost would include the cost of paperwork and billing associated with an order. If the product is made internally rather than ordered from an external source, the cost of labor (and idle time) for setting up and shutting down a machine for a production run would be included in the ordering and setup cost.

Unit Purchasing Cost

This is simply the variable cost associated with purchasing a single unit. Typically, the unit purchasing cost includes the variable labor cost, variable overhead cost, and raw material cost associated with purchasing or producing a single unit. If goods are ordered from an external source, the unit purchase cost must include shipping cost.

Holding or Carrying Cost

This is the cost of carrying one unit of inventory for one time period. If the time period is a year, the carrying cost will be expressed in dollars per unit per year. The holding cost usually includes storage cost, insurance cost, taxes on inventory, and a cost due to the possibility of spoilage, theft, or obsolescence. Usually, however, the most significant component of holding cost is the opportunity cost incurred by tying up capital in inventory. For example, suppose that one unit of a product costs $100 and the company can earn 15% annually on its investments. Then holding one unit in inventory for one year is costing the company $0.15(100) = \$15$. When interest rates are high, most firms assume that their annual holding cost is 20%–40% of the unit purchase cost.

Stockout or Shortage Cost

When a customer demands a product and the demand is not met on time, a stockout, or shortage, is said to occur. If customers will accept delivery at a later date (no matter how late that date may be), we say that demands may be **back-ordered.** The case in which back-ordering is allowed is often referred to as the **backlogged demand** case. If no customer will accept late delivery, we are in the **lost sales** case. Of course, reality lies between these two extremes, but by determining optimal inventory policies for both the backlogged demand and the lost sales cases, we can get a ballpark estimate of what the optimal inventory policy should be.

Many costs are associated with stockouts. If back-ordering is allowed, placement of back orders usually results in an extra cost. Stockouts often cause customers to go elsewhere to meet current and future demands, resulting in lost sales and lost goodwill. Stockouts may also cause a company to fall behind in other aspects of its business and may force a plant to incur the higher cost of overtime production. Usually, the cost of a stockout is harder to measure than ordering, purchasing, or holding costs.

In this chapter, we study several versions of the classic economic order quantity (EOQ) model that was first developed in 1915 by F. W. Harris of Westinghouse Corporation. For the models in this chapter to be valid, certain assumptions must be satisfied.

Assumptions of EOQ Models

Repetitive Ordering

The ordering decision is repetitive, in the sense that it is repeated in a regular fashion. For example, a company that is ordering bearing assemblies will place an order, then see its inventory depleted, then place another order, and so on. This contrasts with one-time orders. For example, when a news vendor decides how many Sunday newspapers to order, only one order (per Sunday) will be placed. Problems where an order is placed just once are referred to as single-period inventory problems; these are discussed in Chapter 4.

Constant Demand

Demand is assumed to occur at a known, constant rate. This implies, for example, that if demand occurs at a rate of 1,000 units per year, the demand during any t-month period will be $\frac{1,000t}{12}$.

Constant Lead Time

The lead time for each order is a known constant, L. By the **lead time** we mean the length of time between the instant when an order is placed and the instant at which the order arrives. For example, if $L = 3$ months, then each order will arrive exactly 3 months after the order is placed.

Continuous Ordering

An order may be placed at any time. Inventory models that allow this are called **continuous review models.** If the amount of on-hand inventory is reviewed periodically and orders may be placed only periodically, we are dealing with a **periodic review model.** For example, if a firm reviews its on-hand inventory only at the end of each month and decides at this time whether an order should be placed, we are dealing with a periodic review model. Periodic review models are discussed in Chapters 4, 5, and 6.

Although the Constant Demand and Constant Lead Time assumptions may seem overly restrictive and unrealistic, there are many situations in which the models of this chapter provide good approximations to reality. Models in which demand is not deterministic are discussed in Chapters 4 and 7. Models in which demand is deterministic but occurs at a nonconstant rate have already been reviewed in our discussion of LP inventory models in Volume 1 and are examined further in Chapter 6.

3.2 The Basic Economic Order Quantity Model

Assumptions of the Basic EOQ Model

For the basic EOQ model to hold, certain assumptions are required (for the sake of definiteness, we assume that the unit of time is one year):

1   Demand is deterministic and occurs at a constant rate.

2   If an order of any size (say, q units) is placed, an ordering and setup cost K is incurred.

3   The lead time for each order is zero.

4   No shortages are allowed.

5   The cost per unit-year of holding inventory is h.

We define D to be the number of units demanded per year. Then assumption 1 implies that during any time interval of length t years, an amount Dt is demanded.

The setup cost K of assumption 2 is in addition to a cost pq of purchasing or producing the q units ordered. Note that we are assuming that the unit purchasing cost p does not depend on the size of the order. This excludes many interesting situations, such as quantity discounts for larger orders. In Section 3.3, we discuss a model that allows quantity discounts.

Assumption 3 implies that each order arrives as soon as it is placed. We relax this assumption later in this section.

Assumption 4 implies that all demands must be met on time; a negative inventory position is not allowed. We relax this assumption in Section 3.5.

Assumption 5 implies that a carrying cost of h dollars will be incurred if 1 unit is held for one year, if 2 units are held for half a year, or if $\frac{1}{4}$ unit is held for four years. In short, if I units are held for T years, a holding cost of ITh is incurred.

Given these five assumptions, the EOQ model determines an ordering policy that minimizes the yearly sum of ordering cost, purchasing cost, and holding cost.

Derivation of Basic EOQ Model

We begin our derivation of the optimal ordering policy by making some simple observations. Since orders arrive instantaneously, we should never place an order when I, the inventory level, is greater than zero; if we place an order when $I > 0$, we are incurring an unnecessary holding cost. On the other hand, if $I = 0$, we must place an order to prevent a shortage from occurring. Together, these observations show that the policy that minimizes yearly costs must place an order whenever $I = 0$. At all instants when an order is placed, we are facing the same situation ($I = 0$). This means that each time we place an order, we should order the same quantity. We let q be the quantity that is ordered each time that $I = 0$.

We now determine the value of q that minimizes annual cost (call it q^*). Let $TC(q)$ be the total annual cost incurred if q units are ordered each time that $I = 0$. Note that

$$TC(q) = \text{annual cost of placing orders} + \text{annual purchasing cost}$$
$$+ \text{ annual holding cost}$$

Since each order is for q units, $\frac{D}{q}$ orders per year will have to be placed so that the annual demand of D units is met. Hence

$$\frac{\text{Ordering cost}}{\text{Year}} = \left(\frac{\text{ordering cost}}{\text{order}}\right)\left(\frac{\text{orders}}{\text{year}}\right) = \frac{KD}{q}$$

For all values of q, the per-unit purchasing cost is p. Since we always purchase D units per year,

$$\frac{\text{Purchasing cost}}{\text{Year}} = \left(\frac{\text{purchasing cost}}{\text{unit}}\right)\left(\frac{\text{units purchased}}{\text{year}}\right) = pD$$

To compute the annual holding cost, note that if we hold I units for a period of one year, we incur a holding cost of (I units)(1 year)(h dollars/unit/year) $= hI$ dollars.

Suppose the inventory level is not constant and varies over time. If the average inventory level during a length of time T is $\bar{I}$, the holding cost for the time period will be $hT\bar{I}$. This idea is illustrated in Figure 1. If we define $I(t)$ to be the inventory level at time t, then during the interval $[0, T]$ the total inventory cost is given by

$$h(\text{area from 0 to } T \text{ under the } I(t) \text{ curve}) = hT\bar{I}$$

The reader may verify that this result holds for the two cases graphed in Figure 1. More formally, $\bar{I}(T)$, the average inventory level from time 0 to time T, is given by

$$\bar{I}(t) = \frac{\int_0^T I(t)dt}{T}$$

and the total holding cost incurred between time 0 and time T is

$$\int_0^T hI(t)dt = hT\bar{I}(T)$$

To determine the annual holding cost, we need to examine the behavior of I over time. Assume that an order of size q has just arrived at time 0. Since demand occurs at a rate of D per year, it will take $\frac{q}{D}$ years for inventory to reach zero again. Since demand during any period of length t is Dt, the inventory level over any time interval will decline along a straight line of slope $-D$. When inventory reaches zero, an order of size q is placed and arrives instantaneously, raising the inventory level back to q. Given these observations, Figure 2 describes the behavior of I over time.

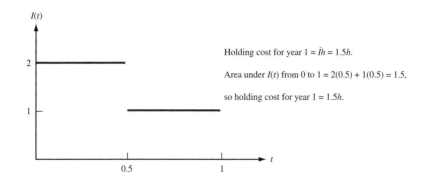

Holding cost for year 1 = $\bar{I}h = 1.5h$.

Area under $I(t)$ from 0 to 1 = 2(0.5) + 1(0.5) = 1.5,

so holding cost for year 1 = 1.5h.

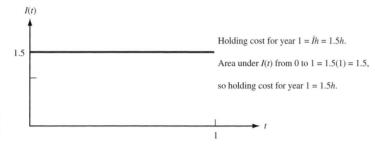

Holding cost for year 1 = $\bar{I}h = 1.5h$.

Area under $I(t)$ from 0 to 1 = 1.5(1) = 1.5,

so holding cost for year 1 = 1.5h.

FIGURE 1
Holding Cost and
Average Inventory Level

A key concept in the study of EOQ models is the idea of a cycle.

DEFINITION ■ Any interval of time that begins with the arrival of an order and ends the instant before the next order is received is called a **cycle.** ■

Observe that Figure 2 simply consists of repeated cycles of length $\frac{q}{D}$. Hence, each year will contain

$$\frac{1}{\frac{q}{D}} = \frac{D}{q}$$

cycles. The average inventory during any cycle is simply half of the maximum inventory level attained during the cycle. This result will hold in any model for which demand occurs at a constant rate and no shortages are allowed. Thus, for our model, the average inventory level during a cycle will be $\frac{q}{2}$ units.

We are now ready to determine the annual holding cost. We write

$$\frac{\text{Holding cost}}{\text{Year}} = \left(\frac{\text{holding cost}}{\text{cycle}}\right)\left(\frac{\text{cycles}}{\text{year}}\right)$$

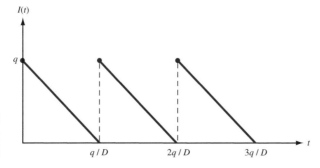

FIGURE 2
Behavior of $I(t)$ in
Basic EOQ Model

Since the average inventory level during each cycle is $\frac{q}{2}$ and each cycle is of length $\frac{q}{D}$,

$$\frac{\text{Holding cost}}{\text{Cycle}} = \frac{q}{2}\left(\frac{q}{D}\right)h = \frac{q^2 h}{2D}$$

Then

$$\frac{\text{Holding cost}}{\text{Year}} = \frac{q^2 h}{2D}\left(\frac{D}{q}\right) = \frac{hq}{2}$$

Combining ordering cost, purchasing cost, and holding cost, we obtain

$$TC(q) = \frac{KD}{q} + pD + \frac{hq}{2}$$

To find the value of q that minimizes $TC(q)$, we set $TC'(q)$ equal to zero. This yields

$$TC'(q) = -\frac{KD}{q^2} + \frac{h}{2} = 0 \tag{1}$$

Equation (1) is satisfied for $q = \pm(2\,KD/h)^{1/2}$. Since $q = -(2\,KD/h)^{1/2}$ makes no sense, let's hope that the **economic order quantity**, or EOQ,

$$q^* = \left(\frac{2KD}{h}\right)^{1/2} \tag{2}$$

minimizes $TC(q)$. Since $TC''(q) = 2KD/q^3 > 0$ for all $q > 0$, we know that $TC(q)$ is a convex function. Then Theorem $1'$ on page 675 of Chapter 12 in Volume 1 implies that any point where $TC'(q) = 0$ will minimize $TC(q)$. Thus, q^* does indeed minimize total annual cost.

REMARKS   **1**   The EOQ does not depend on the unit purchasing price p, because the size of each order does not change the unit purchasing cost. Thus, the total annual purchasing cost is independent of q. In Section 3.3, we discuss models in which the size of the order changes the unit purchasing cost.
2   Since each order is for q^* units, a total of $\frac{D}{q^*}$ orders must be placed during each year.
3   To see whether the EOQ formula is reasonable, let's see how changes in certain parameters change q^*. For example, as K increases, we would expect the number of orders placed each year, $\frac{D}{q^*}$, to decrease. Equivalently, we would expect an increase in K to increase q^*. A glance at (2) shows that this is indeed the case. Analogously, an increase in h makes it more costly to hold inventory, so we would expect an increase in h to reduce the average inventory level, $\frac{q^*}{2}$. Equation (2) shows that an increase in h does reduce q^*; it also shows that the ratio of the ordering cost to the holding cost is the critical factor in determining q^*. For example, if both K and h are doubled, q^* remains unchanged. Also note that q^* is proportional to $D^{1/2}$. Thus, quadrupling demand will only double q^*.
4   It is not difficult to show that if the EOQ is ordered, then

$$\frac{\text{Holding cost}}{\text{Year}} = \frac{\text{ordering cost}}{\text{year}} \tag{3}$$

To prove this, note that

$$\frac{\text{Holding cost}}{\text{Year}} = \frac{hq^*}{2} = \frac{h}{2}\left(\frac{2KD}{h}\right)^{1/2} = \left(\frac{KDh}{2}\right)^{1/2}$$

$$\frac{\text{Ordering cost}}{\text{Year}} = \frac{KD}{q^*} = \frac{KD}{\left(\frac{2KD}{h}\right)^{1/2}} = \left(\frac{KDh}{2}\right)^{1/2}$$

Figure 3 illustrates the trade-off between holding cost and ordering cost. The figure confirms the fact that at q^*, the annual holding and ordering costs are the same.

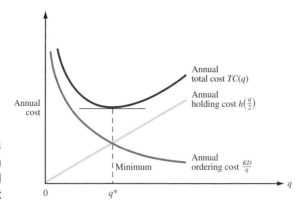

FIGURE 3
Trade-Off between
Holding Cost and
Ordering Cost

We illustrate the use of the EOQ formula with the following example.

EXAMPLE 1 **Braneast Airlines**

Braneast Airlines uses 500 taillights per year. Each time an order for taillights is placed, an ordering cost of $5 is incurred. Each light costs 40¢, and the holding cost is 8¢/light/year. Assume that demand occurs at a constant rate and shortages are not allowed. What is the EOQ? How many orders will be placed each year? How much time will elapse between the placement of orders?

Solution   We are given that $K = \$5$, $h = \$0.08$/light/year, and $D = 500$ lights/year. The EOQ is

$$q^* = \left(\frac{2(5)(500)}{0.08} \right)^{1/2} = 250$$

Hence, the airline should place an order for 250 taillights each time that inventory reaches zero.

$$\frac{\text{Orders}}{\text{Year}} = \frac{D}{q^*} = \frac{500}{250} = \frac{2 \text{ orders}}{\text{year}}$$

The time between placement (or arrival) of orders is simply the length of a cycle. Since the length of each cycle is $\frac{q^*}{D}$, the time between orders will be

$$\frac{q^*}{D} = \frac{250}{500} = \frac{1}{2} \text{ year}$$

Sensitivity of Total Cost to Small Variations in the Order Quantity

In most situations, a slight deviation from the EOQ will result in only a slight increase in costs. For Example 1, let's see how deviations from the EOQ change the total annual cost. Since annual purchasing cost is unaffected by the order quantity, we focus our attention on how the annual holding and ordering costs are affected by changes in the order quantity. Let

$$HC(q) = \text{annual holding cost if the order quantity is } q$$
$$OC(q) = \text{annual ordering cost if the order quantity is } q$$

TABLE 1
Cost Calculations for Figure 4

q	HC(q)	OC(q)	HC(q) + OC(q)
50	2.0	50.00	52.00
100	4.0	25.00	29.00
150	6.0	16.67	22.67
200	8.0	12.50	20.50
220	8.8	11.36	20.16
240	9.6	10.42	20.02
250	10.0	10.00	20.00
260	10.4	9.62	20.02
280	11.2	8.93	20.13
300	12.0	8.33	20.33
350	14.0	7.14	21.14
400	16.0	6.25	22.25

We find that

$$HC(q) = \tfrac{1}{2}(0.08q) = 0.04q \qquad OC(q) = 5\left(\frac{500}{q}\right) = \frac{2,500}{q}$$

Using the information in Table 1, we obtain the sketch of $HC(q) + OC(q)$ given in Figure 4. The figure shows that $HC(q) + OC(q)$ is very flat near q^*. For example, ordering 20% more than the EOQ ($q = 300$) raises $HC(q) + OC(q)$ from 20 to 20.33 (an increase of under 2%).

The flatness of the $HC(q) + OC(q)$ curve is important, because it is often difficult to estimate h and K. Inaccurate estimation of h and K may result in a value of q that differs slightly from the actual EOQ. The flatness of the $HC(q) + OC(q)$ curve indicates that even a moderate error in the determination of the EOQ will only increase costs by a slight amount.

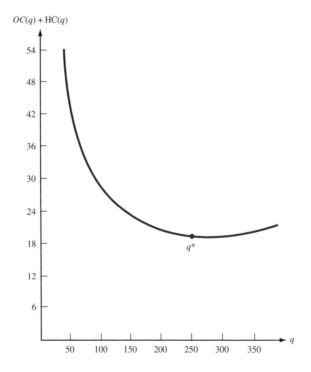

FIGURE 4
$OC(q) + HC(q)$ for
Braneast Example

Determination of EOQ When Holding Cost Is Expressed in Terms of Dollar Value of Inventory

Often, the annual holding cost is expressed in terms of the cost of holding one dollar's worth of inventory for one year. Suppose that h_d = cost of holding one dollar in inventory for one year. Then the cost of holding one unit of inventory for one year will be ph_d, and (2) may be written as

$$q^* = \left(\frac{2KD}{ph_d}\right)^{1/2} \tag{4}$$

EXAMPLE 2   **Ordering Cameras**

A department store sells 10,000 cameras per year. The store orders cameras from a regional warehouse. Each time an order is placed, an ordering cost of $5 is incurred. The store pays $100 for each camera, and the cost of holding $1 worth of inventory for a year is estimated to be the annual capital opportunity cost of 20¢. Determine the EOQ.

Solution   We are given that $K = \$5$, $D = 10,000$ cameras per year, $h_d = 20$¢/dollar/year, and $p = \$100$ per camera. Then

$$q^* = \left(\frac{2(5)(10,000)}{(100)(0.20)}\right)^{1/2} = (5,000)^{1/2} = 70.71 \text{ cameras}$$

Hence, the EOQ recommends that the store order 70.71 cameras each time the inventory level reaches zero. Of course, the number of cameras ordered must be an integer. Since $TC(q)$ is a convex function of q, either $q = 70$ or $q = 71$ must minimize $TC(q)$. (If this seems difficult to believe, look at Figure 4.) Because of the flatness of the $HC(q) + OC(q)$ curve, it doesn't really matter whether the store chooses to order 70 or 71 cameras.

The Effect of a Nonzero Lead Time

We now allow the lead time L to be greater than zero. The introduction of a nonzero lead time leaves the annual holding and ordering costs unchanged. Hence, the EOQ still minimizes total costs. To prevent shortages from occurring and to minimize holding cost, each order must be placed at an inventory level that ensures that when each order arrives, the inventory level will equal zero.

DEFINITION ■   The inventory level at which an order should be placed is the **reorder point.** ■

To determine the reorder point for the basic EOQ model, two cases must be considered.

Case 1

Demand during the lead time does not exceed the EOQ. (This means that $LD \leq$ EOQ.) In this case, the reorder point occurs when the inventory level equals LD. Then the order will arrive L time units later, and upon arrival of the order, the inventory level will equal $LD - LD = 0$. In Example 1, suppose that it takes one month for a shipment of taillights to arrive. Then $L = \frac{1}{12}$ year, and Braneast's reorder point will be $(\frac{1}{12})(500) = 41.67$ taillights. Thus, whenever Braneast has 41.67 taillights on hand, an order should be placed for more taillights.

Case 2

Demand during the lead time exceeds the EOQ. (This means that $LD >$ EOQ.) In this case, the reorder point does not equal LD. Suppose that in Example 1, $L = 15$ months. Then $LD = (15/12)500 = 625$ taillights. Why can't we place an order each time the inventory level reaches 625 taillights? Since the EOQ $= 250$, our inventory level will never reach 625. To determine the correct reorder point, observe that orders are placed every six months. Suppose that an order has just arrived at time 0. Then an order must have been placed $L = 15$ months ago (at $T = -15$ months). Since orders arrive every six months, orders must be placed at $T = -9$ months, $T = -3$ months, $T = 3$ months, and so on. Since at $T = 0$ an order has just arrived, our inventory level at $T = 0$ is 250. Then at $T = 3$ (or any other point when an order is placed), the inventory level will equal $250 - (3/12)(500) = 125$. Thus, the reorder point is 125 taillights.

In general (see Problem 15), it can be shown that the reorder point equals the remainder when LD is divided by the EOQ. Thus, in our example, the reorder point is the remainder when 625 is divided by 250. This again yields a reorder point of 125 taillights.

The determination of the reorder point becomes extremely important when demand is random and stockouts can occur. In Sections 4.6 and 4.7, we discuss the problem of determining the reorder point when demand is random.

We close this section by giving an example of a noninventory problem that can be solved with the reasoning that we used to develop the EOQ.

EXAMPLE 3   Bus Service

Each hour, D students want to ride a bus from the student union to Fraternity Row. The administration places a value of h dollars on each hour that a student is forced to wait for a bus. It costs the university K dollars to send a bus from the student union to Fraternity Row. Assuming that demand occurs at a constant rate, how many buses should be sent each hour from the student union to Fraternity Row?

Solution   Note that

$$\frac{\text{Total cost}}{\text{Hour}} = \frac{\text{cost of sending buses}}{\text{hour}} + \frac{\text{student waiting cost}}{\text{hour}}$$

Since demand occurs at a constant rate, buses should leave at regular intervals. This means that each bus that arrives at the student union will find the same number of students waiting. Let $q =$ number of students present when each bus arrives. Assuming that a bus has just arrived at time 0, "number of students waiting" displays the behavior shown in Figure 5. Then

$$\frac{\text{Cost of sending buses}}{\text{Hour}} = \left(\frac{K \text{ dollars}}{\text{bus}}\right)\left(\frac{\frac{D}{q}\text{buses}}{\text{hour}}\right) = \frac{\frac{KD}{q}\text{ dollars}}{\text{hour}}$$

From Figure 5, the average number of students waiting is $\frac{q}{2}$. Then

$$\frac{\text{Student waiting cost}}{\text{Hour}} = \left(\frac{q}{2}\text{ students}\right)\left(\frac{h \text{ dollars/student}}{\text{hour}}\right) = \frac{\frac{hq}{2}\text{ dollars}}{\text{hour}}$$

These computations show that

$$\frac{\text{Total cost}}{\text{Hour}} = \frac{hq}{2} + \frac{KD}{q}$$

This is identical to $HC(q) + OC(q)$ for the basic EOQ model. Hence, the optimal value of q for our busing problem is simply the EOQ. This means that the optimal value of q is

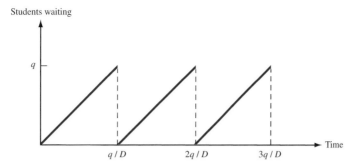

FIGURE 5
Evolution over Time of
Students Waiting
(Example 3)

$q^* = (\frac{2KD}{h})^{1/2}$. Since each bus picks up q^* students, $\frac{D}{q^*}$ buses should be sent each hour. From Figure 5, we see that the time between buses will be $\frac{q^*}{D}$ hours. For example, if $h =$ \$5/student/hour, $D = 100$ students/hour, and $K = $ \$10/bus, we find that

$$q^* = \left(\frac{2(10)(100)}{5}\right)^{1/2} = 20$$

Then $\frac{100}{20} = 5$ buses/hour will leave the student union, and a bus will leave the student union every $\frac{1}{5}$ hour $= 12$ minutes.

Spreadsheet Template for the Basic EOQ Model

EOQ.xls

Figure 6 (file EOQ.xls) illustrates an Excel template for the basic EOQ model. The user inputs the values of K, h (say, per year), lead time (L), and D (again, per year). Cell A5 has been given the range name K; cell B5, H; cell C5, D; and cell A11, L. In A8, the EOQ is determined by the formula (2*K*D/H)^.5. In B8, we compute annual holding costs with the formula .5*A8*H. In D5, we compute orders per year for the EOQ with the formula =D/A8. In C8, we compute annual ordering costs for the EOQ with the formula =K*D5. In D8, we compute total annual cost for the EOQ with the formula =B8+C8. In B11, we compute the reorder point with the formula =**MOD**(L*D,A8). This yields the remainder obtained when L*D is divided by the EOQ. In Figure 6, we have input the data values for Example 1.

A	A	B	C	D
1	SIMPLE			
2	EOQ			
3	MODEL			
4	K	h	D	ORDERS/YR
5	5	0.08	500	2
6				
7	EOQ	HOLDING COSTS	ORDERING COSTS	TOTAL COST
8	250	10	10	20
9				
10	LEADTIME	REORDER POINT		
11	1.25	125		

FIGURE 6
Simple EOQ Model

Power-of-Two Ordering Policies

Suppose a company orders three products, and the EOQs for each product yield times between orders of 3.5 days, 5.6 days, and 9.2 days. It would rarely be the case that orders for different products would arrive on the same day. If we could somehow synchronize our reorder intervals so that orders for different products often arrived on the same day, we could greatly reduce our coordination costs. For example, we would need far fewer trucks to deliver our orders if we could synchronize their arrival. Roundy (1985) devised an elegant and simple method called **power-of-two ordering policies** to ensure that orders for multiple products are well synchronized. Let $q^* =$ EOQ. Then the optimal reorder interval for a product is $t^* = q^*/D$. We assume t^* is at least 1 day. Then for some $m \geq 0$, it must be true that $2^m \leq t^* \leq 2^{m+1}$. If $t^* \leq \sqrt{2} * 2^m$, we choose a reorder quantity corresponding to a reorder interval of 2^m. If $t^* \geq \sqrt{2} * 2^m$, we choose a reorder quantity corresponding to a reorder interval of 2^{m+1}. Roundy proved that using this method (called a *power-of-two* policy) to round the reorder interval to a neighboring power of 2 will increase the sum of fixed and holding costs at most 6%. The virtue of a power-of-two policy is that different products will frequently arrive at the same time. In many circumstances, this will greatly reduce coordination costs. For example, consider our three products with reorder intervals of 3.5 days, 5.6 days, and 9.2 days. Roundy's power-of-two policy would choose order quantities corresponding to reorder periods of 4, 4, and 8 days, respectively. Thus, products 1 and 2 always arrive together; half the time, product 3 arrives with product 2. In most circumstances, this policy will reduce coordination costs by more than the maximum possible 6% increase in total cost. We now give a proof of Roundy's result.

To begin, pick an arbitrary order quantity q' and define the total cost for this order quantity by

$$TC(q') = \frac{hq'}{2} + \frac{KD}{q'}$$

Then

$$\frac{TC(q')}{TC(q^*)} = \frac{\dfrac{hq'}{2} + \dfrac{KD}{q'}}{\sqrt{2KhD}} = \frac{q'}{2}\sqrt{\frac{h^2}{2KDh}} + \frac{1}{q'}\sqrt{\frac{K^2D^2}{2KDh}}$$

$$= \frac{q'}{2}\sqrt{\frac{h}{2KD}} + \frac{1}{2q'}\sqrt{\frac{2KD}{h}}$$

$$= \frac{q'}{2q^*} + \frac{q^*}{2q'}$$

$$= \frac{1}{2}\left(\frac{q'}{q^*} + \frac{q^*}{q'}\right)$$

Since $t^* = \dfrac{q^*}{D}$ and $t' = \dfrac{q'}{D}$, we find that

$$\frac{TC(t')}{TC(t^*)} = \frac{1}{2}\left(\frac{t'}{t^*} + \frac{t^*}{t'}\right) \tag{5}$$

We can now prove Roundy's result. We assume that t^* is at least 1 day. Then for some nonnegative integer m, $2^m \leq t^* \leq 2^{m+1}$.

THEOREM 1

If $t^* \leq 2^m(\sqrt{2})$, then the minimum-cost power-of-two ordering policy is to set $t = 2^m$. If $t^* \geq 2^m(\sqrt{2})$, then the minimum-cost power-of-two ordering policy is 2^{m+1}.

In either case, the total cost of the optimal power-of-two ordering policy will never be more than 6% higher than the total cost of the EOQ.

Proof Since $TC''(q) > 0$, we know that $TC(q)$ is a convex function of q. The convexity of $TC(q)$ implies that the optimal power-of-two reorder time interval is either 2^m or 2^{m+1}. From (5), 2^m will be the optimal power-of-two reorder time interval if and only if

$$\frac{1}{2}\left(\frac{2^m}{t^*} + \frac{t^*}{2^m}\right) \le \frac{1}{2}\left(\frac{2^{m+1}}{t^*} + \frac{t^*}{2^{m+1}}\right) \tag{6}$$

Inequality (6) will hold if and only if

$$\frac{t^*}{2^{m+1}} \le \frac{2^m}{t^*}$$

or $t^* \le \sqrt{2}(2^m)$. We have now shown that if $t^* \le 2^m(\sqrt{2})$, then the minimum-cost power-of-two ordering policy is to set $t = 2^m$. If $t^* \ge 2^m (\sqrt{2})$, then the minimum-cost power-of-two ordering policy is 2^{m+1}. This result shows that the optimal power-of-two ordering policy must choose a reorder time in the interval $\left[\frac{t^*}{\sqrt{2}}, \sqrt{2}t^*\right]$.

From (5), we now find that the maximum discrepancy between the total cost for the power-of-two ordering policy and the total cost for t^* will occur if the power-of-two reorder interval equals either $\sqrt{2}t^*$ or $\frac{t^*}{\sqrt{2}}$. In either case,

$$\frac{TC\left(\sqrt{2}t^* \ or \ \frac{t^*}{\sqrt{2}}\right)}{TC(t^*)} = \frac{1}{2}\left(\frac{1}{\sqrt{2}} + \sqrt{2}\right) = 1.06$$

Thus, a power-of-two policy cannot cause an increase in total cost of more than 6%.

PROBLEMS

Group A

1 Each month, a gas station sells 4,000 gallons of gasoline. Each time the parent company refills the station's tanks, it charges the station $50 plus 70¢, per gallon. The annual cost of holding a gallon of gasoline is 30¢.

  a How large should the station's orders be?

  b How many orders per year will be placed?

  c How long will it be between orders?

  d Would the EOQ assumptions be satisfied in this situation? Why or why not?

  e If the lead time is two weeks, what is the reorder point? If the lead time is ten weeks, what is the reorder point? Assume 1 week $= \frac{1}{52}$ year.

2[†] Money in my savings account earns interest at 10% annual rate. Each time I go to the bank, I waste 15 minutes in line. My time is worth $10 per hour. During each year, I need to withdraw $10,000 to pay my bills.

  a How often should I go to the bank?

  b Each time I go to the bank, how much money should I withdraw?

  c If my need for cash increases, will I go to the bank more often or less often?

  d If interest rates rise, will I go to the bank more often or less often?

  e If the bank adds more tellers, will I go to the bank more often or less often?

3[‡] Father Dominic's Pizza Parlor receives 30 calls per hour for delivery of pizza. It costs Father Dominic's $10 to send out a truck to deliver pizzas. It is estimated that each minute a customer spends waiting for a pizza costs the pizza parlor 20¢ in lost future business.

  a How often should Father Dominic's send out a truck?

  b What would be the answer if a truck could only carry five pizzas?

[†]Based on Baumol (1952).

[‡]Based on Ignall and Kolesar (1972).

4 The efficiency of an inventory system is often measured by the **turnover ratio.** The turnover ratio (TR) is defined by

$$TR = \frac{\text{cost of goods sold during a year}}{\text{average value of on-hand inventory}}$$

a Does a high turnover ratio indicate an efficient inventory system?

b If the EOQ model is being used, determine TR in terms of K, D, h, and q.

c Suppose D is increased. Show that TR will also be increased.

5 Suppose we order three types of appliances for the appliance store Ohm City. The optimal reorder intervals are 9.2 days, 21.2 days, and 38.1 days. What would be the optimal power-of-two ordering policy?

6 Suppose we order three types of clothing for Ceiling Mart. The optimal reorder intervals are 92 days, 21 days, and 60 days. What would be the optimal power-of-two ordering policy?

Group B

7 Suppose we are ordering computer chips. Suppose that in each order, exactly 10% of all chips are defective. As soon as the order arrives, we find out which chips are defective and return them for a complete refund. What would be the optimal ordering policy in this situation?

8 Show that for $q \leq q^*$, an order size of $q + q^*$ will have a lower cost than an order size of $q - q^*$. What is the managerial significance of this result?

9 Suppose that instead of ordering the EOQ q^*, we use the order quantity $0.8q^*$. Use Equation (3) to show that $HC(q) + OC(q)$ will have increased by 2.50%.

10 In terms of K, D, and h, what is the average length of time that an item spends in inventory before being used to meet demand? Explain how this result can be used to characterize a fast- or slow-moving item.

11 A drug store sells 30 bottles of antibiotics per week. Each time it orders antibiotics, there is a fixed ordering cost of $10 and a cost of $10/bottle. Assume that the annual holding cost is 20% of the cost of a bottle of antibiotics, and suppose antibiotics spoil and cannot be sold if they spend more than one week in inventory. When the drug store places an order, how many bottles of antibiotics should be ordered?

12 During each year, CSL Computer Company needs to train 27 service representatives. No matter how many students are trained, it costs $12,000 to run a training program. Since service reps earn a monthly salary of $1,500,

CSL does not want to train them before they are needed. Each training session takes one month.

a State the assumptions needed for the EOQ model to be applicable.

b How many service representatives should be in each training group?

c How many training programs should CSL undertake each year?

d How many trained service reps will be available when each training program begins?

13 A newspaper has 500,000 subscribers who pay $4 per month for the paper. It costs the company $200,000 to bill all its customers. Assume that the company can earn interest at a rate of 20% per year on all revenues. Determine how often the newspaper should bill its customers. (*Hint:* Look at unpaid subscriptions as the inventoried good.)

14 Consider a firm that knows that the price of the product it is ordering is going to increase permanently by X. How much of the product should be ordered before the price increase goes into effect?

Here is one approach to this question: Suppose the firm orders Q units before the price increase goes into effect.

a What extra holding cost is incurred by ordering Q units now?

b How much in purchasing costs is saved by ordering Q units now?

c What value of Q maximizes purchasing cost savings less extra holding costs?

d Suppose that annual demand is 1,000 units, holding cost per unit-year is $7.50, and the price of the item is going to increase by $10. How large an order should be placed before the price increase goes into effect?

15 Show that the reorder point in the EOQ model equals the remainder when LD is divided by the EOQ.

16 The borough of Staten Island has two "sanitation districts." In district 1, street litter piles up at an average rate of 2,000 tons per week, and in district 2 at an average rate of 1,000 tons per week. Each district has 500 miles of streets. Staten Island has 10 sanitation crews and each crew can clean 50 miles per week of streets. To minimize the average level of the total amount of street litter in the two districts, how often should each district be cleaned? Assume that litter in a district grows at a constant rate until it is picked up (assume pickup is instantaneous). (*Hint:* Let p_i equal the average number of times that each district is cleaned per week. Then $p_1 + p_2 = 1$.)[†]

[†]Based on Riccio, Miller, and Little (1986).

3.3 Computing the Optimal Order Quantity When Quantity Discounts Are Allowed

Up to now, we have assumed that the annual purchasing cost does not depend on the order size. In Section 3.2, this assumption allowed us to ignore the annual purchasing cost when we computed the order quantity that minimizes total annual cost. In real life, however, suppliers often reduce the unit purchasing price for large orders. Such price reductions are referred to as *quantity discounts.* If a supplier gives quantity discounts, the annual purchasing cost will depend on the order size. If holding cost is expressed as a percentage of an item's purchasing cost, the annual holding cost will also depend on the order size. Since the annual purchasing cost now depends on the order size, we can no longer ignore purchasing cost while trading off holding cost against setup cost. Thus, the approach used in Section 3.2 to find the optimal quantity is no longer valid, and a new approach is needed.

If we let q be the quantity ordered each time an order is placed, the general quantity discount model analyzed in this section may be described as follows:

$$\text{If } q < b_1, \text{ each item costs } p_1 \text{ dollars.}$$
$$\text{If } b_1 \leq q < b_2, \text{ each item costs } p_2 \text{ dollars.}$$
$$\text{If } b_{k-2} \leq q < b_{k-1}, \text{ each item costs } p_{k-1} \text{ dollars.}$$
$$\text{If } b_{k-1} \leq q < b_k = \infty, \text{ each item costs } p_k \text{ dollars.}$$

Since $b_1, b_2, \ldots, b_{k-1}$ are points where a price change (or break) occurs, we refer to b_1, $b_2, \ldots, b_{k-1}$ as **price break points.** Since larger order quantities should be associated with lower prices, we have $p_k < p_{k-1} < \cdots < p_2 < p_1$. The following example illustrates the quantity discount model.

EXAMPLE 4   Buying Disks

A local accounting firm in Smalltown orders boxes of floppy disks (10 disks to a box) from a store in Megalopolis. The per-box price charged by the store depends on the number of boxes purchased (see Table 2). The accounting firm uses 10,000 disks per year. The cost of placing an order is assumed to be $100. The only holding cost is the opportunity cost of capital, which is assumed to be 20% per year. For this example, $b_1 = 100$, $b_2 = 300$, $p_1 = \$50.00$, $p_2 = \$49.00$, and $p_3 = \$48.50$.

The example is continued later in this section.

TABLE 2
Purchase Costs for Disks

No. of Boxes Ordered (q)	Price per Box
$0 \leq q < 100$	$50.00
$100 \leq q < 300$	$49.00
$q \geq 300$	$48.50

Before explaining how to find the order quantity minimizing total annual costs, we need the following definitions.

1 $TC_i(q)$ = total annual cost (including holding, purchasing, and ordering costs) if each order is for q units at a price p_i.

2 EOQ_i = quantity that minimizes total annual cost if, for any order quantity, the purchasing cost of the item is p_i.

3 EOQ_i is *admissible* if $b_{i-1} \leq EOQ_i < b_i$.

4 $TC(q)$ = actual annual cost if q items are ordered each time an order is placed. (We determine $TC(q)$ by using price p_i if $b_{i-1} \leq q < b_i$.)

Our goal is to find the value of q minimizing $TC(q)$. Figures 7a and 7b illustrate these definitions. Observe that in Figure 7a, EOQ_2 is admissible because $b_1 < EOQ_2 < b_2$, but EOQ_1 and EOQ_3 are not admissible. In each figure, $TC(q)$ is the solid portion of the curve. The dashed portion of each curve represents unattainable costs. For instance, in Figure 7b, $TC_2(q)$ is dotted for $q < b_1$, because the price is not p_2 for $q < b_1$. For $q < b_1$, total annual cost is given by the solid portion of $TC_1(q)$, because for $q < b_1$, the price is p_1, and for $q \geq b_1$, total annual cost is given by the solid portion of $TC_2(q)$.

In general, the value of q minimizing $TC(q)$ can be either a break point (see Figure 7b) or some EOQ_i (see Figure 7a).

The following observations are helpful in determining the point (break point or EOQ_i) that minimizes $TC(q)$.

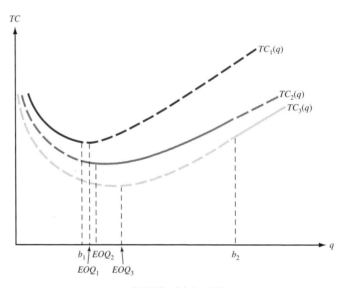

(a) EOQ_2 minimizes TC

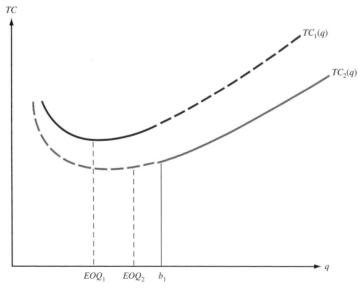

FIGURE 7
Illustrations of
Definitions of $TC_i(q)$
and EOQ_i

(b) b_1 minimizes TC

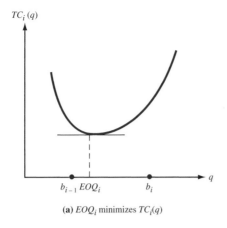

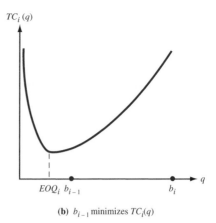

FIGURE 8
For $b_{i-1} \le q < b_i$,
What Value of q
Minimizes $TC_i(q)$?

(a) EOQ_i minimizes $TC_i(q)$ (b) b_{i-1} minimizes $TC_i(q)$

1 For any value of q,

$$TC_k(q) < TC_{k-1}(q) < \cdots < TC_2(q) < TC_1(q)$$

This observation is valid because for any order quantity q, $TC_k(q)$ will have the lowest holding and purchasing costs, since p_k is the lowest available price; $TC_1(q)$ will have the highest holding and purchasing costs, because p_1 is the highest available price. Thus, in Figure 7a, we find that $TC_3(q) < TC_2(q) < TC_1(q)$.

2 If EOQ_i is admissible, then minimum cost for $b_{i-1} \le q < b_i$ occurs for $q = EOQ_i$ (see Figure 8a). If $EOQ_i < b_{i-1}$, the minimum cost for $b_{i-1} \le q < b_i$ occurs for $q = b_{i-1}$ (see Figure 8b). This observation follows from the fact that $TC_i(q)$ decreases for $q < EOQ_i$ and increases for $q > EOQ_i$.

3 If EOQ_i is admissible, then $TC(q)$ cannot be minimized at an order quantity for which the purchasing price per item exceeds p_i. Thus, if EOQ_i is admissible, the optimal order quantity must occur for either price $p_i, p_{i+1}, \ldots,$ or p_k.

To see why observation 3 holds, suppose EOQ_i is admissible. Why can't an order quantity associated with a price $p_j > p_i$ have a lower cost than EOQ_i? Note that EOQ_i minimizes total annual cost if price is p_i and EOQ_j does not minimize total annual cost if price is p_i. Thus,

$$TC_i(EOQ_i) < TC_i(EOQ_j)$$

Since $p_j > p_i$,

$$TC_i(EOQ_j) < TC_j(EOQ_j)$$

The last two inequalities show that

$$TC_i(EOQ_i) < TC_j(EOQ_j)$$

By the definition of EOQ_j, we know that for all q,

$$TC_j(EOQ_j) \le TC_j(q)$$

Thus,

$$TC_i(EOQ_i) < TC_j(EOQ_j) \le TC_j(q)$$

and ordering EOQ_i at price p_i is superior to ordering any quantity at a higher price p_j.

These observations allow us to use the following method to determine the optimal order quantity when quantity discounts are allowed. Beginning with the lowest price, determine for each price the order quantity that minimizes total annual costs for $b_{i-1} \le q < b_i$

(call this order quantity q_i^*). Continue determining q_k^*, q_{k-1}^*, . . . until one of the q_i^*'s (call it q_i^*) is admissible; from observation 2, this will mean that $q_i^* = EOQ_i$. The optimal order quantity will be the member of $\{q_k^*, q_{k-1}^*, \ldots, q_i^*\}$ with the smallest value of $TC(q)$.

EXAMPLE 4   **Buying Disks (Continued)**

Each time an order is placed for disks, how many boxes of disks should be ordered? How many orders will be placed annually? What is the total annual cost of meeting the accounting firm's disk needs?

Solution   Note that $K = \$100$ and $D = 1,000$ boxes per year. We first determine the best order quantity for $p_3 = \$48.50$ and $300 \leq q$. Then

$$EOQ_3 = \left(\frac{2(100)(1,000)}{0.2(48.50)}\right)^{1/2} = 143.59$$

Since $EOQ_3 < 300$, EOQ_3 is not admissible. Therefore, Figure 8b is relevant, and for $q \geq 300$, $TC_3(q)$ is minimized by $q_3^* = 300$.

We next consider $p_2 = \$49.00$ and $100 \leq q < 300$. Then

$$EOQ_2 = \left(\frac{2(100)(1,000)}{9.8}\right)^{1/2} = 142.86$$

Since $100 \leq EOQ_2 < 300$, EOQ_2 is admissible, and for a price $p_2 = \$49.00$, the best we can do is to choose $q_2^* = 142.86$; Figure 8a is relevant. Since q_2^* is admissible, $p_1 = \$50.00$ and $0 \leq q < 100$ cannot yield the order quantity minimizing $TC(q)$ (see observation 3). Thus, either $q_2^* = 142.86$ or $q_3^* = 300$ will minimize $TC(q)$. To determine which of these order quantities minimizes $TC(q)$, we must find the smaller of $TC_3(300)$ and $TC_2(142.86)$. For $q_3^* = 300$, the annual holding cost/item/year is $0.20(48.50) = \$9.70$. Thus, for q_3^*,

$$\text{Annual ordering cost} = 100(\tfrac{1,000}{300}) = \$333.33$$
$$\text{Annual purchasing cost} = 1,000(48.50) = \$48,500$$
$$\text{Annual holding cost} = (\tfrac{1}{2})(300)(9.7) = \$1,455$$
$$TC_3(300) = \$50,288.33$$

For $q_2^* = 142.86$, the annual holding cost/item/year is $0.20(49) = \$9.80$. Thus, for q_2^*,

$$\text{Annual ordering cost} = 100(\tfrac{1,000}{142.86}) = \$699.99$$
$$\text{Annual purchasing cost} = 1,000(49) = \$49,000$$
$$\text{Annual holding cost} = (\tfrac{1}{2})(142.86)(9.8) = \$700.01$$
$$TC_2(142.86) = \$50,400$$

Thus, $q_3^* = 300$ will minimize $TC(q)$.

Our analysis shows that each time an order is placed, 300 boxes of disks should be ordered. Then $\frac{1,000}{300} = 3.33$ orders are placed each year. As we have already seen, the minimum total annual cost is $\$50,288.33$.

A Spreadsheet Template for Quantity Discounts

Qd.xls

Figure 9 (file Qd.xls) illustrates how inventory problems with a quantity discount can be solved on a spreadsheet. In cell B2 (given the range name K), we enter K, the cost per order. In cell C2 (range name D) we enter D, the annual demand. In cell D2 (HD), we enter the annual cost of holding $1 of goods in inventory for one year.

FIGURE 9
Quantity Discount Calculations

A	A	B	C	D	E	F
1	QUANTITY	K	D	hperdollar		
2	DISCOUNT	100	1000	0.2		
3	CALCULATIONS					
4						
5	LEFTENDPOINT	RIGHT ENDPOINT	PRICE	EOQ	MINCOSTOQ	MINIMUM COST
6	0	100	$50.00	141.42135624	99.0000	$51,505.10
7	100	300	$49.00	142.85714286	142.8571	$50,400.00
8	300	10000	$48.50	143.59163172	300.0000	$50,288.33

In the cell range A6:C8, we enter (using the data from Example 4) the left-hand endpoint, right-hand endpoint, and price for each interval. Thus, for an order quantity ≥ 0 and < 100 the per-unit price is $50. Now observe that Figure 8 implies that for each interval the minimum cost in that interval is obtained as follows.

1 If the EOQ for the ith interval's price lies in the interval, then the EOQ for that interval obtains the minimum cost in the ith interval.

2 If the EOQ for the ith interval's price is smaller than the left-hand endpoint of the ith interval (b_{i-1}), then the minimum cost for that interval is attained by an order quantity of b_{i-1}. Here we set $b_0 = 0$.

3 If the EOQ for the ith interval's price is larger than the right-hand endpoint for the ith interval (b_i), then the minimum cost for that interval is attained by an order quantity of $b_i - 1$.

Our spreadsheet incorporates this logic as follows: In D6, we compute the EOQ for the interval $b_0 = 0 \leq$ order quantity $< 100 = b_1$ by entering the formula $(2*K*D/(HD*C6))^\wedge.5$. In E6, we enter the formula

$$=IF(AND(D6>=A6,D6<B6),D6,IF(D6<A6,A6,B6-1))$$

This statement computes the order quantity in the first interval that minimizes annual costs by implementing the logic described in (1)–(3) here. In F6, we compute the annual cost corresponding to the order quantity in E6. This is given by $(K*D/E6)+D*C6+.5*HD*C6*E6$.

In this formula, the first term is the annual cost of placing orders; the second term is the cost of purchasing one year's demand at the price for the first interval; and the third term is the annual holding cost (whose per-unit cost equals the price of item times the annual holding cost per dollar of inventory). Copying from the range D6:F6 to D6:F8 generates the minimum annual cost for the other two intervals. We see that the minimum annual cost is $50,288.33, and it is attained by an order quantity of 300.

PROBLEMS

Group A

1 A consulting firm is trying to determine how to minimize the annual costs associated with purchasing computer paper. Each time an order is placed an ordering cost of $20 is incurred. The price per box of computer paper depends on q, the number of boxes ordered (see Table 3). The annual holding cost is 20% of the dollar value of inventory. During each month, the consulting firm uses 80 boxes of computer paper. Determine the optimal order quantity and the number of orders placed each year.

2 Each year, Shopalot Stores sells 10,000 cases of soda. The company is trying to determine how many cases should be ordered each time. It costs $5 to process each order, and

TABLE 3

No. of Boxes Ordered	Price per Box
$q < 300$	$10.00
$300 \le q < 500$	$9.80
$q \ge 500$	$9.70

TABLE 4

No. of Cases Ordered	Price per Case
$q < 200$	$4.40
$200 \le q < 400$	$4.20
$q \ge 400$	$4.00

the cost of carrying a case of soda in inventory for one year is 20% of the purchase price. The soda supplier offers Shopalot the schedule of quantity discounts shown in Table 4 (q = number of cases ordered per order). Each time an order is placed, how many cases of soda should the company order?

3 A firm buys a product using the price schedule given in Table 5. The company estimates holding costs at 10% of purchase price per year and ordering costs at $40 per order. The firm's annual demand is 460 units.

   a Determine how often the firm should order.

   b Determine the size of each order.

   c At what price should the firm order?

TABLE 5

Order Size	Price per Unit
0–99 units	$20.00
100–199	$19.50
200–499	$19.00
500 or more	$18.75

Group B

4 A hospital orders its thermometers from a hospital supply firm. The cost per thermometer depends on the order size q, as shown in Table 6. The annual holding cost is 25% of the purchasing cost. Let EOQ_{80} be the EOQ if the cost per thermometer is 80¢, and let EOQ_{79} be the EOQ if the cost per thermometer is 79¢.

   a Explain why EOQ_{79} will be larger than EOQ_{80}.

   b Explain why the optimal order quantity must be either EOQ_{79}, EOQ_{80}, or 100.

   c If $EOQ_{80} > 100$, show that the optimal order quantity must be EOQ_{79}.

   d If $EOQ_{80} < 100$ and $EOQ_{79} < 100$, show that the optimal order quantity must be either EOQ_{80} or 100.

   e If $EOQ_{80} < 100$ and $EOQ_{79} > 100$, show that the optimal order quantity must be EOQ_{79}.

5 In Problem 4, suppose the cost per order is $1 and the monthly demand is 50 thermometers. What is the optimal order quantity? How small a discount could the supplier offer and still have the hospital accept the discount?

TABLE 6

Order Size	Price per Thermometer
$q < 100$	80¢
$q \ge 100$	79¢

3.4 The Continuous Rate EOQ Model

Many goods are produced internally rather than purchased from an outside supplier. In this situation, the EOQ assumption that each order arrives at the same instant seems unrealistic; it isn't possible to produce, say, 10,000 cars at the drop of a hard hat. If a company meets demand by making its own products, the continuous rate EOQ model will be more realistic than the traditional EOQ model. Again, we assume that demand is deterministic and occurs at a constant rate; we also assume that shortages are not allowed.

The continuous rate EOQ model assumes that a firm can produce a good at a rate of r units per time period (we again use one year as the time unit). This means that during any time period of length t, the firm can produce rt units. We define

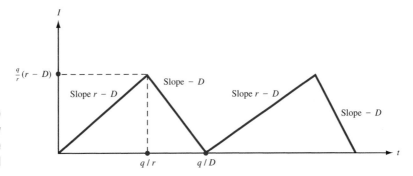

FIGURE 10
Variation of Inventory for Continuous Rate EOQ Model

q = number of units produced during each production run

K = cost of setting up a production run

(often due to idle time that occurs at the beginning or end of a production run)

h = cost of holding one unit in inventory for one year

D = annual demand for the product

Assuming that a production run begins at time 0, the variation of inventory over time is described by Figure 10. At the beginning of a production run, we are producing at a rate of r units per year, and demand is occurring at a rate of D units per year. Thus, until q units are produced, inventory increases at a rate of $r - D$ units per year. (Of course, $r \geq D$ must hold, or else demand could not be met.) At time $\frac{q}{r}$, q units will have been produced. At this time, the production run is complete, and inventory decreases at a rate of D units per year until a zero inventory position is reached. A zero inventory level will occur at time $\frac{q}{D}$. Then another production run begins.

Assuming that per-unit production costs are independent of run size, we must determine the value of q that minimizes

$$\frac{\text{Holding cost}}{\text{Year}} + \frac{\text{setup cost}}{\text{year}}$$

Since demand occurs at a constant rate, we know that (average inventory level) = $(\frac{1}{2})$(maximum inventory level). From Figure 10, we see that the maximum inventory level occurs at time $\frac{q}{r}$. Since between zero and $\frac{q}{r}$, the inventory level is increasing at a rate of $r - D$ units per year, the inventory level at time $\frac{q}{r}$ will be $(\frac{q}{r})(r - D)$. Then (average inventory level) = $(\frac{1}{2})(\frac{q}{r})(r - D)$, and

$$\frac{\text{Holding cost}}{\text{Year}} = h(\text{average inventory})(1 \text{ year}) = \frac{h(r - D)q}{2r}$$

Observe that the annual holding cost for the continuous rate EOQ model is the same as that for a conventional EOQ model in which the unit holding cost is $\frac{h(r - D)}{r}$. As usual,

$$\frac{\text{Ordering cost}}{\text{Year}} = \left(\frac{\text{ordering cost}}{\text{cycle}}\right)\left(\frac{\text{cycles}}{\text{year}}\right) = \frac{KD}{q}$$

The discussion shows that

$$\frac{\text{Holding cost}}{\text{Year}} + \frac{\text{ordering cost}}{\text{year}} = \frac{hq(r - D)}{2r} + \frac{KD}{q}$$

The last equation shows that the problem of minimizing the sum of annual holding and ordering costs for the continuous rate model is equivalent to solving an EOQ model with holding cost $\frac{h(r - D)}{r}$, ordering cost K, and annual demand D. Using this observation and the economic order quantity (or lot size) formula (2), we may immediately deduce that for the continuous rate EOQ model,

$$\text{Optimal run size} = \left(\frac{2KD}{\frac{h(r-D)}{r}}\right)^{1/2} = \left(\frac{2KDr}{h(r-D)}\right)^{1/2} \tag{7}$$

As usual, $\frac{D}{q}$ production runs must be made each year to meet the annual demand of D units. Using the fact that

$$\text{EOQ} = \left(\frac{2KD}{h}\right)^{1/2}$$

we may rewrite (7) as

$$\text{Optimal run size} = \text{EOQ}\left(\frac{r}{r-D}\right)^{1/2} \tag{8}$$

As r increases, production occurs at a more rapid rate. Hence, for large r, the rate model should approach the instantaneous delivery situation of the EOQ model. To see that this is the case, note that for r large, $\frac{r}{(r-D)}$ approaches 1. Then (8) shows that as r increases toward infinity, the optimal run size for the continuous rate model approaches the EOQ.

EXAMPLE 5 **Macho Auto Company**

Macho Auto Company needs to produce 10,000 car chassis per year. Each is valued at $2,000. The plant has the capacity to produce 25,000 chassis per year. It costs $200 to set up a production run, and the annual holding cost is 25¢, per dollar of inventory. Determine the optimal production run size. How many production runs should be made each year?

Solution   We are given that

$$r = 25,000 \text{ chassis per year}$$
$$D = 10,000 \text{ chassis per year}$$
$$h = 0.25(\$2,000)/\text{chassis/year} = \$500/\text{chassis/year}$$
$$K = \$200 \text{ per production run}$$

From (7),

$$\text{Optimal run size} = \left(\frac{2(200)(10,000)(25,000)}{500(25,000 - 10,000)}\right)^{1/2} = 115.47$$

Also, $\frac{10,000}{115.47} = 86.60$ production runs will be made each year.

Spreadsheet Template for the Continuous Rate EOQ Model

Figure 11 (file ConEOQ.xls) illustrates a template for the continuous rate EOQ model. In cell A6, the user inputs K; in B6, h; in C6, D; and in D6, the production rate r. In Figure 11 we have used the parameter values given in Example 5. In A8 (assigned the range name Q), the formula (2*K*D/H)^.5*(R/(R−D))^.5 (again we are using range names) computes the optimal run size. In B8, the formula D/Q computes the number of runs per year. In C8, we compute the annual cost (exclusive of purchasing costs) with the formula (H*Q*(R−D)/(2*R))+K*D/Q. The first term in this formula equals the annual cost of

		A	B	C	D
1	CONTINUOUS				
2	RATE				
3	EOQ				
4	MODEL				
5		K	h	D	r
6		200	500	10000	25000
7	RUN SIZE		RUNS/YR	COST/YR	
8		115.47005383793	86.60254	34641.016	

FIGURE 11
Continuous Rate EOQ Model

holding inventory. This follows because, from Figure 10, the maximum level of inventory during a cycle is $q(r - D)/r$. The second term is the annual cost of placing orders.

PROBLEMS

Group A

1 Show that the optimal run size always exceeds the EOQ. Give an intuitive explanation for this result.

2 A company can produce 100 home computers per day. The setup cost for a production run is $1,000. The cost of holding a computer in inventory for one year is $300. Customers demand 2,000 home computers per month (assume that 1 month = 30 days and 360 days = 1 year). What is the optimal production run size? How many production runs must be made each year?

3 The production process at Father Dominic's Pizza can produce 400 pizza pies per day; the firm operates 250 days per year. Father Dominic's has a cost of $180 per production run and a holding cost of $5 per pizza-year. The pies are frozen immediately after they are produced and stored in a refrigerated warehouse with a current maximum capacity of 2,000 pies.

   a Annual demand is 37,500 pies per year. What production run size should be used?

   b What is the total annual cost incurred in meeting demand?

   c How many days per year will the company be producing pizza pies?

Group B

4 A company has the option of purchasing a good or manufacturing the item. If the item is purchased, the company will be charged $25 per unit plus a cost of $4 per order. If the company manufactures the item, it has a production capacity of 8,000 units per year. It costs $50 to set up a production run, and annual demand is 3,000 units per year. If the annual holding cost is 10% and the cost of manufacturing one unit is $23, determine whether the company should purchase or manufacture the item.

3.5 The EOQ Model with Back Orders Allowed

In many real-life situations, demand is not met on time, and shortages occur. When a shortage occurs, costs are incurred (because of lost business, the cost of placing special orders, loss of future goodwill, and so on). In this section, we modify the EOQ model of Section 3.2 to allow for the possibility of shortages. Let s be the cost of being short one unit for one year. The variables K, D, and h have their usual meanings. In most situations, s is very difficult to measure. We assume that all demand is backlogged and no sales are lost. To determine the order policy that minimizes annual costs, we define

$$q = \text{order quantity}$$

$$q - M = \text{maximum shortage that occurs under an ordering policy}$$

Equivalently (assuming a zero lead time), the firm will be $q - M$ units short each time an order is placed.

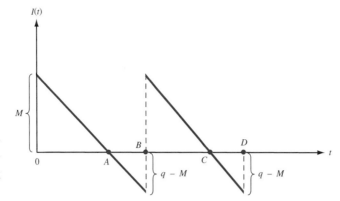

We assume that the lead time for each order is zero. Since an order is placed each time the firm is $q - M$ units short (or when the firm's inventory position is $M - q$), the firm's maximum inventory level will be $M - q + q = M$. For example, if $q = 500$ and $q - M = 100$, we know that an order for 500 units will be used to satisfy the backlogged demand for 100 units and will result in an inventory level of $500 - 100 = 400$ units.

Assuming that an order is placed at time 0, the evolution of the inventory level over time is described by Figure 12. Since purchasing costs do not depend on q and M, we can minimize annual costs by determining the values of q and M that minimize

$$\frac{\text{Holding cost}}{\text{Year}} + \frac{\text{shortage cost}}{\text{year}} + \frac{\text{order cost}}{\text{year}} \tag{9}$$

Notice that what happens between time 0 and time B is identical to what happens between time B and time D. For this reason, we call the time periods $0B$ and BD **cycles.** A cycle may also be thought of as the time interval between placement of orders. To determine holding cost per year and shortage cost per year, we begin by finding holding cost per cycle and shortage cost per cycle. This requires that we find the length of line segments $0A$ and AB in Figure 12. Since a zero inventory level occurs after M units have been demanded, we conclude that $0A = \frac{M}{D}$. Since a cycle ends when q units have been demanded, we conclude that $0B = \frac{q}{D}$. Then

$$\text{Length of } AB = (\text{length of } 0B) - (\text{length of } 0A) = \frac{q - M}{D}$$

Also note that since q units are ordered during each cycle, $\frac{D}{q}$ cycles (and orders) must be placed during each year. We can now express the costs in (9). Recall that

$$\frac{\text{Holding cost}}{\text{Year}} = \left(\frac{\text{holding cost}}{\text{cycle}}\right)\left(\frac{\text{cycles}}{\text{year}}\right)$$

and

$$\frac{\text{Holding cost}}{\text{Cycle}} = \text{holding cost from time 0 to time } A$$

From Figure 12, the average inventory level between time 0 and time A is simply $\frac{M}{2}$. Since $0A$ is of length $\frac{M}{D}$,

$$\frac{\text{Holding cost}}{\text{Cycle}} = \left(\frac{M}{2}\right)\left(\frac{M}{D}\right)h = \frac{M^2 h}{2D}$$

Since there are $\frac{D}{q}$ cycles per year,

$$\frac{\text{Holding cost}}{\text{Year}} = \left(\frac{M^2 h}{2D}\right)\left(\frac{D}{q}\right) = \frac{M^2 h}{2q}$$

Similarly,

$$\frac{\text{Shortage cost}}{\text{Year}} = \left(\frac{\text{shortage cost}}{\text{cycle}}\right)\left(\frac{\text{cycles}}{\text{year}}\right)$$

Also observe that shortage cost per cycle = shortage cost incurred during time AB. Since demand occurs at a constant rate, the average shortage level during time AB is simply half of the maximum shortage. Thus, the average shortage level on the time interval AB is $\frac{q - M}{2}$. Since AB is a time interval of length $\frac{q - M}{D}$,

$$\frac{\text{Shortage cost}}{\text{Cycle}} = \frac{1}{2}(q - M)\left(\frac{q - M}{D}\right)s = \frac{(q - M)^2 s}{2D}$$

Since there are $\frac{D}{q}$ cycles per year,

$$\frac{\text{Shortage cost}}{\text{Year}} = \frac{(q - M)^2 s}{2D}\left(\frac{D}{q}\right) = \frac{(q - M)^2 s}{2q}$$

As usual, ordering cost per year = $\frac{KD}{q}$. Let $TC(q, M)$ be the total annual cost (excluding purchasing cost) if our order policy uses parameters q and M. From our discussion, we must choose q and M to minimize

$$TC(q, M) = \frac{M^2 h}{2q} + \frac{(q - M)^2 s}{2q} + \frac{KD}{q}$$

By using Theorem 3 from p. 677 of Chapter 12 in Volume 1, we can show that $TC(q, M)$ is a convex function of q and M. From Theorem 1′ and Theorem 7 (p. 698) of Chapter 12, the minimum value of $TC(q, M)$ will occur at the point where

$$\frac{\partial TC}{\partial q} = \frac{\partial TC}{\partial M} = 0$$

Some tedious algebra shows that $TC(q, M)$ is minimized for q^* and M^*:

$$q^* = \left[\frac{2KD(h + s)}{hs}\right]^{1/2} = \text{EOQ}\left(\frac{h + s}{s}\right)^{1/2}$$

$$M^* = \left[\frac{2KDs}{h(h + s)}\right]^{1/2} = \text{EOQ}\left(\frac{s}{h + s}\right)^{1/2}$$

Maximum shortage = $q^* - M^*$

As s approaches infinity, q^* and M^* both approach the EOQ, and the maximum shortage approaches zero. This is reasonable, because if s is large, the cost of a shortage is prohibitive, and we would expect the optimal ordering policy to incur very few, if any, shortages. In other words, if s is very large, we are facing (to all intents and purposes) the no-shortages-allowed situation of Section 3.2.

EXAMPLE 6   **Smalltown Optometry Clinic**

Each year, the Smalltown Optometry Clinic sells 10,000 frames for eyeglasses. The clinic orders frames from a regional supplier, which charges $15 per frame. Each order incurs an ordering cost of $50. Smalltown Optometry believes that the demand for frames can be backlogged and that the cost of being short one frame for one year is $15 (because of loss of future business). The annual holding cost for inventory is 30¢ per dollar value of inventory. What is the optimal order quantity? What is the maximum shortage that will occur? What is the maximum inventory level that will occur?

Solution We are given that

$$K = \$50$$
$$D = 10{,}000 \text{ frames per year}$$
$$h = 0.3(15) = \$4.50/\text{frame/year}$$
$$s = \$15/\text{frame/year}$$

Our formula for q^* and M^* now yield

$$q^* = \left(\frac{2(50)(10{,}000)(19.50)}{(4.50)(15)} \right)^{1/2} = 537.48$$

$$M^* = \left(\frac{2(50)(10{,}000)(15)}{(4.50)(19.50)} \right)^{1/2} = 413.45$$

Then the maximum shortage occurring will be $q^* - M^* = 124.03$ frames, and each order should be for 537 or 538 frames. A maximum inventory level of $M^* = 413.45$ frames will occur.

As in Section 3.4, suppose that production is not instantaneous and we can produce at a rate of r units per year. If shortages are allowed, it can be shown that

$$q^* = \left(\frac{2KDr(h + s)}{h(r - D)s} \right)^{1/2}$$

$$M^* = \frac{q^*(r - D)}{r} - \left(\frac{2KD(r - D)h}{sr(h + s)} \right)^{1/2}$$

The maximum shortage occurring in this case (call it S^*) is given by

$$S^* = \left(\frac{2KD(r - D)h}{sr(h + s)} \right)^{1/2}$$

Spreadsheet Template for the EOQ Model with Back Orders

BackEOQ.xls

Figure 13 (file BackEOQ.xls) illustrates a spreadsheet template for the EOQ model with back orders. In cells A6, B6, C6, and D6, we enter the values of K, D, h, and s, respectively, for Example 6. In A8 (given the range name Q), we compute the optimal order quantity with the formula $(2*K*D*(H+S)/(H*S))^{.5}$. In B8 (range name M), we com-

A	A	B	C	D
1	EOQ			
2	MODEL			
3	WITH			
4	BACKORDERS			
5	K	D	h	s
6	50	10000	4.5	15
7	q*	M*	MAX SHORT	ANNUAL COST
8	537.48384989	413.44912	124.03473459	1860.52101884

FIGURE 13
EOQ Model with
Back Orders

pute the optimal value of M with the formula $(2*K*D*S/(H*(H+S)))^\wedge.5$. In C8, we compute the maximum shortage with the formula $Q-M$. In D8, we compute the annual total cost $TC(q, M)$ (exclusive of purchasing costs) with the formula $(M^\wedge2*H)/(2*Q))+((Q-M)^\wedge2*S/2*Q))+(K*D/Q)$.

PROBLEMS

Group A

1 Show that the optimal order quantity for the backlogged demand model is always at least as large as the EOQ but that the maximum inventory level for the backlogged demand model cannot exceed the EOQ.

2 A Mercedes dealer must pay $20,000 for each car purchased. The annual holding cost is estimated to be 25% of the dollar value of inventory. The dealer sells an average of 500 cars per year. He believes that demand is backlogged but estimates that if he is short one car for one year he will lose $20,000 worth of future profits. Each time the dealer places an order for cars, ordering cost amounts to $10,000. Determine the Mercedes dealer's optimal ordering policy. What is the maximum shortage that will occur?

Group B

3 Suppose that instead of measuring shortage in terms of cost per shortage year, a cost of S dollars is incurred for each unit the firm is short. This cost does not depend on the length of time before the backlogged demand is satisfied. Determine a new expression for $TC(q, M)$, and explain how to determine optimal values q^* and M^*.

4 For the model developed in this section, determine

a the average length of time it takes to meet demand for a unit.

b the fraction of all demanded units that are backordered.

3.6 When to Use EOQ Models

Demand is often irregular, or "lumpy." This may be caused by seasonality or other factors. If demand is irregular, the Constant Demand Assumption that was required for all the EOQ models will not be satisfied.

To determine whether the assumption of constant demand is reasonable, suppose that during n periods of time, demands $d_1, d_2, \ldots, d_n$ have been observed. Also, enough is known about future demands to make the assumption of deterministic demand a realistic one. To decide whether demand is sufficiently regular to justify use of EOQ models, Peterson and Silver (1998) recommend that the following computations be done:

1 Determine the estimate $\bar{d}$ of the average demand per period given by

$$\bar{d} = \frac{1}{n} \sum_{i=1}^{i=n} d_i$$

2 Determine an estimate of the variance of the per-period demand **D** from

$$\text{Est. var } \mathbf{D} = \frac{1}{n} \sum_{i=1}^{i=n} d_i^2 - \bar{d}^2$$

3 Determine an estimate of the relative variability of demand (called the **variability coefficient**). This quantity is labeled VC, where

$$VC = \frac{\text{est. var } \mathbf{D}}{\bar{d}^2}$$

Note that if all the d_i are equal, the estimate of the variance of **D** will equal zero. This will also make $VC = 0$. Hence, if VC is small, this indicates that the Constant Demand Assumption is reasonable. Research indicates that the EOQ should be used if $VC < 0.20$;

otherwise, demand is too irregular to justify the use of an EOQ model. (See Peterson and Silver (1998).)

If $VC > 0.20$, dynamic programming methods and the Silver–Meal heuristic, which are discussed in Chapter 6, may be used to determine optimal ordering policies.

As an example of the use of the VC formula, suppose that demands during the four quarters of the past year have been as follows: 80 units, 100 units, 130 units, and 90 units. Assuming that future demand is known to follow a similar pattern, should an EOQ model be used in this situation?

Since $\bar{d} = \frac{400}{4} = 100$ and est. var $\mathbf{D} = (\frac{1}{4})(80^2 + 100^2 + 130^2 + 90^2) - 100^2 = 350$, we have $VC = \frac{350}{(100)^2} = 0.035$. Since VC is smaller than 0.20, an EOQ model can be used in this situation.

In closing, we note that the EOQ models of this chapter require the implicit assumption that demands during different periods of time are independent. In other words, the EOQ models require that any knowledge about demand during one period of time gives no information about demand at any other point in time. If a firm's inventory needs are met through internal production, this is often an unrealistic assumption. For example, suppose a company needs to produce 5 units of product A by December 11 and that each unit of product A requires 2 units of product B and 3 units of product C. Once product B and product C are available, it takes ten days to assemble a unit of product A. Then the fact that there is a demand for 5 units of product A on December 11 *creates* a December 1 demand for 10 units of product B and 15 units of product C. Hence, the December 1 demand for products B and C *depends* on the December 11 product A demand. Our EOQ models do not take into account the dependence of demand that is present in many manufacturing situations. These can best be exploited by using material resource planning (MRP) systems.

REMARK   Use the Excel commands =**AVERAGE,** to estimate the average demand for a given period, and =**VARP,** to estimate the variance in the demand for a given period.

PROBLEM

Group A

1 Observed demand for air conditioners during the last four quarters was as follows: fall, 100; winter, 50; spring, 150; summer, 300. Is it reasonable to use an EOQ model in this situation?

3.7 Multiple-Product EOQ Models

Suppose a company orders several products. Each time an order is received, shipments of some (but perhaps not all) of the products arrive. Each time an order arrives, there is a fixed cost associated with the order (for example, the cost of driving a truck to deliver the order), and there is another fixed cost associated with each product included in the order. How can we minimize the sum of annual holding and fixed costs? An example of this situation would be an appliance store that orders three different types of appliances from a supplier. For a low-demand product, it would be unreasonable to order the product each time a truck arrives. Chopra and Meindl (2001) devised a method to find a nearly optimal solution to this type of problem. To begin, we find the product that is most frequently ordered. Suppose that is product 1; we assume this product will be included in each order. We then set up a Solver model that determines the following changing cells:

- Number of orders received per year. Note that each order is assumed to contain a shipment of product 1.
- For all products other than product 1, the number of orders that need to be received before an order of the product is received. If, for example, product 2 should be contained in every third order, then the changing cell for product 2 will equal 3.

Given trial values of these quantities, we can easily determine the total fixed cost (sum of fixed cost for each product plus fixed cost for each order) and total holding cost for each product. The sum of these costs will be our target cell for Solver. Our model is highly nonlinear. It is necessary to use the Evolutionary Solver (see Chapter 15 of Volume 1) to find the optimal solution. Here is an example of the method.

EXAMPLE 7   Ohm City Appliances

Ohm City Appliances has three types of TVs delivered from Springfield TV. Figure 14 gives the annual demand, unit purchasing cost, annual holding cost (as a percentage of purchase cost), the fixed cost of ordering a product, and the fixed cost of placing an order. Determine an ordering policy that minimizes the sum of fixed and holding costs.

FIGURE 14

	A	B	C	D	E	F
4						
5						
6			Product 1	Product 2	Product 3	
7		annual demand	12000	1200	120	
8		unit cost	$ 500.00	$ 500.00	$ 500.00	
9		holding cost	0.2	0.2	0.2	
10		product order cost	$ 1,000.00	$ 1,000.00	$ 1,000.00	
11		eoq	489.8979486	154.9193338	48.989795	
12		orders per year	24.49489743	7.745966692	2.4494897	
13		overall order cost	$ 4,000.00			
14						
15		Orders per year P1	10.46135741			
16		Orders of P1 per P2	1			
17		Orders of P1 per P3	4			
18		Orders per Year P2	10.46135741			
19		Orders Per Year P3	2.615339354			
20						
21						
22						
23		Main annual order cost	$ 41,845.43			
24						
25			Prod 1	Prod 2	Prod 3	
26		Order quantity	1147.078675	114.7078675	45.883147	
27		Avg. Inventory	573.5393374	57.35393374	22.941573	
28		Annual Holding cost	$ 57,353.93	$ 5,735.39	$ 2,294.16	
29		Annual product ordering cost	$ 10,461.36	$ 10,461.36	$ 2,615.34	
30						
31						
32		Total Annual cost	$ 130,766.97			
33						

Solution   Our work is in file MultipleEOQ.xls. Also see Figure 14.

MultipleEOQ.xls

Step 1   In C11:E11, we compute the EOQ for each product by copying from C11 to D11:E11 the formula

$$=\text{SQRT}(2*C10*C7/(C9*C8))$$

Then, in C12:E12, we compute the number of times each product is ordered during a year by copying from C12 to D12:E12 the formula

$$=C7/C11$$

We find that product 1 is the most frequently ordered.

Step 2   In cell C15, we enter a trial value (not necessarily an integer) for the number of orders placed each year. In C16, we enter a trial value (which must be an integer) for the number of orders with product 1 that must be received before an order of product 2 is received. In C17, we enter a trial value (which must be an integer) for the number of orders with product 1 that must be received before an order of product 3 is obtained.

Step 3   In cell C23, we compute the total fixed cost associated with the orders as (number of orders per year)*(cost per order) with the formula

$$=C15*C13$$

Step 4   In cells C18 and C19, we compute the number of times product I (I = 2 or 3) are ordered each year by computing (orders per year)*(fraction of orders containing product I).

$$=\$C\$15/C16 \text{ (cell C18: orders of product 2 per year)}$$
$$=\$C\$15/C17 \text{ (cell C19: orders of product 3 per year)}$$

Step 5   In cell C26, we compute the size of each product 1 order as (annual product 1 demand)/(orders of product 1 received each year).

$$=C7/C15$$

In a similar fashion, we compute size of product 2 and product 3 orders in cells D26 and E26.

$$=D7/C18 \text{ (cell D26: product 2 order size)}$$
$$=E7/C19 \text{ (cell E26: product 3 order size)}$$

Step 6   In C27:E27, we compute the average inventory level for each product as half the order size. To do this, copy from C27 to D27:E27 the formula

$$=0.5*C26$$

Step 7   In C28:E28, we compute the annual holding cost for each product as (average inventory level for product)*(annual cost of holding one unit of product in inventory). To do this, copy from C28 to D28:E28 the formula

$$=C9*C8*C27$$

Step 8   In C29:E29, we compute the annual ordering cost for each product as (cost per order for product)*(times product is ordered per year). For example, in cell C29, we compute annual ordering cost for product 1 with the formula

$$=C10*C15.$$

Step 9   In cell C32, we compute total annual cost (exclusive of purchasing costs, which do not depend on ordering policy) with the formula

$$=\text{SUM}(C28:E29)+C23$$

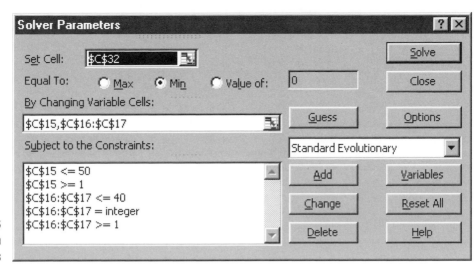

FIGURE 15
Solver Window for Ohm
City Appliances

Step 10 We now use Solver to find the cost-minimizing ordering policy. Figure 15 shows our Solver window.

We minimize total cost (C32) by changing the number of orders per year (C15) and the number of orders that must be placed before orders are placed for less frequently ordered products (C16 and C17). We require an integer for the number of orders before each less frequently ordered product is placed. Included are the lower and upper bounds that Evolutionary Solver requires for each changing cell.

We find that 10.46 truckloads of TVs should be received each year. Each truckload will contain 1,147 type 1 TVs and 114 type 2 TVs. 25% of all orders will include an order of 46 type 3 TVs. Note that the low-demand type 3 TVs are infrequently ordered.

PROBLEMS

Group A

1 Square City Appliance orders four types of washing machines. Table 7 gives the annual demand, purchasing cost, and annual holding cost (as a percentage of purchase cost), and the fixed cost of ordering a product. Determine an ordering policy that minimizes the sum of fixed and holding costs. Each time an order is delivered, a $10,000 cost is incurred. Determine an ordering policy to minimize annual cost of meeting demand.

2 In Problem 1, suppose that Square City manufactures the washing machines. The company can manufacture washing machines at a rate of 30,000 per year. What manufacturing policy will minimize the annual cost of meeting demand?

TABLE 7

	Product 1	Product 2	Product 3	Product 4
Annual demand	10,000	3,000	4,000	500
Unit purchasing cost	$400	$300	$200	$900
Holding cost percentage	.2	.2	.2	.2
Product order cost	$1,000	$1,000	$1,000	$1,000

S U M M A R Y   Notation

$$K = \text{setup or ordering cost}$$
$$h = \text{cost of holding one unit in inventory for one unit of time}$$
$$D = \text{demand rate per unit time}$$
$$r = \text{rate at which firm can make product per unit time } (r > D)$$
$$s = \text{cost of being one unit short for one unit of time}$$

Basic EOQ Model

$$\text{Order quantity} = q^* = \left(\frac{2KD}{h}\right)^{1/2}$$

$\frac{D}{q^*}$ orders are placed each unit of time.

Quantity Discount Model

If $q < b_1$, each item costs p_1 dollars.

If $b_1 \leq q < b_2$, each item costs p_2 dollars.

If $b_{k-2} \leq q < b_{k-1}$, each item costs p_{k-1} dollars.

If $b_{k-1} \leq q < b_k = \infty$, each item costs p_k dollars.

Beginning with the lowest price, determine for each price the order quantity (q_i^*) that minimizes total annual costs for $b_{i-1} \leq q < b_i$. Continue determining $q_k^*, q_{k-1}^*, \ldots$ until one of the q_i^*'s (call it q_i^*) is admissible; from observation 2, this will mean that $q_i^* = EOQ_{i'}$. The optimal order quantity will be the member of $\{q_k^*, q_{k-1}^*, \ldots, q_i^*\}$ with the smallest value of $TC(q)$.

If EOQ_i is admissible, then $q_i^* = EOQ_i$. If $EOQ_i < b_{i-1}$, then $q_i^* = b_{i-1}$.

Continuous Rate Model

$$\text{Optimal run size} = \left[\frac{2KDr}{h(r - D)}\right]^{1/2}$$

EOQ with Back Orders Allowed

$$q^* = \text{optimal order quantity}$$
$$M^* = \text{maximum inventory level under optimal ordering policy}$$
$$q^* - M^* = \text{maximum shortage occurring under optimal ordering policy}$$
$$q^* = \left[\frac{2KD(h + s)}{hs}\right]^{1/2} = EOQ\left(\frac{h + s}{s}\right)^{1/2}$$
$$M^* = \left[\frac{2KDs}{h(h + s)}\right]^{1/2} = EOQ\left(\frac{s}{h + s}\right)^{1/2}$$

REVIEW PROBLEMS

Group A

1 Customers at Joe's Office Supply Store demand an average of 6,000 desks per year. Each time an order is placed, an ordering cost of $300 is incurred. The annual holding cost for a single desk is 25% of the $200 cost of a desk. One week elapses between the placement of an order and the arrival of the order. In parts (a)–(d), assume that no shortages are allowed.

    a Each time an order is placed, how many desks should be ordered?

    b How many orders should be placed each year?

    c Determine the total annual costs (excluding purchasing costs) of meeting the customers' demands for desks.

    d Determine the reorder point. If the lead time were five weeks, what would be the reorder point? (52 weeks = one year.)

    e How would the answers to parts (a) and (b) change if shortages were allowed and a cost of $80 is incurred if Joe's is short one desk for one year?

2 Suppose Joe's is considering manufacturing desks. It costs $250 to set up a production run, and Joe's has the capacity to manufacture up to 10,000 desks per year. What is the optimal production run size? How many production runs will be made each year?

3 A camera store sells an average of 100 cameras per month. The cost of holding a camera in inventory for a year is 30% of the price the camera shop pays for the camera. It costs $120 each time the camera store places an order with its supplier. The price charged per camera depends on the number of cameras ordered (see Table 8). Each time the camera store places an order, how many cameras should be ordered?

Group B

4 A company inventories two items. The relevant data for each item are shown in Table 9. Determine the optimal

inventory policy if no shortages are allowed and if the average investment in inventory is not allowed to exceed $700. If this constraint could be relaxed by $1, by how much would the company's annual costs decrease? (This problem requires knowledge of Section 12.8 in Volume 1.)

5 A company produces three types of items. A single machine is used to produce the three items on a cyclical basis. The company has the policy that every item is produced once during each cycle, and it wants to determine the number of production cycles per year that will minimize the sum of holding and setup costs (no shortages are allowed). The following data are given:

P_i = number of units of product i that could be produced per year if the machine were entirely devoted to producing product i

D_i = annual demand for product i

K_i = cost of setting up production for product i

h_i = cost of holding one unit of product i in inventory for one year

    a Suppose there are N cycles per year. Assuming that during each cycle, a fraction $\frac{1}{N}$ of all demand for each product is met, determine the annual holding cost and the annual setup cost.

    b Let q_i^* be the number of units of product i produced during each cycle. Determine the optimal value of N (call it N^*) and q_i^*.

    c Let $EROQ_i$ be the optimal production run size for product i if the cyclical nature of the problem is ignored. Suppose q_i^* is much smaller than $EROQ_i$. What conclusion could be drawn?

    d Under certain circumstances, it might not be desirable to produce every item during each cycle. Which of the following factors would tend to make it undesirable to produce product i during each cycle: (1) Demand is relatively low. (2) The setup cost is relatively high. (3) The holding cost is relatively high.

TABLE 8

No. of Cameras Ordered	Price per Camera
1–10	$10.00
11–40	$9.00
41–100	$7.00
More than 100	$5.50

TABLE 9

	Item 1	Item 2
Annual demand	6,000	4,000
Per-unit cost	$4.00	$3.50
Annual holding cost	30% per year	25% per year
Price per order	$35	$20

REFERENCES

The following books emphasize applications over theory:

Brown, R. *Decision Rules for Inventory Management.* New York: Holt, Rinehart and Winston, 1967.

Chopra, S., and P. Meindl. *Supply Chain Management.* Englewood Cliffs, N.J.: Prentice Hall, 2001.

McLeavey, D., and S. Narasimhan. *Production Planning and Inventory Control.* Boston: Allyn and Bacon, 1985.

Peterson, R., and E. Silver. *Decision Systems for Inventory Management and Production Planning.* New York: Wiley, 1988.

Tersine, R. *Principles of Inventory and Materials Management.* New York: North-Holland, 1982.

Vollman, T., W. Berry, and C. Whybark. *Manufacturing Planning and Control Systems.* Homewood, Ill.: Irwin, 1998.

Zipkin, P. *Foundations of Inventory Management.* New York: Irwin-McGraw-Hill, 2000.

The following three books contain extensive discussions of inventory theory as well as applications:

Hadley, G., and T. Whitin. *Analysis of Inventory Systems.* Englewood Cliffs, N.J.: Prentice Hall, 1963.

Hax, A., and D. Candea. *Production and Inventory Management.* Englewood Cliffs, N.J.: Prentice Hall, 1984.

Johnson, L., and D. Montgomery. *Operations Research in Production, Planning, Scheduling, and Inventory Control.* New York: Wiley, 1974.

Baumol, W. "The Transactions Demand for Cash: An Inventory Theoretic Approach," *Quarterly Journal of Economics* 16(1952):545–556.

Ignall, E., and P. Kolesar. "Operating Characteristics of a Simple Shuttle under Local Dispatching Rules," *Operations Research* 20(1972):1077–1088.

Riccio, L., J. Miller, and A. Little."Polishing the Big Apple," *Interfaces* 16(no. 1, 1986):83–88.

Roundy, R. "98% Effective Integer Rate Lot-Sizing for One-Warehouse Multi-Retailer Systems," *Management Science* 31(1985):1416–1430.

4

Probabilistic Inventory Models

All the inventory models discussed in Chapter 3 require that demand during any period of time be known with certainty. In this chapter, we consider inventory models in which demand over a given time period is uncertain, or random; single-period inventory models, where a problem is ended once a single ordering decision has been made; single-period bidding models; versions of the EOQ model for uncertain demand that incorporate the important concepts of safety stock and service level; the periodic review (R, S) model; the ABC inventory classification system; and exchange curves.

4.1 Single-Period Decision Models

In many situations, a decision maker is faced with the problem of determining the value q for a variable (q may be the quantity ordered of an inventoried good, for example, or the bid on a contract). After q has been determined, the value d assumed by a random variable $\mathbf{D}$ is observed. Depending on the values of d and q, the decision maker incurs a cost $c(d, q)$. We assume that the person is risk-neutral and wants to choose q to minimize his or her expected cost. Since the decision is made only once, we call a model of this type a *single-period decision model*.

4.2 The Concept of Marginal Analysis

For the single-period model described in Section 4.1, we now assume that $\mathbf{D}$ is an integer-valued discrete random variable with $P(\mathbf{D} = d) = p(d)$. Let $E(q)$ be the decision maker's expected cost if q is chosen. Then

$$E(q) = \sum_{d} p(d)c(d, q)$$

In most practical applications, $E(q)$ is a convex function of q. Let q^* be the value of q that minimizes $E(q)$. If $E(q)$ is a convex function, the graph of $E(q)$ must look something like Figure 1. From the figure, we see that q^* is the smallest value of q for which

$$E(q^* + 1) - E(q^*) \geq 0 \tag{1}$$

Thus, if $E(q)$ is a convex function of q, we can find the value of q minimizing expected cost by finding the smallest value of q that satisfies Inequality (1). Note that $E(q + 1) - E(q)$ is the change in expected cost that occurs if we increase the decision variable q to $q + 1$.

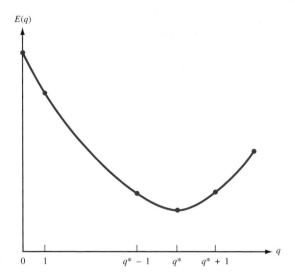

FIGURE 1
Determination of q^* by Marginal Analysis

To determine q^*, we begin with $q = 0$. If $E(1) - E(0) \leq 0$, we can benefit by increasing q from 0 to 1. Now we check to see whether $E(2) - E(1) \leq 0$. If this is true, then increasing q from 1 to 2 will reduce expected cost. Continuing in this fashion, we see that increasing q by 1 will reduce expected costs up to the point where we try to increase q from q^* to $q^* + 1$. In this case, increasing q by 1 will increase expected cost. From Figure 1 (which is the appropriate picture if $E(q)$ is a convex function), we see that if $E(q^* + 1) - E(q^*) \geq 0$, then for $q \geq q^*$, $E(q + 1) - E(q) \geq 0$. Thus, q^* must be the value of q that minimizes $E(q)$. If $E(q)$ is not convex, this argument may not work. (See Problem 1 at the end of this section.)

Our approach determines q^* by repeatedly computing the effect of adding a marginal unit to the value of q. For this reason, it is often called **marginal analysis.** Marginal analysis is very useful if it is easy to determine a simple expression for $E(q + 1) - E(q)$. In the next section, we use marginal analysis to solve the classical news vendor problem.

PROBLEM

Group A

1 Suppose $E(q)$ is $E(0) = 8$, $E(1) = 6$, $E(2) = 5$, $E(3) = 7$, $E(4) = 6$, $E(5) = 5.5$, $E(6) = 4.5$, and $E(7) = 5$.

    a What value of q minimizes $E(q)$?

b If marginal analysis is used to determine the value of q that minimizes $E(q)$, what is the answer?

c Explain why marginal analysis fails to find the value of q that minimizes $E(q)$.

4.3 The News Vendor Problem: Discrete Demand

Organizations often face inventory problems where the following sequence of events occurs:

1 The organization decides how many units to order. We let q be the number of units ordered.

2 With probability $p(d)$, a demand of d units occurs. In this section, we assume that d must be a nonnegative integer. We let **D** be the random variable representing demand.

3 Depending on d and q, a cost $c(d, q)$ is incurred.

Problems that follow this sequence are often called **news vendor problems.** To see why, consider a vendor who must decide how many newspapers should be ordered each day from the newspaper plant. If the vendor orders too many papers, he or she will be left with many worthless newspapers at the end of the day. On the other hand, a vendor who orders too few newspapers will lose profit that could have been earned if enough newspapers to meet customer demand had been ordered, and customers will be disappointed. The news vendor must order the number of papers that properly balances these two costs. We have already encountered a news vendor problem in the discussion of decision theory in Section 2.1.

In this section, we show how marginal analysis can be used to solve news vendor problems when demand is a discrete random variable and $c(d, q)$ has the following form:

$$c(d, q) = c_o q + \text{(terms not involving } q) \qquad (d \leq q) \qquad (2)$$
$$c(d, q) = -c_u q + \text{(terms not involving } q) \qquad (d \geq q + 1) \qquad (2.1)$$

In (2), c_o is the per-unit cost of being overstocked. If $d \leq q$, we have ordered more than was demanded—that is, overstocked. If the size of the order is increased from q to $q + 1$, then (2) shows that the cost increases by c_o. Hence, c_o is the cost due to being overstocked by one extra unit. We refer to c_o as the **overstocking cost.** Similarly, if $d \geq q + 1$, we have understocked (ordered an amount less than demand). If $d \geq q + 1$ and we increase the size of the order by one unit, we are understocked by one less unit. Then (2.1) implies that the cost is reduced by c_u, so c_u is the per-unit cost of being understocked. We call c_u the **understocking cost.**

To derive the optimal order quantity via marginal analysis, let $E(q)$ be the expected cost if an order is placed for q units. We assume that the decision maker's goal is to find the value q^* that minimizes $E(q)$. If $c(d, q)$ can be described by (2) and (2.1), and $E(q)$ is a convex function of q, then marginal analysis can be used to determine q^*.

Following (1), we must determine the smallest value of q for which $E(q + 1) - E(q) \geq 0$. To calculate $E(q + 1) - E(q)$, we must consider two possibilities:

Case 1 $d \leq q$. In this case, ordering $q + 1$ units instead of q units causes us to be overstocked by one more unit. This increases cost by c_o. The probability that Case 1 will occur is simply $P(\mathbf{D} \leq q)$, where $\mathbf{D}$ is the random variable representing demand.

Case 2 $d \geq q + 1$. In this case, ordering $q + 1$ units instead of q units enables us to be short one less unit. This will decrease cost by c_u. The probability that Case 2 will occur is $P(\mathbf{D} \geq q + 1) = 1 - P(\mathbf{D} \leq q)$.

In summary, a fraction $P(\mathbf{D} \leq q)$ of the time, ordering $q + 1$ units will cost c_o more than ordering q units; and a fraction $1 - P(\mathbf{D} \leq q)$ of the time, ordering $q + 1$ units will cost c_u less than ordering q units. Thus, on the average, ordering $q + 1$ units will cost

$$c_o \, P(\mathbf{D} \leq q) - c_u[1 - P(\mathbf{D} \leq q)]$$

more than ordering q units.

More formally, we have shown that

$$E(q + 1) - E(q) = c_o \, P(\mathbf{D} \leq q) - c_u[1 - P(\mathbf{D} \leq q)]$$
$$= (c_o + c_u) \, P(\mathbf{D} \leq q) - c_u{}^{\dagger}$$

Then $E(q + 1) - E(q) \geq 0$ will hold if

$$(c_o + c_u) \, P(\mathbf{D} \leq q) - c_u \geq 0 \qquad \text{or} \qquad P(\mathbf{D} \leq q) \geq \frac{c_u}{c_o + c_u}$$

[†]Since $P(\mathbf{D} \leq q)$ increases as q increases, $E(q + 1) - E(q)$ will increase as q increases. Hence, if $c_o + c_u \geq 0$, $E(q)$ is a convex function of q, and our use of marginal analysis is justified.

Let $F(q) = P(\mathbf{D} \leq q)$ be the demand distribution function. Since marginal analysis is applicable, we have just shown that $E(q)$ will be minimized by the smallest value of q (call it q^*) satisfying

$$F(q^*) \geq \frac{c_u}{c_o + c_u} \tag{3}$$

The following example illustrates the use of (3).

EXAMPLE 1   Walton Bookstore Calendar Sales

In August, Walton Bookstore must decide how many of next year's nature calendars should be ordered. Each calendar costs the bookstore $2 and is sold for $4.50. After January 1, any unsold calendars are returned to the publisher for a refund of 75¢ per calendar. Walton believes that the number of calendars sold by January 1 follows the probability distribution shown in Table 1. Walton wants to maximize the expected net profit from calendar sales. How many calendars should the bookstore order in August?[†]

Solution   Let

$$q = \text{number of calendars ordered in August}$$
$$d = \text{number of calendars demanded by January 1}$$

If $d \leq q$, the costs shown in Table 2 are incurred (revenue is negative cost). From (2), $c_o = 1.25$.

If $d \geq q + 1$, the costs shown in Table 3 are incurred. From (2), $-c_u = -2.5$, or $c_u = 2.50$. Then

$$\frac{c_u}{c_o + c_u} = \frac{2.50}{3.75} = \frac{2}{3}$$

TABLE 1
Probability Mass Function for
Calendar Sales

No. of Calendars Sold	Probability
100	.30
150	.20
200	.30
250	.15
300	.05

TABLE 2
Computation of Total Cost If $d \leq q$

	Cost
Buy q calendars at $2/calendar	$2q$
Sell d calendars at $4.50/calendar	$-4.50d$
Return $q - d$ calendars at 75¢/calendar	$-0.75(q - d)$
Total cost	$1.25q - 3.75d$

[†]Based on Barron (1985).

TABLE 3
Computation of Total Cost If $d \geq q + 1$

	Cost
Buy q calendars at \$2/calendar	$2q$
Sell d calendars at \$4.50/calendar	$-4.50q$
Total cost	$-2.50q$

From (3), Walton should order $q*$ calendars, where $q*$ is the smallest number for which $P(\mathbf{D} \leq q*) \geq \frac{2}{3}$. As a function of q, $P(\mathbf{D} \leq q)$ increases only when $q = 100, 150, 200, 250$, or 300. Also note that $P(\mathbf{D} \leq 100) = .30$, $P(\mathbf{D} \leq 150) = .50$, and $P(\mathbf{D} \leq 200) = .80$. Since $P(\mathbf{D} \leq 200)$ is greater than or equal to $\frac{2}{3}$, $q* = 200$ calendars should be ordered.

REMARKS

1 In terms of marginal analysis, the probability of selling the 200th calendar that is ordered is $P(\mathbf{D} \geq 200) = .50$. This implies that the 200th calendar has a $1 - .50 = .50$ chance of being unsold. Thus, the 200th calendar will increase Walton's expected costs by $.50(-2.50) + .50(1.25) = -\$0.625$. Hence, the 200th calendar should be ordered. On the other hand, the probability that the 201st calendar will be sold is $P(\mathbf{D} \geq 201) = .20$, and the probability that the 201st calendar will not be sold is $1 - .20 = .80$. Therefore, the 201st calendar will increase expected costs by $.20(-2.50) + .80(1.25) = \$0.50$. Thus, the 201st calendar will increase expected costs and should not be ordered.

2 In Example 1, c_o and c_u could easily have been determined without recourse to (2) and (2.1). For example, being one more unit over actual demand increases Walton's costs by $2 - 0.75 = \$1.25$. Thus, $c_o = \$1.25$. Similarly, being one more unit under actual demand will cost Walton $4.50 - 2.00 = \$2.50$ in profit. Hence, $c_u = \$2.50$. If we are able to determine c_o and c_u without using Equations (2) and (2.1), we should do so. In more difficult problems, however, they can be very useful (see Examples 2 and 3).

PROBLEMS

Group A

1 In August 2003, a car dealer is trying to determine how many 2004 models should be ordered. Each car costs the dealer \$10,000. The demand for the dealer's 2004 models has the probability distribution shown in Table 4. Each car is sold for \$15,000. If the demand for 2004 cars exceeds the number of cars ordered in August, the dealer must reorder at a cost of \$12,000 per car. If the demand for 2004 cars falls short, the dealer may dispose of excess cars in an end-of-model-year sale for \$9,000 per car. How many 2004 models should be ordered in August?

2 Each day, a news vendor must determine how many *New York Herald Wonderfuls* to order. She pays 15¢ for each paper and sells each for 30¢. Any leftover papers are a total loss. From past experience, she believes that the number of papers she can sell each day is governed by the probability distribution shown in Table 5. How many papers should she order each day?

3 If c_u is fixed, will an increase in c_o increase or decrease the optimal order quantity?

TABLE 4

No. of Cars Demanded	Probability
20	.30
25	.15
30	.15
35	.20
40	.20

TABLE 5

No. of Papers Demanded	Probability
50	.30
70	.15
90	.25
110	.10
130	.20

TABLE 6

No. of Cells	Probability
50	.20
60	.15
70	.30
80	.10
90	.15
100	.10

TABLE 7

Number Needed	Probability
200	.03
275	.03
350	.03
400	.05
450	.40
500	.30
550	.06
600	.07
650	.03

TABLE 8

Copies Demanded	Probability
5,000	.30
6,000	.20
7,000	.40
8,000	.10

TABLE 9

Week of Birth	Probability
36	.05
37	.15
39	.20
40	.30
41	.15
42	.10
43	.05

4 If c_o is fixed, will an increase in c_u increase or decrease the optimal order quantity?

5 The power at Ice Station Lion is supplied via solar cells. Once a year, a plane flies in and sells solar cells to the ice station at a price of $20 per cell. Because of uncertainty about future power needs, the ice station can only guess the number of cells that will be required during the coming year (see probability distribution in Table 6). If the ice station runs out of solar cells, a special order must be placed at a cost of $30 per cell.

    a Assuming that the news vendor problem is relevant, how many cells should be ordered from the plane?

    b In part (a), what type of cost is being ignored?

6 The daily demand for substitute teachers in the Los Angeles teaching system follows the distribution given in Table 7. Los Angeles wants to know how many teachers to keep in the substitute teacher pool. Whether or not the substitute teacher is needed, it costs $30 per day to keep a substitute teacher in the pool. If not enough substitute teachers are available on a given day, regular teachers are used to cover classes at a cost of $54 per regular teacher. How many teachers should Los Angeles have in the substitute teacher pool?[†]

Group B

7 Every four years, Blockbuster Publishers revises its textbooks. It has been three years since the best-selling book, *The Joy of OR,* has been revised. At present, 2,000

copies of the book are in stock, and Blockbuster must determine how many copies of the book should be printed for the next year. The sales department believes that sales during the next year are governed by the distribution in Table 8. Each copy of *Joy* sold during the next year brings the publisher $35 in revenues. Any copies left at the end of the next year cannot be sold at full price but can be sold for $5 to Bonds Ennoble and Gitano's bookstores. The cost of a printing of the book is $50,000 plus $15 per book printed. How many copies of *Joy* should be printed? Would the answer change if 4,000 copies were currently in stock?

8 Vivian and Wayne are planning on going to Lamaze natural childbirth classes. Lamaze classes meet once a week for five weeks. Each class gives 20% of the knowledge needed for "natural" childbirth. If Vivian and Wayne finish their classes before the birth of their child, they will forget during each week 5% of what they have learned in class. To maximize their expected knowledge at the time of childbirth, during which week of pregnancy should they begin classes? Assume that the number of weeks from conception to childbirth follows the probability distribution given in Table 9.

9[‡] Some universities allow an employee to put an amount q into an account at the beginning of each year, to be used for child-care expenses. The amount q is not subject to federal income tax. Assume that all other income is taxed by the federal government at a 40% rate. If child-care expenses for the year (call them d) are less than q, the employee in effect loses $q - d$ dollars in before-tax income. If child-care expenses exceed q, the employee must pay the excess out of his or her own pocket but may credit 25% of that as a savings on his or her state income tax.

    Suppose Professor Muffy Rabbit believes that there is an equal chance that her child-care expenses for the coming year will be $3,000, $4,000, $5,000, $6,000, or $7,000. At the beginning of the year, how much money should she place in the child-care account?

[†]Based on Bruno (1970).

[‡]Based on Rosenfeld (1986).

4.4 The News Vendor Problem: Continuous Demand

We now consider the news vendor scenario of Section 4.3 when demand $\mathbf{D}$ is a continuous random variable having density function $f(d)$. By modifying our marginal analysis argument of Section 4.3 (or by using Leibniz's rule for differentiating an integral—see Problem 7 at the end of this section), it can be shown that the decision maker's expected cost is minimized by ordering q^* units, where q^* is the smallest number satisfying

$$P(\mathbf{D} \leq q^*) \geq \frac{c_u}{c_o + c_u} \tag{4}$$

Since demand is a continuous random variable, we can find a number q^* for which (4) holds with equality. Hence, in this case, the optimal order quantity can be determined by finding the value of q^* satisfying

$$P(\mathbf{D} \leq q^*) = \frac{c_u}{c_o + c_u} \qquad \text{or} \qquad P(\mathbf{D} \geq q^*) = \frac{c_o}{c_o + c_u} \tag{5}$$

From (5), we see that it is optimal to order units up to the point where the last unit ordered has a chance

$$\frac{c_o}{c_o + c_u}$$

of being sold. Examples 2 and 3 illustrate the use of (5).

EXAMPLE 2   **ABA Room Reservations**

The American Bar Association (ABA) is holding its annual convention in Las Vegas. Six months before the convention begins, the ABA must decide how many rooms should be reserved in the convention hotel. At this time, the ABA can reserve rooms at a cost of $50 per room, but six months before the convention, the ABA does not know with certainty how many people will attend the convention. The ABA believes, however, that the number of rooms required is normally distributed, with a mean of 5,000 rooms and a standard deviation of 2,000 rooms. If the number of rooms required exceeds the number of rooms reserved at the convention hotel, extra rooms will have to be found at neighboring hotels at a cost of $80 per room. It is inconvenient for convention participants to stay at neighboring hotels. We measure this inconvenience by assessing an additional cost of $10 for each room obtained at a neighboring hotel. If the goal is to minimize the expected cost to the ABA and its members, how many rooms should the ABA reserve at the convention hotel?

Solution   Define

$$q = \text{number of rooms reserved}$$
$$d = \text{number of rooms actually required}$$

If $d \leq q$, then the only cost incurred is the cost of the rooms reserved in advance, so if $d \leq q$, the total cost is $50q$. Thus, $c_o = 50$. If $d \geq q + 1$, the following costs are incurred:

Cost of reserving q rooms $= 50q$

Cost of renting $d - q$ rooms in neighboring hotels $= 80(d - q)$

Inconvenience cost to overflow participants $= 10(d - q)$

Total cost $= 90d - 40q \qquad$ and $\qquad c_u = 40$

Since $\dfrac{c_u}{c_u + c_o} = \dfrac{40}{90} = \dfrac{4}{9}$, we see from (5) that the optimal number of rooms to reserve is the number q^* satisfying

$$P(\mathbf{D} \le q^*) = \tfrac{4}{9} \tag{6}$$

The Excel function NORMINV can be used to calculate q^*. Since

$$=\text{NORMINV}(4/9, 5000, 2000)$$

yields 4,720.58, the ABA should reserve 4,720 or 4,721 rooms.

EXAMPLE 3   Airline Overbooking

The ticket price for a New York–Indianapolis flight is \$200. Each plane can hold up to 100 passengers. Usually, some of the passengers who have purchased tickets for a flight fail to show up (no-shows). To protect against no-shows, the airline will try to sell more than 100 tickets for each flight. Federal law states that any ticketed customer who is unable to board the plane is entitled to compensation (say, \$100). Past data indicate that the number of no-shows for each New York–Indianapolis flight is normally distributed, with a mean of 20 and a standard deviation of 5. To maximize expected revenues less compensation costs, how many tickets should the airline sell for each flight? Assume that anybody who doesn't use a ticket receives a \$200 refund.

Solution   Let

$$q = \text{number of tickets sold by airline}$$
$$d = \text{number of no-shows}$$

Observe that $q - d$ will be the number of customers actually showing up for the flight. If $q - d \le 100$, then all customers who show up will board the flight, and the cost to the airline is $-200(q - d) = 200d - 200q$. If $q - d \ge 100$, then 100 passengers will board the plane (paying the airline $200(100) = \$20{,}000$), and $q - d - 100$ customers will be turned away. These $q - d - 100$ customers will receive compensation of $100(q - d - 100)$. Hence, if $q - d \ge 100$, the total cost to the airline is given by $100(q - d - 100) - 200(100) = 100(q - 100) - 100d - 20{,}000$. In summary, the net cost to the airline may be expressed as shown in Table 10.

If $q - 100$ is considered as a decision variable, we have a news vendor problem with $-c_u = -200$ (or $c_u = 200$) and $c_o = 100$. From (5), we should choose $q - 100$ to satisfy

$$P(\mathbf{D} \le q - 100) = \dfrac{c_u}{c_o + c_u} = \dfrac{2}{3} \tag{7}$$

The problem can be solved with the help of Excel. Since

$$=\text{NORMINV}(2/3, 120, 5)$$

TABLE 10
Computation of Total Cost

	Total Cost
$q - d \ge 100$ (or $d \le q - 100$)	$100\,(q - 100) - 100d - 20{,}000$
$q - d \le 100$ (or $d \ge q - 100$)	$200d - 200\,(q - 100) - 200\,(100)$

yields 122.15, we may conclude that the airline should attempt to sell 122 or 123 tickets. This means that once ticket sales have reached 122 (or 123), no more tickets should be sold for the flight. Of course, if fewer than 122 people want to purchase tickets for the flight, the airline should not refuse to sell anybody a ticket for the flight.

PROBLEMS

Group A

1 a In Example 3, why is it unrealistic to assume that the distribution of the number of no-shows is independent of q?

b If the number of no-shows were normally distributed with a mean of $.05q$ and a standard deviation of $.05q$, would we still have a news vendor problem?

2 Condo Construction Company is going to First National Bank for a loan. At the present time, the bank is willing to lend Condo up to $1 million, with interest costs of 10%. Condo believes that the amount of borrowed funds needed during the current year is normally distributed, with a mean of $700,000 and a standard deviation of $300,000. If Condo needs to borrow more money during the year, the company will have to go to Louie the Loan Shark. The cost per dollar borrowed from Louie is 25¢. To minimize expected interest costs for the year, how much money should Condo borrow from the bank?

3 Joe is selling Christmas trees to pay his college tuition. He purchases trees for $10 each and sells them for $25 each. The number of trees he can sell is normally distributed with a mean of 100 and standard deviation of 30. How many trees should Joe purchase?

4 A hot dog vendor at Wrigley Field sells hot dogs for $1.50 each. He buys them for $1.20 each. All the hot dogs he fails to sell at Wrigley Field during the afternoon can be sold that evening at Comiskey Park for $1 each. The daily demand for hot dogs at Wrigley Field is normally distributed with a mean of 40 and a standard deviation of 10.

a If the vendor buys hot dogs once a day, how many should he buy?

b If he buys 52 hot dogs, what is the probability that he will meet all of the day's demand for hot dogs at Wrigley?

Group B

5[†] Motorama TV estimates the annual demand for its TVs is (and will be in the future) normally distributed, with a mean of 6,000 and standard deviation of 2,000. Motorama

[†]Based on Virts and Garrett (1970).

must determine how much production capacity it should have. The cost of building enough production capacity to make 1,000 sets per year is $1,000,000 (equivalent in present value terms to a cost of $100,000 per year forever). Exclusive of the cost of building capacity, each set sold contributes $250 to profits. How much production capacity should Motorama have?

6 I. L. Pea is a well-known mail-order company. During the Christmas rush (from November 1 to December 15), the number of orders that I. L. Pea must fill each day (five days per week) is normally distributed, with a mean of 2,000 and a standard deviation of 500. I. L. Pea must determine how many employees should be working during the Christmas rush. Each employee works five days a week, eight hours a day, can process 50 orders per day, and is paid $10 per hour. If the full-time work force cannot handle the day's orders during regular hours, some employees will have to work overtime. Each employee is paid $15 per hour for overtime work. For example, if 300 orders are received in a day and there are four employees, then $300 - 4(50) = 100$ orders must be processed by employees who are working overtime. Since each employee can fill $\frac{50}{8} = 6.25$ orders per hour, I. L. Pea would need to pay workers $\frac{100}{6.25} = 16$ hours of overtime for that day. To minimize its expected labor costs, how many full-time employees should I. L. Pea employ during the Christmas rush?

7 Suppose demand is a continuous random variable having a probability density function $f(d)$, and $c(d, q)$ is given by Equation (2). Show that if q units are ordered, the expected cost $E(q)$ may be written as

$$E(q) = \int_0^q c_o q f(t) dt + \int_q^\infty (-c_u) q f(t) dt$$
$$+ \text{(terms not involving } q \text{ in integrand)}$$

Now use Leibniz's rule to derive Equation (5).

4.5 Other One-Period Models

Many interesting single-period models in operations research cannot be easily handled by marginal analysis. In such situations, we express the decision maker's objective function (usually expected profit or expected cost) as a function $f(q)$ of the decision variable q.

Then we find a maximum or minimum of $f(q)$ by setting $f'(q) = 0$. In this section, we illustrate this idea by a brief discussion of a bidding model.

EXAMPLE 4 | Condo Construction Company

Condo Construction Company is bidding on an important construction job. The job will cost $2 million to complete. One other company is bidding for the job. Condo believes that the opponent's bid is equally likely to be any amount between $2 million and $4 million. If Condo wants to maximize expected profit, what should its bid be?

Solution   Let

$$B = \text{random variable representing bid of Condo's opponent}$$
$$b = \text{actual bid of Condo's opponent}$$

Then $f(b)$, the density function for B, is given by

$$f(b) = \begin{cases} \dfrac{1}{2,000,000} & (2,000,000 \leq b \leq 4,000,000) \\ 0 & \text{otherwise} \end{cases}$$

Let $q =$ Condo's bid. If $b > q$, Condo outbids the opponent and earns a profit of $q - 2,000,000$. On the other hand, if $b < q$, Condo is outbid by the opponent and earns nothing. The event $b = q$ has a zero probability of occurring and may be ignored. Let $E(q)$ be Condo's expected profit if it bids q. Then

$$E(q) = \int_{2,000,000}^{q} (0)f(b)db + \int_{q}^{4,000,000} (q - 2,000,000)f(b)db$$

Since $f(b) = \frac{1}{2,000,000}$ for $2,000,000 \leq b \leq 4,000,000$, we obtain

$$E(q) = \frac{(q - 2,000,000)(4,000,000 - q)}{2,000,000}$$

To find the value of q maximizing $E(q)$, we find

$$E'(q) = \frac{-(q - 2,000,000) + (4,000,000 - q)}{2,000,000} = \frac{6,000,000 - 2q}{2,000,000}$$

Hence, $E'(q) = 0$ for $q = 3,000,000$. Since $E''(q) = \frac{-2}{2,000,000} < 0$, we know that $E(q)$ is a concave function of q, and $q = 3,000,000$ does indeed maximize $E(q)$. Hence, Condo should bid $3 million. Condo's expected profit will be $E(3,000,000) = \$500,000$.

PROBLEMS

Group A

1 The City of Rulertown consists of the unit interval $[0, 1]$ (see Figure 2). Rulertown needs to determine where to build the city's only fire station. It knows that for small Δx, the probability that a given fire occurs at a location between x and $x + \Delta x$ is $2x(\Delta x)$. Rulertown wants to minimize the average distance between the fire station and a fire. Where should the fire station be located?

FIGURE 2

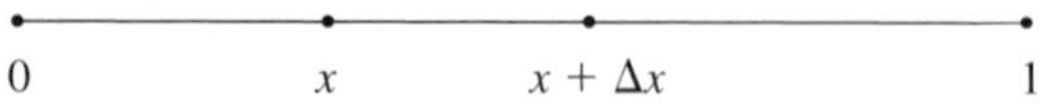

Group B

2 Assume that the Federal Reserve Board can control the growth rate of the U.S. money supply. Also assume that during a year in which the money supply grows by $x\%$, the Gross Domestic Product (GDP) grows by $\mathbf{Z}x\%$, where $\mathbf{Z}$ is a known random variable. The government has decided it wants the GDP to grow by $k\%$ each year. (Too high a growth rate causes excessive inflation, and too low a growth rate causes high unemployment.) To model the government's view, the government assesses a cost of $(d - k)^2$ during a year in which the GDP grows by $d\%$.

a Determine the growth rate of the money supply that should be set by the Federal Reserve Board if the goal is to minimize the expected cost to the government.

b Show that for a given value of $E(\mathbf{Z})$, an increase in var $\mathbf{Z}$ will decrease the optimal growth rate of the money supply found in part (a). (*Hint:* Use the fact that var $\mathbf{Z} = E(\mathbf{Z}^2) - E(\mathbf{Z})^2$.)

4.6 The EOQ with Uncertain Demand: The (r, q) and (s, S) Models

In this section, we discuss a modification of the EOQ that is used when lead time is nonzero and the demand during each lead time is random. We begin by assuming that all demand can be backlogged. As in Chapter 3, we assume a continuous review model, so that orders may be placed at any time, and we define

K = ordering cost

h = holding cost/unit/year

L = lead time for each order (assumed to be known with certainty)

q = quantity ordered each time an order takes place

We also require the following definitions:

D = random variable (assumed continuous) representing annual demand, with mean $E(D)$, variance var D, and standard deviation σ_D

c_B = cost incurred for each unit short, which does not depend on how long it takes to make up stockout

$OHI(t)$ = on-hand inventory (amount of stock on hand) at time t

From Figure 3, we can see that $OHI(1) = 100$, $OHI(0) = 200$, and $OHI(6) = OHI(7) = 0$.

$B(t)$ = number of outstanding back orders at time t

$I(t)$ = net inventory level at time $t = OHI(t) - B(t)$

r = inventory level at which order is placed (reorder point)

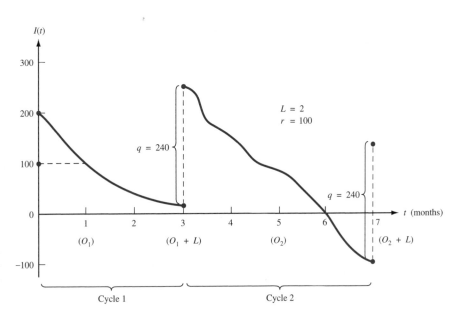

FIGURE 3
Evolution of Inventory over Time in Reorder Point Model

In Figure 3, $B(t) = 0$ for $0 \leq t \leq 6$ and $B(7) = 100$. $I(t)$ agrees with the inventory concept used in Chapter 3; $I(0) = 200 - 0 = 200$, $I(3) = 260 - 0 = 260$, and $I(7) = 0 - 100 = -100$. The reorder point $r = 100$; whenever the inventory level drops to r, an order is placed for q units.

$$X = \text{random variable representing demand during lead time}$$

We assume that X is a continuous random variable having density function $f(x)$ and mean, variance, and standard deviation of $E(X)$, var X, and σ_X, respectively. If we assume that the demands at different points in time are independent, then it can be shown that the random lead time demand X satisfies

$$E(X) = LE(D), \qquad \text{var } X = L(\text{var } D), \qquad \sigma_X = \sigma_D\sqrt{L} \tag{8}$$

We assume that if D is normally distributed, then X will also be normally distributed.

Suppose we allow the lead time L to be a random variable (denoted by L), with mean $E(L)$, variance var L, and standard deviation σ_L. If the length of the lead time is independent of the demand per unit time during the lead time, then

$$E(X) = E(L)E(D) \qquad \text{and} \qquad \text{var } X = E(L)(\text{var } D) + E(D)^2(\text{var } L) \tag{8'}$$

We want to choose q and r to minimize the annual expected total cost (exclusive of purchasing cost). Before showing how optimal values of r and q can be found, we look at an illustration of how inventory evolves over time. Assume that an order of $q = 240$ units has just arrived at time 0. We also assume that $L = 2$. In Figure 3, orders of size q are placed at times $O_1 = 1$ and $O_2 = 5$. These orders are received at times $O_1 + L = 3$ and $O_2 + L = 7$, respectively. A **cycle** is defined to be the time interval between any two instants at which an order is received. Figure 3 contains two complete cycles: cycle 1, from arrival of order at time 0 to the instant before order arrives at time $O_1 + L = 3$; and cycle 2, from arrival of order at time $O_1 + L = 3$ to the instant before order arrives at time $O_2 + L = 7$.

During cycle 1, demand during lead time is less than r, so no shortage occurs. During cycle 2, however, demand during lead time exceeds r, so stockouts do occur between time 6 and time $O_2 + L = 7$. It should be clear that by increasing r, we can reduce the number of stockouts. Unfortunately, increasing r will force us to carry more inventory, thereby resulting in higher holding costs. Thus, an optimal value of r must represent some sort of trade-off between holding and stockout costs.

We now show how the optimal values of q and r may be determined.

Determination of Reorder Point: The Back-Ordered Case

The situation in which all demand must eventually be met and no sales are lost is called the **back-ordered case,** for which we show how to determine the reorder point and order quantity that minimize annual expected cost.

We assume each unit is purchased for the same price, so purchasing costs are fixed. Define $TC(q, r) = $ expected annual cost (excluding purchasing cost) incurred if each order is for q units and is placed when the reorder point is r. Then $TC(q, r) = $ (expected annual holding cost) + (expected annual ordering cost) + (expected annual cost due to shortages). To determine the optimal reorder point and order quantity, we assume that the average number of back orders is small relative to the average on-hand inventory level. In most cases, this assumption is reasonable, because shortages (if they occur at all) usually occur during only a small portion of a cycle. (See Problem 5 at the end of this section.) Then $I(t) = OHI(t) - B(t)$ yields

$$\text{Expected value of } I(t) \cong \text{expected value of } OHI(t) \tag{9}$$

We can now approximate the expected annual holding cost. We know that expected annual holding cost = h(expected value of on-hand inventory level). Then from (9), we can approximate expected annual holding cost by h(expected value of $I(t)$). As in Chapter 3, the expected value of $I(t)$ will equal the expected value of $I(t)$ during a cycle. Since the mean rate at which demand occurs is constant, we may write

$$\text{Expected value of } I(t) \text{ during a cycle}$$
$$= \tfrac{1}{2}[(\text{expected value of } I(t) \text{ at beginning of cycle}) \tag{10}$$
$$+ (\text{expected value of } I(t) \text{ at end of a cycle})]$$

At the end of a cycle (the instant before an order arrives), the inventory level will equal the inventory level at the reorder point (r) less the demand $\mathbf{X}$ during lead time. Thus, expected value of $I(t)$ at end of cycle = $r - E(\mathbf{X})$.

At the beginning of a cycle, the inventory level at the end of the cycle is augmented by the arrival of an order of size q. Thus, expected value of $I(t)$ at beginning of cycle = $r - E(\mathbf{X}) + q$. Now (10) yields

$$\text{Expected value of } I(t) \text{ during cycle} = \tfrac{1}{2}(r - E(\mathbf{X}) + r - E(\mathbf{X}) + q)$$
$$= \tfrac{q}{2} + r - E(\mathbf{X})$$

Thus, expected annual holding cost $\cong h(\tfrac{q}{2} + r - E(\mathbf{X}))$.

To determine the expected annual cost due to stockouts or back orders, we must define

$$\mathbf{B}_r = \text{random variable representing the number of stockouts}$$
$$\text{or back orders during a cycle if the reorder point is } r$$

Now

$$\text{Expected annual shortage cost} = \left(\frac{\text{expected shortage cost}}{\text{cycle}}\right)\left(\frac{\text{expected cycles}}{\text{year}}\right)$$

By the definition of $\mathbf{B}_r$,

$$\frac{\text{Expected shortage cost}}{\text{Cycle}} = c_B E(\mathbf{B}_r)$$

Since all demand will eventually be met, an average of $\frac{E(\mathbf{D})}{q}$ orders will be placed each year. Then

$$\frac{\text{Expected shortage cost}}{\text{Year}} = \frac{c_B E(\mathbf{B}_r)E(\mathbf{D})}{q}$$

Finally,

$$\text{Expected annual order cost} = K\left(\frac{\text{expected orders}}{\text{year}}\right) = \frac{KE(\mathbf{D})}{q}$$

Putting together the expected annual holding, shortage, and ordering costs, we obtain

$$TC(q, r) = h\left(\frac{q}{2} + r - E(\mathbf{X})\right) + \frac{c_B E(\mathbf{B}_r)E(\mathbf{D})}{q} + \frac{KE(\mathbf{D})}{q} \tag{11}$$

Using the method described in Section 12.5 of Volume 1, we could find the values of q and r that minimize (11) by determining values q^* and r^* of q and r satisfying

$$\frac{\partial TC(q^*, r^*)}{\partial q} = \frac{\partial TC(q^*, r^*)}{\partial r} = 0 \tag{12}$$

In Review Problem 7 we show how LINGO can be used to determine values of q and r that exactly satisfy (12). In most cases, however, the value of q^* satisfying (12) is very close to the EOQ[†] of $(\frac{2KE(\mathbf{D})}{h})^{1/2}$. For this reason, we assume that the optimal order quantity q^* may be adequately approximated by the EOQ. Given a value q for the order quantity, we now show how marginal analysis can be used to determine a reorder point r^* that minimizes $TC(q, r)$.

If we assume a given value of q, the expected annual ordering cost is independent of r. Thus, in determining a value of r that minimizes $TC(q, r)$, we may concentrate on minimizing the sum of the expected annual holding and shortage costs. Following the marginal analysis approach of Sections 4.2–4.3, suppose we increase the reorder point (for Δ small) from r to $r + \Delta$ (with q fixed). Will this result in an increase or a decrease in $TC(q, r)$?

If we increase r to $r + \Delta$, the expected annual holding cost will increase by

$$h\left(\frac{q}{2} + r + \Delta - E(\mathbf{X})\right) - h\left(\frac{q}{2} + r - E(\mathbf{X})\right) = h\Delta$$

If we increase the reorder point from r to $r + \Delta$, expected annual stockout costs will be reduced, because of the fact that during any cycle in which lead time demand is at least r, the number of stockouts during the cycle will be reduced by Δ units. In other words, increasing the reorder point from r to $r + \Delta$ will reduce stockout costs by $c_B\Delta$ during a fraction $P(\mathbf{X} \geq r)$ of all cycles. Since there are an average of $\frac{E(\mathbf{D})}{q}$ cycles per year, increasing the reorder point from r to $r + \Delta$ will reduce expected annual stockout cost by

$$\frac{\Delta E(\mathbf{D})c_B P(\mathbf{X} \geq r)}{q}$$

Observe that as r increases, $P(\mathbf{X} \geq r)$ decreases, so as r increases, the expected reduction in expected annual shortage cost resulting from increasing the reorder point by Δ will decrease. This observation allows us to draw Figure 4.

Let r^* be the value of r for which marginal benefit equals marginal cost, or

$$\frac{\Delta E(\mathbf{D})c_B P(\mathbf{X} \geq r^*)}{q} = h\Delta$$

$$P(\mathbf{X} \geq r^*) = \frac{hq}{c_B E(\mathbf{D})}$$

Suppose that $r < r^*$. Then Figure 4 shows that if we increase the reorder point from r to r^*, we can save more in shortage cost than we lose in holding cost. Now suppose that $r > r^*$. Figure 4 shows that by reducing the reorder point from r to r^*, we can save more in holding cost than we lose in increased shortage cost. Thus, r^* does attain the optimal trade-off between shortage and holding costs. In summary, if we assume that the order quantity can be approximated by

$$EOQ = \left(\frac{2KE(\mathbf{D})}{h}\right)^{1/2}$$

then we have the reorder point r^* and the order quantity q^* for the back-ordered case:

$$q^* = \left(\frac{2KE(\mathbf{D})}{h}\right)^{1/2}$$

$$P(\mathbf{X} \geq r^*) = \frac{hq^*}{c_B E(\mathbf{D})}$$

(13)

If

$$\frac{hq^*}{c_B E(\mathbf{D})} > 1$$

[†]Brown (1967) has shown that for approximating the optimal value of q, the EOQ is usually acceptable unless EOQ $\leq \sigma_\mathbf{X}$.

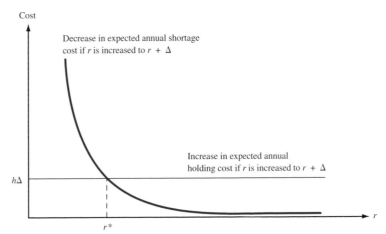

FIGURE 4
Trade-off between
Holding Cost and
Shortage Cost

Cost

Decrease in expected annual shortage
cost if r is increased to $r + \Delta$

Increase in expected annual
holding cost if r is increased to $r + \Delta$

$h\Delta$

r^*

r

then (13) will have no solution, and holding cost is prohibitively high relative to the stock-out cost. Management should set the reorder point at the smallest acceptable level. If (13) yields a negative value of r^*, management should also set the reorder point at the smallest acceptable level.

REMARKS **1** $P(\mathbf{X} \geq r)$ is just the probability that a stockout will occur during a lead time. Also note that for h near zero, (13) yields a stockout probability near zero. For large c_B also, (13) yields a stockout probability near zero. Both of these results should be consistent with intuition.
2 After substituting the EOQ for q in (13), we may easily determine an approximately optimal value of r, the reorder point. Note that $r - E(\mathbf{X})$ is the amount in excess of expected lead time demand that is ordered to protect against the occurrence of stockouts during the lead time. For this reason, $r - E(\mathbf{X})$ is often referred to as **safety stock.**
3 From (11), we find that the expected annual cost of holding safety stock is $h(r - E(\mathbf{X})) = h(\text{safety stock level})$.

The following example illustrates the determination of the reorder point and safety stock level in the back-ordered demand case.

EXAMPLE 5 **Disk Stock**

Each year, a computer store sells an average of 1,000 boxes of disks. Annual demand for boxes of disks is normally distributed with a standard deviation of 40.8 boxes. The store orders disks from a regional distributor. Each order is filled in two weeks. The cost of placing each order is $50, and the annual cost of holding one box of disks in inventory is $10. The per-unit stockout cost (because of loss of goodwill and the cost of placing a special order) is assumed to be $20. The store is willing to assume that all demand is backlogged. Determine the proper order quantity, reorder point, and safety stock level for the computer store. Assume that annual demand is normally distributed. What is the probability that a stockout occurs during the lead time?

Solution We begin by determining the EOQ. Since $h = \$10/\text{box/year}$, $K = \$50$, and $E(\mathbf{D}) = 1,000$, we find that

$$\text{EOQ} = \left(\frac{2(50)(1,000)}{10}\right)^{1/2} = 100$$

We now substitute $q^* = 100$ in (13) and use (13) to determine the reorder point. To do this, we need to determine the probability distribution of **X**, the lead time demand. Since $L = 2$ weeks, **X** will be normally distributed with

$$E(\mathbf{X}) = \frac{E(\mathbf{D})}{26} = \frac{1,000}{26} = 38.46 \quad \text{and} \quad \sigma_{\mathbf{X}} = \frac{\sigma_{\mathbf{D}}}{\sqrt{26}} = \frac{40.8}{\sqrt{26}} = 8$$

Since $c_B = \$20$, (13) now yields

$$P(\mathbf{X} \geq r) = \frac{10(100)}{20(1,000)} = .05 \tag{14}$$

We use the Excel function NORMINV. Since

$$=\text{NORMINV}(0.95, 38.46, 8)$$

yields 51.62, we find that the safety stock level is $r - E(\mathbf{X}) = 51.62 - 38.46 = 13.16$.

To see how the reorder point and safety stock level would be affected by a variable lead time, suppose that the lead time has a mean of two weeks but also has a standard deviation of one week ($\frac{1}{52}$ year). Then (8′) yields

$$\sigma_{\mathbf{X}}^2 = (\tfrac{1}{26})(40.8)^2 + (1,000)^2 (\tfrac{1}{52})^2 = 64.02 + 369.82 = 433.84$$
$$\sigma_{\mathbf{X}} = \sqrt{433.84} = 20.83$$

Assuming that the lead time demand is normally distributed, we would find that $r = 38.46 + 1.65(20.83) = 72.83$, and the safety stock held is $1.65(20.83) = 34.37$. Thus, the variability of the lead time has more than doubled the required safety stock level!

Determination of Reorder Point: The Lost Sales Case

We now assume that all stockouts result in lost sales and that a cost of c_{LS} dollars is incurred for each lost sale. (In addition to penalties for loss of future goodwill, c_{LS} should include profit lost because of a lost sale.)

As in the back-ordered case, we assume that the optimal order quantity can be adequately approximated by the EOQ and attempt to use marginal analysis to determine the optimal reorder point r^* (see Problem 6 at the end of this section). The optimal order quantity q^* and the reorder point r^* for the lost sales case are

$$q^* = \left(\frac{2KE(\mathbf{D})}{h} \right)^{1/2}$$
$$P(\mathbf{X} \geq r^*) = \frac{hq^*}{hq^* + c_{LS}E(\mathbf{D})} \tag{15}$$

The key to the derivation of (15) is to realize that expected inventory in lost sales case = (expected inventory in back-ordered case) + (expected number of shortages per cycle). This equation follows because in the lost sales case, we find that during each cycle, an average of (expected shortages per cycle) fewer orders will be filled from inventory, thereby raising the average inventory level by an amount equal to expected shortages per cycle. Observe that the right-hand side of (15) is smaller than the right-hand side of (13). Thus, the lost sales assumption will yield a lower stockout probability (and a larger reorder point and safety stock level) than the back-ordered assumption.

To illustrate the use of (15), we continue our discussion of Example 5. Suppose that each box of disks sells for $50 and costs the store $30. Assuming that the stockout cost of $20 given in Example 5 represents lost goodwill, we obtain c_{LS} by adding the lost profit ($50 − $30) to the lost goodwill of $20. Thus, $c_{LS} = 20 + 20 = 40$. Recall from Example 5 that $E(\mathbf{D}) = 1,000$ boxes per year, $h = \$10/\text{box/year}$, EOQ $= 100$ boxes, and $K = \$50$. Now (15) yields

$$P(\mathbf{X} \geq r^*) = \frac{10(100)}{10(100) + 40(1,000)} = .024$$

Excel is used to compute r. Since

$$=\text{NORMINV}(.976, 38.46, 8)$$

yields 54.28, we find that $r = 54.28$. Thus, in the lost sales case, the safety stock level is $54.28 − 38.46 = 15.82$.

Continuous Review (r, q) Policies

A continuous review inventory policy, in which we order a quantity q whenever our inventory level reaches a reorder level r, is often called an **(r, q) policy**. An (r, q) policy is also called a **two-bin** policy, because it can easily be implemented by using two bins to store an item. For example, to implement a (30, 500) policy, we fill orders from bin 1 as long as bin 1 contains any items. As soon as bin 1 becomes empty, we know that the reorder point $r = 30$ has been reached, and we place an order for $q = 500$ units. When the order arrives, we bring the number of units in bin 2 up to 30, and place the remainder of the 500 units ordered in bin 1. Thus, whenever bin 1 has been emptied, we know that the reorder point has been reached.

Continuous Review (s, S) Policies

In our derivation of the best (r, q) policy, we assumed that an order could be placed exactly at the point when the inventory level reached the reorder point r. We used this assumption to compute the expected inventory level at the beginning and end of a cycle. Suppose that a demand for more than one unit can arrive at a particular time. Then an order may be triggered when the inventory level is less than r, and our computation of expected inventory level at the end and beginning of a cycle is then incorrect. For example, suppose $r = 30$ and our current inventory level is 35. If an order for 10 units arrives, an order will be placed when the inventory level is 25 (not $r = 30$), and this invalidates the computations that led to (11). From this discussion, we see that it is possible for the inventory level to "undershoot" the reorder point.

Note that this problem could not occur if all demands were for one unit, for then the inventory level would drop from (say) 32 to 31 and then to 30, and each order would be placed when the inventory level equaled the reorder point r. From this example, we see that if demands of size greater than one unit can occur at a point in time, then the (r, q) model may not yield a policy that minimizes expected annual cost.

In such situations, it has been shown that an **(s, S) policy** is optimal. To implement an (s, S) policy, we place an order whenever the inventory level is less than or equal to s. The size of the order is sufficient to raise the inventory level to S (assuming zero lead time). For example, if we were implementing a (5, 40) policy and the inventory level suddenly dropped from 7 to 3, we would immediately place an order for $40 − 3 = 37$ units. Exact computation of the optimal (s, S) policy is difficult. If we neglect the problem of the "undershoots," however, we may approximate the optimal (s, S) policy as follows. Set $S − s$ equal to the economic order quantity q. Then set s equal to the reorder point r obtained from (13) or (15). Finally, we obtain $S = r + q$. Thus, for Example 5 (with back orders allowed), we would set $s = 51.66$ and $S = 51.66 + 100 = 151.66$ and use (assuming that fractional demand is possible) a $(51.66, 151.66)$ policy.

PROBLEMS

Group A

1 A hospital orders its blood from a regional blood bank. Each year, the hospital uses an average of 1,040 pints of Type O blood. Each order placed with the regional blood bank incurs a cost of $20. The lead time for each order is one week. It costs the hospital $20 to hold 1 pint of blood in inventory for a year. The per-pint stockout cost is estimated to be $50. Annual demand for Type O blood is normally distributed, with standard deviation of 43.26 pints. Determine the optimal order quantity, reorder point, and safety stock level. Assume that 52 weeks = 1 year and that all demand is backlogged. To use the techniques of this section, what unrealistic assumptions must be made? What (s, S) policy would be used in this situation?

2 Furnco sells secretarial chairs. Annual demand is normally distributed, with mean of 1,040 chairs and standard deviation of 50.99 chairs. Furnco orders its chairs from its flagship store. It costs $100 to place an order, and the lead time is two weeks. Furnco estimates that each stockout causes a loss of $50 in future goodwill. Furnco pays $60 for each chair and sells it for $100. The annual cost of holding a chair in inventory is 30% of its purchase cost.

   a Assuming that all demand is backlogged, what are the reorder point and the safety stock level?

   b Assuming that all stockouts result in lost sales, determine the optimal reorder point and the safety stock level.

3 We are given the following information for a product:

       Order cost = $50

       Annual demand = N(960, 3,072.49)

       Annual holding cost = $6/item/year

       Shortage cost = $80 per unit

       Lead time = one month

       Sales price = $40 per unit

       Product cost = $30 per unit

   a Determine the order quantity and the reorder point under the assumption that all demands are backordered.

   b Determine the order quantity and reorder point under the lost sales assumption.

4 The lead time demand for bathing suits is governed by the discrete random variable shown in Table 11. The company sells an average of 10,400 suits per year. The cost of placing an order for bathing suits is $30, and the cost of holding one bathing suit in inventory for a year is $3. The stockout cost is $3 per bathing suit. Use marginal analysis to determine the optimal order quantity and the reorder point.

TABLE 11

Lead Time Demand	Probability
180	.30
190	.30
200	.15
210	.10
220	.15

5 In Figure 3, assume that demand occurs at a constant rate during each cycle. Approximate the average level of on-hand inventory between $t = 0$ and $t = 7$. Also approximate the average number of shortages. Does the assumption that the average shortage level is small relative to the average level of on-hand inventory seem valid here?

Group B

6 In this problem, use marginal analysis to determine the optimal reorder point for the lost sales case.

   a Show that the average inventory level for the lost sales case may be written as

$$\tfrac{1}{2}[(r - E(\mathbf{X}) + E(\mathbf{B}_r)) + (r - E(\mathbf{X}) + E(\mathbf{B}_r) + q)]$$
$$= r - E(\mathbf{X}) + E(\mathbf{B}_r) + \tfrac{q}{2}$$

   b Although expected orders per year will no longer equal $\frac{E(\mathbf{D})}{q}$ (why?), we assume that the expected number of lost sales per year is relatively small. Thus, we may still assume that expected orders per year $= \frac{E(\mathbf{D})}{q}$. Now use marginal analysis to derive (15).

7 Suppose that a cost of S dollars (independent of the size of the stockout) is incurred whenever a stockout occurs during a cycle. Under the assumption of backlogged demand, use marginal analysis to determine the reorder point.

8 Explain the following statement: Faster-moving items require larger safety stocks than slower-moving items. (*Hint*: Does $\frac{q}{E(\mathbf{D})}$ large imply that an item is fast-moving or slow-moving?)

9 Suppose annual demand for a product is normally distributed, with a mean of 600 and a variance of 300. Suppose that the lead time for an order is always one month. Show (without using Equation (8)) that the lead time demand has mean 50, variance 25, and standard deviation 5. Assume that the demands during different one-month periods are independent, identically distributed random variables.

4.7 The EOQ with Uncertain Demand: The Service Level Approach to Determining Safety Stock Level

As we have previously stated, it is usually very difficult to determine accurately the cost of being one unit short. For this reason, managers often decide to control shortages by meeting a specified service level. In this section, we discuss two measures of service level:

Service Level Measure 1 SLM_1, the expected fraction (usually expressed as a percentage) of all demand that is met on time.

Service Level Measure 2 SLM_2, the expected number of cycles per year during which a shortage occurs.

Throughout this section, we assume that all shortages are backlogged. The following example illustrates the meaning of the two service level measures.

EXAMPLE 6 ***SLM₁* and *SLM₂***

Suppose that for a given inventory situation, average annual demand is 1,000 and the EOQ is 100. Demand during a lead time is random and is described by the probability distribution in Table 12. For a reorder point of 30 units, determine SLM_1 and SLM_2.

Solution The expected demand during a lead time is $\frac{1}{5}(20) + \frac{1}{5}(30) + \frac{1}{5}(40) + \frac{1}{5}(50) + \frac{1}{5}(60) =$ 40 units. With a reorder point of 30 units, we will reorder during each cycle at the instant when the inventory level hits 30 units. If the lead time demand during a cycle is 20 or 30 units, we will experience no shortage. During a cycle in which lead time demand is 40, a shortage of 10 units will occur; if lead time demand is 50, a shortage of 20 units will occur; if lead time demand is 60, a shortage of 30 units will occur. Hence, the expected number of units short per cycle is given by $\frac{1}{5}(0) + \frac{1}{5}(0) + \frac{1}{5}(10) + \frac{1}{5}(20) + \frac{1}{5}(30) = 12$.

Since the EOQ = 100 and all demand must eventually be met, the average number of orders placed each year will be $\frac{E(\mathbf{D})}{q} = \frac{1,000}{100} = 10$. Then the average number of shortages that occur during a year will equal $10(12) = 120$ units. Thus, each year, on the average, the demand for $1,000 - 120 = 880$ units is met on time. In this case, the $SLM_1 = \frac{880}{1,000} = 0.88$ or 88%. This shows that even if the reorder point is less than the mean lead time demand, a relatively high SLM_1 may result, because stockouts can only occur during the lead time, which is often a small portion of each cycle.

We now determine SLM_2 for a reorder point of 30. With a reorder point of 30, a stockout will occur during any cycle in which lead time demand exceeds 30 units. Thus, the probability of a stockout during a cycle $= P(\mathbf{X} = 40) + P(\mathbf{X} = 50) + P(\mathbf{X} = 60) = \frac{3}{5}$.

TABLE 12
Mass Function for Lead Time Demand

Lead Time Demand	Probability
20	$\frac{1}{5}$
30	$\frac{1}{5}$
40	$\frac{1}{5}$
50	$\frac{1}{5}$
60	$\frac{1}{5}$

Since there are an average of 10 cycles per year, the expected number of cycles per year that will result in shortages is $10\left(\frac{3}{5}\right) = 6$. Thus, a reorder point of 30 yields $SLM_2 = 6$ stockouts per year.

Determination of Reorder Point and Safety Stock Level for SLM_1

Given a desired value of SLM_1, how do we determine a reorder point that provides the desired service level? Suppose we order the EOQ (q) and use a reorder point r. From Section 4.6,

$$\frac{\text{Expected shortages}}{\text{Cycle}} = E(\mathbf{B}_r)$$

$$\frac{\text{Expected shortages}}{\text{Year}} = \frac{E(\mathbf{B}_r)E(\mathbf{D})}{q}$$

Here, $E(\mathbf{D})$ is the average annual demand. Let SLM_1 be the percentage of all demand that is met on time. Then for given values of q (for the order quantity) and r (for the reorder point), we have

$$1 - SLM_1 = \frac{\text{expected shortages per year}}{\text{expected demand per year}} = \frac{E(\mathbf{B}_r)E(\mathbf{D})/q}{E(\mathbf{D})} = \frac{E(\mathbf{B}_r)}{q} \tag{16}$$

Equation (16) can be used to determine the reorder point that yields a desired service level. We now assume that the lead time demand is normally distributed, with mean $E(\mathbf{X})$ and standard deviation $\sigma_\mathbf{X}$. To use (16), we need to determine $E(\mathbf{B}_r)$. If $\mathbf{X}$ is normally distributed, the determination of $E(\mathbf{B}_r)$ requires a knowledge of the normal loss function.

DEFINITION ■ The **normal loss function**, $NL(y)$, is defined by the fact that $\sigma_\mathbf{X}NL(y)$ is the expected number of shortages that will occur during a lead time if (1) lead time demand is normally distributed with mean $E(\mathbf{X})$ and standard deviation $\sigma_\mathbf{X}$ and (2) the reorder point is $E(\mathbf{X}) + y\sigma_\mathbf{X}$. ■

In short, if we hold y standard deviations (in terms of lead time demand) of safety stock, then $NL(y)\sigma_\mathbf{X}$ is the expected number of shortages occurring during a lead time.

Since a larger reorder point leads to fewer shortages, we would expect $NL(y)$ to be a nonincreasing function of y. This is indeed the case. The function $NL(y)$ is tabulated in Table 13. For example, $NL(0) = 0.3989$ means that if the reorder point equals the expected lead time demand, and the standard deviation of lead time demand is $\sigma_\mathbf{X}$, then an average of $0.3989\sigma_\mathbf{X}$ shortages will occur during a lead time. Similarly, $NL(2) = 0.0085$ means that if the reorder point exceeds the mean lead time demand by $2\sigma_\mathbf{X}$, then an average of $0.0085\sigma_\mathbf{X}$ shortages will occur during a given lead time. $NL(y)$ is not tabulated for negative values of y. This is because it can be shown that for $y \le 0$, $NL(y) = NL(-y) - y$. For example, $NL(-2) = NL(2) + 2 = 2.0085$. This means that if the reorder point is $2\sigma_\mathbf{X}$ less than the mean lead time demand, an average of $2.0085\sigma_\mathbf{X}$ shortages will occur during each cycle.

LINGO with the **@PSL** function may be used to compute values of the normal loss function. In LINGO, the program

```
MODEL:
  x = @PSL(2);
END
```

will yield $x = .0085$.

TABLE 13
The Normal Loss Function

x	NL(x)	x	NL(x)	x	NL(x)
0.00	0.3989	0.40	0.2304	0.80	0.1202
0.01	0.3940	0.41	0.2270	0.81	0.1181
0.02	0.3890	0.42	0.2236	0.82	0.1160
0.03	0.3841	0.43	0.2203	0.83	0.1140
0.04	0.3793	0.44	0.2169	0.84	0.1120
0.05	0.3744	0.45	0.2137	0.85	0.1100
0.06	0.3697	0.46	0.2104	0.86	0.1080
0.07	0.3649	0.47	0.2072	0.87	0.1061
0.08	0.3602	0.48	0.2040	0.88	0.1042
0.09	0.3556	0.49	0.2009	0.89	0.1023
0.10	0.3509	0.50	0.1978	0.90	0.1004
0.11	0.3464	0.51	0.1947	0.91	0.09860
0.12	0.3418	0.52	0.1917	0.92	0.09680
0.13	0.3373	0.53	0.1887	0.93	0.09503
0.14	0.3328	0.54	0.1857	0.94	0.09328
0.15	0.3284	0.55	0.1828	0.95	0.09156
0.16	0.3240	0.56	0.1799	0.96	0.08986
0.17	0.3197	0.57	0.1771	0.97	0.08819
0.18	0.3154	0.58	0.1742	0.98	0.08654
0.19	0.3111	0.59	0.1714	0.99	0.08491
0.20	0.3069	0.60	0.1687	1.00	0.08332
0.21	0.3027	0.61	0.1659	1.01	0.08174
0.22	0.2986	0.62	0.1633	1.02	0.08019
0.23	0.2944	0.63	0.1606	1.03	0.07866
0.24	0.2904	0.64	0.1580	1.04	0.07716
0.25	0.2863	0.65	0.1554	1.05	0.07568
0.26	0.2824	0.66	0.1528	1.06	0.07422
0.27	0.2784	0.67	0.1503	1.07	0.07279
0.28	0.2745	0.68	0.1478	1.08	0.07138
0.29	0.2706	0.69	0.1453	1.09	0.06999
0.30	0.2668	0.70	0.1429	1.10	0.06862
0.31	0.2630	0.71	0.1405	1.11	0.06727
0.32	0.2592	0.72	0.1381	1.12	0.06595
0.33	0.2555	0.73	0.1358	1.13	0.06465
0.34	0.2518	0.74	0.1334	1.14	0.02034
0.35	0.2481	0.75	0.1312	1.15	0.06210
0.36	0.2445	0.76	0.1289	1.16	0.06086
0.37	0.2409	0.77	0.1267	1.17	0.05964
0.38	0.2374	0.78	0.1245	1.18	0.05844
0.39	0.2339	0.79	0.1223	1.19	0.05726

(Continued)

TABLE **13**
(Continued)

x	NL(x)	x	NL(x)	x	NL(x)
1.20	0.05610	1.60	0.02324	2.00	0.008491
1.21	0.05496	1.61	0.02270	2.01	0.008266
1.22	0.05384	1.62	0.02217	2.02	0.008046
1.23	0.05274	1.63	0.02165	2.03	0.007832
1.24	0.05165	1.64	0.02114	2.04	0.007623
1.25	0.05059	1.65	0.02064	2.05	0.007418
1.26	0.04954	1.66	0.02015	2.06	0.007219
1.27	0.04851	1.67	0.01967	2.07	0.007024
1.28	0.04750	1.68	0.01920	2.08	0.006835
1.29	0.04650	1.69	0.01874	2.09	0.006649
1.30	0.04553	1.70	0.01829	2.10	0.006468
1.31	0.04457	1.71	0.01785	2.11	0.006292
1.32	0.04363	1.72	0.01742	2.12	0.006120
1.33	0.04270	1.73	0.01699	2.13	0.005952
1.34	0.04179	1.74	0.01658	2.14	0.005788
1.35	0.04090	1.75	0.01617	2.15	0.005628
1.36	0.04002	1.76	0.01578	2.16	0.005472
1.37	0.03916	1.77	0.01539	2.17	0.005320
1.38	0.03831	1.78	0.01501	2.18	0.005172
1.39	0.03748	1.79	0.01464	2.19	0.005028
1.40	0.03667	1.80	0.01428	2.20	0.004887
1.41	0.03587	1.81	0.01392	2.21	0.004750
1.42	0.03508	1.82	0.01357	2.22	0.004616
1.43	0.03431	1.83	0.01323	2.23	0.004486
1.44	0.03356	1.84	0.01290	2.24	0.004358
1.45	0.03281	1.85	0.01257	2.25	0.004235
1.46	0.03208	1.86	0.01226	2.26	0.004114
1.47	0.03137	1.87	0.01195	2.27	0.003996
1.48	0.03067	1.88	0.01164	2.28	0.003882
1.49	0.02998	1.89	0.01134	2.29	0.003770
1.50	0.02931	1.90	0.01105	2.30	0.003662
1.51	0.02865	1.91	0.01077	2.31	0.003556
1.52	0.02800	1.92	0.01049	2.32	0.003453
1.53	0.02736	1.93	0.01022	2.33	0.003352
1.54	0.02674	1.94	0.009957	2.34	0.003255
1.55	0.02612	1.95	0.009698	2.35	0.003159
1.56	0.02552	1.96	0.009445	2.36	0.003067
1.57	0.02494	1.97	0.009198	2.37	0.002977
1.58	0.02436	1.98	0.008957	2.38	0.002889
1.59	0.02380	1.99	0.008721	2.39	0.002804

(Continued)

TABLE 13
(Continued)

x	NL(x)	x	NL(x)	x	NL(x)
2.40	0.002720	2.80	0.0007611	3.20	0.0001852
2.41	0.002640	2.81	0.0007359	3.21	0.0001785
2.42	0.002561	2.82	0.0007115	3.22	0.0001720
2.43	0.002484	2.83	0.0006879	3.23	0.0001657
2.44	0.002410	2.84	0.0006650	3.24	0.0001596
2.45	0.002337	2.85	0.0006428	3.25	0.0001537
2.46	0.002267	2.86	0.0006213	3.26	0.0001480
2.47	0.002199	2.87	0.0006004	3.27	0.0001426
2.48	0.002132	2.88	0.0005802	3.28	0.0001373
2.49	0.002067	2.89	0.0005606	3.29	0.0001322
2.50	0.002004	2.90	0.0005417	3.30	0.0001273
2.51	0.001943	2.91	0.0005233	3.31	0.0001225
2.52	0.001883	2.92	0.0005055	3.32	0.0001179
2.53	0.001826	2.93	0.0004883	3.33	0.0001135
2.54	0.001769	2.94	0.0004716	3.34	0.0001093
2.55	0.001715	2.95	0.0004555	3.35	0.0001051
2.56	0.001662	2.96	0.0004398	3.36	0.0001012
2.57	0.001610	2.97	0.0004247	3.37	0.00009734
2.58	0.001560	2.98	0.0004101	3.38	0.00009365
2.59	0.001511	2.99	0.0003959	3.39	0.00009009
2.60	0.001464	3.00	0.0003822	3.40	0.00008666
2.61	0.001418	3.01	0.0003689	3.41	0.00008335
2.62	0.001373	3.02	0.0003560	3.42	0.00008016
2.63	0.001330	3.03	0.0003436	3.43	0.00007709
2.64	0.001288	3.04	0.0003316	3.44	0.00007413
2.65	0.001247	3.05	0.0003199	3.45	0.00007127
2.66	0.001207	3.06	0.0003087	3.46	0.00006852
2.67	0.001169	3.07	0.0002978	3.47	0.00006587
2.68	0.001132	3.08	0.0002873	3.48	0.00006331
2.69	0.001095	3.09	0.0002771	3.49	0.00006085
2.70	0.001060	3.10	0.0002672	3.50	0.00005848
2.71	0.001026	3.11	0.0002577	3.51	0.00005620
2.72	0.0009928	3.12	0.0002485	3.52	0.00005400
2.73	0.0009607	3.13	0.0002396	3.53	0.00005188
2.74	0.0009295	3.14	0.0002311	3.54	0.00004984
2.75	0.0008992	3.15	0.0002227	3.55	0.00004788
2.76	0.0008699	3.16	0.0002147	3.56	0.00004599
2.77	0.0008414	3.17	0.0002070	3.57	0.00004417
2.78	0.0008138	3.18	0.0001995	3.58	0.00004242
2.79	0.0007870	3.19	0.0001922	3.59	0.00004073

(Continued)

TABLE 13
(Continued)

x	NL(x)	x	NL(x)	x	NL(x)
3.60	0.00003911	3.75	0.00002103	3.90	0.00001108
3.61	0.00003755	3.76	0.00002016	3.91	0.00001061
3.62	0.00003605	3.77	0.00001933	3.92	0.00001016
3.63	0.00003460	3.78	0.00001853	3.93	0.00000972
3.64	0.00003321	3.79	0.00001776	3.94	0.000009307
3.65	0.00003188	3.80	0.00001702	3.95	0.000008908
3.66	0.00003059	3.81	0.00001632	3.96	0.000008525
3.67	0.00002935	3.82	0.00001563	3.97	0.000008158
3.68	0.00002816	3.83	0.00001498	3.98	0.000007806
3.69	0.00002702	3.84	0.00001435	3.99	0.000007469
3.70	0.00002592	3.85	0.00001375	4.00	0.000007145
3.71	0.00002486	3.86	0.00001317		
3.72	0.00002385	3.87	0.00001262		
3.73	0.00002287	3.88	0.00001208		
3.74	0.00002193	3.89	0.00001157		

Source: From R. Peterson and E. Silver, *Decision Systems for Inventory and Production Planning,*
© 1998 John Wiley & Sons, New York. Reprinted with permission.

Assuming normal lead time demand, we now determine the reorder point r that will yield a desired level of SLM_1 (expressed as a fraction). A reorder point of r corresponds to holding

$$y = \frac{r - E(\mathbf{X})}{\sigma_{\mathbf{X}}}$$

standard deviations of safety stock. Now the definition of the normal loss function implies that during a lead time, a reorder point of r will yield an expected number of shortages $E(\mathbf{B}_r)$ given by

$$E(\mathbf{B}_r) = \sigma_{\mathbf{X}} NL\left(\frac{r - E(\mathbf{X})}{\sigma_{\mathbf{X}}}\right) \tag{17}$$

Substituting (17) into (16), we obtain the reorder point for SLM_1 with normal lead time demand:

$$1 - SLM_1 = \frac{\sigma_{\mathbf{X}} NL\left(\frac{r - E(\mathbf{X})}{\sigma_{\mathbf{X}}}\right)}{q}$$

$$NL\left(\frac{r - E(\mathbf{X})}{\sigma_{\mathbf{X}}}\right) = \frac{q(1 - SLM_1)}{\sigma_{\mathbf{X}}} \tag{18}$$

With the exception of r, all quantities in (18) are known. Thus, (18) and Table 13 can be used to determine the reorder point corresponding to a given level of SLM_1.

EXAMPLE 7   Bads, Inc.

Bads, Inc., sells an average of 1,000 food processors each year. Each order for food processors placed by Bads costs $50. The lead time is one month. It costs $10 to hold a food processor in inventory for one year. Annual demand for food processors is normally distributed, with a standard deviation of 69.28. For each of the following values of SLM_1, determine the reorder point: 80%, 90%, 95%, 99%, 99.9%.

Solution Note that $E(\mathbf{D}) = 1,000$, $K = \$50$, and $h = \$10$, so

$$q = \left[\frac{2(50)(1,000)}{10}\right]^{1/2} = 100$$

Also,

$$E(\mathbf{X}) = (\tfrac{1}{12})(1,000) = 83.33 \qquad \text{and} \qquad \sigma_\mathbf{X} = \frac{69.28}{\sqrt{12}} = 20$$

From (18), the reorder point for an 80% value of SLM_1 must satisfy

$$NL\left(\frac{r - 83.33}{20}\right) = \frac{100(1 - 0.80)}{20} = 1$$

From Table 13, we find that 1 exceeds any of the tabulated values of the normal loss function. Thus, the value of r must make $\frac{r - 83.33}{20}$ a negative number. A little trial and error reveals that $NL(-0.9) = NL(0.9) + 0.9 = 1.004$. Hence,

$$\frac{r - 83.33}{20} = -0.9$$

$$r = 83.33 - 20(0.9) = 65.33$$

For $SLM_1 = 0.90$, Equation (18) shows that the reorder point must satisfy

$$NL\left(\frac{r - 83.33}{20}\right) = \frac{(1 - 0.90)100}{20} = 0.5$$

Again, 0.5 exceeds all tabulated values of the normal loss function. Hence, $\frac{r - 83.33}{20}$ must be a negative number. A little trial and error reveals that $N(-0.19) = N(0.19) + 0.19 = 0.5011$. Thus, the reorder point for a 90% service level must satisfy

$$\frac{r - 83.33}{20} = -0.19$$

$$r = 83.33 - 20(0.19) = 79.53$$

A 90% service level can be attained by a reorder point that is less than the expected lead time demand.

To attain a 95% service level, r must satisfy

$$NL\left(\frac{r - 83.33}{20}\right) = \frac{(1 - 0.95)100}{20} = 0.25$$

Since $NL(0.34) = 0.2518$,

$$\frac{r - 83.33}{20} = 0.34$$

$$r = 83.33 + 20(0.34) = 90.13$$

For a 99% service level,

$$NL\left(\frac{r - 83.33}{20}\right) = \frac{(1 - 0.99)100}{20} = 0.05$$

Since $NL(1.25) = 0.0506$, we see that

$$\frac{r - 83.33}{20} = 1.25$$

$$r = 83.33 + 20(1.25) = 108.33$$

TABLE 14
Reorder Points for Various
Service Levels

SLM$_1$	Reorder Point
80%	65.33
90%	79.53
95%	90.13
99%	108.33
99.9%	127.13

Finally, for a 99.9% service level, r must satisfy

$$\frac{r - 83.33}{20} = \frac{(1 - 0.999)100}{20} = 0.005$$

Since $NL(2.19) = 0.005$,

$$\frac{r - 83.33}{20} = 2.19$$

$$r = 83.33 + 20(2.19) = 127.13$$

In summary, the reorder points corresponding to the various values of SLM_1 are given in Table 14. Notice that to go from an 80% to a 90% service level, we must increase the reorder point by 14.20, but to go from a 90% to a 99.9% service level, the reorder point must be increased by 47.60. For higher service levels, a much greater increase in the reorder point is required to cause a commensurate increase in the service level.

Using LINGO to Compute the Reorder Point Level for *SLM*$_1$

Using the **@PSL** function in LINGO, it is a simple matter to compute the reorder point level for SLM_1. For example, to compute the reorder point for Example 7 corresponding to $SLM_1 = .90$ in LINGO, we would use the program

```
MODEL:
1) @PSL((R - 83.33)/20) = 100*(1 - SLM1)/20;
2) SLM1 = .9;
```

This program yields $r = 79.57$. Note that by altering the right-hand side of line 2 we can quickly compute the reorder points for various values of SLM_1.

Using Excel to Compute the Normal Loss Function

It can be shown that

$$NL(y) = \text{(height of normal density at } y) - y*(\text{probability standard normal is greater than or equal to } y)$$

In the file Normalloss.xls, we therefore compute $NL(y)$ with the Excel formula

$$=\text{NORMDIST(D3,0,1,0)-D3*(1-NORMSDIST(D3))}$$

Normalloss.xls

FIGURE 5

	B	C	D	E
1		Computing		
2		Normal Loss Function		
3		y	2	
4		NL(y)	0.008491	
5				
6				

FIGURE 6

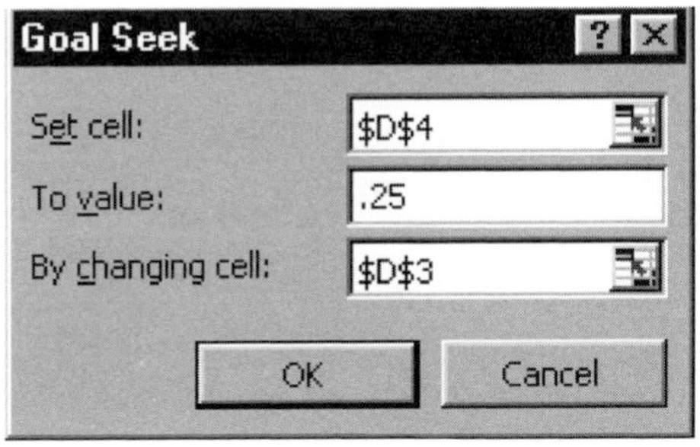

Goal Seek ⸤?⸥⸤✕⸥

Set cell: D4

To value: .25

By changing cell: D3

OK   Cancel

FIGURE 7

	B	C	D	E
1		Computing		
2		Normal Loss Function		
3		y	0.344868	
4		NL(y)	0.25	
5				
6				

Recall that NORMDIST with last argument 0 computes the density function for a normal random variable, and NORMSDIST( ) computes the standardized normal cumulative probability. For example, we see from Figure 5 that (consistent with Table 14) $NL(2) = .008491$.

To illustrate the use of this spreadsheet, recall that in Example 7 we needed to find a value of y such that $NL(y) = .25$. To do this, we use Excel Goal Seek and fill in the Goal Seek dialog box as shown in Figure 6. This tells Excel to change cell D3 until cell D4 (the normal loss value) reaches .25. The result in Figure 7 shows us that $NL(.345) = .25$. Before doing Goal Seek, you should go to Tools Options Calculation Iteration and change the Maximum Change box to a very small number, such as .0000001. This makes Excel force the Set cell within .000001 of its desired value.

Determination of Reorder Point and Safety Stock Level for SLM_2

Suppose that a manager wants to hold sufficient safety stock to ensure that an average of s_0 cycles per year will result in a stockout. Given a reorder point of r, a fraction $P(\mathbf{X} > r)$ of all cycles will lead to a stockout. Since an average of $\frac{E(\mathbf{D})}{q}$ cycles per year will occur (remember we are assuming backlogging), an average of $\frac{P(\mathbf{X} > r)E(\mathbf{D})}{q}$ cycles per year will result in a stockout. Thus, given s_0, the reorder point is the smallest value of r satisfying

$$\frac{P(\mathbf{X} > r)\, E(\mathbf{D})}{q} \leq s_0 \qquad \text{or} \qquad P(\mathbf{X} > r) \leq \frac{s_0 q}{E(\mathbf{D})}$$

If X is a continuous random variable, then $P(X > r) = P(X \geq r)$. Thus, we obtain the re-order point r for SLM_2 for continuous lead time demand,

$$P(X \geq r) = \frac{s_0 q}{E(D)} \qquad (19)$$

and the reorder point for SLM_2 for discrete lead time demand, by choosing the smallest value of r satisfying

$$P(X > r) \leq \frac{s_0 q}{E(D)} \qquad (19')$$

To illustrate the determination of the reorder point for SLM_2, we suppose that Bads, Inc., wants to ensure that stockouts occur during an average of two lead times per year. Recall from Example 7 that EOQ $= 100$, $E(D) = 1,000$ units per year, and X is $N(83.33, 400)$. Now (19) yields $P(X \geq r) = \frac{2(100)}{1,000} = .2$. The reorder point r is calculated using Excel. Since

$$=\text{NORMINV}(.8, 83.33, 20)$$

yields 100.16, we find that $r = 100.16$. The safety stock level yielding an average of two stockouts per year would be $100.16 - E(X) = 16.83$.

PROBLEMS

Group A

1 For Problem 1 of Section 4.6, determine the reorder point that yields 80%, 90%, 95%, and 99% values of SLM_1. What reorder point would yield an average of 0.5 stockout per year?

2 For Problem 2 of Section 4.6, determine the reorder point that yields 80%, 90%, 95%, and 99% values of SLM_1. What reorder point would yield an average of two stockouts per year?

3 Suppose that the EOQ is 100, average annual demand is 1,000 units, and the lead time demand is a random variable having the distribution shown in Table 15.

    a What value of SLM_1 corresponds to a reorder point of 25?

    b If we wanted to attain a 95% value of SLM_1, what reorder point should we choose?

    c If we wanted an average of at most two stockouts per year, what reorder point should we choose?

TABLE 15

Lead Time Demand	Probability
10	$\frac{1}{6}$
15	$\frac{1}{4}$
20	$\frac{1}{4}$
25	$\frac{1}{12}$
30	$\frac{1}{4}$

4 A firm experiences demand with a mean of 100 units per day. Lead time demand is normally distributed, with a mean of 1,000 units and a standard deviation of 200 units. It costs $6 to hold one unit for one year. If the firm wants to meet 90% of all demand on time, what will be the annual cost of holding safety stock? (Assume that each order costs $50.)

4.8 (R, S) Periodic Review Policy[†]

In this section, we describe a widely used periodic review policy: the (R, S) policy. Before describing the operation of this policy, we need to define the concept of **on-order inventory level.** The on-order inventory level is simply the sum of on-hand inventory and inventory on order. Thus, if 30 units of a product are on hand, and we order 70 units (with a lead time of, say, one month), our on-order inventory level is 100.

[†]This section covers topics that may be omitted with no loss of continuity.

We can now describe the operation of the (R, S) inventory policy. Every R units of time (say, years), we review the on-hand inventory level and place an order to bring the on-order inventory level up to S. For example, if we were using a $(.25, 100)$ policy, we would review the inventory level at the end of each quarter. If $i < 100$ units were on hand, an order for $100 - i$ units would be placed. In general, an (R, S) policy will incur higher holding costs than a cost-minimizing (r, q) policy, but an (R, S) policy is usually easier to administer than a continuous review policy. With an (R, S) policy (unlike a continuous review policy), we can predict with certainty the times when an order will be placed. An (R, S) policy also allows a company to coordinate replenishments. For example, a company could use $R = 1$ month for all products ordered from the same supplier and then order all products from that supplier on the first day of each month.

We now assume that the review interval R has been determined and focus on the determination of a value for S that will minimize expected annual costs. Later in this section, we will discuss how to determine an appropriate value for R. We now assume that all shortages are backlogged and demand is a continuous random variable whose distribution remains unchanged over time. Finally, we assume that the per-unit purchase price is constant. This implies that annual purchasing costs do not depend on our choice of R and S. We define

$$R = \text{time (in years) between reviews}$$
$$\mathbf{D} = \text{demand (random) during a one-year period}$$
$$E(\mathbf{D}) = \text{mean demand during a one-year period}$$
$$K = \text{cost of placing an order}$$
$$J = \text{cost of reviewing inventory level}$$
$$h = \text{cost of holding one item in inventory for one year}$$
$$c_B = \text{cost per-unit short in the backlogged case (assumed to be}$$
$$\text{independent of the length of time until the order is filled)}$$
$$L = \text{lead time for each order (assumed constant)}$$
$$\mathbf{D}_{L+R} = \text{demand (random) during a time interval of length } L + R$$
$$E(\mathbf{D}_{L+R}) = \text{mean of } \mathbf{D}_{L+R}$$
$$\sigma_{\mathbf{D}_{L+R}} = \text{standard deviation of } \mathbf{D}_{L+R}$$

Given a value of R, we can now determine a value of S that minimizes expected annual costs. Our derivation mimics the derivation of (13). For a given choice of R and S, our expected costs are given by

(Annual expected purchase costs) + (annual review costs)

+ (annual ordering costs) + (annual expected holding costs)

+ (annual expected shortage costs)

Since $\frac{1}{R}$ reviews per year are placed, annual review costs are given by $\frac{J}{R}$. Also note that whenever an order is placed, the on-order inventory level will equal S. The only way that an order will not be placed at the next review point is if $\mathbf{D}_{L+R} = 0$. Since $\mathbf{D}_{L+R}$ is a continuous random variable, $\mathbf{D}_{L+R} = 0$ will occur with zero probability. Thus, an order is sure to be placed at the next review point (or any review point). This implies that annual ordering cost is given by $K(\frac{1}{R}) = \frac{K}{R}$. Observe that both the annual ordering cost and the review cost are independent of S. Thus, the value of S that minimizes annual expected costs will be the value of S that minimizes (annual expected holding costs) + (annual expected shortage costs).

To determine the annual expected holding cost for a given (R, S) policy, we first define a cycle to be the time interval between the arrival of orders. If we can determine the expected value of the average inventory level over a cycle, then expected annual holding cost is just h(expected value of on-hand inventory level over a cycle). As in our derivation of (11), we now assume that the average number of back orders is small relative to the average on-hand inventory level. Then, as in Section 4.6,

$$\text{Expected value of } I(t) \cong \text{expected value of } OHI(t)$$

Then expected value of $I(t)$ over a cycle may be approximated by 0.5(expected value of $I(t)$ right before an order arrives) + 0.5(expected value of $I(t)$ right after an order arrives).

Right before an order arrives, our maximum on-order inventory level (S) has been reduced by an average of $E(\mathbf{D}_{L+R})$. Thus, expected value of $I(t)$ right before an order arrives = $S - E(\mathbf{D}_{L+R})$.

Since $\frac{1}{R}$ orders are placed each year and an average of $E(\mathbf{D})$ units must be ordered each year, the average size of an order is $E(\mathbf{D})R$. Thus,

$$\text{Expected value of } I(t) \text{ right after an order arrives} = S - E(\mathbf{D}_{L+R}) + E(\mathbf{D})R$$

Then

$$\text{Expected value of } I(t) \text{ during a cycle} = S - E(\mathbf{D}_{L+R}) + \frac{E(\mathbf{D})R}{2}$$

Thus,

$$\text{Expected annual holding cost} = h\left[S - E(\mathbf{D}_{L+R}) + \frac{E(\mathbf{D})R}{2}\right]$$

From this expression, it follows that increasing S to $S + \Delta$ will increase expected annual holding costs by $h\Delta$.

We now focus on how an increase in S to $S + \Delta$ affects expected annual shortage costs. Then we can use marginal analysis to find the value of S that minimizes the sum of annual expected holding and shortage costs. Let's define the shortages "associated" with each order to be the shortages occurring in the time interval between the arrival of the order and the arrival of the next order. For example, an order placed at time 0 arrives at time L, and the next order will not arrive until time $R + L$. Thus, all shortages occurring between L and $R + L$ are associated with the time 0 order. Clearly, the sum of all shortages will equal the sum of the shortages associated with all orders. Let's again focus on the shortages associated with the time 0 order. Since the next order arrives at time $R + L$, and our time 0 order brought the on-order inventory level up to S, a shortage will be associated with the time 0 order if and only if the demand between time 0 and $R + L$ exceeds S. If a shortage occurs, the magnitude of the shortage will equal $\mathbf{D}_{L+R} - S$.

We can now use marginal analysis to determine (for a given R) the value of S that minimizes the sum of annual expected holding and shortage costs. If we increase S to $S + \Delta$, annual expected holding costs increase by $h\Delta$. Increasing S to $S + \Delta$ will decrease shortages associated with an order if $\mathbf{D}_{L+R} \geq S$. Thus, for a fraction $P(\mathbf{D}_{L+R} \geq S)$ of all orders, increasing S to $S + \Delta$ will save $c_B\Delta$ in shortage costs. Since $\frac{1}{R}$ orders are placed each year, increasing S to $S + \Delta$ will reduce expected annual shortage costs by $(\frac{1}{R})c_B\Delta P(\mathbf{D}_{L+R} \geq S)$. Marginal analysis then implies that the value of S minimizing the sum of annual expected holding and shortage costs will occur for the value of S satisfying

$$h\,\Delta = (\tfrac{1}{R})c_B\Delta P(\mathbf{D}_{L+R} \geq S)$$

or

$$P(\mathbf{D}_{L+R} \geq S) = \frac{Rh}{c_B} \tag{20}$$

Suppose that all shortages result in lost sales, and a cost of c_{LS} (including shortage cost plus lost profit) is incurred for each lost sale. Then the value of S minimizing the sum of annual expected holding and shortage costs is given by

$$P(\mathbf{D}_{L+R} \geq S) = \frac{Rh}{Rh + c_{LS}} \qquad (21)$$

The following example illustrates the use of (20).

EXAMPLE 8   **Lowland Appliance**

Lowland Appliance replenishes its stock of color TVs three times a year. Each order takes $\frac{1}{9}$ year to arrive. Annual demand for color TVs is $N(990, 1{,}600)$. The cost of holding one color TV in inventory for one year is \$100. Assume that all shortages are backlogged, with a shortage cost of \$150 per TV. When Lowland places an order, what should the on-order inventory be?

Solution   We are given that $R = \frac{1}{3}$ year, $L = \frac{1}{9}$ year, $R + L = \frac{4}{9}$ year, and $c_B = \$150$. $\mathbf{D}_{L+R}$ is normally distributed, with $E(\mathbf{D}_{L+R}) = \frac{4}{9}(990) = 440$ and $\sigma_{\mathbf{D}_{L+R}} = \sqrt{\frac{4}{9}}\sqrt{1{,}600} = 26.67$. From (20), S should be chosen to satisfy

$$P(\mathbf{D}_{L+R} \geq S) = \frac{\left(\frac{1}{3}\right)100}{150} = .22$$

We use the Excel function NORMINV to compute s. Since

$$=\text{NORMINV}(0.78, 440, 26.67)$$

yields 460.59, when Lowland places an order for TVs, it should order enough to bring the on-order inventory level up to 460.59 (or 461) TVs. For example, if 160 TVs are in stock when a review takes place, $461 - 160 = 301$ TVs should be ordered.

Determination of R

Often, the review interval R is set equal to $\frac{\text{EOQ}}{E(\mathbf{D})}$. This makes the number of orders placed per year equal the number recommended if a simple EOQ model were used to determine the size of orders. Since each order is accompanied by a review, however, we must set the cost per order to $K + J$. This yields

$$\text{EOQ} = \sqrt{\frac{2(K+J)E(\mathbf{D})}{h}}$$

To illustrate the idea, suppose that it costs \$500 to review the inventory level and \$5,000 to place an order for TVs. Then

$$\text{EOQ} = \sqrt{\frac{2(5{,}500)(990)}{100}} = 330$$

This implies a review interval $R = \frac{330}{990} = \frac{1}{3}$ year.

Implementation of an (R, S) System

Retail stores (such as J. C. Penney's) often find an (R, S) policy easy to implement, because the quantity ordered equals the number of sales occurring during the period between reviews. For example, suppose a (1 month, 1,000) policy is being used, and orders are placed on the first day of each month. If 800 items were sold during January, then an order of 800 items must be placed at the beginning of February to bring the on-order inventory level back up to 1,000. By programming a computer to set monthly orders equal to monthly sales, such a policy can easily be implemented.

PROBLEMS

Group A

1 A hospital must order the drug Porapill from Daisy Drug Company. It costs $500 to place an order and $30 to review the hospital's inventory of the drug. Annual demand for the drug is $N(10,000, 640,000)$, and it costs $5 to hold one unit in inventory for one year. Orders arrive one month after being placed. Assume that all shortages are backlogged.

    a Estimate R and the number of orders per year that should be placed.

    b Using the answer in part (a), determine the optimal (R, S) inventory policy. Assume that the shortage cost per unit of the drug is $100.

2 Chicago's Treadway Tires Dealer must order tires from its national warehouse. It costs $10,000 to place an order and $400 to review the inventory level. Annual tire sales are $N(20,000, 4,000,000)$. It costs $10 per year to hold a tire in inventory, and each order arrives two weeks after being placed (52 weeks = 1 year). Assume that all shortages are backlogged.

    a Estimate R and the number of orders per year that should be placed.

    b Using the answer in part (a), determine the optimal (R, S) inventory policy. Assume that the shortage cost is $100 per tire.

3 Suppose we have found the optimal (R, S) policy for the back-ordered case and that $S = 50$. Is the following true or false?

    The optimal S for the lost sales case has $S > 50$.

4.9 The ABC Inventory Classification System

Many companies must develop inventory policies for thousands of items. In such a situation, a company cannot devote a great deal of attention to determining an "optimal" inventory policy for each item. The **ABC classification,** devised at General Electric during the 1950s, helps a company identify a small percentage of its items that account for a large percentage of the dollar value of annual sales. These items are called Type A items. Since most of the firm's inventory investment is in Type A items, concentrating effort on developing effective inventory control policies for these items should produce substantial savings.

Repeated studies have shown that in most companies, 5%–20% of all items stocked account for 55%–65% of sales; these are the Type A items. It has also been found that 20%–30% of all items account for 20%–40% of sales; these are called Type B items. Finally, it is often found that 50%–75% of all items account for only 5%–25% of sales; these are called Type C items. To illustrate how we determine which items are Type A, Type B, and Type C, consider a firm that stocks 100 items. We reorder the items as item 1, item 2, . . . , item 100, where item 1 generates the largest annual sales volume, item 2 generates the second largest annual sales volume, and so on. Then we plot the points (k, percentage of annual sales due to top k% of all items). For example, the point (20, 60) indicates that the top 20 items (from the standpoint of dollar sales) generate 60% of all sales. We then obtain a graph like Figure 8, where items 1–20 are Type A items, items 21–40 are Type B items, and items 41–100 are Type C items.

Since most of our inventory investment is in Type A items, high service levels will result in huge investments in safety stocks. Therefore, Hax and Candea (1984) recommend that SLM_1 be set at only 80%–85% for Type A items. Tight management control of ordering procedures is essential for Type A items; individual demand forecasts should be made for each Type A item. Also, every effort should be made to lower the lead time needed to receive orders or produce the item. If an (R, S) policy is used, R should be small—perhaps one week. This enables us to keep a close watch on inventory levels. Parameters such as estimates of annual mean demand, length of lead time, standard deviation of annual demand, and shortage costs should be reviewed fairly often.

For Type B items, Hax and Candea (1984) recommend that SLM_1 be set at 95%. Inventory policies for Type B items can generally be controlled by computer. Parameters for

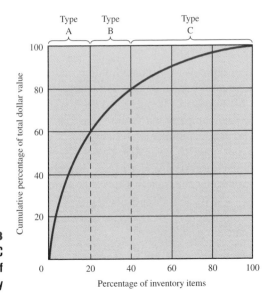

FIGURE **8**
**Example of ABC
Classification of
Inventory**

Type B items should be reviewed less often than for Type A items.

For Type C items, a simple two-bin system is usually adequate. Parameters may be reviewed once or twice a year. Demand for Type C items may be forecast by simple extrapolation methods. A high value of SLM_1 (usually 98%–99%) is recommended. Little extra investment in safety stock will be required to maintain these high service levels.

DEVRO Incorporated, a producer of edible sausage casings, implemented an ABC analysis of its spare parts inventory and found that 2.5% of all items (the Type A items) accounted for 49% of all dollar usage, and 24.7% of all items (the Type B items) accounted for 38% of all dollar usage. By preparing requisition forms in advance for Type A and Type B items, DEVRO was able to substantially reduce the lead time needed to obtain those items. This helped DEVRO effect substantial savings in annual inventory costs. See Flowers and O'Neill (1978) for details.

PROBLEMS

Group A

1 Develop an ABC graph for the data in Table 16. Which items should be classified A, B, and C?

TABLE **16**

Item	Annual Usage	Unit Cost (in dollars)
1	20,000	20
2	23,000	10
3	20,000	3
4	30,000	2
5	5,000	10
6	10,000	7
7	1,000	30
8	2,000	15
9	3,000	10
10	5,000	6

4.10 Exchange Curves

In many situations, it is difficult to estimate holding and shortage costs accurately. **Exchange curves** can be used in such situations to identify "reasonable" inventory policies. Consider a company that stocks two items (1 and 2). Many different ordering policies are possible. For example, the company may order item 1 five times a year and item 2 ten times a year (policy 1), or it may order each item once per year (policy 2). Clearly, policy 1 will result in higher ordering costs than policy 2, but policy 2 will result in higher holding costs and a higher average inventory level than policy 1. An exchange curve enables us to display graphically the trade-off between annual ordering costs and average inventory investment.

To illustrate the construction of an exchange curve, suppose a company stocks two items (item 1 and 2), and suppose that

c_i = cost of purchasing each unit of product i

h = cost of holding \$1 worth of either product in inventory for one year

K_i = order cost for product i

q_i = EOQ for product i

D_i = annual demand for product i

Then

$$q_i = \sqrt{\frac{2K_iD_i}{hc_i}}$$

Suppose the company wants to minimize the sum of annual ordering and holding costs. Then it should follow an EOQ policy for each product and order q_i of product i $\dfrac{D_i}{q_i}$ times per year. Two measures of effectiveness for this (or any other) ordering policy are

AII = average dollar value of inventory cost

AOC = annual ordering cost

If we follow the EOQ policy for each product, then

$$\text{AII} = \left(\frac{q_1}{2}\right)c_1 + \left(\frac{q_2}{2}\right)c_2$$

$$= \left(\frac{1}{2}\right)\left\{c_1\sqrt{\frac{2K_1D_1}{c_1h}} + c_2\sqrt{\frac{2K_2D_2}{c_2h}}\right\}$$

$$= \left(\frac{\sqrt{2}}{2\sqrt{h}}\right)\{\sqrt{K_1D_1c_1} + \sqrt{K_2D_2c_2}\}$$

$$\text{AOC} = K_1\left(\frac{D_1}{q_1}\right) + K_2\left(\frac{D_2}{q_2}\right)$$

$$= K_1D_1\sqrt{\frac{c_1h}{2K_1D_1}} + K_2D_2\sqrt{\frac{c_2h}{2K_2D_2}}$$

$$= \left(\frac{\sqrt{2h}}{2}\right)\{\sqrt{K_1D_1c_1} + \sqrt{K_2D_2c_2}\}$$

The expression for AII follows from the fact that the average inventory level of an item equals half the order quantity. The expression for AOC follows from the fact that $\dfrac{D_i}{q_i}$ orders per year are placed for item i.

Since h is often hard to estimate, let's suppose h is unknown and look at how a change in h affects AII and AOC. A plot of the points (AOC, AII) associated with each value of h is known as an **exchange curve.** For any point on the exchange curve, we see that

$$\text{AII(AOC)} = (\tfrac{1}{2})\{\sqrt{K_1 D_1 c_1} + \sqrt{K_2 D_2 c_2}\}^2 \tag{21}$$

This shows that the exchange curve is a hyperbola. Also, any point on the exchange curve satisfies $\frac{\text{AII}}{\text{AOC}} = \frac{1}{h}$ or $\frac{\text{AOC}}{\text{AII}} = h$. Thus, for any point on the exchange curve, the annual holding cost per dollar of inventory is the ratio of the x-coordinate to the y-coordinate. This shows how each point on the exchange curve can be identified with a value of h.

We now illustrate the computation of an exchange curve and show how the exchange curve can be used as an aid in decision making.

EXAMPLE 9   Exchange Curve

A company stocks two products. Relevant information is given in Table 17.

1 Draw an exchange curve.

2 Currently, the company is ordering each product ten times per year. Use the exchange curve to demonstrate to management that this is an unsatisfactory ordering policy.

3 Suppose that management limits the company's average inventory investment to $10,000. Use the exchange curve to determine an appropriate ordering policy.

Solution **1** From (21), we find the equation of the exchange curve to be

$$(\text{AII})(\text{AOC}) = (\tfrac{1}{2})\{\sqrt{50(10,000)(200)} + \sqrt{80(20,000)(2.5)}\}^2$$
$$= 72,000,000$$

Some representative points on the exchange curve, along with the associated value of h, are given in Table 18. The exchange curve is graphed in Figure 9.

2 If the company orders each product ten times per year,

$$\text{AOC} = 10(\$50) + 10(\$80) = \$1,300$$
$$\text{AII} = \tfrac{1}{2}(1,000)(\$200) + \tfrac{1}{2}(2,000)(\$2.50) = \$102,500$$

TABLE 17
Relevant Information for Example 9

	K_i	D_i	c_i
Product 1	$50	10,000	$200
Product 2	$80	20,000	$2.50

TABLE 18
Points on Exchange Curve

AOC	AII	h
$2,000	$36,000	.06
$3,000	$24,000	.13
$4,000	$18,000	.22
$5,000	$14,400	.35
$6,000	$12,000	.50
$8,000	$9,000	.89

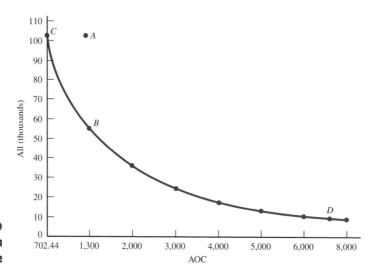

FIGURE 9
Example of an
Exchange Curve

This is point A in Figure 9. Observe that point $B = (1{,}300, 55{,}385)$, corresponding to $h = .02$, yields the same AOC as the current policy, but a much lower AII. Also, point $C = (702.44, 102{,}500)$, corresponding to $h = .01$, yields the same AII as the current policy, but a much lower AOC. Thus, we can use the exchange curve to show the manager how to improve on the current ordering policy.

3   From the exchange curve, we find that $D = (7{,}200, 10{,}000)$ is on the exchange curve. Thus, for a \$10,000 AII, the best we can do is to hold ordering costs to \$7,200. Of course, the manager could opt for AII = \$9,000 and AOC = \$8,000 or one of many other possibilities. The point is that the exchange curve clarifies many of the options available to management.

Exchange Curves for Stockouts

Exchange curves can also be used to assess the trade-offs between average inventory investment (AII) and the expected number of lead times per year resulting in stockouts. To illustrate, consider a company stocking a single item for which

$$c = \text{purchase cost per unit}$$
$$K = \text{setup cost}$$
$$h = \text{annual cost of holding one unit in inventory}$$
$$c_B = \text{cost of a stockout (we assume all items are back-ordered)}$$
$$E(\mathbf{D}) = \text{mean annual demand}$$
$$q = \text{economic order quantity}$$
$$\mathbf{X} = \text{lead time demand}$$
$$E(\mathbf{X}) = \text{mean lead time demand}$$
$$\sigma_{\mathbf{X}} = \text{standard deviation of lead time demand}$$
$$r = \text{reorder point (determined from Equation (13))}$$

From (13), a fraction

$$\frac{qh}{c_B E(\mathbf{D})}$$

of all lead times will have a stockout. Since there are an average of $\frac{E(\mathbf{D})}{q}$ orders placed per year, an average of

$$\left(\frac{E(\mathbf{D})}{q}\right)\left(\frac{qh}{c_B E(\mathbf{D})}\right) = \frac{h}{c_B}$$

lead times per year will result in stockouts. We let SY = expected number of lead times per year resulting in stockouts. From (11), we know that the average inventory level is $(\frac{q}{2} + r - E(\mathbf{X}))$. Thus, we have AII $= c(\frac{q}{2} + r - E(\mathbf{X}))$.

An exchange curve for this situation is a graph of the points (AII, SY) corresponding to different values of c_B. To illustrate the construction of an exchange curve, let $E(\mathbf{X}) = 200$, $\sigma_X = 50$, $E(\mathbf{D}) = 100{,}000$, $K = \$12.50$, $h = \$10$, and $c = \$100$. We will find four points on the exchange curve by setting $c_B = \$1, \$5, \$10$, and $\$20$. First we find that

$$q = \sqrt{\frac{2(12.5)(100{,}000)}{10}} = 500$$

The stockout probabilities and SY are given in Table 19.

Using Table 4 in Chapter 1 or the Excel NORMSDIST( ) function, we can calculate the reorder point r for each value of c_B. Then we determine the average inventory level and AII = average inventory investment. These calculations are given in Table 20.

The exchange curve (based on the four points we have computed) is graphed in Figure 10. For example, the exchange curve shows us that if current AII is \$33,250, then for a \$3,400 increase in AII, we can reduce SY from 10 to 2, but an additional increase in AII of \$3,400 would decrease SY by less than 2.

Exchange Surfaces

Using more sophisticated techniques (see Gardner and Dannenbring (1979)), an **exchange surface** involving three or more quantities can be derived. The exchange surface in Figure 11 was derived from a sample of 500 items in a military distribution system. The

TABLE 19
Computation of SY

c_B	Stockout Probability $= \dfrac{qh}{c_B E(\mathbf{D})}$	$SY = \dfrac{h}{c_B}$
\$1	$\frac{500(10)}{1(100{,}000)} = .05$	$\frac{10}{1} = 10$
\$5	$\frac{500(10)}{5(100{,}000)} = .01$	$\frac{10}{5} = 2$
\$10	$\frac{500(10)}{10(100{,}000)} = .005$	$\frac{10}{10} = 1$
\$20	$\frac{500(10)}{20(100{,}000)} = .0025$	$\frac{10}{20} = 0.50$

TABLE 20
Calculation of AII

c_B	Reorder Point	Average Inventory Level	AII
\$1	$200 + 50(1.65) = 282.5$	$250 + 282.5 - 200 = 332.5$	\$33,250
\$5	$200 + 50(2.33) = 316.5$	$250 + 316.5 - 200 = 366.5$	\$36,650
\$10	$200 + 50(2.58) = 329$	$250 + 329 \ \ - 200 = 379$	\$37,900
\$20	$200 + 50(2.81) = 340.5$	$250 + 340.5 - 200 = 390.5$	\$39,050

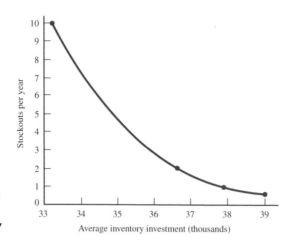

FIGURE 10
Exchange Curve for All
and _SY_

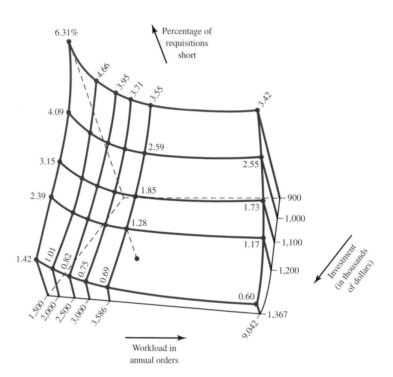

FIGURE 11
Example of an
Exchange Surface[†]

x-coordinate is the annual number of orders placed, the *y*-coordinate is the average inventory investment (in thousands of dollars), and the *z*-coordinate is the percentage of requests that yield shortages. For example, suppose the military has fixed a $900,000 average inventory investment. By varying the number of orders per year between 1,500 and 9,042, the military can vary the percentage of requests that yield shortages between 6.31% and 3.42%. Also, if annual orders are fixed at 3,000, then the percentage of requests yielding shortages can vary between 0.75% and 3.71%. An exchange surface makes it easy to identify the trade-offs involved between improving service, increased inventory investment, and increased work load (orders per year).

[†]Reprinted by permission of E. Gardner and D. Dannenbring, "Using Optimal Policy Surfaces to Analyze Aggregate Inventory Tradeoffs," *Management Science,* Vol. 25, No. 8, August 1979. Copyright 1979, the Institute of Management Sciences.

PROBLEMS

Group A

1 Consider a two-item inventory system with the attributes in Table 21.

   a Draw an exchange curve for these products (use AOC and AII as the x- and y-coordinates).

   b Currently, management is ordering each product twice a year. How can it improve on this strategy?

   c The order costs correspond to machine setup times. Machine time is valued at $50 per hour. If management wants to limit machine setup time to 500 hours per year, what strategies are available?

2 Explain how to draw an exchange curve where the x-coordinate is AII and the y-coordinate is percentage of all requests for stock that result in shortages.

3 Consider the exchange surface in Figure 11. The current inventory policy has yielded 3,586 orders per year, an AII of $1,367,000, and 0.89% shortages.

   a Without changing orders per year and AII, by how much can shortages be improved?

   b If AII and shortages are maintained at current levels, by how much can orders per year be reduced?

   c If shortages and orders per year are maintained at current levels, by how much can AII be reduced?

TABLE 21

	K_i	D_i	c_i
Product 1	$500	10,000	$2,000
Product 2	$800	20,000	$250

SUMMARY   Single-Period Decision Models

A decision maker begins by choosing a value q of a decision variable. Then a random variable $\mathbf{D}$ assumes a value d. Finally, a cost $c(d, q)$ is incurred. The decision maker's goal is to choose q to minimize expected cost.

News Vendor Problem

If $c(d, q)$ has the structure

$$c(d, q) = c_o q + \text{(terms not involving } q) \qquad (d \leq q) \qquad (2)$$
$$c(d, q) = -c_u q + \text{(terms not involving } q) \qquad (d \geq q + 1) \qquad (2.1)$$

the single-period decision model is a **news vendor problem.** Here

$$c_o = \text{per-unit overstocking cost}$$
$$c_u = \text{per-unit understocking cost}$$

If $\mathbf{D}$ is a discrete random variable, the optimal decision is given by the smallest value of q (q^*) satisfying

$$F(q^*) \geq \frac{c_u}{c_o + c_u} \qquad (3)$$

If $\mathbf{D}$ is a continuous random variable, the optimal decision is the value of q (q^*) satisfying

$$P(\mathbf{D} \leq q^*) = \frac{c_u}{c_o + c_u} \qquad (5)$$

Determination of Reorder Point and Order Quantity with Uncertain Demand: Minimizing Annual Expected Cost

Let

K = ordering cost

h = holding cost/unit/year

L = lead time for each order (assumed to be known with certainty)

q = order quantity

$\mathbf{D}$ = random variable representing annual demand, with mean $E(\mathbf{D})$, variance var $\mathbf{D}$, and standard deviation $\sigma_{\mathbf{D}}$

c_B = cost incurred for each unit short if shortages are backlogged

c_{LS} = cost (including lost profits, lost goodwill) incurred for each lost sale if each shortage results in a lost sale

$\mathbf{X}$ = random variable representing lead time demand

Then

$$E(\mathbf{X}) = LE(\mathbf{D}), \qquad \text{var } \mathbf{X} = L \text{ (var } \mathbf{D}), \qquad \sigma_{\mathbf{X}} = \sqrt{L}\sigma_{\mathbf{D}}$$

and r is the reorder point, or inventory level at which an order should be placed. Safety stock, $r - E(\mathbf{X})$, is the amount of inventory held in excess of lead time demand to meet shortages that may occur before an order arrives.

Assume that the optimal order quantity can be reasonably approximated by the EOQ, $\mathbf{D}$ is a continuous random variable, and all shortages are backlogged. Then annual expected cost is minimized by q^* and r^* given by

$$q^* = \left(\frac{2KE(\mathbf{D})}{h} \right)^{1/2}$$

$$P(\mathbf{X} \geq r^*) = \frac{hq^*}{c_B E(\mathbf{D})}$$

(13)

Assume that the optimal order quantity can be reasonably approximated by the EOQ, $\mathbf{D}$ is a continuous random variable, and all shortages result in lost sales. Then annual expected cost is minimized by q^* and r^* satisfying

$$q^* = \left(\frac{2KE(\mathbf{D})}{h} \right)^{1/2}$$

$$P(\mathbf{X} \geq r^*) = \frac{hq^*}{hq^* + c_{LS}E(\mathbf{D})}$$

(15)

Determination of Reorder Point: The Service Level Approach

Since it may be difficult to determine the exact cost of a shortage or lost sale, it is often desirable to choose a reorder point that meets a desired service level. Two common measures of service level are

Service Level Measure 1 SLM_1, the expected fraction (usually expressed as a percentage) of all demand that is met on time.

Service Level Measure 2 SLM_2, the expected number of cycles per year during which a shortage occurs.

If lead time is normally distributed, then for a desired value SLM_1, the reorder point r is found from

$$NL\left(\frac{r - E(\mathbf{X})}{\sigma_{\mathbf{X}}}\right) = \frac{q(1 - SLM_1)}{\sigma_{\mathbf{X}}} \qquad (18)$$

where $NL(y)$ is the **normal loss function,** tabulated in Table 13, and q is the EOQ.

If lead time demand is a continuous random variable, and we desire $SLM_2 = s_0$ shortages per year, the reorder point r is given by

$$P(\mathbf{X} \geq r) = \frac{s_0 q}{E(\mathbf{D})} \qquad (19)$$

Again, q is the EOQ.

If lead time demand is a discrete random variable, and we desire $SLM_2 = s_0$ shortages per year, the reorder point is the smallest value of r satisfying

$$P(\mathbf{X} > r) \leq \frac{s_0 q}{E(\mathbf{D})} \qquad (19')$$

Again, q is the EOQ.

(R, S) Periodic Review Policy

Every R units of time, we review the inventory level and place an order to bring our on-hand inventory level up to S. Given a value of R, we determine the value of S from

$$P(\mathbf{D}_{L+R} \geq S) = \frac{Rh}{c_B}$$

ABC Classification

The 5%–20% of all items accounting for 55%–65% of sales are Type A items; the 20%–30% of all items accounting for 20%–40% of sales are Type B items; and the 50%–75% of all items that account for 5%–25% of all sales are Type C items. By concentrating effort on Type A (and possibly Type B) items, we can achieve substantial cost reductions.

Exchange Curves

Exchange curves (and exchange surfaces) are used to display trade-offs between various objectives. For example, an exchange curve may display the trade-off between annual ordering costs and average dollar level of inventory. An exchange curve can be used to compare how various ordering policies compare with respect to several objectives.

REVIEW PROBLEMS

Group A

1 The Chocochip Cookie Store bakes its cookies every morning before opening. It costs the store 15¢ to bake each cookie, and each cookie is sold for 35¢. At the end of the day, leftover cookies may be sold to a thrift bakery for 5¢ per cookie. The number of cookies sold each day is described by the discrete random variable in Table 22.

TABLE 22

Demand (dozens)	Probability
20	.30
30	.20
40	.20
50	.15
60	.15

TABLE 23

Cash Needs	Probability
$4,000	.30
$5,000	.20
$6,000	.10
$7,000	.30
$8,000	.10

a How many dozen cookies should be baked before the store opens?

b If the daily demand (in dozens) for cookies is $N(50, 400)$, how many dozen cookies should be baked? A description of the $N(\mu, \sigma^2)$ notation can be found in Section 1.7.

c If the daily demand (in dozens) for cookies has a density function

$$f(d) = \frac{e^{-d/50}}{50} \qquad (d \geq 0)$$

how many dozen cookies should be baked?

2 An optometrist orders eyeglass frames at a cost of $40 per frame and sells each frame for $70. Annual holding cost is 20% of the optometrist's cost of purchasing a frame. Each time frames are ordered, a cost of $200 is incurred. Because of lost goodwill, a cost of $50 is incurred each time a customer wants a frame that is not in stock. Frames are delivered one week after an order is placed. Annual demand for frames is $N(1,040, 15.73)$.

a Assuming all shortages are backlogged, determine the order quantity and reorder point.

b Assuming all shortages result in lost sales, determine the order quantity and reorder point.

c To meet 95% of all orders from stock, what should be the reorder point?

d To have shortages occur during an average of two lead times per year, what should be the reorder point?

3 We are given the following information about a product:

Cost of placing an order = $100
Cost per item = $5
Sale price per item = $8
Annual holding cost = 40% of cost of item
Annual demand = 5,000 units
Lead time demand = $N(20, 900)$

a If the reorder point that minimizes expected cost is 80, what is the shortage cost? (Assume backlogging.)

b If the reorder point that minimizes expected cost is 80, what is the shortage cost? (Assume lost sales.)

c What reorder point would meet 90% of all demand on time?

d What reorder point would result in a stockout occurring during an average of 0.5 lead time per year?

Group B

4 A business believes that its needs for cash during the next month are described by the random variable shown in Table 23. At the beginning of the month, the business has $10,000 available, and the business manager must determine how much of the money should be placed in an account bearing 24% annual interest. If any money must be withdrawn before the end of the month, all interest on the withdrawn money is forfeited, and a penalty equal to 2% of the withdrawn money must be paid. How much money should be placed in the 24% annual interest account?

5 A fur dealer buys fur coats for $100 each and sells them for $200 each. He believes that the demand for coats is $N(100, 100)$. Any coat not sold can be sold to a discount house for $100, but the fur dealer believes he must charge himself a cost of 10¢ per dollar invested in a fur coat that is sold at discount. How many coats should the dealer order? If the price at which the dealer sold his coats increased (assuming demand is unchanged), would he buy more or fewer coats?

6 A company currently has two warehouses. Each warehouse services half the company's demand, and the annual demand serviced by each warehouse is $N(10,000, 1,000,000)$. The lead time for meeting demand is $\frac{1}{10}$ year. The company wants to meet 95% of all demand on time. Assume that the EOQ at each warehouse is 2,000.

a How much safety stock must be held?

b Show that, if the company had only one warehouse, it would hold less safety stock than it does when it has two warehouses.

c A young MBA argues, "By having one central warehouse, I can reduce the total amount of safety stock needed to meet 95% of all customer demands on time. Therefore, we can save money by having only one central warehouse instead of several branch warehouses." How might this argument be rebutted?

7 Use LINGO to determine the values of q and r that minimize expected annual cost for Example 5. How close are your answers to those given in the text?

REFERENCES

The following references emphasize applications over theory:

Brown, R. *Decision Rules for Inventory Management.* New York: Holt, Rinehart and Winston, 1967.

Peterson, R., and E. Silver. *Decision Systems for Inventory Management and Production Planning.* New York: Wiley, 1998.

Tersine, R. *Principles of Inventory and Materials Management.* New York: North-Holland, 1982.

Vollman, T., W. Berry, and C. Whybark. *Manufacturing Planning and Control Systems.* Homewood, Ill.: Irwin, 1997.

The following references contain extensive theoretical discussions as well as applications:

Hadley, G., and T. Whitin. *Analysis of Inventory Systems.* Englewood Cliffs, N.J.: Prentice Hall, 1963.

Hax, A., and D. Candea. *Production and Inventory Management.* Englewood Cliffs, N.J.: Prentice Hall, 1984.

Johnson, L., and D. Montgomery. *Operations Research in Production, Scheduling, and Inventory Control.* New York: Wiley, 1974.

Barron, H. "Payoff Matrices Pay Off at Hallmark," *Interfaces* 15(no. 4, 1985):20–25.

Bruno, J. "The Use of Monte-Carlo Techniques for Determining the Size of Substitute Teacher Pools," *Socio-Economic Planning Science* 4(1970):415–428.

Flowers, D., and J. O'Neill. "An Application of Classical Inventory Analysis to a Spare Parts Inventory," *Interfaces* 8(no. 2, 1978):76–79.

Rosenfeld, D. "Optimal Management of Tax-Sheltered Employment Reimbursement Programs," *Interfaces* 16(no. 3, 1986):68–72.

Virts, J., and R. Garrett. "Weighting Risk in Capacity Expansion," *Harvard Business Review* 48(1970).

For a discussion of exchange curves and surfaces, see:

Gardner, E., and D. Dannenbring. "Using Optimal Policy Surfaces to Analyze Aggregate Inventory Tradeoffs," *Management Science* 25(1979):709–720.

5

Markov Chains

Sometimes we are interested in how a random variable changes over time. For example, we may want to know how the price of a share of stock or a firm's market share evolves. The study of how a random variable changes over time includes stochastic processes, which are explained in this chapter. In particular, we focus on a type of stochastic process known as a Markov chain. Markov chains have been applied in areas such as education, marketing, health services, finance, accounting, and production. We begin by defining the concept of a stochastic process. In the rest of the chapter, we will discuss the basic ideas needed for an understanding of Markov chains.

5.1 What Is a Stochastic Process?

Suppose we observe some characteristic of a system at discrete points in time (labeled 0, 1, 2, ...). Let X_t be the value of the system characteristic at time t. In most situations, X_t is not known with certainty before time t and may be viewed as a random variable. A **discrete-time stochastic process** is simply a description of the relation between the random variables X_0, X_1, X_2, ... Some examples of discrete-time stochastic processes follow.

EXAMPLE 1   The Gambler's Ruin

At time 0, I have \$2. At times 1, 2, ..., I play a game in which I bet \$1. With probability p, I win the game, and with probability $1 - p$, I lose the game. My goal is to increase my capital to \$4, and as soon as I do, the game is over. The game is also over if my capital is reduced to \$0. If we define X_t to be my capital position after the time t game (if any) is played, then X_0, X_1, ..., X_t may be viewed as a discrete-time stochastic process. Note that $X_0 = 2$ is a known constant, but X_1 and later X_t's are random. For example, with probability p, $X_1 = 3$, and with probability $1 - p$, $X_1 = 1$. Note that if $X_t = 4$, then X_{t+1} and all later X_t's will also equal 4. Similarly, if $X_t = 0$, then X_{t+1} and all later X_t's will also equal 0. For obvious reasons, this type of situation is called a *gambler's ruin* problem.

EXAMPLE 2   Choosing Balls from an Urn

An urn contains two unpainted balls at present. We choose a ball at random and flip a coin. If the chosen ball is unpainted and the coin comes up heads, we paint the chosen unpainted ball red; if the chosen ball is unpainted and the coin comes up tails, we paint the chosen unpainted ball black. If the ball has already been painted, then (whether heads or tails has been tossed) we change the color of the ball (from red to black or from black to red). To model this situation as a stochastic process, we define time t to be the time after the coin has been flipped for the tth time and the chosen ball has been painted. The state at any time may be described by the vector $[u \quad r \quad b]$, where u is the number of unpainted balls in the urn, r is the number of red balls in the urn, and b is the number of black balls in the urn. We are given that $\mathbf{X}_0 = [2 \quad 0 \quad 0]$. After the first coin toss, one ball will have been painted either red or black, and the state will be either $[1 \quad 1 \quad 0]$ or $[1 \quad 0 \quad 1]$. Hence, we can be sure that $\mathbf{X}_1 = [1 \quad 1 \quad 0]$ or $\mathbf{X}_1 = [1 \quad 0 \quad 1]$. Clearly, there must be some sort of relation between the $\mathbf{X}_t$'s. For example, if $\mathbf{X}_t = [0 \quad 2 \quad 0]$, we can be sure that $\mathbf{X}_{t+1}$ will be $[0 \quad 1 \quad 1]$.

EXAMPLE 3 **CSL Computer Stock**

Let $\mathbf{X}_0$ be the price of a share of CSL Computer stock at the beginning of the current trading day. Also, let $\mathbf{X}_t$ be the price of a share of CSL stock at the beginning of the tth trading day in the future. Clearly, knowing the values of $\mathbf{X}_0, \mathbf{X}_1, \ldots, \mathbf{X}_t$ tells us something about the probability distribution of $\mathbf{X}_{t+1}$; the question is, what does the past (stock prices up to time t) tell us about $\mathbf{X}_{t+1}$? The answer to this question is of critical importance in finance. (See Section 5.2 for more details.)

We close this section with a brief discussion of continuous-time stochastic processes. A **continuous-time stochastic process** is simply a stochastic process in which the state of the system can be viewed at any time, not just at discrete instants in time. For example, the number of people in a supermarket t minutes after the store opens for business may be viewed as a continuous-time stochastic process. (Models involving continuous-time stochastic processes are studied in Chapter 8.) Since the price of a share of stock can be observed at any time (not just the beginning of each trading day), it may be viewed as a continuous-time stochastic process. Viewing the price of a share of stock as a continuous-time stochastic process has led to many important results in the theory of finance, including the famous Black–Scholes option pricing formula.

5.2 What Is a Markov Chain?

One special type of discrete-time stochastic process is called a *Markov chain*. To simplify our exposition, we assume that at any time, the discrete-time stochastic process can be in one of a finite number of states labeled $1, 2, \ldots, s$.

DEFINITION ■ A discrete-time stochastic process is a **Markov chain** if, for $t = 0, 1, 2, \ldots$ and all states,

$$P(\mathbf{X}_{t+1} = i_{t+1} | \mathbf{X}_t = i_t, \mathbf{X}_{t-1} = i_{t-1}, \ldots, \mathbf{X}_1 = i_1, \mathbf{X}_0 = i_0)$$
$$= P(\mathbf{X}_{t+1} = i_{t+1} | \mathbf{X}_t = i_t) \quad ■ \tag{1}$$

Essentially, (1) says that the probability distribution of the state at time $t + 1$ depends on the state at time t (i_t) and does not depend on the states the chain passed through on the way to i_t at time t.

In our study of Markov chains, we make the further assumption that for all states i and j and all t, $P(\mathbf{X}_{t+1} = j | \mathbf{X}_t = i)$ is independent of t. This assumption allows us to write

$$P(\mathbf{X}_{t+1} = j | \mathbf{X}_t = i) = p_{ij} \tag{2}$$

where p_{ij} is the probability that given the system is in state i at time t, it will be in a state j at time $t + 1$. If the system moves from state i during one period to state j during the next period, we say that a **transition** from i to j has occurred. The p_{ij}'s are often referred to as the **transition probabilities** for the Markov chain.

Equation (2) implies that the probability law relating the next period's state to the current state does not change (or remains stationary) over time. For this reason, (2) is often called the **Stationarity Assumption.** Any Markov chain that satisfies (2) is called a **stationary Markov chain.**

Our study of Markov chains also requires us to define q_i to be the probability that the chain is in state i at time 0; in other words, $P(X_0 = i) = q_i$. We call the vector $\mathbf{q} = [q_1 \ q_2 \ \cdots \ q_s]$ the **initial probability distribution** for the Markov chain. In most applications, the transition probabilities are displayed as an $s \times s$ **transition probability matrix** P. The transition probability matrix P may be written as

$$P = \begin{bmatrix} p_{11} & p_{12} & \cdots & p_{1s} \\ p_{21} & p_{22} & \cdots & p_{2s} \\ \vdots & \vdots & & \vdots \\ p_{s1} & p_{s2} & \cdots & p_{ss} \end{bmatrix}$$

Given that the state at time t is i, the process must be somewhere at time $t + 1$. This means that for each i,

$$\sum_{j=1}^{j=s} P(X_{t+1} = j | P(X_t = i)) = 1$$

$$\sum_{j=1}^{j=s} p_{ij} = 1$$

We also know that each entry in the P matrix must be nonnegative. Hence, all entries in the transition probability matrix are nonnegative, and the entries in each row must sum to 1.

EXAMPLE 1   The Gambler's Ruin (Continued)

Find the transition matrix for Example 1.

Solution   Since the amount of money I have after $t + 1$ plays of the game depends on the past history of the game only through the amount of money I have after t plays, we definitely have a Markov chain. Since the rules of the game don't change over time, we also have a stationary Markov chain. The transition matrix is as follows (state i means that we have i dollars):

State

	\$0	\$1	\$2	\$3	\$4
0	1	0	0	0	0
1	$1-p$	0	p	0	0
$P = 2$	0	$1-p$	0	p	0
3	0	0	$1-p$	0	p
4	0	0	0	0	1

If the state is \$0 or \$4, I don't play the game anymore, so the state cannot change; hence, $p_{00} = p_{44} = 1$. For all other states, we know that with probability p, the next period's state will exceed the current state by 1, and with probability $1 - p$, the next period's state will be 1 less than the current state.

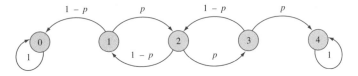

A transition matrix may be represented by a graph in which each node represents a state and arc (i, j) represents the transition probability p_{ij}. Figure 1 gives a graphical representation of Example 1's transition probability matrix.

EXAMPLE 2 Choosing Balls (Continued)

Find the transition matrix for Example 2.

Solution Since the state of the urn after the next coin toss only depends on the past history of the process through the state of the urn after the current coin toss, we have a Markov chain. Since the rules don't change over time, we have a stationary Markov chain. The transition matrix for Example 2 is as follows:

$$
\begin{array}{c}
 & & \text{State} \\
 & \begin{array}{cccccc} [0\ \ 1\ \ 1] & [0\ \ 2\ \ 0] & [0\ \ 0\ \ 2] & [2\ \ 0\ \ 0] & [1\ \ 1\ \ 0] & [1\ \ 0\ \ 1] \end{array} \\
P = \begin{array}{c} [0\ \ 1\ \ 1] \\ [0\ \ 2\ \ 0] \\ [0\ \ 0\ \ 2] \\ [2\ \ 0\ \ 0] \\ [1\ \ 1\ \ 0] \\ [1\ \ 0\ \ 1] \end{array}
\left[\begin{array}{cccccc}
0 & \frac{1}{2} & \frac{1}{2} & 0 & 0 & 0 \\
1 & 0 & 0 & 0 & 0 & 0 \\
1 & 0 & 0 & 0 & 0 & 0 \\
0 & 0 & 0 & 0 & \frac{1}{2} & \frac{1}{2} \\
\frac{1}{4} & \frac{1}{4} & 0 & 0 & 0 & \frac{1}{2} \\
\frac{1}{4} & 0 & \frac{1}{4} & 0 & \frac{1}{2} & 0
\end{array} \right]
\end{array}
$$

To illustrate the determination of the transition matrix, we determine the [1 1 0] row of this transition matrix. If the current state is [1 1 0], then one of the events shown in Table 1 must occur. Thus, the next state will be [1 0 1] with probability $\frac{1}{2}$, [0 2 0] with probability $\frac{1}{4}$, and [0 1 1] with probability $\frac{1}{4}$. Figure 2 gives a graphical representation of this transition matrix.

TABLE 1
Computations of Transition Probabilities If Current State Is [1 1 0]

Event	Probability	New State
Flip heads and choose unpainted ball	$\frac{1}{4}$	[0 2 0]
Choose red ball	$\frac{1}{2}$	[1 0 1]
Flip tails and choose unpainted ball	$\frac{1}{4}$	[0 1 1]

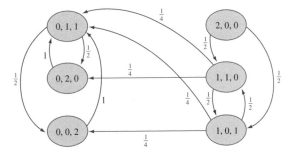

FIGURE 2
Graphical
Representation of
Transition Matrix
for Urn

EXAMPLE 3   **CSL Computer Stock (Continued)**

In recent years, students of finance have devoted much effort to answering the question of whether the daily price of a stock share can be described by a Markov chain. Suppose the daily price of a stock share (such as CSL Computer stock) can be described by a Markov chain. What does that tell us? Simply that the probability distribution of tomorrow's price for one share of CSL stock depends *only* on today's price of CSL stock, *not* on the past prices of CSL stock. If the price of a stock share can be described by a Markov chain, the "chartists" who attempt to predict future stock prices on the basis of the patterns followed by past stock prices are barking up the wrong tree. For example, suppose the daily price of a share of CSL stock follows a Markov chain, and today's price for a share of CSL stock is $50. Then to predict tomorrow's price of a share of CSL stock, it does not matter whether the price has increased or decreased during each of the last 30 days. In either situation (or any other situation that might have led to today's $50 price), a prediction of tomorrow's stock price should be based only on the fact that today's price of CSL stock is $50. At this time, the consensus is that for most stocks the daily price of the stock can be described as a Markov chain. This idea is often referred to as the **efficient market hypothesis.**

PROBLEMS

Group A

1 In Smalltown, 90% of all sunny days are followed by sunny days, and 80% of all cloudy days are followed by cloudy days. Use this information to model Smalltown's weather as a Markov chain.

2 Consider an inventory system in which the sequence of events during each period is as follows. (1) We observe the inventory level (call it i) at the beginning of the period. (2) If $i \leq 1$, $4 - i$ units are ordered. If $i \geq 2$, 0 units are ordered. Delivery of all ordered units is immediate. (3) With probability $\frac{1}{3}$, 0 units are demanded during the period; with probability $\frac{1}{3}$, 1 unit is demanded during the period; and with probability $\frac{1}{3}$, 2 units are demanded during the period. (4) We observe the inventory level at the beginning of the next period.

Define a period's state to be the period's beginning inventory level. Determine the transition matrix that could be used to model this inventory system as a Markov chain.

3 A company has two machines. During any day, each machine that is working at the beginning of the day has a $\frac{1}{3}$ chance of breaking down. If a machine breaks down during the day, it is sent to a repair facility and will be working two days after it breaks down. (Thus, if a machine breaks down during day 3, it will be working at the beginning of day 5.) Letting the state of the system be the number of machines working at the beginning of the day, formulate a transition probability matrix for this situation.

Group B

4 Referring to Problem 1, suppose that tomorrow's Smalltown weather depends on the last two days of Smalltown weather, as follows: (1) If the last two days have been sunny, then 95% of the time, tomorrow will be sunny. (2) If yesterday was cloudy and today is sunny, then 70% of the time, tomorrow will be sunny. (3) If yesterday was sunny and today is cloudy, then 60% of the time, tomorrow will be cloudy. (4) If the last two days have been cloudy, then 80% of the time, tomorrow will be cloudy.

Using this information, model Smalltown's weather as a Markov chain. If tomorrow's weather depended on the last three days of Smalltown weather, how many states will be needed to model Smalltown's weather as a Markov chain? (*Note:* The approach used in this problem can be used to model a discrete-time stochastic process as a Markov chain even if X_{t+1} depends on states prior to X_t, such as X_{t-1} in the current example.)

5 Let X_t be the location of your token on the Monopoly board after t dice rolls. Can X_t be modeled as a Markov chain? If not, how can we modify the definition of the state at time t so that $X_0, X_1, \ldots, X_t, \ldots$ would be a Markov chain? (*Hint:* How does a player go to Jail? In this problem, assume that players who are sent to Jail stay there until they roll doubles or until they have spent three turns there, whichever comes first.)

6 In Problem 3, suppose a machine that breaks down returns to service three days later (for instance, a machine that breaks down during day 3 would be back in working order at the beginning of day 6). Determine a transition probability matrix for this situation.

5.3 n-Step Transition Probabilities

Suppose we are studying a Markov chain with a known transition probability matrix P. (Since all chains that we will deal with are stationary, we will not bother to label our Markov chains as stationary.) A question of interest is: If a Markov chain is in state i at time m, what is the probability that n periods later the Markov chain will be in state j? Since we are dealing with a stationary Markov chain, this probability will be independent of m, so we may write

$$P(X_{m+n} = j | X_m = i) = P(X_n = j | X_0 = i) = P_{ij}(n)$$

where $P_{ij}(n)$ is called the **n-step probability** of a transition from state i to state j.

Clearly, $P_{ij}(1) = p_{ij}$. To determine $P_{ij}(2)$, note that if the system is now in state i, then for the system to end up in state j two periods from now, we must go from state i to some state k and then go from state k to state j (see Figure 3). This reasoning allows us to write

$$P_{ij}(2) = \sum_{k=1}^{k=s} (\text{probability of transition from } i \text{ to } k)$$
$$\times (\text{probability of transition from } k \text{ to } j)$$

Using the definition of P, the transition probability matrix, we rewrite the last equation as

$$P_{ij}(2) = \sum_{k=1}^{k=s} p_{ik}p_{kj} \tag{3}$$

The right-hand side of (3) is just the scalar product of row i of the P matrix with column j of the P matrix. Hence, $P_{ij}(2)$ is the ijth element of the matrix P^2. By extending this reasoning, it can be shown that for $n > 1$,

$$P_{ij}(n) = ij\text{th element of } P^n \tag{4}$$

Of course, for $n = 0$, $P_{ij}(0) = P(X_0 = j | X_0 = i)$, so we must write

$$P_{ij}(0) = \begin{cases} 1 & \text{if } j = i \\ 0 & \text{if } j \neq i \end{cases}$$

We illustrate the use of Equation (4) in Example 4.

FIGURE 3
$P_{ij}(2) = p_{i1}p_{1j} + p_{i2}p_{2j} + \cdots + p_{is}p_{sj}$

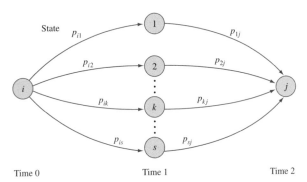

EXAMPLE 4   The Cola Example

Suppose the entire cola industry produces only two colas. Given that a person last purchased cola 1, there is a 90% chance that her next purchase will be cola 1. Given that a person last purchased cola 2, there is an 80% chance that her next purchase will be cola 2.

1 If a person is currently a cola 2 purchaser, what is the probability that she will purchase cola 1 two purchases from now?

2 If a person is currently a cola 1 purchaser, what is the probability that she will purchase cola 1 three purchases from now?

Solution We view each person's purchases as a Markov chain with the state at any given time being the type of cola the person last purchased. Hence, each person's cola purchases may be represented by a two-state Markov chain, where

State 1 = person has last purchased cola 1

State 2 = person has last purchased cola 2

If we define X_n to be the type of cola purchased by a person on her nth future cola purchase (present cola purchase $= X_0$), then $X_0, X_1, \ldots$ may be described as the Markov chain with the following transition matrix:

$$P = \begin{array}{c} \\ \text{Cola 1} \\ \text{Cola 2} \end{array} \begin{array}{cc} \text{Cola 1} & \text{Cola 2} \\ \begin{bmatrix} .90 & .10 \\ .20 & .80 \end{bmatrix} \end{array}$$

We can now answer questions 1 and 2.

1 We seek $P(X_2 = 1 | X_0 = 2) = P_{21}(2) = $ element 21 of P^2:

$$P^2 = \begin{bmatrix} .90 & .10 \\ .20 & .80 \end{bmatrix} \begin{bmatrix} .90 & .10 \\ .20 & .80 \end{bmatrix} = \begin{bmatrix} .83 & .17 \\ .34 & .66 \end{bmatrix}$$

Hence, $P_{21}(2) = .34$. This means that the probability is .34 that two purchases in the future a cola 2 drinker will purchase cola 1. By using basic probability theory, we may obtain this answer in a different way (see Figure 4). Note that $P_{21}(2) = $ (probability that next purchase is cola 1 and second purchase is cola 1) + (probability that next purchase is cola 2 and second purchase is cola 1) $= p_{21}p_{11} + p_{22}p_{21} = (.20)(.90) + (.80)(.20) = .34$.

2 We seek $P_{11}(3) = $ element 11 of P^3:

$$P^3 = P(P^2) = \begin{bmatrix} .90 & .10 \\ .20 & .80 \end{bmatrix} \begin{bmatrix} .83 & .17 \\ .34 & .66 \end{bmatrix} = \begin{bmatrix} .781 & .219 \\ .438 & .562 \end{bmatrix}$$

Therefore, $P_{11}(3) = .781$.

FIGURE 4
Probability That Two Periods from Now, a Cola 2 Purchaser Will Purchase Cola 1 Is .20(.90) + .80(.20) = .34

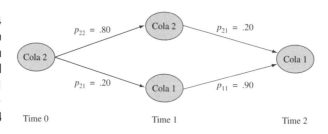

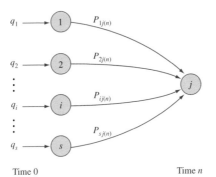

FIGURE 5
Determination of Probability of Being in State _j_ at Time _n_ When Initial State Is Unknown

Time 0   Time _n_

In many situations, we do not know the state of the Markov chain at time 0. As defined in Section 5.2, let q_i be the probability that the chain is in state i at time 0. Then we can determine the probability that the system is in state i at time n by using the following reasoning (see Figure 5).

Probability of being in state j at time n

$$= \sum_{i=1}^{i=s} (\text{probability that state is originally } i)$$

$$\times (\text{probability of going from } i \text{ to } j \text{ in } n \text{ transitions})$$

$$= \sum_{i=1}^{i=s} q_i P_{ij}(n)$$ (5)

$$= \mathbf{q}(\text{column } j \text{ of } P^n)$$

where $\mathbf{q} = [q_1 \quad q_2 \quad \cdots \quad q_s]$.

To illustrate the use of (5), we answer the following question: Suppose 60% of all people now drink cola 1, and 40% now drink cola 2. Three purchases from now, what fraction of all purchasers will be drinking cola 1? Since $\mathbf{q} = [.60 \quad .40]$ and $\mathbf{q}(\text{column 1 of } P^3) =$ probability that three purchases from now a person drinks cola 1, the desired probability is

$$[.60 \quad .40] \begin{bmatrix} .781 \\ .438 \end{bmatrix} = .6438$$

Hence, three purchases from now, 64% of all purchasers will be purchasing cola 1.

To illustrate the behavior of the n-step transition probabilities for large values of n, we have computed several of the n-step transition probabilities for the Cola example in Table 2.

TABLE 2
_n_-Step Transition Probabilities for Cola Drinkers

n	$P_{11}(n)$	$P_{12}(n)$	$P_{21}(n)$	$P_{22}(n)$
1	.90	.10	.20	.80
2	.83	.17	.34	.66
3	.78	.22	.44	.56
4	.75	.25	.51	.49
5	.72	.28	.56	.44
10	.68	.32	.65	.35
20	.67	.33	.67	.33
30	.67	.33	.67	.33
40	.67	.33	.67	.33

For large n, both $P_{11}(n)$ and $P_{21}(n)$ are nearly constant and approach .67. This means that for large n, no matter what the initial state, there is a .67 chance that a person will be a cola 1 purchaser. Similarly, we see that for large n, both $P_{12}(n)$ and $P_{22}(n)$ are nearly constant and approach .33. This means that for large n, no matter what the initial state, there is a .33 chance that a person will be a cola 2 purchaser. In Section 5.5, we make a thorough study of this settling down of the n-step transition probabilities.

REMARK We can easily multiply matrices on a spreadsheet using the MMULT command, as discussed in Section 2.7.

PROBLEMS

Group A

1 Each American family is classified as living in an urban, rural, or suburban location. During a given year, 15% of all urban families move to a suburban location, and 5% move to a rural location; also, 6% of all suburban families move to an urban location, and 4% move to a rural location; finally, 4% of all rural families move to an urban location, and 6% move to a suburban location.

a If a family now lives in an urban location, what is the probability that it will live in an urban area two years from now? A suburban area? A rural area?

b Suppose that at present, 40% of all families live in an urban area, 35% live in a suburban area, and 25% live in a rural area. Two years from now, what percentage of American families will live in an urban area?

c What problems might occur if this model were used to predict the future population distribution of the United States?

2 The following questions refer to Example 1.

a After playing the game twice, what is the probability that I will have $3? How about $2?

b After playing the game three times, what is the probability that I will have $2?

3 In Example 2, determine the following n-step transition probabilities:

a After two balls are painted, what is the probability that the state is [0   2   0]?

b After three balls are painted, what is the probability that the state is [0   1   1]? (Draw a diagram like Figure 4.)

5.4 Classification of States in a Markov Chain

In Section 5.3, we mentioned the fact that after many transitions, the n-step transition probabilities tend to settle down. Before we can discuss this in more detail, we need to study how mathematicians classify the states of a Markov chain. We use the following transition matrix to illustrate most of the following definitions (see Figure 6).

$$P = \begin{bmatrix} .4 & .6 & 0 & 0 & 0 \\ .5 & .5 & 0 & 0 & 0 \\ 0 & 0 & .3 & .7 & 0 \\ 0 & 0 & .5 & .4 & .1 \\ 0 & 0 & 0 & .8 & .2 \end{bmatrix}$$

DEFINITION ■ Given two states i and j, a **path** from i to j is a sequence of transitions that begins in i and ends in j, such that each transition in the sequence has a positive probability of occurring. ■

A state j is **reachable** from state i if there is a path leading from i to j. ■

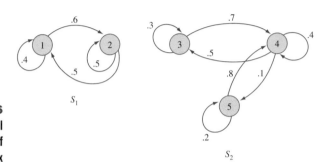

FIGURE **6**
Graphical
Representation of
Transition Matrix

DEFINITION ■ Two states i and j are said to **communicate** if j is reachable from i, and i is reachable from j. ■

For the transition probability matrix P represented in Figure 6, state 5 is reachable from state 3 (via the path 3–4–5), but state 5 is not reachable from state 1 (there is no path from 1 to 5 in Figure 6). Also, states 1 and 2 communicate (we can go from 1 to 2 and from 2 to 1).

DEFINITION ■ A set of states S in a Markov chain is a **closed set** if no state outside of S is reachable from any state in S. ■

From the Markov chain with transition matrix P in Figure 6, $S_1 = \{1, 2\}$ and $S_2 = \{3, 4, 5\}$ are both closed sets. Observe that once we enter a closed set, we can never leave the closed set (in Figure 6, no arc begins in S_1 and ends in S_2 or begins in S_2 and ends in S_1).

DEFINITION ■ A state i is an **absorbing state** if $p_{ii} = 1$. ■

Whenever we enter an absorbing state, we never leave the state. In Example 1, the gambler's ruin, states 0 and 4 are absorbing states. Of course, an absorbing state is a closed set containing only one state.

DEFINITION ■ A state i is a **transient state** if there exists a state j that is reachable from i, but the state i is not reachable from state j. ■

In other words, a state i is transient if there is a way to leave state i that never returns to state i. In the gambler's ruin example, states 1, 2, and 3 are transient states. For example (see Figure 1), from state 2, it is possible to go along the path 2–3–4, but there is no way to return to state 2 from state 4. Similarly, in Example 2, [2 0 0], [1 1 0], and [1 0 1] are all transient states (in Figure 2, there is a path from [1 0 1] to [0 0 2], but once both balls are painted, there is no way to return to [1 0 1]).

After a large number of periods, the probability of being in any transient state i is zero. Each time we enter a transient state i, there is a positive probability that we will leave i forever and end up in the state j described in the definition of a transient state. Thus, eventually we are sure to enter state j (and then we will never return to state i). To illustrate, in Example 2, suppose we are in the transient state [1 0 1]. With probability 1, the unpainted ball will eventually be painted, and we will never reenter state [1 0 1] (see Figure 2).

DEFINITION ■ If a state is not transient, it is called a **recurrent state.** ■

In Example 1, states 0 and 4 are recurrent states (and also absorbing states), and in Example 2, [0 2 0], [0 0 2], and [0 1 1] are recurrent states. For the transition matrix P in Figure 6, all states are recurrent.

DEFINITION ■ A state i is **periodic** with period $k > 1$ if k is the smallest number such that all paths leading from state i back to state i have a length that is a multiple of k. If a recurrent state is not periodic, it is referred to as **aperiodic.** ■

For the Markov chain with transition matrix

$$Q = \begin{bmatrix} 0 & 1 & 0 \\ 0 & 0 & 1 \\ 1 & 0 & 0 \end{bmatrix}$$

each state has period 3. For example, if we begin in state 1, the only way to return to state 1 is to follow the path 1–2–3–1 for some number of times (say, m). (See Figure 7.) Hence, any return to state 1 will take $3m$ transitions, so state 1 has period 3. Wherever we are, we are sure to return three periods later.

DEFINITION ■ If all states in a chain are recurrent, aperiodic, and communicate with each other, the chain is said to be **ergodic.** ■

The gambler's ruin example is not an ergodic chain, because (for example) states 3 and 4 do not communicate. Example 2 is also not an ergodic chain, because (for example) [2 0 0] and [0 1 1] do not communicate. Example 4, the cola example, is an ergodic Markov chain. Of the following three Markov chains, P_1 and P_3 are ergodic, and P_2 is not ergodic.

$$P_1 = \begin{bmatrix} \frac{1}{3} & \frac{2}{3} & 0 \\ \frac{1}{2} & 0 & \frac{1}{2} \\ 0 & \frac{1}{4} & \frac{3}{4} \end{bmatrix} \qquad \text{Ergodic}$$

$$P_2 = \begin{bmatrix} \frac{1}{2} & \frac{1}{2} & 0 & 0 \\ \frac{1}{2} & \frac{1}{2} & 0 & 0 \\ 0 & 0 & \frac{2}{3} & \frac{1}{3} \\ 0 & 0 & \frac{1}{4} & \frac{3}{4} \end{bmatrix} \qquad \text{Nonergodic}$$

$$P_3 = \begin{bmatrix} \frac{1}{4} & \frac{1}{2} & \frac{1}{4} \\ \frac{2}{3} & \frac{1}{3} & 0 \\ 0 & \frac{2}{3} & \frac{1}{3} \end{bmatrix} \qquad \text{Ergodic}$$

FIGURE 7
A Periodic Markov Chain $k = 3$

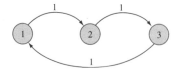

P_2 is not ergodic because there are two closed classes of states (class $1 = \{1, 2\}$ and class $2 = \{3, 4\}$), and the states in different classes do not communicate with each other.

After the next two sections, the importance of the concepts introduced in this section will become clear.

PROBLEMS

Group A

1 In Example 1, what is the period of states 1 and 3?

2 Is the Markov chain of Section 5.3, Problem 1, an ergodic Markov chain?

3 Consider the following transition matrix:

$$P = \begin{bmatrix} 0 & 0 & 1 & 0 & 0 & 0 \\ 0 & 0 & 0 & 0 & 0 & 1 \\ 0 & 0 & 0 & 0 & 1 & 0 \\ \frac{1}{4} & \frac{1}{4} & 0 & \frac{1}{2} & 0 & 0 \\ 1 & 0 & 0 & 0 & 0 & 0 \\ 0 & \frac{1}{3} & 0 & 0 & 0 & \frac{2}{3} \end{bmatrix}$$

a Which states are transient?

b Which states are recurrent?

c Identify all closed sets of states.

d Is this chain ergodic?

4 For each of the following chains, determine whether the Markov chain is ergodic. Also, for each chain, determine the recurrent, transient, and absorbing states.

$$P_1 = \begin{bmatrix} 0 & .8 & .2 \\ .3 & .7 & 0 \\ .4 & .5 & .1 \end{bmatrix} \quad P_2 = \begin{bmatrix} .2 & .8 & 0 & 0 \\ 0 & 0 & .9 & .1 \\ .4 & .5 & .1 & 0 \\ 0 & 0 & 0 & 1 \end{bmatrix}$$

5 Fifty-four players (including Gabe Kaplan and James Garner) participated in the 1980 World Series of Poker. Each player began with $10,000. Play continued until one player had won everybody else's money. If the World Series of Poker were to be modeled as a Markov chain, how many absorbing states would the chain have?

6 Which of the following chains is ergodic?

$$P_1 = \begin{bmatrix} .4 & 0 & .6 \\ .3 & .3 & .4 \\ 0 & .5 & .5 \end{bmatrix} \quad P_2 = \begin{bmatrix} .7 & 0 & 0 & .3 \\ .2 & .2 & .4 & .2 \\ .6 & .1 & .1 & .2 \\ .2 & 0 & 0 & .8 \end{bmatrix}$$

5.5 Steady-State Probabilities and Mean First Passage Times

In our discussion of the cola example (Example 4), we found that after a long time, the probability that a person's next cola purchase would be cola 1 approached .67 and .33 that it would be cola 2 (see Table 2). These probabilities *did not* depend on whether the person was initially a cola 1 or a cola 2 drinker. In this section, we discuss the important concept of steady-state probabilities, which can be used to describe the long-run behavior of a Markov chain.

The following result is vital to an understanding of steady-state probabilities and the long-run behavior of Markov chains.

THEOREM 1

Let P be the transition matrix for an s-state ergodic chain.[†] Then there exists a vector $\pi = [\pi_1 \quad \pi_2 \quad \cdots \quad \pi_s]$ such that

$$\lim_{n \to \infty} P^n = \begin{bmatrix} \pi_1 & \pi_2 & \cdots & \pi_s \\ \pi_1 & \pi_2 & \cdots & \pi_s \\ \vdots & \vdots & & \vdots \\ \pi_1 & \pi_2 & \cdots & \pi_s \end{bmatrix}$$

[†]To see why Theorem 1 fails to hold for a nonergodic chain, see Problems 11 and 12 at the end of this section. For a proof of this theorem, see Isaacson and Madsen (1976, Chapter 3).

Recall that the ijth element of P^n is $P_{ij}(n)$. Theorem 1 tells us that for any initial state i,

$$\lim_{n \to \infty} P_{ij}(n) = \pi_j$$

Observe that for large n, P^n approaches a matrix with identical rows. This means that after a long time, the Markov chain settles down, and (independent of the initial state i) there is a probability π_j that we are in state j.

The vector $\pi = [\pi_1 \quad \pi_2 \quad \cdots \quad \pi_s]$ is often called the **steady-state distribution,** or **equilibrium distribution,** for the Markov chain. For a given chain with transition matrix P, how can we find the steady-state probability distribution? From Theorem 1, observe that for large n and all i,

$$P_{ij}(n + 1) \cong P_{ij}(n) \cong \pi_j \tag{6}$$

Since $P_{ij}(n + 1) = $ (row i of P^n) (column j of P), we may write

$$P_{ij}(n + 1) = \sum_{k=1}^{k=s} P_{ik}(n) p_{kj} \tag{7}$$

If n is large, substituting (6) into (7) yields

$$\pi_j = \sum_{k=1}^{k=s} \pi_k p_{kj} \tag{8}$$

In matrix form, (8) may be written as

$$\pi = \pi P \tag{8'}$$

Unfortunately, the system of equations specified in (8) has an infinite number of solutions, because the rank of the P matrix always turns out to be $\leq s - 1$ (see Volume 1, Chapter 2, Review Problem 21). To obtain unique values of the steady-state probabilities, note that for any n and any i,

$$P_{i1}(n) + P_{i2}(n) + \cdots + P_{is}(n) = 1 \tag{9}$$

Letting n approach infinity in (9), we obtain

$$\pi_1 + \pi_2 + \cdots + \pi_s = 1 \tag{10}$$

Thus, after replacing any of the equations in (8) with (10), we may use (8) to solve for the steady-state probabilities.

To illustrate how to find the steady-state probabilities, we find the steady-state probabilities for Example 4, the cola example. Recall that the transition matrix for Example 4 was

$$P = \begin{bmatrix} .90 & .10 \\ .20 & .80 \end{bmatrix}$$

Then (8) or (8') yields

$$[\pi_1 \quad \pi_2] = [\pi_1 \quad \pi_2] \begin{bmatrix} .90 & .10 \\ .20 & .80 \end{bmatrix}$$

$$\pi_1 = .90\pi_1 + .20\pi_2$$
$$\pi_2 = .10\pi_1 + .80\pi_2$$

Replacing the second equation with the condition $\pi_1 + \pi_2 = 1$, we obtain the system

$$\pi_1 = .90\pi_1 + .20\pi_2$$
$$1 = \pi_1 + \pi_2$$

Solving for π_1 and π_2 we obtain $\pi_1 = \frac{2}{3}$ and $\pi_2 = \frac{1}{3}$. Hence, after a long time, there is a $\frac{2}{3}$ probability that a given person will purchase cola 1 and a $\frac{1}{3}$ probability that a given person will purchase cola 2.

Transient Analysis

A glance at Table 2 shows that for Example 4, the steady state is reached (to two decimal places) after only ten transitions. No general rule can be given about how quickly a Markov chain reaches the steady state, but if P contains very few entries that are near 0 or near 1, the steady state is usually reached very quickly. The behavior of a Markov chain before the steady state is reached is often called **transient** (or short-run) **behavior.** To study the transient behavior of a Markov chain, one simply uses the formulas for $P_{ij}(n)$ given in (4) and (5). It's nice to know, however, that for large n, the steady-state probabilities accurately describe the probability of being in any state.

Intuitive Interpretation of Steady-State Probabilities

An intuitive interpretation can be given to the steady-state probability equations (8). By subtracting $\pi_j p_{jj}$ from both sides of (8), we obtain

$$\pi_j(1 - p_{jj}) = \sum_{k \neq j} \pi_k p_{kj} \tag{11}$$

Equation (11) states that in the steady state,

$$\text{Probability that a particular transition leaves state } j \tag{12}$$
$$= \text{probability that a particular transition enters state } j$$

Recall that in the steady state, the probability that the system is in state j is π_j. From this observation, it follows that

Probability that a particular transition leaves state j
$$= (\text{probability that the current period begins in } j)$$
$$\times (\text{probability that the current transition leaves } j)$$
$$= \pi_j(1 - p_{jj})$$

and

Probability that a particular transition enters state j
$$= \sum_{k} (\text{probability that the current period begins in } k \neq j)$$
$$\times (\text{probability that the current transition enters } j)$$
$$= \sum_{k \neq j} \pi_k p_{kj}$$

Equation (11) is reasonable; if (11) were violated for any state, then for some state j, the right-hand side of (11) would exceed the left-hand side of (11). This would result in probability "piling up" at state j, and a steady-state distribution would not exist. Equation (11) may be viewed as saying that in the steady state, the "flow" of probability into each state must equal the flow of probability out of each state. This explains why steady-state probabilities are often called equilibrium probabilities.

Use of Steady-State Probabilities in Decision Making

EXAMPLE 5   The Cola Example (Continued)

In Example 4, suppose that each customer makes one purchase of cola during any week (52 weeks = 1 year). Suppose there are 100 million cola customers. One selling unit of cola costs the company $1 to produce and is sold for $2. For $500 million per year, an advertising firm guarantees to decrease from 10% to 5% the fraction of cola 1 customers who switch to cola 2 after a purchase. Should the company that makes cola 1 hire the advertising firm?

Solution   At present, a fraction $\pi_1 = \frac{2}{3}$ of all purchases are cola 1 purchases. Each purchase of cola 1 earns the company a $1 profit. Since there are a total of 52(100,000,000), or 5.2 billion, cola purchases each year, the cola 1 company's current annual profit is

$$\tfrac{2}{3}(5,200,000,000) = \$3,466,666,667$$

The advertising firm is offering to change the P matrix to

$$P_1 = \begin{bmatrix} .95 & .05 \\ .20 & .80 \end{bmatrix}$$

For P_1, the steady-state equations become

$$\pi_1 = .95\pi_1 + .20\pi_2$$
$$\pi_2 = .05\pi_1 + .80\pi_2$$

Replacing the second equation by $\pi_1 + \pi_2 = 1$ and solving, we obtain $\pi_1 = .8$ and $\pi_2 = .2$. Now the cola 1 company's annual profit will be

$$(.80)(5,200,000,000) - 500,000,000 = \$3,660,000,000$$

Hence, the cola 1 company should hire the ad agency.

EXAMPLE 6   Playing Monopoly

With the assumption that each Monopoly player who goes to Jail stays until he or she rolls doubles or has spent three turns in Jail, the steady-state probability of a player landing on any Monopoly square has been determined by Ash and Bishop (1972) (see Table 3).[†] These steady-state probabilities can be used to measure the cost-effectiveness of various monopolies. For example, it costs $1,500 to build hotels on the Orange monopoly. Each time a player lands on a Tennessee Ave. or a St. James Place hotel, the owner of the monopoly receives $950, and each time a player lands on a New York Ave. hotel, the owner receives $1,000. From Table 3, we can compute the expected rent per dice roll earned by the Orange monopoly:

$$950(.0335) + 950(.0318) + 1,000(.0334) = \$95.44$$

Thus, per dollar invested, the Orange monopoly yields $\frac{95.44}{1,500} = \$0.064$ per dice roll.

Now let's consider the Green monopoly. To put hotels on the Green monopoly costs $3,000. If a player lands on a North Carolina Ave. or a Pacific Ave. hotel, the owner receives $1,275. If a player lands on a Pennsylvania Ave. hotel, the owner receives $1,400. From Table 3, the average revenue per dice roll earned from hotels on the Green monopoly is

$$1,275(.0294) + 1,275(.0300) + 1,400(.0279) = \$114.80$$

[†]This example is based on Ash and Bishop (1972).

TABLE **3**

Steady-State Probabilities for Monopoly

n	Position	Steady-State Probability
0	Go	.0346
1	Mediterranean Ave.	.0237
2	Community Chest 1	.0218
3	Baltic Ave.	.0241
4	Income tax	.0261
5	Reading RR	.0332
6	Oriental Ave.	.0253
7	Chance 1	.0096
8	Vermont Ave.	.0258
9	Connecticut Ave.	.0237
10	Visiting jail	.0254
11	St. Charles Place	.0304
12	Electric Co.	.0311
13	State Ave.	.0258
14	Virginia Ave.	.0288
15	Pennsylvania RR	.0313
16	St. James Place	.0318
17	Community Chest 2	.0272
18	Tennessee Ave.	.0335
19	New York Ave.	.0334
20	Free parking	.0336
21	Kentucky Ave.	.0310
22	Chance 2	.0125
23	Indiana Ave.	.0305
24	Illinois Ave.	.0355
25	B and O RR	.0344
26	Atlantic Ave.	.0301
27	Ventnor Ave.	.0299
28	Water works	.0315
29	Marvin Gardens	.0289
30	Jail	.1123
31	Pacific Ave.	.0300
32	North Carolina Ave.	.0294
33	Community Chest 3	.0263
34	Pennsylvania Ave.	.0279
35	Short Line RR	.0272
36	Chance 3	.0096
37	Park Place	.0245
38	Luxury tax	.0295
39	Boardwalk	.0295

Source: Reprinted by permission from R. Ash and
R. Bishop, "Monopoly as a Markov Process," *Mathematics
Magazine* 45(1972):26–29. Copryright © 1972
Mathematical Association of America.

Thus, per dollar invested, the Green monopoly yields only $\frac{114.80}{3,000} = \$0.038$ per dice roll.

This analysis shows that the Orange monopoly is superior to the Green monopoly. By the way, why does the Orange get landed on so often?

Mean First Passage Times

For an ergodic chain, let m_{ij} = expected number of transitions before we first reach state j, given that we are currently in state i; m_{ij} is called the **mean first passage time** from state i to state j. In Example 4, m_{12} would be the expected number of bottles of cola purchased by a person who just bought cola 1 before first buying a bottle of cola 2. Assume that we are currently in state i. Then with probability p_{ij}, it will take one transition to go from state i to state j. For $k \neq j$, we next go with probability p_{ik} to state k. In this case, it will take an average of $1 + m_{kj}$ transitions to go from i to j. This reasoning implies that

$$m_{ij} = p_{ij}(1) + \sum_{k \neq j} p_{ik}(1 + m_{kj})$$

Since

$$p_{ij} + \sum_{k \neq j} p_{ik} = 1$$

we may rewrite the last equation as

$$m_{ij} = 1 + \sum_{k \neq j} p_{ik}m_{kj} \tag{13}$$

By solving the linear equations given in (13), we may find all the mean first passage times. It can be shown that

$$m_{ii} = \frac{1}{\pi_i}$$

This can simplify the use of (13).

To illustrate the use of (13), let's solve for the mean first passage times in Example 4. Recall that $\pi_1 = \frac{2}{3}$ and $\pi_2 = \frac{1}{3}$. Then

$$m_{11} = \frac{1}{\frac{2}{3}} = 1.5 \quad \text{and} \quad m_{22} = \frac{1}{\frac{1}{3}} = 3$$

Now (13) yields the following two equations:

$$m_{12} = 1 + p_{11}m_{12} = 1 + 0.9m_{12}, \qquad m_{21} = 1 + p_{22}m_{21} = 1 + 0.8m_{21}$$

Solving these two equations, we find that $m_{12} = 10$ and $m_{21} = 5$. This means, for example, that a person who last drank cola 1 will drink an average of ten bottles of soda before switching to cola 2.

Solving for Steady-State Probabilities and Mean First Passage Times on the Computer

Since we solve for steady-state probabilities and mean first passage times by solving a system of linear equations, we may use LINDO to determine them. Simply type in an objective function of 0, and type the equations you need to solve as your constraints.

Markov.lng

Alternatively, you may use the following LINGO model (file Markov.lng) to determine steady-state probabilities and mean first passage times for an ergodic chain.

```
MODEL:
 1]
 2]SETS:
 3]STATE/1..2/:PI;
```

```
 4]SXS(STATE,STATE):TPROB,MFP;
 5]ENDSETS
 6]DATA:
 7]TPROB = .9,.1,
 8].2,.8;
 9]ENDDATA
10]@FOR(STATE(J)|J #LT# @SIZE(STATE):
11]PI(J) = @SUM(SXS(I,J): PI(I) * TPROB(I,J)););
12]@SUM(STATE:PI) = 1;
13]@FOR(SXS(I,J):MFP(I,J)=
14]1+@SUM(STATE(K)|K#NE#J:TPROB(I,K)*MFP(K,J)););
   END
```

In line 3, we define the set of states and associate a steady-state probability ($PI(I)$) with each state I. In line 4, we create for each pairing of states (I, J) a transition probability ($TPROB(I, J)$) which equals p_{ij} and $MFP(I, J)$ which equals m_{ij}. The transition probabilities for the cola example are input in lines 7 and 8. In lines 10 and 11, we create (for each state except the highest-numbered state) the steady-state equation

$$PI(J) = \sum_I PI(I) * TPROB(I, J)$$

In line 12, we ensure that the steady-state probabilities sum to 1. In lines 13 and 14, we create the equations that must be solved to compute the mean first passage times. For each (I, J), lines 13–14 create the equation

$$MFP(I, J) = 1 + \sum_{K \neq J} TPROB(I, K) * MFP(K, J)$$

which is needed to compute the mean first passage times.

PROBLEMS

Group A

1 Find the steady-state probabilities for Problem 1 of Section 5.3.

2 For the gambler's ruin problem (Example 1), why is it unreasonable to talk about steady-state probabilities?

3 For each of the following Markov chains, determine the long-run fraction of the time that each state will be occupied.

a $\begin{bmatrix} \frac{2}{3} & \frac{1}{3} \\ \frac{1}{2} & \frac{1}{2} \end{bmatrix}$ **b** $\begin{bmatrix} .8 & .2 & 0 \\ 0 & .2 & .8 \\ .8 & .2 & 0 \end{bmatrix}$

c Find all mean first passage times for part (b).

4 At the beginning of each year, my car is in good, fair, or broken-down condition. A good car will be good at the beginning of next year with probability .85; fair with probability .10; or broken-down with probability .05. A fair car will be fair at the beginning of the next year with probability .70 or broken-down with probability .30. It costs $6,000 to purchase a good car; a fair car can be traded in for $2,000; and a broken-down car has no trade-in value and must immediately be replaced by a good car. It costs $1,000 per year to operate a good car and $1,500 to operate a fair car. Should I replace my car as soon as it becomes a fair car, or should I drive my car until it breaks down? Assume that the cost of operating a car during a year depends on the type of car on hand at the beginning of the year (after a new car, if any, arrives).

5 A square matrix is said to be doubly stochastic if its entries are all nonnegative and the entries in each row and each column sum to 1. For any ergodic, doubly stochastic matrix, show that all states have the same steady-state probability.

6 This problem will show why steady-state probabilities are sometimes referred to as stationary probabilities. Let $\pi_1, \pi_2, \ldots, \pi_s$ be the steady-state probabilities for an ergodic chain with transition matrix P. Also suppose that with probability π_i, the Markov chain begins in state i.

a What is the probability that after one transition, the system will be in state i? (*Hint:* Use Equation (8).)

b For any value of $n(n = 1, 2, \ldots)$, what is the probability that a Markov chain will be in state i after n transitions?

c Why are steady-state probabilities sometimes called stationary probabilities?

7 Consider two stocks. Stock 1 always sells for $10 or $20. If stock 1 is selling for $10 today, there is a .80 chance that it will sell for $10 tomorrow. If it is selling for $20 today, there is a .90 chance that it will sell for $20 tomorrow.

Stock 2 always sells for $10 or $25. If stock 2 sells today for $10, there is a .90 chance that it will sell tomorrow for $10. If it sells today for $25, there is a .85 chance that it will sell tomorrow for $25. On the average, which stock will sell for a higher price? Find and interpret all mean first passage times.

8 Three balls are divided between two containers. During each period a ball is randomly chosen and switched to the other container.

    a Find (in the steady state) the fraction of the time that a container will contain 0, 1, 2, or 3 balls.

    b If container 1 contains no balls, on the average how many periods will go by before it again contains no balls? (*Note:* This is a special case of the Ehrenfest Diffusion model, which is used in biology to model diffusion through a membrane.)

9 Two types of squirrels—gray and black—have been seen in Pine Valley. At the beginning of each year, we determine which of the following is true:

There are only gray squirrels in Pine Valley.

There are only black squirrels in Pine Valley.

There are both gray and black squirrels in Pine Valley.

There are no squirrels in Pine Valley.

Over the course of many years, the following transition matrix has been estimated.

	Gray	Black	Both	Neither
Gray	.7	.2	.05	.05
Black	.2	.6	.1	.1
Both	.1	.1	.8	0
Neither	.05	.05	.1	.8

    a During what fraction of years will gray squirrels be living in Pine Valley?

    b During what fraction of years will black squirrels be living in Pine Valley?

Group B

10 Payoff Insurance Company charges a customer according to his or her accident history. A customer who has had no accident during the last two years is charged a $100 annual premium. Any customer who has had an accident during each of the last two years is charged a $400 annual premium. A customer who has had an accident during only one of the last two years is charged an annual premium of $300. A customer who has had an accident during the last year has a 10% chance of having an accident during the current year. If a customer has not had an accident during the last year, there is only a 3% chance that he or she will have an accident during the current year. During a given year, what is the average premium paid by a Payoff customer? (*Hint:* In case of difficulty, try a four-state Markov chain.)

11 Consider the following nonergodic chain:

$$P = \begin{bmatrix} \frac{1}{2} & \frac{1}{2} & 0 & 0 \\ \frac{1}{2} & \frac{1}{2} & 0 & 0 \\ 0 & 0 & \frac{1}{3} & \frac{2}{3} \\ 0 & 0 & \frac{2}{3} & \frac{1}{3} \end{bmatrix}.$$

a Why is the chain nonergodic?

b Explain why Theorem 1 fails for this chain. *Hint:* Find out if the following equation is true:

$$\lim_{n \to \infty} P_{12}(n) = \lim_{n \to \infty} P_{32}(n)$$

c Despite the fact that Theorem 1 fails, determine

$$\lim_{n \to \infty} P_{13}(n), \quad \lim_{n \to \infty} P_{21}(n),$$

$$\lim_{n \to \infty} P_{43}(n), \quad \lim_{n \to \infty} P_{41}(n)$$

12 Consider the following nonergodic chain:

$$P = \begin{bmatrix} 0 & 1 & 0 \\ 0 & 0 & 1 \\ 1 & 0 & 0 \end{bmatrix}$$

a Why is this chain nonergodic?

b Explain why Theorem 1 fails for this chain. (*Hint:* Show that $\lim_{n \to \infty} P_{11}(n)$ does not exist by listing the pattern that $P_{11}(n)$ follows as n increases.)

13 An important machine is known to never last more than four months. During its first month of operation, it fails 10% of the time. If the machine completes its first month, then it fails during its second month 20% of the time. If the machine completes its second month of operation, then it will fail during its third month 50% of the time. If the machine completes its third month, then it is sure to fail by the end of the fourth month. At the beginning of each month, we must decide whether or not to replace our machine with a new machine. It costs $500 to purchase a new machine, but if a machine fails during a month, we incur a cost of $1,000 (due to factory downtime) and must replace the machine (at the beginning of the next month) with a new machine. Three maintenance policies are under consideration:

Policy 1 Plan to replace a machine at the beginning of its fourth month of operation.
Policy 2 Plan to replace a machine at the beginning of its third month of operation.
Policy 3 Plan to replace a machine at the beginning of its second month of operation.

Which policy will give the lowest average monthly cost?

14 Each month, customers are equally likely to demand 1 or 2 computers from a Pearco dealer. All orders must be met from current stock. Two ordering policies are under consideration:

Policy 1 If ending inventory is 2 units or less, order enough to bring next month's beginning inventory to 4 units.
Policy 2 If ending inventory is 1 unit or less, order enough to bring next month's beginning inventory up to 3 units.

    The following costs are incurred by Pearco:

It costs $4,000 to order a computer.

It costs $100 to hold a computer in inventory for a month.

It costs $500 to place an order for computers. This is in addition to the per-customer cost of $4,000.

Which ordering policy has a lower expected monthly cost?

15 The Gotham City Maternity Ward contains 2 beds. Admissions are made only at the beginning of the day. Each day, there is a .5 probability that a potential admission will arrive. A patient can be admitted only if there is an open bed at the beginning of the day. Half of all patients are discharged after one day, and all patients that have stayed one day are discharged at the end of their second day.

a What is the fraction of days where all beds are utilized?

b On the average, what percentage of the beds are utilized?

5.6 Absorbing Chains

Many interesting applications of Markov chains involve chains in which some of the states are absorbing and the rest are transient states. Such a chain is called an **absorbing chain.** Consider an absorbing Markov chain: If we begin in a transient state, then eventually we are sure to leave the transient state and end up in one of the absorbing states. To see why we are interested in absorbing chains, we consider the following two absorbing chains.

EXAMPLE 7 **Accounts Receivable**

The accounts receivable situation of a firm is often modeled as an absorbing Markov chain.[†] Suppose a firm assumes that an account is uncollectable if the account is more than three months overdue. Then at the beginning of each month, each account may be classified into one of the following states:

State 1 New account

State 2 Payment on account is one month overdue.

State 3 Payment on account is two months overdue.

State 4 Payment on account is three months overdue.

State 5 Account has been paid.

State 6 Account is written off as bad debt.

Suppose that past data indicate that the following Markov chain describes how the status of an account changes from one month to the next month:

	New	1 month	2 months	3 months	Paid	Bad debt
New	0	.6	0	0	.4	0
1 month	0	0	.5	0	.5	0
2 months	0	0	0	.4	.6	0
3 months	0	0	0	0	.7	.3
Paid	0	0	0	0	1	0
Bad debt	0	0	0	0	0	1

For example, if an account is two months overdue at the beginning of a month, there is a 40% chance that at the beginning of next month, the account will not be paid up (and therefore be three months overdue) and a 60% chance that the account will be paid up. To simplify our example, we assume that after three months, a debt is either collected or written off as a bad debt.

Once a debt is paid up or written off as a bad debt, the account is closed, and no further transitions occur. Hence, Paid and Bad Debt are absorbing states. Since every account

[†]This example is based on Cyert, Davidson, and Thompson (1963).

will eventually be paid up or written off as a bad debt, New, 1 Month, 2 Months, and 3 Months are transient states. For example, a two-month overdue account can follow the path 2 Months–Collected, but there is no return path from Collected to 2 Months.

A typical new account will be absorbed as either a collected debt or a bad debt. A question of major interest is: What is the probability that a new account will eventually be collected? The answer is worked out later in this section.

<table>
<tr><td>**EXAMPLE 8**</td><td>**Work-Force Planning**</td></tr>
</table>

The law firm of Mason and Burger employs three types of lawyers: junior lawyers, senior lawyers, and partners. During a given year, there is a .15 probability that a junior lawyer will be promoted to senior lawyer and a .05 probability that he or she will leave the firm. Also, there is a .20 probability that a senior lawyer will be promoted to partner and a .10 probability that he or she will leave the firm. There is a .05 probability that a partner will leave the firm. The firm never demotes a lawyer.

There are many interesting questions the law firm might want to answer. For example, what is the probability that a newly hired junior lawyer will leave the firm before becoming a partner? On the average, how long does a newly hired junior lawyer stay with the firm? The answers are worked out later in this section.

We model the career path of a lawyer through Mason and Burger as an absorbing Markov chain with the following transition probability matrix:

	Junior	Senior	Partner	Leave as NP	Leave as P
Junior	.80	.15	0	.05	0
Senior	0	.70	.20	.10	0
Partner	0	0	.95	0	.05
Leave as nonpartner	0	0	0	1	0
Leave as partner	0	0	0	0	1

The last two states are absorbing states, and all other states are transient. For example, Senior is a transient state, because there is a path from Senior to Leave as Nonpartner, but there is no path returning from Leave as Nonpartner to Senior (we assume that once a lawyer leaves the firm, he or she never returns).

For any absorbing chain, one might want to know certain things. (1) If the chain begins in a given transient state, and before we reach an absorbing state, what is the expected number of times that each state will be entered? How many periods do we expect to spend in a given transient state before absorption takes place? (2) If a chain begins in a given transient state, what is the probability that we end up in each absorbing state?

To answer these questions, we need to write the transition matrix with the states listed in the following order: transient states first, then absorbing states. For the sake of definiteness, let's assume that there are $s - m$ transient states $(t_1, t_2, \ldots, t_{s-m})$ and m absorbing states $(a_1, a_2, \ldots, a_m)$. Then the transition matrix for the absorbing chain may be written as follows:

$$P = \begin{array}{c} \\ s - m \text{ rows} \\ m \text{ rows} \end{array} \begin{array}{c} \overset{s-m}{\underset{\text{columns}}{}} \quad \overset{m}{\underset{\text{columns}}{}} \\ \left[\begin{array}{c|c} Q & R \\ \hline 0 & I \end{array} \right] \end{array}$$

In this format, the rows and column of P correspond (in order) to the states $t_1, t_2, \ldots,$ $t_{s-m}, a_1, a_2, \ldots, a_m$. Here, I is an $m \times m$ identity matrix reflecting the fact that we can never leave an absorbing state: Q is an $(s - m) \times (s - m)$ matrix that represents transitions between transient states; R is an $(s - m) \times m$ matrix representing transitions from transient states to absorbing states; 0 is an $m \times (s - m)$ matrix consisting entirely of zeros. This reflects the fact that it is impossible to go from an absorbing state to a transient state.

Applying this notation to Example 7, we let

$$t_1 = \text{New}$$
$$t_2 = 1 \text{ Month}$$
$$t_3 = 2 \text{ Months}$$
$$t_4 = 3 \text{ Months}$$
$$a_1 = \text{Paid}$$
$$a_2 = \text{Bad Debt}$$

Then for Example 7, the transition probability matrix may be written as

	New	1 month	2 months	3 months	Paid	Bad debt
New	0	.6	0	0	.4	0
1 month	0	0	.5	0	.5	0
2 months	0	0	0	.4	.6	0
3 months	0	0	0	0	.7	.3
Paid	0	0	0	0	1	0
Bad debt	0	0	0	0	0	1

Then $s = 6$, $m = 2$, and

$$Q = \begin{bmatrix} 0 & .6 & 0 & 0 \\ 0 & 0 & .5 & 0 \\ 0 & 0 & 0 & .4 \\ 0 & 0 & 0 & 0 \end{bmatrix}_{4\times4} \qquad R = \begin{bmatrix} .4 & 0 \\ .5 & 0 \\ .6 & 0 \\ .7 & .3 \end{bmatrix}_{4\times2}$$

For Example 8, we let

$$t_1 = \text{Junior}$$
$$t_2 = \text{Senior}$$
$$t_3 = \text{Partner}$$
$$a_1 = \text{Leave as nonpartner}$$
$$a_2 = \text{Leave as partner}$$

and we may write the transition probability matrix as

	Junior	Senior	Partner	Leave as NP	Leave as P
Junior	.80	.15	0	.05	0
Senior	0	.70	.20	.10	0
Partner	0	0	.95	0	.05
Leave as nonpartner	0	0	0	1	0
Leave as partner	0	0	0	0	1

Then $s = 5$, $m = 2$, and

$$Q = \begin{bmatrix} .80 & .15 & 0 \\ 0 & .70 & .20 \\ 0 & 0 & .95 \end{bmatrix}_{3\times3} \qquad R = \begin{bmatrix} .05 & 0 \\ .10 & 0 \\ 0 & .05 \end{bmatrix}_{3\times2}$$

We can now find out some facts about absorbing chains (see Kemeny and Snell (1960). (1) If the chain begins in a given transient state, and before we reach an absorbing state, what is the expected number of times that each state will be entered? How many periods do we expect to spend in a given transient state before absorption takes place? *Answer:* If we are at present in transient state t_i, the expected number of periods that will be spent in transient state t_j before absorption is the ijth element of the matrix $(I - Q)^{-1}$. (See Problem 12 at the end of this section for a proof.) (2) If a chain begins in a given transient state, what is the probability that we end up in each absorbing state? *Answer:* If we are at present in transient state t_i, the probability that we will eventually be absorbed in absorbing state a_j is the ijth element of the matrix $(I - Q)^{-1} R$. (See Problem 13 at the end of this section for a proof.)

The matrix $(I - Q)^{-1}$ is often referred to as the **Markov chain's fundamental matrix.** The reader interested in further study of absorbing chains is referred to Kemeny and Snell (1960).

EXAMPLE 7   **Accounts Receivable (Continued)**

1 What is the probability that a new account will eventually be collected?

2 What is the probability that a one-month-overdue account will eventually become a bad debt?

3 If the firm's sales average $100,000 per month, how much money per year will go uncollected?

Solution From our previous discussion, recall that

$$Q = \begin{bmatrix} 0 & .6 & 0 & 0 \\ 0 & 0 & .5 & 0 \\ 0 & 0 & 0 & .4 \\ 0 & 0 & 0 & 0 \end{bmatrix} \qquad R = \begin{bmatrix} .4 & 0 \\ .5 & 0 \\ .6 & 0 \\ .7 & .3 \end{bmatrix}$$

Then

$$I - Q = \begin{bmatrix} 1 & -.6 & 0 & 0 \\ 0 & 1 & -.5 & 0 \\ 0 & 0 & 1 & -.4 \\ 0 & 0 & 0 & 1 \end{bmatrix}$$

By using the Gauss–Jordan method of Volume 1, Chapter 2, we find that

$$(I - Q)^{-1} = \begin{array}{c} \\ t_1 \\ t_2 \\ t_3 \\ t_4 \end{array} \begin{array}{c} \begin{matrix} t_1 & t_2 & t_3 & t_4 \end{matrix} \\ \begin{bmatrix} 1 & .60 & .30 & .12 \\ 0 & 1 & .50 & .20 \\ 0 & 0 & 1 & .40 \\ 0 & 0 & 0 & 1 \end{bmatrix} \end{array}$$

To answer questions 1–3, we need to compute

$$(I - Q)^{-1}R = \begin{array}{c} \\ t_1 \\ t_2 \\ t_3 \\ t_4 \end{array} \begin{array}{cc} a_1 & a_2 \\ \left[\begin{array}{cc} .964 & .036 \\ .940 & .060 \\ .880 & .120 \\ .700 & .300 \end{array} \right] \end{array}$$

Then

1 t_1 = New, a_1 = Paid. Thus, the probability that a new account is eventually collected is element 11 of $(I - Q)^{-1}R = .964$.

2 t_2 = 1 Month, a_2 = Bad Debt. Thus, the probability that a one-month overdue account turns into a bad debt is element 22 of $(I - Q)^{-1}R = .06$.

3 From answer 1, only 3.6% of all debts are uncollected. Since yearly accounts payable are $1,200,000, on the average, $(.036)(1,200,000) = \$43,200$ per year will be uncollected.

EXAMPLE 8   **Work-Force Planning (Continued)**

1 What is the average length of time that a newly hired junior lawyer spends working for the firm?

2 What is the probability that a junior lawyer makes it to partner?

3 What is the average length of time that a partner spends with the firm (as a partner)?

Solution   Recall that for Example 8,

$$Q = \begin{bmatrix} .80 & .15 & 0 \\ 0 & .70 & .20 \\ 0 & 0 & .95 \end{bmatrix} \qquad R = \begin{bmatrix} .05 & 0 \\ .10 & 0 \\ 0 & .05 \end{bmatrix}$$

Then

$$I - Q = \begin{bmatrix} .20 & -.15 & 0 \\ 0 & .30 & -.20 \\ 0 & 0 & .05 \end{bmatrix}$$

By using the Gauss–Jordan method of Volume 1, Chapter 2, we find that

$$(I - Q)^{-1} = \begin{array}{c} \\ t_1 \\ t_2 \\ t_3 \end{array} \begin{array}{ccc} t_1 & t_2 & t_3 \\ \left[\begin{array}{ccc} 5 & 2.5 & 10 \\ 0 & \frac{10}{3} & \frac{40}{3} \\ 0 & 0 & 20 \end{array} \right] \end{array}$$

Then

$$(I - Q)^{-1}R = \begin{array}{c} \\ t_1 \\ t_2 \\ t_3 \end{array} \begin{array}{cc} a_1 & a_2 \\ \left[\begin{array}{cc} .50 & .50 \\ \frac{1}{3} & \frac{2}{3} \\ 0 & 1 \end{array} \right] \end{array}$$

Then

1 Expected time junior lawyer stays with firm = (expected time junior lawyer stays with firm as junior) + (expected time junior lawyer stays with firm as senior) + (expected time junior lawyer stays with firm as partner). Now

$$\text{Expected time as junior} = (I - Q)_{11}^{-1} = 5$$
$$\text{Expected time as senior} = (I - Q)_{12}^{-1} = 2.5$$
$$\text{Expected time as partner} = (I - Q)_{13}^{-1} = 10$$

Hence, the total expected time that a junior lawyer spends with the firm is $5 + 2.5 + 10 = 17.5$ years.

2 The probability that a new junior lawyer makes it to partner is just the probability that he or she leaves the firm as a partner. Since $t_1 =$ Junior Lawyer and $a_2 =$ Leave as Partner, the answer is element 12 of $(I - Q)^{-1}R = .50$.

3 Since $t_3 =$ Partner, we seek the expected number of years that are spent in t_3, given that we begin in t_3. This is just element 33 of $(I - Q)^{-1} = 20$ years. This is reasonable, because during each year, there is 1 chance in 20 that a partner will leave the firm, so it should take an average of 20 years before a partner leaves the firm.

REMARKS

Computations with absorbing chains are greatly facilitated if we multiply matrices on a spreadsheet with the MMULT command and find the inverse of $(I - Q)$ with the MINVERSE function.

IQinverse.xls

To use the Excel MINVERSE command to find $(I - Q)^{-1}$, we enter $(I - Q)$ into a spreadsheet (see cell range C4:E6 of file IQinverse.xls) and select the range (C8:E10) where we want to compute $(I - Q)^{-1}$. Next we type the formula

$$=\text{MINVERSE(C4:E6)}$$

in the upper left-hand corner (cell C8) of the output range C8:E10. Finally, we select **CONTROL SHIFT ENTER** (not just ENTER) to complete the computation of the desired inverse. The MINVERSE function must be entered with CONTROL SHIFT ENTER because it is an array function. We cannot edit or delete any part of a range computed by an array function. See Figure 8.

	B	C	D	E	F
2					
3					
4			0.2	-0.15	0	
5	I-Q		0	0.3	-0.2	
6			0	0	0.05	
7						
8			5	2.5	10	
9	(I-Q)$^{-1}$		0	3.333333	13.33333	
10			0	0	20	
11						

FIGURE 8

PROBLEMS

Group A

1[†] The State College admissions office has modeled the path of a student through State College as a Markov chain:

	F.	So.	J.	Sen.	Q.	G.
Freshman	.10	.80	0	0	.10	0
Sophmore	0	.10	.85	0	.05	0
Junior	0	0	.15	.80	.05	0
Senior	0	0	0	.10	.05	.85
Quits	0	0	0	0	1	0
Graduates	0	0	0	0	0	1

[†]Based on Bessent and Bessent (1980).

Each student's state is observed at the beginning of each fall semester. For example, if a student is a junior at the beginning of the current fall semester, there is an 80% chance that he will be a senior at the beginning of the next fall semester, a 15% chance that he will still be a junior, and a 5% chance that he will have quit. (We assume that once a student quits, he never reenrolls.)

a If a student enters State College as a freshman, how many years can he expect to spend as a student at State?

b What is the probability that a freshman graduates?

2[†] The *Herald Tribble* has obtained the following information about its subscribers: During the first year as subscribers, 20% of all subscribers cancel their subscriptions. Of those who have subscribed for one year, 10% cancel during the second year. Of those who have been subscribing for more than two years, 4% will cancel during any given year. On the average, how long does a subscriber subscribe to the *Herald Tribble?*

3 A forest consists of two types of trees: those that are 0–5 ft and those that are taller than 5 ft. Each year, 40% of all 0–5-ft tall trees die, 10% are sold for $20 each, 30% stay between 0 and 5 ft, and 20% grow to be more than 5 ft. Each year, 50% of all trees taller than 5 ft are sold for $50, 20% are sold for $30, and 30% remain in the forest.

    a What is the probability that a 0–5-ft tall tree will die before being sold?

    b If a tree (less than 5 ft) is planted, what is the expected revenue earned from that tree?

4[‡] Absorbing Markov chains are used in marketing to model the probability that a customer who is contacted by telephone will eventually buy a product. Consider a prospective customer who has never been called about purchasing a product. After one call, there is a 60% chance that the customer will express a low degree of interest in the product, a 30% chance of a high degree of interest, and a 10% chance the customer will be deleted from the company's list of prospective customers. Consider a customer who currently expresses a low degree of interest in the product. After another call, there is a 30% chance that the customer will purchase the product, a 20% chance the person will be deleted from the list, a 30% chance that the customer will still possess a low degree of interest, and a 20% chance that the customer will express a high degree of interest. Consider a customer who currently expresses a high degree of interest in the product. After another call, there is a 50% chance that the customer will have purchased the product, a 40% chance that the customer will still have a high degree of interest, and a 10% chance that the customer will have a low degree of interest.

    a What is the probability that a new prospective customer will eventually purchase the product?

    b What is the probability that a low-interest prospective customer will ever be deleted from the list?

    c On the average, how many times will a new prospective customer be called before either purchasing the product or being deleted from the list?

5 Each week, the number of acceptable-quality units of a drug that are processed by a machine is observed: >100, 50–100, 1–50, 0 (indicating that the machine was broken during the week). Given last week's observation, the probability distribution of next week's observation is as follows.

	>100	50–100	1–50	0
>100	.8	.1	.05	.05
50–100	.1	.6	.1	.2
1–50	.1	.1	.5	.3
0	0	0	0	1

[†]Based on Deming and Glasser (1968).
[‡]Based on Thompson and McNeal (1967).

For example, if we observe a week in which more than 100 units are produced, then there is a .10 chance that during the next week 50–100 units are produced.

    a Suppose last week the machine produced 200 units. On average, how many weeks will elapse before the machine breaks down?

    b Suppose last week the machine produced 50 units. On average, how many weeks will elapse before the machine breaks down?

6 I now have $2, and my goal is to have $6. I will repeatedly flip a coin that has a .4 chance of coming up heads. If the coin comes up heads, I win the amount I bet. If the coin comes up tails, I lose the amount of my bet. Let us suppose I follow the **bold strategy** of betting Min($6 − current asset position, current asset position). This strategy (see Section 7.3) maximizes my chance of reaching my goal. What is the probability that I reach my goal?

7 Suppose I toss a fair coin, and the first toss comes up heads. If I keep tossing the coin until I either see two consecutive heads or two consecutive tails, what is the probability that I will see two consecutive heads before I see two consecutive tails?

8 Suppose each box of Corn Snaps cereal contains one of five different Harry Potter trading cards. On the average, how many boxes of cereal will I have to buy to obtain a complete set of trading cards?

Group B

9 In the gambler's ruin problem (Example 1), assume $p = .60$.

    a What is the probability that I reach $4?

    b What is the probability that I am wiped out?

    c What is the expected duration of the game?

10[§] In caring for elderly patients at a mental hospital, a major goal of the hospital is successful placement of the patients in boarding homes or nursing homes. The movement of patients between the hospital, outside homes, and the absorbing state (death) may be described by the following Markov chain (the unit of time is one month):

	Hospital	Homes	Death
Hospital	.991	.003	.006
Homes	.025	.969	.006
Death	0	0	1

Each month that a patient spends in the hospital costs the state $655, and each month that a patient spends in a home costs the state $226. To improve the success rate of the placement of patients in homes, the state has recently begun a "geriatric resocialization program" (GRP) to prepare the patients for functioning in the homes. Some patients are placed in the GRP and then released to homes. These patients presumably are less likely to fail to adjust in the homes. Other patients continue to go directly from the hospital to homes without taking part in the GRP. The state pays $680 for each month that a patient spends in the GRP. The

[§]Based on Meredith (1973).

movement of the patients through various states is governed by the following Markov chain:

	GRP	Hos.	Homes (GRP)	Homes (Direct)	Dead
GRP	.854	.028	.112	0	.006
Hospital	.013	.978	0	.003	.006
Homes (GRP)	.025	0	.969	0	.006
Homes (Direct)	0	.025	0	.969	.006
Dead	0	0	0	0	1

a Does the GRP save the state money?

b Under the old system and under the GRP, compute the expected number of months that a patient spends in the hospital.

11 Freezco, Inc., sells refrigerators. The company has issued a warranty on all refrigerators that requires free replacement of any refrigerator that fails before it is three years old. We are given the following information: (1) 3% of all new refrigerators fail during their first year of operation; (2) 5% of all one-year-old refrigerators fail during their second year of operation; and (3) 7% of all two-year-old refrigerators fail during their third year of operation. A replacement refrigerator is not covered by the warranty.

a Use Markov chain theory to predict the fraction of all refrigerators that Freezco will have to replace.

b Suppose that it costs Freezco $500 to replace a refrigerator and that Freezco sells 10,000 refrigerators per year. If the company reduced the warranty period to two years, how much money in replacement costs would be saved?

12 For a Q matrix representing the transitions between transient states in an absorbing Markov chain, it can be shown that

$$(I - Q)^{-1} = I + Q + Q^2 + \cdots + Q^n + \cdots$$

a Explain why this expression for $(I - Q)^{-1}$ is plausible.

b Define m_{ij} = expected number of periods spent in transient state t_j before absorption, given that we begin in state t_i. (Assume that the initial period is spent in state t_i.) Explain why m_{ij} = (probability that we are in state t_j initially) + (probability that we are in state t_j after first transition) + (probability that we are in state t_j after second transition) + $\cdots$ + (probability that we are in state t_j after nth transition) + $\cdots$.

c Explain why the probability that we are in state t_j initially = ijth entry of the $(s - m) \times (s - m)$ identity matrix. Explain why the probability that we are in state t_j after nth transition = ijth entry of Q^n.

d Now explain why m_{ij} = ijth entry of $(I - Q)^{-1}$.

13 Define

b_{ij} = probability of ending up in absorbing state a_j given that we begin in transient state t_i

r_{ij} = ijth entry of R

q_{ik} = ikth entry of Q

B = $(s - m) \times m$ matrix whose ijth entry is b_{ij}

Suppose we begin in state t_i. On our first transition, three types of events may happen:

Event 1 We go to absorbing state a_j (with probability r_{ij}).

Event 2 We go to an absorbing state other than a_j (with probability $\sum_{k \neq j} r_{ik}$).

Event 3 We go to transient state t_k (with probability q_{ik}).

    a Explain why

$$b_{ij} = r_{ij} + \sum_{k=1}^{k=s-m} q_{ik}b_{kj}$$

    b Now show that b_{ij} = ijth entry of $(R + QB)$ and that $B = R + QB$.

    c Show that $B = (I - Q)^{-1}R$ and that b_{ij} = ijth entry of $B = (I - Q)^{-1}R$.

14 Consider an LP with five basic feasible solutions and a unique optimal solution. Assume that the simplex method begins at the worst basic feasible solution, and on each pivot the simplex is equally likely to move to any better basic feasible solution. On the average, how many pivots will be required to find the optimal solution to the LP?

Group C

15 General Motors has three auto divisions (1, 2, and 3). It also has an accounting division and a management consulting division. The question is: What fraction of the cost of the accounting and management consulting divisions should be allocated to each auto division? We assume that the entire cost of the accounting and management consulting departments must be allocated to the three auto divisions. During a given year, the work of the accounting division and management consulting division is allocated as shown in Table 4.

For example, accounting spends 10% of its time on problems generated by the accounting department, 20% of its time on work generated by division 3, and so forth. Each year, it costs $63 million to run the accounting department and $210 million to run the management consulting department. What fraction of these costs should be allocated to each auto division? Think of $1 in costs incurred in accounting work. There is a .20 chance that this dollar should be allocated to each auto division, a .30 chance it should be allocated to consulting, and a .10 chance to accounting. If the dollar is allocated to an auto division, we know which division should be charged for that dollar. If the dollar is charged to consulting (for example), we repeat the process until the dollar is eventually charged to an auto division. Use knowledge of absorbing chains to figure out how to allocate the costs of running the accounting and management consulting departments among the three auto divisions.

16 A telephone sales force can model its contact with customers as a Markov chain. The six states of the chain are as follows:

State 1 Sale completed during most recent call
State 2 Sale lost during most recent call
State 3 New customer with no history
State 4 During most recent call, customer's interest level low

TABLE 4

	Accounting	Management Consulting	Division 1	Division 2	Division 3
Accounting	10%	30%	20%	20%	20%
Management	30%	20%	30%	0%	20%

State 5 During most recent call, customer's interest level medium

State 6 During most recent call, customer's interest level high

Based on past phone calls, the following transition matrix has been estimated:

$$
\begin{array}{c c}
 & \begin{array}{cccccc} 1 & 2 & 3 & 4 & 5 & 6 \end{array} \\
\begin{array}{c} 1 \\ 2 \\ 3 \\ 4 \\ 5 \\ 6 \end{array} &
\left[\begin{array}{cccccc}
1 & 0 & 0 & 0 & 0 & 0 \\
0 & 1 & 0 & 0 & 0 & 0 \\
.10 & .30 & 0 & .25 & .20 & .15 \\
.05 & .45 & 0 & .20 & .20 & .10 \\
.15 & .10 & 0 & .15 & .25 & .35 \\
.20 & .05 & 0 & .15 & .30 & .30
\end{array} \right]
\end{array}
$$

a For a new customer, determine the average number of calls made before the customer buys the product or the sale is lost.

b What fraction of new customers will buy the product?

c What fraction of customers currently having a low degree of interest will buy the product?

d Suppose a call costs $15 and a sale earns $190 in revenue. Determine the "value" of each type of customer.

17 Seas Beginning sells clothing by mail order. An important question is: When should the company strike a customer from its mailing list? At present, the company does so if a customer fails to order from six consecutive catalogs. Management wants to know if striking a customer after failure to order from four consecutive catalogs will result in a higher profit per customer.

The following data are available: Six percent of all customers who receive a catalog for the first time place an order. If a customer placed an order from the last-received catalog, then there is a 20% chance he or she will order from the next catalog. If a customer last placed an order one catalog ago, there is a 16% chance he or she will order from the next catalog received. If a customer last placed an order two catalogs ago, there is a 12% chance he or she will place an order from the next catalog received. If a customer last placed an order three catalogs ago, there is an 8% chance he or she will place an order from the next catalog received. If a customer last placed an order four catalogs ago, there is a 4% chance he or she will place an order from the next catalog received. If a customer last placed an order five catalogs ago, there is a 2% chance he or she will place an order from the next catalog received.

It costs $1 to send a catalog, and the average profit per order is $15. To maximize expected profit per customer, should Seas Beginning cancel customers after six nonorders or four nonorders?

Hint: Model each customer's evolution as a Markov chain with possible states New, 0, 1, 2, 3, 4, 5, Canceled. A customer's state represents the number of catalogs received since the customer last placed an order. "New" means the customer received a catalog for the first time. "Canceled" means that the customer has failed to order from six consecutive catalogs. For example, suppose a customer placed the following sequence of orders (O) and nonorders (NO):

NO NO O NO NO O O NO NO O NO NO NO NO NO NO Canceled

Here we are assuming a customer is stricken from the mailing list after six consecutive nonorders. For this sequence of orders and nonorders, the states are (*i*th listed state occurs right before *i*th catalog is received)

New 1 2 0 1 2 0 0 1 2 0 1 2 3 4 5 Canceled

You should be able to figure (for each cancellation policy) the expected number of orders a customer will place before cancellation and the expected number of catalogs a customer will receive before cancellation. This will enable you to compute expected profit per customer.

5.7 Work-Force Planning Models[†]

Many organizations, like the Mason and Burger law firm of Example 8, employ several categories of workers. For long-term planning purposes, it is often useful to be able to predict the number of employees of each type who will (if present trends continue) be available in the steady state. Such predictions can be made via an analysis similar to the one in Section 5.5 of steady-state probabilities for Markov chains.

More formally, consider an organization whose members are classified at any point in time into one of s groups (labeled $1, 2, \ldots, s$). During every time period, a fraction p_{ij} of

[†]This section covers topics that may be omitted with no loss of continuity.

those who begin a time period in group i begin the next time period in group j. Also, during every time period, a fraction $p_{i,s+1}$ of all group i members leave the organization. Let P be the $s \times (s + 1)$ matrix whose ijth entry is p_{ij}. At the beginning of each time period, the organization hires H_i group i members. Let $N_i(t)$ be the number of group i members at the beginning of period t. A question of natural interest is whether $N_i(t)$ approaches a limit as t grows large (call the limit, if it exists, N_i). If each $N_i(t)$ does not approach a limit, we call $\mathbf{N} = (N_1, N_2, \ldots, N_s)$ the **steady-state census** of the organization.

If a steady-state census exists, we can find it by solving a system of s equations that is derived as follows: Simply note that for a steady-state census to exist, it must be true that in the steady state, for $i = 1, 2, \ldots, s$,

$$\begin{array}{r} \text{Number of people entering group } i \text{ during each period} \\ = \text{number of people leaving group } i \text{ during each period} \end{array} \tag{14}$$

After all, if (14) did not hold for all groups, then the number of people in at least one group would pile up as time progressed. We note that

$$\text{Number of people entering state } i \text{ during each period} = H_i + \sum_{k \neq i} N_k p_{ki}$$

$$\text{Number of people leaving state } i \text{ during each period} = N_i \sum_{k \neq i} p_{ik}$$

Then the equation used to compute the steady-state census is

$$H_i + \sum_{k \neq i} N_k p_{ki} = N_i \sum_{k \neq i} p_{ik} \ (i = 1, 2, \ldots, s) \tag{14$'$}$$

Note that $\sum_{k \neq i} p_{ik} = 1 - p_{ii}$. This can be used to simplify (14$'$).

If a steady-state census does not exist, then (14$'$) will have no solution. See Problem 6 for an example of this. Given the values of the p_{ij}'s and the H_i's, (14$'$) can be used to solve for the steady-state census. Conversely, given the p_{ij}'s and a desired steady-state census, (14$'$) can be used to determine a hiring policy (specified by values of $H_1, H_2, \ldots, H_s$) that attains the desired steady-state census. Some steady-state censuses may be impossible to maintain unless some of the H_i's are negative (corresponding to firing employees).

The following two examples illustrate the use of the steady-state census equation.

EXAMPLE 9   **Steady-State Census**

Suppose that each American can be classified into one of three groups: children, working adults, or retired people. During a one-year period, .959 of all children remain children, .04 of all children become working adults, and .001 of all children die. During any given year, .96 of all working adults remain working adults, .03 of all working adults retire, and .01 of all working adults die. Also, .95 of all retired people remain retired, and .05 of all retired people die. One thousand children are born each year.

1 Determine the steady-state census.

2 Each retired person receives a pension of $5,000 per year. The pension fund is funded by payments from working adults. How much money must each working adult contribute annually to the pension fund?

Solution   **1**   Let

$$\text{Group 1} = \text{children}$$
$$\text{Group 2} = \text{working adults}$$
$$\text{Group 3} = \text{retired people}$$
$$\text{Group 4} = \text{died}$$

We are given that $H_1 = 1{,}000$, $H_2 = H_3 = 0$, and

$$P = \begin{bmatrix} .959 & .040 & 0 & .001 \\ 0 & .960 & .030 & .010 \\ 0 & 0 & .950 & .050 \end{bmatrix}$$

Now (14) or (14′) yields

Number entering group i each year = number leaving group i each year

$$1{,}000 = (.04 + .001)N_1 \qquad \text{(Children)}$$
$$.04N_1 = (.03 + .01)N_2 \qquad \text{(Working adults)}$$
$$.03N_2 = .05N_3 \qquad \text{(Retired people)}$$

Solving this system of equations, we find that $N_1 = 24{,}390$, $N_2 = 24{,}390.24$, and $N_3 = 14{,}634.14$.

2 Since in the steady state, there are 14,634.14 retired people, in the steady state they receive 14,634.14(5,000) dollars per year. Hence, each working adult must pay

$$\frac{14{,}634.14(5{,}000)}{24{,}390.24} = \$3{,}000 \text{ per year}$$

This result is reasonable, because in the steady state, there are $\frac{5}{3}$ as many working adults as there are retired people.

EXAMPLE 10 **The Mason and Burger Law Firm (continued)**

Let's return to the law firm of Mason and Burger (Example 8). Suppose the firm's long-term goal is to employ 50 junior lawyers, 30 senior lawyers, and 10 partners. To achieve this steady-state census, how many lawyers of each type should Mason and Burger hire each year?

Solution Let

Group 1 = junior lawyers

Group 2 = senior lawyers

Group 3 = partners

Group 4 = lawyers who have left firm

Mason and Burger want to obtain $N_1 = 50$, $N_2 = 30$, and $N_3 = 10$. Recall from Example 8 that

$$P = \begin{bmatrix} .80 & .15 & 0 & .05 \\ 0 & .70 & .20 & .10 \\ 0 & 0 & .95 & .05 \end{bmatrix}$$

Then (14) or (14′) yields

Number entering group i = number leaving group i

$$H_1 = (.15 + .05)50 \qquad \text{(Junior lawyers)}$$
$$(.15)50 + H_2 = (.20 + .10)30 \qquad \text{(Senior lawyers)}$$
$$(.20)30 + H_3 = (.05)10 \qquad \text{(Partners)}$$

The unique solution to this system of equations is $H_1 = 10$, $H_2 = 1.5$, $H_3 = -5.5$. This means that to maintain the desired steady-state census, Mason and Burger would have to fire 5.5 partners each year. This is reasonable, because an average of .20(30) = 6 senior lawyers become partners every year, and once a senior lawyer becomes a partner, he or she stays a partner for an average of 20 years. This shows that to keep the number of partners down to 10, several partners must be released each year. An alternative solution might be to reduce (below its current value of .20) the fraction of senior lawyers who become partners during each year.

For more information on work-force planning models, the interested reader should consult the excellent book by Grinold and Marshall (1977).

Using LINGO to Solve for the Steady-State Census

Census.lng

The following LINGO model (file Census.lng) can be used to determine the steady-state census for a work-force planning problem:

```
MODEL:
1]SETS:
2]STATE/1..3/:N,H;
3]SXS(STATE,STATE):TPROB;
4]ENDSETS
5]DATA:
6]H=1000,0,0;
7]TPROB=.959,.04,0,
8]0,.96,.03,
9]0,0,.95;
10]ENDDATA
11]@FOR(STATE(I):H(I)
12]+@SUM(STATE(K)|K#NE#I:N(K)*TPROB(K,I))=
13]N(I)*(1-TPROB(I,I)););
END
```

In line 2, we create the possible states and define for each state I the steady-state census level and number hired, $N(I)$ and $H(I)$, respectively. In line 3, we create for each pair (I, J) of states the probability TPROB (I, J) of going from state I in one period to state J during the next period. In line 6, we input the value of $H(I)$ for each state I. In lines 7 through 9, we input the TPROB (I, J) for Example 9. In lines 11 through 13, we create for each state I the equation $(14)'$. Note that we use the fact that

$$\sum_{K \neq I} \text{TPROB}(I, K) = 1 - \text{TPROB}(I, I)$$

Entering the **GO** command will yield the steady-state census level $N(I)$ for state I. Note that by modifying the DATA portion of the program, we could also enter a desired steady-state census ($N(I)$) and have LINGO solve for a set of hiring levels ($H(I)$) which yield the desired steady-state census.

PROBLEMS

Group A

1 Refer to Problem 1 of Section 5.6. Suppose that each year, State College admits 7,000 freshmen, 500 sophomore transfers, and 500 junior transfers. In the long run, what will be the composition of the State College student body?

2 In Example 9, suppose that advances in medical science have reduced the annual death rate for retired people from 5% to 3%. By how much would this increase the annual pension contribution that a working adult would have to make to the pension fund?

3 New York City produces 1,000 tons of air pollution per day, Jersey City 100 tons, and Newark 50 tons. Each day, $\frac{1}{3}$ of New York's pollution is blown to Newark, $\frac{1}{3}$ dissipates, and $\frac{1}{3}$ remains in New York. Each day, $\frac{1}{3}$ of Jersey City's pollution is blown to New York, $\frac{1}{3}$ stays in Jersey City, and $\frac{1}{3}$ is blown to Newark. Each day, $\frac{1}{3}$ of Newark's pollution stays in Newark, and the rest is blown to Jersey City. On a typical day, which city will be the most polluted?

4 Money circulates among the Federation's three "capital" planets: Vulcan, Romulanville, and Klingonville. Ideally, the Federation would like to have $5 billion in circulation at each planet. Each month, $\frac{1}{3}$ of all the money at Vulcan leaves circulation, $\frac{1}{3}$ stays at Vulcan, and $\frac{1}{3}$ ends up in Klingonville. Each month, $\frac{1}{3}$ of the money at Romulanville remains in Romulanville, $\frac{1}{3}$ ends up in Klingonville, and $\frac{1}{3}$ ends up at Vulcan. Each month, $\frac{2}{3}$ of the money in Klingonville ends up in Romulanville, and $\frac{1}{3}$ stays in Klingonville. The Federation introduces money into the system at Vulcan. Is there any way to have a steady-state level of $5 billion in circulation at each planet?

Group B

5 All State University Business School faculty members are classified as tenured or untenured. Each year, 10% of the untenured faculty are granted tenure and 10% leave State University; 95% of the tenured faculty remain and 5% leave. The business school wants to maintain a faculty with 100 members, of which x% are untenured. Determine a hiring policy that will achieve this goal. For what values of x does this goal require firing tenured faculty members? Describe a hiring policy that maintains a faculty that is 10% untenured. Describe a hiring policy that maintains a faculty that is 40% untenured.

6 In the world of Never-Ever Land, one child is born at the beginning of each year. During each year, 90% of the children alive at the beginning of the year remain children, and 10% become adults. During each year, 90% of the adults alive at the beginning of the year remain adults and 10% of the adults become children.

   a Explain why no steady-state census exists.

   b Show that equation (14′) has no solution.

Group C

7[†] For simplicity, suppose that fresh blood obtained by a hospital will spoil if it is not transfused within five days. The hospital receives 100 pints of fresh blood daily from a local blood bank. Two policies are possible for determining the order in which blood is transfused (see Table 5). For example, under policy 1, blood has a 10% chance of being transfused during its first day at the hospital. Under policy 2, four-day-old blood has a 10% chance of being transfused.

   a A FIFO (first in, first out) blood-issuing policy issues "old" blood first, whereas a LIFO (last in, first out) policy issues "young" blood first. Which policy represents a LIFO policy, and which represents a FIFO policy?

TABLE 5

	Age of Blood (beginning of day)				
Chance of transfusion	0	1	2	3	4
Policy 1	.10	.20	.30	.40	.50
Policy 2	.50	.40	.30	.20	.10

   b For each policy, determine the probability that a new pint of blood will spoil.

   c For each policy, determine the average number of pints of blood in inventory.

   d For each policy, find the average age of transfused blood.

   e Comment on the relative merits of a FIFO policy and a LIFO policy.

8 Suppose that each week every American family buys a gallon of orange juice from company A or B or C. Let p_i = probability that a gallon produced by company i is of unsatisfactory quality. If the last gallon of juice purchased by a family is satisfactory, then the next week they will purchase a gallon of juice from the same company. If the last gallon of juice purchased by a family is not satisfactory, then the family will purchase a gallon from a competitor. Consider a week in which A families have purchased juice A, B families have purchased juice B, and C families have purchased juice C. Assume that families that switch brands during a period are allocated to the remaining brands in a manner proportionate to the current market shares of the other brands. Thus, if a family switches from brand A, there is a chance $B/(B + C)$ that they will switch to B and a chance $C/(B + C)$ that they will switch to C. Suppose that 1 million gallons of orange juice are purchased each week.

   a After a long time, what will be the market share for each firm? *Hint:* Show that for some k in the steady state, brand A will sell $k(p_B + p_C - p_A)$ gallons of juice each week, and conjecture the number of gallons of brands B and C that will be sold each week.

   b Suppose a 1% increase in market share is worth $10,000 per week to firm A. Also suppose that currently $p_A = .10$, $p_B = .15$, and $p_C = .20$. Firm A believes that for a cost of $1 million per year, it can cut the percentage of unsatisfactory juice cartons in half. Is this worthwhile?[‡]

9 The age-based probability that an American dies during a given year is shown in Table 6. For example, a fraction

TABLE 6

Age	Death Probability
0	0.007557
1–4	0.000383
5–9	0.000217
10–14	0.000896
15–24	0.001267
25–34	0.002213
35–44	0.004459
45–54	0.010941
55–64	0.025384
65–84	0.058031
85+	0.15327

[†]Based on Pegels and Jelmert (1970).

[‡]Based on Babich (1992).

.007557 of all babies die during their first year of life. Suppose 100 babies are born each year, and nobody lives to be older than 110.

a What is the average age of people in the United States?

b Suppose all people ages 21–65 work, and all people over age 65 are retired. If we want to pay each retiree $20,000 per year, how much money must each worker pay in to ensure that during each year, the retirement plan is self-financing?

SUMMARY

Let X_t be the value of a system's characteristic at time t. A **discrete-time stochastic process** is simply a description of the relation between the random variables X_0, X_1, X_2, ... A discrete-time stochastic process is a **Markov chain** if, for $t = 0, 1, 2, \ldots$ and all states,

$$P(X_{t+1} = i_{t+1}|X_t = i_t, X_{t-1} = i_{t-1}, \ldots, X_1 = i_1, X_0 = i_0)$$
$$= P(X_{t+1} = i_{t+1}|X_t = i_t)$$

For a stationary Markov chain, the **transition probability** p_{ij} is the probability that given the system is in state i at time t, the system will be in state j at time $t + 1$.

The vector $\mathbf{q} = [q_1 \quad q_2 \quad \cdots \quad q_s]$ is the **initial probability distribution** for the Markov chain. $P(X_0 = i)$ is given by q_i.

n-Step Transition Probabilities

The **n-step transition probability,** $p_{ij}(n)$, is the probability that n periods from now, the state will be j, given that the current state is i. $P_{ij}(n) = ij$th element of P^n.

Given the intial probability vector $\mathbf{q}$, the probability of being in state j at time n is given by $\mathbf{q}(\text{column } j \text{ of } P^n)$.

Classification of States in a Markov Chain

Given two states i and j, a **path** from i to j is a sequence of transitions that begins in i and ends in j, such that each transition in the sequence has a positive probability of occurring. A state j is **reachable** from a state i if there is a path leading from i to j. Two states i and j are said to **communicate** if j is reachable from i, and i is reachable from j.

A set of states S in a Markov chain is a **closed set** if no state outside of S is reachable from any state in S.

A state i is an **absorbing state** if $p_{ii} = 1$. A state i is a **transient state** if there exists a state j that is reachable from i, but the state i is not reachable from state j.

If a state is not transient, it is a **recurrent state**. A state i is **periodic** with period $k > 1$ if all paths leading from state i back to state i have a length that is a multiple of k. If a recurrent state is not periodic, it is **aperiodic.** If all states in a chain are recurrent, aperiodic, and communicate with each other, the chain is said to be **ergodic.**

Steady-State Probabilities

Let P be the transition probability matrix for an ergodic Markov chain with states 1, 2, ..., s (with ijth element p_{ij}). After a large number of periods have elapsed, the probability (call it π_j) that the Markov chain is in state j is independent of the initial state. The long-run, or **steady-state,** probability π_j may be found by solving the following set of linear equations:

$$\pi_j = \sum_{k=1}^{k=s} \pi_k p_{kj} \qquad (j = 1, 2, \ldots, s; \text{ omit one of these equations})$$

$$\pi_1 + \pi_2 + \cdots + \pi_s = 1$$

Absorbing Chains

A Markov chain in which one or more states is an absorbing state is an **absorbing Markov chain.** To answer important questions about an absorbing Markov chain, we list the states in the following order: transient states first, then absorbing states. Assume there are $s - m$ transient states $(t_1, t_2, \ldots, t_{s-m})$ and m absorbing states $(a_1, a_2, \ldots, a_m)$. Write the transition probability matrix P as follows:

$$
P = \begin{array}{c} \\ s - m \text{ rows} \\ m \text{ rows} \end{array}
\begin{array}{cc}
\overset{\substack{s - m \\ \text{columns}}}{} & \overset{\substack{m \\ \text{columns}}}{} \\
\left[\begin{array}{c|c} Q & R \\ \hline 0 & I \end{array} \right]
\end{array}
$$

The following questions may now be answered. (1) If the chain begins in a given transient state, and before we reach an absorbing state, what is the expected number of times that each state will be entered? How many periods do we expect to spend in a given transient state before absorption takes place? *Answer:* If we are at present in transient state t_i, the expected number of periods that will be spent in transient state t_j before absorption is the ijth element of the matrix $(I - Q)^{-1}$. (2) If a chain begins in a given transient state, what is the probability that we will end up in each absorbing state? *Answer:* If we are at present in transient state t_i, the probability that we will eventually be absorbed in absorbing state a_j is the ijth element of the matrix $(I - Q)^{-1}R$.

Work-Force Planning Models

For an organization in which each member is classified into one of s groups,

$\quad p_{ij} = $ fraction of members beginning a time period in group i
$\qquad\qquad$ who begin the next time period in group j

$p_{i,s+1} = $ fraction of all group i members
$\qquad\qquad$ who leave the organization during a period

$\quad P = s \times (s + 1)$ matrix whose ijth entry is p_{ij}

$\quad H_i = $ number of group i members
$\qquad\qquad$ hired at the beginning of each period

$\quad N_i = $ limiting number (if it exists) of group i members

N_i may be found by equating the number of people per period who enter group i with the number of people per period who leave group i. Thus, $(N_1, N_2, \ldots, N_s)$ may be found by solving

$$H_i + \sum_{k \neq i} N_k p_{ki} = N_i \sum_{k \neq i} p_{ik} \qquad (i = 1, 2, \ldots, s)$$

REVIEW PROBLEMS

Group A

1 A machine is used to produce precision tools. If the machine is in good condition today, then 90% of the time, it will be in good condition tomorrow. If the machine is in bad condition today, then 80% of the time, it will be in bad condition tomorrow. If the machine is in good condition, it produces 100 tools per day. If the machine is in bad condition, it produces 60 tools per day. On the average, how many tools per day are produced?

2 Customers buy cars from three auto companies. Given the company from which a customer last bought a car, the probability that she will buy her next car from each company is as follows:

	Will Buy Next from		
Last Bought from	Co. 1	Co. 2	Co. 3
Co. 1	.80	.10	.10
Co. 2	.05	.85	.10
Co. 3	.10	.20	.70

a If someone currently owns a company 1 car, what is the probability that at least one of the next two cars she buys will be a company 1 car?

b At present, it costs company 1 an average of $5,000 to produce a car, and the average price a customer pays for one is $8,000. Company 1 is considering instituting a five-year warranty. It estimates that this will increase the cost per car by $300, but a market research survey indicates that the probabilities will change as follows:

	Will Buy Next from		
Last Bought from	Co. 1	Co. 2	Co. 3
Co. 1	.85	.10	.05
Co. 2	.10	.80	.10
Co. 3	.15	.10	.75

Should company 1 institute the five-year warranty?

3[†] A baseball team consists of 2 stars, 13 starters, and 10 substitutes. For tax purposes, the team owner must value the players. The value of each player is defined to be the total value of the salary he will earn until retirement. At the beginning of each season, the players are classified into one of four categories:

Category 1 Star (earns $1 million per year)
Category 2 Starter (earns $400,000 per year)
Category 3 Substitute (earns $100,000 per year)
Category 4 Retired (earns no more salary)

Given that a player is a star, starter, or substitute at the beginning of the current season, the probabilities that he will be a star, starter, substitute, or retired at the beginning of the next season are as follows:

	Next Season			
This Season	Star	Starter	Substitute	Retired
Star	.50	.30	.15	.05
Starter	.20	.50	.20	.10
Substitute	.05	.15	.50	.30
Retired	0	0	0	1

Determine the value of the team's players.

[†]Based on Flamholtz, Geis, and Perle (1984).

4 The best-selling college statistics text, *The Thrill of Statistics,* sells 5 million copies every fall. Some users keep the book, and some sell it back to the bookstore. Suppose that 90% of all students who buy a new book sell it back, 80% of all students who buy a once-used book sell it back, and 60% of all students who buy a twice-used book sell it back. If a book has been used four or more times, the cover falls off, and it cannot be sold back.

a In the steady state, how many new copies of the book will the publisher be able to sell each year?

b Suppose that a bookstore's profit on each type of book is as follows:

> New book: $6
> Once-used book: $3
> Twice-used book: $2
> Thrice-used book: $1

If the steady-state census is representative of the bookstore's sales, what will be its average profit per book?

5 Hearts Dog Food and Corporal Dog Food are battling tooth and nail for the nation's dog biscuit market. A dog owner buys one box of dog biscuits per month. If a dog owner's last purchase was a Hearts box of biscuits, there is a .8 chance that his next purchase will also be Hearts. If a dog owner's last purchase was a Corporal box of biscuits, there is a .9 chance that his next purchase will also be Corporal. It cost Hearts 80¢ to produce a box of biscuits, which sells for $1.

a If there are 40 million dog owners in the United States, what is Hearts' annual expected profit?

b If Hearts sells each box of biscuits for $100 - x$ cents ($0 \le x \le 20$), then a fraction $.8 + \frac{x}{100}$ of all dog owners whose last purchase was from Hearts will purchase their next box of biscuits from Hearts. How can Hearts maximize profit?

6 A small video store tracks the number of times per week a video is rented and estimates the following transition probabilities:

	5 times	4 times	3 times	2 times	1 time	0 time
5 times	.8	.1	.1	0	0	0
4 times	0	.7	.2	.1	0	0
3 times	0	0	.6	.3	.1	0
2 times	0	0	.5	.4	.1	0
1 time	0	0	0	0	.6	.4
0 time	0	0	0	0	0	1

For example, if a video was rented 5 times this week, then there is an 80% chance it will be rented 5 times next week, a 10% chance it will be rented 4 times, and a 10% chance it will be rented 3 times.

a Suppose a video was rented 5 times this week. On the average, how many times will it be rented during the next 2 weeks?

b Suppose a video was rented 5 times this week. On the average, how many more weeks will it be rented at least once?

c Suppose a video was rented 5 times this week. On the average, how many more times will it be rented?

7 Ross and Rachel have just tied the knot. The probability that they are happy each day depends on whether they were happy or sad during the last two days, in the following fashion:

Last two days	Happy	Sad
HH	.8	.2
HS	.5	.5
SH	.7	.3
SS	.4	.6

For example, if the newlyweds were sad two days ago and yesterday they were happy, then there is a 70% chance they will be happy tomorrow and a 30% chance they will be sad tomorrow. On what fraction of days will Ross and Rachel be happy?

8 Suppose that during a given year, 15% of all untenured processors leave a university (they are fired or find another job), and 15% are given tenure. Also assume that during each year, 5% of all tenured professors leave the university (via retirement or finding another job). If the university wants to have a faculty consisting of 200 untenured and 500 tenured professors, how many tenured and untenured professors should be hired each year?

Group B

9 At the beginning of a period, a company observes its inventory level. Then an order may be placed (and is instantaneously received). Finally, the period's demand is observed. We are given the following information: (1) A $2 cost is assessed against each unit of inventory on hand at the end of a period. (2) A $3 penalty is assessed against each unit of demand not met on time. Assume that all shortages result in lost sales. (3) Placing an order costs 50¢ per unit plus a $5 ordering cost. (4) During each period, demand is equally likely to equal 1, 2, or 3 units.

The company is considering the following ordering policy: At the end of any period, if the on-hand inventory is 1 unit or less, order sufficient units to bring the on-hand inventory level at the beginning of the next period up to 4 units.

a What fraction of the time will the on-hand inventory level at the end of each period be 0 unit? 1 unit? 2 units? 3 units? 4 units?

b Determine the average cost per period incurred by this ordering policy.

c Answer parts (a) and (b) if all shortages are backlogged. Assume that the cost for each unit backlogged is $3.

10[†] In problem 3, suppose that in evaluating a player's value, the owner must discount future salaries. Assume that $1 paid out in salary during the next season is equivalent to 90¢ paid out during the current season. Can you still determine the value of the team's players? (*Hint:* Modify

the probabilities in the transition probability matrix to account for the discounting of future salaries, or look at Problem 8 of Section 5.6.)

11[‡] During any month, Cashco has a .5 chance of receiving a $1,000 cash inflow and a .5 chance that there will be a $1,000 cash outflow. For every $1,000 in cash on hand at the end of a month, Cashco incurs a $15 cost (due to lost interest). At the beginning of each month, Cashco can adjust its on-hand cash balance upward or downward with the cost per transaction being $20. Cashco can never let the on-hand balance become negative. The company is considering the following two cash management policies:

Policy 1 At the beginning of a month in which the on-hand cash balance is $3,000, immediately reduce the cash balance to $1,000. At the beginning of a month in which the on-hand cash balance is $0, immediately bring the on-hand cash balance up to $1,000.

Policy 2 At the beginning of a month in which the on-hand cash balance is $4,000, immediately reduce the cash balance to $2,000. At the beginning of a month in which the on-hand cash balance is $0, immediately bring the on-hand cash balance up to $2,000.

Which policy will incur a smaller expected monthly cost (opportunity plus transaction)? The sequence of events during each month is as follows:

a Observe beginning cash balance

b Adjust (if desired) cash balance

c Cash balance changes

d Opportunity cost is assessed

12 In the game of craps, we roll a pair of six-sided dice. On the first throw, if we roll a 7 or an 11, we win right away. If we roll a 2, a 3, or a 12, we lose right away. If we first roll a total of 4, 5, 6, 8, 9, or 10, we keep rolling the dice until we get either a 7 or the total rolled on the first throw. If we get a 7, we lose. If we roll the same total as the first throw, we win. Use knowledge of Markov chains to determine our probability of winning at craps.

13 At the beginning of each day, a patient in a hospital is classified into one of three conditions: good, fair, or critical. At the beginning of the next day, the patient will either still be in the hospital and be in good, fair, or critical condition or will be discharged in one of three conditions: improved, unimproved, or dead. The transition probabilities for this situation are as follows:

	Good	Fair	Critical
Good	.65	.20	.05
Fair	.50	.30	.12
Critical	.51	.25	.20

	Improved	Unimproved	Dead
Good	.06	.03	.01
Fair	.03	.02	.03
Critical	.01	.01	.02

For example, a patient who begins the day in fair condition has a 12% chance of being in critical condition the next day

[†]Based on Flamholtz, Geis, and Perle (1984).

[‡]Based on Eppen and Fama (1970).

and a 3% chance of being discharged the next day in improved condition.

a Consider a patient who enters the hospital in good condition. On the average, how many days does this patient spend in the hospital?

b This morning there were 500 patients in good condition, 300 in fair condition, and 200 patients in critical condition in the hospital. Tomorrow morning the following admissions will be made: good condition, 50; fair condition, 40; critical condition, 30. Predict tomorrow morning's hospital census.

c The hospital's daily admissions are as follows: 20 patients in good condition, 10 patients in fair condition, and 10 patients in critical condition. On the average, how many patients of each type would you expect to see in the hospital?

d What fraction of patients who enter the hospital in good condition will leave the hospital in improved condition?

14 A major problem for a hospital is managing the database containing patient records. Blair General Hospital is considering two policies:

Policy 1 Dispose of a patient's records if he or she has not reentered the hospital in the last five years.
Policy 1 Dispose of a patient's records if he or she has not reentered the hospital in the last ten years.

The following information is available: If a patient has been hospitalized, there is a 30% chance he or she will reenter the hospital during the next year. If a patient has not been hospitalized during the last year, there is a 20% chance he or she will be hospitalized during the next year. If a patient has not been hospitalized during the last two years, there is a 10% chance he or she will be hospitalized during the next year. If a patient has not been hospitalized during

the last three years, there is a 5% chance he or she will be hospitalized during the next year. If a patient has not been hospitalized during the last four years, there is a 3% chance he or she will be hospitalized during the next year. If a patient has not been hospitalized during the last five years, there is a 2% chance he or she will be hospitalized during the next year. If a patient has not been hospitalized for at least six years, there is a 1% chance he or she will be hospitalized during the next year.

Assume that the hospital admits an average of 10,000 new patients each year. For each policy, estimate the number of patient records that will be in the system.[†]

15 Consider an n-state Markov chain in which each transition probability is positive and the transition matrix is symmetric; the entry in row I and column J of the transition matrix is identical to the entry in row J and column I.

a Why do we know that steady-state probabilities exist for this situation?

b What are the steady-state probabilities?

16[‡] The Euro was introduced on January 1, 2002 as the common currency for 15 European countries. Each Euro has a marking on the coin indicating the country of origin. For example, Euros minted in Portugal have a different marking than Euros minted in Spain. European politicians are interested in determining what fraction of Euros will eventually end up circulating in each country. For example, will 30% of all Euros circulate in France? How could Markov chains be used to answer this question? What parameters must be known before using Markov chain theory to solve this problem?

[†]Based on Liu, Wang, and Guh (1991).
[‡]Based on "Statisticians Count Euros and Find More Than Money," *New York Times,* July 2, 2002.

REFERENCES

Grinold, R., and K. Marshall. *Manpower Planning Models.* New York: North-Holland, 1977. Contains an extensive discussion of manpower planning models.

The following two references are the sources for two applications of Markov chains discussed in the chapter:

Meredith, J. "A Markovian Analysis of a Geriatric Ward," *Management Science* 19(1973):604–612.
Walker, J., and J. Lehmann. *100 Ways to Win at Monopoly.* New York: Dell, 1975.

The following books give both the theoretical aspects and many interesting applications of Markov chains:

Bhat, N. *Elements of Applied Stochastic Processes,* 2d ed. New York: Wiley, 1985.
Isaacson, D., and R. Madsen. *Markov Chains: Theory and Applications.* New York: Wiley, 1976.

Karlin, S., and H. Taylor. *A First Course in Stochastic Processes,* 2d ed. Orlando, Fla.: Academic Press, 1975.
Kemeny, J., and L. Snell. *Finite Markov Chains.* Princeton, N.J.: Van Nostrand, 1960.
Ash, R. and R. Bishop. "Monopoly as a Markov Process," *Mathematics Magazine* 45(1972):26–29.
Babich, P. "Customer Satisfaction: How Good Is Good Enough," *Quality Progress* (December 1992): 65–68.
Bessent, W., and A. Bessent. "Student Education Flow in a University Department: Results of a Markov Analysis," *Interfaces* 10(1980):52–59.
Cyert, R., M. Davidson, and G. Thompson. "Estimation of the Allowance for Doubtful Accounts by Markov Chains," *Management Science* 8(1963):287–303.
Deming, E., and G. Glasser. "A Markovian Analysis of the Life of Newspaper Subscriptions," *Management Science* 14(1968):B283–B294.
Eppen, G., and E. Fama. "Three Asset Cash Balance and Dynamic Portfolio Problems," *Management Science* 17(1970):311–319.

Flamholtz, E., G. Geis, and R. Perle. "A Markovian Model for the Valuation of Human Assets Acquired by an Organizational Purchase," *Interfaces* 14(1984):11–15.

Liu, C., K. Wang, and Y. Guh. "A Markov Chain Model for Medical Record Analysis," *Operations Research Quarterly* 42(no. 5, 1991):357–364.

Pegels, C., and A. Jelmert. "An Evaluation of Blood-Inventory Policies: A Markov Chain Application," *Operations Research* 18(1970):1097–1098.

Thompson, W., and J. McNeal. "Sales Planning and Control Using Absorbing Markov Chains," *Journal of Marketing Research* 4(1967):62–66.

Deterministic Dynamic Programming

Dynamic programming is a technique that can be used to solve many optimization problems. In most applications, dynamic programming obtains solutions by working backward from the end of a problem toward the beginning, thus breaking up a large, unwieldy problem into a series of smaller, more tractable problems.

We introduce the idea of working backward by solving two well-known puzzles and then show how dynamic programming can be used to solve network, inventory, and resource-allocation problems. We close the chapter by showing how to use spreadsheets to solve dynamic programming problems.

6.1 Two Puzzles[†]

In this section, we show how working backward can make a seemingly difficult problem almost trivial to solve.

EXAMPLE 1   **Match Puzzle**

Suppose there are 30 matches on a table. I begin by picking up 1, 2, or 3 matches. Then my opponent must pick up 1, 2, or 3 matches. We continue in this fashion until the last match is picked up. The player who picks up the last match is the loser. How can I (the first player) be sure of winning the game?

Solution   If I can ensure that it will be my opponent's turn when 1 match remains, I will certainly win. Working backward one step, if I can ensure that it will be my opponent's turn when 5 matches remain, I will win. The reason for this is that no matter what he does when 5 matches remain, I can make sure that when he has his next turn, only 1 match will remain. For example, suppose it is my opponent's turn when 5 matches remain. If my opponent picks up 2 matches, I will pick up 2 matches, leaving him with 1 match and sure defeat. Similarly, if I can force my opponent to play when 5, 9, 13, 17, 21, 25, or 29 matches remain, I am sure of victory. Thus, I cannot lose if I pick up $30 - 29 = 1$ match on my first turn. Then I simply make sure that my opponent will always be left with 29, 25, 21, 17, 13, 9, or 5 matches on his turn. Notice that we have solved this puzzle by working backward from the end of the problem toward the beginning. Try solving this problem without working backward!

EXAMPLE 2   **Milk**

I have a 9-oz cup and a 4-oz cup. My mother has ordered me to bring home exactly 6 oz of milk. How can I accomplish this goal?

[†]This section covers topics that may be omitted with no loss of continuity.

TABLE 1
Moves in the Cup-and-Milk Problem

No. of Ounces in 9-oz Cup	No. of Ounces in 4-oz Cup
6	0
6	4
9	1
0	1
1	0
1	4
5	0
5	4
9	0
0	0

Solution By starting near the end of the problem, I cleverly realize that the problem can easily be solved if I can somehow get 1 oz of milk into the 4-oz cup. Then I can fill the 9-oz cup and empty 3 oz from the 9-oz cup into the partially filled 4-oz cup. At this point, I will be left with 6 oz of milk. After I have this flash of insight, the solution to the problem may easily be described as in Table 1 (the initial situation is written last, and the final situation is written first).

PROBLEMS

Group A

1 Suppose there are 40 matches on a table. I begin by picking up 1, 2, 3, or 4 matches. Then my opponent must pick up 1, 2, 3, or 4 matches. We continue until the last match is picked up. The player who picks up the last match is the loser. Can I be sure of victory? If so, how?

2 Three players have played three rounds of a gambling game. Each round has one loser and two winners. The losing player must pay each winner the amount of money that the winning player had at the beginning of the round. At the end of the three rounds each player has $10. You are told that each player has won one round. By working backward, determine the original stakes of the three players. [*Note:* If the answer turns out to be (for example) 5, 15, 10, don't worry about which player had which stake; we can't really tell which player ends up with how much, but we can determine the numerical values of the original stakes.]

Group B

3 We have 21 coins and are told that one is heavier than any of the other coins. How many weighings on a balance will it take to find the heaviest coin? (*Hint:* If the heaviest coin is in a group of three coins, we can find it in one weighing. Then work backward to two weighings, and so on.)

4 Given a 7-oz cup and a 3-oz cup, explain how we can return from a well with 5 oz of water.

6.2 A Network Problem

Many applications of dynamic programming reduce to finding the shortest (or longest) path that joins two points in a given network. The following example illustrates how dynamic programming (working backward) can be used to find the shortest path in a network.

EXAMPLE 3   **Shortest Path**

Joe Cougar lives in New York City, but he plans to drive to Los Angeles to seek fame and fortune. Joe's funds are limited, so he has decided to spend each night on his trip at a friend's house. Joe has friends in Columbus, Nashville, Louisville, Kansas City, Omaha, Dallas, San Antonio, and Denver. Joe knows that after one day's drive he can reach Columbus, Nashville, or Louisville. After two days of driving, he can reach Kansas City, Omaha, or Dallas. After three days of driving, he can reach San Antonio or Denver. Finally, after four days of driving, he can reach Los Angeles. To minimize the number of miles traveled, where should Joe spend each night of the trip? The actual road mileages between cities are given in Figure 1.

Solution   Joe needs to know the shortest path between New York and Los Angeles in Figure 1. We will find it by working backward. We have classified all the cities that Joe can be in at the beginning of the nth day of his trip as stage n cities. For example, because Joe can only be in San Antonio or Denver at the beginning of the fourth day (day 1 begins when Joe leaves New York), we classify San Antonio and Denver as stage 4 cities. The reason for classifying cities according to stages will become apparent later.

The idea of working backward implies that we should begin by solving an easy problem that will eventually help us to solve a complex problem. Hence, we begin by finding the shortest path to Los Angeles from each city in which there is only one day of driving left (stage 4 cities). Then we use this information to find the shortest path to Los Angeles from each city for which only two days of driving remain (stage 3 cities). With this information in hand, we are able to find the shortest path to Los Angeles from each city that is three days distant (stage 2 cities). Finally, we find the shortest path to Los Angeles from each city (there is only one: New York) that is four days away.

To simplify the exposition, we use the numbers 1, 2, . . . , 10 given in Figure 1 to label the 10 cities. We also define c_{ij} to be the road mileage between city i and city j. For example, $c_{35} = 580$ is the road mileage between Nashville and Kansas City. We let $f_t(i)$ be the length of the shortest path from city i to Los Angeles, given that city i is a stage t city.[†]

Stage 4 Computations

We first determine the shortest path to Los Angeles from each stage 4 city. Since there is only one path from each stage 4 city to Los Angeles, we immediately see that $f_4(8) = 1{,}030$, the shortest path from Denver to Los Angeles simply being the *only* path from Denver to Los Angeles. Similarly, $f_4(9) = 1{,}390$, the shortest (and only) path from San Antonio to Los Angeles.

Stage 3 Computations

We now work backward one stage (to stage 3 cities) and find the shortest path to Los Angeles from each stage 3 city. For example, to determine $f_3(5)$, we note that the shortest path from city 5 to Los Angeles must be one of the following:

Path 1   Go from city 5 to city 8 and then take the shortest path from city 8 to city 10.

Path 2   Go from city 5 to city 9 and then take the shortest path from city 9 to city 10.

The length of path 1 may be written as $c_{58} + f_4(8)$, and the length of path 2 may be written as $c_{59} + f_4(9)$. Hence, the shortest distance from city 5 to city 10 may be written as

---

[†]In this example, keeping track of the stages is unnecessary; to be consistent with later examples, however, we do keep track.

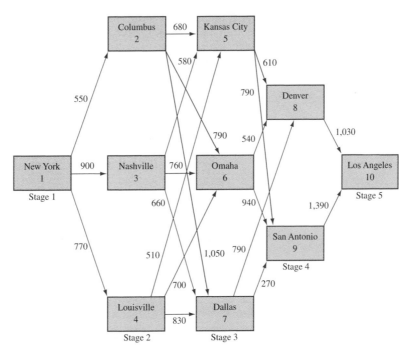

FIGURE 1
Joe's Trip Across the
United States

$$f_3(5) = \min \begin{cases} c_{58} + f_4(8) = 610 + 1{,}030 = 1{,}640^* \\ c_{59} + f_4(9) = 790 + 1{,}390 = 2{,}180 \end{cases}$$

[the * indicates the choice of arc that attains the $f_3(5)$]. Thus, we have shown that the shortest path from city 5 to city 10 is the path 5–8–10. Note that to obtain this result, we made use of our knowledge of $f_4(8)$ and $f_4(9)$.

Similarly, to find $f_3(6)$, we note that the shortest path to Los Angeles from city 6 must begin by going to city 8 or to city 9. This leads us to the following equation:

$$f_3(6) = \min \begin{cases} c_{68} + f_4(8) = 540 + 1{,}030 = 1{,}570^* \\ c_{69} + f_4(9) = 940 + 1{,}390 = 2{,}330 \end{cases}$$

Thus, $f_3(6) = 1{,}570$, and the shortest path from city 6 to city 10 is the path 6–8–10.

To find $f_3(7)$, we note that

$$f_3(7) = \min \begin{cases} c_{78} + f_4(8) = 790 + 1{,}030 = 1{,}820 \\ c_{79} + f_4(9) = 270 + 1{,}390 = 1{,}660^* \end{cases}$$

Therefore, $f_3(7) = 1{,}660$, and the shortest path from city 7 to city 10 is the path 7–9–10.

Stage 2 Computations

Given our knowledge of $f_3(5)$, $f_3(6)$, and $f_3(7)$, it is now easy to work backward one more stage and compute $f_2(2)$, $f_2(3)$, and $f_2(4)$ and thus the shortest paths to Los Angeles from city 2, city 3, and city 4. To illustrate how this is done, we find the shortest path (and its length) from city 2 to city 10. The shortest path from city 2 to city 10 must begin by going from city 2 to city 5, city 6, or city 7. Once this shortest path gets to city 5, city 6, or city 7, then it must follow a shortest path from that city to Los Angeles. This reasoning shows that the shortest path from city 2 to city 10 must be one of the following:

Path 1 Go from city 2 to city 5. Then follow a shortest path from city 5 to city 10. A path of this type has a total length of $c_{25} + f_3(5)$.

Path 2 Go from city 2 to city 6. Then follow a shortest path from city 6 to city 10. A path of this type has a total length of $c_{26} + f_3(6)$.

Path 3 Go from city 2 to city 7. Then follow a shortest path from city 7 to city 10. This path has a total length of $c_{27} + f_3(7)$. We may now conclude that

$$f_2(2) = \min \begin{cases} c_{25} + f_3(5) = 680 + 1{,}640 = 2{,}320^* \\ c_{26} + f_3(6) = 790 + 1{,}570 = 2{,}360 \\ c_{27} + f_3(7) = 1{,}050 + 1{,}660 = 2{,}710 \end{cases}$$

Thus, $f_2(2) = 2{,}320$, and the shortest path from city 2 to city 10 is to go from city 2 to city 5 and then follow the shortest path from city 5 to city 10 (5–8–10).

Similarly,

$$f_2(3) = \min \begin{cases} c_{35} + f_3(5) = 580 + 1{,}640 = 2{,}220^* \\ c_{36} + f_3(6) = 760 + 1{,}570 = 2{,}330 \\ c_{37} + f_3(7) = 660 + 1{,}660 = 2{,}320 \end{cases}$$

Thus, $f_2(3) = 2{,}220$, and the shortest path from city 3 to city 10 consists of arc 3–5 and the shortest path from city 5 to city 10 (5–8–10).

In similar fashion,

$$f_2(4) = \min \begin{cases} c_{45} + f_3(5) = 510 + 1{,}640 = 2{,}150^* \\ c_{46} + f_3(6) = 700 + 1{,}570 = 2{,}270 \\ c_{47} + f_3(7) = 830 + 1{,}660 = 2{,}490 \end{cases}$$

Thus, $f_2(4) = 2{,}150$, and the shortest path from city 4 to city 10 consists of arc 4–5 and the shortest path from city 5 to city 10 (5–8–10).

Stage 1 Computations

We can now use our knowledge of $f_2(2), f_2(3),$ and $f_2(4)$ to work backward one more stage to find $f_1(1)$ and the shortest path from city 1 to city 10. Note that the shortest path from city 1 to city 10 must begin by going to city 2, city 3, or city 4. This means that the shortest path from city 1 to city 10 must be one of the following:

Path 1 Go from city 1 to city 2 and then follow a shortest path from city 2 to city 10. The length of such a path is $c_{12} + f_2(2)$.

Path 2 Go from city 1 to city 3 and then follow a shortest path from city 3 to city 10. The length of such a path is $c_{13} + f_2(3)$.

Path 3 Go from city 1 to city 4 and then follow a shortest path from city 4 to city 10. The length of such a path is $c_{14} + f_2(4)$. It now follows that

$$f_1(1) = \min \begin{cases} c_{12} + f_2(2) = 550 + 2{,}320 = 2{,}870^* \\ c_{13} + f_2(3) = 900 + 2{,}220 = 3{,}120 \\ c_{14} + f_2(4) = 770 + 2{,}150 = 2{,}920 \end{cases}$$

Determination of the Optimal Path

Thus, $f_1(1) = 2{,}870$, and the shortest path from city 1 to city 10 goes from city 1 to city 2 and then follows the shortest path from city 2 to city 10. Checking back to the $f_2(2)$ calculations, we see that the shortest path from city 2 to city 10 is 2–5–8–10. Translating the numerical labels into real cities, we see that the shortest path from New York to Los Angeles passes through New York, Columbus, Kansas City, Denver, and Los Angeles. This path has a length of $f_1(1) = 2{,}870$ miles.

Computational Efficiency of Dynamic Programming

For Example 3, it would have been an easy matter to determine the shortest path from New York to Los Angeles by enumerating all the possible paths [after all, there are only $3(3)(2) = 18$ paths]. Thus, in this problem, the use of dynamic programming did not really serve much purpose. For larger networks, however, dynamic programming is much more efficient for determining a shortest path than the explicit enumeration of all paths. To see this, consider the network in Figure 2. In this network, it is possible to travel from any node in stage k to any node in stage $k + 1$. Let the distance between node i and node j be c_{ij}. Suppose we want to determine the shortest path from node 1 to node 27. One way to solve this problem is explicit enumeration of all paths. There are 5^5 possible paths from node 1 to node 27. It takes five additions to determine the length of each path. Thus, explicitly enumerating the length of all paths requires $5^5(5) = 5^6 = 15{,}625$ additions.

Suppose we use dynamic programming to determine the shortest path from node 1 to node 27. Let $f_t(i)$ be the length of the shortest path from node i to node 27, given that node i is in stage t. To determine the shortest path from node 1 to node 27, we begin by finding $f_6(22)$, $f_6(23)$, $f_6(24)$, $f_6(25)$, and $f_6(26)$. This does not require any additions. Then we find $f_5(17)$, $f_5(18)$, $f_5(19)$, $f_5(20)$, $f_5(21)$. For example, to find $f_5(21)$ we use the following equation:

$$f_5(21) = \min_j \{c_{21,j} + f_6(j)\} \qquad (j = 22, 23, 24, 25, 26)$$

Determining $f_5(21)$ in this manner requires five additions. Thus, the calculation of all the $f_5(\cdot)$'s requires $5(5) = 25$ additions. Similarly, the calculation of all the $f_4(\cdot)$'s requires 25 additions, and the calculation of all the $f_3(\cdot)$'s requires 25 additions. The determination of all the $f_2(\cdot)$'s also requires 25 additions, and the determination of $f_1(1)$ requires 5 additions. Thus, in total, dynamic programming requires $4(25) + 5 = 105$ additions to find

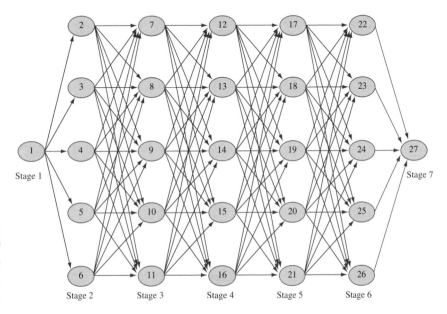

FIGURE 2
Illustration of
Computational Efficiency
of Dynamic
Programming

the shortest path from node 1 to node 27. Because explicit enumeration requires 15,625 additions, we see that dynamic programming requires only 0.007 times as many additions as explicit enumeration. For larger networks, the computational savings effected by dynamic programming are even more dramatic.

Besides additions, determination of the shortest path in a network requires comparisons between the lengths of paths. If explicit enumeration is used, then $5^5 - 1 = 3,124$ comparisons must be made (that is, compare the length of the first two paths, then compare the length of the third path with the shortest of the first two paths, and so on). If dynamic programming is used, then for $t = 2, 3, 4, 5$, determination of each $f_t(i)$ requires $5 - 1 = 4$ comparisons. Then to compute $f_1(1)$, $5 - 1 = 4$ comparisons are required. Thus, to find the shortest path from node 1 to node 27, dynamic programming requires a total of $20(5 - 1) + 4 = 84$ comparisons. Again, dynamic programming comes out far superior to explicit enumeration.

Characteristics of Dynamic Programming Applications

We close this section with a discussion of the characteristics of Example 3 that are common to most applications of dynamic programming.

Characteristic 1

The problem can be divided into stages with a decision required at each stage. In Example 3, stage t consisted of those cities where Joe could be at the beginning of day t of his trip. As we will see, in many dynamic programming problems, the stage is the amount of time that has elapsed since the beginning of the problem. We note that in some situations, decisions are not required at every stage (see Section 6.5).

Characteristic 2

Each stage has a number of states associated with it. By a **state,** we mean the information that is needed at any stage to make an optimal decision. In Example 3, the state at stage t is simply the city where Joe is at the beginning of day t. For example, in stage 3, the possible states are Kansas City, Omaha, and Dallas. Note that to make the correct decision at any stage, Joe doesn't need to know how he got to his current location. For example, if Joe is in Kansas City, then his remaining decisions don't depend on how he goes to Kansas City; his future decisions just depend on the fact that he is now in Kansas City.

Characteristic 3

The decision chosen at any stage describes how the state at the current stage is transformed into the state at the next stage. In Example 3, Joe's decision at any stage is simply the next city to visit. This determines the state at the next stage in an obvious fashion. In many problems, however, a decision does not determine the next stage's state with certainty; instead, the current decision only determines the probability distribution of the state at the next stage.

Characteristic 4

Given the current state, the optimal decision for each of the remaining stages must not depend on previously reached states or previously chosen decisions. This idea is known as the **principle of optimality.** In the context of Example 3, the principle of optimality reduces to the following: Suppose the shortest path (call it R) from city 1 to city 10 is known to pass through city i. Then the portion of R that goes from city i to city 10 must be a shortest path from city i to city 10. If this were not the case, then we could create a path from city 1 to city 10 that was shorter than R by appending a shortest path from city i to city 10 to the portion of R leading from city 1 to city i. This would create a path from city 1 to city 10 that is shorter than R, thereby contradicting the fact that R is a shortest path from city 1 to city 10. For example, if the shortest path from city 1 to city 10 is known to pass through city 2, then the shortest path from city 1 to city 10 must include a shortest path from city 2 to city 10 (2–5–8–10). This follows because any path from city 1 to city 10 that passes through city 2 and does not contain a shortest path from city 2 to city 10 will have a length of c_{12} + [something bigger than $f_2(2)$]. Of course, such a path cannot be a shortest path from city 1 to city 10.

Characteristic 5

If the states for the problem have been classified into one of T stages, there must be a recursion that relates the cost or reward earned during stages t, t + 1, . . . , T to the cost or reward earned from stages t + 1, t + 2, . . . , T. In essence, the recursion formalizes the working-backward procedure. In Example 3, our recursion could have been written as

$$f_t(i) = \min_j \{c_{ij} + f_{t+1}(j)\}$$

where j must be a stage $t + 1$ city and $f_5(10) = 0$.

We can now describe how to make optimal decisions. Let's assume that the initial state during stage 1 is i_1. To use the recursion, we begin by finding the optimal decision for each state associated with the last stage. Then we use the recursion described in characteristic 5 to determine $f_{T-1}(\cdot)$ (along with the optimal decision) for every stage $T - 1$ state. Then we use the recursion to determine $f_{T-2}(\cdot)$ (along with the optimal decision) for every stage $T - 2$ state. We continue in this fashion until we have computed $f_1(i_1)$ and the optimal decision when we are in stage 1 and state i_1. Then our optimal decision in stage 1 is chosen from the set of decisions attaining $f_1(i_1)$. Choosing this decision at stage 1 will lead us to some stage 2 state (call it state i_2) at stage 2. Then at stage 2, we choose any decision attaining $f_2(i_2)$. We continue in this fashion until a decision has been chosen for each stage.

In the rest of this chapter, we discuss many applications of dynamic programming. The presentation will seem easier if the reader attempts to determine how each problem fits into the network context introduced in Example 3. In the next section, we begin by studying how dynamic programming can be used to solve inventory problems.

PROBLEMS

Group A

1 Find the shortest path from node 1 to node 10 in the network shown in Figure 3. Also, find the shortest path from node 3 to node 10.

2 A sales representative lives in Bloomington and must be in Indianapolis next Thursday. On each of the days Monday, Tuesday, and Wednesday, he can sell his wares in Indianapolis, Bloomington, or Chicago. From past experience, he believes that he can earn $12 from spending a day in Indianapolis, $16 from spending a day in Bloomington, and $17 from spending a day in Chicago. Where should he spend the first three days

FIGURE 3

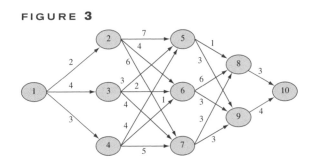

TABLE 2

	To		
From	Indianapolis	Bloomington	Chicago
Indianapolis	—	5	2
Bloomington	5	—	7
Chicago	2	7	—

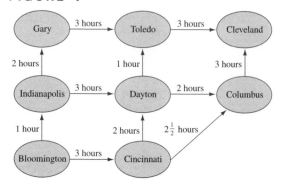

FIGURE 4

and nights of the week to maximize his sales income less travel costs? Travel costs are shown in Table 2.

Group B

3 I must drive from Bloomington to Cleveland. Several paths are available (see Figure 4). The number on each arc is the length of time it takes to drive between the two cities. For example, it takes 3 hours to drive from Bloomington to Cincinnati. By working backward, determine the shortest path (in terms of time) from Bloomington to Cleveland. [*Hint:* Work backward and don't worry about stages—only about states.]

6.3 An Inventory Problem

In this section, we illustrate how dynamic programming can be used to solve an inventory problem with the following characteristics:

1 Time is broken up into periods, the present period being period 1, the next period 2, and the final period T. At the beginning of period 1, the demand during each period is known.

2 At the beginning of each period, the firm must determine how many units should be produced. Production capacity during each period is limited.

3 Each period's demand must be met on time from inventory or current production. During any period in which production takes place, a fixed cost of production as well as a variable per-unit cost is incurred.

4 The firm has limited storage capacity. This is reflected by a limit on end-of-period inventory. A per-unit holding cost is incurred on each period's ending inventory.

5 The firm's goal is to minimize the total cost of meeting on time the demands for periods 1, 2, . . . , T.

In this model, the firm's inventory position is reviewed at the end of each period (say, at the end of each month), and then the production decision is made. Such a model is called a **periodic review model.** This model is in contrast to the continuous review models in which the firm knows its inventory position at all times and may place an order or begin production at any time.

If we exclude the setup cost for producing any units, the inventory problem just described is similar to the Sailco inventory problem that we solved by linear programming in Section 3.10 of Volume 1. Here, we illustrate how dynamic programming can be used to determine a production schedule that minimizes the total cost incurred in an inventory problem that meets the preceding description.

EXAMPLE 4   **Inventory**

A company knows that the demand for its product during each of the next four months will be as follows: month 1, 1 unit; month 2, 3 units; month 3, 2 units; month 4, 4 units. At the beginning of each month, the company must determine how many units should be produced during the current month. During a month in which any units are produced, a setup cost of $3 is incurred. In addition, there is a variable cost of $1 for every unit produced. At the end of each month, a holding cost of 50¢ per unit on hand is incurred. Capacity limitations allow a maximum of 5 units to be produced during each month. The size of the company's warehouse restricts the ending inventory for each month to 4 units at most. The company wants to determine a production schedule that will meet all demands on time and will minimize the sum of production and holding costs during the four months. Assume that 0 units are on hand at the beginning of the first month.

Solution   Recall from Section 3.10 of Volume 1 that we can ensure that all demands are met on time by restricting each month's ending inventory to be nonnegative. To use dynamic programming to solve this problem, we need to identify the appropriate state, stage, and decision. The stage should be defined so that when one stage remains, the problem will be trivial to solve. If we are at the beginning of month 4, then the firm would meet demand at minimum cost by simply producing just enough units to ensure that (month 4 production) + (month 3 ending inventory) = (month 4 demand). Thus, when one month remains, the firm's problem is easy to solve. Hence, we let time represent the stage. In most dynamic programming problems, the stage has something to do with time.

At each stage (or month), the company must decide how many units to produce. To make this decision, the company need only know the inventory level at the beginning of the current month (or the end of the previous month). Therefore, we let the state at any stage be the beginning inventory level.

Before writing a recursive relation that can be used to "build up" the optimal production schedule, we must define $f_t(i)$ to be the minimum cost of meeting demands for months $t, t + 1, \ldots, 4$ if i units are on hand at the beginning of month t. We define $c(x)$ to be the cost of producing x units during a period. Then $c(0) = 0$, and for $x > 0$, $c(x) = 3 + x$. Because of the limited storage capacity and the fact that all demand must be met on time, the possible states during each period are 0, 1, 2, 3, and 4. Thus, we begin by determining $f_4(0)$, $f_4(1)$, $f_4(2)$, $f_4(3)$, and $f_4(4)$. Then we use this information to determine $f_3(0)$, $f_3(1)$, $f_3(2)$, $f_3(3)$, and $f_3(4)$. Then we determine $f_2(0)$, $f_2(1)$, $f_2(2)$, $f_2(3)$, and $f_2(4)$. Finally, we determine $f_1(0)$. Then we determine an optimal production level for each month. We define $x_t(i)$ to be a production level during month t that minimizes the total cost during months $t, t + 1, \ldots, 4$ if i units are on hand at the beginning of month t. We now begin to work backward.

Month 4 Computations

During month 4, the firm will produce just enough units to ensure that the month 4 demand of 4 units is met. This yields

$$f_4(0) = \text{cost of producing } 4 - 0 \text{ units} = c(4) = 3 + 4 = \$7 \text{ and } x_4(0) = 4 - 0 = 4$$
$$f_4(1) = \text{cost of producing } 4 - 1 \text{ units} = c(3) = 3 + 3 = \$6 \text{ and } x_4(1) = 4 - 1 = 3$$
$$f_4(2) = \text{cost of producing } 4 - 2 \text{ units} = c(2) = 3 + 2 = \$5 \text{ and } x_4(2) = 4 - 2 = 2$$
$$f_4(3) = \text{cost of producing } 4 - 3 \text{ units} = c(1) = 3 + 1 = \$4 \text{ and } x_4(3) = 4 - 3 = 1$$
$$f_4(4) = \text{cost of producing } 4 - 4 \text{ units} = c(0) = \$0 \quad \text{and} \quad x_4(4) = 4 - 4 = 0$$

Month 3 Computations

How can we now determine $f_3(i)$ for $i = 0, 1, 2, 3, 4$? The cost $f_3(i)$ is the minimum cost incurred during months 3 and 4 if the inventory at the beginning of month 3 is i. For each possible production level x during month 3, the total cost during months 3 and 4 is

$$(\tfrac{1}{2})(i + x - 2) + c(x) + f_4(i + x - 2) \tag{1}$$

This follows because if x units are produced during month 3, the ending inventory for month 3 will be $i + x - 2$. Then the month 3 holding cost will be $(\tfrac{1}{2})(i + x - 2)$, and the month 3 production cost will be $c(x)$. Then we enter month 4 with $i + x - 2$ units on hand. Since we proceed optimally from this point onward (remember the principle of optimality), the cost for month 4 will be $f_4(i + x - 2)$. We want to choose the month 3 production level to minimize (1), so we write

$$f_3(i) = \min_x \{(\tfrac{1}{2})(i + x - 2) + c(x) + f_4(i + x - 2)\} \tag{2}$$

In (2), x must be a member of $\{0, 1, 2, 3, 4, 5\}$, and x must satisfy $4 \geq i + x - 2 \geq 0$. This reflects the fact that the current month's demand must be met ($i + x - 2 \geq 0$), and ending inventory cannot exceed the capacity of $4(i + x - 2 \leq 4)$. Recall that $x_3(i)$ is any value of x attaining $f_3(i)$. The computations for $f_3(0), f_3(1), f_3(2), f_3(3),$ and $f_3(4)$ are given in Table 3.

Month 2 Computations

We can now determine $f_2(i)$, the minimum cost incurred during months 2, 3, and 4 given that at the beginning of month 2, the on-hand inventory is i units. Suppose that month 2 production $= x$. Because month 2 demand is 3 units, a holding cost of $(\tfrac{1}{2})(i + x - 3)$ is

TABLE 3
Computations for $f_3(i)$

i	x	$(\tfrac{1}{2})(i + x - 2) + c(x)$	$f_4(i + x - 2)$	Total Cost Months 3, 4	$f_3(i)$ $x_3(i)$
0	2	$0 + 5 = 5$	7	$5 + 7 = 12*$	$f_3(0) = 12$
0	3	$\tfrac{1}{2} + 6 = \tfrac{13}{2}$	6	$\tfrac{13}{2} + 6 = \tfrac{25}{2}$	$x_3(0) = 2$
0	4	$1 + 7 = 8$	5	$8 + 5 = 13$	
0	5	$\tfrac{3}{2} + 8 = \tfrac{19}{2}$	4	$\tfrac{19}{2} + 4 = \tfrac{27}{2}$	
1	1	$0 + 4 = 4$	7	$4 + 7 = 11$	$f_3(1) = 10$
1	2	$\tfrac{1}{2} + 5 = \tfrac{11}{2}$	6	$\tfrac{11}{2} + 6 = \tfrac{23}{2}$	$x_3(1) = 5$
1	3	$1 + 6 = 7$	5	$7 + 5 = 12$	
1	4	$\tfrac{3}{2} + 7 = \tfrac{17}{2}$	4	$\tfrac{17}{2} + 4 = \tfrac{25}{2}$	
1	5	$2 + 8 = 10$	0	$10 + 0 = 10*$	
2	0	$0 + 0 = 0$	7	$0 + 7 = 7*$	$f_3(2) = 7$
2	1	$\tfrac{1}{2} + 4 = \tfrac{9}{2}$	6	$\tfrac{9}{2} + 6 = \tfrac{21}{2}$	$x_3(2) = 0$
2	2	$1 + 5 = 6$	5	$6 + 5 = 11$	
2	3	$\tfrac{3}{2} + 6 = \tfrac{15}{2}$	4	$\tfrac{15}{2} + 4 = \tfrac{23}{2}$	
2	4	$2 + 7 = 9$	0	$9 + 0 = 9$	
3	0	$\tfrac{1}{2} + 0 = \tfrac{1}{2}$	6	$\tfrac{1}{2} + 6 = \tfrac{13}{2}*$	$f_3(3) = \tfrac{13}{2}$
3	1	$1 + 4 = 5$	5	$5 + 5 = 10$	$x_3(3) = 0$
3	2	$\tfrac{3}{2} + 5 = \tfrac{13}{2}$	4	$\tfrac{13}{2} + 4 = \tfrac{21}{2}$	
3	3	$2 + 6 = 8$	0	$8 + 0 = 8$	
4	0	$1 + 0 = 1$	5	$1 + 5 = 6*$	$f_3(4) = 6$
4	1	$\tfrac{3}{2} + 4 = \tfrac{11}{2}$	4	$\tfrac{11}{2} + 4 = \tfrac{19}{2}$	$x_3(4) = 0$
4	2	$2 + 5 = 7$	0	$7 + 0 = 7$	

incurred at the end of month 2. Thus, the total cost incurred during month 2 is $(\frac{1}{2})(i + x - 3) + c(x)$. During months 3 and 4, we follow an optimal policy. Since month 3 begins with an inventory of $i + x - 3$, the cost incurred during months 3 and 4 is $f_3(i + x - 3)$. In analogy to (2), we now write

$$f_2(i) = \min_x \{(\tfrac{1}{2})(i + x - 3) + c(x) + f_3(i + x - 3)\} \tag{3}$$

where x must be a member of $\{0, 1, 2, 3, 4, 5\}$ and x must also satisfy $0 \le i + x - 3 \le 4$. The computations for $f_2(0)$, $f_2(1)$, $f_2(2)$, $f_2(3)$, and $f_2(4)$ are given in Table 4.

Month 1 Computations

The reader should now be able to show that the $f_1(i)$'s can be determined via the following recursive relation:

$$f_1(i) = \min_x \{(\tfrac{1}{2})(i + x - 1) + c(x) + f_2(i + x - 1)\} \tag{4}$$

where x must be a member of $\{0, 1, 2, 3, 4, 5\}$ and x must satisfy $0 \le i + x - 1 \le 4$. Since the inventory at the beginning of month 1 is 0 units, we actually need only determine $f_1(0)$ and $x_1(0)$. To give the reader more practice, however, the computations for $f_1(1)$, $f_1(2)$, $f_1(3)$, and $f_1(4)$ are given in Table 5.

Determination of the Optimal Production Schedule

We can now determine a production schedule that minimizes the total cost of meeting the demand for all four months on time. Since our initial inventory is 0 units, the minimum cost for the four months will be $f_1(0) = \$20$. To attain $f_1(0)$, we must produce $x_1(0) = 1$

TABLE 4
Computations for $f_2(i)$

i	x	$(\tfrac{1}{2})(i + x - 3) + c(x)$	$f_3(i + x - 3)$	Total Cost Months 2–4	$f_2(i)$ $x_2(i)$
0	3	$0 + 6 = 6$	12	$6 + 12 = 18$	$f_2(0) = 16$
0	4	$\frac{1}{2} + 7 = \frac{15}{2}$	10	$\frac{15}{2} + 10 = \frac{35}{2}$	$x_2(0) = 5$
0	5	$1 + 8 = 9$	7	$9 + 7 = 16*$	
1	2	$0 + 5 = 5$	12	$5 + 12 = 17$	$f_2(1) = 15$
1	3	$\frac{1}{2} + 6 = \frac{13}{2}$	10	$\frac{13}{2} + 10 = \frac{33}{2}$	$x_2(1) = 4$
1	4	$1 + 7 = 8$	7	$8 + 7 = 15*$	
1	5	$\frac{3}{2} + 8 = \frac{19}{2}$	$\frac{13}{2}$	$\frac{19}{2} + \frac{13}{2} = 16$	
2	1	$0 + 4 = 4$	12	$4 + 12 = 16$	$f_2(2) = 14$
2	2	$\frac{1}{2} + 5 = \frac{11}{2}$	10	$\frac{11}{2} + 10 = \frac{31}{2}*$	$x_2(2) = 3$
2	3	$1 + 6 = 7$	7	$7 + 7 = 14*$	
2	4	$\frac{3}{2} + 7 = \frac{17}{2}$	$\frac{13}{2}$	$\frac{17}{2} + \frac{13}{2} = 15$	
2	5	$2 + 8 = 10$	6	$10 + 6 = 16$	
3	0	$0 + 0 = 0$	12	$0 + 12 = 12*$	$f_2(3) = 12$
3	1	$\frac{1}{2} + 4 = \frac{9}{2}$	10	$\frac{9}{2} + 10 = \frac{29}{2}$	$x_2(3) = 0$
3	2	$1 + 5 = 6$	7	$6 + 7 = 13$	
3	3	$\frac{3}{2} + 6 = \frac{15}{2}$	$\frac{13}{2}$	$\frac{15}{2} + \frac{13}{2} = 14$	
3	4	$2 + 7 = 9$	6	$9 + 6 = 15$	
4	0	$\frac{1}{2} + 0 = \frac{1}{2}$	10	$\frac{1}{2} + 10 = \frac{21}{2}*$	$f_2(4) = \frac{21}{2}$
4	1	$1 + 4 = 5$	7	$5 + 7 = 12$	$x_2(4) = 0$
4	2	$\frac{3}{2} + 5 = \frac{13}{2}$	$\frac{13}{2}$	$\frac{13}{2} + \frac{13}{2} = 13$	
4	3	$2 + 6 = 8$	6	$8 + 6 = 14$	

TABLE 5
Computations for $f_1(i)$

i	x	$(\frac{1}{2})(i + x - 1) + c(x)$	$f_2(i + x - 1)$	Total Cost	$f_1(i)$ $x_1(i)$
0	1	$0 + 4 = 4$	16	$4 + 16 = 20^*$	$f_1(0) = 20$
0	2	$\frac{1}{2} + 5 = \frac{11}{2}$	15	$\frac{11}{2} + 15 = \frac{41}{2}$	$x_1(0) = 1$
0	3	$1 + 6 = 7$	14	$7 + 14 = 21$	
0	4	$\frac{3}{2} + 7 = \frac{17}{2}$	12	$\frac{17}{2} + 12 = \frac{41}{2}$	
0	5	$2 + 8 = 10$	$\frac{21}{2}$	$10 + \frac{21}{2} = \frac{41}{2}$	
1	0	$0 + 0 = 0$	16	$0 + 16 = 16^*$	$f_1(1) = 16$
1	1	$\frac{1}{2} + 4 = \frac{9}{2}$	15	$\frac{9}{2} + 15 = \frac{39}{2}$	$x_1(1) = 0$
1	2	$1 + 5 = 6$	14	20	
1	3	$\frac{3}{2} + 6 = \frac{15}{2}$	12	$\frac{15}{2} + 12 = \frac{39}{2}$	
1	4	$2 + 7 = 9$	$\frac{21}{2}$	$9 + \frac{21}{2} = \frac{39}{2}$	
2	0	$\frac{1}{2} + 0 = \frac{1}{2}$	15	$\frac{1}{2} + 15 = \frac{31}{2}^*$	$f_1(2) = \frac{31}{2}$
2	1	$1 + 4 = 5$	14	$5 + 14 = 19$	$x_1(2) = 0$
2	2	$\frac{3}{2} + 5 = \frac{13}{2}$	12	$\frac{13}{2} + 12 = \frac{37}{2}$	
2	3	$2 + 6 = 8$	$\frac{21}{2}$	$8 + \frac{21}{2} = \frac{37}{2}$	
3	0	$1 + 0 = 1$	14	$1 + 14 = 15^*$	$f_1(3) = 15$
3	1	$\frac{3}{2} + 4 = \frac{11}{2}$	12	$\frac{11}{2} + 12 = \frac{35}{2}$	$x_1(3) = 0$
3	2	$2 + 5 = 7$	$\frac{21}{2}$	$7 + \frac{21}{2} = \frac{35}{2}$	
4	0	$\frac{3}{2} + 0 = \frac{3}{2}$	12	$\frac{3}{2} + 12 = \frac{27}{2}^*$	$f_1(4) = \frac{27}{2}$
4	1	$2 + 4 = 6$	$\frac{21}{2}$	$6 + \frac{21}{2} = \frac{33}{2}$	$x_1(4) = 0$

unit during month 1. Then the inventory at the beginning of month 2 will be $0 + 1 - 1 = 0$. Thus, in month 2, we should produce $x_2(0) = 5$ units. Then at the beginning of month 3, our beginning inventory will be $0 + 5 - 3 = 2$. Hence, during month 3, we need to produce $x_3(2) = 0$ units. Then month 4 will begin with $2 - 2 + 0 = 0$ units on hand. Thus, $x_4(0) = 4$ units should be produced during month 4. In summary, the optimal production schedule incurs a total cost of \$20 and produces 1 unit during month 1, 5 units during month 2, 0 units during month 3, and 4 units during month 4.

Note that finding the solution to Example 4 is equivalent to finding the shortest route joining the node (1, 0) to the node (5, 0) in Figure 5. Each node in Figure 5 corresponds to a state, and each column of nodes corresponds to all the possible states associated with a given stage. For example, if we are at node (2, 3), then we are at the beginning of month 2, and the inventory at the beginning of month 2 is 3 units. Each arc in the network represents the way in which a decision (how much to produce during the current month) transforms the current state into next month's state. For example, the arc joining nodes (1, 0) and (2, 2) (call it arc 1) corresponds to producing 3 units during month 1. To see this, note that if 3 units are produced during month 1, then we begin month 2 with $0 + 3 - 1 = 2$ units. The length of each arc is simply the sum of production and inventory costs during the current period, given the current state and the decision associated with the chosen arc. For example, the cost associated with arc 1 would be $6 + (\frac{1}{2})2 = 7$. Note that some nodes in adjacent stages are not joined by an arc. For example, node (2, 4) is not joined to node (3, 0). The reason for this is that if we begin month 2 with 4 units, then at the beginning of month 3, we will have at least $4 - 3 = 1$ unit on hand. Also note that we have drawn arcs joining all month 4 states to the node (5, 0), since having a positive inventory at the end of month 4 would clearly be suboptimal.

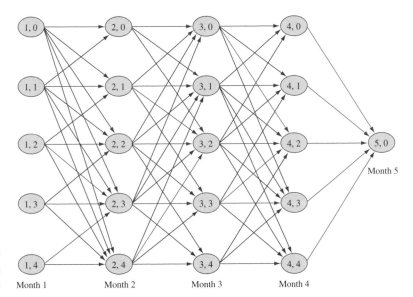

FIGURE 5
Network Representation
of Inventory Example

Month 1   Month 2   Month 3   Month 4

Returning to Example 4, the minimum-cost production schedule corresponds to the shortest path joining (1, 0) and (5, 0). As we have already seen, this would be the path corresponding to production levels of 1, 5, 0, and 4. In Figure 5, this would correspond to the path beginning at (1, 0), then going to (2, 0 + 1 − 1) = (2, 0), then to (3, 0 + 5 − 3) = (3, 2), then to (4, 2 + 0 − 2) = (4, 0), and finally to (5, 0 + 4 − 4) = (5, 0). Thus, our optimal production schedule corresponds to the path (1, 0)–(2, 0)–(3, 2)–(4, 0)–(5, 0) in Figure 5.

PROBLEMS

Group A

1 In Example 4, determine the optimal production schedule if the initial inventory is 3 units.

2 An electronics firm has a contract to deliver the following number of radios during the next three months; month 1, 200 radios; month 2, 300 radios; month 3, 300 radios. For each radio produced during months 1 and 2, a $10 variable cost is incurred; for each radio produced during month 3, a $12 variable cost is incurred. The inventory cost is $1.50 for each radio in stock at the end of a month. The cost of setting up for production during a month is $250.

Radios made during a month may be used to meet demand for that month or any future month. Assume that production during each month must be a multiple of 100. Given that the initial inventory level is 0 units, use dynamic programming to determine an optimal production schedule.

3 In Figure 5, determine the production level and cost associated with each of the following arcs:

   a (2, 3)–(3, 1)

   b (4, 2)–(5, 0)

6.4 Resource-Allocation Problems

Resource-allocation problems, in which limited resources must be allocated among several activities, are often solved by dynamic programming. Recall that we have solved such problems by linear programming (for instance, the Giapetto problem). To use linear programming to do resource allocation, three assumptions must be made:

Assumption 1 The amount of a resource assigned to an activity may be any nonnegative number.

Assumption 2 The benefit obtained from each activity is proportional to the amount of the resource assigned to the activity.

Assumption 3 The benefit obtained from more than one activity is the sum of the benefits obtained from the individual activities.

Even if assumptions 1 and 2 do not hold, dynamic programming can be used to solve resource-allocation problems efficiently when assumption 3 is valid and when the amount of the resource allocated to each activity is a member of a finite set.

EXAMPLE 5   Resource Allocation

Finco has $6,000 to invest, and three investments are available. If d_j dollars (in thousands) are invested in investment j, then a net present value (in thousands) of $r_j(d_j)$ is obtained, where the $r_j(d_j)$'s are as follows:

$$r_1(d_1) = 7d_1 + 2 \qquad (d_1 > 0)$$
$$r_2(d_2) = 3d_2 + 7 \qquad (d_2 > 0)$$
$$r_3(d_3) = 4d_3 + 5 \qquad (d_3 > 0)$$
$$r_1(0) = r_2(0) = r_3(0) = 0$$

The amount placed in each investment must be an exact multiple of $1,000. To maximize the net present value obtained from the investments, how should Finco allocate the $6,000?

Solution The return on each investment is not proportional to the amount invested in it [for example, $16 = r_1(2) \neq 2r_1(1) = 18$]. Thus, linear programming cannot be used to find an optimal solution to this problem.[†]

Mathematically, Finco's problem may be expressed as

$$\max\{r_1(d_1) + r_2(d_2) + r_3(d_3)\}$$
$$\text{s.t.} \qquad d_1 + d_2 + d_3 = 6$$
$$d_j \text{ nonnegative integer} \qquad (j = 1, 2, 3)$$

Of course, if the $r_j(d_j)$'s were linear, then we would have a knapsack problem like those we studied in Section 9.5 of Volume 1.

To formulate Finco's problem as a dynamic programming problem, we begin by identifying the stage. As in the inventory and shortest-route examples, the stage should be chosen so that when one stage remains the problem is easy to solve. Then, given that the problem has been solved for the case where one stage remains, it should be easy to solve the problem where two stages remain, and so forth. Clearly, it would be easy to solve when only one investment was available, so we define stage t to represent a case where funds must be allocated to investments $t, t + 1, \ldots, 3$.

For a given stage, what must we know to determine the optimal investment amount? Simply how much money is available for investments $t, t + 1, \ldots, 3$. Thus, we define the state at any stage to be the amount of money (in thousands) available for investments $t, t + 1, \ldots, 3$. We can never have more than $6,000 available, so the possible states at any stage are 0, 1, 2, 3, 4, 5, and 6. We define $f_t(d_t)$ to be the maximum net present value (NPV) that can be obtained by investing d_t thousand dollars in investments $t, t + 1, \ldots, 3$. Also define $x_t(d_t)$ to be the amount that should be invested in investment t to attain $f_t(d_t)$. We start to work backward by computing $f_3(0), f_3(1), \ldots, f_3(6)$ and then determine $f_2(0), f_2(1), \ldots, f_2(6)$. Since $6,000 is available for investment in investments 1, 2, and 3, we

[†]The fixed-charge approach described in Section 9.2 of Volume 1 could be used to solve this problem.

terminate our computations by computing $f_1(6)$. Then we retrace our steps and determine the amount that should be allocated to each investment (just as we retraced our steps to determine the optimal production level for each month in Example 4).

Stage 3 Computations

We first determine $f_3(0), f_3(1), \ldots, f_3(6)$. We see that $f_3(d_3)$ is attained by investing all available money (d_3) in investment 3. Thus,

$$f_3(0) = 0 \qquad x_3(0) = 0$$
$$f_3(1) = 9 \qquad x_3(1) = 1$$
$$f_3(2) = 13 \qquad x_3(2) = 2$$
$$f_3(3) = 17 \qquad x_3(3) = 3$$
$$f_3(4) = 21 \qquad x_3(4) = 4$$
$$f_3(5) = 25 \qquad x_3(5) = 5$$
$$f_3(6) = 29 \qquad x_3(6) = 6$$

TABLE 6
Computations for $f_2(0), f_2(1), \ldots, f_2(6)$

d_2	x_2	$r_2(x_2)$	$f_3(d_2 - x_2)$	NPV from Investments 2, 3	$f_2(d_2)$ $x_2(d_2)$
0	0	0	0	0*	$f_2(0) = 0$
					$x_2(0) = 0$
1	0	0	9	9	$f_2(1) = 10$
1	1	10	0	10*	$x_2(1) = 1$
2	0	0	13	13	$f_2(2) = 19$
2	1	10	9	19*	$x_2(2) = 1$
2	2	13	0	13	
3	0	0	17	17	$f_2(3) = 23$
3	1	10	13	23*	$x_2(3) = 1$
3	2	13	9	22	
3	3	16	0	16	
4	0	0	21	21	$f_2(4) = 27$
4	1	10	17	27*	$x_2(4) = 1$
4	2	13	13	26	
4	3	16	9	25	
4	4	19	0	19	
5	0	0	25	25	$f_2(5) = 31$
5	1	10	21	31*	$x_2(5) = 1$
5	2	13	17	30	
5	3	16	13	29	
5	4	19	9	28	
5	5	22	0	22	
6	0	0	29	29	$f_2(6) = 35$
6	1	10	25	35*	$x_2(6) = 1$
6	2	13	21	34	
6	3	16	17	33	
6	4	19	13	32	
6	5	22	9	31	
6	6	25	0	25	

TABLE 7
Computations for $f_1(6)$

d_1	x_1	$r_1(x_1)$	$f_2(6 - x_1)$	NPV from Investments 1-3	$f_1(6)$ $x_1(6)$
6	0	0	35	35	$f_1(6) = 49$
6	1	9	31	40	$x_1(6) = 4$
6	2	16	27	43	
6	3	23	23	46	
6	4	30	19	49*	
6	5	37	10	47	
6	6	44	0	44	

Stage 2 Computations

To determine $f_2(0), f_2(1), \ldots, f_2(6)$, we look at all possible amounts that can be placed in investment 2. To find $f_2(d_2)$, let x_2 be the amount invested in investment 2. Then an NPV of $r_2(x_2)$ will be obtained from investment 2, and an NPV of $f_3(d_2 - x_2)$ will be obtained from investment 3 (remember the principle of optimality). Since x_2 should be chosen to maximize the net present value earned from investments 2 and 3, we write

$$f_2(d_2) = \max_{x_2} \{r_2(x_2) + f_3(d_2 - x_2)\} \tag{5}$$

where x_2 must be a member of $\{0, 1, \ldots, d_2\}$. The computations for $f_2(0), f_2(1), \ldots, f_2(6)$ and $x_2(0), x_2(1), \ldots, x_2(6)$ are given in Table 6.

Stage 1 Computations

Following (5), we write

$$f_1(6) = \max_{x_1} \{r_1(x_1) + f_2(6 - x_1)\}$$

where x_1 must be a member of $\{0, 1, 2, 3, 4, 5, 6\}$. The computations for $f_1(6)$ are given in Table 7.

Determination of Optimal Resource Allocation

Since $x_1(6) = 4$, Finco invests \$4,000 in investment 1. This leaves $6,000 - 4,000 = $ \$2,000 for investments 2 and 3. Hence, Finco should invest $x_2(2) = $ \$1,000 in investment 2. Then \$1,000 is left for investment 3, so Finco chooses to invest $x_3(1) = $ \$1,000 in investment 3. Therefore, Finco can attain a maximum net present value of $f_1(6) = $ \$49,000 by investing \$4,000 in investment 1, \$1,000 in investment 2, and \$1,000 in investment 3.

Network Representation of Resource Example

As with the inventory example of Section 6.3, Finco's problem has a network representation, equivalent to finding the *longest route* from (1, 6) to (4, 0) in Figure 6. In the figure, the node (t, d) represents the situation in which d thousand dollars is available for investments $t, t + 1, \ldots, 3$. The arc joining the nodes (t, d) and $(t + 1, d - x)$ has a length $r_t(x)$ corresponding to the net present value obtained by investing x thousand dollars in investment t. For example, the arc joining nodes (2, 4) and (3, 1) has a length $r_2(3) = $ \$16,000, corresponding to the \$16,000 net present value that can be obtained by invest-

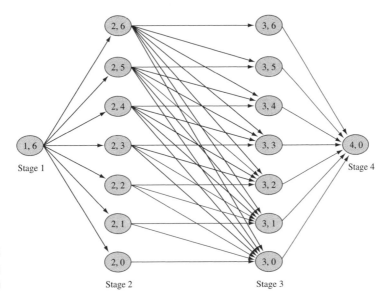

FIGURE 6
Network Representation
of Finco

ing \$3,000 in investment 2. Note that not all pairs of nodes in adjacent stages are joined by arcs. For example, there is no arc joining the nodes (2, 4) and (3, 5); after all, if you have only \$4,000 available for investments 2 and 3, how can you have \$5,000 available for investment 3? From our computations, we see that the longest path from (1, 6) to (4, 0) is (1, 6)–(2, 2)–(3, 1)–(4, 0).

Generalized Resource Allocation Problem

We now consider a generalized version of Example 5. Suppose we have w units of a resource available and T activities to which the resource can be allocated. If activity t is implemented at a level x_t (we assume x_t must be a nonnegative integer), then $g_t(x_t)$ units of the resource are used by activity t, and a benefit $r_t(x_t)$ is obtained. The problem of determining the allocation of resources that maximizes total benefit subject to the limited resource availability may be written as

$$\max \sum_{t=1}^{t=T} r_t(x_t)$$

$$\text{s.t.} \quad \sum_{t=1}^{t=T} g_t(x_t) \leq w \tag{6}$$

where x_t must be a member of $\{0, 1, 2, \dots \}$. Some possible interpretations of $r_t(x_t)$, $g_t(x_t)$, and w are given in Table 8.

To solve (6) by dynamic programming, define $f_t(d)$ to be the maximum benefit that can be obtained from activities $t, t + 1, \dots, T$ if d units of the resource may be allocated to activities $t, t + 1, \dots, T$. We may generalize the recursions of Example 5 to this situation by writing

$$f_{T+1}(d) = 0 \quad \text{for all } d$$

$$f_t(d) = \max_{x_t} \{r_t(x_t) + f_{t+1}[d - g_t(x_t)]\} \tag{7}$$

where x_t must be a nonnegative integer satisfying $g_t(x_t) \leq d$. Let $x_t(d)$ be any value of x_t that attains $f_t(d)$. To use (7) to determine an optimal allocation of resources to activities $1, 2, \dots, T$, we begin by determining all $f_T(\cdot)$ and $x_T(\cdot)$. Then we use (7) to determine all $f_{T-1}(\cdot)$ and $x_{T-1}(\cdot)$, continuing to work backward in this fashion until all $f_2(\cdot)$ and $x_2(\cdot)$

TABLE 8
Examples of a Generalized Resource Allocation Problem

Interpretation of $r_t(x_t)$	Interpretation of $g_t(x_t)$	Interpretation of w
Benefit from placing x_t type t items in a knapsack	Weight of x_t type t items	Maximum weight that knapsack can hold
Grade obtained in course t if we study course t for x_t hours per week	Number of hours per week x_t spent studying course t	Total number of study hours available each week
Sales of a product in region t if x_t sales reps are assigned to region t	Cost of assigning x_t sales reps to region t	Total sales force budget
Number of fire alarms per week responded to within one minute if precinct t is assigned x_t engines	Cost per week of maintaining x_t fire engines in precinct t	Total weekly budget for maintaining fire engines

have been determined. To wind things up, we now calculate $f_1(w)$ and $x_1(w)$. Then we implement activity 1 at a level $x_1(w)$. At this point, we have $w - g_1[x_1(w)]$ units of the resource available for activities 2, 3, . . . , T. Then activity 2 should be implemented at a level of $x_2\{w - g_1[x_1(w)]\}$. We continue in this fashion until we have determined the level at which all activities should be implemented.

Solution of Knapsack Problems by Dynamic Programming

We illustrate the use of (7) by solving a simple knapsack problem (see Section 9.5 of Volume 1). Then we develop an alternative recursion that can be used to solve knapsack problems.

EXAMPLE 6 Knapsack

Suppose a 10-lb knapsack is to be filled with the items listed in Table 9. To maximize total benefit, how should the knapsack be filled?

Solution We have $r_1(x_1) = 11x_1$, $r_2(x_2) = 7x_2$, $r_3(x_3) = 12x_3$, $g_1(x_1) = 4x_1$, $g_2(x_2) = 3x_2$, and $g_3(x_3) = 5x_3$. Define $f_t(d)$ to be the maximum benefit that can be earned from a d-pound knapsack that is filled with items of Type t, $t + 1$, . . . , 3.

Stage 3 Computations

Now (7) yields

$$f_3(d) = \max_{x_3}\{12x_3\}$$

TABLE 9
Weights and Benefits for Knapsack

Item	Weight (lb)	Benefit
1	4	11
2	3	7
3	5	12

where $5x_3 \le d$ and x_3 is a nonnegative integer. This yields

$$f_3(10) = 24$$
$$f_3(5) = f_3(6) = f_3(7) = f_3(8) = f_3(9) = 12$$
$$f_3(0) = f_3(1) = f_3(2) = f_3(3) = f_3(4) = 0$$
$$x_3(10) = 2$$
$$x_3(9) = x_3(8) = x_3(7) = x_3(6) = x_3(5) = 1$$
$$x_3(0) = x_3(1) = x_3(2) = x_3(3) = x_3(4) = 0$$

Stage 2 Computations

Now (7) yields

$$f_2(d) = \max_{x_2} \{7x_2 + f_3(d - 3x_2)\}$$

where x_2 must be a nonnegative integer satisfying $3x_2 \le d$. We now obtain

$$f_2(10) = \max \begin{cases} 7(0) + f_3(10) = 24^* & x_2 = 0 \\ 7(1) + f_3(7) = 19 & x_2 = 1 \\ 7(2) + f_3(4) = 14 & x_2 = 2 \\ 7(3) + f_3(1) = 21 & x_2 = 3 \end{cases}$$

Thus, $f_2(10) = 24$ and $x_2(10) = 0$.

$$f_2(9) = \max \begin{cases} 7(0) + f_3(9) = 12 & x_2 = 0 \\ 7(1) + f_3(6) = 19 & x_2 = 1 \\ 7(2) + f_3(3) = 14 & x_2 = 2 \\ 7(3) + f_3(0) = 21^* & x_2 = 3 \end{cases}$$

Thus, $f_2(9) = 21$ and $x_2(9) = 3$.

$$f_2(8) = \max \begin{cases} 7(0) + f_3(8) = 12 & x_2 = 0 \\ 7(1) + f_3(5) = 19^* & x_2 = 1 \\ 7(2) + f_3(2) = 14 & x_2 = 2 \end{cases}$$

Thus, $f_2(8) = 19$ and $x_2(8) = 1$.

$$f_2(7) = \max \begin{cases} 7(0) + f_3(7) = 12 & x_2 = 0 \\ 7(1) + f_3(4) = 7 & x_2 = 1 \\ 7(2) + f_3(1) = 14^* & x_2 = 2 \end{cases}$$

Thus, $f_2(7) = 14$ and $x_2(7) = 2$.

$$f_2(6) = \max \begin{cases} 7(0) + f_3(6) = 12 & x_2 = 0 \\ 7(1) + f_3(3) = 7 & x_2 = 1 \\ 7(2) + f_3(0) = 14^* & x_2 = 2 \end{cases}$$

Thus, $f_2(6) = 14$ and $x_2(6) = 2$.

$$f_2(5) = \max \begin{cases} 7(0) + f_3(5) = 12^* & x_2 = 0 \\ 7(1) + f_3(2) = 7 & x_2 = 1 \end{cases}$$

Thus, $f_2(5) = 12$ and $x_2(5) = 0$.

$$f_2(4) = \max \begin{cases} 7(0) + f_3(4) = 0 & x_2 = 0 \\ 7(1) + f_3(1) = 7^* & x_2 = 1 \end{cases}$$

Thus, $f_2(4) = 7$ and $x_2(4) = 1$.

$$f_2(3) = \max \begin{cases} 7(0) + f_3(3) = 0 & x_2 = 0 \\ 7(1) + f_3(0) = 7^* & x_2 = 1 \end{cases}$$

Thus, $f_2(3) = 7$ and $x_2(3) = 1$.

$$f_2(2) = 7(0) + f_3(2) = 0 \qquad x_2 = 0$$

Thus, $f_2(2) = 0$ and $x_2(2) = 0$.

$$f_2(1) = 7(0) + f_3(1) = 0 \qquad x_2 = 0$$

Thus, $f_2(1) = 0$ and $x_2(1) = 0$.

$$f_2(0) = 7(0) + f_3(0) = 0 \qquad x_2 = 0$$

Thus, $f_2(0) = 0$ and $x_2(0) = 0$.

Stage 1 Computations

Finally, we determine $f_1(10)$ from

$$f_1(10) = \max \begin{cases} 11(0) + f_2(10) = 24 & x_1 = 0 \\ 11(1) + f_2(6) \;\; = 25^* & x_1 = 1 \\ 11(2) + f_2(2) \;\; = 22 & x_1 = 2 \end{cases}$$

Determination of the Optimal Solution to Knapsack Problem

We have $f_1(10) = 25$ and $x_1(10) = 1$. Hence, we should include one Type 1 item in the knapsack. Then we have $10 - 4 = 6$ lb left for Type 2 and Type 3 items, so we should include $x_2(6) = 2$ Type 2 items. Finally, we have $6 - 2(3) = 0$ lb left for Type 3 items, and we include $x_3(0) = 0$ Type 3 items. In summary, the maximum benefit that can be gained from a 10-lb knapsack is $f_3(10) = 25$. To obtain a benefit of 25, one Type 1 and two Type 2 items should be included.

Network Representation of Knapsack Problem

Finding the optimal solution to Example 6 is equivalent to finding the longest path in Figure 7 from node (10, 1) to some stage 4 node. In Figure 7, for $t \leq 3$, the node (d, t) represents a situation in which d pounds of space may be allocated to items of Type $t, t + 1$, ..., 3. The node $(d, 4)$ represents d pounds of unused space. Each arc from a stage t node to a stage $t + 1$ node represents a decision of how many Type t items are placed in the knapsack. For example, the arc from (10, 1) to (6, 2) represents placing one Type 1 item in the knapsack. This leaves $10 - 4 = 6$ lb for items of Types 2 and 3. This arc has a length of 11, representing the benefit obtained by placing one Type 1 item in the knapsack. Our solution to Example 6 shows that the longest path in Figure 7 from node (10, 1) to a stage 4 node is (10, 1)–(6, 2)–(0, 3)–(0, 4). We note that the optimal solution to a knapsack problem does not always use all the available weight. For example, the reader should verify that if a Type 1 item earned 16 units of benefit, the optimal solution would be to include two type 1 items, corresponding to the path (10, 1)–(2, 2)–(2, 3)–(2, 4). This solution leaves 2 lb of space unused.

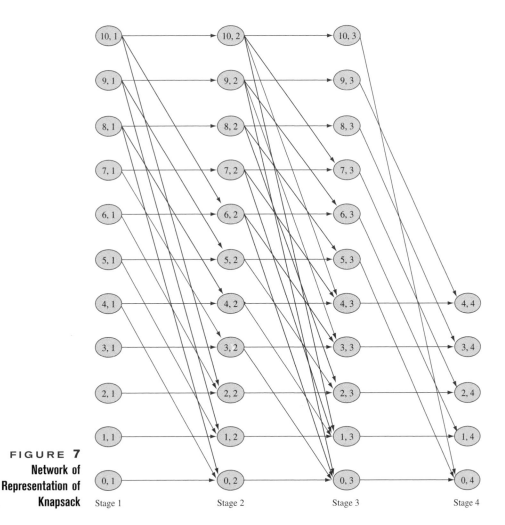

FIGURE 7
Network of
Representation of
Knapsack

Stage 1   Stage 2   Stage 3   Stage 4

An Alternative Recursion for Knapsack Problems

Other approaches can be used to solve knapsack problems by dynamic programming. The approach we now discuss builds up the optimal knapsack by first determining how to fill a small knapsack optimally and then, using this information, how to fill a larger knapsack optimally. We define $g(w)$ to be the maximum benefit that can be gained from a w-lb knapsack. In what follows, b_j is the benefit earned from a single Type j item, and w_j is the weight of a single Type j item. Clearly, $g(0) = 0$, and for $w > 0$,

$$g(w) = \max_j \{b_j + g(w - w_j)\} \tag{8}$$

where j must be a member of $\{1, 2, 3\}$, and j must satisfy $w_j \leq w$. The reasoning behind (8) is as follows: To fill a w-lb knapsack optimally, we must begin by putting some type of item into the knapsack. If we begin by putting a Type j item into a w-lb knapsack, the best we can do is earn $b_j +$ [best we can do from a $(w - w_j)$-lb knapsack]. After noting that a Type j item can be placed into a w-lb knapsack only if $w_j \leq w$, we obtain (8). We define $x(w)$ to be any type of item that attains the maximum in (8) and $x(w) = 0$ to mean that no item can fit into a w-lb knapsack.

To illustrate the use of (8), we re-solve Example 6. Because no item can fit in a 0-, 1-, or 2-lb knapsack, we have $g(0) = g(1) = g(2) = 0$ and $x(0) = x(1) = x(2) = 0$. Only a Type 2 item fits into a 3-lb knapsack, so we have that $g(3) = 7$ and $x(3) = 2$. Continuing, we find that

$$g(4) = \max \begin{cases} 11 + g(0) = 11^* & \text{(Type 1 item)} \\ 7 + g(1) = 7 & \text{(Type 2 item)} \end{cases}$$

Thus, $g(4) = 11$ and $x(4) = 1$.

$$g(5) = \max \begin{cases} 11 + g(1) = 11 & \text{(Type 1 item)} \\ 7 + g(2) = 7 & \text{(Type 2 item)} \\ 12 + g(0) = 12^* & \text{(Type 3 item)} \end{cases}$$

Thus, $g(5) = 12$ and $x(5) = 3$.

$$g(6) = \max \begin{cases} 11 + g(2) = 11 & \text{(Type 1 item)} \\ 7 + g(3) = 14^* & \text{(Type 2 item)} \\ 12 + g(1) = 12 & \text{(Type 3 item)} \end{cases}$$

Thus, $g(6) = 14$ and $x(6) = 2$.

$$g(7) = \max \begin{cases} 11 + g(3) = 18^* & \text{(Type 1 item)} \\ 7 + g(4) = 18^* & \text{(Type 2 item)} \\ 12 + g(2) = 12 & \text{(Type 3 item)} \end{cases}$$

Thus, $g(7) = 18$ and $x(7) = 1$ or $x(7) = 2$.

$$g(8) = \max \begin{cases} 11 + g(4) = 22^* & \text{(Type 1 item)} \\ 7 + g(5) = 19 & \text{(Type 2 item)} \\ 12 + g(3) = 19 & \text{(Type 3 item)} \end{cases}$$

Thus, $g(8) = 22$ and $x(8) = 1$.

$$g(9) = \max \begin{cases} 11 + g(5) = 23^* & \text{(Type 1 item)} \\ 7 + g(6) = 21 & \text{(Type 2 item)} \\ 12 + g(4) = 23^* & \text{(Type 3 item)} \end{cases}$$

Thus, $g(9) = 23$ and $x(9) = 1$ or $x(9) = 3$.

$$g(10) = \max \begin{cases} 11 + g(6) = 25^* & \text{(Type 1 item)} \\ 7 + g(7) = 25^* & \text{(Type 2 item)} \\ 12 + g(5) = 24 & \text{(Type 3 item)} \end{cases}$$

Thus, $g(10) = 25$ and $x(10) = 1$ or $x(10) = 2$. To fill the knapsack optimally, we begin by putting any $x(10)$ item in the knapsack. Let's arbitrarily choose a Type 1 item. This leaves us with $10 - 4 = 6$ lb to fill, so we now put an $x(10 - 4) = 2$ (Type 2) item in the knapsack. This leaves us with $6 - 3 = 3$ lb to fill, which we do with an $x(6 - 3) = 2$ (Type 2) item. Hence, we may attain the maximum benefit of $g(10) = 25$ by filling the knapsack with two Type 2 items and one Type 1 item.

A Turnpike Theorem

For a knapsack problem, let

$$c_j = \text{benefit obtained from each type } j \text{ item}$$
$$w_j = \text{weight of each type } j \text{ item}$$

In terms of benefit per unit weight, the best item is the item with the largest value of $\frac{c_j}{w_j}$. Assume there are n types of items that have been ordered, so that

$$\frac{c_1}{w_1} \geq \frac{c_2}{w_2} \geq \cdots \geq \frac{c_n}{w_n}$$

Thus, Type 1 items are the best, Type 2 items are the second best, and so on. Recall from Volume 1, Section 9.5 that it is possible for the optimal solution to a knapsack problem to use none of the best item. For example, the optimal solution to the knapsack problem

$$\max z = 16x_1 + 22x_2 + 12x_3 + 8x_4$$
$$\text{s.t.} \quad 5x_1 + 7x_2 + 5x_3 + 4x_4 \leq 14$$
$$x_i \text{ nonnegative integer}$$

is $z = 44$, $x_2 = 2$, $x_1 = x_3 = x_4 = 0$, and this solution does not use any of the best (Type 1) item. Assume that

$$\frac{c_1}{w_1} > \frac{c_2}{w_2}$$

Thus, there is a unique best item type. It can be shown that for some number w^*, it is optimal to use at least one Type 1 item if the knapsack is allowed to hold w pounds, where $w \geq w^*$. In Problem 6 at the end of this section, you will show that this result holds for

$$w^* = \frac{c_1 w_1}{c_1 - w_1 \left(\frac{c_2}{w_2}\right)}$$

Thus, for the knapsack problem

$$\max z = 16x_1 + 22x_2 + 12x_3 + 8x_4$$
$$\text{s.t.} \quad 5x_1 + 7x_2 + 5x_3 + 4x_4 \leq w$$
$$x_i \text{ nonnegative integer}$$

at least one Type 1 item will be used if

$$w \geq \frac{16(5)}{16 - 5(\frac{22}{7})} = 280$$

This result can greatly reduce the computation needed to solve a knapsack problem. For example, suppose that $w = 4,000$. We know that for $w \geq 280$, the optimal solution will use at least one Type 1 item, so we can conclude that the optimal way to fill a 4,000-lb knapsack will consist of one Type 1 item plus the optimal way to fill a knapsack of $4,000 - 5 = 3,995$ lb. Repeating this reasoning shows that the optimal way to fill a 4,000-lb knapsack will consist of $\frac{4,000-280}{5} = 744$ Type 1 items plus the optimal way to fill a knapsack of 280 lb. This reasoning substantially reduces the computation needed to determine how to fill a 4,000-lb knapsack. (Actually, the 280-lb knapsack will use at least one Type 1 item, so we know that to fill a 4,000-lb knapsack optimally, we can use 745 Type 1 items and then optimally fill a 275-lb knapsack.)

Why is this result referred to as a **turnpike theorem**? Think about taking an automobile trip in which our goal is to minimize the time needed to complete the trip. For a long enough trip, it may be advantageous to go slightly out of our way so that most of the trip will be spent on a turnpike, on which we can travel at the greatest speed. For a short trip, it may not be worth our while to go out of our way to get on the turnpike.

Similarly, in a long (large-weight) knapsack problem, it is always optimal to use some of the best items, but this may not be the case in a short knapsack problem. Turnpike results abound in the dynamic programming literature [see Morton (1979)].

PROBLEMS

Group A

1 J. R. Carrington has $4 million to invest in three oil well sites. The amount of revenue earned from site $i (i = 1, 2, 3)$ depends on the amount of money invested in site i (see Table 10). Assuming that the amount invested in a site must be an exact multiple of $1 million, use dynamic programming to determine an investment policy that will maximize the revenue J. R. will earn from his three oil wells.

2 Use either of the approaches outlined in this section to solve the following knapsack problem:

$$\max z = 5x_1 + 4x_2 + 2x_3$$
$$\text{s.t.} \quad 4x_1 + 3x_2 + 2x_3 \leq 8$$
$$x_1, x_2, x_3 \geq 0; x_1, x_2, x_3 \text{ integer}$$

3 The knapsack problem of Problem 2 can be viewed as finding the longest route in a particular network.

    a Draw the network corresponding to the recursion derived from (7).

    b Draw the network corresponding to the recursion derived from (8).

4 The number of crimes in each of a city's three police precincts depends on the number of patrol cars assigned to each precinct (see Table 11). Five patrol cars are available. Use dynamic programming to determine how many patrol cars should be assigned to each precinct.

TABLE 11

	No. of Patrol Cars Assigned to Precinct					
Precinct	0	1	2	3	4	5
1	14	10	7	4	1	0
2	25	19	16	14	12	11
3	20	14	11	8	6	5

5 Use dynamic programming to solve a knapsack problem in which the knapsack can hold up to 13 lb (see Table 12).

Group B

6 Consider a knapsack problem for which

$$\frac{c_1}{w_1} > \frac{c_2}{w_2}$$

Show that if the knapsack can hold w pounds, and $w \geq w^*$, where

$$w^* = \frac{c_1 w_1}{c_1 - w_1 \left(\dfrac{c_2}{w_2} \right)}$$

then the optimal solution to the knapsack problem must use at least one Type 1 item.

TABLE 10

Amount Invested ($ Millions)	Revenue ($ Millions)		
	Site 1	Site 2	Site 3
0	4	3	3
1	7	6	7
2	8	10	8
3	9	12	13
4	11	14	15

TABLE 12

Item	Weight (lb)	Benefit
1	3	12
2	5	25
3	7	50

6.5 Equipment-Replacement Problems

Many companies and customers face the problem of determining how long a machine should be utilized before it should be traded in for a new one. Problems of this type are called **equipment-replacement problems** and can often be solved by dynamic programming.

EXAMPLE 7   **Equipment Replacement**

An auto repair shop always needs to have an engine analyzer available. A new engine an lyzer costs $1,000. The cost m_i of maintaining an engine analyzer during its ith year of operation is as follows: $m_1 = \$60$, $m_2 = \$80$, $m_3 = \$120$. An analyzer may be kept for

FIGURE 8
Time Horizon for
Equipment
Replacement

1, 2, or 3 years; after i years of use ($i = 1, 2, 3$), it may be traded in for a new one. If an i-year-old engine analyzer is traded in, a salvage value s_i is obtained, where $s_1 = \$800$, $s_2 = \$600$, and $s_3 = \$500$. Given that a new machine must be purchased now (time 0; see Figure 8), the shop wants to determine a replacement and trade-in policy that minimizes net costs = (maintenance costs) + (replacement costs) – (salvage value received) during the next 5 years.

Solution We note that after a new machine is purchased, the firm must decide when the newly purchased machine should be traded in for a new one. With this in mind, we define $g(t)$ to be the minimum net cost incurred from time t until time 5 (including the purchase cost and salvage value for the newly purchased machine) given that a new machine has been purchased at time t. We also define c_{tx} to be the net cost (including purchase cost and salvage value) of purchasing a machine at time t and operating it until time x. Then the appropriate recursion is

$$g(t) = \min_{x} \{c_{tx} + g(x)\} \qquad (t = 0, 1, 2, 3, 4) \tag{9}$$

where x must satisfy the inequalities $t + 1 \le x \le t + 3$ and $x \le 5$. Because the problem is over at time 5, no cost is incurred from time 5 onward, so we may write $g(5) = 0$.

To justify (9), note that after a new machine is purchased at time t, we must decide when to replace the machine. Let x be the time at which the replacement occurs. The replacement must be after time t but within 3 years of time t. This explains the restriction that $t + 1 \le x \le t + 3$. Since the problem ends at time 5, we must also have $x \le 5$. If we choose to replace the machine at time x, then what will be the cost from time t to time 5? Simply the sum of the cost incurred from the purchase of the machine to the sale of the machine at time x (which is by definition c_{tx}) and the total cost incurred from time x to time 5 (given that a new machine has just been purchased at time x). By the principle of optimality, the latter cost is, of course, $g(x)$. Hence, if we keep the machine that was purchased at time t until time x, then from time t to time 5, we incur a cost of $c_{tx} + g(x)$. Thus, x should be chosen to minimize this sum, and this is exactly what (9) does. We have assumed that maintenance costs, salvage value, and purchase price remain unchanged over time, so each c_{tx} will depend only on how long the machine is kept; that is, each c_{tx} depends only on $x - t$. More specifically,

$$c_{tx} = \$1,000 + m_1 + \cdots + m_{x-t} - s_{x-t}$$

This yields

$$c_{01} = c_{12} = c_{23} = c_{34} = c_{45} = 1,000 + 60 - 800 = \$260$$
$$c_{02} = c_{13} = c_{24} = c_{35} = 1,000 + 60 + 80 - 600 = \$540$$
$$c_{03} = c_{14} = c_{25} = 1,000 + 60 + 80 + 120 - 500 = \$760$$

We begin by computing $g(4)$ and work backward until we have computed $g(0)$. Then we use our knowledge of the values of x attaining $g(0)$, $g(1)$, $g(2)$, $g(3)$, and $g(4)$ to determine the optimal replacement strategy. The calculations follow.

At time 4, there is only one sensible decision (keep the machine until time 5 and sell it for its salvage value), so we find

$$g(4) = c_{45} + g(5) = 260 + 0 = \$260^*$$

Thus, if a new machine is purchased at time 4, it should be traded in at time 5.

If a new machine is purchased at time 3, we keep it until time 4 or time 5. Hence,

$$g(3) = \min \begin{cases} c_{34} + g(4) = 260 + 260 = \$520^* & \text{(Trade at time 4)} \\ c_{35} + g(5) = 540 + 0 = \$540 & \text{(Trade at time 5)} \end{cases}$$

Thus, if a new machine is purchased at time 3, we should trade it in at time 4.

If a new machine is purchased at time 2, we trade it in at time 3, time 4, or time 5. This yields

$$g(2) = \min \begin{cases} c_{23} + g(3) = 260 + 520 = \$780 & \text{(Trade at time 3)} \\ c_{24} + g(4) = 540 + 260 = \$800 & \text{(Trade at time 4)} \\ c_{25} + g(5) = \$760^* & \text{(Trade at time 5)} \end{cases}$$

Thus, if we purchase a new machine at time 2, we should keep it until time 5 and then trade it in.

If a new machine is purchased at time 1, we trade it in at time 2, time 3, or time 4. Then

$$g(1) = \min \begin{cases} c_{12} + g(2) = 260 + 760 = \$1,020^* & \text{(Trade at time 2)} \\ c_{13} + g(3) = 540 + 520 = \$1,060 & \text{(Trade at time 3)} \\ c_{14} + g(4) = 760 + 260 = \$1,020^* & \text{(Trade at time 4)} \end{cases}$$

Thus, if a new machine is purchased at time 1, it should be traded in at time 2 or time 4.

The new machine that was purchased at time 0 may be traded in at time 1, time 2, or time 3. Thus,

$$g(0) = \min \begin{cases} c_{01} + g(1) = 260 + 1,020 = \$1,280^* & \text{(Trade at time 1)} \\ c_{02} + g(2) = 540 + 760 = \$1,300 & \text{(Trade at time 2)} \\ c_{03} + g(3) = 760 + 520 = \$1,280^* & \text{(Trade at time 3)} \end{cases}$$

Thus, the new machine purchased at time 0 should be replaced at time 1 or time 3. Let's arbitrarily choose to replace the time 0 machine at time 1. Then the new time 1 machine may be traded in at time 2 or time 4. Again we make an arbitrary choice and replace the time 1 machine at time 2. Then the time 2 machine should be kept until time 5, when it is sold for salvage value. With this replacement policy, we will incur a net cost of $g(0) = \$1,280$. The reader should verify that the following replacement policies are also optimal: (1) trading in at times 1, 4, and 5 and (2) trading in at times 3, 4, and 5.

We have assumed that all costs remain stationary over time. This assumption was made solely to simplify the computation of the c_{tx}'s. If we had relaxed the assumption of stationary costs, then the only complication would have been that the c_{tx}'s would have been messier to compute. We also note that if a short planning horizon is used, the optimal replacement policy may be extremely sensitive to the length of the planning horizon. Thus, more meaningful results can be obtained by using a longer planning horizon.

An equipment-replacement model was actually used by Phillips Petroleum to reduce costs associated with maintaining the company's stock of trucks (see Waddell (1983)).

Network Representation of Equipment-Replacement Problem

The reader should verify that our solution to Example 7 was equivalent to finding the shortest path from node 0 to node 5 in the network in Figure 9. The length of the arc joining nodes i and j is c_{ij}.

FIGURE 9
Network Representation
of Equipment
Replacement

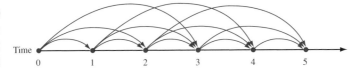

An Alternative Recursion

There is another dynamic programming formulation of the equipment-replacement model. If we define the stage to be the time t and the state at any stage to be the age of the engine analyzer at time t, then an alternative dynamic programming recursion can be developed. Define $f_t(x)$ to be the minimum cost incurred from time t to time 5, given that at time t the shop has an x-year-old analyzer. The problem is over at time 5, so we sell the machine at time 5 and receive $-s_x$. Then $f_5(x) = -s_x$, and for $t = 0, 1, 2, 3, 4$,

$$f_t(3) = -500 + 1{,}000 + 60 + f_{t+1}(1) \qquad \text{(Trade)} \qquad (10)$$

$$f_t(2) = \min \begin{cases} -600 + 1{,}000 + 60 + f_{t+1}(1) & \text{(Trade)} \\ 120 + f_{t+1}(3) & \text{(Keep)} \end{cases} \qquad (10.1)$$

$$f_t(1) = \min \begin{cases} -800 + 1{,}000 + 60 + f_{t+1}(1) & \text{(Trade)} \\ 80 + f_{t+1}(2) & \text{(Keep)} \end{cases} \qquad (10.2)$$

$$f_0(0) = 1{,}000 + 60 + f_1(1) \qquad \text{(Keep)} \qquad (10.3)$$

The rationale behind Equations (10)–(10.3) is that if we have a 1- or 2-year-old analyzer, then we must decide between replacing the machine or keeping it another year. In (10.1) and (10.2), we compare the costs of these two options. For any option, the total cost from t until time 5 is the sum of the cost during the current year plus costs from time $t + 1$ to time 5. If we have a 3-year-old analyzer, then we must replace it, so there is no choice. The way we have defined the state means that it is only possible to be in state 0 at time 0. In this case, we must keep the analyzer for the first year (incurring a cost of $1,060). From this point on, a total cost of $f_1(1)$ is incurred. Thus, (10.3) follows. Since we know that $f_5(1) = -800$, $f_5(2) = -600$, and $f_5(3) = -500$, we can immediately compute all the $f_4(\cdot)$'s. Then we can compute the $f_3(\cdot)$'s. We continue in this fashion until $f_0(0)$ is determined (remember that we begin with a new machine). Then we follow our usual method for determining an optimal policy. That is, if $f_0(0)$ is attained by keeping the machine, then we keep the machine for a year and then, during year 1, we choose the action that attains $f_1(1)$. Continuing in this fashion, we can determine for each time whether or not the machine should be replaced. (See Problem 1 below.)

PROBLEMS

Group A

1 Use Equations (10)–(10.3) to determine an optimal replacement policy for the engine analyzer example.

2 Suppose that a new car costs $10,000 and that the annual operating cost and resale value of the car are as shown in Table 13. If I have a new car now, determine a replacement policy that minimizes the net cost of owning and operating a car for the next six years.

3 It costs $40 to buy a telephone from a department store. The estimated maintenance cost for each year of operation is shown in Table 14. (I can keep a telephone for at most five years.) I have just purchased a new telephone, and my old telephone has no salvage value. Determine how to minimize the total cost of purchasing and operating a telephone for the next six years.

TABLE 13

Age of Car (Years)	Resale Value ($)	Operating Cost ($)	
1	7,000	300	(year 1)
2	6,000	500	(year 2)
3	4,000	800	(year 3)
4	3,000	1,200	(year 4)
5	2,000	1,600	(year 5)
6	1,000	2,200	(year 6)

TABLE 14

Year	Maintenance Cost ($)
1	20
2	30
3	40
4	60
5	70

6.6 Formulating Dynamic Programming Recursions

In many dynamic programming problems (such as the inventory and shortest path examples), a given stage simply consists of all the possible states that the system can occupy at that stage. If this is the case, then the dynamic programming recursion (for a min problem) can often be written in the following form:

$$f_t(i) = \min\{(\text{cost during stage } t) + f_{t+1} (\text{new state at stage } t + 1)\} \tag{11}$$

where the minimum in (11) is over all decisions that are allowable, or feasible, when the state at stage t is i. In (11), $f_t(i)$ is the minimum cost incurred from stage t to the end of the problem (say, the problem ends after stage T), given that at stage t the state is i.

Equation (11) reflects the fact that the minimum cost incurred from stage t to the end of the problem must be attained by choosing at stage t an allowable decision that minimizes the sum of the costs incurred during the current stage (stage t) plus the minimum cost that can be incurred from stage $t + 1$ to the end of the problem. Correct formulation of a recursion of the form (11) requires that we identify three important aspects of the problem:

Aspect 1 *The set of decisions that is allowable, or feasible, for the given state and stage.* Often, the set of feasible decisions depends on both t and i. For instance, in the inventory example of Section 6.3, let

$$d_t = \text{demand during month } t$$

$$i_t = \text{inventory at beginning of month } t$$

In this case, the set of allowable month t decisions (let x_t represent an allowable production level) consists of the members of $\{0, 1, 2, 3, 4, 5\}$ that satisfy $0 \le (i_t + x_t - d_t) \le 4$. Note how the set of allowable decisions at time t depends on the stage t and the state at time t, which is i_t.

Aspect 2 *We must specify how the cost during the current time period (stage t) depends on the value of t, the current state, and the decision chosen at stage t.* For instance, in the inventory example of Section 6.3, suppose a production level x_t is chosen during month t. Then the cost during month t is given by $c(x_t) + (\frac{1}{2})(i_t + x_t - d_t)$.

Aspect 3 *We must specify how the state at stage t + 1 depends on the value of t, the state at stage t, and the decision chosen at stage t.* Again referring to the inventory example, the month $t + 1$ state is $i_t + x_t - d_t$.

If you have properly identified the state, stage, and decision, then aspects 1–3 shouldn't be too hard to handle. A word of caution, however: Not all recursions are of the form (11). For instance, our first equipment-replacement recursion skipped over time $t + 1$.

This often occurs when the stage alone supplies sufficient information to make an optimal decision. We now work through several examples that illustrate the art of formulating dynamic programming recursions.

EXAMPLE 8   A Fishery

The owner of a lake must decide how many bass to catch and sell each year. If she sells x bass during year t, then a revenue $r(x)$ is earned. The cost of catching x bass during a year is a function $c(x, b)$ of the number of bass caught during the year and of b, the number of bass in the lake at the beginning of the year. Of course, bass do reproduce. To model this, we assume that the number of bass in the lake at the beginning of a year is 20% more than the number of bass left in the lake at the end of the previous year. Assume that there are 10,000 bass in the lake at the beginning of the first year. Develop a dynamic programming recursion that can be used to maximize the owner's net profits over a T-year horizon.

Solution   In problems where decisions must be made at several points in time, there is often a trade-off of current benefits against future benefits. For example, we could catch many bass early in the problem, but then the lake would be depleted in later years, and there would be very few bass to catch. On the other hand, if we catch very few bass now, we won't make much money early, but we can make a lot of money near the end of the horizon. In intertemporal optimization problems, dynamic programming is often used to analyze these complex trade-offs.

At the beginning of year T, the owner of the lake need not worry about the effect that the capture of bass will have on the future population of the lake. (At time T, there is no future!) So at the beginning of year T, the problem is relatively easy to solve. For this reason, we let time be the stage. At each stage, the owner of the lake must decide how many bass to catch. We define x_t to be the number of bass caught during year t. To determine an optimal value of x_t, the owner of the lake need only know the number of bass (call it b_t) in the lake at the beginning of year t. Therefore, the state at the beginning of year t is b_t.

We define $f_t(b_t)$ to be the maximum net profit that can be earned from bass caught during years $t, t + 1, \ldots, T$ given that b_t bass are in the lake at the beginning of year t. We may now dispose of aspects 1–3 of the recursion.

Aspect 1   What are the allowable decisions? During any year, we can't catch more bass than there are in the lake. Thus, in each state and for all t, $0 \leq x_t \leq b_t$ must hold.

Aspect 2   What is the net profit earned during year t? If x_t bass are caught during a year that begins with b_t bass in the lake, then the net profit is $r(x_t) - c(x_t, b_t)$.

Aspect 3   What will be the state during year $t + 1$? At the end of year t, there will be $b_t - x_t$ bass in the lake. By the beginning of year $t + 1$, these bass will have multiplied by 20%. This implies that at the beginning of year $t + 1$, $1.2(b_t - x_t)$ bass will be in the lake. Thus, the year $t + 1$ state will be $1.2(b_t - x_t)$.

We can now use (11) to develop the appropriate recursion. After year T, there are no future profits to consider, so

$$f_T(b_T) = \max_{x_T}\{r_T(x_T) - c(x_T, b_T)\}$$

where $0 \leq x_T \leq b_T$. Applying (11), we obtain

$$f_t(b_t) = \max\{r(x_t) - c(x_t, b_t) + f_{t+1}[1.2(b_t - x_t)]\} \tag{12}$$

where $0 \leq x_t \leq b_t$. To begin the computations, we first determine $f_T(b_T)$ for all values of b_T that might occur [b_T could be up to $10{,}000(1.2)^{T-1}$; why?]. Then we use (12) to work backward until $f_1(10,000)$ has been computed. Then, to determine an optimal fishing policy, we begin by choosing x_1 to be any value attaining the maximum in the (12) equation for $f_1(10,000)$. Then year 2 will begin with $1.2(10,000 - x_1)$ bass in the lake. This means that x_2 should be chosen to be any value attaining the maximum in the (12) equation for $f_2(1.2(10,000 - x_1))$. Continue in this fashion until the optimal values of $x_3, x_4, \ldots, x_T$ have been determined.

Incorporating the Time Value of Money into Dynamic Programming Formulations

A weakness of the current formulation is that profits received during later years are weighted the same as profits received during earlier years. As mentioned in the discussion of discounting (Volume 1, Chapter 3), later profits should be weighted less than earlier profits. Suppose that for some $\beta < 1$, \$1 received at the beginning of year $t + 1$ is equivalent to β dollars received at the beginning of year t. We can incorporate this idea into the dynamic programming recursion by replacing (12) with

$$f_t(b_t) = \max_{x_t} \{r(x_t) - c(x_t, b_t) + \beta f_{t+1}[1.2(b_t - x_t)]\} \tag{12'}$$

where $0 \le x_t \le b_t$. Then we redefine $f_t(b_t)$ to be the maximum net profit *(in year t dollars)* that can be earned during years $t, t + 1, \ldots, T$. Since f_{t+1} is measured in year $t + 1$ dollars, multiplying it by β converts $f_{t+1}(\cdot)$ to year t dollars, which is just what we want. In Example 8, once we have worked backward and determined $f_1(10,000)$, an optimal fishing policy is found by using the same method that was previously described. This approach can be used to account for the time value of money in any dynamic programming formulation.

EXAMPLE 9   **Power Plant**

An electric power utility forecasts that r_t kilowatt-hours (kwh) of generating capacity will be needed during year t (the current year is year 1). Each year, the utility must decide by how much generating capacity should be expanded. It costs $c_t(x)$ dollars to increase generating capacity by x kwh during year t. It may be desirable to reduce capacity, so x need not be nonnegative. During each year, 10% of the old generating capacity becomes obsolete and unusable (capacity does not become obsolete during its first year of operation). It costs the utility $m_t(i)$ dollars to maintain i units of capacity during year t. At the beginning of year 1, 100,000 kwh of generating capacity are available. Formulate a dynamic programming recursion that will enable the utility to minimize the total cost of meeting power requirements for the next T years.

Solution   Again, we let time be the stage. At the beginning of year t, the utility must determine the amount of capacity (call it x_t) to add during year t. To choose x_t properly, all the utility needs to know is the amount of available capacity at the beginning of year t (call it i_t). Hence, we define the state at the beginning of year t to be the current capacity level. We may now dispose of aspects 1–3 of the formulation.

Aspect 1   What values of x_t are feasible? To meet year t's requirement of r_t, we must have $i_t + x_t \ge r_t$, or $x_t \ge r_t - i_t$. So the feasible x_t's are those values of x_t satisfying $x_t \ge r_t - i_t$.

Aspect 2   What cost is incurred during year t? If x_t kwh are added during a year that begins with i_t kwh of available capacity, then during year t, a cost $c_t(x_t) + m_t(i_t + x_t)$ is incurred.

Aspect 3   What will be the state at the beginning of year $t + 1$? At the beginning of year $t + 1$, the utility will have $0.9i_t$ kwh of old capacity plus the x_t kwh that have been added during year t. Thus, the state at the beginning of year $t + 1$ will be $0.9i_t + x_t$.

We can now use (11) to develop the appropriate recursion. Define $f_t(i_t)$ to be the minimum cost incurred by the utility during years $t, t + 1, \ldots, T$, given that i_t kwh of capacity are available at the beginning of year t. At the beginning of year T, there are no future costs to consider, so

$$f_T(i_T) = \min_{x_T} \{c_T(x_T) + m_T(i_T + x_T)\} \tag{13}$$

where x_T must satisfy $x_T \geq r_T - i_T$. For $t < T$,

$$f_t(i_t) = \min_{x_T} \{c_t(x_t) + m_t(i_t + x_t) + f_{t+1}(0.9i_t + x_t)\} \tag{14}$$

where x_t must satisfy $x_t \geq r_t - i_t$. If the utility does not start with any excess capacity, then we can safely assume that the capacity level would never exceed $r_{\text{MAX}} = \max_{t=1, 2, \ldots, T} \{r_t\}$. This means that we need consider only states $0, 1, 2, \ldots, r_{\text{MAX}}$. To begin computations, we use (13) to compute $f_T(0), f_T(1), \ldots, f_T(r_{\text{MAX}})$. Then we use (14) to work backward until $f_1(100,000)$ has been determined. To determine the optimal amount of capacity that should be added during each year, proceed as follows. During year 1, add an amount of capacity x_1 that attains the minimum in the (14) equation for $f_1(100,000)$. Then the utility will begin year 2 with $90,000 + x_1$ kwh of capacity. Then, during year 2, x_2 kwh of capacity should be added, where x_2 attains the minimum in the (14) equation for $f_2(90,000 + x_1)$. Continue in this fashion until the optimal value of x_T has been determined.

EXAMPLE 10   **Wheat Sale**

Farmer Jones now possesses $5,000 in cash and 1,000 bushels of wheat. During month t, the price of wheat is p_t. During each month, he must decide how many bushels of wheat to buy (or sell). There are three restrictions on each month's wheat transactions: (1) During any month, the amount of money spent on wheat cannot exceed the cash on hand at the beginning of the month; (2) during any month, he cannot sell more wheat than he has at the beginning of the month; and (3) because of limited warehouse capacity, the ending inventory of wheat for each month cannot exceed 1,000 bushels.

Show how dynamic programming can be used to maximize the amount of cash that farmer Jones has on hand at the end of six months.

Solution   Again, we let time be the stage. At the beginning of month t (the present is the beginning of month 1), farmer Jones must decide by how much to change the amount of wheat on hand. We define Δw_t to be the change in farmer Jones's wheat position during month t: $\Delta w_t \geq 0$ corresponds to a month t wheat purchase, and $\Delta w_t \leq 0$ corresponds to a month t sale of wheat. To determine an optimal value for Δw_t, we must know two things: the amount of wheat on hand at the beginning of month t (call it w_t) and the cash on hand at the beginning of month t, (call this c_t). We define $f_t(c_t, w_t)$ to be the maximum cash that farmer Jones can obtain at the end of month 6, given that farmer Jones has c_t dollars and w_t bushels of wheat at the beginning of month t. We now discuss aspects 1–3 of the formulation.

Aspect 1   What are the allowable decisions? If the state at time t is (c_t, w_t), then restrictions 1–3 limit Δw_t in the following manner:

$$p_t(\Delta w_t) \leq c_t \quad \text{or} \quad \Delta w_t \leq \frac{c_t}{p_t}$$

ensures that we won't run out of money at the end of month t. The inequality $\Delta w_t \geq -w_t$ ensures that during month t, we will not sell more wheat than we had at the beginning of month t; and $w_t + \Delta w_t \leq 1,000$, or $\Delta w_t \leq 1,000 - w_t$, ensures that we will end month t with at most 1,000 bushels of wheat. Putting these three restrictions together, we see that

$$-w_t \leq \Delta w_t \leq \min \left\{ \frac{c_t}{p_t}, 1{,}000 - w_t \right\}$$

will ensure that restrictions 1–3 are satisfied during month t.

Aspect 2 Since farmer Jones wants to maximize his cash on hand at the end of month 6, no benefit is earned during months 1 through 5. In effect, during months 1–5, we are doing bookkeeping to keep track of farmer Jones's position. Then, during month 6, we turn all of farmer Jones's assets into cash.

Aspect 3 If the current state is (c_t, w_t) and farmer Jones changes his month t wheat position by an amount Δw_t, what will be the new state at the beginning of month $t + 1$? Cash on hand will increase by $-(\Delta w_t)p_t$, and farmer Jones's wheat position will increase by Δw_t. Hence, the month $t + 1$ state will be $[c_t - (\Delta w_t)p_t, w_t + \Delta w_t]$.

We may now use (11) to develop the appropriate recursion. To maximize his cash position at the end of month 6, farmer Jones should convert his month 6 wheat into cash by selling all of it. This means that $\Delta w_6 = -w_6$. This leads to the following relation:

$$f_6(c_6, w_6) = c_6 + w_6 p_6 \tag{15}$$

Using (11), we obtain for $t < 6$

$$f_t(c_t, w_t) = \max_{\Delta w_t} \{0 + f_{t+1}[c_t - (\Delta w_t)p_t, w_t + \Delta w_t]\} \tag{16}$$

where Δw_t must satisfy

$$-w_t \leq \Delta w_t \leq \min \left\{ \frac{c_t}{p_t}, 1{,}000 - w_t \right\}$$

We begin our calculations by determining $f_6(c_6, w_6)$ for all states that can possibly occur during month 6. Then we use (16) to work backward until $f_1(5{,}000, 1{,}000)$ has been computed. Next, farmer Jones should choose Δw_1 to attain the maximum value in the (16) equation for $f_1(5{,}000, 1{,}000)$, and a month 2 state of $[5{,}000 - p_1(\Delta w_1), 1{,}000 + \Delta w_1]$ will ensue. Farmer Jones should next choose Δw_2 to attain the maximum value in the (16) equation for $f_2[5{,}000 - p_1(\Delta w_1), 1{,}000 + \Delta w_1]$. We continue in this manner until the optimal value of Δw_6 has been determined.

EXAMPLE 11 **Refinery Capacity**

Sunco Oil needs to build enough refinery capacity to refine 5,000 barrels of oil per day and 10,000 barrels of gasoline per day. Sunco can build refinery capacity at four locations. The cost of building a refinery at site t that has the capacity to refine x barrels of oil per day and y barrels of gasoline per day is $c_t(x, y)$. Use dynamic programming to determine how much capacity should be located at each site.

Solution If Sunco had only one possible refinery site, then the problem would be easy to solve. Sunco could solve a problem in which there were two possible refinery sites, and finally, a problem in which there were four refinery sites. For this reason, we let the stage represent the number of available oil sites. At any stage, Sunco must determine how much oil and gas capacity should be built at the given site. To do this, the company must know how much refinery capacity of each type must be built at the available sites. We now define $f_t(o_t, g_t)$ to be the minimum cost of building o_t barrels per day of oil refinery capacity and g_t barrels per day of gasoline refinery capacity at sites $t, t + 1, \ldots, 4$.

To determine $f_4(o_4, g_4)$, note that if only site 4 is available, Sunco must build a refinery at site 4 with o_4 barrels of oil capacity and g_4 barrels of gasoline capacity. This implies that $f_4(o_4, g_4) = c_4(o_4, g_4)$. For $t = 1, 2, 3$, we can determine $f_t(o_t, g_t)$ by noting that if we build a refinery at site t that can refine x_t barrels of oil per day and y_t barrels of gasoline per day, then we incur a cost of $c_t(x_t, y_t)$ at site t. Then we will need to build a total oil refinery capacity of $o_t - x_t$ and a gas refinery capacity of $g_t - y_t$ at sites $t + 1$, $t + 2, \ldots, 4$. By the principle of optimality, the cost of doing this will be $f_{t+1}(o_t - x_t, g_t - y)$. Since $0 \le x_t \le o_t$ and $0 \le y_t \le g_t$ must hold, we obtain the following recursion:

$$f_t(o_t, g_t) = \min \{c_t(o_t, g_t) + f_{t+1}(o_t - x_t, g_t - y_t)\} \tag{17}$$

where $0 \le x_t \le o_t$ and $0 \le y_t \le g_t$. As usual, we work backward until $f_1(5,000, 10,000)$ has been determined. Then Sunco chooses x_1 and y_1 to attain the minimum in the (17) equation for f_1 (5,000, 10,000). Then Sunco should choose x_2 and y_2 that attain the minimum in the (17) equation for $f_2(5,000 - x_1, 10,000 - y_1)$. Sunco continues in this fashion until optimal values of x_4 and y_4 are determined.

EXAMPLE 12   **Traveling Salesperson**

The traveling salesperson problem (see Section 9.6 of Volume 1) can be solved by using dynamic programming. As an example, we solve the following traveling salesperson problem: It's the last weekend of the 2004 election campaign, and candidate Walter Glenn is in New York City. Before election day, Walter must visit Miami, Dallas, and Chicago and then return to his New York City headquarters. Walter wants to minimize the total distance he must travel. In what order should he visit the cities? The distances in miles between the four cities are given in Table 15.

Solution   We know that Walter must visit each city exactly once, the last city he visits must be New York, and his tour originates in New York. When Walter has only one city left to visit, his problem is trivial: simply go from his current location to New York. Then we can work backward to a problem in which he is in some city and has only two cities left to visit, and finally we can find the shortest tour that originates in New York and has four cities left to visit. We therefore let the stage be indexed by the number of cities that Walter has already visited. At any stage, to determine which city should next be visited, we need to know two things: Walter's current location and the cities he has already visited. The state at any stage consists of the last city visited and the set of cities that have already been visited. We define $f_t(i, S)$ to be the minimum distance that must be traveled to complete a tour if the $t - 1$ cities in the set S have been visited and city i was the last city visited. We let c_{ij} be the distance between cities i and j.

Stage 4 Computations

We note that, at stage 4, it must be the case that $S = \{2, 3, 4\}$ (why?), and the only possible states are $(2, \{2, 3, 4\})$, $(3, \{2, 3, 4\})$, and $(4, \{2, 3, 4\})$. In stage 4, we must go from the current location to New York. This observation yields

TABLE 15
Distances for a Traveling Salesperson

	City			
	New York	Miami	Dallas	Chicago
1 New York	—	1,334	1,559	809
2 Miami	1,334	—	1,343	1,397
3 Dallas	1,559	1,343	—	921
4 Chicago	809	1,397	921	—

$$f_4(2, \{2, 3, 4\}) = c_{21} = 1,334^* \qquad \text{(Go from city 2 to city 1)}$$
$$f_4(3, \{2, 3, 4\}) = c_{31} = 1,559^* \qquad \text{(Go from city 3 to city 1)}$$
$$f_4(4, \{2, 3, 4\}) = c_{41} = 809^* \qquad \text{(Go from city 4 to city 1)}$$

Stage 3 Computations

Working backward to stage 3, we write

$$f_3(i, S) = \min_{\substack{j \notin S \\ \text{and } j \neq 1}} \{c_{ij} + f_4[j, S \cup \{j\}]\} \tag{18}$$

This result follows, because if Walter is now at city i and he travels to city j, he travels a distance c_{ij}. Then he is at stage 4, has last visited city j, and has visited the cities in $S \cup \{j\}$. Hence, the length of the rest of his tour must be $f_4(j, S \cup \{j\})$. To use (18), note that at stage 3, Walter must have visited $\{2, 3\}$, $\{2, 4\}$, or $\{3, 4\}$ and must next visit the nonmember of S that is not equal to 1. We can use (18) to determine $f_3(\cdot)$ for all possible states:

$$f_3(2, \{2, 3\}) = c_{24} + f_4(4, \{2, 3, 4\}) = 1,397 + 809 = 2,206^* \qquad \text{(Go from 2 to 4)}$$
$$f_3(3, \{2, 3\}) = c_{34} + f_4(4, \{2, 3, 4\}) = 921 + 809 = 1,730^* \qquad \text{(Go from 3 to 4)}$$
$$f_3(2, \{2, 4\}) = c_{23} + f_4(3, \{2, 3, 4\}) = 1,343 + 1,559 = 2,902^* \qquad \text{(Go from 2 to 3)}$$
$$f_3(4, \{2, 4\}) = c_{43} + f_4(3, \{2, 3, 4\}) = 921 + 1,559 = 2,480^* \qquad \text{(Go from 4 to 3)}$$
$$f_3(3, \{3, 4\}) = c_{32} + f_4(2, \{2, 3, 4\}) = 1,343 + 1,334 = 2,677^* \qquad \text{(Go from 3 to 2)}$$
$$f_3(4, \{3, 4\}) = c_{42} + f_4(2, \{2, 3, 4\}) = 1,397 + 1,334 = 2,731^* \qquad \text{(Go from 4 to 2)}$$

In general, we write, for $t = 1, 2, 3$,

$$f_t(i, S) = \min_{\substack{j \notin S \\ \text{and } j \neq 1}} \{c_{ij} + f_{t+1}[j, S \cup \{j\}]\} \tag{19}$$

This result follows, because if Walter is at present in city i and he next visits city j, then he travels a distance c_{ij}. The remainder of his tour will originate from city j, and he will have visited the cities in $S \cup \{j\}$. Hence, the length of the remainder of his tour must be $f_{t+1}(j, S \cup \{j\})$. Equation (19) now follows.

Stage 2 Computations

At stage 2, Walter has visited only one city, so the only possible states are $(2, \{2\})$, $(3, \{3\})$, and $(4, \{4\})$. Applying (19), we obtain

$$f_2(2, \{2\}) = \min \begin{cases} c_{23} + f_3(3, \{2, 3\}) = 1,343 + 1,730 = 3,073^* \\ \text{(Go from 2 to 3)} \\ c_{24} + f_3(4, \{2, 4\}) = 1,397 + 2,480 = 3,877 \\ \text{(Go from 2 to 4)} \end{cases}$$

$$f_2(3, \{3\}) = \min \begin{cases} c_{34} + f_3(4, \{3, 4\}) = 921 + 2,731 = 3,652 \\ \text{(Go from 3 to 4)} \\ c_{32} + f_3(2, \{2, 3\}) = 1,343 + 2,206 = 3,549^* \\ \text{(Go from 3 to 2)} \end{cases}$$

$$f_2(4, \{4\}) = \min \begin{cases} c_{42} + f_3(2, \{2, 4\}) = 1,397 + 2,902 = 4,299 \\ \text{(Go from 4 to 2)} \\ c_{43} + f_3(3, \{3, 4\}) = 921 + 2,677 = 3,598^* \\ \text{(Go from 4 to 3)} \end{cases}$$

Stage 1 Computations

Finally, we are back to stage 1 (where no cities have been visited). Since Walter is currently in New York and has visited no cities, the stage 1 state must be $f_1(1, \{\cdot\})$. Applying (19),

$$f_1(1, \{\cdot\}) = \min \begin{cases} c_{12} + f_2(2, \{2\}) = 1{,}334 + 3{,}073 = 4{,}407^* \\ \text{(Go from 1 to 2)} \\ c_{13} + f_2(3, \{3\}) = 1{,}559 + 3{,}549 = 5{,}108 \\ \text{(Go from 1 to 3)} \\ c_{14} + f_2(4, \{4\}) = 809 + 3{,}598 = 4{,}407^* \\ \text{(Go from 1 to 4)} \end{cases}$$

So from city 1 (New York), Walter may go to city 2 (Miami) or city 4 (Chicago). We arbitrarily have him choose to go to city 4. Then he must choose to visit the city that attains $f_2(4, \{4\})$, which requires that he next visit city 3 (Dallas). Then he must visit the city attaining $f_3(3, \{3, 4\})$, which requires that he next visit city 2 (Miami). Then Walter must visit the city attaining $f_4(2, \{2, 3, 4\})$, which means, of course, that he must next visit city 1 (New York). The optimal tour (1–4–3–2–1, or New York–Chicago–Dallas–Miami–New York) is now complete. The length of this tour is $f_1(1, \{\cdot\}) = 4{,}407$. As a check, note that

$$\text{New York to Chicago distance} = 809 \text{ miles}$$
$$\text{Chicago to Dallas distance} = 921 \text{ miles}$$
$$\text{Dallas to Miami distance} = 1{,}343 \text{ miles}$$
$$\text{Miami to New York distance} = 1{,}334 \text{ miles}$$

so the total distance that Walter travels is $809 + 921 + 1{,}343 + 1{,}334 = 4{,}407$ miles. Of course, if we had first sent him to city 2, we would have obtained another optimal tour (1–2–3–4–1) that would simply be a reversal of the original optimal tour.

Computational Difficulties in Using Dynamic Programming

For traveling salesperson problems that are large, the state space becomes very large, and the branch-and-bound approach outlined in Volume 1, Chapter 9 (along with other branch-and-bound approaches) is much more efficient than the dynamic programming approach outlined here. For example, for a 30-city problem, suppose we are at stage 16 (this means that 15 cities have been visited). Then it can be shown that there are more than 1 billion possible states. This brings up a problem that limits the practical application of dynamic programming. In many problems, *the state space becomes so large that excessive computational time is required to solve the problem by dynamic programming.* For instance, in Example 8, suppose that $T = 20$. It is possible that if no bass were caught during the first 20 years, then the lake might contain $10{,}000(1.2)^{20} = 383{,}376$ bass at the beginning of year 21. If we view this example as a network in which we need to find the longest route from the node (1, 10,000) (representing year 1 and 10,000 bass in the lake) to some stage 21 node, then stage 21 would have 383,377 nodes. Even a powerful computer would have difficulty solving this problem. Techniques to make problems with large state spaces computationally tractable are discussed in Bersetkas (1987) and Denardo (1982).

Nonadditive Recursions

The last two examples in this section differ from the previous ones in that the recursion does not represent $f_t(i)$ as the sum of the cost (or reward) incurred during the current period and future costs (or rewards) incurred during future periods.

Joe Cougar needs to drive from city 1 to city 10. He is no longer interested in minimizing the length of his trip, but he is interested in minimizing the maximum altitude above sea level that he will encounter during his drive. To get from city 1 to city 10, he must follow a path in Figure 10. The length c_{ij} of the arc connecting city i and city j represents the maximum altitude (in thousands of feet above sea level) encountered when driving from city i to city j. Use dynamic programming to determine how Joe should proceed from city 1 to city 10.

Solution To solve this problem by dynamic programming, note that for a trip that begins in city i and goes through stages $t, t + 1, \ldots, 5$, the maximum altitude that Joe encounters will be the maximum of the following two quantities: (1) the maximum altitude encountered on stages $t + 1, t + 2, \ldots, 5$ or (2) the altitude encountered when traversing the arc that begins in stage t. Of course, if we are in a stage 4 state, quantity 1 does not exist.

After defining $f_t(i)$ as the smallest maximum altitude that Joe can encounter in a trip from city i in stage t to city 10, this reasoning leads us to the following recursion:

$$f_4(i) = c_{i,10} \tag{20}$$
$$f_t(i) = \min_j \{\max[c_{ij}, f_{t+1}(j)]\} \qquad (t = 1, 2, 3)$$

where j may be any city such that there is an arc connecting city i and city j.

We first compute $f_4(7)$, $f_4(8)$, and $f_4(9)$ and then use (20) to work backward until $f_1(1)$ has been computed. We obtain the following results:

$$f_4(7) = 13^* \qquad\qquad \text{(Go from 7 to 10)}$$
$$f_4(8) = 8^* \qquad\qquad \text{(Go from 8 to 10)}$$
$$f_4(9) = 9^* \qquad\qquad \text{(Go from 9 to 10)}$$

$$f_3(5) = \min \begin{cases} \max\,[c_{57}, f_4(7)] = 13 & \text{(Go from 5 to 7)} \\ \max\,[c_{58}, f_4(8)] = 8^* & \text{(Go from 5 to 8)} \\ \max\,[c_{59}, f_4(9)] = 10 & \text{(Go from 5 to 9)} \end{cases}$$

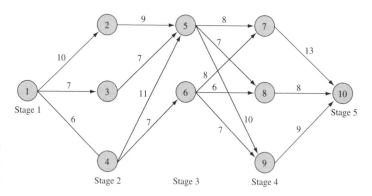

FIGURE 10
Joe's Trip
(Altitudes Given)

$$f_3(6) = \min \begin{cases} \max\ [c_{67}, f_4(7)] = 13 & \text{(Go from 6 to 7)} \\ \max\ [c_{68}, f_4(8)] = 8^* & \text{(Go from 6 to 8)} \\ \max\ [c_{69}, f_4(9)] = 9 & \text{(Go from 6 to 9)} \end{cases}$$

$$f_2(2) = \max\ [c_{25}, f_3(5)] = 9^* \qquad \text{(Go from 2 to 5)}$$

$$f_2(3) = \max\ [c_{35}, f_3(5)] = 8^* \qquad \text{(Go from 3 to 5)}$$

$$f_2(4) = \min \begin{cases} \max\ [c_{45}, f_3(5)] = 11 & \text{(Go from 4 to 5)} \\ \max\ [c_{46}, f_3(6)] = 8^* & \text{(Go from 4 to 6)} \end{cases}$$

$$f_1(1) = \min \begin{cases} \max\ [c_{12}, f_2(2)] = 10 & \text{(Go from 1 to 2)} \\ \max\ [c_{13}, f_2(3)] = 8^* & \text{(Go from 1 to 3)} \\ \max\ [c_{14}, f_2(4)] = 8^* & \text{(Go from 1 to 4)} \end{cases}$$

To determine the optimal strategy, note that Joe can begin by going from city 1 to city 3 or from city 1 to city 4. Suppose Joe begins by traveling to city 3. Then he should choose the arc attaining $f_2(3)$, which means he should next travel to city 5. Then Joe must choose the arc that attains $f_3(5)$, driving next to city 8. Then, of course, he must drive to city 10. Thus, the path 1–3–5–8–10 is optimal, and Joe will encounter a maximum altitude equal to $f_1(1) = 8{,}000$ ft. The reader should verify that the path 1–4–6–8–10 is also optimal.

EXAMPLE 14   **Sales Allocation**

Glueco is planning to introduce a new product in three different regions. Current estimates are that the product will sell well in each region with respective probabilities .6, .5, and .3. The firm has available two top sales representatives that it can send to any of the three regions. The estimated probabilities that the product will sell well in each region when 0, 1, or 2 additional sales reps are sent to a region are given in Table 16. If Glueco wants to maximize the probability that its new product will sell well in all three regions, then where should it assign sales representatives? You may assume that sales in the three regions are independent.

Solution   If Glueco had just one region to worry about and wanted to maximize the probability that the new product would sell in that region, then the proper strategy would be clear: Assign both sales reps to the region. We could then work backward and solve a problem in which Glueco's goal is to maximize the probability that the product will sell in two regions. Finally, we could work backward and solve a problem with three regions. We define $f_t(s)$ as the probability that the new product will sell in regions $t, t + 1, \ldots, 3$ if s sales reps are optimally assigned to these regions. Then

$$f_3(2) = .7 \qquad \text{(Assign 2 sales reps to region 3)}$$
$$f_3(1) = .55 \qquad \text{(Assign 1 sales rep to region 3)}$$
$$f_3(0) = .3 \qquad \text{(Assign 0 sales reps to region 3)}$$

TABLE 16
Relation between Regional Sales and Sales Representatives

No. of Additional Sales Representatives	Probability of Selling Well		
	Region 1	Region 2	Region 3
0	.6	.5	.3
1	.8	.7	.55
2	.85	.85	.7

Also, $f_1(2)$ will be the maximum probability that the product will sell well in all three regions. To develop a recursion for $f_2(\cdot)$ and $f_1(\cdot)$, we define p_{tx} to be the probability that the new product sells well in region t if x sales reps are assigned to region t. For example, $p_{21} = .7$. For $t = 1$ and $t = 2$, we then write

$$f_t(s) = \max_x \{p_{tx} f_{t+1}(s - x)\} \tag{21}$$

where x must be a member of $\{0, 1, \ldots, s\}$. To justify (21), observe that if s sales reps are available for regions $t, t + 1, \ldots, 3$ and x sales reps are assigned to region t, then

$$p_{tx} = \text{probability that product sells in region } t$$
$$f_{t+1}(s - x) = \text{probability that product sells well in regions } t + 1, \ldots, 3$$

Note that the sales in each region are independent. This implies that if x sales reps are assigned to region t, then the probability that the new product sells well in regions t, $t + 1, \ldots, 3$ is $p_{tx} f_{t+1}(s - x)$. We want to maximize this probability, so we obtain (21). Applying (21) yields the following results:

$$f_2(2) = \max \begin{cases} (.5)f_3(2 - 0) = .35 \\ (\text{Assign 0 sales reps to region 2}) \\ (.7)f_3(2 - 1) = .385^* \\ (\text{Assign 1 sales rep to region 2}) \\ (.85)f_3(2 - 2) = .255 \\ (\text{Assign 2 sales reps to region 2}) \end{cases}$$

Thus, $f_2(2) = .385$, and 1 sales rep should be assigned to region 2.

$$f_2(1) = \max \begin{cases} (.5)f_3(1 - 0) = .275^* \\ (\text{Assign 0 sales reps to region 2}) \\ (.7)f_3(1 - 1) = .21 \\ (\text{Assign 1 sales rep to region 2}) \end{cases}$$

Thus, $f_2(1) = .275$, and no sales reps should be assigned to region 2.

$$f_2(0) = (.5)f_3(0 - 0) = .15^*$$
$$(\text{Assign 0 sales reps to region 2})$$

Finally, we are back to the original problem, which is to find $f_1(2)$. Equation (21) yields

$$f_1(2) = \max \begin{cases} (.6)f_2(2 - 0) = .231^* \\ (\text{Assign 0 sales reps to region 1}) \\ (.8)f_2(2 - 1) = .220 \\ (\text{Assign 1 sales rep to region 1}) \\ (.85)f_2(2 - 2) = .1275 \\ (\text{Assign 2 sales reps to region 1}) \end{cases}$$

Thus, $f_1(2) = .231$, and no sales reps should be assigned to region 1. Then Glueco needs to attain $f_2(2 - 0)$, which requires that 1 sales rep be assigned to region 2. Glueco must next attain $f_3(2 - 1)$, which requires that 1 sales rep be assigned to region 3. In summary, Glueco can obtain a .231 probability of the new product selling well in all three regions by assigning 1 sales rep to region 2 and 1 sales rep to region 3.

PROBLEMS

Group A

1 At the beginning of year 1, Sunco Oil owns i_0 barrels of oil reserves. During year $t(t = 1, 2, \ldots, 10)$, the following events occur in the order listed: (1) Sunco extracts and refines x barrels of oil reserves and incurs a cost $c(x)$: (2) Sunco sells year t's extracted and refined oil at a price of p_t dollars per barrel; and (3) exploration for new reserves results in a discovery of b_t barrels of new reserves.

Sunco wants to maximize sales revenues less costs over the next 10 years. Formulate a dynamic programming recursion that will help Sunco accomplish its goal. If Sunco felt that cash flows in later years should be discounted, how should the formulation be modified?

2 At the beginning of year 1, Julie Ripe has D dollars (this includes year 1 income). During each year, Julie earns i dollars and must determine how much money she should consume and how much she should invest in Treasury bills. During a year in which Julie consumes d dollars, she earns a utility of $\ln d$. Each dollar invested in Treasury bills yields $1.10 in cash at the beginning of the next year. Julie's goal is to maximize the total utility she earns during the next 10 years.

   a Why might $\ln d$ be a better indicator of Julie's utility than a function such as d^2?

   b Formulate a dynamic programming recursion that will enable Julie to maximize the total utility she receives during the next 10 years. Assume that year t revenue is received at the beginning of year t.

3 Assume that during minute t (the current minute is minute 1), the following sequence of events occurs: (1) At the beginning of the minute, x_t customers arrive at the cash register; (2) the store manager decides how many cash registers should be operated during the current minute; (3) if s cash registers are operated and i customers are present (including the current minute's arrivals), $c(s, i)$ customers complete service; and (4) the next minute begins.

A cost of 10¢ is assessed for each minute a customer spends waiting to check out (this time includes checkout time). Assume that it costs $c(s)$ cents to operate s cash registers for 1 minute. Formulate a dynamic programming recursion that minimizes the sum of holding and service costs during the next 60 minutes. Assume that before the first minute's arrivals, no customers are present and that holding cost is assessed at the end of each minute.

4 Develop a dynamic programming formulation of the CSL Computer problem of Volume 1, Section 3.12.

5 To graduate from State University, Angie Warner needs to pass at least one of the three subjects she is taking this semester. She is now enrolled in French, German, and statistics. Angie's busy schedule of extracurricular activities allows her to spend only 4 hours per week on studying. Angie's probability of passing each course depends on the number of hours she spends studying for the course (see Table 17). Use dynamic programming to determine how many hours per week Angie should spend studying each subject. (*Hint:* Explain why maximizing the probability of

TABLE 17

Hours of Study per Week	Probability of Passing Course		
	French	German	Statistics
0	.20	.25	.10
1	.30	.30	.30
2	.35	.33	.40
3	.38	.35	.44
4	.40	.38	.50

TABLE 18

Component	No. of Actors Assigned to Component			
	0	1	2	3
Warp drive	.30	.55	.65	.95
Solar relay	.40	.50	.70	.90
Candy maker	.45	.55	.80	.98

passing at least one course is equivalent to minimizing the probability of failing all three courses.)

6 E.T. is about to fly home. For the trip to be successful, the ship's solar relay, warp drive, and candy maker must all function properly. E.T. has found three unemployed actors who are willing to help get the ship ready for takeoff. Table 18 gives, as a function of the number of actors assigned to repair each component, the probability that each component will function properly during the trip home. Use dynamic programming to help E.T. maximize the probability of having a successful trip home.

7 Farmer Jones is trying to raise a prize steer for the Bloomington 4-H show. The steer now weighs w_0 pounds. Each week, farmer Jones must determine how much food to feed the steer. If the steer weighs w pounds at the beginning of a week and is fed p pounds of food during a week, then at the beginning of the next week, the steer will weigh $g(w, p)$ pounds. It costs farmer Jones $c(p)$ dollars to feed the steer p pounds of food during a week. At the end of the 10th week (or equivalently, the beginning of the 11th week), the steer may be sold for $10/lb. Formulate a dynamic programming recursion that can be used to determine how farmer Jones can maximize profit from the steer.

Group B

8 MacBurger has just opened a fast-food restaurant in Bloomington. Currently, i_0 customers frequent MacBurger (we call these loyal customers), and $N - i_0$ customers frequent other fast-food establishments (we call these nonloyal customers). At the beginning of each month, MacBurger must decide how much money to spend on advertising. At the end of a month in which MacBurger spends d dollars on advertising, a fraction $p(d)$ of the loyal customers become nonloyal customers, and a fraction $q(d)$ of the nonloyal customers become loyal customers. During the next 12 months, MacBurger wants to spend D dollars on advertising. Develop a dynamic programming recursion that will enable MacBurger to maximize the number of loyal customers the company will have at the end of month 12. (Ignore the possibility of a fractional number of loyal customers.)

9 Public Service Indiana (PSI) is considering five possible locations to build power plants during the next 20 years. It will cost c_i dollars to build a plant at site i and h_i dollars to operate a site i plant for a year. A plant at site i can supply k_i kilowatt-hours (kwh) of generating capacity. During year t, d_t kwh of generating capacity are required. Suppose that at most one plant can be built during a year, and if it is decided to build a plant at site i during year t, then the site i plant can be used to meet the year t (and later) generating requirements. Initially, PSI has 500,000 kwh of generating capacity available. Formulate a recursion that PSI could use to minimize the sum of building and operating costs during the next 20 years.

10 During month t, a firm faces a demand for d_t units of a product. The firm's production cost during month t consists of two components. First, for each unit produced during month t, the firm incurs a variable production cost of c_t. Second, if the firm's production level during month $t - 1$ is x_{t-1} and the firm's production level during month t is x_t, then during month t, a smoothing cost of $5|x_t - x_{t-1}|$ will be incurred (see Section 4.12 for an explanation of smoothing costs). At the end of each month, a holding cost of h_t per unit is incurred. Formulate a recursion that will enable the firm to meet (on time) its demands over the next 12 months. Assume that at the beginning of the first month, 20 units are in inventory and that last month's production was 20 units. (*Hint:* The state during each month must consist of two quantities.)

11 The state of Transylvania consists of three cities with the following populations: city 1, 1.2 million people; city 2, 1.4 million people; city 3, 400,000 people. The Transylvania House of Representatives consists of three representatives. Given proportional representation, city 1 should have $d_1 = (\frac{1.2}{3}) = 1.2$ representatives; city 2 should have $d_2 = 1.4$ representatives; and city 3 should have $d_3 = 0.40$ representative. Each city must receive an integral number of representatives, so this is impossible. Transylvania has therefore decided to allocate x_i representatives to city i, where the allocation x_1, x_2, x_3 minimizes the maximum discrepancy between the desired and actual number of representatives received by a city. In short, Transylvania must determine x_1, x_2, and x_3 to minimize the largest of the following three numbers: $|x_1 - d_1|$, $|x_2 - d_2|$, $|x_3 - d_3|$. Use dynamic programming to solve Transylvania's problem.

12 A job shop has four jobs that must be processed on a single machine. The due date and processing time for each job are given in Table 19. Use dynamic programming to determine the order in which the jobs should be done so as to minimize the total lateness of the jobs. (The lateness of a job is simply how long after the job's due date the job is completed; for example, if the jobs are processed in the given order, then job 3 will be 2 days late, job 4 will be 4 days late, and jobs 1 and 2 will not be late.)

TABLE 19

Job	Processing Time (Days)	Due Date (Days from Now)
1	2	4
2	4	14
3	6	10
4	8	16

6.7 The Wagner–Whitin Algorithm and the Silver–Meal Heuristic[†]

The inventory example of Section 6.3 is a special case of the *dynamic lot-size model.*

Description of Dynamic Lot-Size Model

1 Demand d_t during period $t(t = 1, 2, \ldots, T)$ is known at the beginning of period 1.

2 Demand for period t must be met on time from inventory or from period t production. The cost $c(x)$ of producing x units during any period is given by $c(0) = 0$, and for $x > 0$, $c(x) = K + cx$, where K is a fixed cost for setting up production during a period, and c is the variable per-unit cost of production.

3 At the end of period t, the inventory level i_t is observed, and a holding cost of hi_t is incurred. We let i_0 denote the inventory level before period 1 production occurs.

4 The goal is to determine a production level x_i for each period t that minimizes the total cost of meeting (on time) the demands for periods $1, 2, \ldots, T$.

[†]This section covers topics that may be omitted with no loss of continuity.

5 There is a limit c_t placed on period t's ending inventory.

6 There is a limit r_t placed on period t's production.

In this section, we consider these first four points. We let x_t = period t production. Period t production can be used to meet period t demand.

EXAMPLE 15   **Dynamic Lot-Size Model**

We now determine an optimal production schedule for a five-period dynamic lot-size model with $K = \$250$, $c = \$2$, $h = \$1$, $d_1 = 220$, $d_2 = 280$, $d_3 = 360$, $d_4 = 140$, and $d_5 = 270$. We assume that the initial inventory level is zero. The solution to this example is given later in this section.

Discussion of the Wagner–Whitin Algorithm

If the dynamic programming approach outlined in Section 6.3 were used to find an optimal production policy for Example 15, we would have to consider the possibility of producing any amount between 0 and $d_1 + d_2 + d_3 + d_4 + d_5 = 1{,}270$ units during period 1. Thus, it would be possible for the period 2 state (period 2's entering inventory) to be $0, 1, \ldots, 1{,}270 - d_1 = 1{,}050$, and we would have to determine $f_2(0), f_2(1), \ldots, f_2(1{,}050)$. Using the dynamic programming approach of Section 6.3 to find an optimal production schedule for Example 15 would therefore require a great deal of computational effort. Fortunately, however, Wagner and Whitin (1958) have developed a method that greatly simplifies the computation of optimal production schedules for dynamic lot-size models. Lemmas 1 and 2 are necessary for the development of the Wagner–Whitin algorithm.

LEMMA 1

Suppose it is optimal to produce a positive quantity during a period t. Then for some $j = 0, 1, \ldots, T - t$, the amount produced during period t must be such that after period t's production, a quantity $d_t + d_{t+1} + \cdots + d_{t+j}$ will be in stock. In other words, if production occurs during period t, we must (for some j) produce an amount that exactly suffices to meet the demands for periods $t, t + 1, \ldots, t + j$.

Proof   If the lemma is false, then for some t, some $j = 0, 1, \ldots, T - t - 1$, and some x satisfying $0 < x < d_{t+j+1}$, period t production must bring the stock level to $d_t + d_{t+1} + \cdots + d_{t+j} + x$, and at the beginning of period $t + j + 1$, our inventory level would be $x < d_{t+j+1}$. Thus, production must occur during period $t + j + 1$. By deferring production of x units from period t to period $t + j + 1$ (with all other production levels unchanged), we save $h(j + 1)x$ in holding costs while incurring no additional setup costs (because production is already occurring during period $t + j + 1$). Thus, it cannot have been optimal to bring our period t stock level to $d_t + d_{t+1} + \cdots + d_{t+j} + x$. This contradiction proves the lemma.

LEMMA 2

If it is optimal to produce anything during period t, then $i_{t-1} < d_t$. In other words, production cannot occur during period t unless there is insufficient stock to meet period t demand.

Proof If the lemma is false, there must be an optimal policy that (for some t) has $x_t > 0$ and $i_{t-1} \geq d_t$. If this is the case, then by deferring the period t production of x_t units to period $t + 1$, we save hx_t in holding costs and possibly K (if the optimal policy produces during period $t + 1$) in setup costs. Thus, any production schedule having $x_t > 0$ and $i_{t-1} \geq d_t$ cannot be optimal.

Lemma 2 shows that no production will occur until the first period t for which $i_{t-1} < d_t$, so production must occur during period t (or else period t's demand would not be met on time). Lemma 1 now implies that for some $j = 0, 1, \ldots, T - t$, period t production will be such that after period t's production, on-hand stock will equal $d_t + d_{t+1} + \cdots + d_{t+j}$. Then Lemma 2 implies that no production can occur until period $t + j + 1$. Since the entering inventory level for period $t + j + 1$ will equal zero, production must occur during period $t + j + 1$. During period $t + j + 1$, Lemma 1 implies that period $t + j + 1$ production will (for some k) equal $d_{t+j+1} + d_{t+j+2} + \cdots + d_{t+j+k}$ units. Then period $t + j + k + 1$ will begin with zero inventory, and production again occurs, and so on. *With the possible exception of the first period, production will occur only during periods in which beginning inventory is zero, and during each period in which beginning inventory is zero (and $d_t \neq 0$), production must occur.*

Using this insight, Wagner and Whitin developed a recursion that can be used to determine an optimal production policy. We assume that the initial inventory level is zero. (See Problem 1 at the end of this section if this is not the case.) Define f_t as the minimum cost incurred during periods $t, t + 1, \ldots, T$, given that at the beginning of period t, the inventory level is zero. Then $f_1, f_2, \ldots, f_T$ must satisfy

$$f_t = \min_{j=0, 1, 2, \ldots, T-t} (c_{tj} + f_{t+j+1}) \tag{22}$$

where $f_{T+1} = 0$ and c_{tj} is the total cost incurred during periods $t, t + 1, \ldots, t + j$ if production during period t is exactly sufficient to meet demands for periods $t, t + 1, \ldots, t + j$. Thus,

$$c_{tj} = K + c(d_t + d_{t+1} + \cdots + d_{t+j}) + h[jd_{t+j} + (j - 1)d_{t+j-1} + \cdots + d_{t+1}]$$

where K is the setup cost incurred during period t, $c(d_t + d_{t+1} + \cdots + d_{t+j})$ is the variable production cost incurred during period t, and $h[jd_{t+j} + (j - 1)d_{t+j-1} + \cdots + d_{t+1}]$ is the holding cost incurred during periods $t, t + 1, \ldots, t + j$. For example, an amount d_{t+j} of period t production will be held in inventory for j periods (during periods $t, t + 1, \ldots, t + j - 1$), thereby incurring a holding cost of hjd_{t+j}.

To find an optimal production schedule by the Wagner–Whitin algorithm, begin by using (22) to find f_T. Then use (22) to compute $f_{T-1}, f_{T-2}, \ldots, f_1$. Once f_1 has been determined, an optimal production schedule may be easily obtained.

EXAMPLE 15   Dynamic Lot-Size Model (continued)

Solution To illustrate the Wagner–Whitin algorithm, we find an optimal production schedule for Example 15. The computations follow.

$$f_6 = 0$$
$$f_5 = 250 + 2(270) + f_6 = 790^* \qquad \text{(Produce for period 5)}$$

If we begin period 5 with zero inventory, we should produce enough during period 5 to meet period 5 demand.

$$f_4 = \min \begin{cases} 250 + 2(140) + f_5 = 1{,}320^* \\ \text{(Produce for period 4)} \\ 250 + 2(140 + 270) + 270 + f_6 = 1{,}340 \\ \text{(Produce for periods 4, 5)} \end{cases}$$

If we begin period 4 with zero inventory, we should produce enough during period 4 to meet the demand for period 4.

$$f_3 = \min \begin{cases} 250 + 2(360) + f_4 = 2{,}290 \\ \text{(Produce for period 3)} \\ 250 + 2(360 + 140) + 140 + f_5 = 2{,}180^* \\ \text{(Produce for periods 3, 4)} \\ 250 + 2(360 + 140 + 270) + 140 + 2(270) + f_6 = 2{,}470 \\ \text{(Produce for periods 3, 4, 5)} \end{cases}$$

If we begin period 3 with zero inventory, we should produce enough during period 3 to meet the demand for periods 3 and 4.

$$f_2 = \min \begin{cases} 250 + 2(280) + f_3 = 2{,}990^* \\ \text{(Produce for period 2)} \\ 250 + 2(280 + 360) + 360 + f_4 = 3{,}210 \\ \text{(Produce for periods 2, 3)} \\ 250 + 2(280 + 360 + 140) + 360 + 2(140) + f_5 = 3{,}240 \\ \text{(Produce for periods 2, 3, 4)} \\ 250 + 2(280 + 360 + 140 + 270) + 360 + 2(140) + 3(270) + f_6 = 3{,}800 \\ \text{(Produce for periods 2, 3, 4, 5)} \end{cases}$$

If we begin period 2 with zero inventory, we should produce enough during period 2 to meet the demand for period 2.

$$f_1 = \min \begin{cases} 250 + 2(220) + f_2 = 3{,}680^* \\ \text{(Produce for period 1)} \\ 250 + 2(220 + 280) + 280 + f_3 = 3{,}710 \\ \text{(Produce for periods 1, 2)} \\ 250 + 2(220 + 280 + 360) + 280 + 2(360) + f_4 = 4{,}290 \\ \text{(Produce for periods 1, 2, 3)} \\ 250 + 2(220 + 280 + 360 + 140) + 280 + 2(360) + 3(140) + f_5 = 4{,}460 \\ \text{(Produce for periods 1, 2, 3, 4)} \\ 250 + 2(220 + 280 + 360 + 140 + 270) + 280 \\ \quad + 2(360) + 3(140) + 4(270) + f_6 = 5{,}290 \\ \text{(Produce for periods 1, 2, 3, 4, 5)} \end{cases}$$

If we begin period 1 with zero inventory, it is optimal to produce $d_1 = 220$ units during period 1; then we begin period 2 with zero inventory. Since f_2 is attained by producing period 2's demand, we should produce $d_2 = 280$ units during period 2; then we enter period 3 with zero inventory. Since f_3 is attained by meeting the demands for periods 3 and 4, we produce $d_3 + d_4 = 500$ units during period 3; then we enter period 5 with zero inventory and produce $d_5 = 270$ units during period 5. The optimal production schedule will incur at total cost of $f_1 = \$3{,}680$.

For Example 15, any optimal production schedule must produce exactly $d_1 + d_2 + d_3 + d_4 + d_5 = 1,270$ units, incurring variable production costs of $2(1,270) = \$2,540$. Thus, in computing the optimal production schedule, we may always ignore the variable production costs. This substantially simplifies the calculations.

The Silver–Meal Heuristic

The Silver–Meal (S–M) heuristic involves less work than the Wagner–Whitin algorithm and can be used to find a near-optimal production schedule. The S–M heuristic is based on the fact that our goal is to minimize average cost per period (for the reasons stated, variable production costs may be ignored). Suppose we are at the beginning of period 1 and are trying to determine how many periods of demand should be satisfied by period 1's production. During period 1, if we produce an amount sufficient to meet demand for the next t periods, then a cost of $TC(t) = K + HC(t)$ will be incurred (ignoring variable production costs). Here, $HC(t)$ is the holding cost incurred during the next t periods (including the current period) if production during the current period is sufficient to meet demand for the next t periods.

Let $AC(t) = \frac{TC(t)}{t}$ be the average per-period cost incurred during the next t periods. Since $\frac{1}{t}$ is a decreasing convex function of t, as t increases, $\frac{K}{t}$ decreases at a decreasing rate. In most cases, $\frac{HC(t)}{t}$ tends to be an increasing function of t (see Problem 4 at the end of this section). Thus, in most situations, an integer t^* can be found such that for $t < t^*$, $AC(t + 1) \leq AC(t)$ and $AC(t^* + 1) \geq AC(t^*)$. The S–M heuristic recommends that period 1's production be sufficient to meet the demands for periods $1, 2, \ldots, t^*$ (if no t^* exists, period 1 production should satisfy the demand for periods $1, 2, \ldots, T$). Since t^* is a local (and perhaps a global) minimum for $AC(t)$, it seems reasonable that producing $d_1 + d_2 + \cdots + d_{t^*}$ units during period 1 will come close to minimizing the average per-period cost incurred during periods $1, 2, \ldots, t^*$. Next we apply the S–M heuristic while considering period $t^* + 1$ as the initial period. We find that during period $t^* + 1$, the demand for the next t_1^* periods should be produced. Continue in this fashion until the demand for period T has been produced.

To illustrate, we apply the S–M heuristic to Example 15. We have

$$TC(1) = 250 \qquad\qquad AC(1) = \frac{250}{1} = 250$$

$$TC(2) = 250 + 280 = 530 \qquad AC(2) = \frac{530}{2} = 265$$

Since $AC(2) \geq AC(1)$, $t^* = 1$, and the S–M heuristic dictates that we produce $d_1 = 220$ units during period 1. Then

$$TC(1) = 250 \qquad\qquad AC(1) = \frac{250}{1} = 250$$

$$TC(2) = 250 + 360 = 610 \qquad AC(2) = \frac{610}{2} = 305$$

Since $AC(2) \geq AC(1)$, the S–M heuristic recommends producing $d_2 = 280$ units during period 2. Then

$$TC(1) = 250 \qquad\qquad\qquad AC(1) = \frac{250}{1} = 250$$

$$TC(2) = 250 + 140 = 390 \qquad\qquad AC(2) = \frac{390}{2} = 195$$

$$TC(3) = 250 + 2(270) + 140 = 930 \qquad AC(3) = \frac{930}{3} = 310$$

Since $AC(3) \geq AC(2)$, period 3 production should meet the demand for the next two periods (periods 3 and 4). During period 3, we should produce $d_3 + d_4 = 500$ units. This brings us to period 5. Period 5 is the final period, so $d_5 = 270$ units should be produced during period 5.

For Example 15 (and many other dynamic lot-size problems), the S–M heuristic yields an optimal production schedule. In extensive testing, the S–M heuristic usually yielded a production schedule costing less than 1% above the optimal policy obtained by the Wagner–Whitin algorithm (see Peterson and Silver (1998)).

PROBLEMS

Group A

1 For Example 15, suppose we had an inventory of 200 units. What would be the optimal production schedule? What if the initial inventory were 400 units?

2 Use the Wagner–Whitin and Silver–Meal methods to find production schedules for the following dynamic lot-size problem: $K = \$50$, $h = \$0.40$, $d_1 = 10$, $d_2 = 60$, $d_3 = 20$, $d_4 = 140$, $d_5 = 90$.

3 Use the Wagner–Whitin and Silver–Meal methods to find production schedules for the following dynamic lot-

size problem: $K = \$30$, $h = \$1$, $d_1 = 40$, $d_2 = 60$, $d_3 = 10$, $d_4 = 70$, $d_5 = 20$.

Group B

4 Explain why $HC(t)/t$ tends to be an increasing function of t.

6.8 Using Excel to Solve Dynamic Programming Problems[†]

In earlier chapters, we have seen that any LP problem can be solved with LINDO or LINGO, and any NLP can be solved with LINGO. Unfortunately, no similarly user-friendly package can be used to solve dynamic programming problems. LINGO can be used to solve DP problems, but student LINGO can only handle a very small problem. Fortunately, Excel can often be used to solve DP problems. Our three illustrations solve a knapsack problem (Example 6), a resource-allocation problem (Example 5), and an inventory problem (Example 4).

Solving Knapsack Problems on a Spreadsheet

Recall the knapsack problem of Example 6. The question is how to (using three types of items) fill a 10-lb knapsack and obtain the maximum possible benefit. Recall that $g(w) =$ maximum benefit that can be obtained from a w-lb knapsack. Recall that

$$g(w) = \max_{j} \{b_j + g(w - w_j)\} \tag{8}$$

where b_j = benefit from a type j item and w_j = weight of a type j item.

Dpknap.xls

In each row of the spreadsheet (see Figure 11 or file Dpknap.xls) we compute $g(w)$ for various values of w. We begin by entering $g(0) = g(1) = g(2) = 0$ and $g(3) = 7$; [$g(3) = 7$ follows because a 3-lb item is the only item that will fit in a 3-lb knapsack]. The

[†]This section covers topics that may be omitted with no loss of continuity.

A	A	B	C	D	E	F	G
1	KNAPSACK	ITEM1	ITEM2	ITEM3	g(SIZE)		FIGURE11
2	SIZE						KNAPSACK
3	0				0		PROBLEM
4	1				0		
5	2				0		
6	3				7		
7	4	11	7	-10000	11		
8	5	11	7	12	12		
9	6	11	14	12	14		
10	7	18	18	12	18		
11	8	22	19	19	22		
12	9	23	21	23	23		
13	10	25	25	24	25		
14							
15							
16							
17							
18							
19							
20							
21							
22							
23							
24							
25							
26							
27							
28							
29							
30							

FIGURE 11
Knapsack Problem

columns labeled ITEM1, ITEM2, and ITEM3 correspond to the terms $j = 1, 2, 3$, respectively, in (8). Thus, in the ITEM1 column, we should enter a formula to compute $b_1 + g(w - w_1)$; in the ITEM2 column, we should enter a formula to compute $b_2 + g(w - w_2)$; in the ITEM3 column, we should enter a formula to compute $b_3 + g(w - w_3)$. The only exception to this occurs when a w_j-lb item will not fit in a w-lb knapsack. In this situation, we enter a very negative number (such as 10,000) to ensure that a w_j-lb item will not be considered.

More specifically, in row 7, we want to compute $g(4)$. To do this, we enter the following formulas:

> B7: 11 + E3   [This is $b_1 + g(4 - w_1)$]
>
> C7: 7 + E4   [This is $b_2 + g(4 - w_2)$]
>
> D7: −10,000   (This is because a 5-lb item will not fit in a 4-lb knapsack)

In E7, we compute $g(4)$ by entering the formula =MAX(B7:D7). In row 8, we compute $g(5)$ by entering the following formulas:

> B8: 11 + E4
>
> C8: 7 + E5
>
> D8: 12 + E3

To compute $g(5)$, we enter =MAX(B8:D8) in E8. Now comes the fun part! Simply copy the formulas from the range B8:E8 to B8:E13. Then $g(10)$ will be computed in E13. We see that $g(10) = 25$. Because both item 1 and item 2 attain $g(10)$, we may begin filling a knapsack with a Type 1 or Type 2 item. We choose to begin with a Type 1 item. This leaves us with $10 - 4 = 6$ lb to fill. From row 9 we find that $g(6) = 14$ is attained by a Type 2 item. This leaves us with $6 - 3 = 3$ lb to fill. We also use a Type 2 item to attain $g(3) = 7$. This leaves us with 0 lb. Thus, we conclude that we can obtain 25 units of benefit by filling a 10-lb knapsack with two Type 2 items and one Type 1 item.

By the way, if we had been interested in filling a 100-lb knapsack, we would have copied the formulas from B8:E8 to B8:E103.

Solving a General Resource-Allocation Problem on a Spreadsheet

Solving a nonknapsack resource-allocation problem on a spreadsheet is more difficult. To illustrate, consider Example 5 in which we have \$6,000 to allocate between three investments. Define $f_t(d)$ = maximum NPV obtained from investments $t, \ldots, 3$ given that d (in thousands) dollars are available for investments $t, \ldots, 3$. Then we may write

$$f_t(d) = \max_{0 \le x \le d} \{r_t(x) + f_{t+1}(d - x)\} \tag{10}$$

where $f_4(d) = 0 (d = 0, 1, 2, 3, 4, 5, 6)$, $r_t(x)$ = NPV obtained if x (in thousands) dollars are invested in investment t, and the maximization in (10) is only taken over integral values for d. Our subsequent discussion will be simplified if we define $J_t(d, x) = r_t(x) + f_{t+1}(d - x)$ and rewrite (10) as

$$f_t(d) = \max_{0 \le x \le d} \{J_t(d, x)\} \tag{10'}$$

Dpresour.xls

We begin the construction of the spreadsheet (Figure 12 and file Dpresour.xls) by entering the $r_t(x)$ in A1:H4. For example, $r_2(3) = 16$ is entered in E3. In rows 18–20, we have set up the computations to compute the $J_t(d, x)$. These computations require using the Excel =**HLOOKUP** command to look up the values of $r_t(x)$ (in rows 2–4) and $f_{t+1}(d - x)$ (in rows 11–14). For example, to compute $J_3(3, 1)$, we enter the following formula in I18:

=HLOOKUP(I\$17,\$B\$1:\$H\$4,\$A18+1)

+ HLOOKUP(I\$16-I\$17,\$B\$10:\$H\$14,\$A18+1)

The portion =HLOOKUP(I\$17,\$B\$1:\$H\$4,\$A18+1) of the formula in cell I18 finds the column in B1:H4 whose first entry matches I17. Then we pick off the entry in row A18 + 1 of that column. This returns $r_3(1) = 9$. Note that H stands for horizontal lookup. The portion HLOOKUP(I\$16-I\$17,\$b\$10:\$h\$14,\$A18+1) finds the column in B10:H14 whose first entry matches I16-I17. Then we pick off the entry in row A18 + 1 of that column. This yields $f_4(3 - 1) = 0$.

FIGURE 12
Resource Allocation

A	A	B	C	D	E	F	G	H	I	J	K	L	M
1	REWARD	0	1	2	3	4	5	6					
2	PERIOD3	0	9	13	17	21	25	29					
3	PERIOD2	0	10	13	16	19	22	25					
4	PERIOD1	0	9	16	23	30	37	44					
5													
6													
7	FIGURE 12												
8	RESOURCE	ALLOCATION											
9													
10	VALUE	0	1	2	3	4	5	6					
11	PERIOD4	0	0	0	0	0	0	0					
12	PERIOD3	0	9	13	17	21	25	29					
13	PERIOD2	0	10	19	23	27	31	35					
14	PERIOD1	0	10	19	28	35	42	49					
15													
16	d	0	1	1	2	2	2	3	3	3	3	4	4
17	x	0	0	1	0	1	2	0	1	2	3	0	1
18	1	0	0	9	0	9	13	0	9	13	17	0	9
19	2	0	9	10	13	19	13	17	23	22	16	21	27
20	3	0	10	9	19	19	16	23	28	26	23	27	32

FIGURE **12**
(Continued)

A	N	O	P	Q	R	S	T	U	V	W	X	Y	Z
1													
2													
3													
4													
5													
6													
7													
8													
9													
10													
11													
12													
13													
14													
15													
16	4	4	4	5	5	5	5	5	5	6	6	6	6
17	2	3	4	0	1	2	3	4	5	0	1	2	3
18	13	17	21	0	9	13	17	21	25	0	9	13	17
19	26	25	19	25	31	30	29	28	22	29	35	34	33
20	35	33	30	31	36	39	42	40	37	35	40	43	46

A	AA	AB	AC	AD	AE	AF	AG	AH	AI	AJ	AK
1											
2											
3											
4											
5											
6											
7											
8											
9											
10											
11											
12											
13											
14											
15											
16	6	6	6	0	1	2	3	4	5	6	
17	4	5	6	ft(0)	ft(1)	ft(2)	ft(3)	ft(4)	ft(5)	ft(6)	t
18	21	25	29	0	9	13	17	21	25	29	3
19	32	31	25	0	10	19	23	27	31	35	2
20	49	47	44	0	10	19	28	35	42	49	1

We now copy any of the $J_t(d, x)$ formulas (such as the one in I18) to the range B18:AC20.

The $f_t(d)$ are computed in AD18:AJ20. We begin by manually entering in AD18:AJ18 the formulas used to compute $f_3(0), f_3(1), \ldots, f_3(6)$. These formulas are as follows:

AD18:	0	(Computes $f_3(0)$)
AE18:	=MAX(C18:D18)	(Computes $f_3(1)$)
AF18:	=MAX(E18:G18)	(Computes $f_3(2)$)
AG18:	=MAX(H18:K18)	(Computes $f_3(3)$)
AH18:	=MAX(L18:P18)	(Computes $f_3(4)$)
AI18:	=MAX(Q18:V18)	(Computes $f_3(5)$)
AJ18:	=MAX(W18:AC18)	(Computes $f_3(6)$)

We now copy these formulas from the range AD18:AJ18 to the range AD18:AJ20.

For our spreadsheet to work we must be able to compute the $J_t(d, x)$ by looking up the appropriate value of $f_t(d)$ in rows 11–14. Thus, in B11:H11, we enter a zero in each cell [because $f_4(d) = 0$ for all d]. In B12, we enter =AD18 [this is the cell in which $f_3(0)$ is computed]. We now copy this formula to the range B12:H14.

Note that rows 11–14 of our spreadsheet are defined in terms of rows 18–20, and rows 18–20 are defined in terms of rows 11–14. This creates **circularity** or **circular references** in our spreadsheet. To resolve the circular references in this (or any) spreadsheet, simply select Tools, Options, Calculations and select the Iteration box. This will cause Excel to resolve all circular references until the circularity is resolved.

To determine how $6,000 should be allocated to the three investments, note that $f_1(6) = 49$. Because $f_1(6) = J_1(6, 4)$, we allocate $4,000 to investment 1. Then we must find $f_2(6 - 4) = 19 = J_2(2, 1)$. We allocate $1,000 to investment 2. Finally, we find that $f_3(2 - 1) = J_3(1, 1)$ and allocate $1,000 to investment 3.

Solving an Inventory Problem on a Spreadsheet

We now show how to determine an optimal production policy for Example 4. An important aspect of this production problem is that each month's ending inventory must be between 0 and 4 units. We can ensure that this occurs by manually determining the allowable actions in each state. We will design our spreadsheet to ensure that the ending inventory for each month must be between 0 and 4 inclusive.

Dpinv.xls

Our first step in setting up the spreadsheet (Figure 13, file Dpinv.xls) is to enter the production cost for each possible production level (0, 1, 2, 3, 4, 5) in B1:G2. Then we define $f_t(i)$ to be the minimum cost incurred in meeting demands for months $t, t + 1, \ldots,$ 4 when i units are on hand at the beginning of month t. If d_t is month t's demand, then for $t = 1, 2, 3, 4$ we may write

$$f_t(i) = \min_{x|0\leq i+x-d_t\leq 4} \{.5(i + x - d_t) + c(x) + f_{t+1}(i + x - d_t)\} \qquad (23)$$

where $c(x) = $ cost of producing x units during a month, and $f_5(i) = 0$ for $(i = 0, 1, 2, 3, 4)$.

If we define $J_t(i, x) = .5(i + x - d_t) + c(x) + f_{t+1}(i + x - d_t)$ we may write

$$f_t(i) = \min_{x|0\leq i+x-d_t\leq 4} \{J_t(i, x)\}$$

FIGURE 13
Inventory Example

A	A	B	C	D	E	F	G	H	I	J	K	L	M
1	PROD COST	0	1	2	3	4	5						
2		0	4	5	6	7	8						
3													
4	VALUE	-5	0	1	2	3	4	5					
5	M5	10000	0	0	0	0	0	10000					
6	M4	10000	7	6	5	4	0	10000					
7	M3	10000	12	10	7	6.5	6	10000					
8	M2	10000	16	15	14	12	10.5	10000					
9													
10		STATE	0	0	0	0	0	0	1	1	1	1	1
11		ACTION	0	1	2	3	4	5	0	1	2	3	4
12	DEMAND												
13	4		10000	10004	10005	10006	7	8.5	10000	10004	10005	6	7.5
14	2		10000	10004	12	12.5	13	13.5	10000	11	11.5	12	12.5
15	3		10000	10004	10005	18	17.5	16	10000	10004	17	16.5	15
16	1		10000	20	20.5	21	20.5	20.5	16	19.5	20	19.5	19.5
17													

A	N	O	P	Q	R	S	T	U	V	W	X	Y	Z
1													
2													
3													
4													
5													
6													
7													
8													
9													
10	1	2	2	2	2	2	2	3	3	3	3	3	3
11	5	0	1	2	3	4	5	0	1	2	3	4	5
12													
13	9	10000	10004	5	6.5	8	9.5	10000	4	5.5	7	8.5	10
14	10	7	10.5	11	11.5	9	10010.5	6.5	10	10.5	8	10009.5	10011
15	16	10000	16	15.5	14	15	16	12	14.5	13	14	15	10010.5
16	10010.5	15.5	19	18.5	18.5	10009.5	10011	15	17.5	17.5	10008.5	10010	10011.5
17													

FIGURE **13**
(Continued)

A	AA	AB	AC	AD	AE	AF	AG	AH	AI	AJ	AK	AL
1												
2												
3												
4												
5												
6												
7												
8												
9												
10	4	4	4	4	4	4						
11	0	1	2	3	4	5	F(0)	F(1)	F(2)	F(3)	F(4)	
12												
13	0	4.5	6	7.5	9	10010.5	7	6	5	4	0	1
14	6	9.5	7	10008.5	10010	10011.5	12	10	7	6.5	6	2
15	10.5	12	13	14	10009.5	10011	16	15	14	12	10.5	3
16	13.5	16.5	10007.5	10009	10010.5	10012	20	16	15.5	15	13.5	4
17												

Next we compute $J_t(i, x)$ in A13:AF16. For example, to compute $J_4(0, 2)$, we enter the following formula in E13:

$$=\text{HLOOKUP(E\$11,\$B\$1:\$G\$2,2)}$$
$$+.5*+1\text{MAX(E\$10+E\$11}-\text{\$A13,0)}$$
$$+\text{HLOOKUP(E\$10+E\$11}-\text{\$A13,\$B\$4:\$H\$8,1}+\text{\$AL13)}$$

The first term in this sum yields $c(x)$ (this is because E\$11 is the production level). The second term gives the holding cost for the month (this is because E\$10+E\$11−\$A13 gives the month's ending inventory). The final term yields $f_{t+1}(i + x - d_t)$. This is because E\$10+E\$11−\$A13 is the beginning inventory for month $t + 1$. The reference to 1+\$AL13 in the final term ensures that we look up the value of $f_{t+1}(i + x - d_t)$ in the correct row [the values of the $f_{t+1}(\ )$ will be tabulated in C5:G8]. Copying the formula in E13 to the range C13:AF16 computes all the $J_t(i, x)$.

In AG13:AK16, we compute the $f_t(d)$. To begin, we enter the following formulas in cells AG13:AK13:

AG13:	=MIN(C13:H13)	[Computes $f_4(0)$]
AH13:	=MIN(I13:N13)	[Computes $f_4(1)$]
AI13:	=MIN(O13:T13)	[Computes $f_4(2)$]
AJ13:	=MIN(U13:Z13)	[Computes $f_4(3)$]
AK13:	=MIN(AA13:AF13)	[Computes $f_4(4)$]

To compute all the $f_t(i)$, we now copy from the range AG13:AK13 to the range AG13:AK16. For this to be successful, we need to have the correct values of the $f_t(i)$ in B5:H8. In columns B and H of rows 5–8, we enter 10,000 (or any large positive number). This ensures that it is very costly to end a month with an inventory that is negative or that exceeds 4. This will ensure that each month's ending inventory is between 0 and 4 inclusive. In the range C5:G5, we enter a 0 in each cell. This is because $f_5(i) = 0$ for $i = 0$, 1, 2, 3, 4. In cell C6, we enter +AG13; this enters the value of $f_1(0)$. By copying this formula to the range C6:G8, we have created a table of the $f_t(d)$, which can be used (in rows 13–16) to look up the $f_t(d)$.

As with the spreadsheet we used to solve Example 5, our current spreadsheet exhibits circular references. This is because rows 6–8 refer to rows 13–16, and rows 13–16 refer to rows 6–8. Pressing F9 several times, however, resolves the circular references. You also can resolve circular references by selecting Tools, Options, Calculations and checking the Iterations box.

For any initial inventory level, we can now compute the optimal production schedule. For example, suppose the inventory at the beginning of month 1 is 0. Then $f_1(0) = 20 = J_1(0, 1)$. Thus, it is optimal to produce 1 unit during month 1. Now we seek $f_2(0 + 1 - 1) = 16 = J_2(0, 5)$, so we produce 5 units during month 2. Then we seek $f_3(0 + 5 - 3) = 7 = J_3(2, 0)$, so we produce 0 units during month 3. Solving $f_4(2 + 0 - 2) = J_4(0, 4)$, we produce 4 units during month 4.

PROBLEMS

Group A

1 Use a spreadsheet to solve Problem 2 of Section 6.3.

2 Use a spreadsheet to solve Problem 4 of Section 6.4.

3 Use a spreadsheet to solve Problem 5 of Section 6.4.

SUMMARY

Dynamic programming solves a relatively complex problem by decomposing the problem into a series of simpler problems. First we solve a one-stage problem, then a two-stage problem, and finally a T-stage problem (T = total number of stages in the original problem).

In most applications, a decision is made at each stage (t = current stage), a reward is earned (or a cost is incurred) at each stage, and we go on to the stage $t + 1$ state.

Working Backward

In formulating dynamic programming recursions by working backward, it is helpful to remember that in most cases:

1 The **stage** is the mechanism by which we build up the problem.

2 The **state** at any stage gives the information needed to make the correct decision at the current stage.

3 In most cases, we must determine how the reward received (or cost incurred) during the current stage depends on the stage t decision, the stage t state, and the value of t.

4 We must also determine how the stage $t + 1$ state depends on the stage t decision, the stage t state, and the value of t.

5 If we define (for a minimization problem) $f_t(i)$ as the minimum cost incurred during stages $t, t + 1, \ldots, T$, given that the stage t state is i, then (in many cases) we may write $f_t(i) = \min \{(\text{cost during stage } t) + f_{t+1}(\text{new state at stage } t + 1)\}$, where the minimum is over all decisions allowable in state i during stage t.

6 We begin by determining all the $f_T(\cdot)$'s, then all the $f_{T-1}(\cdot)$'s, and finally f_1 (the initial state).

7 We then determine the optimal stage 1 decision. This leads us to a stage 2 state, at which we determine the optimal stage 2 decision. We continue in this fashion until the optimal stage T decision is found.

Wagner–Whitin Algorithm and Silver–Meal Heuristic for Dynamic Lot-Size Model

A periodic review inventory model in which each period's demand is known at the beginning of the problem is a **dynamic lot-size model.** A cost-minimizing production or ordering policy may be found via a backward recursion, a forward recursion, the Wagner–Whitin algorithm, or the Silver–Meal heuristic.

The Wagner–Whitin algorithm uses the fact that production occurs during a period if and only if the period's beginning inventory is zero. The decision during such a period is the number of consecutive periods of demand that production should meet.

During a period in which beginning inventory is zero, the Silver–Meal heuristic computes the average cost per period (setup plus holding) incurred in meeting the demand during the next k periods. If k^* minimizes this average cost, then the next k^* periods of demand should be met by the current period's production.

Computational Considerations

Dynamic programming is much more efficient than explicit enumeration of the total cost associated with each possible set of decisions that may be chosen during the T stages. Unfortunately, however, many practical applications of dynamic programming involve very large state spaces, and in these situations, considerable computational effort is required to determine optimal decisions.

REVIEW PROBLEMS

Group A

1 In the network in Figure 14, find the shortest path from node 1 to node 10 and the shortest path from node 2 to node 10.

2 A company must meet the following demands on time: month 1, 1 unit; month 2, 1 unit; month 3, 2 units; month 4, 2 units. It costs $4 to place an order, and a $2 per-unit holding cost is assessed against each month's ending inventory. At the beginning of month 1, 1 unit is available. Orders are delivered instantaneously.

    a Use a backward recursion to determine an optimal ordering policy.

    b Use the Wagner–Whitin method to determine an optimal ordering policy.

    c Use the Silver–Meal heuristic to determine an ordering policy.

3 Reconsider Problem 2, but now suppose that demands need not be met on time. Assume that all lost demand is backlogged and that a $1 per-unit shortage cost is assessed against the number of shortages incurred during each month. All demand must be met by the end of month 4. Use dynamic programming to determine an ordering policy that minimizes total cost.

4 Indianapolis Airlines has been told that it may schedule six flights per day departing from Indianapolis. The destination of each flight may be New York, Los Angeles,

FIGURE **14**

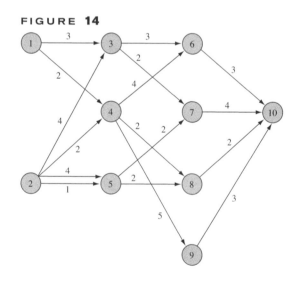

or Miami. Table 20 shows the contribution to the company's profit from any given number of daily flights from Indianapolis to each possible destination. Find the optimal number of flights that should depart Indianapolis for each destination. How would the answer change if the airline were restricted to only four daily flights?

TABLE 20

| Destination | Profit per Flight ($) | | | | | |
| | Number of Planes | | | | | |
	1	2	3	4	5	6
New York	80	150	210	250	270	280
Los Angeles	100	195	275	325	300	250
Miami	90	180	265	310	350	320

5 I am working as a cashier at the local convenience store. A customer's bill is $1.09, and he gives me $2.00. I want to give him change using the smallest possible number of coins. Use dynamic programming to determine how to give the customer his change. Does the answer suggest a general result about giving change? Resolve the problem if a 20¢ piece (in addition to other United States coins) were available.

6 A company needs to have a working machine during each of the next six years. Currently, it has a new machine. At the beginning of each year, the company may keep the machine or sell it and buy a new one. A machine cannot be kept for more than three years. A new machine costs $5,000. The revenues earned by a machine, the cost of maintaining it, and the salvage value that can be obtained by selling it at the end of a year depend on the age of the machine (see Table 21). Use dynamic programming to maximize the net profit earned during the next six years.

7 A company needs the following number of workers during each of the next five years: year 1, 15; year 2, 30; year 3, 10; year 4, 30; year 5, 20. At present, the company has 20 workers. Each worker is paid $30,000 per year. At the beginning of each year, workers may be hired or fired. It costs $10,000 to hire a worker and $20,000 to fire a worker. A newly hired worker can be used to meet the current year's worker requirement. During each year, 10% of all workers quit (workers who quit do not incur any firing cost).

a With dynamic programming, formulate a recursion that can be used to minimize the total cost incurred in meeting the worker requirements of the next five years.

b How would the recursion be modified if hired workers cannot be used to meet worker requirements until the year following the year in which they are hired?

8 At the beginning of each year, Barnes Carr Oil sets the world oil price. If a price p is set, then $D(p)$ barrels of oil will be demanded by world customers. We assume that

during any year, each oil company sells the same number of barrels of oil. It costs Barnes Carr Oil c dollars to extract and refine each barrel of oil. Barnes Carr cannot set too high a price, however, because if a price p is set and there are currently N oil companies, then $g(p, N)$ oil companies will enter the oil business [$g(p, N)$ could be negative]. Setting too high a price will dilute future profits because of the entrance of new companies. Barnes Carr wants to maximize the discounted profit the company will earn over the next 20 years. Formulate a recursion that will aid Barnes Carr in meeting its goal. Initially, there are 10 oil companies.

9 For a computer to work properly, three subsystems of the computer must all function properly. To increase the reliability of the computer, spare units may be added to each system. It costs $100 to add a spare unit to system 1, $300 to system 2, and $200 to system 3. As a function of the number of added spares (a maximum of two spares may be added to each system), the probability that each system will work is given in Table 22. Use dynamic programming to maximize the probability that the computer will work properly, given that $600 is available for spare units.

Group B

10 During any year, I can consume any amount that does not exceed my current wealth. If I consume c dollars during a year, I earn c^a units of happiness. By the beginning of the next year, the previous year's ending wealth grows by a factor k.

a Formulate a recursion that can be used to maximize total utility earned during the next T years. Assume I originally have w_0 dollars.

b Let $f_t(w)$ be the maximum utility earned during years $t, t + 1, \ldots, T$, given that I have w dollars at the beginning of year t; and $c_t(w)$ be the amount that should be consumed during year t to attain $f_t(w)$. By working backward, show that for appropriately chosen constants a_t and b_t,

$$f_t(w) = b_t w^a \quad \text{and} \quad c_t(w) = a_t w$$

Interpret these results.

11 At the beginning of month t, farmer Smith has x_t bushels of wheat in his warehouse. He has the opportunity to sell wheat at a price s_t dollars per bushel and can buy wheat at p_t dollars per bushel. Farmer Smith's warehouse can hold at most C units at the end of each month.

a Formulate a recursion that can be used to maximize the total profit earned during the next T months.

b Let $f_t(x_t)$ be the maximum profit that can be earned during months $t, t + 1, \ldots, T$, given that x_t bushels of

TABLE 21

| | Age of Machine at Beginning of Year | | |
	0 Year	1 Year	2 Years
Revenues ($)	4,500	3,000	1,500
Operating Costs ($)	500	700	1,100
Salvage Value at End of Year ($)	3,000	1,800	500

TABLE 22

| Number of Spares | Probability That a System Works | | |
	System 1	System 2	System 3
0	.85	.60	.70
1	.90	.85	.90
2	.95	.95	.98

wheat are in the warehouse at the beginning of month t. By working backward, show that for appropriately chosen constants a_t and b_t,

$$f_t(x_t) = a_t + b_t x_t$$

c During any given month, show that the profit-maximizing policy has the following properties: (1) The amount sold during month t will equal either x_t or zero. (2) The amount purchased during a given month will be either zero or sufficient to bring the month's ending stock to C bushels.

REFERENCES

The following references are oriented toward applications and are written at an intermediate level:

Dreyfus, S., and A. Law. *The Art and Theory of Dynamic Programming.* Orlando, Fla.: Academic Press, 1977.

Nemhauser, G. *Introduction to Dynamic Programming.* New York: Wiley, 1966.

Wagner, H. *Principles of Operations Research,* 2d ed. Englewood Cliffs, N.J.: Prentice Hall, 1975.

The following five references are oriented toward theory and are written at a more advanced level:

Bellman, R. *Dynamic Programming.* Princeton, N.J.: Princeton University Press, 1957.

Bellman, R., and S. Dreyfus. *Applied Dynamic Programming.* Princeton, N.J.: Princeton University Press, 1962.

Bersetkas, D. *Dynamic Programming and Optimal Control,* vol. 1. Cambridge, Mass.: Athena Scientific, 2000.

Denardo, E. *Dynamic Programming: Theory and Applications.* Englewood Cliffs, N.J.: Prentice Hall, 1982.

Whittle, P. *Optimization Over Time: Dynamic Programming and Stochastic Control,* vol. 1. New York: Wiley, 1982.

Morton, T. "Planning Horizons for Dynamic Programs," *Operations Research* 27(1979):730–743. A discussion of turnpike theorems.

Peterson, R., and E. Silver. *Decision Systems for Inventory Management and Production Planning.* New York: Wiley, 1998. Discusses the Silver–Meal method.

Waddell, R. "A Model for Equipment Replacement Decisions and Policies," *Interfaces* 13(1983):1–8. An application of the equipment replacement model.

Wagner, H., and T. Whitin. "Dynamic Version of the Economic Lot Size Model," *Management Science* 5(1958):89–96. Discusses Wagner–Whitin method.

7 Probabilistic Dynamic Programming

Recall from our study of deterministic dynamic programming that many recursions were of the following form:

$$f_t \text{ (current state)} = \min_{\substack{\text{all feasible} \\ \text{decisions}}} \text{ (or max)}\{\text{costs during current stage} + f_{t+1} \text{ (new state)}\}$$

For all the examples in Chapter 6, a specification of the current state and current decision was enough to tell us *with certainty* the new state and the costs during the current stage. In many practical problems, these factors may not be known with certainty, even if the current state and decision are known. For example, in the inventory model of Section 6.3, we assumed that each period's demand was known at the beginning of the problem. In most situations, it would be more realistic to assume that period t's demand is a random variable whose value is not known until after period t's production decision is made. Even if we know the current period's state (beginning inventory level) and decision (production during the current period), the next period's state and the current period's cost will be random variables whose values are not known until the value of period t's demand is known. The Chapter 6 discussion simply does not apply to this problem.

In this chapter, we explain how to use dynamic programming to solve problems in which the current period's cost or the next period's state are random. We call these problems *probabilistic dynamic programming problems* (or PDPs). In a PDP, the decision maker's goal is usually to minimize expected (or expected discounted) cost incurred or to maximize expected (or expected discounted) reward earned over a given time horizon. Chapter 7 concludes with a brief study of *Markov decision processes*. A Markov decision process is just a probabilistic dynamic programming problem in which the decision maker faces an infinite horizon.

7.1 When Current Stage Costs Are Uncertain, but the Next Period's State Is Certain

For problems in this section, the next period's state is known with certainty, but the reward earned during the current stage is not known with certainty (given the current state and decision).

EXAMPLE 1   Milk Distribution

For a price of $1/gallon, the Safeco Supermarket chain has purchased 6 gallons of milk from a local dairy. Each gallon of milk is sold in the chain's three stores for $2/gallon. The dairy must buy back for 50¢/gallon any milk that is left at the end of the day. Unfortunately for Safeco, demand for each of the chain's three stores is uncertain. Past data indicate that the daily demand at each store is as shown in Table 1. Safeco wants to allo-

TABLE 1
Probability Distributions for Daily Milk Demand

	Daily Demand (gallons)	Probability
Store 1	1	.60
	2	0
	3	.40
Store 2	1	.50
	2	.10
	3	.40
Store 3	1	.40
	2	.30
	3	.30

cate the 6 gallons of milk to the three stores so as to maximize the expected net daily profit (revenues less costs) earned from milk. Use dynamic programming to determine how Safeco should allocate the 6 gallons of milk among the three stores.

Solution   With the exception of the fact that the demand (and therefore the revenue) is uncertain, this problem is very similar to the resource allocation problems studied in Section 6.4.

Observe that since Safeco's daily purchase costs are always $6, we may concentrate our attention on the problem of allocating the milk to maximize daily expected revenue earned from the 6 gallons.

Define

$$r_t(g_t) = \text{expected revenue earned from } g_t \text{ gallons assigned to store } t$$
$$f_t(x) = \text{maximum expected revenue earned from } x \text{ gallons assigned}$$
$$\text{to stores } t, t + 1, \ldots, 3$$

Since $f_3(x)$ must by definition be the expected revenue earned from assigning x gallons of milk to store 3, we see that $f_3(x) = r_3(x)$. For $t = 1, 2$, we may write

$$f_t(x) = \max_{g_t} \{r_t(g_t) + f_{t+1}(x - g_t)\} \qquad (1)$$

where g_t must be a member of $\{0, 1, \ldots, x\}$. Equation (1) follows, because for any choice of g_t (the number of gallons assigned to store t), the expected revenue earned from store $t, t + 1, \ldots, 3$ will be the sum of the expected revenue earned from store t if g_t gallons are assigned to store t plus the maximum expected revenue that can be earned from the stores $t + 1, t + 2, \ldots, 3$ when $x - g_t$ gallons are assigned to these stores. To compute the optimal allocation of milk to the stores, we begin by computing $f_3(0), f_3(1), \ldots, f_3(6)$. Then we use Equation (1) to compute $f_2(0), f_2(1), \ldots, f_2(6)$. Finally we determine $f_1(6)$.

We begin by computing the $r_t(g_t)$'s. Note that it would be foolish to assign more than 3 gallons to any store. For this reason, we compute the $r_t(g_t)$'s only for $g_t = 0, 1, 2,$ or 3. As an example, we compute $r_3(2)$, the expected revenue earned if 2 gallons are assigned to store 3. If the demand at store 3 is for 2 or more gallons, both gallons assigned to store 3 will be sold, and $4 in revenue will be earned. If the demand at store 3 is 1 gallon, 1 gallon will be sold for $2, and 1 gallon will be returned for 50¢. Hence, if demand at store 3 is for 1 gallon, a revenue of $2.50 will be earned. Since there is a .60 chance that demand at store 3 will be for 2 or more gallons and a .40 chance that store 3 demand will be for 1 gallon, it follows that $r_3(2) = (.30 + .30)(4.00) + .40(2.50) = \$3.40$. Similar computations yield the following results:

$$r_3(0) = \$0 \qquad r_2(0) = \$0 \qquad r_1(0) = \$0$$
$$r_3(1) = \$2.00 \qquad r_2(1) = \$2.00 \qquad r_1(1) = \$2.00$$
$$r_3(2) = \$3.40 \qquad r_2(2) = \$3.25 \qquad r_1(2) = \$3.10$$
$$r_3(3) = \$4.35 \qquad r_2(3) = \$4.35 \qquad r_1(3) = \$4.20$$

We now use (1) to determine an optimal allocation of milk to stores. Let $g_t(x)$ be an allocation of milk to store t that attains $f_t(x)$. Then

$$f_3(0) = r_3(0) = 0 \qquad g_3(0) = 0$$
$$f_3(1) = r_3(1) = 2.00 \qquad g_3(1) = 1$$
$$f_3(2) = r_3(2) = 3.40 \qquad g_3(2) = 2$$
$$f_3(3) = r_3(3) = 4.35 \qquad g_3(3) = 3$$

We need not compute $f_3(4)$, $f_3(5)$, and $f_3(6)$, because an optimal allocation will never have more than 3 gallons to allocate to a single store (demand at any store is never more than 3 gallons).

Using (1) to work backward, we obtain

$$f_2(0) = r_2(0) + f_3(0 - 0) = 0 \qquad g_2(0) = 0$$

$$f_2(1) = \max \begin{cases} r_2(0) + f_3(1 - 0) = 2.00^* \\ r_2(1) + f_3(1 - 1) = 2.00^* \end{cases} \qquad g_2(1) = 0 \text{ or } 1$$

$$f_2(2) = \max \begin{cases} r_2(0) + f_3(2 - 0) = 0 + 3.40 = 3.40 \\ r_2(1) + f_3(2 - 1) = 2.00 + 2.00 = 4.00^* \\ r_2(2) + f_3(2 - 2) = 3.25 + 0 = 3.25 \end{cases} \qquad g_2(2) = 1$$

$$f_2(3) = \max \begin{cases} r_2(0) + f_3(3 - 0) = 0 + 4.35 = 4.35 \\ r_2(1) + f_3(3 - 1) = 2.00 + 3.40 = 5.40^* \\ r_2(2) + f_3(3 - 2) = 3.25 + 2.00 = 5.25 \\ r_2(3) + f_3(3 - 3) = 4.35 + 0 = 4.35 \end{cases} \qquad g_2(3) = 1$$

Note that in computing $f_2(4)$, $f_2(5)$, and $f_2(6)$, we need not consider any allocation for more than 3 gallons to store 2 or any that leaves more than 3 gallons for store 3.

$$f_2(4) = \max \begin{cases} r_2(1) + f_3(4 - 1) = 2.00 + 4.35 = 6.35 \\ r_2(2) + f_3(4 - 2) = 3.25 + 3.40 = 6.65^* \\ r_2(3) + f_3(4 - 3) = 4.35 + 2.00 = 6.35 \end{cases} \qquad g_2(4) = 2$$

$$f_2(5) = \max \begin{cases} r_2(2) + f_3(5 - 2) = 3.25 + 4.35 = 7.60 \\ r_2(3) + f_3(5 - 3) = 4.35 + 3.40 = 7.75^* \end{cases} \qquad g_2(5) = 3$$

$$f_2(6) = r_2(3) + f_3(6 - 3) = 4.35 + 4.35 = 8.70^* \qquad g_2(6) = 3$$

Finally,

$$f_1(6) = \max \begin{cases} r_1(0) + f_2(6 - 0) = 0 + 8.70 \\ r_1(1) + f_2(6 - 1) = 2.00 + 7.75 = 9.75^* \\ r_1(2) + f_2(6 - 2) = 3.10 + 6.65 = 9.75^* \\ r_1(3) + f_2(6 - 3) = 4.20 + 5.40 = 9.60 \end{cases} \qquad g_1(6) = 1 \text{ or } 2$$

Thus, we can either assign 1 or 2 gallons to store 1. Suppose we arbitrarily choose to assign 1 gallon to store 1. Then we have $6 - 1 = 5$ gallons for stores 2 and 3. Since $f_2(5)$ is attained by $g_2(5) = 3$, we assign 3 gallons to store 2. Then $5 - 3 = 2$ gallons are available for store 3. Since $g_3(2) = 2$, we assign 2 gallons to store 3. Note that although this policy obtains the maximum expected revenue, $f_1(6) = \$9.75$, the total revenue actually received on a given day may be more or less than \$9.75. For example, if demand at each store were 1 gallon, total revenue would be $3(2.00) + 3(0.50) = \$7.50$, whereas if demand at each store were 3 gallons, all the milk would be sold at \$2/gallon, and the total revenue would be $6(2.00) = \$12.00$.

PROBLEMS

Group A

1 In Example 1, find another allocation of milk that maximizes expected daily revenue.

2 Suppose that \$4 million is available for investment in three projects. The probability distribution of the net present value earned from each project depends on how much is invested in each project. Let $\mathbf{I}_t$ be the random variable

denoting the net present value earned by project t. The distribution of $\mathbf{I}_t$ depends on the amount of money invested in project t, as shown in Table 2 (a zero investment in a project always earns a zero NPV). Use dynamic programming to determine an investment allocation that maximizes the expected NPV obtained from the three investments.

TABLE 2
Investment Probability for Problem 2

	Investment (millions)	Probability		
Project 1	\$1	$P(\mathbf{I}_1 = 2) = .6$	$P(\mathbf{I}_1 = 4) = .3$	$P(\mathbf{I}_1 = 5) = .1$
	\$2	$P(\mathbf{I}_1 = 4) = .5$	$P(\mathbf{I}_1 = 6) = .3$	$P(\mathbf{I}_1 = 8) = .2$
	\$3	$P(\mathbf{I}_1 = 6) = .4$	$P(\mathbf{I}_1 = 7) = .5$	$P(\mathbf{I}_1 = 10) = .1$
	\$4	$P(\mathbf{I}_1 = 7) = .2$	$P(\mathbf{I}_1 = 9) = .4$	$P(\mathbf{I}_1 = 10) = .4$
Project 2	\$1	$P(\mathbf{I}_2 = 1) = .5$	$P(\mathbf{I}_2 = 2) = .4$	$P(\mathbf{I}_2 = 4) = .1$
	\$2	$P(\mathbf{I}_2 = 3) = .4$	$P(\mathbf{I}_2 = 5) = .4$	$P(\mathbf{I}_2 = 6) = .2$
	\$3	$P(\mathbf{I}_2 = 4) = .3$	$P(\mathbf{I}_2 = 6) = .3$	$P(\mathbf{I}_2 = 8) = .4$
	\$4	$P(\mathbf{I}_2 = 3) = .4$	$P(\mathbf{I}_2 = 8) = .3$	$P(\mathbf{I}_2 = 9) = .3$
Project 3	\$1	$P(\mathbf{I}_3 = 0) = .2$	$P(\mathbf{I}_3 = 4) = .6$	$P(\mathbf{I}_3 = 5) = .2$
	\$2	$P(\mathbf{I}_3 = 4) = .4$	$P(\mathbf{I}_3 = 6) = .4$	$P(\mathbf{I}_3 = 7) = .2$
	\$3	$P(\mathbf{I}_3 = 5) = .3$	$P(\mathbf{I}_3 = 7) = .4$	$P(\mathbf{I}_3 = 8) = .3$
	\$4	$P(\mathbf{I}_3 = 6) = .1$	$P(\mathbf{I}_3 = 8) = .5$	$P(\mathbf{I}_3 = 9) = .4$

7.2 A Probabilistic Inventory Model

In this section, we modify the inventory model of Section 6.3 to allow for uncertain demand. This will illustrate the difficulties involved in solving a PDP for which the state during the next period is uncertain (given the current state and current decision).

EXAMPLE 2   Three-Period Production Policy

Consider the following three-period inventory problem. At the beginning of each period, a firm must determine how many units should be produced during the current period. During a period in which x units are produced, a production cost $c(x)$ is incurred, where $c(0) =$

0, and for $x > 0$, $c(x) = 3 + 2x$. Production during each period is limited to at most 4 units. After production occurs, the period's random demand is observed. Each period's demand is equally likely to be 1 or 2 units. After meeting the current period's demand out of current production and inventory, the firm's end-of-period inventory is evaluated, and a holding cost of $1 per unit is assessed. Because of limited capacity, the inventory at the end of each period cannot exceed 3 units. It is required that all demand be met on time. Any inventory on hand at the end of period 3 can be sold at $2 per unit. At the beginning of period 1, the firm has 1 unit of inventory. Use dynamic programming to determine a production policy that minimizes the expected net cost incurred during the three periods.

Solution Define $f_t(i)$ to be the minimum expected net cost incurred during the periods $t, t + 1,$ ..., 3 when the inventory at the beginning of period t is i units. Then

$$f_3(i) = \min_x \{c(x) + (\tfrac{1}{2})(i + x - 1) + (\tfrac{1}{2})(i + x - 2)$$
$$- (\tfrac{1}{2})2(i + x - 1) - (\tfrac{1}{2})2(i + x - 2)\} \tag{2}$$

where x must be a member of $\{0, 1, 2, 3, 4\}$ and x must satisfy $(2 - i) \leq x \leq (4 - i)$.

Equation (2) follows, because if x units are produced during period 3, the net cost during period 3 is (expected production cost) + (expected holding cost) − (expected salvage value). If x units are produced, the expected production cost is $c(x)$, and there is a $\tfrac{1}{2}$ chance that the period 3 holding cost will be $i + x - 1$ and a $\tfrac{1}{2}$ chance that it will be $i + x - 2$. Hence, the period 3 expected holding cost will be $(\tfrac{1}{2})(i + x - 1) + (\tfrac{1}{2})(i + x - 2) = i + x - \tfrac{3}{2}$. Similar reasoning shows that the expected salvage value (a negative cost) at the end of period 3 will be $(\tfrac{1}{2})2(i + x - 1) + (\tfrac{1}{2})2(i + x - 2) = 2i + 2x - 3$. To ensure that period 3 demand is met, we must have $i + x \geq 2$, or $x \geq 2 - i$. Similarly, to ensure that ending period three inventory does not exceed 3 units, we must have $i + x - 1 \leq 3$, or $x \leq 4 - i$.

For $t = 1, 2$, we can derive the recursive relation for $f_t(i)$ by noting that for any month t production level x, the expected costs incurred during periods $t, t + 1, \ldots, 3$ are the sum of the expected costs incurred during period t and the expected costs incurred during periods $t + 1, t + 2, \ldots, 3$. As before, if x units are produced during month t, the expected cost during month t will be $c(x) + (\tfrac{1}{2})(i + x - 1) + (\tfrac{1}{2})(i + x - 2)$. (Note that during periods 1 and 2, no salvage value is received.) If x units are produced during month t, the expected cost during periods $t + 1, t + 2, \ldots, 3$ is computed as follows. Half of the time, the demand during period t will be 1 unit, and the inventory at the beginning of period $t + 1$ will be $i + x - 1$. In this situation, the expected costs incurred during periods $t + 1, t + 2, \ldots, 3$ (assuming we act optimally during these periods) is $f_{t+1}(i + x - 1)$. Similarly, there is a $\tfrac{1}{2}$ chance that the inventory at the beginning of period $t + 1$ will be $i + x - 2$. In this case, the expected cost incurred during periods $t + 1, t + 2, \ldots, 3$ will be $f_{t+1}(i + x - 2)$. In summary, the expected cost during periods $t + 1, t + 2, \ldots, 3$ will be $(\tfrac{1}{2})f_{t+1}(i + x - 1) + (\tfrac{1}{2})f_{t+1}(i + x - 2)$. With this in mind, we may write for $t = 1, 2$,

$$f_t(i) = \min_x [c(x) + (\tfrac{1}{2})(i + x - 1) + (\tfrac{1}{2})(i + x - 2)$$
$$+ (\tfrac{1}{2})f_{t+1}(i + x - 1) + (\tfrac{1}{2})f_{t+1}(i + x - 2)] \tag{3}$$

where x must be a member of $\{0, 1, 2, 3, 4\}$ and x must satisfy $(2 - i) \leq x \leq (4 - i)$.

Generalizing the reasoning that led to (3) yields the following important observation concerning the formulation of PDPs. Suppose the possible states during period $t + 1$ are $s_1, s_2, \ldots, s_n$ and the probability that the period $t + 1$ state will be s_i is p_i. Then the minimum expected cost incurred during periods $t + 1, t + 2, \ldots,$ end of the problem is

$$\sum_{i=1}^{i=n} p_i f_{t+1}(s_i)$$

where $f_{t+1}(s_i)$ is the minimum expected cost incurred from period $t + 1$ to the end of the problem, given that the state during period $t + 1$ is s_i.

We define $x_t(i)$ to be a period t production level attaining the minimum in (3) for $f_t(i)$. We now work backward until $f_1(1)$ is determined. The relevant computations are summarized in Tables 3, 4, and 5. Since each period's ending inventory must be nonnegative and cannot exceed 3 units, the state during each period must be 0, 1, 2, or 3.

As in Section 6.3, we begin by producing $x_1(1) = 3$ units during period 1. We cannot, however, determine period 2's production level until period 1's demand is observed. Also, period 3's production level cannot be determined until period 2's demand is observed. To illustrate the idea, we determine the optimal production schedule if period 1 and period

TABLE 3
Computations for $f_3(i)$

i	x	$c(x)$	Expected Holding Cost $(i + x - \frac{3}{2})$	Expected Salvage Value $(2i + 2x - 3)$	Total Expected Cost	$f_3(i)$ $x_3(i)$
3	0	0	$\frac{3}{2}$	3	$-\frac{3}{2}*$	$f_3(3) = -\frac{3}{2}$
3	1	5	$\frac{5}{2}$	5	$\frac{5}{2}$	$x_3(3) = 0$
2	0	0	$\frac{1}{2}$	1	$-\frac{1}{2}*$	$f_3(2) = -\frac{1}{2}$
2	1	5	$\frac{3}{2}$	3	$\frac{7}{2}$	$x_3(2) = 0$
2	2	7	$\frac{5}{2}$	5	$\frac{9}{2}$	
1	1	5	$\frac{1}{2}$	1	$\frac{9}{2}*$	$f_3(1) = \frac{9}{2}$
1	2	7	$\frac{3}{2}$	3	$\frac{11}{2}$	$x_3(1) = 1$
1	3	9	$\frac{5}{2}$	5	$\frac{13}{2}$	
0	2	7	$\frac{1}{2}$	1	$\frac{13}{2}*$	$f_3(0) = \frac{13}{2}$
0	3	9	$\frac{3}{2}$	3	$\frac{15}{2}$	$x_3(0) = 2$
0	4	11	$\frac{5}{2}$	5	$\frac{17}{2}$	

TABLE 4
Computations for $f_2(i)$

i	x	$c(x)$	Expected Holding Cost $(i + x - \frac{3}{2})$	Expected Future Cost $(\frac{1}{2})f_3(i + x - 1)$ $+ (\frac{1}{2})f_3 (i + x - 2))$	Total Expected Cost Periods 2,3	$f_2(i)$ $x_2(i)$
3	0	0	$\frac{3}{2}$	2	$\frac{7}{2}*$	$f_2(3) = \frac{7}{2}$
3	1	5	$\frac{5}{2}$	-1	$\frac{13}{2}$	$x_2(3) = 0$
2	0	0	$\frac{1}{2}$	$\frac{11}{2}$	$6*$	$f_2(2) = 6$
2	1	5	$\frac{3}{2}$	2	$\frac{17}{2}$	$x_2(2) = 0$
2	2	7	$\frac{5}{2}$	-1	$\frac{17}{2}$	
1	1	5	$\frac{1}{2}$	$\frac{11}{2}$	11	$f_2(1) = \frac{21}{2}$
1	2	7	$\frac{3}{2}$	2	$\frac{21}{2}*$	$x_2(1) = 2$ or 3
1	3	9	$\frac{5}{2}$	-1	$\frac{21}{2}*$	
0	2	7	$\frac{1}{2}$	$\frac{11}{2}$	13	$f_2(0) = \frac{25}{2}$
0	3	9	$\frac{3}{2}$	2	$\frac{25}{2}*$	$x_2(0) = 3$ or 4
0	4	11	$\frac{5}{2}$	-1	$\frac{25}{2}*$	

TABLE 5
Computations for $f_1(1)$

x	$c(x)$	Expected Holding Cost $(i + x - \frac{3}{2})$	Expected Future Cost $((\frac{1}{2})f_2(i + x - 1)$ $+ (\frac{1}{2})f_2(i + x - 2))$	Total Expected Cost Periods 1–3	$f_1(1)$ $x_1(1)$
1	5	$\frac{1}{2}$	$\frac{23}{2}$	17	$f_1(1) = \frac{65}{4}$
2	7	$\frac{3}{2}$	$\frac{33}{4}$	$\frac{67}{4}$	$x_1(1) = 3$
3	9	$\frac{5}{2}$	$\frac{19}{4}$	$\frac{65}{4}*$	

2 demands are both 2 units. Since $x_1(1) = 3$, 3 units will be produced during period 1. Then period 2 will begin with an inventory of $1 + 3 - 2 = 2$ units, so $x_2(2) = 0$ units should be produced. After period 2's demand of 2 units is met, period 3 will begin with $2 - 2 = 0$ units on hand. Thus, $x_3(0) = 2$ units will be produced during period 3.

In contrast, suppose that period 1 and period 2 demands are both 1 unit. As before, $x_1(1) = 3$ units will be produced during period 1. Then period 2 will begin with $1 + 3 - 1 = 3$ units, and $x_2(3) = 0$ units will be produced during period 2. Then period 3 will begin with $3 - 1 = 2$ units on hand, and $x_3(2) = 0$ units will be produced during period 3. Note that the optimal production policy has adapted to the low demand by reducing period 3 production. This example illustrates an important aspect of dynamic programming solutions for problems in which future states are not known with certainty at the beginning of the problem: *If a random factor (such as random demand) influences transitions from the period t state to the period t + 1 state, the optimal action for period t cannot be determined until period t's state is known.*

(s, S) Policies

Consider the following modification of the dynamic lot-size model of Section 6.7, for which there exists an optimal production policy called an (s, S) inventory policy:

1 The cost of producing $x > 0$ units during a period consists of a fixed cost K and a per-unit variable production cost c.

2 With a probability $p(x)$, the demand during a given period will be x.

3 A holding cost of h per unit is assessed on each period's ending inventory. If we are short, a per-unit shortage cost of d is incurred. (The case where no shortages are allowed may be obtained by letting d be very large.)

4 The goal is to minimize the total expected cost incurred during periods $1, 2, \ldots, T$.

5 All demands must be met by the end of period T.

For such an inventory problem, Scarf (1960) used dynamic programming to prove that there exists an optimal production policy of the following form: For each t ($t = 1, 2, \ldots,$ T) there exists a pair of numbers (s_t, S_t) such that if i_{t-1}, the entering inventory for period t, is less than s_t, then an amount $S_t - i_{t-1}$ is produced; if $i_{t-1} \geq s_t$, then it is optimal not to produce during period t. Such a policy is called an **(s, S) policy.**

For Example 2, our calculations show that $s_2 = 2$, $S_2 = 3$ or 4, $s_3 = 2$, and $S_3 = 2$. Thus, if we enter period 2 with 1 or 0 units, we produce enough to bring our stock level (before meeting period 2 demand) up to 3 or 4 units. If we enter period 2 with more than 1 unit, then no production should take place during period 2.

PROBLEMS

Group A

1 For Example 2, suppose that the period 1 demand is 1 unit, and the period 2 demand is 2 units. What would be the optimal production schedule?

2 Re-solve Example 2 if the end-of-period holding cost is $2 per unit.

3 In Example 2, suppose that shortages are allowed, and each shortage results in a lost sale and a cost incurred of $3. Now re-solve Example 2.

Group B

4 Chip Bilton sells sweatshirts at State U football games. He is equally likely to sell 200 or 400 sweatshirts at each game. Each time Chip places an order, he pays $500 plus $5 for each sweatshirt he orders. Each sweatshirt sells for $8. A holding cost of $2 per shirt (because of the opportunity cost for capital tied up in sweatshirts as well as storage costs) is assessed against each shirt left at the end of a game. Chip can store at most 400 shirts after each game. Assuming that the number of shirts ordered by Chip must be a multiple of 100, determine an ordering policy that maximizes expected profits earned during the first three games of the season. Assume that any leftover sweatshirts have a value of $6.

7.3 How to Maximize the Probability of a Favorable Event Occurring[†]

There are many occasions on which the decision maker's goal is to maximize the probability of a favorable event occurring. For instance, a company may want to maximize its probability of reaching a specified level of annual profits. To solve such a problem, we assign a reward of 1 if the favorable event occurs and a reward of 0 if it does not occur. Then the maximization of expected reward will be equivalent to maximizing the probability that the favorable event will occur. Also, the maximum expected reward will equal the maximum probability of the favorable event occurring. The following two examples illustrate how this idea may be used to solve some fairly complex problems.

EXAMPLE 3   Gambling Game

A gambler has $2. She is allowed to play a game of chance four times, and her goal is to maximize her probability of ending up with a least $6. If the gambler bets b dollars on a play of the game, then with probability .40, she wins the game and increases her capital position by b dollars; with probability .60, she loses the game and decreases her capital by b dollars. On any play of the game, the gambler may not bet more money than she has available. Determine a betting strategy that will maximize the gambler's probability of attaining a wealth of at least $6 by the end of the fourth game. We assume that bets of zero dollars (that is, not betting) are permissible.

Solution Define $f_t(d)$ to be the probability that by the end of game 4, the gambler will have at least $6, given that she acts optimally and has d dollars immediately before the game is played for the tth time. If we give the gambler a reward of 1 when her ending wealth is at least $6 and a reward of 0 if it is less, then $f_t(d)$ will equal the maximum expected reward that can

[†]This section covers topics that may be omitted with no loss of continuity.

be earned during games $t, t + 1, \ldots, 4$ if the gambler has d dollars immediately before the tth play of the game. As usual, we define $b_t(d)$ dollars to be a bet size that attains $f_t(d)$.

If the gambler is playing the game for the fourth and final time, her optimal strategy is clear: If she has \$6 or more, don't bet anything, but if she has less than \$6, bet enough money to ensure (if possible) that she will have \$6 if she wins the last game. Note that if she begins game 4 with \$0, \$1, or \$2, there is no way to win (no way to earn a reward of 1). This reasoning yields the following results:

$$f_4(0) = 0 \qquad b_4(0) = \$0$$
$$f_4(1) = 0 \qquad b_4(1) = \$0 \text{ or } \$1$$
$$f_4(2) = 0 \qquad b_4(2) = \$0, \$1, \text{ or } \$2$$
$$f_4(3) = .40 \qquad b_4(3) = \$3$$
$$f_4(4) = .40 \qquad b_4(4) = \$2, \$3, \text{ or } \$4$$
$$f_4(5) = .40 \qquad b_4(5) = \$1, \$2, \$3, \$4, \text{ or } \$5$$

For $d \geq 6$,

$$f_4(d) = 1 \qquad b_4(d) = \$0, \$1, \ldots, \$(d - 6)$$

For $t \leq 3$, we can find a recursion for $f_t(d)$ by noting that if the gambler has d dollars, is about to play the game for the tth time, and bets b dollars, then the following diagram summarizes what can occur:

$$\text{With probability .40 win game } t \qquad f_{t+1}(d + b)$$
$$\text{(Expected reward)}$$
$$\text{With probability .60 lose game } t \qquad f_{t+1}(d - b)$$

Thus, if the gambler has d dollars at the beginning of game t and bets b dollars, the expected reward (or expected probability of reaching \$6) will be $.4 f_{t+1}(d + b) + .6 f_{t+1}(d - b)$. This leads to the following recursion:

$$f_t(d) = \max_b (.4 f_{t+1}(d + b) + .6 f_{t+1}(d - b)) \qquad (4)$$

where b must be a member of $\{0, 1, \ldots, d\}$. Then $b_t(d)$ is any bet size that attains the maximum in (4) for $f_t(d)$. Using (4), we work backward until $f_1(2)$ has been determined.

Stage 3 Computations

$$f_3(0) = 0 \qquad b_3(0) = \$0$$

$$f_3(1) = \max \begin{cases} .4 f_4(1) + .6 f_4(1) = 0^* & \text{(Bet \$0)} \\ .4 f_4(2) + .6 f_4(0) = 0^* & \text{(Bet \$1)} \end{cases}$$

Thus, $f_3(1) = 0$, and $b_3(1) = \$0$ or \$1.

$$f_3(2) = \max \begin{cases} .4 f_4(2) + .6 f_4(2) = 0 & \text{(Bet \$0)} \\ .4 f_4(3) + .6 f_4(1) = .16^* & \text{(Bet \$1)} \\ .4 f_4(4) + .6 f_4(0) = .16^* & \text{(Bet \$2)} \end{cases}$$

Thus, $f_3(2) = .16$, and $b_3(2) = \$1$ or \$2.

$$f_3(3) = \max \begin{cases} .4 f_4(3) + .6 f_4(3) = .40^* & \text{(Bet \$0)} \\ .4 f_4(4) + .6 f_4(2) = .16 & \text{(Bet \$1)} \\ .4 f_4(5) + .6 f_4(1) = .16 & \text{(Bet \$2)} \\ .4 f_4(6) + .6 f_4(0) = .40^* & \text{(Bet \$3)} \end{cases}$$

Thus, $f_3(3) = .40$, and $b_3(3) = \$0$ or $\$3$.

$$f_3(4) = \max \begin{cases} .4f_4(4) + .6f_4(4) = .40^* & \text{(Bet \$0)} \\ .4f_4(5) + .6f_4(3) = .40^* & \text{(Bet \$1)} \\ .4f_4(6) + .6f_4(2) = .40^* & \text{(Bet \$2)} \\ .4f_4(7) + .6f_4(1) = .40^* & \text{(Bet \$3)} \\ .4f_4(8) + .6f_4(0) = .40^* & \text{(Bet \$4)} \end{cases}$$

Thus, $f_3(4) = .40$, and $b_3(4) = \$0, \$1, \$2, \$3$, or $\$4$.

$$f_3(5) = \max \begin{cases} .4f_4(5) + .6f_4(5) = .40 & \text{(Bet \$0)} \\ .4f_4(6) + .6f_4(4) = .64^* & \text{(Bet \$1)} \\ .4f_4(7) + .6f_4(3) = .64^* & \text{(Bet \$2)} \\ .4f_4(8) + .6f_4(2) = .40 & \text{(Bet \$3)} \\ .4f_4(9) + .6f_4(1) = .40 & \text{(Bet \$4)} \\ .4f_4(10) + .6f_4(0) = .40 & \text{(Bet \$5)} \end{cases}$$

Thus, $f_3(5) = .64$, and $b_3(5) = \$1$ or $\$2$. For $d \geq 6$, $f_3(d) = 1$, and $b_3(d) = \$0, \$1, \ldots,$ $\$(d - 6)$.

Stage 2 Computations

$$f_2(0) = 0 \qquad b_2(0) = \$0$$

$$f_2(1) = \max \begin{cases} .4f_3(1) + .6f_3(1) = 0 & \text{(Bet \$0)} \\ .4f_3(2) + .6f_3(0) = .064^* & \text{(Bet \$1)} \end{cases}$$

Thus, $f_2(1) = .064$, and $b_2(1) = \$1$.

$$f_2(2) = \max \begin{cases} .4f_3(2) + .6f_3(2) = .16^* & \text{(Bet \$0)} \\ .4f_3(3) + .6f_3(1) = .16^* & \text{(Bet \$1)} \\ .4f_3(4) + .6f_3(0) = .16^* & \text{(Bet \$2)} \end{cases}$$

Thus, $f_2(2) = .16$, and $b_2(2) = \$0, \$1$, or $\$2$.

$$f_2(3) = \max \begin{cases} .4f_3(3) + .6f_3(3) = .40^* & \text{(Bet \$0)} \\ .4f_3(4) + .6f_3(2) = .256 & \text{(Bet \$1)} \\ .4f_3(5) + .6f_3(1) = .256 & \text{(Bet \$2)} \\ .4f_3(6) + .6f_3(0) = .40^* & \text{(Bet \$3)} \end{cases}$$

Thus, $f_2(3) = .40$, and $b_2(3) = \$0$ or $\$3$.

$$f_2(4) = \max \begin{cases} .4f_3(4) + .6f_3(4) = .40 & \text{(Bet \$0)} \\ .4f_3(5) + .6f_3(3) = .496^* & \text{(Bet \$1)} \\ .4f_3(6) + .6f_3(2) = .496^* & \text{(Bet \$2)} \\ .4f_3(7) + .6f_3(1) = .40 & \text{(Bet \$3)} \\ .4f_3(8) + .6f_3(0) = .40 & \text{(Bet \$4)} \end{cases}$$

Thus, $f_2(4) = .496$, and $b_2(4) = \$1$ or $\$2$.

$$f_2(5) = \max \begin{cases} .4f_3(5) + .6f_3(5) = .64^* & \text{(Bet \$0)} \\ .4f_3(6) + .6f_3(4) = .64^* & \text{(Bet \$1)} \\ .4f_3(7) + .6f_3(3) = .64^* & \text{(Bet \$2)} \\ .4f_3(8) + .6f_3(2) = .496 & \text{(Bet \$3)} \\ .4f_3(9) + .6f_3(1) = .40 & \text{(Bet \$4)} \\ .4f_3(10) + .6f_3(0) = .40 & \text{(Bet \$5)} \end{cases}$$

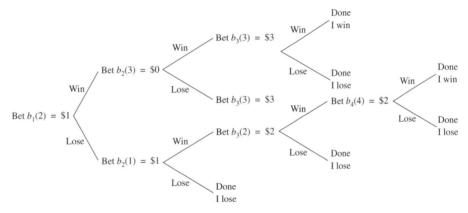

FIGURE 1
Ways Gambler Can
Reach $6

Thus, $f_2(5) = .64$, and $b_2(5) = \$0, \$1$, or $\$2$. For $d \geq 6$, $f_2(d) = 1$ and $b_2(d) = \$0$, $\$1, \ldots, \$(d - 6)$.

Stage 1 Computations

$$f_1(2) = \max \begin{cases} .4f_2(2) + .6f_2(2) = .16 & (\text{Bet } \$0) \\ .4f_2(3) + .6f_2(1) = .1984^* & (\text{Bet } \$1) \\ .4f_2(4) + .6f_2(0) = .1984^* & (\text{Bet } \$2) \end{cases}$$

Thus, $f_1(2) = .1984$, and $b_1(2) = \$1$ or $\$2$. Hence, the gambler has a .1984 chance of reaching $6. Suppose the gambler begins by betting $b_1(2) = \$1$. Then Figure 1 indicates the various possibilities that can occur. By following the strategy outlined in the figure, the gambler can reach her goal of $6 in two different ways. First, she can win game 1 and game 3. This will occur with probability $(.4)^2 = .16$. Second, the gambler can win if she loses the first game but wins the next three games. This will occur with probability $.6(.4)^3 = .0384$. Hence, the gambler's probability of reaching $6 is $.16 + .0384 = .1984 = f_1(2)$.

EXAMPLE 4   Tennis Serves

Martina McEnroe has two types of serves: a hard serve (H) and a soft serve (S).[†] The probability that Martina's hard serve will land in bounds is p_H, and the probability that her soft serve will land in bounds is p_S. If Martina's hard serve lands in bounds, there is a probability w_H that Martina will win the point. If Martina's soft serve lands in bounds, there is a probability w_S that Martina will win the point. We assume that $p_H < p_S$ and $w_H > w_S$. Martina's goal is to maximize the probability of winning a point on which she serves. Use dynamic programming to help Martina select an optimal serving strategy. Remember that if both serves are out of bounds, Martina loses the point.

Solution   To maximize Martina's probability of winning the point, we give her a reward of 1 if she wins the point and a reward of 0 if she loses the point. We also define $f_t(t = 1, 2)$ to be the probability that Martina wins a point if she plays optimally and is about to take her tth serve. To determine the optimal serving strategy, we work backward, beginning with f_2. If Martina serves hard on the second serve, she will win the point (and earn a reward

[†]Based on material by E. V. Denardo, personal communication.

of 1) with probability $p_H w_H$. Similarly, if she serves soft on the second serve, her expected reward is $p_S w_S$. Thus, we have

$$f_2 = \max \begin{cases} p_H w_H & \text{(Serve hard)} \\ p_S w_S & \text{(Serve soft)} \end{cases}$$

For the moment, let's assume that

$$p_S w_S > p_H w_H \tag{5}$$

If (5) holds, then Martina should serve soft on the second serve. In this situation, $f_2 = p_S w_S$.

To determine f_1, we need to look at what happens on the first serve. If Martina serves hard on the first serve, the events in Table 6 can occur, and Martina earns an expected reward of $p_H w_H + (1 - p_H) f_2$. If Martina serves soft on the first serve, then the events in Table 7 can occur, and Martina's expected reward is $p_S w_S + (1 - p_S) f_2$. We now write the following recursion for f_1:

$$f_1 = \max \begin{cases} p_H w_H + (1 - p_H) f_2 & \text{(Serve hard)} \\ p_S w_S + (1 - p_S) f_2 & \text{(Serve soft)} \end{cases}$$

From this equation, we see that Martina should serve hard on the first serve if

$$p_H w_H + (1 - p_H) f_2 \geq p_S w_S + (1 - p_S) f_2 \tag{6}$$

(If (6) is not satisfied, Martina should serve soft on the first serve.)

Continuing with the assumption that $p_S w_S > w_H p_H$ (which implies that $f_2 = p_S w_S$), we may substitute $f_2 = p_S w_S$ into (6) to obtain the result that Martina should serve hard on the first serve if

$$p_H w_H + (1 - p_H) p_S w_S \geq p_S w_S + (1 - p_S) p_S w_S$$

TABLE 6
Computation of Expected Reward If First Serve Is Hard

Event	Probability of Event	Expected Reward for Given Event
First serve in and Martina wins point	$p_H w_H$	1
First serve in and Martina loses point	$p_H (1 - w_H)$	0
First serve out of bounds	$1 - p_H$	f_2

TABLE 7
Computation of Expected Reward If First Serve Is Soft

Event	Probability of Event	Expected Reward for Given Event
First serve in and Martina wins point	$p_s w_s$	1
First serve in and Martina loses point	$p_s (1 - w_s)$	0
First serve out of bounds	$1 - p_s$	f_2

or

$$p_H w_H \geq p_S w_S (1 + p_H - p_S) \tag{7}$$

For example, if $p_H = .60$, $p_S = .90$, $w_H = .55$, and $w_S = .50$, then (5) and (7) are both satisfied, and Martina should serve hard on her first serve and soft on her second serve. On the other hand, if $p_H = .25$, $p_S = .80$, $w_H = .60$, and $w_S = .45$, then both serves should be soft. The reason for this is that in this case, the hard serve's advantage from the fact that w_H exceeds w_S is outweighed by the fact that a hard serve on the first serve greatly increases the chances of a double fault.

To complete our analysis, we must consider the situation where (5) does not hold. We now show that if

$$p_H w_H \geq p_S w_S \tag{8}$$

Martina should serve hard on both serves. Note that if (8) holds, then $f_2 = \max \{p_H w_H, p_S w_S\} = p_H w_H$, and Martina should serve hard on the second serve. Now (6) implies that Martina should serve hard on the first serve if

$$p_H w_H + (1 - p_H) p_H w_H \geq p_S w_S + (1 - p_S) p_H w_H$$

Upon rearrangement, the last inequality becomes

$$p_H w_H (1 + p_S - p_H) \geq p_S w_S$$

Dividing both sides of the last inequality by $p_S w_S$ shows that Martina should serve hard on the first serve if

$$\frac{p_H w_H}{p_S w_S} (1 + p_S - p_H) \geq 1$$

After noting that $p_H w_H \geq p_S w_S$ and $(1 + p_S - p_H) > 1$ (because $p_S > p_H$), we see that the last inequality holds. Thus, we have shown that if $p_H w_H \geq p_S w_S$, Martina should serve hard on both serves. This is reasonable, because if it is optimal to serve hard on the second (and this requires $p_H w_H \geq p_S w_S$), then it should be optimal to serve hard on the first serve, because the danger of double-faulting (which is the drawback to the hard serve) is less immediate on the first serve. Of course, Example 4 could have been solved using a decision tree; see Problem 10 of Section 2.4.

In our solution to Example 4, we have shown how Martina's optimal strategy depends on the values of the parameters defining the problem. This is a kind of sensitivity analysis like the one applied to linear programming problems in Chapters 5 and 6 of Volume 1.

PROBLEMS

Group A

1 Vladimir Ulanowsky is playing Keith Smithson in a two-game chess match. Winning a game scores 1 match point, and drawing a game scores $\frac{1}{2}$ match point. After the two games are played, the player with more match points is declared the champion. If the two players are tied after two games, they continue playing until someone wins a game (the winner of that game will be the champion). During each game, Ulanowsky can play one of two ways: boldly or conservatively. If he plays boldly, he has a 45% chance of winning the game and a 55% chance of losing the game. If

he plays conservatively, he has a 90% chance of drawing the game and a 10% chance of losing the game. Ulanowsky's goal is to maximize his probability of winning the match. Use dynamic programming to help him accomplish this goal. If this problem is solved correctly, even though Ulanowsky is the inferior player, his chance of winning the match is over $\frac{1}{2}$. Explain this anomalous result.

2 Dickie Hustler has \$2 and is going to toss an unfair coin (probability .4 of heads) three times. Before each toss, he can bet any amount of money (up to what he now has). If heads comes up, Dickie wins the number of dollars he bets; if tails comes up, he loses the number of dollars he bets. Use dynamic programming to determine a strategy that maximizes Dickie's probability of having at least $5 after the third coin toss.

Group B

3 Supppose that Army trails by 14 points in the Army–Navy football game. Army's guardian angel has assured the Army coach that his team will have the ball two more times during the game and will score a touchdown (worth 6 points) each time it has the ball. The Army coach has also been assured

that Navy will not score any more points. Suppose a win is assigned a value of 1, a tie is .3, and a loss is 0. Army's problem is to determine whether to go for 1 or 2 points after each touchdown. A 1-point conversion is always successful, and a 2-point conversion is successful only 40% of the time. The Army coach wants to maximize the expected reward earned from the outcome of the game. Use dynamic programming to determine an optimal strategy. Then prove the following result: *No matter what value is assigned to a tie, it is never optimal to use the following strategy: Go for a 1-point conversion after the first touchdown and go for a 2-point conversion after the second touchdown.* Note that this (suboptimal) strategy is the one most coaches follow!

7.4 Further Examples of Probabilistic Dynamic Programming Formulations

Many probabilistic dynamic programming problems can be solved using recursions of the following form (for max problems):

$$f_t(i) = \max_a \left\{ (\text{expected reward during stage } t|i, a) + \sum_j p(j|i, a, t)f_{t+1}(j) \right\} \qquad (9)$$

In (9), $f_t(i)$ is the maximum expected reward that can be earned during stages $t, t + 1, \ldots$ end of the problem, given that the state at the beginning of stage t is i. The max in (9) is taken over all actions a that are feasible when the state at the beginning of stage t is i. In (9), $p(j|i, a, t)$ is the probability that the next period's state will be j, given that the current (stage t) state is i and action a is chosen. Hence, the summation in (9) represents the expected reward from stage $t + 1$ to the end of the problem. By choosing a to maximize the right-hand side of (9), we are choosing a to maximize the expected reward earned from stage t to the end of the problem, and this is what we want to do. The following are six examples of probabilistic dynamic programming formulations.

EXAMPLE 5 Sunco Oil Drilling

Sunco Oil has D dollars to allocate for drilling at sites $1, 2, \ldots, T$. If x dollars are allocated to site t, the probability is $q_t(x)$ that oil will be found on site t. Sunco estimates that if site t has any oil, it is worth r_t dollars. Formulate a recursion that could be used to enable Sunco to maximize the expected value of all oil found on sites $1, 2, \ldots, T$.

Solution This is a typical resource allocation problem (see Example 1). Therefore, the stage should represent the number of sites, the decision for site t is how many dollars to allocate to site t, and the state is the number of dollars available to allocate to sites $t, t + 1, \ldots, T$. We therefore define $f_t(d)$ to be the maximum expected value of the oil that can be found on sites $t, t + 1, \ldots, T$ if d dollars are available to allocate to sites $t, t + 1, \ldots, T$.

We make the reasonable assumption that $q_T(x)$ is a nondecreasing function of x. If this is the case, then at stage T, all the money should be allocated to site T. This yields

$$f_T(d) = r_T q_T(d) + (1 - q_T(d))0 = r_T q_T(d)$$

For $t < T$,

$$f_t(d) = \max_x \{ r_t q_t(x) + f_{t+1}(d - x) \}$$

where x must satisfy $0 \le x \le d$. The last recursion follows, because $r_t q_t(x)$ is the expected value of the reward for stage t, and since Sunco will have $d - x$ dollars available for sites

$t + 1, t + 2, \ldots, T, f_{t+1}(d - x)$ is the expected value of the oil that can be found by optimally drilling at sites $t + 1, t + 2, \ldots, T$. To solve the problem, we would work backward until $f_1(D)$ had been determined.

EXAMPLE 6   **Bass Fishing**

Each year, the owner of a lake must determine how many bass to capture and sell. During year t, a price p_t will be received for each bass that is caught. If the lake contains b bass at the beginning of year t, the cost of capturing x bass is $c_t(x|b)$. Between the time that year t's bass are caught and year $t + 1$ begins, the bass in the lake multiply by a random factor $\mathbf{D}$, where $P(\mathbf{D} = d) = q(d)$.

Formulate a dynamic programming recursion that can be used to determine a bass-catching strategy that will maximize the owner's net profit over the next ten years. At present, the lake contains 10,000 bass.

Solution   As in Example 8 of Chapter 6, the stage is the year, the state is the number of bass in the lake at the beginning of the year, and the decision is how many bass to catch during each year. We define $f_t(b)$ to be the maximum expected net profit that can be earned during the years $t, t + 1, \ldots, 10$ if the lake contains b bass at the beginning of year t. Then

$$f_{10}(b) = \max_x \{xp_{10} - c_{10}(x|b)\}$$

where $0 \leq x \leq b$, and for $t < 10$

$$f_t(b) = \max_x \left\{ xp_t - c_t(x|b) + \sum_d q(d) f_{t+1}(d(b - x)) \right\}$$

In this recursion, x must satisfy $0 \leq x \leq b$. To justify the recursion for $t < 10$, first note that the profits during year t are (with certainty) $xp_t - c_t(x|b)$. Then with probability $q(d)$, year $t + 1$'s state will be $d(b - x)$. It then follows that if x bass are caught during year t, the maximum expected net profit that can be earned during the years $t + 1, t + 2, \ldots,$ 10 will be

$$\sum_d q(d) f_{t+1}(d(b - x))$$

Hence, the recursion chooses the number of bass during year t to maximize the sum of year t profits and future profits. To use this recursion, we work backward until $f_1(10,000)$ is computed. Then, after the number of bass in the lake at the beginning of year t is observed, we use the recursion to determine the number of bass that should be caught during year t.

EXAMPLE 7   **Waiting in Line**

When Sally Mutton arrives at the bank, 30 minutes remain on her lunch break. If Sally makes it to the head of the line and enters service before the end of her lunch break, she earns reward r. However, Sally does not enjoy waiting in lines, so to reflect her dislike for waiting in line, she incurs a cost c for each minute she waits. During a minute in which n people are ahead of Sally, there is a probability $p(x|n)$ that x people will complete their transactions. Suppose that when Sally arrives, 20 people are ahead of her in line. Use dynamic programming to determine a strategy for Sally that will maximize her expected net revenue (reward − waiting costs).

Solution   When Sally arrives at the bank, she must decide whether to join the line or to give up and leave. At any later time, she may also decide to leave if it is unlikely that she will be served by the end of her lunch break. If 1 minute remained, Sally's decision would be simple: She should stay in line if and only if her expected reward exceeds the cost of waiting for 1 minute (c). Then we can work backward to a problem with 2 minutes left, and so on. We define $f_t(n)$ to be the maximum expected net reward that Sally can receive from time t to the end of her lunch break if at time t, n people are ahead of her. We let $t = 0$ be the present and $t = 30$ be the end of the problem. Since $t = 29$ is the beginning of the last minute of the problem, we write

$$f_{29}(n) = \max \begin{cases} 0 & \text{(Leave)} \\ rp(n|n) - c & \text{(Stay)} \end{cases}$$

This follows because if Sally chooses to leave at time 29, she earns no reward and incurs no more costs. On the other hand, if she stays at time 29, she will incur a waiting cost of c (a revenue of $-c$) and with probability $p(n|n)$ will enter service and receive a reward r. Thus, if Sally stays, her expected net reward is $rp(n|n) - c$.

For $t < 29$, we write

$$f_t(n) = \max \begin{cases} 0 & \text{(Leave)} \\ rp(n|n) - c + \sum_{k < n} p(k|n)f_{t+1}(n - k) & \text{(Stay)} \end{cases}$$

The last recursion follows, because if Sally stays, she will earn an expected reward (as in the $t = 29$ case) of $rp(n|n) - c$ during the current minute, and with probability $p(k|n)$, there will be $n - k$ people ahead of her; in this case, her expected net reward from time $t + 1$ to time 30 will be $f_{t+1}(n - k)$. If Sally stays, her overall expected reward received from time $t + 1, t + 2, \ldots, 30$ will be

$$\sum_{k < n} p(k|n)f_{t+1}(n - k)$$

Of course, if n people complete their transactions during the current minute, the problem ends, and Sally's future net revenue will be zero.

To determine Sally's optimal waiting policy, we work backward until $f_0(20)$ is computed. If $f_0(20)$ is attained by "stay," Sally stays and sees how many people are ahead of her at time 1. She continues to stay until a situation arises for which the optimal action is "leave" or she begins to be served. In either case, the problem terminates.

Problems in which the decision maker can terminate the problem by choosing a particular action are known as **stopping rule problems;** they often have a special structure that simplifies the determination of optimal policies. See Ross (1983) for more information on stopping rule problems.

| EXAMPLE 8 | *Cash Management Policy* |

E. J. Korvair Department Store is trying to determine an optimal cash management policy. During each day, the demand for cash may be described by a random variable **D**, where $p(\mathbf{D} = d) = p(d)$. At the beginning of each day, the store sends an employee to the bank to deposit or withdraw funds. Each bank transaction costs K dollars. Then E. J.'s demand for cash is met by cash left from the previous day plus money withdrawn (or minus money deposited). At the end of the day, the store determines its cash balance at the store. If the cash balance is negative, a shortage cost of s dollars per dollar short is incurred. If the ending balance is positive, a cost of i dollars per dollar held is incurred (because of loss of interest that could have been earned by depositing cash in the bank). At the beginning of day 1, the store has $10,000 cash on hand and a bank balance of $100,000. Formulate a dynamic programming model that can be used to minimize the expected cost of filling the store's cash needs for the next 30 days.

Solution To determine how much money should be withdrawn or deposited, E. J. needs to know its cash on hand and bank balance at the beginning of the day. As usual, we let time be the stage. At the beginning of each stage (or day), E. J. must decide how much to withdraw from or deposit in the bank. We let $f_t(c, b)$ be the minimum expected cost incurred by the store during days $t, t + 1, \ldots, 30$, given that at the beginning of day t, the store has c dollars cash at the store and b dollars in the bank.

We observe that

$$f_{30}(c, b) = \min_x \left\{ K\delta(x) + \sum_{d \le c+x} p(d)(c + x - d)i + \sum_{d \ge c+x} p(d)(d - c - x)s \right\} \quad (10)$$

Here, x is the amount of money transferred from the bank to the store (if $x < 0$ money is transferred from the store to the bank). Since the store cannot withdraw more than b dollars from the bank or deposit more than c dollars in the bank, x must satisfy $b \ge x \ge -c$. Also, in (10), $\delta(0) = 0$ and $\delta(x) = 1$ for $x \ne 0$. In short, $K\delta(x)$ picks up the transaction cost (if there is a transaction). If $d \le c + x$, the store will end the day with $c + x - d$ dollars, so a cost of $i(c + x - d)$ is incurred (because of lost interest). Since this occurs with probability $p(d)$, the first sum in (10) represents the expected interest costs incurred during day 30. Also note that if $d \ge c + x$, the store will be $d - c - x$ dollars short, and a shortage cost of $s(d - c - x)$ will be incurred. Again, this cost is incurred with probability $p(d)$. Hence, the second sum in (10) is the expected shortage cost incurred during day 30.

For $t < 30$, we write

$$f_t(c, b) = \min_x \left\{ K\delta(x) + \sum_{d \le c+x} p(d)(c + x - d)i \right.$$

$$\left. + \sum_{d \ge c+x} p(d)(d - c - x)s + \sum_d p(d)f_{t+1}(c + x - d, b - x) \right\} \quad (11)$$

As in (10), x must satisfy $b \ge x \ge -c$. Also, the term $K\delta(x)$ and the first two summations yield the expected cost incurred during day t. If day t demand is d, then at the beginning of day $t + 1$, the store will have $c + x - d$ dollars cash on hand and a bank balance of $b - x$. Thus, with probability $p(d)$, the store's expected cost during days $t + 1$, $t + 2, \ldots, 30$ will be $f_{t+1}(c + x - d, b - x)$. Weighting $f_{t+1}(c + x - d, b - x)$ by the probability that day t demand will be d, we see that the last sum in (11) is the expected cost incurred during days $t + 1, t + 2, \ldots, 30$. Hence, (11) is correct. To determine the optimal cash management policy, we would use (10) and (11) to work backward until $f_1(10,000, 100,000)$ has been computed.

EXAMPLE 9　　**Parking Spaces**

Robert Blue is trying to find a parking place near his favorite restaurant. He is approaching the restaurant from the west, and his goal is to park as nearby as possible. The available parking places are pictured in Figure 2. Robert is nearsighted and cannot see ahead; he can only see whether the space he is at now is empty. When Robert arrives at an empty space, he must decide whether to park there or to continue to look for a closer space. Once he passes a space, he cannot return to it. Robert estimates that the probability that space t is empty is p_t. If he does not end up with a parking space, he is embarrassed and incurs a cost M (M is a big positive number). If he does park in space t, he incurs a cost $|t|$. Show how Robert can use dynamic programming to develop a parking strategy that minimizes his expected cost.

Solution　　If Robert is at space T, his problem is easy to solve: park in space T if it is empty; otherwise, incur a cost of M. Then Robert can work backward until he determines what to do at space $-T$. For this reason, we let the space Robert is at represent the stage. In order to make a decision at any stage, all Robert must know is whether or not the space is empty (if a space is not empty, he must continue). Thus, the state at any stage is whether

FIGURE 2
Location of
Parking Places

$$\rightarrow \boxed{-T}\;\boxed{1-T}\;\boxed{2-T}\;\cdots\;\boxed{-2}\;\boxed{-1}\;\boxed{0}\;\boxed{1}\;\boxed{2}\;\cdots\;\boxed{T}$$

$$0 = \text{Restaurant}$$

or not the space is empty. Of course, if the space is empty, Robert's decision is whether to take the space or to continue.

We define

$f_t(o)$ = minimum expected cost if Robert is at space t and space t is occupied

$f_t(e)$ = minimum expected cost if Robert is at space t and space t is empty

If Robert is at space T, he will park in the space if it is empty (incurring a cost T) or incur a cost M if the space is occupied. Thus, we have $f_T(o) = M$ and $f_T(e) = T$.

For $t < T$, we write

$$f_t(o) = p_{t+1}f_{t+1}(e) + (1 - p_{t+1})f_{t+1}(o) \tag{12}$$

$$f_t(e) = \min \begin{cases} |t| & \text{(Take space } t\text{)} \\ p_{t+1}f_{t+1}(e) + (1 - p_{t+1})f_{t+1}(o) & \text{(Don't take space } t\text{)} \end{cases} \tag{13}$$

To justify (12), note that if space t is occupied, Robert must next look at space $t + 1$. With probability p_{t+1}, space $t + 1$ will be empty; in this case, Robert's expected cost will be $f_{t+1}(e)$. Similarly, with probability $(1 - p_{t+1})$, space $t + 1$ will be occupied, and Robert will incur an expected cost of $f_{t+1}(o)$. Thus, Robert's expected cost is

$$p_{t+1}f_{t+1}(e) + (1 - p_{t+1})f_{t+1}(o)$$

To justify (13), note that Robert can either take space t (incurring a cost of $|t|$) or continue. Thus, if Robert continues, his expected cost will be

$$p_{t+1}f_{t+1}(e) + (1 - p_{t+1})f_{t+1}(o)$$

Since Robert wants to minimize his expected cost, (13) follows. By using (12) and (13), Robert can work backward to compute $f_{-T}(e)$ and $f_{-T}(o)$. Then he will continue until he reaches an empty space at some location t for which the minimum in (13) is attained by taking space t. If no such empty space is reached, Robert will not find a space, and he will incur a cost M.

EXAMPLE 10 Safecracker

During month $t(t = 1, 2, \ldots, 60)$, expert safecracker Dirk Stack knows that he will be offered a role in a bank job that will pay him d_t dollars. There is, however, a probability p_t that month t's job will result in his capture. If Dirk is captured, all his money will be lost. Dirk's goal is to maximize his expected asset position at the end of month 60. Formulate a dynamic programming recursion that will help Dirk accomplish his goal. At the beginning of month 1, Dirk has $50,000.

Solution At the beginning of month 60, Dirk has no future to consider and his problem is easy to solve, so we let time represent the stage. At the beginning of each month, Dirk must decide whether or not to take the current month's job offer. In order to make this decision, Dirk must know how much money he has at the beginning of the month. We define $f_t(d)$ to be Dirk's maximum expected asset position at the end of month 60, given that at the beginning of month t, Dirk has d dollars. Then

$$f_{60}(d) = \max \begin{cases} p_{60}(0) + (1 - p_{60})(d + d_{60}) & \text{(Accept month 60 job)} \\ d & \text{(Reject month 60 job)} \end{cases}$$

This result follows, because if Dirk takes the job during month 60, there is a probability p_{60} that he will be caught and end up with zero dollars and a probability $(1 - p_{60})$ that he will not be caught and end up with $d + d_{60}$ dollars. Of course, if Dirk does not take the month 60 job, he ends month 60 with d dollars.

Extending this reasoning yields, for $t < 60$,

$$f_t(d) = \max \begin{cases} p_t(0) + (1 - p_t)f_{t+1}(d + d_t) & \text{(Accept month } t \text{ job)} \\ f_{t+1}(d) & \text{(Reject month } t \text{ job)} \end{cases}$$

Note that if Dirk accepts month t's job, there is a probability p_t that he will be caught (and end up with zero) and a probability $(1 - p_t)$ that he will successfully complete month t's job and earn d_t dollars. In this case, Dirk will begin month $t + 1$ with $d + d_t$ dollars, and his expected final cash position will be $f_{t+1}(d + d_t)$. Of course, if Dirk rejects the month t job, he begins month $t + 1$ with d dollars, and his expected final cash position will be $f_{t+1}(d)$. Since Dirk wants to maximize his expected cash position at the end of month 60, the recursion follows. By using the recursion, Dirk can work backward to compute $f_1(50,000)$. Then he can decide whether to accept the month 1 job. Assuming he has not been caught, he can then determine whether to accept the month 2 job, and so on.

As described in Section 6.8, spreadsheets can be used to solve dynamic programming recursions. See Problems 14 and 15 for some examples of how spreadsheets can be used to solve PDPs.

PROBLEMS

Group A

1 The space shuttle is about to go up on another flight. With probability $p_t(z)$, it will use z type t fuel cells during the flight. The shuttle has room for at most W fuel cells. If at any time during the flight, all the type t fuel cells burn out, a cost c_t will be incurred. Assuming the goal is to minimize the expected cost due to fuel cell shortages, set up a dynamic programming model that could be used to determine how to stock the space shuttle with fuel cells. There are T different types of fuel cells.

2 At the beginning of each year, a firm observes its asset position (call it d) and may invest any amount x ($0 \le x \le d$) in a risky investment. During each year, the money invested doubles with probability p and is completely lost with probability $1 - p$. Independently of this investment, the firm's asset position increases by an amount y with probability q_y (y may be negative). If the firm's asset position is negative at the beginning of a year, it cannot invest any money during that year. The firm initially has \$10,000 in assets and wants to maximize its expected asset position ten years from now. Formulate a dynamic programming recursion that will help accomplish this goal.

3 Consider a machine that may be in any one of the states $0, 1, 2, \ldots$. At the beginning of each month, the state of the machine is observed, and it is decided whether to replace or keep the machine. If the machine is replaced, a new state 0 machine arrives instantaneously. It costs R dollars to replace

a machine. Each month that a state i machine is in operation, a maintenance cost of $c(i)$ is incurred. If a machine is in state i at the beginning of a month, then with probability p_{ij}, the machine will begin the next month in state j. At the beginning of the first month, we own a state i_0 machine. Assuming that the interest rate is 12% per year, formulate a dynamic programming recursion that could be used to minimize the expected discounted cost incurred during the next T months. Note that if we replace a machine at the beginning of a month, we incur a maintenance cost of $c(0)$ during the month, and with probability p_{0i}, we begin the next month with a state i machine.

4 In the time interval between t and $t - 1$ seconds before the departure of Braneast Airlines Flight 313, there is a probability p_t that the airline will receive a reservation for the flight and a probability $1 - p_t$ that the airline will receive no reservation. The flight can seat up to 100 passengers. At departure time, if r reservations have been accepted by the airline, there is a probability $q(y|r)$ that y passengers will show up for the flight. Each passenger who boards the flight adds \$500 to Braneast's revenues, but each passenger who shows up for the flight and cannot be seated receives \$200 in compensation. Formulate a dynamic programming recursion to enable the airline to maximize its expected revenue from Flight 313. Assume that no reservations are received more than 100,000 seconds before flight time.

5 At the beginning of each week, a machine is either running or broken down. If the machine runs throughout the week, it earns revenues of $100. If the machine breaks down during a week, it earns no revenue for that week. If the machine is running at the beginning of the week, we may perform maintenance on it to lessen the chance of a breakdown. If the maintenance is performed, a running machine has a .4 chance of breaking down during the week; if maintenance is not performed, a running machine has a .7 chance of breaking down during the week. Maintenance costs $20 per week. If the machine is broken down at the beginning of the week, it must be replaced or repaired. Both repair and replacement occur instantaneously. Repairing a machine costs $40, and there is a .4 chance that the repaired machine will break down during the week. Replacing a broken machine costs $90, but the new machine is guaranteed to run throughout the next week of operation. Use dynamic programming to determine a repair, replacement, and maintenance policy that maximizes the expected net profit earned over a four-week period. Assume that the machine is running at the beginning of the first week.

6 I own a single share of Wivco stock. I must sell my share at the beginning of one of the next 30 days. Each day, the price of the stock changes. With probability $q(x)$, the price tomorrow will increase by $x\%$ over today's stock price (x can be negative). For example, with probability $q(5)$, tomorrow's stock price will be 5% higher than today's. Show how dynamic programming can be used to determine a strategy that maximizes the expected revenue earned from selling the share of Wivco stock. Assume that at the beginning of the first day, the stock sells for $10 per share.

Group B

7 The National Cat Foundling Home encourages people to adopt its cats, but (because of limited funds) it allows each prospective owner to inspect only four cats before choosing one of them to take home. Ten-year-old Sara is eager to adopt a cat and agrees to abide by the following rules. A randomly selected cat is brought for Sara to see, and then Sara must either choose the cat or reject it. If the first cat is rejected, Sara sees another randomly selected cat and must accept or reject it. This procedure continues until Sara has selected her cat. Once Sara rejects a cat, she cannot go back later and choose it as her pet. Determine a strategy for Sara that will maximize her probability of ending up with the cat she actually prefers.

8 Consider the following probabilistic inventory model:

a At the beginning of each period, a firm observes its inventory position.

b Then the firm decides how many units to produce during the current period. It costs $c(x)$ dollars to produce x units during a period.

c With probability $q(d)$, d units are demanded during the period. From units on hand (including the current period's production), the firm satisfies as much of the demand as possible. The firm receives r dollars for each unit sold. For each unit of demand that is unsatisfied, a penalty cost p is incurred. All unsatisfied demand is assumed to be lost. For example, if the firm has 20 units available and current demand is 30, a revenue of $20r$

would be received, and a penalty of $10p$ would be incurred.

d If ending inventory is positive, a holding cost of $1 per unit is incurred.

e The next period now begins.

The firm's inital inventory is zero, and its goal is to minimize the expected cost over a 100-period horizon. Formulate a dynamic programming recursion that will help the firm accomplish its goal.

9 Martha and Ken Allen want to sell their house. At the beginning of each day, they receive an offer. We assume that from day to day, the sizes of the offers are independent random variables and that the probability that a given day's offer is for j dollars is p_j. An offer may be accepted during the day it is made or at any later date. For each day the house remains unsold, a maintenance cost of c dollars is incurred. The house must be sold within 30 days. Formulate a dynamic programming recursion that Martha and Ken can use to maximize their expected net profit (selling price − maintenance cost). Assume that the maintenance cost for a day is incurred before the current day's offer is received and that each offer is for an integer number of dollars.

10 An advertising firm has D dollars to spend on reaching customers in T separate markets. Market t consists of k_t people. If x dollars are spent on advertising in market t, the probability that a given person in market t will be reached is $p_t(x)$. Each person in market t who is reached will buy c_t units of the product. A person who is not reached will not buy any of the product. Formulate a dynamic programming recursion that could be used to maximize the expected number of units sold in T markets.

11 Georgia Stein is the new owner of the New York Yankees. Each season, Georgia must decide how much money to spend on the free agent draft. During each season, Georgia can spend any amount of money on free agents up to the team's capital position at the beginning of the season. If the Yankees finish in ith place during the season, their capital position increases by $R(i)$ dollars less the amount of money spent in the free agent draft. If the Yankees finished in ith place last season and spend d dollars on free agents during the off-season, the probability that the Yankees will finish in place j during the next season is $p_{ij}(d)(j = 1, 2, \ldots, 7)$. Last season, the Yankees finished in first place, and at the end of the season, they had a capital position of D dollars. Formulate a dynamic programming recursion that will enable the Yankees to maximize their expected cash position at the end of T seasons.

12 Bailey Bliss is the campaign manager for Walter Glenn's presidential campaign. He has D dollars to allocate to T winner-take-all primaries. If x_t dollars are allocated to primary t, then with probability $p_t(x_t)$, Glenn will win primary t and obtain v_t delegates. With probability $1 - p_t(x_t)$, Glenn loses primary t and obtains no delegates. Glenn needs K delegates to be nominated. Use dynamic programming to help Bliss maximize Glenn's probability of being nominated. What aspect of a real campaign does the present formulation ignore?

13 At 7 A.M., eight people leave their cars for repair at Harry's Auto Repair Shop. If person i's car is ready by time

t (7 A.M. = time 0, and so on), he will pay Harry $r_i(t)$ dollars. For example, if person 2's car must be ready by 2 P.M., we may have $r_2(8) = 0$. Harry estimates that with probability $p_i(t)$, it will take t hours to repair person i's car. Formulate a dynamic programming recursion that will enable Harry to maximize his expected revenue for the day. His workday ends at 5 P.M. = time 10.

14 In Example 10 suppose $p_t = t/60$ and $d_t = t$. Using a spreadsheet, solve for Dirk's optimal strategy. (*Hint:* The possible states are 50, 51, ..., 1,880 (thousands).)

15 In Example 9, assume $T = 10$ and $p_t = |t|/10$. Using a spreadsheet, solve for Robert's optimal strategy.

7.5 Markov Decision Processes[†]

To use dynamic programming in a problem for which the stage is represented by time, one must determine the value of T, the number of time periods over which expected revenue or expected profit is maximized (or expected costs are minimized). T is referred to as the **horizon length.** For instance, in the equipment replacement problem of Section 6.5, if our goal is to minimize costs over a 30-year period, then $T = 30$. Of course, it may be difficult for a decision maker to determine exactly the most suitable horizon length. In fact, when a decision maker is facing a long horizon and is not sure of the horizon length, it is more convenient to assume that the horizon length is infinite.

Suppose a decision maker's goal is to maximize the expected reward earned over an infinite horizon. In many situations, the expected reward earned over an infinite horizon may be unbounded. For example, if for any state and decision, the reward earned during a period is at least \$3, then the expected reward earned during an infinite number of periods will, no matter what decisions are chosen, be unbounded. In this situation, it is not clear how a decision maker should choose a decision. Two approaches are commonly used to resolve the problem of unbounded expected rewards over an infinite horizon.

1 We can discount rewards (or costs) by assuming that a \$1 reward received during the next period will have the same value as a reward of β dollars ($0 < \beta < 1$) received during the current period. This is equivalent to assuming that the decision maker wants to maximize expected discounted reward. Let M be the maximum reward (over all possible states and choices of decisions) that can be received during a single period. Then the maximum expected discounted reward (measured in terms of current period dollars) that can be received over an infinite period horizon is

$$M + M\beta + M\beta^2 + \cdots = \frac{M}{1 - \beta} < \infty$$

Thus, discounting rewards (or costs) resolves the problem of an infinite expected reward.

2 The decision maker can choose to maximize the expected reward earned per period. Then he or she would choose a decision during each period in an attempt to maximize the average reward per period as given by

$$E\left(\lim_{n \to \infty} \frac{\text{reward earned during periods } 1, 2, \ldots, n}{n}\right)$$

Thus, if a \$3 reward were earned each period, the total reward earned during an infinite number of periods would be unbounded, but the average reward per period would equal \$3.

In our discussion of infinite horizon problems, we choose to resolve the problem of unbounded expected rewards by discounting rewards by a factor β per period. A brief discussion of the criterion of average reward per period is also included. Infinite horizon probabilistic dynamic programming problems are called **Markov decision processes** (or MDPs).

[†]This section covers topics that may be omitted with no loss of continuity.

Description of an MDP

An MDP is described by four types of information:

1 State space

2 Decision set

3 Transition probabilities

4 Expected rewards

State Space

At the beginning of each period, the MDP is in some state i, where i is a member of $S = \{1, 2, \ldots, N\}$. S is referred to as the MDP's **state space.**

Decision Set

For each state i, there is a finite set of allowable decisions, $D(i)$.

Transition Probabilities

Suppose a period begins in state i, and a decision $d \in D(i)$ is chosen. Then with probability $p(j|i, d)$, the next period's state will be j. The next period's state depends only on the current period's state and on the decision chosen during the current period (not on previous states and decisions). This is why we use the term *Markov* decision process.

Expected Rewards

During a period in which the state is i and a decision $d \in D(i)$ is chosen, an expected reward of r_{id} is received.

EXAMPLE 11   **Machine Replacement**

At the beginning of each week, a machine is in one of four conditions (states): excellent (E), good (G), average (A), or bad (B). The weekly revenue earned by a machine in each type of condition is as follows: excellent, $100; good, $80; average, $50; bad, $10. After observing the condition of a machine at the beginning of the week, we have the option of instantaneously replacing it with an excellent machine, which costs $200. The quality of a machine deteriorates over time, as shown in Table 8. For this situation, determine the state space, decision sets, transition probabilities, and expected rewards.

TABLE 8
Next Period's States of Machines

Present State of Machine	Probability That Machine Begins Next Week As			
	Excellent	Good	Average	Bad
Excellent	.7	.3	—	—
Good	—	.7	.3	—
Average	—	—	.6	.4
Bad	—	—	—	1.0
				until replaced

Solution　The set of possible states is $S = \{E, G, A, B\}$. Let

$$R = \text{replace at beginning of current period}$$
$$NR = \text{do not replace during current period}$$

Since it is absurd to replace an excellent machine, we write

$$D(E) = \{NR\} \qquad D(G) = D(A) = D(B) = \{R, NR\}$$

We are given the following transition probabilities:

$$
\begin{array}{llll}
p(E|NR, E) = .7 & p(G|NR, E) = .3 & p(A|NR, E) = 0 & p(B|NR, E) = 0 \\
p(E|NR, G) = 0 & p(G|NR, G) = .7 & p(A|NR, G) = .3 & p(B|NR, G) = 0 \\
p(E|NR, A) = 0 & p(G|NR, A) = 0 & p(A|NR, A) = .6 & p(B|NR, A) = .4 \\
p(E|NR, B) = 0 & p(G|NR, B) = 0 & p(A|NR, B) = 0 & p(B|NR, B) = 1
\end{array}
$$

If we replace a machine with an excellent machine, the transition probabilities will be the same as if we had begun the week with an excellent machine. Thus,

$$
\begin{aligned}
p(E|G, R) &= p(E|A, R) = p(E|B, R) = .7 \\
p(G|G, R) &= p(G|A, R) = p(G|B, R) = .3 \\
p(A|G, R) &= p(A|A, R) = p(A|B, R) = 0 \\
p(B|G, R) &= p(B|A, R) = p(B|B, R) = 0
\end{aligned}
$$

If the machine is not replaced, then during the week, we receive the revenues given in the problem. Therefore, $r_{E,NR} = \$100$, $r_{G,NR} = \$80$, $r_{A,NR} = \$50$, and $r_{B,NR} = \$10$. If we replace a machine with an excellent machine, then no matter what type of machine we had at the beginning of the week, we receive $\$100$ and pay a cost of $\$200$. Thus, $r_{E,R} = r_{G,R} = r_{A,R} = r_{B,R} = -\$100$.

In an MDP, what criterion should be used to determine the correct decision? Answering this question requires that we discuss the idea of an **optimal policy** for an MDP.

DEFINITION ■　A **policy** is a rule that specifies how each period's decision is chosen.　■

Period t's decision may depend on the prior history of the process. Thus, period t's decision can depend on the state during periods $1, 2, \ldots, t$ and the decisions chosen during periods $1, 2, \ldots, t - 1$.

DEFINITION ■　A policy δ is a **stationary policy** if whenever the state is i, the policy δ chooses (independently of the period) the same decision (call this decision $\delta(i)$).　■

We let δ represent an arbitrary policy and Δ represent the set of all policies. Then

$\mathbf{X}_t$ = random variable for the state of MDP at the beginning of period t (for example, $\mathbf{X}_2, \mathbf{X}_3, \ldots, \mathbf{X}_n$)

X_1 = given state of the process at beginning of period 1 (initial state)

d_t = decision chosen during period t

$V_\delta(i)$ = expected discounted reward earned during an infinite number of periods, given that at beginning of period 1, state is i and stationary policy will be δ

Then

$$V_\delta(i) = E_\delta \left(\sum_{t=1}^{t=\infty} \beta^{t-1} r_{\mathbf{X}_t d_t} | X_1 = i \right)$$

where $E_\delta(\beta^{t-1} r_{\mathbf{X}_t d_t} | X_1 = i)$ is the expected discounted reward earned during period t, given that at the beginning of period 1, the state is i and stationary policy δ is followed.

In a maximization problem, we define

$$V(i) = \max_{\delta \in \Delta} V_\delta(i) \tag{14}$$

In a minimization problem, we define

$$V(i) = \min_{\delta \in \Delta} V_\delta(i)$$

DEFINITION ■ If a policy δ^* has the property that for all $i \in S$

$$V(i) = V_{\delta^*}(i)$$

then δ^* is an **optimal policy.** ■

The existence of a single policy δ^* that simultaneously attains all N maxima in (14) is not obvious. If the r_{id}'s are bounded, Blackwell (1962) has shown that an optimal policy exists, and there is always a stationary policy that is optimal. (Even if the r_{id}'s are not bounded, an optimal policy may exist.)

We now consider three methods that can be used to determine an optimal stationary policy:

1 Policy iteration

2 Linear programming

3 Value iteration, or successive approximations

Policy Iteration

Value Determination Equations

Before we can explain the policy iteration method, we need to determine a system of linear equations that can be used to find $V_\delta(i)$ for $i \in S$ and any stationary policy δ. Let $\delta(i)$ be the decision chosen by the stationary policy δ whenever the process begins a period in state i. Then $V_\delta(i)$ can be found by solving the following system of N linear equations, the value determination equations:

$$V_\delta(i) = r_{i,\delta(i)} + \beta \sum_{j=1}^{j=N} p(j|i, \delta(i)) V_\delta(j) \qquad (i = 1, 2, \ldots, N) \tag{15}$$

To justify (15), suppose we are in state i and we follow a stationary policy δ. The current period is period 1. Then the expected discounted reward earned during an infinite number of periods consists of $r_{i,\delta(i)}$ (the expected reward received during the current period) plus β (expected discounted reward, to beginning of period 2, earned from period 2 onward). But with probability $p(j|i,\delta(i))$, we will begin period 2 in state j and earn an expected discounted reward, back to period 2, of $V_\delta(j)$. Thus, the expected discounted reward, discounted back to the beginning of period 2 and earned from the beginning of period 2 onward, is given by

$$\sum_{j=1}^{j=N} p(j|i,\ \delta(i))V_\delta(j)$$

Equation (15) now follows.

To illustrate the use of the value determination equations, we consider the following stationary policy for the machine replacement example:

$$\delta(E) = \delta(G) = NR \qquad \delta(A) = \delta(B) = R$$

This policy replaces a bad or average machine and does not replace a good or excellent machine. For this policy, (15) yields the following four equations:

$$V_\delta(E) = 100 + .9(.7V_\delta(E) + .3V_\delta(G))$$
$$V_\delta(G) = 80 + .9(.7V_\delta(G) + .3V_\delta(A))$$
$$V_\delta(A) = -100 + .9(.7V_\delta(E) + .3V_\delta(G))$$
$$V_\delta(B) = -100 + .9(.7V_\delta(E) + .3V_\delta(G))$$

Solving these equations yields $V_\delta(E) = 687.81$, $V_\delta(G) = 572.19$, $V_\delta(A) = 487.81$, and $V_\delta(B) = 487.81$.

Howard's Policy Iteration Method

We now describe Howard's (1960) policy iteration method for finding an optimal stationary policy for an MDP (max problem).

Step 1 Policy evaluation—Choose a stationary policy δ and use the value determination equations to find $V_\delta(i)(i = 1, 2, \ldots, N)$.

Step 2 Policy improvement—For all states $i = 1, 2, \ldots, N$, compute

$$T_\delta(i) = \max_{d \in D(i)} \left(r_{id} + \beta \sum_{j=1}^{j=N} p(j|i, d)V_\delta(j) \right) \tag{16}$$

Since we can choose $d = \delta(i)$ for $i = 1, 2, \ldots, N$, $T_\delta(i) \geq V_\delta(i)$. If $T_\delta(i) = V_\delta(i)$ for $i = 1, 2, \ldots N$, then δ is an optimal policy. If $T_\delta(i) > V_\delta(i)$ for at least one state, then δ is not an optimal policy. In this case, modify δ so that the decision in each state i is the decision attaining the maximum in (16) for $T_\delta(i)$. This yields a new stationary policy δ' for which $V_{\delta'}(i) \geq V_\delta(i)$ for $i = 1, 2, \ldots N$, and for at least one state i', $V_{\delta'}(i') > V_\delta(i')$. Return to step 1, with policy δ' replacing policy δ.

In a minimization problem, we replace max in (16) with min. If $T_\delta(i) = V_\delta(i)$ for $i = 1, 2, \ldots, N$, then δ is an optimal policy. If $T_\delta(i) < V_\delta(i)$ for at least one state, then δ is not an optimal policy. In this case, modify δ so that the decision in each state i is the decision attaining the minimum in (16) for $T_\delta(i)$. This yields a new stationary policy δ' for which $V_{\delta'}(i) \leq V_\delta(i)$ for $i = 1, 2, \ldots N$, and for at least one state i', $V_{\delta'}(i') < V_\delta(i')$. Return to step 1, with policy δ' replacing policy δ.

The policy iteration method is guaranteed to find an optimal policy for the machine replacement example after evaluating a finite number of policies. We begin with the following stationary policy:

$$\delta(E) = \delta(G) = NR \qquad \delta(A) = \delta(B) = R$$

For this policy, we have already found that $V_\delta(E) = 687.81$, $V_\delta(G) = 572.19$, $V_\delta(A) = 487.81$, and $V_\delta(B) = 487.81$. We now compute $T_\delta(E)$, $T_\delta(G)$, $T_\delta(A)$, and $T_\delta(B)$. Since NR is the only possible decision in E,

$$T_\delta(E) = V_\delta(E) = 687.81$$

and $T_\delta(E)$ is attained by the decision NR.

$$T_\delta(G) = \max \begin{cases} -100 + .9(.7V_\delta(E) + .3V_\delta(G)) = 487.81 & (R) \\ 80 + .9(.7V_\delta(G) + .3V_\delta(A)) = V_\delta(G) = 572.19* & (NR) \end{cases}$$

Thus, $T_\delta(G) = 572.19$ is attained by the decision NR.

$$T_\delta(A) = \max \begin{cases} -100 + .9(.7V_\delta(E) + .3V_\delta(G)) = 487.81 & (R) \\ 50 + .9(.6V_\delta(A) + .4V_\delta(B)) = 489.03* & (NR) \end{cases}$$

Thus, $T_\delta(A) = 489.03$ is attained by the decision NR.

$$T_\delta(B) = \max \begin{cases} -100 + .9(.7V_\delta(E) + .3V_\delta(G)) = V_\delta(B) = 487.81* & (R) \\ 10 + .9V_\delta(B) = 449.03 & (NR) \end{cases}$$

Thus, $T_\delta(B) = V_\delta(B) = 487.81$. We have found that $T_\delta(E) = V_\delta(E)$, $T_\delta(G) = V_\delta(G)$, $T_\delta(B) = V_\delta(B)$, and $T_\delta(A) > V_\delta(A)$. Thus, the policy δ is not optimal, and the policy δ' given by $\delta'(E) = \delta'(G) = \delta'(A) = NR$, $\delta'(B) = R$, is an improvement over δ. We now return to step 1 and solve the value determination equations for δ'. From (15), the value determination equations for δ' are

$$V_{\delta'}(E) = 100 + .9(.7V_{\delta'}(E) + .3V_{\delta'}(G))$$
$$V_{\delta'}(G) = 80 + .9(.7V_{\delta'}(G) + .3V_{\delta'}(A))$$
$$V_{\delta'}(A) = 50 + .9(.6V_{\delta'}(A) + .4V_{\delta'}(B))$$
$$V_{\delta'}(B) = -100 + .9(.7V_{\delta'}(E) + .3V_{\delta'}(G))$$

Solving these equations, we obtain $V_{\delta'}(E) = 690.23$, $V_{\delta'}(G) = 575.50$, $V_{\delta'}(A) = 492.35$, and $V_{\delta'}(B) = 490.23$. Observe that in each state i, $V_{\delta'}(i) > V_\delta(i)$. We now apply the policy iteration procedure to δ'. We compute

$$T_{\delta'}(E) = V_{\delta'}(E) = 690.23$$

$$T_{\delta'}(G) = \max \begin{cases} -100 + .9(.7V_{\delta'}(E) + .3V_{\delta'}(G)) = 490.23 & (R) \\ 80 + .9(.7V_{\delta'}(G) + .3V_{\delta'}(A)) = V_{\delta'}(G) = 575.50* & (NR) \end{cases}$$

Thus, $T_{\delta'}(G) = V_{\delta'}(G) = 575.50$ is attained by NR.

$$T_{\delta'}(A) = \max \begin{cases} -100 + .9(.7V_{\delta'}(E) + .3V_{\delta'}(G)) = 490.23 & (R) \\ 50 + .9(.6V_{\delta'}(A) + .4V_{\delta'}(B)) = V_{\delta'}(A) = 492.35* & (NR) \end{cases}$$

Thus, $T_{\delta'}(A) = V_{\delta'}(A) = 492.35$ is attained by NR.

$$T_{\delta'}(B) = \max \begin{cases} -100 + .9(.7V_{\delta'}(E) + .3V_{\delta'}(G)) = V_\delta(B) = 490.23* & (R) \\ 10 + .9V_{\delta'}(B) = 451.21 & (NR) \end{cases}$$

Thus, $T_{\delta'}(B) = V_{\delta'}(B) = 490.23$ is attained by R.

For each state i, $T_{\delta'}(i) = V_{\delta'}(i)$. Thus, δ' is an optimal stationary policy. To maximize expected discounted rewards (profits), a bad machine should be replaced, but an excellent, good, or average machine should not be replaced. If we began period 1 with an excellent machine, an expected discounted reward of $690.23 could be earned.

Linear Programming

It can be shown (see Ross (1983)) that an optimal stationary policy for a maximization problem can be found by solving the following LP:

$$\min z = V_1 + V_2 + \cdots + V_N$$

$$\text{s.t. } V_i - \beta \sum_{j=1}^{j=N} p(j|i, d)V_j \geq r_{id} \qquad \text{(For each state } i \text{ and each } d \in d(i))$$

All variables urs

For a minimization problem, we solve the following LP:

$$\max z = V_1 + V_2 + \cdots + V_N$$

$$\text{s.t. } V_i - \beta \sum_{j=1}^{j=N} p(j|i, d)V_j \leq r_{id} \qquad \text{(For each state } i \text{ and each } d \in d(i))$$

All variables urs

The optimal solution to these LPs will have $V_i = V(i)$. Also, if a constraint for state i and decision d is binding (has no slack or excess), then decision d is optimal in state i.

REMARKS 1 In the objective function, the coefficient of each V_i may be any positive number.
2 If all the V_i's are nonnegative (this will surely be the case if all the r_{id}'s are nonnegative), we may assume that all variables are nonnegative. If it is possible for some state to have $V(i)$ negative, then we must replace each variable V_i by $V_i' - V_i''$, where both V_i' and V_i'' are nonnegative.
3 With LINDO, we may allow $V(i)$ to be negative with the statement **FREE** Vi. With LINGO, use the **@FREE** statement to allow a variable to assume a negative value.

Our machine replacement example yields the following LP:

$$\min z = V_E + V_G + V_A + V_B$$

s.t.	$V_E \geq 100 + .9(.7V_E + .3V_G)$	(*NR* in *E*)
	$V_G \geq 80 + .9(.7V_G + .3V_A)$	(*NR* in *G*)
	$V_G \geq -100 + .9(.7V_E + .3V_G)$	(*R* in *G*)
	$V_A \geq 50 + .9(.6V_A + .4V_B)$	(*NR* in *A*)
	$V_A \geq -100 + .9(.7V_E + .3V_G)$	(*R* in *A*)
	$V_B \geq 10 + .9V_B$	(*NR* in *B*)
	$V_B \geq -100 + .9(.7V_E + .3V_G)$	(*R* in *B*)

All variables urs

The LINDO output for this LP yields $V_E = 690.23$, $V_G = 575.50$, $V_A = 492.35$, and $V_B = 490.23$. These values agree with those found via the policy iteration method. The LINDO output also indicates that the first, second, fourth, and seventh constraints have no slack. Thus, the optimal policy is to replace a bad machine and not to replace an excellent, good, or average machine.

Value Iteration

There are several versions of value iteration (see Denardo (1982)). We discuss for a maximization problem the simplest value iteration scheme, also known as *successive approx-*

imations. Let $V_t(i)$ be the maximum expected discounted reward that can be earned during t periods if the state at the beginning of the current period is i. Then

$$V_t(i) = \max_{d \in D(i)} \left\{ r_{id} + \beta \sum_{j=1}^{j=N} p(j|i, d)V_{t-1}(j) \right\} \qquad (t \geq 1)$$

$$V_0(i) = 0$$

This result follows, because during the current period, we earn an expected reward (in current dollars) of r_{id}, and during the next $t - 1$ periods, our expected discounted reward (in terms of period 2 dollars) is

$$\sum_{j=1}^{j=N} p(j|i, d)V_{t-1}(j)$$

Let $d_t(i)$ be the decision that must be chosen during period 1 in state i to attain $V_t(i)$. For an MDP with a finite state space and each $D(i)$ containing a finite number of elements, the most basic result in successive approximations states that for $i = 1, 2, \ldots, N$,

$$|V_t(i) - V(i)| \leq \frac{\beta^t}{1 - \beta} \max_{i,d} |r_{id}|$$

Recall that $V(i)$ is the maximum expected discounted reward earned during an infinite number of periods if the state is i at the beginning of the current period. Then

$$\lim_{t \to \infty} d_t(i) = \delta^*(i)$$

where $\delta^*(i)$ defines an optimal stationary policy. Since $\beta < 1$, for t sufficiently large, $V_t(i)$ will come arbitrarily close to $V(i)$. For instance, in the machine replacement example, $\beta = .9$ and $\max |r_{id}| = 100$. Thus, for all states, $V_{50}(i)$ would differ by at most $(.9)^{50}(\frac{100}{.10}) = \$5.15$ from $V(i)$. The equation

$$\lim_{t \to \infty} d_t(i) = \delta^*(i)$$

implies that for t sufficiently large, the decision that is optimal in state i for a t-period problem is also optimal in state i for an infinite horizon problem. This result is reminiscent of the turnpike theorem result for the knapsack problem that was discussed in Chapter 6.

Unfortunately, there is usually no easy way to determine a t^* such that for all i and $t \geq t^*$, $d_t(i) = \delta^*(i)$. (See Denardo (1982) for a partial result in this direction.) Despite this fact, value iteration methods usually obtain a satisfactory approximation to the $V(i)$ and $\delta^*(i)$ with less computational effort than is needed by the policy iteration method or by linear programming. Again, see Denardo (1982) for a discussion of this matter.

We illustrate the computation of V_1 and V_2 for the machine replacement example:

$$V_1(E) = 100 \qquad (NR)$$

$$V_1(G) = \max \begin{cases} 80^* & (NR) \\ -100 & (R) \end{cases} = 80$$

$$V_1(A) = \max \begin{cases} 50^* & (NR) \\ -100 & (R) \end{cases} = 50$$

$$V_1(B) = \max \begin{cases} 10^* & (NR) \\ -100 & (R) \end{cases} = 10$$

The * indicates the action attaining $V_1(i)$. Then

$$V_2(E) = 100 + .9(.7V_1(E) + .3V_1(G)) = 184.6 \qquad (NR)$$

$$V_2(G) = \max \begin{cases} 80 + .9(.7V_1(G) + .3V_1(A)) = 143.9^* & (NR) \\ -100 + .9(.7V_1(E) + .3V_1(G)) = -15.4 & (R) \end{cases}$$

$$V_2(A) = \max \begin{cases} 50 + .9(.6V_1(A) + .4V_1(B)) = 80.6^* & (NR) \\ -100 + .9(.7V_1(E) + .3V_1(G)) = -15.4 & (R) \end{cases}$$

$$V_2(B) = \max \begin{cases} 10 + .9V_1(B) = 19^* & (NR) \\ -100 + .9(.7V_1(E) + .3V_1(G)) = -15.4 & (R) \end{cases}$$

The * now indicates the decision $d_2(i)$ attaining $V_2(i)$. Observe that after two iterations of successive aproximations, we have not yet come close to the actual values of $V(i)$ and have not found it optimal to replace even a bad machine.

In general, if we want to ensure that all the $V_t(i)$'s are within ϵ of the corresponding $V(i)$, we would perform t^* iterations of successive approximations, where

$$\frac{\beta^{t^*}}{1 - \beta} \max_{i,d} |r_{id}| < \epsilon$$

There is no guarantee, however, that after t^* iterations of successive approximations, the optimal stationary policy will have been found.

Maximizing Average Reward per Period

We now briefly discuss how linear programming can be used to find a stationary policy that maximizes the expected per-period reward earned over an infinite horizon. Consider a decision rule or policy Q that chooses decision $d \in D(i)$ with probability $q_i(d)$ during a period in which the state is i. A policy Q will be a stationary policy if each $q_i(d)$ equals 0 or 1. To find a policy that maximizes expected reward per period over an infinite horizon, let π_{id} be the fraction of all periods in which the state is i and the decision $d \in D(i)$ is chosen. Then the expected reward per period may be written as

$$\sum_{i=1}^{i=N} \sum_{d \in D(i)} \pi_{id} r_{id} \qquad (17)$$

What constraints must be satisfied by the π_{id}? First, all π_{id}'s must be nonnegative. Second,

$$\sum_{i=1}^{i=N} \sum_{d \in D(i)} \pi_{id} = 1$$

must hold. Finally, the fraction of all periods during which a transition occurs out of state j must equal the fraction of all periods during which a transition occurs into state j. This is identical to the restriction on steady-state probabilities for Markov chains discussed in Section 5.5. This yields (for $j = 1, 2, \ldots, n$),

$$\sum_{d \in D(j)} \pi_{jd}(1 - p(j|j, d)) = \sum_{d \in D(i)} \sum_{i \neq j} \pi_{id} p(j|i, d)$$

Rearranging the last equality yields (for $j = 1, 2, \ldots, N$)

$$\sum_{d \in D(j)} \pi_{jd} = \sum_{d \in D(i)} \sum_{i=1}^{i=N} \pi_{id} p(j|i, d)$$

Putting together our objective function (17) and all the constraints yields the following LP:

$$\max z = \sum_{i=1}^{i=N} \sum_{d \in D(i)} \pi_{id} r_{id}$$

$$\text{s.t.} \quad \sum_{i=1}^{i=N} \sum_{d \in D(i)} \pi_{id} = 1$$

$$\sum_{d \in D(j)} \pi_{jd} = \sum_{d \in D(i)} \sum_{i=1}^{i=N} \pi_{id} p(j|i, d)$$

$$(j = 1, 2, \ldots, N)$$

$$\text{All } \pi_{id's} \geq 0$$

(18)

It can be shown that this LP has an optimal solution in which for each i, at most one $\pi_{id} > 0$. This optimal solution implies that expected reward per period is maximized by a solution in which each $q_i(d)$ equals 0 or 1. Thus, the optimal solution to (18) will occur for a stationary policy. For states having $\pi_{id} = 0$, any decision may be chosen without affecting the expected reward per period.

We illustrate the use of (18) for Example 11 (machine replacement). For this example, (18) yields

$$\max z = 100\pi_{ENR} + 80\pi_{GNR} + 50\pi_{ANR} + 10\pi_{BNR} - 100(\pi_{GR} + \pi_{AR} + \pi_{BR})$$

$$\text{s.t.} \quad \pi_{ENR} + \pi_{GNR} + \pi_{ANR} + \pi_{BNR} + \pi_{GR} + \pi_{AR} + \pi_{BR} = 1$$

$$\pi_{ENR} = .7(\pi_{ENR} + \pi_{GR} + \pi_{AR} + \pi_{BR})$$

$$\pi_{GNR} + \pi_{GR} = .3(\pi_{GR} + \pi_{AR} + \pi_{BR} + \pi_{ENR}) + .7\pi_{GNR}$$

$$\pi_{AR} + \pi_{ANR} = .3\pi_{GNR} + .6\pi_{ANR}$$

$$\pi_{BR} + \pi_{BNR} = \pi_{BNR} + .4\pi_{ANR}$$

Using LINDO, we find the optimal objective function value for this LP to be $z = 60$. The only nonzero decision variables are $\pi_{ENR} = .35$, $\pi_{GNR} = .50$, $\pi_{AR} = .15$. Thus, an average of $60 profit per period can be earned by not replacing an excellent or good machine but replacing an average machine. Since we are replacing an average machine, the action chosen during a period in which a machine is in bad condition is of no importance.

PROBLEMS

Group A

1 A warehouse has an end-of-period capacity of 3 units. During a period in which production takes place, a setup cost of $4 is incurred. A $1 holding cost is assessed against each unit of a period's ending inventory. Also, a variable production cost of $1 per unit is incurred. During each period, demand is equally likely to be 1 or 2 units. All demand must be met on time, and $\beta = .8$. The goal is to minimize expected discounted costs over an infinite horizon.

a Use the policy iteration method to determine an optimal stationary policy.

b Use linear programming to determine an optimal stationary policy.

c Perform two iterations of value iteration.

2 Priceler Auto Corporation must determine whether or not to give consumers 8% or 11% financing on new cars. If Priceler gives 8% financing during the current month, the probability distribution of sales during the current month will be as shown in Table 9. If Priceler gives 11% financing during the current month, the probability distribution of sales during the current month will be as shown in Table 10. "Good" sales represents 400,000 sales per month, "bad" sales represents 300,000 sales per month. For example, if last month's sales were bad and Priceler gives 8% financing during the current month, there is a .40 chance that sales will be good during the current month. At 11% financing rates, Priceler earns $1,000 per car, and at 8% financing, Priceler earns $800 per car. Priceler's goal is to maximize

TABLE 9

Last Month's Sales	Current Month's Sales	
	Good	Bad
Good	.95	.05
Bad	.40	.60

TABLE 10

Last Month's Sales	Current Month's Sales	
	Good	Bad
Good	.80	.20
Bad	.20	.80

expected discounted profit over an infinite horizon (use $\beta = .98$).

a Use the policy iteration method to determine an optimal stationary policy.

b Use linear programming to determine an optimal stationary policy.

c Perform two iterations of value iteration.

d Find a policy that maximizes average profit per month.

3 Suppose you are using the policy iteration method to determine an optimal policy for an MDP. How might you use LINDO to solve the value determination equations?

Group B

4 During any day, I may own either 0 or 1 share of a stock. The price of the stock is governed by the Markov chain shown in Table 11. At the beginning of a day in which I own a share of stock, I may either sell it at today's price or keep it. At the beginning of a day in which I don't own a share of stock, I may either buy a share of stock at today's price

TABLE 11

Today's Price	Tomorrow's Price			
	$0	$1	$2	$3
$0	.5	.3	.1	.1
$1	.1	.5	.2	.2
$2	.2	.1	.5	.2
$3	.1	.1	.3	.5

or not buy a share. My goal is to maximize my expected discounted profit over an infinite horizon (use $\beta = .95$).

a Use the policy iteration method to determine an optimal stationary policy.

b Use linear programming to determine an optimal stationary policy.

c Perform two iterations of value iteration.

d Find a policy that maximizes average daily profit.

5 Ethan Sherwood owns two printing presses, on which he prints two types of jobs. At the beginning of each day, there is a .5 probability that a type 1 job will arrive, a .1 probability that a type 2 job will arrive, and a .4 probability that no job will arrive. Ethan receives $400 for completing a type 1 job and $200 for completing a type 2 job. (Payment for each job is received in advance.) Each type of job takes an average of three days to complete. To model this, we assume that each day a job is in press there is a $\frac{1}{3}$ probability that its printing will be completed at the end of the day. If both presses are busy at the beginning of the day, any arriving job is lost to the system. The crucial decision is when (if ever) Ethan should accept the less profitable type 2 job. Ethan's goal is to maximize expected discounted profit (use $\beta = .90$).

a Use the policy iteration method to determine an optimal stationary policy.

b Use linear programming to determine an optimal stationary policy.

c Perform two iterations of value iteration.

SUMMARY

Key to Formulating Probabilistic Dynamic Programming Problems (PDPs)

Suppose the possible states during period $t + 1$ are $s_1, s_2, \ldots s_n$, and the probability that the period $t + 1$ state will be s_i is p_i. Then the minimum expected cost incurred during periods $t + 1, t + 2, \ldots$, end of the problem is

$$\sum_{i=1}^{i=n} p_i f_{t+1}(s_i)$$

where $f_{t+1}(s_i)$ is the minimum expected cost incurred from period $t + 1$ to the end of the problem, given that the state during period $t + 1$ is s_i.

Maximizing the Probability of a Favorable Event Occurring

To maximize the probability that a favorable event will occur, assign a reward of 1 if the favorable event occurs and a reward of 0 if it does not occur.

Markov Decision Processes

A **Markov decision process** (MDP) is simply an infinite-horizon PDP. Let $V_\delta(i)$ be the expected discounted reward earned during an infinite number of periods, given that at the beginning of period 1, the state is i and the stationary policy δ is followed.

For a maximization problem, we define

$$V(i) = \max_{\delta \in \Delta} V_\delta(i)$$

For a minimization problem, we define

$$V(i) = \min_{\delta \in \Delta} V_\delta(i)$$

If a policy δ^* has the property that for all $i \in S$,

$$V(i) = V_{\delta^*}(i)$$

then δ^* is an **optimal policy.** We can use the **value determination equations** to determine $V_\delta(i)$:

$$V_\delta(i) = r_{i,\delta(i)} + \beta \sum_{j=1}^{j=N} p(j|i, \delta(i))V_\delta(j) \qquad (i = 1, 2, \ldots, N) \tag{15}$$

An optimal policy for an MDP may be determined by one of three methods:

1 Policy iteration

2 Linear programming

3 Value iteration, or successive approximations

Policy Iteration

A summary of Howard's policy iteration method for a maximization problem follows.

Step 1 Policy evaluation—Choose a stationary policy δ and use the value determination equations to find $V_\delta(i)(i = 1, 2, \ldots, N)$.

Step 2 Policy improvement—For all states $i = 1, 2, \ldots, N$, compute

$$T_\delta(i) = \max_{d \in D(i)} \left\{ r_{id} + \beta \sum_{j=1}^{j=N} p(j|i, d)V_\delta(j) \right\} \tag{16}$$

Since we can choose $d = \delta(i)$ for $i = 1, 2, \ldots, N$, $T_\delta(i) \geq V_\delta(i)$. If $T_\delta(i) = V_\delta(i)$ for $i = 1, 2, \ldots, N$, then δ is an optimal policy. If $T_\delta(i) > V_\delta(i)$ for at least one state, then δ is not an optimal policy. In this case, modify δ so that the decision in each state i is the decision attaining the maximum in (16) for $T_\delta(i)$. This yields a new stationary policy δ' for which $V_{\delta'}(i) \geq V_\delta(i)$ for $i = 1, 2, \ldots, N$, and for at least one state i', $V_{\delta'}(i') > V_\delta(i')$. Return to step 1, with policy δ' replacing policy δ.

Linear Programming

In a maximization problem, $V(i)$ for each state may be determined by solving the following LP:

$$\min z = V_1 + V_2 + \cdots + V_N$$

$$\text{s.t.} \quad V_i - \beta \sum_{j=1}^{j=N} p(j|i, d)V_j \geq r_{id} \qquad \text{(For each state } i \text{ and each } d \in D(i))$$

All variables urs

If the constraint for state i and decision d has no slack, then decision d is optimal in state i.

Value Iteration, or Successive Approximations

Let $V_t(i)$ be the maximum expected discounted reward that can be earned during t periods if the state at the beginning of the current period is i. Then

$$V_t(i) = \max_{d \in D(i)} \left\{ r_{id} + \beta \sum_{j=1}^{j=N} p(j|i, d)V_{t-1}(j) \right\} \qquad (t \geq 1)$$

$$V_0(i) = 0$$

As t grows large, $V_t(i)$ will approach $V(i)$. For t sufficiently large, the decision that is optimal in state i for a t-period problem is also optimal in state i for an infinite-horizon problem.

REVIEW PROBLEMS

Group A

1 A company has five sales representatives available for assignment to three sales districts. The sales in each district during the current year depend on the number of sales representatives assigned to the district and on whether the national economy has a bad or good year (see Table 12). In the Sales column for each district, the first number represents sales if the national economy had a bad year, and the second number represents sales if the economy had a good year. There is a .3 chance that the national economy will have a good year and a .7 chance that the national economy will have a bad year. Use dynamic programming to determine an assignment of sales representatives to districts that maximizes the company's expected sales.

TABLE 12

No. of Sales Reps Assigned to District	Sales (millions)		
	District 1	District 2	District 3
0	$1, $4	$2, $5	$3, $4
1	$2, $6	$4, $6	$5, $5
2	$3, $7	$5, $6	$6, $7
3	$4, $8	$6, $6	$7, $7

2 At the beginning of each period, a company must determine how many units to produce. A setup cost of $5 is incurred during each period in which production takes place. The production of each unit also incurs a $2 variable cost. All demand must be met on time, and there is a $1 per-unit holding cost on each period's ending inventory. During each period, it is equally likely that demand will equal 0 or 1 unit. Assume that each period's ending inventory cannot exceed 2 units.

a Use dynamic programming to minimize the expected costs incurred during three periods. Assume that the initial inventory is 0 units.

b Now suppose that each unit demanded can be sold for $4. If the demand is not met on time, the sale is lost. Use dynamic programming to maximize the expected profit earned during three periods. Assume that the initial inventory is 0 units.

c In parts (a) and (b), is an (s, S) policy optimal?

3 At Hot Dog Queen Restaurant, the following sequence of events occurs during each minute:

a With probability p, a customer arrives and waits in line.

b Hot Dog Queen determines the rate s at which customers are served. If any customers are in the restaurant, then with probability s, one of the customers completes service and leaves the restaurant. It costs $c(s)$ dollars per period to serve customers at a rate s. Each customer spends R dollars, and the customer's food costs Hot Dog Queen $R - 1$ dollars to prepare.

c For each customer in line at the end of the minute, a cost of h dollars is assessed (because of customer inconvenience).

d The next minute begins.

Formulate a recursion that could be used to maximize expected revenues less costs (including customer inconvenience costs) incurred during the next T minutes. Assume that initially there are no customers present.

4 At the beginning of 2004, the United States has B barrels of oil. If x barrels of oil are consumed during a year, then consumers earn a benefit (measured in dollars) of $u(x)$. The United States may spend money on oil exploration. If d dollars are spent during a year on oil exploration, then there is a probability $p(d)$ that an oil field (containing 500,000 barrels of oil) will be found. Formulate a recursion that can be used to maximize the expected discounted benefits less exploration expenditures earned from the beginning of 2004 to the end of the year 2539.

5 I am a contestant on the popular TV show "Tired of Fortune." During the bonus round, I will be asked up to four questions. For each question that is correctly answered, I win a certain amount of money. One incorrect answer, however, means that I lose all the money I have previously won, and the game is over. If I elect to pass, or not answer a question, the game is over, but I may keep what I have already won. The amount of money I win for each correct question and the probability that I will answer each question correctly are shown in Table 13.

a My goal is to maximize the expected amount of money won. Use dynamic programming to accomplish this goal.

b Suppose that I am allowed to pass, or not answer a question, and still go on to the next question. Now determine how to maximize the amount of money won.

6 A machine in excellent condition earns $100 profit per week, a machine in good condition earns $70 per week, and a machine in bad condition earns $20 per week. At the beginning of any week, a machine may be sent out for repairs at a cost of $90. A machine that is sent out for repairs returns in excellent condition at the beginning of the next week. If a machine is not repaired, the condition of the machine evolves in accordance with the Markov chain shown in Table 14. The company wants to maximize its expected discounted profit over an infinite horizon ($\beta = .9$).

TABLE 13

Question	Probability of Correct Answer	Money Won
1	.6	$10,000
2	.5	$20,000
3	.4	$30,000
4	.3	$40,000

TABLE 14

This Week	Next Week		
	Excellent	Good	Bad
Excellent	.7	.2	.1
Good	0	.7	.3
Bad	0	.1	.9

a Use policy iteration to determine an optimal stationary policy.

b Use linear programming to detemine an optimal stationary policy.

c Perform two iterations of value iteration.

7 A country now has 10 units of capital. Each year, it may consume any amount of the available capital and invest the rest. Invested capital has a 50% chance of doubling and a 50% chance of losing half its value. For example, if the country invests 6 units of capital, there is a 50% chance that the 6 units will turn into 12 capital units and a 50% chance that the invested capital will turn into 3 units. What strategy should be used to maximize total expected consumption over a four-year period?

8 The Dallas Mavericks trail by two points and have the ball with 10 seconds remaining. They must decide whether to take a two- or a three-point shot. Assume that once the Mavericks take their shot, time expires. The probability that a two-point shot is successful is TWO, and the probability that a three-point shot is successful is THREE. If the game is tied, an overtime period will be played. Assume that there is a .5 chance the Mavericks will win in overtime. (*Note:* This problem is often used on Microsoft job interviews.)

a Give a rule based on the values of TWO and THREE that tells Dallas what to do.

b Typical values for an NBA team are TWO = .45 and THREE = .35. Based on this information, what strategy should most NBA teams follow?

9 At any time, the size of a tree is 0, 1, 2, or 3. We must decide when to harvest the tree. Each year, it costs $1 to maintain the tree. It costs $5 to harvest a tree. The sales price for a tree of each size is as follows:

Tree Size	Sales Price
0	$20
1	$30
2	$45
3	$49

The transition probability matrix for the size of the tree is as follows:

$$
\begin{array}{c@{\quad}cccc}
 & 0 & 1 & 2 & 3 \\
0 & \begin{bmatrix} .8 & .2 & 0 & 0 \\ 0 & .9 & .1 & 0 \\ 0 & 0 & .7 & .3 \\ 0 & 0 & 0 & 1 \end{bmatrix} \\
1 \\
2 \\
3
\end{array}
$$

For example, 80% of all size 0 trees begin the next year as size 0 trees, and 20% of all size 0 trees begin the next year as size 1 trees. Assuming the discount factor for cash flows is .9 per year, determine an optimal harvesting strategy.

10 For \$50, we can enter a raffle. We draw a certificate containing a number 100, 200, 300, ... , 1,000. Each number is equally likely. At any time, we can redeem the highest-numbered certificate we have obtained so far for the face value of the certificate. We may enter the raffle as many times as we wish. Assuming no discounting, what strategy would maximize our expected profit? How does this model relate to the problem faced by an unemployed person who is searching for a job?

11 At the beginning of each year, an aircraft engine is in good, fair, or poor condition. It costs \$500,000 to run a good engine for a year, \$1 million to run a fair engine for a year, and \$2 million to run a poor engine for a year. A fair engine can be overhauled for \$2 million, and it immediately becomes a good engine. A poor engine can be replaced for \$3 million, and it immediately becomes a good engine. The transition probability matrix for an engine is as follows:

	Good	Fair	Poor
Good	.7	.2	.1
Fair	0	.6	.4
Poor	0	0	1

The discount factor for costs is .9. What strategy minimizes expected discounted cost over an infinite horizon?

Group B

12 A syndicate of college students spends weekends gambling in Las Vegas. They begin week 1 with W dollars.

At the beginning of each week, they may wager any amount of their money at the gambling tables. If they wager d dollars, then with probability p, their wealth increases by d dollars, and with probability $1 - p$, their wealth decreases by d dollars. Their goal is to maximize their expected wealth at the end of T weeks.

a Show that if $p \geq \frac{1}{2}$, the students should bet all their money.

b Show that if $p < \frac{1}{2}$, the students should bet no money. (*Hint:* Define $f_t(w)$ as the maximum expected wealth at the end of week T, given that wealth is w dollars at the beginning of week t; by working backward, find an expression for $f_t(w)$.)

Group C

13 You have invented a new product: the HAL DVD player. Each of 1,000 potential customers places a different value on this product. A consumer's valuation is equally likely to be any number between \$0 and \$1,000. It costs \$100 to produce the HAL player. During a year in which we set a price p for the product, all customers valuing the product at \$$p$ or more will purchase the product. Each year, we set a price for the product. What pricing strategy will maximize our expected profit over three years? What commonly observed phenomenon does this problem illustrate?

REFERENCES

The following books contain elementary discussions of Markov decision processes and probabilistic dynamic programming:

Howard, R. *Dynamic Programming and Markov Processes.* Cambridge, Mass.: MIT Press, 1960.
Wagner, H. *Principles of Operations Research,* 2d ed. Englewood Cliffs, N.J.: Prentice Hall, 1975.

The following books treat Markov decision processes and probabilistic dynamic programming at a more advanced level:

Bersetkas, D. *Dynamic Programming and Optimal Control,* vols. 1 & 2. Cambridge, Mass.: Athena Publishing, 2000.
Heyman, D., and M. Sobel. *Stochastic Models in Operations Research,* vol. 2. New York: McGraw-Hill, 1984.
Kohlas, S. *Stochastic Methods of Operations Research.* Cambridge, U.K.: Cambridge University Press, 1982.
Puterman, M. *Markov Decision Processes: Discrete Stochastic Dynamic Programming.* New York: John Wiley, 1994.
Ross, S. *Introduction to Stochastic Dynamic Programming.* Orlando, Fla.: Academic Press, 1983.

Whittle, P. *Optimization Over Time: Dynamic Programming and Stochastic Control.* New York: Wiley, 1982.
White, D.J. *Markov Decision Processes.* New York: John Wiley, 1993.

Excellent one-chapter introductions to Markov decision processes are given in the following two books:

Denardo, E. *Dynamic Programming Theory and Applications.* Englewood Cliffs, N.J.: Prentice Hall, 1982.
Shapiro, J. *Mathematical Programming: Structures and Algorithms.* New York: Wiley, 1979.

Blackwell, D. "Discrete Dynamic Programming," *Annals of Mathematical Statistics* 33(1962):719–726. Indicates how one proves that a stationary policy is optimal for a Markov decision process.
Scarf, H. "The Optimality of (*s, S*) Policies for the Dynamic Inventory Problem," *Proceedings of the First Stanford Symposium on Mathematical Methods in the Social Sciences.* Stanford, Calif.: Stanford University Press, 1960. A proof of the optimality of (*s, S*) policies.

8 Queuing Theory

Each of us has spent a great deal of time waiting in lines. In this chapter, we develop mathematical models for waiting lines, or queues. In Section 8.1, we begin by discussing some terminology that is often used to describe queues. In Section 8.2, we look at some distributions (the exponential and the Erlang distributions) that are needed to describe queuing models. In Section 8.3, we introduce the idea of a birth–death process, which is basic to many queuing models involving the exponential distribution. The remainder of the chapter examines several models of queuing systems that can be used to answer questions like the following:

1 What fraction of the time is each server idle?

2 What is the expected number of customers present in the queue?

3 What is the expected time that a customer spends in the queue?

4 What is the probability distribution of the number of customers present in the queue?

5 What is the probability distribution of a customer's waiting time?

6 If a bank manager wants to ensure that only 1% of all customers will have to wait more than 5 minutes for a teller, how many tellers should be employed?

8.1 Some Queuing Terminology

To describe a queuing system, an input process and an output process must be specified. Some examples of input and output processes are given in Table 1.

The Input or Arrival Process

The input process is usually called the **arrival process.** Arrivals are called **customers.** In all models that we will discuss, we assume that no more than one arrival can occur at a given instant. For a case like a restaurant, this is a very unrealistic assumption. If more than one arrival can occur at a given instant, we say that **bulk arrivals** are allowed.

Usually, we assume that the arrival process is unaffected by the number of customers present in the system. In the context of a bank, this would imply that whether there are 500 or 5 people at the bank, the process governing arrivals remains unchanged.

There are two common situations in which the arrival process may depend on the number of customers present. The first occurs when arrivals are drawn from a small population. Suppose that there are only four ships in a naval shipyard. If all four ships are being repaired, then no ship can break down in the near future. On the other hand, if all four ships are at sea, a breakdown has a relatively high probability of occurring in the near fu-

TABLE **1**

Examples of Queuing Systems

Situation	Input Process	Output Process
Bank	Customers arrive at bank	Tellers serve the customers
Pizza parlor	Requests for pizza delivery are received	Pizza parlor sends out truck to deliver pizzas
Hospital blood bank	Pints of blood arrive	Patients use up pints of blood
Naval shipyard	Ships at sea break down and are sent to shipyard for repairs	Ships are repaired and return to sea

ture. Models in which arrivals are drawn from a small population are called **finite source models.** Another situation in which the arrival process depends on the number of customers present occurs when the rate at which customers arrive at the facility decreases when the facility becomes too crowded. For example, if you see that the bank parking lot is full, you might pass by and come another day. If a customer arrives but fails to enter the system, we say that the customer has **balked.** The phenomenon of balking was described by Yogi Berra when he said, "Nobody goes to that restaurant anymore; it's too crowded."

If the arrival process is unaffected by the number of customers present, we usually describe it by specifying a probability distribution that governs the time between successive arrivals.

The Output or Service Process

To describe the output process (often called the service process) of a queuing system, we usually specify a probability distribution—the **service time distribution**—which governs a customer's service time. In most cases, we assume that the service time distribution is independent of the number of customers present. This implies, for example, that the server does not work faster when more customers are present.

In this chapter, we study two arrangements of servers: **servers in parallel** and **servers in series.** Servers are in parallel if all servers provide the same type of service and a customer need only pass through one server to complete service. For example, the tellers in a bank are usually arranged in parallel; any customer need only be serviced by one teller, and any teller can perform the desired service. Servers are in series if a customer must pass through several servers before completing service. An assembly line is an example of a series queuing system.

Queue Discipline

To describe a queuing system completely, we must also describe the queue discipline and the manner in which customers join lines.

The **queue discipline** describes the method used to determine the order in which customers are served. The most common queue discipline is the **FCFS discipline** (first come, first served), in which customers are served in the order of their arrival. Under the **LCFS discipline** (last come, first served), the most recent arrivals are the first to enter service. If we consider exiting from an elevator to be service, then a crowded elevator illustrates an LCFS discipline. Sometimes the order in which customers arrive has no effect on the order in which they are served. This would be the case if the next customer to enter service is randomly chosen from those customers waiting for service. Such a situation is referred to as the **SIRO discipline** (service in random order). When callers to an airline are put on hold, the luck of the draw often determines the next caller serviced by an operator.

Finally, we consider **priority queuing disciplines.** A priority discipline classifies each arrival into one of several categories. Each category is then given a priority level, and within each priority level, customers enter service on an FCFS basis. Priority disciplines are often used in emergency rooms to determine the order in which customers receive treatment, and in copying and computer time-sharing facilities, where priority is usually given to jobs with shorter processing times.

Method Used by Arrivals to Join Queue

Another factor that has an important effect on the behavior of a queuing system is the method that customers use to determine which line to join. For example, in some banks, customers must join a single line, but in other banks, customers may choose the line they want to join. When there are several lines, customers often join the shortest line. Unfortunately, in many situations (such as a supermarket), it is difficult to define the shortest line. If there are several lines at a queuing facility, it is important to know whether or not customers are allowed to switch, or jockey, between lines. In most queuing systems with multiple lines, jockeying is permitted, but jockeying at a toll booth plaza is not recommended.

8.2 Modeling Arrival and Service Processes

Modeling the Arrival Process

As previously mentioned, we assume that at most one arrival can occur at a given instant of time. We define t_i to be the time at which the ith customer arrives. To illustrate this, consider Figure 1. For $i \geq 1$, we define $T_i = t_{i+1} - t_i$ to be the ith interarrival time. Thus, in the figure, $T_1 = 8 - 3 = 5$, and $T_2 = 15 - 8 = 7$. In modeling the arrival process, we assume that the T_i's are independent, continuous random variables described by the random variable $\mathbf{A}$. The independence assumption means, for example, that the value of T_2 has no effect on the value of T_3, T_4, or any later T_i. The assumption that each T_i is continuous is usually a good approximation of reality. After all, an interarrival time need not be exactly 1 minute or 2 minutes; it could just as easily be, say, 1.55892 minutes. The assumption that each interarrival time is governed by the same random variable implies that the distribution of arrivals is independent of the time of day or the day of the week. This is the assumption of stationary interarrival times. Because of phenomena such as rush hours, the assumption of stationary interarrival times is often unrealistic, but we may often approximate reality by breaking the time of day into segments. For example, if we were modeling traffic flow, we might break the day up into three segments: a morning rush hour segment, a midday segment, and an afternoon rush hour segment. During each of these segments, interarrival times may be stationary.

We assume that $\mathbf{A}$ has a density function $a(t)$. Recall from Section 1.6 that for small Δt, $P(t \leq \mathbf{A} \leq t + \Delta t)$ is approximately $\Delta t a(t)$. Of course, a negative interarrival time is impossible. This allows us to write

$$P(\mathbf{A} \leq c) = \int_0^c a(t)dt \quad \text{and} \quad P(\mathbf{A} > c) = \int_c^\infty a(t)dt \tag{1}$$

FIGURE 1
Definition of
Interarrival Times

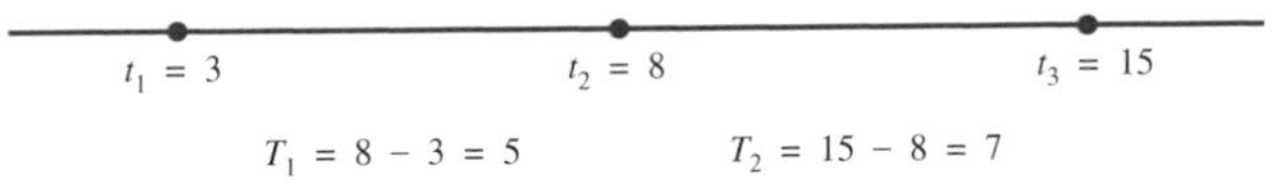

We define $\frac{1}{\lambda}$ to be the mean or average interarrival time. Without loss of generality, we assume that time is measured in units of hours. Then $\frac{1}{\lambda}$ will have units of hours per arrival. From Section 1.6, we may compute $\frac{1}{\lambda}$ from $a(t)$ by using the following equation:

$$\frac{1}{\lambda} = \int_0^\infty ta(t)dt \tag{2}$$

We define λ to be the **arrival rate,** which will have units of arrivals per hour.

In most applications of queuing, an important question is how to choose **A** to reflect reality and still be computationally tractable. The most common choice for **A** is the **exponential distribution.** An exponential distribution with parameter λ has a density $a(t) = \lambda e^{-\lambda t}$. Figure 2 shows the density function for an exponential distribution. We see that $a(t)$ decreases very rapidly for t small. This indicates that very long interarrival times are unlikely. Using Equation (2) and integration by parts, we can show that the average or mean interarrival time (call it $E(\mathbf{A})$) is given by

$$E(\mathbf{A}) = \frac{1}{\lambda} \tag{3}$$

Using the fact that var $\mathbf{A} = E(\mathbf{A}^2) - E(\mathbf{A})^2$, we can show that

$$\text{var } \mathbf{A} = \frac{1}{\lambda^2} \tag{4}$$

No-Memory Property of the Exponential Distribution

The reason the exponential distribution is often used to model interarrival times is embodied in the following lemma.

LEMMA 1

If **A** has an exponential distribution, then for all nonnegative values of t and h,

$$P(\mathbf{A} > t + h | \mathbf{A} \geq t) = P(\mathbf{A} > h) \tag{5}$$

Proof First note that from Equation (1), we have

$$P(\mathbf{A} > h) = \int_h^\infty \lambda e^{-\lambda t} = [-e^{-\lambda t}]_h^\infty = e^{-\lambda h} \tag{6}$$

Then

$$P(\mathbf{A} > t + h | \mathbf{A} \geq t) = \frac{P(\mathbf{A} > t + h \cap \mathbf{A} \geq t)}{P(\mathbf{A} \geq t)}$$

From (6),

$$P(\mathbf{A} > t + h \cap \mathbf{A} \geq t) = e^{-\lambda(t+h)} \quad \text{and} \quad P(\mathbf{A} \geq t) = e^{-\lambda t}$$

Thus,

$$P(\mathbf{A} > t + h | \mathbf{A} \geq t) = \frac{e^{-\lambda(t+h)}}{e^{-\lambda t}} = e^{-\lambda h} = P(\mathbf{A} > h)$$

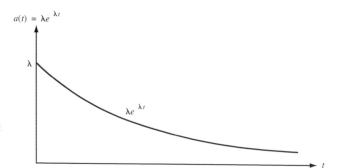

FIGURE 2
Density Function for
Exponential Distribution

It can be shown that no other density function can satisfy (5) (see Feller (1957)). For reasons that become apparent, a density that satisfies (5) is said to have the **no-memory property.** Suppose we are told that there has been no arrival for the last t hours (this is equivalent to being told that $\mathbf{A} \geq t$) and are asked what the probability is that there will be no arrival during the next h hours (that is, $\mathbf{A} > t + h$). Then (5) implies that this probability *does not depend on the value of* t, and for all values of t, this probability equals $P(\mathbf{A} > h)$. In short, if we know that at least t time units have elapsed since the last arrival occurred, then the distribution of the remaining time until the next arrival (h) does not depend on t. For example, if $h = 4$, then (5) yields, for $t = 5$, $t = 3$, $t = 2$, and $t = 0$,

$$P(\mathbf{A} > 9|\mathbf{A} \geq 5) = P(\mathbf{A} > 7|\mathbf{A} \geq 3) = P(\mathbf{A} > 6|\mathbf{A} \geq 2)$$
$$= P(\mathbf{A} > 4|\mathbf{A} \geq 0) = e^{-4\lambda}$$

The no-memory property of the exponential distribution is important, because it implies that if we want to know the probability distribution of the time until the next arrival, then *it does not matter how long it has been since the last arrival.* To put it in concrete terms, suppose interarrival times are exponentially distributed with $\lambda = 6$. Then the no-memory property implies that no matter how long it has been since the last arrival, the probability distribution governing the time until the next arrival has the density function $6e^{-6t}$. This means that to predict future arrival patterns, we need not keep track of how long it has been since the last arrival. This observation can appreciably simplify analysis of a queuing system.

To see that knowledge of the time since the last arrival does affect the distribution of time until the next arrival in most situations, suppose that $\mathbf{A}$ is discrete with $P(\mathbf{A} = 5) = P(\mathbf{A} = 100) = \frac{1}{2}$. If we are told that there has been no arrival during the last 6 time units, we *know with certainty* that it will be $100 - 6 = 94$ time units until the next arrival. On the other hand, if we are told that no arrival has occurred during the last time unit, then there is some chance that the time until the next arrival will be $5 - 1 = 4$ time units and some chance that it will be $100 - 1 = 99$ time units. Hence, in this situation, the distribution of the next interarrival time cannot easily be predicted with knowledge of the time that has elapsed since the last arrival.

Relation Between Poisson Distribution and Exponential Distribution

If interarrival times are exponential, the probability distribution of the number of arrivals occurring in any time interval of length t is given by the following important theorem.

THEOREM 1

Interarrival times are exponential with parameter λ if and only if the number of arrivals to occur in an interval of length t follows a Poisson distribution with parameter λt.

A discrete random variable **N** has a Poisson distribution with parameter λ if, for $n = 0, 1, 2, \ldots,$

$$P(\mathbf{N} = n) = \frac{e^{-\lambda}\lambda^n}{n!} \qquad (n = 0, 1, 2, \ldots) \tag{7}$$

If **N** is a Poisson random variable, it can be shown that $E(\mathbf{N}) = \text{var } \mathbf{N} = \lambda$. If we define $\mathbf{N}_t$ to be the number of arrivals to occur during any time interval of length t, Theorem 1 states that

$$P(\mathbf{N}_t = n) = \frac{e^{-\lambda t}(\lambda t)^n}{n!} \qquad (n = 0, 1, 2, \ldots)$$

Since $\mathbf{N}_t$ is Poisson with parameter λt, $E(\mathbf{N}_t) = \text{var } \mathbf{N}_t = \lambda t$. An average of λt arrivals occur during a time interval of length t, so λ may be thought of as the average number of arrivals per unit time, or the arrival rate.

What assumptions are required for interarrival times to be exponential? Theorem 2 provides a partial answer. Consider the following two assumptions:

1 Arrivals defined on nonoverlapping time intervals are independent (for example, the number of arrivals occurring between times 1 and 10 does not give us any information about the number of arrivals occurring between times 30 and 50).

2 For small Δt (and any value of t), the probability of one arrival occurring between times t and $t + \Delta t$ is $\lambda \Delta t + o(\Delta t)$, where $o(\Delta t)$ refers to any quantity satisfying

$$\lim_{\Delta t \to 0} \frac{o(\Delta t)}{\Delta t} = 0$$

Also, the probability of no arrival during the interval between t and $t + \Delta t$ is $1 - \lambda \Delta t + o(\Delta t)$, and the probability of more than one arrival occurring between t and $t + \Delta t$ is $o(\Delta t)$.

THEOREM 2

If assumptions 1 and 2 hold, then $\mathbf{N}_t$ follows a Poisson distribution with parameter λt, and interarrival times are exponential with parameter λ; that is, $a(t) = \lambda e^{-\lambda t}$.

In essence, Theorem 2 states that if the arrival rate is stationary, if bulk arrivals cannot occur, and if past arrivals do not affect future arrivals, then interarrival times will follow an exponential distribution with parameter λ, and the number of arrivals in any interval of length t is Poisson with parameter λt. The assumptions of Theorem 2 may appear to be very restrictive, but interarrival times are often exponential even if the assumptions of Theorem 2 are not satisfied (see Denardo (1982)). In Section 8.12, we discuss how to use data to test whether the hypothesis of exponential interarrival times is reasonable. In many applications, the assumption of exponential interarrival times turns out to be a fairly good approximation of reality.

Using Excel to Compute Poisson and Exponential Probabilities

Excel contains functions that facilitate the computation of probabilities concerning the Poisson and exponential random variables.

The syntax of the Excel POISSON function is as follows:

■ =POISSON(x,MEAN,TRUE) gives the probability that a Poisson random variable with mean = Mean is less than or equal to x.

	A	B	C	D	E
3					
4	Poisson	Lambda			
5	P(X=40)	40	0.541918		
6					
7	P(X<=40) ·	40	0.062947		
8					
9	Exponential				
10		Lambda			
11	P(X<=10)	0.1	0.632121		
12	Density for X = 10	0.1	0.036788		
13					
14					
15					
16					

FIGURE 3

- =POISSON(x,MEAN,FALSE) gives probability that a Poisson random variable with mean = Mean is equal to x.

For example, if an average of 40 customers arrive per hour and arrivals follow a Poisson distribution then the function =POISSON(40,40,TRUE) yields the probability .542 that 40 or fewer customers arrive during an hour. The function =POISSON(40,40,FALSE) yields the probability .063 that *exactly* 40 customers arrive during an hour.

The syntax of the Excel EXPONDIST function is as follows:

- =EXPONDIST(x,LAMBDA,TRUE) gives the probability that an exponential random variable with parameter λ assumes a value less than or equal to x.

- =EXPONDIST(x,LAMBDA,FALSE) gives the value of the density function for an exponential random variable with parameter λ.

For example, suppose the average time between arrivals follows an exponential distribution with mean 10. Then $\lambda = .1$, and =EXPONDIST(10,0.1,TRUE) yields the probability .632 that the time between arrivals is 10 minutes or less.

Poissexp.xls

The function =EXPONDIST(10,.1,FALSE) yields the height .037 of the density function for $x = 10$ and $\lambda = .1$. See file Poissexp.xls and Figure 3.

Example 1 illustrates the relation between the exponential and Poisson distributions.

EXAMPLE 1   Beer Orders

The number of glasses of beer ordered per hour at Dick's Pub follows a Poisson distribution, with an average of 30 beers per hour being ordered.

1 Find the probability that exactly 60 beers are ordered between 10 P.M. and 12 midnight.

2 Find the mean and standard deviation of the number of beers ordered between 9 P.M. and 1 A.M.

3 Find the probability that the time between two consecutive orders is between 1 and 3 minutes.

Solution   **1** The number of beers ordered between 10 P.M. and 12 midnight will follow a Poisson distribution with parameter $2(30) = 60$. From Equation (7), the probability that 60 beers are ordered between 10 P.M. and 12 midnight is

$$\frac{e^{-60}60^{60}}{60!}$$

Alternatively, we can find the answer with the Excel function =POISSON(60,60,FALSE). This yields .051.

2   We have $\lambda = 30$ beers per hour; $t = 4$ hours. Thus, the mean number of beers ordered between 9 P.M. and 1 A.M. is $4(30) = 120$ beers. The standard deviation of the number of beers ordered between 10 P.M. and 1 A.M. is $(120)^{1/2} = 10.95$.

3   Let **X** be the time (in minutes) between successive beer orders. The mean number of orders per minute is exponential with parameter or rate $\frac{30}{60} = 0.5$ beer per minute. Thus, the probability density function of the time between beer orders is $0.5e^{-0.5t}$. Then

$$P(1 \leq \mathbf{X} \leq 3) = \int_1^3 (0.5e^{-0.5t})dt = e^{-0.5} - e^{-1.5} = .38$$

Alternatively, we can use Excel to find the answer with the formula

$$=\text{EXPONDIST}(3,.5,\text{TRUE}) - \text{EXPONDIST}(1,.5,\text{TRUE})$$

This yields a probability of .383.

The Erlang Distribution

If interarrival times do not appear to be exponential, they are often modeled by an Erlang distribution. An Erlang distribution is a continuous random variable (call it **T**) whose density function $f(t)$ is specified by two parameters: a rate parameter R and a shape parameter k (k must be a positive integer). Given values of R and k, the Erlang density has the following probability density function:

$$f(t) = \frac{R(Rt)^{k-1}e^{-Rt}}{(k-1)!} \qquad (t \geq 0) \tag{8}$$

Using integration by parts, we can show that if **T** is an Erlang distribution with rate parameter R and shape parameter k, then

$$E(\mathbf{T}) = \frac{k}{R} \quad \text{and} \quad \text{var } \mathbf{T} = \frac{k}{R^2} \tag{9}$$

To see how varying the shape parameter changes the shape of the Erlang distribution, we consider for a given value of λ, a family of Erlang distributions with rate parameter $k\lambda$ and shape parameter k. By (9), each of these Erlangs has a mean of $\frac{1}{\lambda}$. As k varies, the Erlang distribution takes on many shapes. For example, Figure 4 shows, for a given value of λ, the density functions for Erlang distributions having shape parameters 1, 2, 4, 6, and 20. For $k = 1$, the Erlang density looks similar to an exponential distribution; in fact, if we set $k = 1$ in (8), we find that for $k = 1$, the Erlang distribution is an exponential distribution with parameter R. As k increases, the Erlang distribution behaves more and more like a normal distribution. For extremely large values of k, the Erlang distribution approaches a random variable with zero variance (that is, a constant interarrival time). Thus, by varying k, we may approximate both skewed and symmetric distributions.

It can be shown that an Erlang distribution with shape parameter k and rate parameter $k\lambda$ has the same distribution as the random variable $\mathbf{A}_1 + \mathbf{A}_2 + \cdots + \mathbf{A}_k$, where each $\mathbf{A}_i$ is an exponential random variable with parameter $k\lambda$, and the $\mathbf{A}_i$'s are independent random variables.

If we model interarrival times as an Erlang distribution with shape parameter k, we are really saying that the interarrival process is equivalent to a customer going through k phases (each of which has the no-memory property) before arriving. For this reason, the shape parameter is often referred to as the *number of phases* of the Erlang distribution.

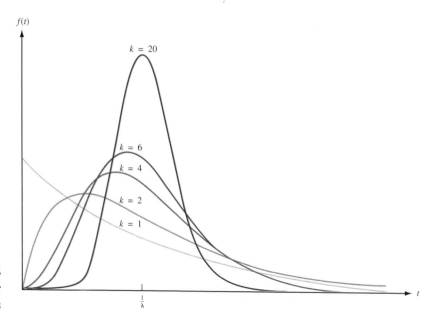

FIGURE 4
Density Functions for
Erlang Distributions

Modeling the Service Process

We now turn our attention to modeling the service process. We assume that the service times of different customers are independent random variables and that each customer's service time is governed by a random variable **S** having a density function $s(t)$. We let $\frac{1}{\mu}$ be the mean service time for a customer. Of course,

$$\frac{1}{\mu} = \int_0^\infty ts(t)dt$$

The variable $\frac{1}{\mu}$ will have units of hours per customer, so μ has units of customers per hour. For this reason, we call μ the service rate. For example, $\mu = 5$ means that if customers were always present, the server could serve an average of 5 customers per hour, and the average service time of each customer would be $\frac{1}{5}$ hour. As with interarrival times, we hope that service times can be accurately modeled as exponential random variables. If we can model a customer's service time as an exponential random variable, we can determine the distribution of a customer's remaining service time without having to keep track of how long the customer has been in service. Also note that if service times follow an exponential density $s(t) = \mu e^{-\mu t}$, then a customer's mean service time will be $\frac{1}{\mu}$.

As an example of how the assumption of exponential service times can simplify computations, consider a three-server system in which each customer's service time is governed by an exponential distribution $s(t) = \mu e^{-\mu t}$. Suppose all three servers are busy, and a customer is waiting (see Figure 5). What is the probability that the customer who is waiting will be the last of the four customers to complete service? From Figure 5, it is clear that the following will occur. One of customers 1–3 (say, customer 3) will be the first to complete service. Then customer 4 will enter service. By the no-memory property, customer 4's service time has the same distribution as the remaining service times of customers 1 and 2. Thus, by symmetry, customers 4, 1, and 2 will have the same chance of being the last customer to complete service. This implies that customer 4 has a $\frac{1}{3}$ chance of being the last customer to complete service. Without the no-memory property, this problem would be hard to solve, because it would be very difficult to determine the prob-

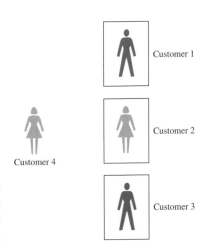

FIGURE 5
Example of Usefulness of Exponential Distribution

Customer 4

Customer 1

Customer 2

Customer 3

ability distribution of the remaining service time (after customer 3 completes service) of customers 1 and 2.

Unfortunately, actual service times may not be consistent with the no-memory property. For this reason, we often assume that $s(t)$ is an Erlang distribution with shape parameter k and rate parameter $k\mu$. From (9), this yields a mean service time of $\frac{1}{\mu}$. Modeling service times as an Erlang distribution with shape parameter k also implies that a customer's service time may be considered to consist of passage through k phases of service, in which the time to complete each phase has the no-memory property and a mean of $\frac{1}{k\mu}$ (see Figure 6). In many situations, an Erlang distribution can be closely fitted to observed service times.

In certain situations, interarrival or service times may be modeled as having zero variance; in this case, interarrival or service times are considered to be **deterministic.** For example, if interarrival times are deterministic, then each interarrival time will be exactly $\frac{1}{\lambda}$, and if service times are deterministic, each customer's service time will be exactly $\frac{1}{\mu}$.

The Kendall–Lee Notation for Queuing Systems

We have now developed enough terminology to describe the standard notation used to describe many queuing systems. The notation that we discuss in this section is used to describe a queuing system in which all arrivals wait in a single line until one of s identical parallel servers is free. Then the first customer in line enters service, and so on (see Figure 7). If, for example, the customer in server 3 is the next customer to complete service, then (assuming an FCFS discipline) the first customer in line would enter server 3. The next customer in line would enter service after the next service completion, and so on.

To describe such a queuing system, Kendall (1951) devised the following notation. Each queuing system is described by six characteristics:

$$1/2/3/4/5/6$$

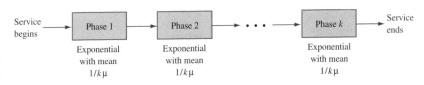

FIGURE 6
Representation of Erlang Service Time

Service begins → Phase 1 → Phase 2 → ··· → Phase k → Service ends

Exponential with mean $1/k\mu$ Exponential with mean $1/k\mu$ Exponential with mean $1/k\mu$

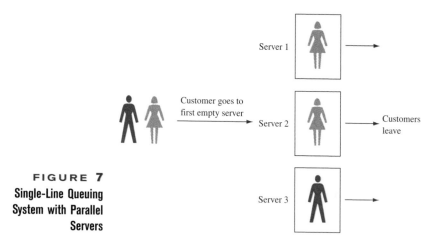

FIGURE 7
Single-Line Queuing System with Parallel Servers

The first characteristic specifies the nature of the arrival process. The following standard abbreviations are used:

M = Interarrival times are independent, identically distributed (iid) random variables having an exponential distribution.

D = Interarrival times are iid and deterministic.

E_k = Interarrival times are iid Erlangs with shape parameter k.

GI = Interarrival times are iid and governed by some general distribution.

The second characteristic specifies the nature of the service times:

M = Service times are iid and exponentially distributed.

D = Service times are iid and deterministic.

E_k = Service times are iid Erlangs with shape parameter k.

G = Service times are iid and follow some general distribution.

The third characteristic is the number of parallel servers. The fourth characteristic describes the queue discipline:

FCFS = First come, first served

LCFS = Last come, first served

SIRO = Service in random order

GD = General queue discipline

The fifth characteristic specifies the maximum allowable number of customers in the system (including customers who are waiting and customers who are in service). The sixth characteristic gives the size of the population from which customers are drawn. Unless the number of potential customers is of the same order of magnitude as the number of servers, the population size is considered to be infinite. In many important models 4/5/6 is $GD/\infty/\infty$. If this is the case, then 4/5/6 is often omitted.

As an illustration of this notation, $M/E_2/8/FCFS/10/\infty$ might represent a health clinic with 8 doctors, exponential interarrival times, two-phase Erlang service times, an FCFS queue discipline, and a total capacity of 10 patients.

The Waiting Time Paradox

We close this section with a brief discussion of an interesting paradox known as the waiting time paradox.

FIGURE 8
The Waiting
Time Paradox

Suppose the time between the arrival of buses at the student center is exponentially distributed, with a mean of 60 minutes. If we arrive at the student center at a randomly chosen instant, what is the average amount of time that we will have to wait for a bus?

The no-memory property of the exponential distribution implies that no matter how long it has been since the last bus arrived, we would still expect to wait an average of 60 minutes until the next bus arrived. This answer is indeed correct, but it appears to be contradicted by the following argument. On the average, somebody who arrives at a random time should arrive in the middle of a typical interval between arrivals of successive buses. If we arrive at the midpoint of a typical interval, and the average time between buses is 60 minutes, then we should have to wait, on the average, $(\frac{1}{2})60 = 30$ minutes for the next bus. Why is this argument incorrect? Simply because the typical interval between buses is *longer* than 60 minutes. The reason for this anomaly is that we are more likely to arrive during a longer interval than a shorter interval. Let's simplify the situation by assuming that half of all buses run 30 minutes apart and half of all buses run 90 minutes apart. One might think that since the average time between buses is 60 minutes, the average wait for a bus would be $(\frac{1}{2})60 = 30$ minutes, but this is incorrect. Look at a typical sequence of bus interarrival times (see Figure 8). Half of the interarrival times are 30 minutes, and half are 90 minutes. Clearly, there is a $\frac{90}{30+90} = \frac{3}{4}$ chance that one will arrive during a 90-minute interarrival time and a $\frac{30}{30+90} = \frac{1}{4}$ chance that one will arrive during a 30-minute interarrival time. Thus, the average-size interarrival time into which a customer arrives is $(\frac{3}{4})(90) + (\frac{1}{4})(30) = 75$ minutes. Since we do arrive, on the average, in the middle of an interarrival time, our average wait will be $(\frac{3}{4})(\frac{1}{2})90 + (\frac{1}{4})(\frac{1}{2})30 = 37.5$ minutes, which is longer than 30 minutes.

Returning to the case where interarrival times are exponential with mean 60 minutes, the average size of a typical interarrival time turns out to be 120 minutes. Thus, the average time that we will have to wait for a bus is $(\frac{1}{2})(120) = 60$ minutes. Note that if buses *always* arrived 60 minutes apart, then the average time a person would have to wait for a bus would be $(\frac{1}{2})(60) = 30$ minutes. In general, it can be shown that if **A** is the random variable for the time between buses, then the average time until the next bus (as seen by an arrival who is equally likely to come at any time) is given by

$$\frac{1}{2}\left(E(\mathbf{A}) + \frac{\text{var } \mathbf{A}}{E(\mathbf{A})}\right)$$

For our bus example, $\lambda = \frac{1}{60}$, so Equations (3) and (4) show that $E(\mathbf{A}) = 60$ minutes and var $\mathbf{A} = 3,600$ minutes^2. Substituting into this formula yields

$$\text{Expected waiting time} = \frac{1}{2}(60 + \frac{3,600}{60}) = 60 \text{ minutes}$$

PROBLEMS

Group A

1 Suppose I arrive at an $M/M/7/FCFS/8/\infty$ queuing system when all servers are busy. What is the probability that I will complete service before at least one of the seven customers in service?

2 The time between buses follows the mass function shown in Table 2. What is the average length of time one must wait for a bus?

TABLE 2

Time Between Buses	Probability
30 minutes	$\frac{1}{4}$
1 hour	$\frac{1}{4}$
2 hours	$\frac{1}{2}$

3 There are four sections of the third grade at Jefferson Elementary School. The number in each section is as follows: section 1, 20 students; section 2, 25 students; section 3, 35 students; section 4, 40 students. What is the average size of a third-grade section? Suppose the board of education randomly selects a Jefferson third-grader. On the average, how many students will be in her class?

4 The time between arrivals of buses follows an exponential distribution, with a mean of 60 minutes.

  a What is the probability that exactly four buses will arrive during the next 2 hours?

  b That at least two buses will arrive during the next 2 hours?

  c That no buses will arrive during the next 2 hours?

d A bus has just arrived. What is the probability that it will be between 30 and 90 minutes before the next bus arrives?

5 During the year 2000, there was an average of .022 car accident per person in the United States. Using your knowledge of the Poisson random variable, explain the truth in the statement, "Most drivers are better than average."

6 Suppose it is equally likely that a plane flight is 50%, 60%, 70%, 80%, or 90% full.

  a What fraction of seats on a typical flight are full? This is known as the *flight load factor*.

  b We are always complaining that there are never empty seats on our plane flights. Given the previous information, what is the average load factor on a plane trip I take?

7 An average of 12 jobs per hour arrive at our departmental printer.

  a Use two different computations (one involving the Poisson and another the exponential random variable) to determine the probability that no job will arrive during the next 15 minutes.

  b What is the probability that 5 or fewer jobs will arrive during the next 30 minutes?

8.3 Birth–Death Processes

In this section, we discuss the important idea of a birth–death process. We subsequently use birth–death processes to answer questions about several different types of queuing systems.

We define the number of people present in any queuing system at time t to be the **state** of the queuing system at time t. For $t = 0$, the state of the system will equal the number of people initially present in the system. Of great interest to us is the quantity $P_{ij}(t)$ which is defined as the probability that j people will be present in the queuing system at time t, given that at time 0, i people are present. Note that $P_{ij}(t)$ is analogous to the n-step transition probability $P_{ij}(n)$ (the probability that after n transitions, a Markov chain will be in state j, given that the chain began in state i), discussed in Chapter 5. Recall that for most Markov chains, the $P_{ij}(n)$ approached a limit π_j, which was independent of the initial state i. Similarly, it turns out that for many queuing systems, $P_{ij}(t)$ will, for large t, approach a limit π_j, which is independent of the initial state i. We call π_j the **steady state,** or equilibrium probability, of state j.

For the queuing systems that we will discuss, π_j may be thought of as the probability that at an instant in the distant future, j customers will be present. Alternatively, π_j may be thought of (for time in the distant future) as the fraction of the time that j customers are present. In most queuing systems, the value of $P_{ij}(t)$ for small t will critically depend on i, the number of customers initially present. For example, if t is small, then we would expect that $P_{50,1}(t)$ and $P_{1,1}(t)$ would differ substantially. However, if steady-state probabilities exist, then for large t, both $P_{50,1}(t)$ and $P_{1,1}(t)$ will be near π_1. The question of how large t must be before the steady state is approximately reached is difficult to answer. The behavior of $P_{ij}(t)$ before the steady state is reached is called the **transient behavior** of the queuing system. Analysis of the system's transient behavior will be discussed in Section 8.16. For now, when we analyze the behavior of a queuing system, we assume that the steady state has been reached. This allows us to work with the π_j's instead of the $P_{ij}(t)$'s.

We now discuss a certain class of continuous-time stochastic processes, called birth–death processes, which includes many interesting queuing systems. For a birth–death process, it is easy to determine the steady-state probabilities (if they exist).

A **birth–death process** is a continuous-time stochastic process for which the system's state at any time is a nonnegative integer (see Section 5.1 for a definition of a continuous-time stochastic process). If a birth–death process is in state j at time t, then the motion of the process is governed by the following laws.

Laws of Motion for Birth–Death Processes

Law 1 With probability $\lambda_j\Delta t + o(\Delta t)$, a birth occurs between time t and time $t + \Delta t$.[†] A birth increases the system state by 1, to $j + 1$. The variable λ_j is called the **birth rate** in state j. In most queuing systems, a birth is simply an arrival.

Law 2 With probability $\mu_j\Delta t + o(\Delta t)$, a death occurs between time t and time $t + \Delta t$. A death decreases the system state by 1, to $j - 1$. The variable μ_j is the **death rate** in state j. In most queuing systems, a death is a service completion. Note that $\mu_0 = 0$ must hold, or a negative state could occur.

Law 3 Births and deaths are independent of each other.

Laws 1–3 can be used to show that the probability that more than one event (birth or death) occurs between t and $t + \Delta t$ is $o(\Delta t)$. Note that any birth–death process is completely specified by knowledge of the birth rates λ_j and the death rates μ_j. Since a negative state cannot occur, any birth–death process must have $\mu_0 = 0$.

Relation of Exponential Distribution to Birth–Death Processes

Most queuing systems with exponential interarrival times and exponential service times may be modeled as birth–death processes. To illustrate why this is so, consider an $M/M/1/FCFS/\infty/\infty$ queuing system in which interarrival times are exponential with parameter λ and service times are exponentially distributed with parameter μ. If the state (number of people present) at time t is j, then the no-memory property of the exponential distribution implies that the probability of a birth during the time interval $[t, t + \Delta t]$ will not depend on how long the system has been in state j. This means that the probability of a birth occurring during $[t, t + \Delta t]$ will not depend on how long the system has been in state j and thus may be determined as if an arrival had just occurred at time t. Then the probability of a birth occurring during $[t, t + \Delta t]$ is

$$\int_0^{\Delta t} \lambda e^{-\lambda t} dt = 1 - e^{-\lambda \Delta t}$$

By the Taylor series expansion given in Section 1.1,

$$e^{-\lambda \Delta t} = 1 - \lambda \Delta t + o(\Delta t)$$

This means that the probability of a birth occurring during $[t, t + \Delta t]$ is $\lambda \Delta t + o(\Delta t)$. From this we may conclude that the birth rate in state j is simply the arrival rate λ.

To determine the death rate at time t, note that if the state is zero at time t, then nobody is in service, so no service completion can occur between t and $t + \Delta t$. Thus, $\mu_0 = 0$.

[†]Recall from Section 8.2 that $o(\Delta t)$ means that $\lim\limits_{\Delta t \to 0} \dfrac{o(\Delta t)}{\Delta t} = 0$.

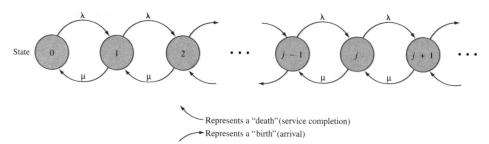

FIGURE 9

Rate Diagram for
M/*M*/1/FCFS/∞/∞
Queuing System

Represents a "death" (service completion)
Represents a "birth" (arrival)

If the state at time t is $j \geq 1$, then we know (since there is only one server) that exactly one customer will be in service. The no-memory property of the exponential distribution then implies that the probability that a customer will complete service between t and $t + \Delta t$ is given by

$$\int_0^{\Delta t} \mu e^{-\mu t} dt = 1 - e^{-\mu \Delta t} = \mu \Delta t + o(\Delta t)$$

Thus, for $j \geq 1$, $\mu_j = \mu$. In summary, if we assume that service completions and arrivals occur independently, then an $M/M/1/\text{FCFS}/\infty/\infty$ queuing system is a birth–death process. The birth and death rates for the $M/M/1/\text{FCFS}/\infty/\infty$ queuing system may be represented in a rate diagram (see Figure 9).

More complicated queuing systems with exponential interarrival times and exponential service times may often be modeled as birth–death processes by adding the service rates for occupied servers and adding the arrival rates for different arrival streams. For example, consider an $M/M/3/\text{FCFS}/\infty/\infty$ queuing system in which interarrival times are exponential with $\lambda = 4$ and service times are exponential with $\mu = 5$. To model this system as a birth–death process, we would use the following parameters (see Figure 10):

$$\lambda_j = 4 \qquad\qquad\qquad\qquad\qquad\qquad\qquad (j = 0, 1, 2, \ldots)$$
$$\mu_0 = 0, \quad \mu_1 = 5, \quad \mu_2 = 5 + 5 = 10, \quad \mu_j = 5 + 5 + 5 = 15 \quad (j = 3, 4, 5, \ldots)$$

If either interarrival times or service times are nonexponential, then the birth–death process model is not appropriate.[†] Suppose, for example, that service times are not exponential and we are considering an $M/G/1/\text{FCFS}/\infty/\infty$ queuing system. Since the service times for an $M/G/1/\text{FCFS}/\infty/\infty$ system may be nonexponential, the probability that a death (service completion) occurs between t and $t + \Delta t$ will depend on the time since the last service completion. This violates law 2, so we cannot model an $M/G/1/\text{FCFS}/\infty/\infty$ system as a birth–death process.

FIGURE 10

Rate Diagram for
M/*M*/3/FCFS/∞/∞
Queuing System

$\lambda = 4 \quad \mu = 5$

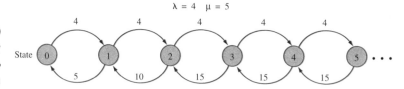

[†]A modified birth–death model can be developed if service times and interarrival times are Erlang distributions.

Derivation of Steady-State Probabilities for Birth–Death Processes

We now show how the π_j's may be determined for an arbitrary birth–death process. The key is to relate (for small Δt) $P_{ij}(t + \Delta t)$ to $P_{ij}(t)$. The way to do this is to note that there are four ways for the state at time $t + \Delta t$ to be j. For $j \geq 1$, the four ways are shown in Table 3. For $j \geq 1$, the probability that the state of the system will be $j - 1$ at time t and j at time $t + \Delta t$ is (see Figure 11)

$$P_{i,j-1}(t)(\lambda_{j-1}\Delta t + o(\Delta t))$$

Similar arguments yield (II) and (III). (IV) follows, because if the system is in a state other than j, $j - 1$, or $j + 1$ at time t, then to end up in state j at time $t + \Delta t$, more than one event (birth or death) must occur between t and $t + \Delta t$. By law 3, this has probability $o(\Delta t)$. Thus,

$$P_{ij}(t + \Delta t) = (\text{I}) + (\text{II}) + (\text{III}) + (\text{IV})$$

After regrouping terms in this equation, we obtain

$$\begin{aligned} P_{ij}(t + \Delta t) = {} & P_{ij}(t) \\ & + \Delta t(\lambda_{j-1}P_{i,j-1}(t) + \mu_{j+1}P_{i,j+1}(t) - P_{ij}(t)\mu_j - P_{ij}(t)\lambda_j) \\ & + o(\Delta t)(P_{i,j-1}(t) + P_{i,j+1}(t) + 1 - 2P_{ij}(t)) \end{aligned} \qquad (10)$$

Since the underlined term may be written as $o(\Delta t)$, we rewrite (10) as

$$P_{ij}(t + \Delta t) - P_{ij}(t) = \Delta t(\lambda_{j-1}P_{i,j-1}(t) + \mu_{j+1}P_{i,j+1}(t) - P_{ij}(t)\mu_j - P_{ij}(t)\lambda_j) + o(\Delta t)$$

Dividing both sides of this equation by Δt and letting Δt approach zero, we see that for all i and $j \geq 1$,

$$P'_{ij}(t) = \lambda_{j-1}P_{i,j-1}(t) + \mu_{j+1}P_{i,j+1}(t) - P_{ij}(t)\mu_j - P_{ij}(t)\lambda_j \qquad (10')$$

Since for $j = 0$, $P_{i,j-1}(t) = 0$ and $\mu_j = 0$, we obtain, for $j = 0$,

$$P'_{i,0}(t) = \mu_1 P_{i,1}(t) - \lambda_0 P_{i,0}(t)$$

This is an infinite system of differential equations. (A differential equation is simply an equation in which a derivative appears.) In theory, these equations may be solved for the $P_{ij}(t)$. In reality, however, this system of equations is usually extremely difficult to solve. All is not lost, however. We can use (10') to obtain the steady-state probabilities π_j ($j = 0, 1, 2, \ldots$). As with Markov chains, we define the steady-state probability π_j to be

$$\lim_{t \to \infty} P_{ij}(t)$$

Then for large t and any initial state i, $P_{ij}(t)$ will not change very much and may be thought of as a constant. Thus, in the steady state (t large), $P'_{ij}(t) = 0$. In the steady state,

TABLE 3
Computations of Probability That State at Time $t + \Delta t$ Is j

State at Time t	State at Time $t + \Delta t$	Probability of This Sequence of Events
$j - 1$	j	$P_{i,j-1}(t)\,(\lambda_{j-1}\Delta t + o(\Delta t)) = (\text{I})$
$j + 1$	j	$P_{i,j+1}(t)\,(\mu_{j+1}\Delta t + o(\Delta t)) = (\text{II})$
j	j	$P_{i,j}(t)\,(1 - \mu_j \Delta t - \lambda_j \Delta t - 2o(\Delta t)) = (\text{III})$
Any other state	j	$o(\Delta t) = (\text{IV})$

FIGURE **11**
Probability That State
Is $j - 1$ at Time t and
j at Time $t + \Delta t$ Is
$P_{i,j-1}(t)(\lambda_{j-1}(\Delta t) + o(\Delta t))$

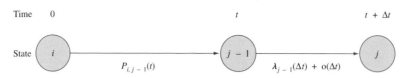

also, $P_{i,j-1}(t) = \pi_{j-1}$, $P_{i,j+1}(t) = \pi_{j+1}$, and $P_{ij}(t) = \pi_j$ will all hold. Substituting these relations into (10'), we obtain, for $j \geq 1$,

$$\lambda_{j-1}\pi_{j-1} + \mu_{j+1}\pi_{j+1} - \pi_j\mu_j - \pi_j\lambda_j = 0 \qquad (10'')$$

$$\lambda_{j-1}\pi_{j-1} + \mu_{j+1}\pi_{j+1} = \pi_j(\lambda_j + \mu_j) \qquad (j = 1, 2, \ldots)$$

For $j = 0$, we obtain

$$\mu_1\pi_1 = \pi_0\lambda_0$$

Equations (10'') are an infinite system of *linear* equations that can be easily solved for the π_j's. Before discussing how to solve (10''), we give an intuitive derivation of (10''), based on the following observation: *At any time t that we observe a birth–death process, it must be true that for each state j, the number of times we have entered state j differs by at most 1 from the number of times we have left state j.*

Suppose that by time t, we have entered state 6 three times. Then one of the cases in Table 4 must have occurred. For example, if Case 2 occurs, we begin in state 6 and end up in some other state. Since we have observed three transitions into state 6 by time t, the following events (among others) must have occurred:

Start in state 6	Enter state 6 (second time)
Leave state 6 (first time)	Leave state 6 (third time)
Enter state 6 (first time)	Enter state 6 (third time)
Leave state 6 (second time)	Leave state 6 (fourth time)

Hence, if Case 2 occurs, then by time t, we must have left state 6 four times.

This observation suggests that for large t and for $j = 0, 1, 2, \ldots$ (and for any initial conditions), it will be true that

$$\frac{\text{Expected no. of departures from state } j}{\text{Unit time}}$$
$$= \frac{\text{Expected no. of entrances into state } j}{\text{Unit time}} \qquad (11)$$

Assuming the system has settled down into the steady state, we know that the system spends a fraction π_j of its time in state j. We can now use (11) to determine the steady-state probabilities π_j. For $j \geq 1$, we can only leave state j by going to state $j + 1$ or state $j - 1$, so for $j \geq 1$, we obtain

$$\frac{\text{Expected no. of departures from state } j}{\text{Unit time}} = \pi_j(\lambda_j + \mu_j) \qquad (12)$$

Since for $j \geq 1$ we can only enter state j from state $j - 1$ or state $j + 1$,

$$\frac{\text{Expected no. of entrances into state } j}{\text{Unit time}} = \pi_{j-1}\lambda_{j-1} + \pi_{j+1}\mu_{j+1} \qquad (13)$$

Substituting (12) and (13) into (11) yields

$$\pi_{j-1}\lambda_{j-1} + \pi_{j+1}\mu_{j+1} = \pi_j(\lambda_j + \mu_j) \qquad (j = 1, 2, \ldots) \qquad (14)$$

TABLE 4
Relation between Number of Transitions into and out of a State by Time t

Initial State	State of Time t	Number of Transitions Out of State 6 by Time t
Case 1: state 6	State 6	3
Case 2: state 6	Any state except 6	4
Case 3: any state except state 6	State 6	2
Case 4: any state except state 6	Any state except 6	3

For $j = 0$, we know that $\mu_0 = \pi_{-1} = 0$, so we also have

$$\pi_1\mu_1 = \pi_0\lambda_0 \tag{14'}$$

Equations (14) and (14′) are often called the **flow balance equations,** or **conservation of flow equations,** for a birth–death process. Note that (14) expresses the fact that in the steady state, the rate at which transitions occur into any state i must equal the rate at which transitions occur out of state i. If (14) did not hold for all states, then probability would "pile up" at some state, and a steady state would not exist.

Writing out the equations for (14) and (14′), we obtain the flow balance equations for a birth–death process:

$$
\begin{array}{lll}
(j = 0) & \pi_0\lambda_0 = \pi_1\mu_1 \\
(j = 1) & (\lambda_1 + \mu_1)\pi_1 = \lambda_0\pi_0 + \mu_2\pi_2 \\
(j = 2) & (\lambda_2 + \mu_2)\pi_2 = \lambda_1\pi_1 + \mu_3\pi_3 \\
& \quad\vdots \\
(j\text{th equation}) & (\lambda_j + \mu_j)\pi_j = \lambda_{j-1}\pi_{j-1} + \mu_{j+1}\pi_{j+1}
\end{array}
\tag{15}
$$

Solution of Birth–Death Flow Balance Equations

To solve (15), we begin by expressing all the π_j's in terms of π_0. From the ($j = 0$) equation, we obtain

$$\pi_1 = \frac{\pi_0\lambda_0}{\mu_1}$$

Substituting this result into the ($j = 1$) equation yields

$$\lambda_0\pi_0 + \mu_2\pi_2 = \frac{(\lambda_1 + \mu_1)\pi_0\lambda_0}{\mu_1}$$

$$\mu_2\pi_2 = \frac{\pi_0(\lambda_0\lambda_1)}{\mu_1}$$

Thus,

$$\pi_2 = \frac{\pi_0(\lambda_0\lambda_1)}{\mu_1\mu_2}$$

We could now use the ($j = 3$) equation to solve for π_3 in terms of π_0 and so on. If we define

$$c_j = \frac{\lambda_0\lambda_1 \cdots \lambda_{j-1}}{\mu_1\mu_2 \cdots \mu_j}$$

then it can be shown that

$$\pi_j = \pi_0 c_j \tag{16}$$

(See Problem 1 at the end of this section.) Since at any given time, we must be in some state, the steady-state probabilities must sum to 1:

$$\sum_{j=0}^{j=\infty} \pi_j = 1 \tag{17}$$

Substituting (16) into (17) yields

$$\pi_0 \left(1 + \sum_{j=1}^{j=\infty} c_j \right) = 1 \tag{18}$$

If $\sum_{j=1}^{j=\infty} c_j$ is finite, we can use (18) to solve for π_0:

$$\pi_0 = \frac{1}{1 + \sum_{j=1}^{j=\infty} c_j} \tag{19}$$

Then (16) can be used to determine $\pi_1, \pi_2, \ldots$. It can be shown that if $\sum_{j=1}^{j=\infty} c_j$ is infinite, then no steady-state distribution exists. The most common reason for a steady-state failing to exist is that the arrival rate is at least as large as the maximum rate at which customers can be served.

Using a Spreadsheet to Compute Steady-State Probabilities

The following example illustrates how a spreadsheet can be used to compute steady-state probabilities for a birth–death process.

EXAMPLE 2   Indiana Bell

Indiana Bell customer service representatives receive an average of 1,700 calls per hour. The time between calls follows an exponential distribution. A customer service representative can handle an average of 30 calls per hour. The time required to handle a call is also exponentially distributed. Indiana Bell can put up to 25 people on hold. If 25 people are on hold, a call is lost to the system. Indiana Bell has 75 service representatives.

1   What fraction of the time are all operators busy?

2   What fraction of all calls are lost to the system?

Bell.xls   Solution   In Figure 12 (file Bell.xls), we set up a spreadsheet to compute the steady-state probabilities for this birth–death process. We let the state i at any time equal the number of callers whose calls are being processed or are on hold. We have that for $i = 0, 1, 2, \ldots, 99$, $\lambda_i = 1,700$. The fact that any calls received when $75 + 25 = 100$ calls are in the system are lost to the system implies that $\lambda_{100} = 0$. Then no state $i > 100$ can occur (why?). We have $u_0 = 0$ and for $i = 1, 2, \ldots, 75$, $\mu_i = 30i$. For $i > 75$, $\mu_i = 30(75) = 2,250$.

To answer parts (1) and (2), we need to compute the steady-state probabilities $\pi_i =$ fraction of the time the state is i. In cells A4:A104, we enter the possible states of the system (0–100). To do this, enter 0 in cell A4 and 1 in A5. Then select the range A4:A5 and drag the cursor to A6:A104. In B4, type the arrival rate of 1,700 and just drag the cursor

	A	B	C	D	E	F
1						Prob(i>=75)
2	INDIANA	BELL	EXAMPLE		0	.012759326
3	STATE	LAMBDA	MU	CJ	PROB	
4	0	1700	0	1	2.451E-25	
5	1	1700	30	56.6666667	1.3889E-23	
6	2	1700	60	1605.55556	3.9352E-22	
7	3	1700	90	30327.1605	7.4332E-21	
8	4	1700	120	429634.774	1.053E-19	
9	5	1700	150	4869194.1	1.1934E-18	
10	6	1700	180	45986833.2	1.1271E-17	
11	7	1700	210	372274364	9.1244E-17	
12	8	1700	240	2636943411	6.4631E-16	
13	9	1700	270	1.6603E+10	4.0694E-15	
14	10	1700	300	9.4084E+10	2.306E-14	
15	11	1700	330	4.8467E+11	1.1879E-13	
16	12	1700	360	2.2887E+12	5.6097E-13	
17	13	1700	390	9.9765E+12	2.4452E-12	
18	14	1700	420	4.0381E+13	9.8974E-12	
19	15	1700	450	1.5255E+14	3.739E-11	
20	16	1700	480	5.4029E+14	1.3242E-10	
21	17	1700	510	1.801E+15	4.4141E-10	
22	18	1700	540	5.6697E+15	1.3896E-09	
23	19	1700	570	1.691E+16	4.1445E-09	
24	20	1700	600	4.791E+16	1.1743E-08	
25	21	1700	630	1.2928E+17	3.1687E-08	
26	22	1700	660	3.33E+17	8.1618E-08	
27	23	1700	690	8.2043E+17	2.0109E-07	
28	24	1700	720	1.9371E+18	4.7479E-07	
29	25	1700	750	4.3908E+18	1.0762E-06	
30	26	1700	780	9.5697E+18	2.3455E-06	
31	27	1700	810	2.0085E+19	4.9227E-06	
32	28	1700	840	4.0648E+19	9.9627E-06	
33	29	1700	870	7.9426E+19	1.9467E-05	
34	30	1700	900	1.5003E+20	3.6772E-05	
35	31	1700	930	2.7424E+20	6.7217E-05	
36	32	1700	960	4.8564E+20	0.00011903	
37	33	1700	990	8.3393E+20	0.00020439	
38	34	1700	1020	1.3899E+21	0.00034066	
39	35	1700	1050	2.2503E+21	0.00055154	
40	36	1700	1080	3.5421E+21	0.00086817	
41	37	1700	1110	5.4248E+21	0.00132962	
42	38	1700	1140	8.0897E+21	0.00198277	
43	39	1700	1170	1.1754E+22	0.00288095	
44	40	1700	1200	1.6652E+22	0.00408134	
45	41	1700	1230	2.3015E+22	0.00564088	
46	42	1700	1260	3.1052E+22	0.00761072	
47	43	1700	1290	4.0921E+22	0.01002963	
48	44	1700	1320	5.2701E+22	0.01291694	
49	45	1700	1350	6.6364E+22	0.01626578	
50	46	1700	1380	8.1753E+22	0.02003755	
51	47	1700	1410	9.8567E+22	0.02415875	
52	48	1700	1440	1.1636E+23	0.02852075	
53	49	1700	1470	1.3457E+23	0.03298318	
54	50	1700	1500	1.5251E+23	0.03738094	
55	51	1700	1530	1.6946E+23	0.04153437	
56	52	1700	1560	1.8467E+23	0.04526182	
57	53	1700	1590	1.9744E+23	0.04839314	
58	54	1700	1620	2.0719E+23	0.05078292	
59	55	1700	1650	2.1347E+23	0.0523218	
60	56	1700	1680	2.1601E+23	0.05294468	

FIGURE 12
Indiana Bell

A	A	B	C	D	E	F
6 1	57	1700	1710	2.1E+23	0.0526351	
6 2	58	1700	1740	2.1E+23	0.0514251	
6 3	59	1700	1770	2.0E+23	0.0493913	
6 4	60	1700	1800	1.9E+23	0.0466473	
6 5	61	1700	1830	1.8E+23	0.0433336	
6 6	62	1700	1860	1.6E+23	0.039606	
6 7	63	1700	1890	1.5E+23	0.0356244	
6 8	64	1700	1920	1.3E+23	0.0315425	
6 9	65	1700	1950	1.1E+23	0.0274985	
7 0	66	1700	1980	9.6E+22	0.0236099	
7 1	67	1700	2010	8.1E+22	0.0199685	
7 2	68	1700	2040	6.8E+22	0.0166405	
7 3	69	1700	2070	5.6E+22	0.0136661	
7 4	70	1700	2100	4.5E+22	0.011063	
7 5	71	1700	2130	3.6E+22	0.0088296	
7 6	72	1700	2160	2.8E+22	0.0069492	
7 7	73	1700	2190	2.2E+22	0.0053944	
7 8	74	1700	2220	1.7E+22	0.0041308	
7 9	75	1700	2250	1.3E+22	0.0031211	
8 0	76	1700	2250	9.6E+21	0.0023581	
8 1	77	1700	2250	7.3E+21	0.0017817	
8 2	78	1700	2250	5.5E+21	0.0013462	
8 3	79	1700	2250	4.1E+21	0.0010171	
8 4	80	1700	2250	3.1E+21	0.0007685	
8 5	81	1700	2250	2.4E+21	0.0005806	
8 6	82	1700	2250	1.8E+21	0.0004387	
8 7	83	1700	2250	1.4E+21	0.0003315	
8 8	84	1700	2250	1.0E+21	0.0002504	
8 9	85	1700	2250	7.7E+20	0.0001892	
9 0	86	1700	2250	5.8E+20	0.000143	
9 1	87	1700	2250	4.4E+20	0.000108	
9 2	88	1700	2250	3.3E+20	0.0000816	
9 3	89	1700	2250	2.5E+20	0.0000617	
9 4	90	1700	2250	1.9E+20	0.0000466	
9 5	91	1700	2250	1.4E+20	0.0000352	
9 6	92	1700	2250	1.1E+20	0.0000266	
9 7	93	1700	2250	8.2E+19	0.0000201	
9 8	94	1700	2250	6.2E+19	0.0000152	
9 9	95	1700	2250	4.7E+19	0.0000115	
1 0 0	96	1700	2250	3.5E+19	0.0000087	
1 0 1	97	1700	2250	2.7E+19	0.0000065	
1 0 2	98	1700	2250	2.0E+19	0.0000049	
1 0 3	99	1700	2250	1.5E+19	0.0000037	
1 0 4	100	0	2250	1.2E+19	0.0000028	

FIGURE 12
(Continued)

down to B5:B104 to create the arrival rates for all states. To create the service rates, enter 0 in cell C4. Then enter 30 in C5 and 60 in cell C6. Then select the range C5:C6 and drag the cursor down to C79. This creates the service rates for states 0–75. In C80, enter 2,250 and drag that result down to C81:C104. This creates the service rate (2,250) for states 76–100. In the cell range D4:D104, we calculate the c_j's that are needed to compute the steady-state probabilities. To begin, we enter a 1 in D4. Since $c_1 = \lambda_0/\mu_1$, we enter =B4/C5 in cell D5. Since $c_2 = c_1\lambda_1/\mu_2$, we enter =D5*B5/C6 into D6. Copying from D6 to D7:D104 now generates the rest of the c_j's. In E4, we compute π_0 by entering =SUM(D$4:D$104). In E5, we compute π_1 by entering =D5*E$4. Copying from the range E5 to the range E5:E104 generates the rest of the steady-state probabilities. We can now answer questions (1) and (2).

1 We seek $\pi_{75} + \pi_{76} + \cdots + \pi_{100}$. To obtain this, we enter the command =SUM(E79:E104) in cell F2 and obtain .013.

2 An arriving call is turned away if the state equals 100. A fraction $\pi_{100} = .0000028$ of all arrivals will be turned away. Thus, the phone company is providing very good service!

In Sections 8.4–8.6 and 8.9–8.10, we apply the theory of birth–death processes to determine the steady-state probability distributions for a variety of queuing systems. Then we use the steady-state probability distributions to determine other quantities of interest (such as expected waiting time and expected number of customers in the system).

Birth–death models have been used to model phenomena other than queuing systems. For example, the number of firms in an industry can be modeled as a birth–death process: The state of the industry at any given time is the number of firms that are in business; a birth corresponds to a firm entering the industry; and a death corresponds to a firm going out of business.

PROBLEMS

Group A

1 Show that the values of the π_j's given in (16) do indeed satisfy the flow balance equations (14) and (14′).

2 My home uses two light bulbs. On average, a light bulb lasts for 22 days (exponentially distributed). When a light bulb burns out, it takes an average of 2 days (exponentially distributed) before I replace the bulb.

a Formulate a three-state birth–death model of this situation.

b Determine the fraction of the time that both light bulbs are working.

c Determine the fraction of the time that no light bulbs are working.

Group B

3 You are doing an industry analysis of the Bloomington pizza industry. The rate (per year) at which pizza restaurants enter the industry is given by p, where p = price of a pizza in dollars. The price of a pizza is assumed to be max(0, $16 - .5F$), where F = number of pizza restaurants in Bloomington. During a given year, the probability that a pizza restaurant fails is $1/(10 + p)$. Create a birth–death model of this situation.

a In the steady state, estimate the average number of pizza restaurants in Bloomington.

b What fraction of the time will there be more than 20 pizza restaurants in Bloomington?

8.4 The *M*/*M*/1/*GD*/∞/∞ Queuing System and the Queuing Formula $L = \lambda W$

We now use the birth–death methodology explained in the previous section to analyze the properties of the *M*/*M*/1/*GD*/∞/∞ queuing system. Recall that the *M*/*M*/1/*GD*/∞/∞ queuing system has exponential interarrival times (we assume that the arrival rate per unit time is λ) and a single server with exponential service times (we assume that each customer's service time is exponential with rate μ). In Section 8.3, we showed that an *M*/*M*/1/*GD*/∞/∞ queuing system may be modeled as a birth–death process with the following parameters:

$$\lambda_j = \lambda \quad (j = 0, 1, 2, \ldots)$$
$$\mu_0 = 0 \tag{20}$$
$$\mu_j = \mu \quad (j = 1, 2, 3, \ldots)$$

Derivation of Steady-State Probabilities

We can use Equations (15)–(19) to solve for π_j, the steady-state probability that j customers will be present. Substituting (20) into (16) yields

$$\pi_1 = \frac{\lambda \pi_0}{\mu}, \qquad \pi_2 = \frac{\lambda^2 \pi_0}{\mu^2}, \qquad \cdots, \qquad \pi_j = \frac{\lambda^j \pi_0}{\mu^j} \qquad (21)$$

We define $\rho = \frac{\lambda}{\mu}$. For reasons that will become apparent later, we call ρ the **traffic intensity** of the queuing system. Substituting (21) into (17) yields

$$\pi_0(1 + \rho + \rho^2 + \cdots) = 1 \qquad (22)$$

We now assume that $0 \leq \rho < 1$. Then we evaluate the sum $S = 1 + \rho + \rho^2 + \cdots$ as follows: Multiplying S by ρ yields $\rho S = \rho + \rho^2 + \rho^3 + \cdots$. Then $S - \rho S = 1$, and

$$S = \frac{1}{1 - \rho} \qquad (23)$$

Substituting (23) into (22) yields

$$\pi_0 = 1 - \rho \qquad (0 \leq \rho < 1) \qquad (24)$$

Substituting (24) into (21) yields

$$\pi_j = \rho^j(1 - \rho) \qquad (0 \leq \rho < 1) \qquad (25)$$

If $\rho \geq 1$, however, the infinite sum in (22) "blows up" (try $\rho = 1$, for example, and you get $1 + 1 + 1 + \cdots$). Thus, if $\rho \geq 1$, no steady-state distribution exists. Since $\rho = \frac{\lambda}{\mu}$, we see that if $\lambda \geq \mu$ (that is, the arrival rate is at least as large as the service rate), then no steady-state distribution exists.

If $\rho > 1$, it is easy to see why no steady-state distribution can exist. Suppose $\lambda = 6$ customers per hour and $\mu = 4$ customers per hour. Even if the server were working all the time, she could only serve an average of 4 people per hour. Thus, the average number of customers in the system would grow by at least $6 - 4 = 2$ customers per hour. This means that after a long time, the number of customers present would "blow up," and no steady-state distribution could exist. If $\rho = 1$, the nonexistence of a steady state is not quite so obvious, but our analysis does indicate that no steady state exists.

Derivation of L

Throughout the rest of this section, we assume that $\rho < 1$, ensuring that a steady-state probability distribution, as given in (25), does exist. We now use the steady-state probability distribution in (25) to determine several quantities of interest. For example, assuming that the steady state has been reached, the average number of customers present in the queuing system (call it L) is given by

$$L = \sum_{j=0}^{j=\infty} j\pi_j = \sum_{j=0}^{j=\infty} j\rho^j(1 - \rho)$$

$$= (1 - \rho) \sum_{j=0}^{j=\infty} j\rho^j$$

Defining

$$S' = \sum_{j=0}^{j=\infty} j\rho^j = \rho + 2\rho^2 + 3\rho^3 + \cdots$$

we see that $\rho S' = \rho^2 + 2\rho^3 + 3\rho^4 + \cdots$. Subtracting yields

$$S' - \rho S' = \rho + \rho^2 + \cdots = \frac{\rho}{1 - \rho}$$

Thus,

$$S' = \frac{\rho}{(1 - \rho)^2}$$

and

$$L = (1 - \rho)\frac{\rho}{(1 - \rho)^2} = \frac{\rho}{1 - \rho} = \frac{\lambda}{\mu - \lambda} \tag{26}$$

Derivation of L_q

In some circumstances, we are interested in the expected number of people waiting in line (or in the queue). We denote this number by L_q. Note that if 0 or 1 customer is present in the system, then nobody is waiting in line, but if j people are present ($j \geq 1$), there will be $j - 1$ people waiting in line. Thus, if we are in the steady state,

$$L_q = \sum_{j=1}^{j=\infty} (j - 1)\pi_j = \sum_{j=1}^{j=\infty} j\pi_j - \sum_{j=1}^{j=\infty} \pi_j$$
$$= L - (1 - \pi_0) = L - \rho$$

where the last equation follows from (24). Since $L = \frac{\rho}{1-\rho}$, we write

$$L_q = \frac{\rho}{1 - \rho} - \rho = \frac{\rho^2}{1 - \rho} = \frac{\lambda^2}{\mu(\mu - \lambda)} \tag{27}$$

Derivation of L_s

Also of interest is L_s, the expected number of customers in service. For an $M/M/1/GD/\infty/\infty$ queuing system,

$$L_s = 0\pi_0 + 1(\pi_1 + \pi_2 + \cdots) = 1 - \pi_0 = 1 - (1 - \rho) = \rho$$

Since every customer who is present is either in line or in service, it follows that for any queuing system (not just an $M/M/1/GD/\infty/\infty$ system), $L = L_s + L_q$. Thus, using our formulas for L and L_s, we could have determined L_q from

$$L_q = L - L_s = \frac{\rho}{1 - \rho} - \rho = \frac{\rho^2}{1 - \rho}$$

The Queuing Formula $L = \lambda W$

Often we are interested in the amount of time that a typical customer spends in a queuing system. We define W as the expected time a customer spends in the queuing system, including time in line plus time in service, and W_q as the expected time a customer spends waiting in line. Both W and W_q are computed under the assumption that the steady state has been reached. By using a powerful result known as **Little's queuing formula,** W and

W_q may be easily computed from L and L_q. We first define (for any queuing system or any subset of a queuing system) the following quantities:

λ = average number of arrivals *entering* the system per unit time

L = average number of customers present in the queuing system

L_q = average number of customers waiting in line

L_s = average number of customers in service

W = average time a customer spends in the system

W_q = average time a customer spends in line

W_s = average time a customer spends in service

In these definitions, all averages are steady-state averages. For most queuing systems, Little's queuing formula may be summarized as in Theorem 3.

THEOREM 3

For *any* queuing system in which a steady-state distribution exists, the following relations hold:

$$L = \lambda W \tag{28}$$

$$L_q = \lambda W_q \tag{29}$$

$$L_s = \lambda W_s \tag{30}$$

Before using these important results, we present an intuitive justification of (28). First note that both sides of (28) have the same units (we assume the unit of time is hours). This follows, because L is expressed in terms of number of customers, λ is expressed in terms of customers per hour, and W is expressed in hours. Thus, λW has the same units (customers) as L. For a rigorous proof of Little's theorem, see Ross (1970). We content ourselves with the following heuristic discussion.

Consider a queuing system in which customers are served on a first come, first served basis. An arbitrary arrival enters the system (assume that the steady state has been reached). This customer stays in the system until he completes service, and upon his departure, there will be (on the average) L customers present in the system. But when this customer leaves, who will be left in the system? Only those customers who arrive during the time the initial customer spends in the system. Since the initial customer spends an average of W hours in the system, an average of λW customers will arrive during his stay in the system. Hence, $L = \lambda W$. The "real" proof of $L = \lambda W$ is virtually independent of the number of servers, the interarrival time distribution, the service discipline, and the service time distribution. Thus, as long as a steady state exists, we may apply Equations (28)–(30) to any queuing system.

To illustrate the use of (28) and (29), we determine W and W_q for an $M/M/1/GD/\infty/\infty$ queuing system. From (26),

$$L = \frac{\rho}{1 - \rho}$$

Then (28) yields

$$W = \frac{L}{\lambda} = \frac{\rho}{\lambda(1 - \rho)} = \frac{1}{\mu - \lambda} \tag{31}$$

From (27), we obtain

$$L_q = \frac{\lambda^2}{\mu(\mu - \lambda)}$$

and (29) implies

$$W_q = \frac{L_q}{\lambda} = \frac{\lambda}{\mu(\mu - \lambda)} \qquad (32)$$

Notice that (as expected) as ρ approaches 1, both W and W_q become very large. For ρ near zero, W_q approaches zero, but for small ρ, W approaches $\frac{1}{\mu}$, the mean service time.

The following three examples show applications of the formulas we have developed.

EXAMPLE 3   **Drive-in Banking**

An average of 10 cars per hour arrive at a single-server drive-in teller. Assume that the average service time for each customer is 4 minutes, and both interarrival times and service times are exponential. Answer the following questions:

1   What is the probability that the teller is idle?

2   What is the average number of cars waiting in line for the teller? (A car that is being served is not considered to be waiting in line.)

3   What is the average amount of time a drive-in customer spends in the bank parking lot (including time in service)?

4   On the average, how many customers per hour will be served by the teller?

Solution   By assumption, we are dealing with an $M/M/1/GD/\infty/\infty$ queuing system for which $\lambda = 10$ cars per hour and $\mu = 15$ cars per hour. Thus, $\rho = \frac{10}{15} = \frac{2}{3}$.

1   From (24), $\pi_0 = 1 - \rho = 1 - \frac{2}{3} = \frac{1}{3}$. Thus, the teller will be idle an average of one-third of the time.

2   We seek L_q. From (27),

$$L_q = \frac{\rho^2}{1 - \rho} = \frac{(\frac{2}{3})^2}{1 - \frac{2}{3}} = \frac{4}{3} \quad \text{customers}$$

3   We seek W. From (28), $W = \frac{L}{\lambda}$. Then from (26).

$$L = \frac{\rho}{1 - \rho} = \frac{\frac{2}{3}}{1 - \frac{2}{3}} = 2 \quad \text{customers}$$

Thus, $W = \frac{2}{10} = \frac{1}{5}$ hour $= 12$ minutes (W will have the same units as λ).

4   If the teller were always busy, he would serve an average of $\mu = 15$ customers per hour. From part (1), we know that the teller is only busy two-thirds of the time. Thus, during each hour, the teller will serve an average of $(\frac{2}{3})(15) = 10$ customers. This must be the case, because in the steady state, 10 customers are arriving each hour, so each hour, 10 customers must leave the system.

EXAMPLE 4   **Service Station**

Suppose that all car owners fill up when their tanks are exactly half full.[†] At the present time, an average of 7.5 customers per hour arrive at a single-pump gas station. It takes an

---

[†]This example is based on Erickson (1973).

average of 4 minutes to service a car. Assume that interarrival times and service times are both exponential.

1 For the present situation, compute L and W.

2 Suppose that a gas shortage occurs and panic buying takes place. To model this phenomenon, suppose that all car owners now purchase gas when their tanks are exactly three-quarters full. Since each car owner is now putting less gas into the tank during each visit to the station, we assume that the average service time has been reduced to $3\frac{1}{3}$ minutes. How has panic buying affected L and W?

Solution **1** We have an $M/M/1/GD/\infty/\infty$ system with $\lambda = 7.5$ cars per hour and $\mu = 15$ cars per hour. Thus, $\rho = \frac{7.5}{15} = .50$. From (26), $L = \frac{.50}{1-.50} = 1$, and from (28), $W = \frac{L}{\lambda} = \frac{1}{7.5} = 0.13$ hour. Hence, in this situation, everything is under control, and long lines appear to be unlikely.

2 We now have an $M/M/1/GD/\infty/\infty$ system with $\lambda = 2(7.5) = 15$ cars per hour. (This follows because each car owner will fill up twice as often.) Now $\mu = \frac{60}{3.333} = 18$ cars per hour, and $\rho = \frac{15}{18} = \frac{5}{6}$. Then

$$L = \frac{\frac{5}{6}}{1 - \frac{5}{6}} = 5 \text{ cars} \quad \text{and} \quad W = \frac{L}{\lambda} = \frac{5}{15} = \frac{1}{3} \text{ hours} = 20 \text{ minutes}$$

Thus, panic buying has caused long lines.

Example 4 illustrates the fact that as ρ approaches 1, L and therefore W increase rapidly. Table 5 illustrates this fact.

A Queuing Optimization Model

Example 5 shows how queuing theory can be used as an aid in decision making.

TABLE 5
Relation between ρ and L for an
$M/M/1/GD/\infty/\infty$ System

ρ	L for an $M/M/1/GD/\infty/\infty$ System
0.30	0.43
0.40	0.67
0.50	1.00
0.60	1.50
0.70	2.33
0.80	4.00
0.90	9.00
0.95	19.00
0.99	99.00

EXAMPLE 5   **Tool Center**

Machinists who work at a tool-and-die plant must check out tools from a tool center.[†] An average of ten machinists per hour arrive seeking tools. At present, the tool center is staffed by a clerk who is paid $6 per hour and who takes an average of 5 minutes to handle each request for tools. Since each machinist produces $10 worth of goods per hour, each hour that a machinist spends at the tool center costs the company $10. The company is deciding whether or not it is worthwhile to hire (at $4 per hour) a helper for the clerk. If the helper is hired, the clerk will take an average of only 4 minutes to process requests for tools. Assume that service and interarrival times are exponential. Should the helper be hired?

Solution Problems in which a decision maker must choose between alternative queuing systems are called **queuing optimization problems.** In the current problem, the company's goal is to minimize the sum of the hourly service cost and the expected hourly cost due to the idle times of machinists. In queuing optimization problems, the component of cost due to customers waiting in line is referred to as the *delay cost.* Thus, the firm wants to minimize

$$\frac{\text{Expected cost}}{\text{Hour}} = \frac{\text{service cost}}{\text{hour}} + \frac{\text{expected delay cost}}{\text{hour}}$$

The computation of the hourly service cost is usually simple. The easiest way to compute the hourly delay cost is to note that

$$\frac{\text{Expected delay cost}}{\text{Hour}} = \left(\frac{\text{expected delay cost}}{\text{customer}}\right)\left(\frac{\text{expected customers}}{\text{hour}}\right)$$

In our problem,

$$\frac{\text{Expected delay cost}}{\text{Customer}} = \left(\frac{\$10}{\text{machinist-hour}}\right)\left(\begin{matrix}\text{average hours machinist}\\ \text{spends in system}\end{matrix}\right)$$

Thus,

$$\frac{\text{Expected delay cost}}{\text{Customer}} = 10W \quad \text{and} \quad \frac{\text{expected delay cost}}{\text{hour}} = 10W\lambda$$

We can now compare the expected cost per hour if the helper is not hired to the expected cost per hour if the helper is hired. If the helper is not hired, $\lambda = 10$ machinists per hour and $\mu = 12$ machinists per hour. From (31), $W = \frac{1}{12-10} = \frac{1}{2}$ hour. Since the clerk is paid $6 per hour, we have that

$$\frac{\text{Service cost}}{\text{Hour}} = \$6 \quad \text{and} \quad \frac{\text{expected delay cost}}{\text{hour}} = 10(\tfrac{1}{2})10 = \$50$$

[†]This example is based on Brigham (1955).

Thus, without the helper, the expected hourly cost is $6 + 50 = \$56$. With the helper, $\mu = 15$ customers per hour. Then $W = \frac{1}{15-10} = \frac{1}{5}$ hour and

$$\frac{\text{Expected delay cost}}{\text{Hour}} = 10(\tfrac{1}{5})(10) = \$20$$

Since the hourly service cost is now $6 + 4 = \$10$ per hour, the expected hourly cost with the helper is $20 + 10 = \$30$. Thus, the helper should be hired, because he saves $50 - 20 = \$30$ per hour in delay costs, which more than makes up for his $4-per-hour salary.

The queuing formula $L = \lambda W$ is very general and can be applied to many situations that do not seem to be queuing problems. Think of any situation where a quantity (such as mortgage loan applications, potatoes at McDonald's, revenues from computer sales) flows through a system. If we let

L = average amount of quantity present

λ = rate at which quantity arrives at system

W = average time a unit of quantity spends in system

then $L = \lambda W$ or $W = L/\lambda$.

Here are some examples of $L = \lambda W$ in non-queuing situations.

EXAMPLE 6 Potatoes at McDonald's

Our local MacDonald's uses an average of 10,000 pounds of potatoes per week. The average number of pounds of potatoes on hand is 5,000. On the average, how long do potatoes stay in the restaurant before being used?

Solution We are given that $L = 5,000$ pounds and $\lambda = 10,000$ pounds/week. Therefore, $W = 5,000$ pounds/(10,000 pounds/week) $= .5$ week.

EXAMPLE 7 Accounts Receivable

A local computer store sells $300,000 worth of computers per year. On average accounts receivable are $45,000. On average, how long does it take from the time a customer is billed until the store receives payment?

Solution We are given that $L = \$45,000$ and $\lambda = \$300,000$/year. Therefore $W = \$45,000/(\$300,000/\text{year}) = .15$ year.

A Spreadsheet for the *M/M/1/GD/∞/∞* Queuing System

MM1.xls

Figure 13 (file MM1.xls) gives a template that can be used to compute important quantities for the $M/M/1/GD/\infty/\infty$ queuing system. Simply input λ in cell A4 and μ in cell B4. L, L_q, L_s, W, W_q, and W_s are computed in rows 6 and 8. Column B prints out the steady-state probabilities (computed from (24) and (25)). We are assuming that λ and μ are such that the probability that more than 1,000 customers will be present is very small. In Figure 13, we have input the values of λ and μ for Example 3.

	A	B	C
1	M/M/1	QUEUE	
2			
3	LAMBDA?	MU?	RO
4	10	15	0.66666667
5	L	LQ	LS
6	2	1.33333333	0.66666667
7	W	WQ	WS
8	0.2	0.13333333	0.06666667
9	J	PI(J)	
10	0	0.33333333	
11	1	0.22222222	
12	2	0.14814815	
13	3	0.09876543	
14	4	0.06584362	
15	5	0.04389575	
16	6	0.02926383	
17	7	0.01950922	
18	8	0.01300615	
19	9	0.00867076	
20	10	0.00578051	
21	11	0.00385367	
22	12	0.00256912	
23	13	0.00171274	
24	14	0.00114183	
25	15	0.00076122	
26	16	0.00050748	
27	17	0.00033832	
28	18	0.00022555	
29	19	0.00015036	
30	20	0.00010024	
31	21	6.6829E-05	
32	22	4.4552E-05	
33	23	2.9702E-05	
34	24	1.9801E-05	
35	25	1.3201E-05	
36	26	8.8005E-06	
37	27	5.867E-06	
38	28	3.9113E-06	
39	29	2.6075E-06	
40	30	1.7384E-06	

FIGURE 13
$M/M/1$ Queue

A	A	B	C
4 1	31	0.0000012	
4 2	32	0.0000008	
4 3	33	0.0000005	
4 4	34	0.0000003	
4 5	35	0.0000002	
4 6	36	0.0000002	
4 7	37	0.0000001	
4 8	38	6.8E-08	
4 9	39	4.5E-08	
5 0	40	3.0E-08	
5 1	41	2.0E-08	
5 2	42	1.3E-08	
5 3	43	8.9E-09	
5 4	44	6.0E-09	
5 5	45	4.0E-09	
5 6	46	2.6E-09	
5 7	47	1.8E-09	
5 8	48	1.2E-09	
5 9	49	7.8E-10	
6 0	50	5.2E-10	
6 1	51	3.5E-10	
6 2	52	2.3E-10	
6 3	53	1.5E-10	
6 4	54	1.0E-10	
6 5	55	6.9E-11	
6 6	56	4.6E-11	
6 7	57	3.1E-11	
6 8	58	2.0E-11	
6 9	59	1.4E-11	
7 0	60	9.1E-12	
7 1	61	6.0E-12	
7 2	62	4.0E-12	
7 3	63	2.7E-12	
7 4	64	1.8E-12	
7 5	65	1.2E-12	
7 6	66	8.0E-13	
7 7	67	5.3E-13	
7 8	68	3.5E-13	
7 9	69	2.4E-13	
8 0	70	1.6E-13	

FIGURE 13
(Continued)

PROBLEMS

Group A

1[†] Each airline passenger and his or her luggage must be checked to determine whether he or she is carrying weapons onto the airplane. Suppose that at Gotham City Airport, an average of 10 passengers per minute arrive (interarrival times are exponential). To check passengers for weapons, the airport must have a checkpoint consisting of a metal detector and baggage X-ray machine. Whenever a check-point is in operation, two employees are required. A checkpoint can check an average of 12 passengers per minute (the time to check a passenger is exponential). Under the assumption that the airport has only one checkpoint, answer the following questions:

a What is the probability that a passenger will have to wait before being checked for weapons?

b On the average, how many passengers are waiting in line to enter the checkpoint?

[†]Based on Gilliam (1979).

c On the average, how long will a passenger spend at the checkpoint?

2 The Decision Sciences Department is trying to determine whether to rent a slow or a fast copier. The department believes that an employee's time is worth $15 per hour. The slow copier rents for $4 per hour and it takes an employee an average of 10 minutes to complete copying (exponentially distributed). The fast copier rents for $15 per hour and it takes an employee an average of 6 minutes to complete copying. An average of 4 employees per hour need to use the copying machine (interarrival times are exponential). Which machine should the department rent?

3 For an $M/M/1/GD/\infty/\infty$ queuing system, suppose that both λ and μ are doubled.

a How is L changed?

b How is W changed?

c How is the steady-state probability distribution changed?

4 A fast-food restaurant has one drive-through window. An average of 40 customers per hour arrive at the window. It takes an average of 1 minute to serve a customer. Assume that interarrival and service times are exponential.

a On the average, how many customers are waiting in line?

b On the average, how long does a customer spend at the restaurant (from time of arrival to time service is completed)?

c What fraction of the time are more than 3 cars waiting for service (this includes the car (if any) at the window)?

5 On a typical Saturday, Red Lobster serves 1,000 customers. The restaurant is open for 12 hours. On average, 150 customers are present. How long does an average customer spend in the restaurant?

6 Our local maternity ward delivers 1,500 babies per year. On the average, 5 beds in the maternity ward are filled. How long does the average mother stay in the maternity ward?

7 Assume that an average of 125 packets per second of information arrive to a router and that it takes an average of .002 second to process each packet. Assuming exponential interarrival and service times, answer the following questions.

a What is the average number of packets waiting for entry into the router?

b What is the probability that 10 or more packets are present?

Group B

8 Referring to Problem 1, suppose the airline wants to determine how many checkpoints to operate to minimize operating costs and delay costs over a ten-year period. Assume that the cost of delaying a passenger for 1 hour is $10 and that the airport is open every day for 16 hours per day. It costs $1 million to purchase, staff, and maintain a metal detector and baggage X-ray machine for a ten-year period. Finally, assume that each passenger is equally likely to enter a given checkpoint.

9[†] Each machine on Widgetco's assembly line gets out of whack an average of once a minute. Laborers are assigned to reset a machine that gets out of whack. The company pays each laborer c_s dollars per hour and estimates that each hour of idle machine time costs the company c_m dollars in lost production. Data indicate that the time between successive breakdowns of a machine and the time to reset a machine are exponential. Widgetco plans to assign each worker a certain number of machines to watch over and repair. Let M = total number of Widgetco machines, w = number of laborers hired by Widgetco, and $R = \frac{M}{w}$ = machines assigned to each laborer.

a Express Widgetco's hourly cost in terms of R and M.

b Show that the optimal value of R does not depend on the value of M.

c Use calculus to show that costs are minimized by choosing

$$R = \frac{\frac{\mu}{60}}{1 + \left(\frac{c_m}{c_s}\right)^{1/2}}$$

d Suppose c_m = 78¢ and c_s = $2.75. Widgetco has 200 machines, and a laborer can reset a machine in an average of 7.8 seconds. How can Widgetco minimize costs?

e In parts (a)–(d), we have tacitly assumed that at any point in time, the rate at which the machines assigned to a worker break down does not depend on the number of his or her assigned machines that are currently working properly. Does this assumption seem reasonable?

10 Consider an airport where taxis and customers arrive (exponential interarrival times) with respective rates of 1 and 2 per minute. No matter how many other taxis are present, a taxi will wait. If an arriving customer does not find a taxi, the customer immediately leaves.

a Model this system as a birth–death process (*Hint:* Determine what the state of the system is at any given time and draw a rate diagram.)

b Find the average number of taxis that are waiting for a customer.

c Suppose all customers who use a taxi pay a $2 fare. During a typical hour, how much revenue will the taxis receive?

11 A bank is trying to determine which of two machines should be rented to process checks. Machine 1 rents for $10,000 per year and processes 1,000 checks per hour. Machine 2 rents for $15,000 per year and processes 1,600 checks per hour. Assume that the machines work 8 hours a day, 5 days a week, 50 weeks a year. The bank must process an average of 800 checks per hour, and the average check processed is for $100. Assume an annual interest rate of 20%. Then determine the cost to the bank (in lost interest) for each hour that a check spends waiting for and undergoing processing. Assuming that interarrival times and service times are exponential, which machine should the bank rent?

12[‡] A tire plant must produce an average of 100 tires per day. The plant produces tires in a batch of size x. The plant

[†]Based on Vogel (1979).
[‡]Based on Karmarkar (1985).

manager must determine the batch size x that minimizes the time a batch spends in the plant. From the time a batch of tires arrives, it takes an average of $\frac{1}{20}$ of a day to set up the plant for production of tires. Once the plant is set up, it takes an average of $\frac{1}{150}$ day to produce each tire. Assume that the time to produce a batch of tires is exponentially distributed and that the time for a batch of tires to "arrive" is also exponentially distributed. Determine the batch size that minimizes the expected time a batch spends in the plant (from arrival of batch to time production of batch is completed).

13 A worker at the State Unemployment Office is responsible for processing a company's forms when it opens for business. The worker can process an average of 4 forms per week. In 2002, an average of 1.8 companies per week submitted forms for processing, and the worker had a backlog of .45 week. In 2003, an average of 3.9 companies per week submitted forms for processing, and the worker had a 5-week backlog. The poor worker was fired and sued to get his job back. The court said that since the amount of work submitted to the worker had approximately doubled, the worker's backlog should have also doubled. Since his backlog increased by more than a factor of 10, he must have been slacking off, so the state was justified in firing him. Use queuing theory to defend the worker (based on an actual case!).

14 For the $M/M/1/GD/\infty/\infty$ queuing model, show that the following results hold:

a $W = (L + 1)W_s.$

b $W_q = LW_s.$

c Interpret the results in (a) and (b).

15 From the time a request for data is submitted until the request is fulfilled, a database takes an average of 3 seconds to respond to a request for data. We find that the database is idle around 20% of the time. Answer the following questions, assuming that the database can be modeled as an $M/M/1$ system.

a What is the average service time per database query?

b What is the average number of queries in the system?

c What is the probability that 5 or more queries are present?

8.5 The $M/M/1/GD/c/\infty$ Queuing System

In this section, we analyze the $M/M/1/GD/c/\infty$ queuing system. Recall that this queuing system is an $M/M/1/GD/\infty/\infty$ system with a total capacity of c customers. The $M/M/1/GD/c/\infty$ system is identical to the $M/M/1/GD/\infty/\infty$ system except for the fact that when c customers are present, all arrivals are turned away and are forever lost to the system. As in Section 8.4, we assume that interarrival times are exponential with rate λ, and service times are exponential with rate μ. Then the $M/M/1/GD/c/\infty$ system may be modeled (see Figure 14) as a birth–death process with the following parameters:

$$\lambda_j = \lambda \quad (j = 0, 1, \ldots, c - 1)$$
$$\lambda_c = 0$$
$$\mu_0 = 0$$
$$\mu_j = \mu \quad (j = 1, 2, \ldots, c)$$

(33)

Since $\lambda_c = 0$, the system will never reach state $c + 1$ (or any higher-numbered state). As in Section 8.4, it is convenient to define $\rho = \frac{\lambda}{\mu}$. Then we can apply Equations (16)–(19) to find that if $\lambda \neq \mu$, the steady-state probabilities for the $M/M/1/GD/c/\infty$ model are given by

$$\pi_0 = \frac{1 - \rho}{1 - \rho^{c+1}}$$
$$\pi_j = \rho^j \pi_0 \quad (j = 1, 2, \ldots, c)$$
$$\pi_j = 0 \quad (j = c + 1, c + 2, \ldots)$$

(34)

FIGURE 14
Rate Diagram for
$M/M/1/GD/c/\infty$
Queuing System

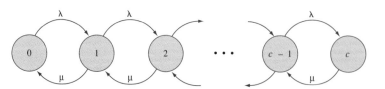

Combining (34) with the fact that $L = \sum_{j=0}^{j=c} j\pi_j$, we can show that when $\lambda \neq \mu$,

$$L = \frac{\rho[1 - (c + 1)\rho^c + c\rho^{c+1}]}{(1 - \rho^{c+1})(1 - \rho)} \quad (35)$$

If $\lambda = \mu$, then all the c_j's in (16) equal 1, and all the π_j's must be equal. Hence, if $\lambda = \mu$, the steady-state probabilities for the $M/M/1/GD/c/\infty$ system are

$$\pi_j = \frac{1}{c + 1} \quad (j = 0, 1, \ldots, c) \quad (36)$$

$$L = \frac{c}{2}$$

As with the $M/M/1/GD/\infty/\infty$ system, $L_s = 0\pi_0 + 1(\pi_1 + \pi_2 + \cdots) = 1 - \pi_0$. As before, we may determine L_q from $L_q = L - L_s$.

Determination of W and W_q from (28) and (29) is a tricky matter. Recall that in (28) and (29), λ represents the average number of customers per unit time who *actually enter* the system. In our finite capacity model, an average of λ arrivals per unit time arrive, but $\lambda\pi_c$ of these arrivals find the system filled to capacity and leave. Thus, an average of $\lambda - \lambda\pi_c = \lambda(1 - \pi_c)$ arrivals per unit time will actually enter the system. Combining this fact with (28) and (29) yields

$$W = \frac{L}{\lambda(1 - \pi_c)} \quad \text{and} \quad W_q = \frac{L_q}{\lambda(1 - \pi_c)} \quad (37)$$

For an $M/M/1/GD/c/\infty$ system, a steady state will exist even if $\lambda \geq \mu$. This is because, even if $\lambda \geq \mu$, the finite capacity of the system prevents the number of people in the system from "blowing up."

EXAMPLE 8   **Barber Shop**

A one-man barber shop has a total of 10 seats. Interarrival times are exponentially distributed, and an average of 20 prospective customers arrive each hour at the shop. Those customers who find the shop full do not enter. The barber takes an average of 12 minutes to cut each customer's hair. Haircut times are exponentially distributed.

1 On the average, how many haircuts per hour will the barber complete?

2 On the average, how much time will be spent in the shop by a customer who enters?

Solution   **1** A fraction π_{10} of all arrivals will find the shop is full. Thus, an average of $\lambda(1 - \pi_{10})$ will enter the shop each hour. All entering customers will receive a haircut, so the barber will give an average of $\lambda(1 - \pi_{10})$ haircuts per hour. From our problem, $c = 10$, $\lambda = 20$ customers per hour, and $\mu = 5$ customers per hour. Then $\rho = \frac{20}{5} = 4$, and (34) yields

$$\pi_0 = \frac{1 - 4}{1 - 4^{11}}$$

and

$$\pi_{10} = 4^{10}\left(\frac{1 - 4}{1 - 4^{11}}\right) = \frac{-3(4^{10})}{1 - 4^{11}} = .75$$

Thus, an average of $20(1 - \frac{3}{4}) = 5$ customers per hour will receive haircuts. This means that an average of $20 - 5 = 15$ prospective customers per hour will not enter the shop.

2 To determine W, we use (35) and (37). From (35),

$$L = \frac{4[1 - 11(4^{10}) + 10(4^{11})]}{(1 - 4^{11})(1 - 4)} = 9.67 \text{ customers}$$

Then (37) yields

$$W = \frac{9.67}{20(1 - \frac{3}{4})} = 1.93 \text{ hours}$$

This barber shop is crowded, and the barber would be well advised to hire at least one more barber!

A Spreadsheet for the *M/M/1/GD/c/∞* Queuing System

MM1CAP.xls

Figure 15 (file MM1CAP.xls) gives a template that can be used to compute important quantities for the *M/M/1/GD/c/∞* queuing system. Input λ in cell B2, μ in cell C2, and c (we assume $c \leq 1,000$) in cell D2. In cell F2, the steady-state probability that the state is c is given. This is the fraction of all arrivals who find the system full. In row 4, the quantities L, L_s, L_q, W, W_s, and W_q are computed. In column E, the steady-state probabilities are computed from equations (16)–(18). In Figure 15, we have input the data from Example 8.

PROBLEMS

Group A

1 A service facility consists of one server who can serve an average of 2 customers per hour (service times are exponential). An average of 3 customers per hour arrive at the facility (interarrival times are assumed exponential). The system capacity is 3 customers.

  a On the average, how many potential customers enter the system each hour?

  b What is the probability that the server will be busy?

2 An average of 40 cars per hour (interarrival times are exponentially distributed) are tempted to use the drive-in window at the Hot Dog King Restaurant. If a total of more than 4 cars are in line (including the car at the window) a car will not enter the line. It takes an average of 4 minutes (exponentially distributed) to serve a car.

  a What is the average number of cars waiting for the drive-in window (not including a car at the window)?

  b On the average, how many cars will be served per hour?

  c I have just joined the line at the drive-in window. On the average, how long will it be before I have received my food?

3 An average of 125 packets of information per minute arrive at an internet router. It takes an average of .002 second to process a packet of information. The router is designed to have a limited buffer to store waiting messages. Any message that arrives when the buffer is full is lost to the system. Assuming that interarrival and service times are

exponentially distributed, how big a buffer size is needed to ensure that at most 1 in a million messages is lost?

Group B

4 Show that if $\rho \neq 1$

$$1 + \rho + \rho^2 + \cdots + \rho^c = \frac{1 - \rho^{c+1}}{1 - \rho}$$

(*Hint:* Recall how we evaluated $1 + \rho + \rho^2 + \cdots$.)

5 Use the answer to Problem 3 to derive the steady-state probabilities for the *M/M/1/GD/c/∞* system given in Equation (34).

6 Two one-man barber shops sit side by side in Dunkirk Square. Each can hold a maximum of 4 people, and any potential customer who finds a shop full will not wait for a haircut. Barber 1 charges $11 per haircut and takes an average of 12 minutes to complete a haircut. Barber 2 charges $5 per haircut and takes an average of 6 minutes to complete a haircut. An average of 10 potential customers per hour arrive at each barber shop. Of course, a potential customer becomes an actual customer only if he finds that the shop is not full. Assuming that interarrival times and haircut times are exponential, which barber will earn more money?

7 A small mail order firm Seas Beginnings has one phone line. An average of 60 people per hour call in orders, and it takes an average of 1 minute to handle a call. Time between

FIGURE 15

	A	B	C	D	E	F	G
1	M/M/1/GD/c	LAMBDA?	MU?	c?	RO	PI(c)	TURNED AWAY
2		20	5	10	4	0.75000018	15.00000358
3		L	LS	LQ	W	WS	WQ
4		9.66666929	0.99999928	8.66667	1.93333524	0.2	1.733335241
5							
6							
7							
8							
9							
10							
11							
12	STATE	LAMBDA(J)	MU(J)	CJ	PROB	#IN QUEUE	COLA*COLE
13	0	20	0	1	7.1526E-07	0	0
14	1	20	5	4	2.861E-06	0	2.86102E-06
15	2	20	5	16	1.1444E-05	1	2.28882E-05
16	3	20	5	64	4.5776E-05	2	0.000137329
17	4	20	5	256	0.00018311	3	0.000732422
18	5	20	5	1024	0.00073242	4	0.00366211
19	6	20	5	4096	0.00292969	5	0.017578129
20	7	20	5	16384	0.01171875	6	0.08203127
21	8	20	5	65536	0.04687501	7	0.375000089
22	9	20	5	262144	0.18750004	8	1.687500402
23	10	0	5	1048576	0.75000018	9	7.500001788
24	11	0	5	0	0	10	0
25	12	0	5	0	0	11	0
26	13	0	5	0	0	12	0
27	14	0	5	0	0	13	0
28	15	0	5	0	0	14	0
29	16	0	5	0	0	15	0
30	17	0	5	0	0	16	0
31	18	0	5	0	0	17	0
32	19	0	5	0	0	18	0
33	20	0	5	0	0	19	0
34	21	0	5	0	0	20	0
35	22	0	5	0	0	21	0
36	23	0	5	0	0	22	0
37	24	0	5	0	0	23	0
38	25	0	5	0	0	24	0
39	26	0	5	0	0	25	0
40	27	0	5	0	0	26	0
41	28	0	5	0	0	27	0

calls and time to handle calls are exponentially distributed. If the phone line is busy, Seas Beginnings can put up to $c - 1$ people on hold. If $c - 1$ people are on hold, a caller gets a busy signal and calls a competitor (Air End). Seas Beginnings wants only 1% of all callers to get a busy signal.

How many people should the company be able to put on hold?

8.6 The *M*/*M*/*s*/*GD*/∞/∞ Queuing System

We now consider the *M*/*M*/*s*/*GD*/∞/∞ system. We assume that interarrival times are exponential (with rate λ), service times are exponential (with rate μ), and there is a single line of customers waiting to be served at one of s parallel servers. If $j \leq s$ customers are present, then all j customers are in service; if $j > s$ customers are present, then all s servers are occupied, and $j - s$ customers are waiting in line. Any arrival who finds an idle server enters service immediately, but an arrival who does not find an idle server joins the queue of customers awaiting service. Banks and post office branches in which all customers wait in a single line for service can often be modeled as *M*/*M*/*s*/*GD*/∞/∞ queuing systems.

To describe the *M*/*M*/*s*/*GD*/∞/∞ system as a birth–death model, note that (as in the *M*/*M*/1/*GD*/∞/∞ model) $\lambda_j = \lambda$ ($j = 0, 1, 2, \ldots$). If j servers are occupied, then service completions occur at a rate

$$\underbrace{\mu + \mu + \cdots = j\mu}_{j\mu\text{'s}}$$

Whenever j customers are present, min (j, s) servers will be occupied. Thus, $\mu_j = $ min $(j, s)\mu$. Summarizing, we find that the *M*/*M*/*s*/*GD*/∞/∞ system can be modeled as a birth–death process (see Figure 16) with parameters

$$
\begin{aligned}
\lambda_j &= \lambda & (j = 0, 1, \ldots) \\
\mu_j &= j\mu & (j = 0, 1, \ldots, s) \\
\mu_j &= s\mu & (j = s + 1, s + 2, \ldots)
\end{aligned}
\tag{38}
$$

we define $\rho = \frac{\lambda}{s\mu}$. For $\rho < 1$, substituting (38) into (16)–(19) yields the following steady-state probabilities:

$$
\pi_0 = \frac{1}{\displaystyle\sum_{i=0}^{i=(s-1)} \frac{(s\rho)^i}{i!} + \frac{(s\rho)^s}{s!(1 - \rho)}}
\tag{39}
$$

$$
\pi_j = \frac{(s\rho)^j \pi_0}{j!} \qquad (j = 1, 2, \ldots, s)
\tag{39.1}
$$

$$
\pi_j = \frac{(s\rho)^j \pi_0}{s! s^{j-s}} \qquad (j = s, s + 1, s + 2, \ldots)
\tag{39.2}
$$

If $\rho \geq 1$, no steady state exists. In other words, if the arrival rate is at least as large as the maximum possible service rate ($\lambda \geq s\mu$), the system "blows up."

From (39.2) it can be shown that the steady-state probability that all servers are busy is given by

$$
P(j \geq s) = \frac{(s\rho)^s \pi_0}{s!(1 - \rho)}
\tag{40}
$$

FIGURE 16
Rate Diagram for
M/*M*/*s*/*GD*/∞/∞
Queuing System

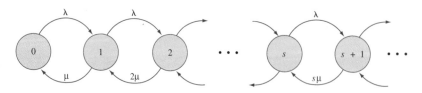

TABLE 6

$P(j \geq s)$ for the $M/M/s/GD/\infty/\infty$ Queuing System

ρ	$s = 2$	$s = 3$	$s = 4$	$s = 5$	$s = 6$	$s = 7$
.10	.02	.00	.00	.00	.00	.00
.20	.07	.02	.00	.00	.00	.00
.30	.14	.07	.04	.02	.01	.00
.40	.23	.14	.09	.06	.04	.03
.50	.33	.24	.17	.13	.10	.08
.55	.39	.29	.23	.18	.14	.11
.60	.45	.35	.29	.24	.20	.17
.65	.51	.42	.35	.30	.26	.21
.70	.57	.51	.43	.38	.34	.30
.75	.64	.57	.51	.46	.42	.39
.80	.71	.65	.60	.55	.52	.49
.85	.78	.73	.69	.65	.62	.60
.90	.85	.83	.79	.76	.74	.72
.95	.92	.91	.89	.88	.87	.85

Table 6 tabulates $P(j \geq s)$ for a variety of situations. It can also be shown that

$$L_q = \frac{P(j \geq s)\rho}{1 - \rho} \tag{41}$$

Then (28) yields

$$W_q = \frac{L_q}{\lambda} = \frac{P(j \geq s)}{s\mu - \lambda} \tag{42}$$

To determine L (and then W), we use the fact that $L = L_q + L_s$. Since $W_s = \frac{1}{\mu}$, Equation (30) shows that $L_s = \frac{\lambda}{\mu}$. Then

$$L = L_q + \frac{\lambda}{\mu} \tag{43}$$

Also,

$$\begin{aligned} W &= \frac{L}{\lambda} \\ &= \frac{L_q}{\lambda} + \frac{1}{\mu} \\ &= W_q + \frac{1}{\mu} \\ &= \frac{P(j \geq s)}{s\mu - \lambda} + \frac{1}{\mu} \end{aligned} \tag{44}$$

When we need to determine L, L_q, W, or W_q, we begin by looking up $P(j \geq s)$ in Table 6. Then we use (41)–(44) to calculate the quantity we want. If we are interested in the steady-state probability distribution, we find $P(j \geq s)$ in Table 6 and then use (40) to obtain π_0. Then (39.1) and (39.2) yield the entire steady-state distribution. The following two examples illustrate the use of the preceding formulas.

EXAMPLE 9   Bank Tellers

Consider a bank with two tellers. An average of 80 customers per hour arrive at the bank and wait in a single line for an idle teller. The average time it takes to serve a customer is 1.2 minutes. Assume that interarrival times and service times are exponential. Determine

1 The expected number of customers present in the bank

2 The expected length of time a customer spends in the bank

3 The fraction of time that a particular teller is idle

Solution **1** We have an $M/M/2/GD/\infty/\infty$ system with $\lambda = 80$ customers per hour and $\mu = 50$ customers per hour. Thus $\rho = \frac{80}{2(50)} = 0.80 < 1$, so a steady state does exist. (For $\lambda \geq 100$, no steady state would exist.) From Table 6, $P(j \geq 2) = .71$. Then (41) yields

$$L_q = \frac{.80(.71)}{1 - .80} = 2.84 \text{ customers}$$

and from (43), $L = 2.84 + \frac{80}{50} = 4.44$ customers.

2 Since $W = \frac{L}{\lambda}$, $W = \frac{4.44}{80} = 0.055$ hour $= 3.3$ minutes.

3 To determine the fraction of time that a particular server is idle, note that he or she is idle during the entire time that $j = 0$ and half the time (by symmetry) that $j = 1$. The probability that a server is idle is given by $\pi_0 + 0.5\pi_1$. Using the fact that $P(j \geq 2) = .71$, we obtain π_0 from (40):

$$\pi_0 = \frac{s!P(j \geq s)(1 - \rho)}{(s\rho)^2} = \frac{2!(.71)(1 - .80)}{(1.6)^2} = .11$$

Now (39.1) yields

$$\pi_1 = \frac{(1.6)^1 \pi_0}{1!} = .176$$

Thus, the probability that a particular teller is idle is $\pi_0 + 0.5\pi_1 = .11 + 0.5(.176) = .198$. We could have determined π_0 directly from (39):

$$\pi_0 = \frac{1}{1 + \frac{[2(.80)]^1}{1!} + \frac{[2(.80)]^2}{2!(1 - .80)}} = \frac{1}{1 + 1.6 + 6.4} = \frac{1}{9}$$

This is consistent with our computation of $\pi_0 = .11$.

EXAMPLE 10   Bank Staffing

The manager of a bank must determine how many tellers should work on Fridays. For every minute a customer stands in line, the manager believes that a delay cost of 5¢ is incurred. An average of 2 customers per minute arrive at the bank. On the average, it takes a teller 2 minutes to complete a customer's transaction. It cost the bank $9 per hour to hire a teller. Interarrival times and service times are exponential. To minimize the sum of service costs and delay costs, how many tellers should the bank have working on Fridays?

Solution Since $\lambda = 2$ customers per minute and $\mu = 0.5$ customer per minute, $\frac{\lambda}{s\mu} < 1$ requires that $\frac{4}{s} < 1$ or $s \geq 5$. Thus, there must be at least 5 tellers, or the number of customers present will "blow up." We now compute, for $s = 5, 6, \ldots,$

$$\frac{\text{Expected service cost}}{\text{Minute}} + \frac{\text{expected delay cost}}{\text{minute}}$$

Since each teller is paid $\frac{9}{60} = 15\cancel{c}$ per minute,

$$\frac{\text{Expected service cost}}{\text{Minute}} = 0.15s$$

As in Example 4,

$$\frac{\text{Expected delay cost}}{\text{Minute}} = \left(\frac{\text{expected customers}}{\text{minute}}\right)\left(\frac{\text{expected delay cost}}{\text{customer}}\right)$$

But

$$\frac{\text{Expected delay cost}}{\text{Customer}} = 0.05 W_q$$

Since an average of 2 customers arrive per minute,

$$\frac{\text{Expected delay cost}}{\text{Minute}} = 2(0.05 W_q) = 0.10 W_q$$

For $s = 5$, $\rho = \frac{2}{.5(5)} = .80$ and $P(j \geq 5) = .55$. From (42),

$$W_q = \frac{.55}{5(.5) - 2} = 1.1 \text{ minutes}$$

Thus, for $s = 5$,

$$\frac{\text{Expected delay cost}}{\text{Minute}} = 0.10(1.1) = 11\cancel{c}$$

and, for $s = 5$,

$$\frac{\text{Total expected cost}}{\text{Minute}} = 0.15(5) + 0.11 = 86\cancel{c}$$

Since $s = 6$ has a service cost per minute of $6(0.15) = 90\cancel{c}$, 6 tellers cannot have a lower total cost than 5 tellers. Hence, having 5 tellers serve is optimal. Putting it another way, adding an additional teller can save the bank at most $11\cancel{c}$ per minute in delay costs. Since an additional teller cost $15\cancel{c}$ per minute, it cannot be optimal to hire more than 5 tellers.

In addition to a customer's expected time in the system, the distribution of a customer's waiting time is of interest. For example, if all customers who have to wait more than 5 minutes at a supermarket checkout counter decide to switch to another store, the probability that a given customer will switch to another store equals $P(\mathbf{W} > 5)$. To determine this probability, we need to know the distribution of a customer's waiting time. For an $M/M/s/FCFS/\infty/\infty$ queuing system, it can be shown that

$$P(\mathbf{W} > t) = e^{-\mu t}\left\{1 + P(j \geq s)\frac{1 - \exp[-\mu t(s - 1 - s\rho)]}{s - 1 - s\rho}\right\}^\dagger \qquad (45)$$

$$P(\mathbf{W}_q > t) = P(j \geq s) \exp[-s\mu(1 - \rho)t] \qquad (46)$$

To illustrate the use of (45) and (46), suppose that in Example 7 (for $s = 5$), the bank manager wants to know the probability that a customer will have to wait in line for more than 10 minutes. For $s = 5$, $\rho = .80$, $P(j \geq 5) = .55$, and $\mu = 0.5$ customer per minute. (46) yields

$$P(\mathbf{W}_q > 10) = .55 \exp[-5(0.5)(1 - .80)(10)] = .55\, e^{-5} = .004$$

†If $s - 1 = s\rho$, then $P(\mathbf{W} > t) = e^{-\mu t}(1 + P(j \geq s)\mu t)$.

Thus, the bank manager can be sure that the chance of a customer's having to wait more than 10 minutes is quite small.

A Spreadsheet for the *M/M/s/GD/∞/∞* Queuing System

Multiple.xls

Figure 17 (file Multiple.xls) gives a template that can be used to compute the important quantities for the *M/M/s/GD/∞/∞* queuing system. In cell B2 we input λ, in cell C2 we input μ, and in cell D2 we input s. In cell B6, we compute $P(j \geq s)$. In row 4, the quantities L, L_s, L_q, W, W_s, and W_q are computed. In A8, we compute $P(W_q > t)$ for the value of t input in cell B8. In cell C8, we compute $P(W > t)$ for the value of t input in cell B8. Steady-state probabilities are computed in column E (we are assuming that there is a small probability that more than 1,000 customers are present). In Figure 17, we have input data for Example 10 (with 5 servers).

Having a spreadsheet to compute quantities of interest for the *M/M/s* system enables us to use spreadsheet techniques such as Data Tables and Goal Seek to answer questions of interest. For example, reconsider Example 10. To determine the number of servers that minimizes expected cost per minute, we would like to vary the number of servers (starting with 5) and compute expected cost per minute for different numbers of servers. This is easily done with a **one-way data table.** (See Figure 18.)

Step 1 Enter the possible number of servers (5–8) in cells J5:J8.

Step 2 Enter the formula for expected cost per minute one column over to the right and one row above where the possible number of servers are listed. This is in cell K4.

$$=0.15*D2+B2*G4*0.05$$

Step 3 Highlight the *table range*. This includes the inputs, the calculated formula, and the range where values of the calculated formula are placed. In our example, the table range is J4:K8.

Step 4 Select Data Table and choose One-Way Table (because we are changing only one input, the number of servers).

Step 5 Fill in the dialog box as shown in Figure 19. This instructs Excel to repeatedly place the input values in the left-hand column of the table range in cell D2 (number of servers) and recalculate our formula (expected cost per minute, which is entered in cell K4). We then obtain the expected cost per minute for 5–8 servers. As before, we find that 5 servers yield the lowest expected cost per minute.

As another example of how we can use powerful spreadsheet tools to answer important queuing questions, suppose we want to know (for 5 servers) the 90th percentile of a customer's time in the system. That is, we wish to know the value of t that makes $P(W > t)$ equal to .10. This may easily be determined with the Excel Goal Seek feature. Goal Seek enables us to find what value of one cell (the *changing* cell) causes a formula in another cell (the *set* cell) to assume a desired value (called the *to value*).

To use Goal Seek to find the 90th percentile of a customer's time in the system, we select Tools Goal Seek and fill in the dialog box as shown in Figure 20. This dialog box finds the value for t in B8 that makes $P(W > t)$ (computed in C8) equal to .1. We find that with 5 servers, 10% of all customers will spend at least 6.7 minutes in the bank. See Figure 21.

We note that the precision of Goal Seek may be improved by selecting Tools Options Calculation and setting Maximum Change to a smaller number than the default value of .001. For example, a Maximum Change of .000001 ensures that upon completion of the Goal Seek operation, $P(W > q)$ will be within .000001 of .10.

FIGURE 17

	A	B	C	D	E	F	G	H
1	M/M/s/GD	LAMBDA?	MU?	s?	RO			
2		2	0.5	5	0.8			
3		L	LS	LQ	W	WS	WQ	
4		6.21645022	4	2.21645022	3.10822511	1.999999999	1.108225109	
5	STATE	P(j>=s)						
6	1	0.55411255						
7	P(Wq>t)	t?	P(W>t)					
8	0.019390014	6.70521931	0.10000006					
9								
10								
11								
12	STATE	LAMBDA(J)	MU(J)	CJ	PROB	#IN QUEUE	COLA*COLE	COLE*COL
13	0	2	0	1	0.01298701	0	0	0
14	1	2	0.5	4	0.05194805	0	0.051948052	0
15	2	2	1	8	0.1038961	0	0.207792208	0
16	3	2	1.5	10.6666667	0.13852814	0	0.415584416	0
17	4	2	2	10.6666667	0.13852814	0	0.554112554	0
18	5	2	2.5	8.53333333	0.11082251	0	0.554112554	0
19	6	2	2.5	6.82666667	0.08865801	1	0.531948052	0.08865801
20	7	2	2.5	5.46133333	0.07092641	2	0.496484848	0.14185281
21	8	2	2.5	4.36906667	0.05674113	3	0.453929004	0.17022338
22	9	2	2.5	3.49525333	0.0453929	4	0.408536104	0.1815716
23	10	2	2.5	2.79620267	0.03631432	5	0.363143203	0.1815716
24	11	2	2.5	2.23696213	0.02905146	6	0.319566019	0.17430874
25	12	2	2.5	1.78956971	0.02324117	7	0.27889398	0.16268816
26	13	2	2.5	1.43165577	0.01859293	8	0.241708116	0.14874346
27	14	2	2.5	1.14532461	0.01487435	9	0.208240839	0.13386911
28	15	2	2.5	0.91625969	0.01189948	10	0.178492147	0.11899476
29	16	2	2.5	0.73300775	0.00951958	11	0.152313299	0.10471539
30	17	2	2.5	0.5864062	0.00761566	12	0.129466304	0.09138798
31	18	2	2.5	0.46912496	0.00609253	13	0.109665575	0.07920292
32	19	2	2.5	0.37529997	0.00487403	14	0.092606486	0.06823636
33	20	2	2.5	0.30023998	0.00389922	15	0.077984409	0.05848831
34	21	2	2.5	0.24019198	0.00311938	16	0.065506904	0.04991002
35	22	2	2.5	0.19215358	0.0024955	17	0.054901024	0.04242352
36	23	2	2.5	0.15372287	0.0019964	18	0.04591722	0.03593522
37	24	2	2.5	0.12297829	0.00159712	19	0.038330897	0.03034529
38	25	2	2.5	0.09838264	0.0012777	20	0.031942414	0.02555393
39	26	2	2.5	0.07870611	0.00102216	21	0.026576088	0.0214653
40	27	2	2.5	0.06296489	0.00081773	22	0.022078597	0.01798997
41	28	2	2.5	0.05037191	0.00065418	23	0.018317058	0.01504615
42	29	2	2.5	0.04029753	0.00052334	24	0.015176991	0.01256027
43	30	2	2.5	0.03223802	0.00041868	25	0.012560268	0.01046689
44	31	2	2.5	0.02579042	0.00033494	26	0.010383155	0.00870845
45	32	2	2.5	0.02063233	0.00026795	27	0.008574476	0.00723471
46	33	2	2.5	0.01650587	0.00021436	28	0.007073943	0.00600213
47	34	2	2.5	0.01320469	0.00017149	29	0.005830644	0.0049732
48	35	2	2.5	0.01056376	0.00013719	30	0.004801707	0.00411575
49	36	2	2.5	0.008451	0.00010975	31	0.003951119	0.00340235
50	37	2	2.5	0.0067608	8.7803E-05	32	0.003248698	0.00280968
51	38	2	2.5	0.00540864	7.0242E-05	33	0.0026692	0.00231799

	J	K
2		
3		
4	Servers	0.86082251
5	5	0.86082251
6	6	0.92847608
7	7	1.05900734
8	8	1.2029522

FIGURE 18

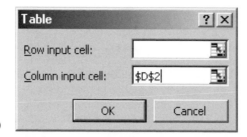

FIGURE 19

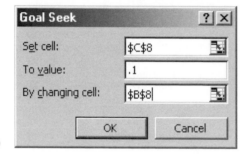

FIGURE 20

FIGURE 21

	A	B	C
7	P(Wq>t)	t?	P(W>t)
8	0.019390014	6.70521931	0.10000006

Using LINGO for *M/M/s/GD/∞/∞* Computations

The LINGO function **@PEB( )** yields the probability that all servers are busy ($P(j \geq s)$) for an *M/M/s/GD/∞/∞* system. The **@PEB** function has two arguments: the first is the value of λ/μ and the second is the number of servers. Thus, for Example 9, **@PEB** (80/50,2) = .711111 yields $P(j \geq 2)$.

The **@PEB** function can be used to solve queuing optimization problems with LINGO. For instance, to determine the cost-minimizing number of servers in Example 10, we would input the following problem into LINGO:

```
MODEL:
1)  MIN=.10*@PEB(4,S)/(.5*S-2) + .15*S;
2)  S>5;
END
```

In line 1 .10*@PEB(4,S)/(.5*S−2) is the expected cost per minute due to customers waiting in line, while .15*S is the per-minute service cost. Line 2 follows, because we need at least 5 servers for a steady state to exist. LINGO outputs $S = 5$ with an objective function value of .860823 (this is expected cost per minute).

PROBLEMS

Group A

1 A supermarket is trying to decide how many cash registers to keep open. Suppose an average of 18 customers arrive each hour, and the average checkout time for a customer is 4 minutes. Interarrival times and service times are exponential, and the system may be modeled as an $M/M/s/GD/\infty/\infty$ queuing system. It costs $20 per hour to operate a cash register, and a cost of 25¢ is assessed for each minute the customer spends in the cash register area. How many registers should the store open?

2 A small bank is trying to determine how many tellers to employ. The total cost of employing a teller is $100 per day, and a teller can serve an average of 60 customers per day. An average of 50 customers per day arrive at the bank, and both service times and interarrival times are exponential. If the delay cost per customer-day is $100, how many tellers should the bank hire?

3 In this problem, all interarrival and service times are exponential.

a At present, the finance department and the marketing department each have one typist. Each typist can type 25 letters per day. Finance requires that an average of 20 letters per day be typed, and marketing requires that an average of 15 letters per day be typed. For each department, determine the average length of time elapsing between a request for a letter and completion of the letter.

b Suppose that the two typists were grouped into a typing pool; that is, each typist would be available to type letters for either department. For this arrangement, calculate the average length of time between a request for a letter and completion of the letter.

c Comment on the results of parts (a) and (b).

d Under the pooled arrangement, what is the probability that more than .200 day will elapse between a request for a letter and completion of the letter?

4 MacBurger's is attempting to determine how many servers (or lines) should be available during the breakfast shift. During each hour, an average of 100 customers arrive at the restaurant. Each line or server can handle an average of 50 customers per hour. A server costs $5 per hour, and the cost of a customer waiting in line for 1 hour is $20. Assuming that an $M/M/s/GD/\infty/\infty$ model is applicable, determine the number of lines that minimizes the sum of delay and service costs.

5 An average of 100 customers arrive each hour at the Gotham City Bank. The average service time for each customer is 1 minute. Service times and interarrival times are exponential. The manager wants to ensure that no more than 1% of all customers will have to wait in line for more than 5 minutes. If the bank follows the policy of having all customers join a single line, how many tellers must the bank hire?

6 An average of 90 patrons per hour arrive at a hotel lobby (interarrival times are exponential), waiting to check in. At present, there are 5 clerks, and patrons are waiting in a single line for the first available clerk. The average time for a clerk to service a patron is 3 minutes (exponentially distributed). Clerks earn $10 per hour, and the hotel assesses a waiting time cost of $20 for each hour that a patron waits in line.

a Compute the expected cost per hour of the current system.

b The hotel is considering replacing one clerk with an Automatic Clerk Machine (ACM). Management estimates that 20% of all patrons will use an ACM. An ACM takes an average of 1 minute to service a patron. It costs $48 per day (1 day = 8 hours) to operate an ACM. Should the hotel install the ACM? Assume that all customers who are willing to use the ACM wait in a single queue.

7 An average of 50 customers per hour arrive at a small post office. Interarrival times are exponentially distributed. Each window can serve an average of 25 customers per hour. Service times are exponentially distributed. It costs $25 per hour to open a window, and the post office values the time a customer spends waiting in line at $15 per customer-hour. To minimize expected hourly costs, how many postal windows should be opened?

8 An average of 300 customers per hour arrive at a huge branch of bank 2. It takes an average of 2 minutes to serve each customer. It costs $10 per hour to keep open a teller window, and the bank estimates that it will lose $50 in future profits for each hour that a customer waits in line. How many teller windows should bank 2 open?

9 An average of 40 students per hour arrive at the MBA computing lab. The average student uses a computer for 20 minutes. Assume exponential interarrival and service times.

a If we want the average time a student waits for a PC to be at most 10 minutes, how many computers should the lab have?

b If we want 95% of all students to spend 5 minutes or less waiting for a PC, how many PCs should the lab have?

10 A data storage system consists of 3 disk drives sharing a common queue. An average of 50 storage requests arrive per second. The average time required to service a request is .03 second. Assuming that interarrival times and service times are exponential, determine:

a The probability that a given disk drive is busy

b The probability that no disk drives are busy

c The probability that a job will have to wait

d The average number of jobs present in the storage system

11 A Northwest Airlines ticket counter forecasts that 200 people per hour will need to check in. It takes an average of two minutes to service a customer. Assume that interarrival times and service times are exponential and that all customers wait in a single line for the first available agent.

a If we want the average time a customer spends in line and in service to be 30 minutes or less, how many ticket agents should be on duty?

b If we want 95% of all customers to wait 45 minutes or less in line, how many ticket agents should be on duty?

Group B

12 An average of 100 customers per hour arrive at Gotham City Bank. It takes a teller an average of 2 minutes to serve a customer. Interarrival and service times are exponential. The bank currently has four tellers working. The bank manager wants to compare the following two systems with regard to average number of customers present in the bank and the probability that a customer will spend more than 8 minutes in the bank:

System 1 Each teller has her own line, and no jockeying between lines is permitted.

System 2 All customers wait in a single line for the first available teller.

If you were the bank manager, which system would you prefer?

13 A muffler shop has three mechanics. Each mechanic takes an average of 45 minutes to install a new muffler.

Suppose an average of 1 customer per hour arrives. What is the expected number of mechanics that are busy at any given time? Answer this question without assuming that service times and interarrival times are exponential.

14 Consider the following two queuing systems:

System 1 An $M/M/1$ system with arrival rate λ and service rate 3μ.

System 2 An $M/M/3$ system with arrival rate λ and each server working at rate μ.

Without doing extensive calculations, which system will have the smaller W and L? (*Hint:* Write down the birth–death parameters for each system. Then determine which system is more efficient.)

15 (Requires the use of a spreadsheet or LINGO) The Carco plant in Bedford produces windshield wipers for Fords. In a given day, each machine in the plant can produce 1,000 wipers. The plant operates 250 days per year, and Ford will need 3 million wipers per year. It costs $50,000 per year to operate a machine. For each day that a wiper is delayed, a cost of $100 (due to production downtime at other plants) is incurred. How many machines should the Ford plant have? Assume that interarrival times and service times are exponential.

8.7 The $M/G/\infty/GD/\infty/\infty$ and $GI/G/\infty/GD/\infty/\infty$ Models

There are many examples of systems in which a customer never has to wait for service to begin. In such a system, the customer's entire stay in the system may be thought of as his or her service time. Since a customer never has to wait for service, there is, in essence, a server available for each arrival, and we may think of such a system as an **infinite-server** (or self-service) system. Two examples of an infinite-server system are given in Table 7.

Using the Kendall–Lee notation, an infinite-server system in which interarrival and service times may follow arbitrary probability distributions may be written as $GI/G/\infty/GD/\infty/\infty$ queuing system. Such a system operates as follows:

1 Interarrival times are iid with common distribution **A**. Define $E(\mathbf{A}) = \frac{1}{\lambda}$. Thus, λ is the arrival rate.

2 When a customer arrives, he or she immediately enters service. Each customer's time in the system is governed by a distribution **S** having $E(\mathbf{S}) = \frac{1}{\mu}$.

TABLE 7
Examples of Infinite-Server Queuing Systems

Situation	Arrival	Service Time (time in system)	State of System
Industry	Firm enters industry	Time until firm leaves industry	Number of firms in industry
College program	Student enters program	Time student remains in program	Number of students in program

Let L be the expected number of customers in the system in the steady state, and W be the expected time that a customer spends in the system. By definition, $W = \frac{1}{\mu}$. Then Equation (30) implies that

$$L = \frac{\lambda}{\mu} \qquad (47)$$

Equation (47) does not require any assumptions of exponentiality. If interarrival times are exponential, it can be shown (even for an arbitrary service time distribution) that the steady-state probability that j customers are present (call it π_j) follows a Poisson distribution with mean $\frac{\lambda}{\mu}$. This implies that

$$\pi_j = \frac{(\frac{\lambda}{\mu})^j e^{-\lambda/\mu}}{j!}$$

The following example is a typical application of a $GI/G/\infty/GD/\infty/\infty$ system.

EXAMPLE 11   Smalltown Ice Cream Shops

During each year, an average of 3 ice cream shops open up in Smalltown. The average time that an ice cream shop stays in business is 10 years. On January 1, 2525, what is the average number of ice cream shops that you would find in Smalltown? If the time between the opening of ice cream shops is exponential, what is the probability that on January 1, 2525, there will be 25 ice cream shops in Smalltown?

Solution   We are given that $\lambda = 3$ shops per year and $\frac{1}{\mu} = 10$ years per shop. Assuming that the steady state has been reached, there will be an average of $L = \lambda(\frac{1}{\mu}) = 3(10) = 30$ shops in Smalltown. If interarrivals of ice cream shops are exponential, then

$$\pi_{25} = \frac{(30)^{25}e^{-30}}{25!} = .05$$

Of course, we could also compute the probability that there are 25 ice cream shops with the Excel formula

$$=\text{POISSON}(30,25,0)$$

This yields .045.

PROBLEMS

Group A

1 Each week, the Columbus Record Club attracts 100 new members. Members remain members for an average of one year (1 year = 52 weeks). On the average, how many members will the record club have?

2 The State U doctoral program in business admits an average of 25 doctoral students each year. If a doctoral student spends an average of 4 years in residence at State U, how many doctoral students would one expect to find there?

3 There are at present 40 solar energy construction firms in the state of Indiana. An average of 20 solar energy construction firms open each year in the state. The average firm stays in business for 10 years. If present trends continue, what is the expected number of solar energy construction firms that will be found in Indiana? If the time between the entries of firms into the industry is exponentially distributed, what is the probability that (in the steady state) there will be more than 300 solar energy firms in business? (*Hint:* For large λ, the Poisson distribution can be approximated by a normal distribution.)

8.8 The $M/G/1/GD/\infty/\infty$ Queuing System

In this section, we consider a single-server queuing system in which interarrival times are exponential, but the service time distribution (**S**) need not be exponential. Let λ be the arrival rate (assumed to be measured in arrivals per hour). Also define $\frac{1}{\mu} = E(\mathbf{S})$ and $\sigma^2 = \text{var } \mathbf{S}$.

In Kendall's notation, such a queuing system is described as an $M/G/1/GD/\infty/\infty$ queuing system. An $M/G/1/GD/\infty/\infty$ system is *not* a birth–death process, because the probability that a service completion occurs between t and $t + \Delta t$ when the state of the system at time t is j depends on the length of time since the last service completion (because service times no longer have the no-memory property). Thus, we cannot write the probability of a service completion between t and $t + \Delta t$ in the form $\mu \Delta t$, and a birth–death model is not appropriate.

Determination of the steady-state probabilities for an $M/G/1/GD/\infty/\infty$ queuing system is a difficult matter. Since the birth–death steady-state equations are no longer valid, a different approach must be taken. Markov chain theory is used to determine π_i', the probability that after the system has operated for a long time, i customers will be present at the instant immediately after a service completion occurs (see Problem 5 at the end of this section). It can be shown that $\pi_i' = \pi_i$, where π_i is the fraction of the time after the system has operated for a long time that i customers are present (see Kleinrock (1975)).

Fortunately, however, utilizing the results of Pollaczek and Khinchin, we may determine L_q, L, L_s, W_q, W, and W_s. Pollaczek and Khinchin showed that for the $M/G/1/GD/\infty/\infty$ queuing system,

$$L_q = \frac{\lambda^2 \sigma^2 + \rho^2}{2(1 - \rho)} \tag{48}$$

where $\rho = \frac{\lambda}{\mu}$. Since $W_s = \frac{1}{\mu}$, (30) implies that $L_s = \lambda(\frac{1}{\mu}) = \rho$. Since $L = L_s + L_q$, we obtain

$$L = L_q + \rho \tag{49}$$

Then (29) and (28) imply that

$$W_q = \frac{L_q}{\lambda} \tag{50}$$

$$W = W_q + \frac{1}{\mu} \tag{51}$$

It can also be shown that π_0, the fraction of the time that the server is idle, is $1 - \rho$. (See Problem 2 at the end of this section.) This result is similar to the one for the $M/M/1/GD/\infty/\infty$ system.

To illustrate the use of (48)–(51), consider an $M/M/1/GD/\infty/\infty$ system with $\lambda = 5$ customers per hour and $\mu = 8$ customers per hour. From our study of the $M/M/1/GD/\infty/\infty$ model, we know that

$$L = \frac{\lambda}{\mu - \lambda} = \frac{5}{8 - 5} = \frac{5}{3} \text{ customers}$$

$$L_q = L - \rho = \frac{5}{3} - \frac{5}{8} = \frac{25}{24} \text{ customers}$$

$$W = \frac{L}{\lambda} = \frac{\frac{5}{3}}{5} = \frac{1}{3} \text{ hour}$$

$$W_q = \frac{L_q}{\lambda} = \frac{\frac{25}{24}}{5} = \frac{5}{24} \text{ hour}$$

From (3) and (4), we know that $E(\mathbf{S}) = \frac{1}{8}$ hour and var $\mathbf{S} = \frac{1}{64}$ hour^2. Then (48) yields

$$L_q = \frac{\frac{(5)^2}{64} + \left(\frac{5}{8}\right)^2}{2\left(1 - \frac{5}{8}\right)} = \frac{25}{24} \text{ customers}$$

$$L = L_q + \rho = \frac{25}{24} + \frac{5}{8} = \frac{40}{24} = \frac{5}{3} \text{ customers}$$

$$W_q = \frac{L_q}{\lambda} = \frac{\frac{25}{24}}{5} = \frac{5}{24} \text{ hour}$$

$$W = \frac{L}{\lambda} = \frac{\frac{5}{3}}{5} = \frac{1}{3} \text{ hour}$$

To demonstrate how the variance of the service time can significantly affect the efficiency of a queuing system, we consider an $M/D/1/GD/\infty/\infty$ queuing system having λ and μ identical to the $M/M/1/GD/\infty/\infty$ system that we have just analyzed. For this $M/D/1/GD/\infty/\infty$ model, $E(\mathbf{S}) = \frac{1}{8}$ hour and var $\mathbf{S} = 0$. Then

$$L_q = \frac{\left(\frac{5}{8}\right)^2}{2\left(1 - \frac{5}{8}\right)} = \frac{25}{48} \text{ customer}$$

$$W_q = \frac{L_q}{\lambda} = \frac{\frac{25}{48}}{5} = \frac{5}{48} \text{ hour}$$

In this $M/D/1/GD/\infty/\infty$ system, a typical customer will spend only half as much time in line as in an $M/M/1/GD/\infty/\infty$ queuing system with identical arrival and service rates. As this example shows, even if mean service times are not decreased, a decrease in the variability of service times can substantially reduce queue size and customer waiting time.

PROBLEMS

Group A

1 An average of 20 cars per hour arrive at the drive-in window of a fast-food restaurant. If each car's service time is 2 minutes, how many cars (on the average) will be waiting in line? Assume exponential interarrival times.

2 Using the fact that $L_s = \frac{\lambda}{\mu}$, demonstrate that for an $M/G/1/GD/\infty/\infty$ queuing system, the probability that the server is busy is $\rho = \frac{\lambda}{\mu}$.

3 An average of 40 cars per hour arrive to be painted at a single-server GM painting facility. 95% of the cars require 1 minute to paint; 5% must be painted twice and require 2.5 minutes to paint. Assume that interarrival times are exponential.

a On the average, how long does a car wait before being painted?

b If cars never had to be repainted, how would your answer to part (a) change?

Group B

4 Consider an $M/G/1/GD/\infty/\infty$ queuing system in which an average of 10 arrivals occur each hour. Suppose that each customer's service time follows an Erlang distribution, with rate parameter 1 customer per minute and shape parameter 4.

a Find the expected number of customers waiting in line.

b Find the expected time that a customer will spend in the system.

c What fraction of the time will the server be idle?

5 Consider an $M/G/1/GD/\infty/\infty$ queuing system in which interarrival times are exponentially distributed with parameter λ and service times have a probability density function $s(t)$. Let X_i be the number of customers present an instant after the ith customer completes service.

a Explain why $X_1, X_2, \ldots, X_k, \ldots$ is a Markov chain.

b Explain why $P_{ij} = P(X_{k+1} = j | X_k = i)$ is zero for $j < i - 1$.

c Explain why for $i > 0$, $P_{i,i-1} = $ (probability that no arrival occurs during a service time); $P_{ii} = $ (probability that one arrival occurs during a service time); and for j

$\geq i$, $P_{ij} = $ (probability that $j - i + 1$ arrivals occur during a service time).

d Explain why, for $j \geq i - 1$ and $i > 0$,

$$P_{ij} = \int_0^\infty \frac{s(x)e^{-\lambda x}(\lambda x)^{j-i+1}}{(j - i + 1)!} \, dx$$

Hint: The probability that a service time is between x and $x + \Delta x$ is $\Delta x s(x)$. Given that the service time equals x, the probability that $j - i + 1$ arrivals will occur during the service time is

$$\frac{e^{-\lambda x}(\lambda x)^{j-i+1}}{(j - i + 1)!}$$

8.9 Finite Source Models: The Machine Repair Model

With the exception of the $M/M/1/GD/c/\infty$ model, all the models we have studied have displayed arrival rates that were independent of the state of the system. As discussed previously, there are two situations where the assumption of the state-independent arrival rates may be invalid:

1 If customers do not want to buck long lines, the arrival rate may be a decreasing function of the number of people present in the queuing system. For an illustration of this situation, see Problems 4 and 5 at the end of this section.

2 If arrivals to a system are drawn from a small population, the arrival rate may greatly depend on the state of the system. For example, if a bank has only 10 depositors, then at an instant when all depositors are in the bank, the arrival rate must be zero, while if fewer than 10 people are in bank, the arrival rate will be positive.

Models in which arrivals are drawn from a small population are called **finite source models.** We now analyze an important finite source model known as the *machine repair* (or *machine interference*) model.

In the machine repair problem, the system consists of K machines and R repair people. At any instant in time, a particular machine is in either good or bad condition. The length of time that a machine remains in good condition follows an exponential distribution with rate λ. Whenever a machine breaks down, the machine is sent to a repair center consisting of R repair people. The repair center services the broken machines as if they were arriving at an $M/M/R/GD/\infty/\infty$ system.

Thus, if $j \leq R$ machines are in bad condition, a machine that has just broken will immediately be assigned for repair; if $j > R$ machines are broken, $j - R$ machines will be waiting in a single line for a repair worker to become idle. The time it takes to complete repairs on a broken machine is assumed exponential with rate μ (or mean repair time is $\frac{1}{\mu}$). Once a machine is repaired, it returns to good condition and is again susceptible to breakdown. The machine repair model may be modeled as a birth–death process, where the state j at any time is the number of machines in bad condition. Using the Kendall–Lee notation, the model just described may be expressed as an $M/M/R/GD/K/K$ model. The first K indicates that at any time, no more than K customers (or machines) may be present, and the second K indicates that arrivals are drawn from a finite source of size K.

Table 8 exhibits the interpretation of each state for a machine repair model having $K = 5$ and $R = 2$ ($G = $ machine in good condition; $B = $ broken machine). To find the birth–death parameters for the machine repair model (see Figure 22), note that a birth cor-

TABLE 8
Possible States in a Machine Repair Problem When $K = 5$ and $R = 2$

State	No. of Good Machines	Repair Queue	No. of Repair Workers Busy
0	$G\,G\,G\,G\,G$		0
1	$G\,G\,G\,G$		1
2	$G\,G\,G$		2
3	$G\,G$	B	2
4	G	$B\,B$	2
5		$B\,B\,B$	2

FIGURE 22
Rate Diagram for $M/M/R/GD/K/K$ Queuing System When $R = 2$, $K = 5$

State is number of machines in bad condition

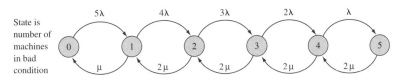

responds to a machine breaking down and a death corresponds to a machine having just been repaired. To figure out the birth rate in state j, we must determine the rate at which machines break down when the state of the system is j. When the state is j, there are $K - j$ machines in good condition. Since each machine breaks down at rate λ, the total rate at which breakdowns occur when the state is j is

$$\lambda_j = \underbrace{\lambda + \lambda + \cdots + \lambda}_{(K-j)\lambda\text{'s}} = (K - j)\lambda$$

To determine the death rate for the machine repair model, we proceed as we did in our discussion of the $M/M/s/GD/\infty/\infty$ queuing model. When the state is j, min (j, R) repair people will be busy. Since each occupied repair worker completes repairs at rate μ, the death rate μ_j is given by

$$\mu_j = j\mu \qquad (j = 0, 1, \dots, R)$$
$$\mu_j = R\mu \qquad (j = R + 1, R + 2, \dots, K)$$

If we define $\rho = \frac{\lambda}{\mu}$, an application of (16)–(18) yields the following steady-state probability distribution:

$$\pi_j = \binom{K}{j} \rho^j \pi_0 \qquad (j = 0, 1, \dots, R)$$

$$= \frac{\binom{K}{j} \rho^j j! \pi_0}{R! R^{j-R}} \qquad (j = R + 1, R + 2, \dots, K) \tag{52}$$

In (52),

$$\binom{K}{j} = \frac{K!}{j!(K - j)!}$$

where $0! = 1$, and for $n \geq 1$, $n! = n(n-1)\cdots(2)(1)$. To use (52), begin by finding π_0 from the fact that $\pi_0 + \pi_1 + \cdots + \pi_k = 1$. Using the steady-state probabilities in (52), we can determine the following quantities of interest:

$$L = \text{expected number of broken machines}$$
$$L_q = \text{expected number of machines waiting for service}$$
$$W = \text{average time a machine spends broken (down time)}$$
$$W_q = \text{average time a machine spends waiting for service}$$

Unfortunately, there are no simple formulas for L, L_q, W, and W_q. The best we can do is to express these quantities in terms of the π_j's:

$$L = \sum_{j=0}^{j=K} j\pi_j \tag{53}$$

$$L_q = \sum_{j=R}^{j=K} (j - R)\pi_j \tag{54}$$

We can now use (28) and (29) to obtain W and W_q. Since the arrival rate is state-dependent, the average number of arrivals per unit time is given by $\bar{\lambda}$, where

$$\bar{\lambda} = \sum_{j=0}^{j=K} \pi_j\lambda_j = \sum_{j=0}^{j=K} \lambda(K - j)\pi_j = \lambda(K - L) \tag{55}$$

If (28) is applied to the machines being repaired and to those machines awaiting repairs, we obtain

$$W = \frac{L}{\bar{\lambda}} \tag{56}$$

Applying (29) to the machines awaiting repair, we obtain

$$W_q = \frac{L_q}{\bar{\lambda}} \tag{57}$$

The following example illustrates the use of these formulas.

EXAMPLE 12   **Patrol Cars**

The Gotham Township Police Department has 5 patrol cars. A patrol car breaks down and requires service once every 30 days. The police department has two repair workers, each of whom takes an average of 3 days to repair a car. Breakdown times and repair times are exponential.

1 Determine the average number of police cars in good condition.

2 Find the average down time for a police car that needs repairs.

3 Find the fraction of the time a particular repair worker is idle.

Solution   This is a machine repair problem with $K = 5$, $R = 2$, $\lambda = \frac{1}{30}$ car per day, and $\mu = \frac{1}{3}$ car per day. Then

$$\rho = \frac{\frac{1}{30}}{\frac{1}{3}} = \frac{1}{10}$$

From (52),

$$\pi_1 = \binom{5}{1}\left(\frac{1}{10}\right)\pi_0 = .5\pi_0$$

$$\pi_2 = \binom{5}{2}\left(\frac{1}{10}\right)^2\pi_0 = .1\pi_0$$

$$\pi_3 = \binom{5}{3}\left(\frac{1}{10}\right)^3\frac{3!}{2!2}\pi_0 = .015\pi_0 \tag{58}$$

$$\pi_4 = \binom{5}{4}\left(\frac{1}{10}\right)^4\frac{4!}{2!(2)^2}\pi_0 = .0015\pi_0$$

$$\pi_5 = \binom{5}{5}\left(\frac{1}{10}\right)^5\frac{5!}{2!(2)^3}\pi_0 = .000075\pi_0$$

Then $\pi_0(1 + .5 + .1 + .015 + .0015 + .000075) = 1$, or $\pi_0 = .619$. Now (58) yields $\pi_1 = .310$, $\pi_2 = .062$, $\pi_3 = .009$, $\pi_4 = .001$, and $\pi_5 = 0$.

1 The expected number of cars in good condition is $K - L$, which is given by

$$K - \sum_{j=0}^{j=5} j\pi_j = 5 - [0(.619) + 1(.310) + 2(.062) + 3(.009) + 4(.001) + 5(0)]$$

$$= 5 - .465 = 4.535 \text{ cars in good condition}$$

2 We seek $W = \frac{L}{\bar{\lambda}}$. From (55),

$$\bar{\lambda} = \sum_{j=0}^{j=5} \lambda(5 - j)\pi_j = \frac{1}{30}(5\pi_0 + 4\pi_1 + 3\pi_2 + 2\pi_3 + \pi_4 + 0\pi_5)$$

$$= \frac{1}{30}[5(.619) + 4(.310) + 3(.062) + 2(.009) + 1(.001) + 0(0)]$$

$$= 0.151 \text{ car per day}$$

or

$$\bar{\lambda} = \lambda(K - L) = \frac{4.535}{30} = 0.151 \text{ car per day}$$

Since $L = 0.465$ car, we find that $W = \frac{0.465}{.0151} = 3.08$ days.

3 The fraction of the time that a particular repair worker will be idle is $\pi_0 + 0.5\pi_1 = .619 + .5(.310) = .774$.

If there were three repair people, the fraction of the time that a particular server would be idle would be $\pi_0 + (\frac{2}{3})\pi_1 + (\frac{1}{3})\pi_2$, and for a repair staff of R people, the probability that a particular server would be idle is given by

$$\pi_0 + \frac{(R-1)\pi_1}{R} + \frac{(R-2)\pi_2}{R} + \cdots + \frac{\pi_{R-1}}{R}$$

A Spreadsheet for the Machine Repair Problem

Machrep.xls

Figure 23 (file Machrep.xls) gives a spreadsheet template for the machine repair model. In cell B2, we input λ; in cell C2, μ; in cell D2, the number of repairers; and in cell F2,

	A	B	C	D	E	F	G	H
1	MACHINE	LAMBDA?	MU?	R?	RO	K?		
2	REPAIR	0.03333333	0.33333333	3	0.1	5		
3	MODEL	L	LS	LQ	W	WS	WQ	
4		0.45494681	0.45450532	0.0004415	3.00291412	3	0.002914124	
5								
6								
7								
8								
9								
10								
11								
12	STATE	LAMBDA(J)	MU(J)	CJ	PROB	#IN QUEUE	COLA*COLE	COLE*COL
13	0	0.16666667	0	1	0.62085236	0	0	0
14	1	0.13333333	0.33333333	0.5	0.31042618	0	0.310426181	0
15	2	0.1	0.66666667	0.1	0.06208524	0	0.124170472	0
16	3	0.06666667	1	0.01	0.00620852	0	0.018625571	0
17	4	0.03333333	1	0.00066667	0.0004139	1	0.001655606	0.0004139
18	5	0	1	2.2222E-05	1.3797E-05	2	6.89836E-05	2.7593E-05
19	6	0	1	0	0	3	0	0
20	7	0	1	0	0	4	0	0
21	8	0	1	0	0	5	0	0

FIGURE 23

the number of machines. In row 4, L, L_q, L_s, W, W_q, and W_s are computed. L_s equals the expected number of machines (in the steady state) being repaired and W_s equals the expected time that a broken machine spends being repaired. In column E, the steady-state probabilities are computed. We are assuming that $K \leq 1,000$. In Figure 23, we have input the information for Example 12.

Using LINGO for Machine Repair Model Computations

The LINGO function $@\mathbf{PFS}(K*\lambda/\mu,R,K)$ will yield L, the expected number (in the steady state) of machines in bad condition. The FS stands for Finite Source. Thus, for Example 12, $@\mathbf{PFS}(5*(1/30)/(1/3),2,5)$ will yield .465.

PROBLEMS

Group A

1 A laundromat has 5 washing machines. A typical machine breaks down once every 5 days. A repairer can repair a machine in an average of 2.5 days. Currently, three repairers are on duty. The owner of the laundromat has the option of replacing them with a superworker, who can repair a machine in an average of $\frac{5}{6}$ day. The salary of the superworker equals the pay of the three regular employees. Breakdown and service times are exponential. Should the laundromat replace the three repairers with the superworker?

2 My dog just had 3 frisky puppies who jump in and out of their whelping box. A puppy spends an average of 10 minutes (exponentially distributed) in the whelping box before jumping out. Once out of the box, a puppy spends an average of 15 minutes (exponentially distributed) before jumping back into the box.

   a At any given time, what is the probability that more puppies will be out of the box than will be in the box?

   b On the average, how many puppies will be in the box?

Group B

3[†] Gotham City has 10,000 streetlights. City investigators have determined that at any given time, an average of 1,000 lights are burned out. A streetlight burns out after an average of 100 days of use. The city has hired Mafia, Inc., to replace burned-out lamps. Mafia, Inc.'s contract states that the company is supposed to replace a burned-out street lamp in an average of 7 days. Do you think that Mafia, Inc. is living up to the contract?

4 This problem illustrates balking. The Oryo Cookie Ice Cream Shop in Dunkirk Square has three competitors. Since

[†]Based on Kolesar (1979).

people don't like to wait in long lines for ice cream, the arrival rate to the Oryo Cookie Ice Cream Shop depends on the number of people in the shop. More specifically, while $j \leq 4$ customers are present in the Oryo shop, customers arrive at a rate of $(20 - 5j)$ customers per hour. If more than 4 people are in the Oryo shop, the arrival rate is zero. For each customer, revenues less raw material costs are 50¢. Each server is paid $3 per hour. A server can serve an average of 10 customers per hour. To maximize expected profits (revenues less raw material and labor costs), how many servers should Oryo hire? Assume that interarrival and service times are exponential.

5 Suppose that interarrival times to a single-server system are exponential, but when n customers are present, there is a probability $\frac{n}{n+1}$ that an arrival will balk and leave the system before entering service. Also assume exponential service times.

a Find the probability distribution of the number of people present in the steady state.

b Find the expected number of people present in the steady state. (*Hint:* The fact that

$$e^x = 1 + x + \frac{x^2}{2!} + \frac{x^3}{3!} + \cdots$$

may be useful.)

6 For the machine repair model, show that $W = K/\bar{\lambda} - (1/\lambda)$.

7 (Requires use of a spreadsheet or LINGO) The machine repair model may often be used to approximate the behavior of a computer's CPU (central processing unit). Suppose that 20 terminals (assumed to always be busy) feed the CPU. After the CPU responds to a user, he or she takes an average of 80 seconds before sending another request to the CPU (this is called the *think time*). The CPU takes an average of

2 seconds to respond to any request. On the average, how long will a user have to wait before the CPU acts on his or her request? How will your answer change if there are 30 terminals? 40 terminals? Of course, you must make appropriate assumptions about exponentiality to answer this question.

8 Allbest airlines has 100 planes. Planes break down an average of twice a year and take one week to fix. Assuming the times between breakdowns and repairs are exponential, how many repairmen are needed to ensure that there is at least a 95% chance that 90 or more planes are available? (*Hint:* Use a one-way data table.)

9 An army has 200 tanks. Tanks need maintenance 10 times per year, and maintenance takes an average of 2 days. The army would like to have an average of at least 180 tanks working. How many repairmen are needed? Assume exponential interarrival and service times. (*Hint:* Use a one-way data table.)

Group C

10 Bectol, Inc. is building a dam. A total of 10 million cu ft of dirt is needed to construct the dam. A bulldozer is used to collect dirt for the dam. Then the dirt is moved via dumpers to the dam site. Only one bulldozer is available, and it rents for $100 per hour. Bectol can rent, at $40 per hour, as many dumpers as desired. Each dumper can hold 1,000 cu ft of dirt. It takes an average of 12 minutes for the bulldozer to load a dumper with dirt, and each dumper an average of five minutes to deliver the dirt to the dam and return to the bulldozer. Making appropriate assumptions about exponentiality, determine how Bectol can minimize the total expected cost of moving the dirt needed to build the dam. (*Hint:* There is a machine repair problem somewhere!)

8.10 Exponential Queues in Series and Open Queuing Networks

In the queuing models that we have studied so far, a customer's entire service time is spent with a single server. In many situations (such as the production of an item on an assembly line), the customer's service is not complete until the customer has been served by more than one server (see, for example, Figure 24).

Upon entering the system in Figure 24, the arrival undergoes stage 1 service (after waiting in line if all stage 1 servers are busy on arrival). After completing stage 1 service, the customer waits for and undergoes stage 2 service. This process continues until the customer completes stage k service. A system like Figure 24 is called a **k-stage series** (or tandem) **queuing system.** A remarkable theorem due to Jackson (1957) is as follows (see Heyman and Sobel (1984) for a proof).

THEOREM 4

If (1) interarrival times for a series queuing system are exponential with rate λ, (2) service times for each stage i server are exponential, and (3) each stage has an infinite-capacity waiting room, then interarrival times for arrivals to each stage of the queuing system are exponential with rate λ.

FIGURE **24**
Exponential Queues in Series

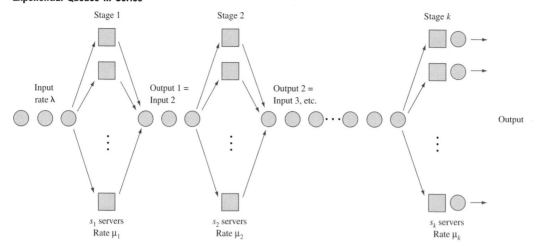

Stage 1 Stage 2 Stage k

Input rate λ

Output 1 = Input 2

Output 2 = Input 3, etc.

Output

s_1 servers
Rate μ_1

s_2 servers
Rate μ_2

s_k servers
Rate μ_k

For this result to be valid, each stage must have sufficient capacity to service a stream of arrivals that arrives at rate λ; otherwise, the queue will "blow up" at the stage with insufficient capacity. From our discussion of the $M/M/s/GD/\infty/\infty$ queuing system in Section 8.6, we see that each stage will have sufficient capacity to handle an arrival stream of rate λ if and only if, for $j = 1, 2, \ldots, k$, $\lambda > s_j\mu_j$. If $\lambda < s_j\mu_j$, Jackson's result implies that stage j of the system in Figure 24 may be analyzed as an $M/M/s_j/GD/\infty/\infty$ system with exponential interarrival times having rate λ and exponential service times with a mean service time of $\frac{1}{\mu_j}$. The usefulness of Jackson's result is illustrated by the following example.

EXAMPLE 13   **Auto Assembly**

The last two things that are done to a car before its manufacture is complete are installing the engine and putting on the tires. An average of 54 cars per hour arrive requiring these two tasks. One worker is available to install the engine and can service an average of 60 cars per hour. After the engine is installed, the car goes to the tire station and waits for its tires to be attached. Three workers serve at the tire station. Each works on one car at a time and can put tires on a car in an average of 3 minutes. Both interarrival times and service times are exponential.

1   Determine the mean queue length at each work station.

2   Determine the total expected time that a car spends waiting for service.

Solution   This is a series queuing system with $\lambda = 54$ cars per hour, $s_1 = 1$, $\mu_1 = 60$ cars per hour, $s_2 = 3$, and $\mu_2 = 20$ cars per hour (see Figure 25). Since $\lambda < \mu_1$ and $\lambda < 3\mu_2$, neither queue will "blow up," and Jackson's theorem is applicable. For stage 1 (engine), $\rho = \frac{54}{60} = .90$. Then (27) yields

$$L_q \text{ (for engine)} = \left(\frac{\rho^2}{1 - \rho}\right) = \left[\frac{(.90)^2}{1 - .90}\right] = 8.1 \text{ cars}$$

Now (32) yields

$$W_q \text{ (for engine)} = \frac{L_q}{\lambda} = \frac{8.1}{54} = 0.15 \text{ hour}$$

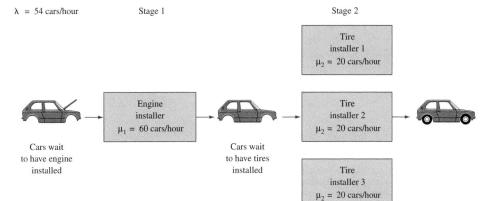

$\lambda = 54$ cars/hour   Stage 1   Stage 2

FIGURE 25
Series Queuing System
for Automobile

For stage 2 (tires), $\rho = \frac{54}{3(20)} = .90$. Table 6 yields $P(j \geq 3) = .83$. Now (41) yields

$$L_q \text{ (for tires)} = \frac{.83(.90)}{1 - .90} = 7.47 \text{ cars}$$

Then

$$W_q \text{ (for tires)} = \frac{L_q}{\lambda} = \frac{7.47}{54} = 0.138 \text{ hour}$$

Thus, the total expected time a car spends waiting for engine installation and tires is $0.15 + 0.138 = 0.288$ hour.

Open Queuing Networks

We now describe **open queuing networks,** a generalization of queues in series. As in Figure 24, assume that station j consists of s_j exponential servers, each operating at rate μ_j. Customers are assumed to arrive at station j from outside the queuing system at rate r_j. These interarrival times are assumed to be exponentially distributed. Once completing service at station i, a customer joins the queue at station j with probability p_{ij} and completes service with probability

$$1 - \sum_{j=1}^{j=k} p_{ij}$$

Define λ_j, the rate at which customers arrive at station j (this includes arrivals at station j from outside the system *and* from other stations). $\lambda_1, \lambda_2, \ldots, \lambda_k$ can be found by solving the following system of linear equations:

$$\lambda_j = r_j + \sum_{i=1}^{i=k} p_{ij}\lambda_i \qquad (j = 1, 2, \ldots, k)$$

This follows, because a fraction p_{ij} of the λ_i arrivals to station i will next go to station j. Suppose $s_j\mu_j > \lambda_j$ holds for all stations. Then it can be shown that the probability distribution of the number of customers present at station j and the expected number of customers present at station j can be found by treating station j as an $M/M/s_j/GD/\infty/\infty$ system with arrival rate λ_j and service rate μ_j. If for some j, $s_j\mu_j \leq \lambda_j$, then no steady-state distribution of customers exists. Remarkably, the numbers of customers present at each station are independent random variables. That is, knowledge of the number of people at all stations other than station j tells us nothing about the distribution of the number of people at station j! This result does not hold, however, if either interarrival or service times are not exponential.

To find L, the expected number of customers in the queuing system, simply add up the expected number of customers present at each station. To find W, the average time a customer spends in the system, simply apply the formula $L = \lambda W$ to the entire system. Here, $\lambda = r_1 + r_2 + \cdots + r_k$, because this represents the average number of customers per unit time arriving at the system. The following example illustrates the analysis of open queuing networks.

EXAMPLE 14	Open Queuing Network Example

Consider two servers. An average of 8 customers per hour arrive from outside at server 1, and an average of 17 customers per hour arrive from outside at server 2. Interarrival times are exponential. Server 1 can serve at an exponential rate of 20 customers per hour, and server 2 can serve at an exponential rate of 30 customers per hour. After completing service at server 1, half of the customers leave the system, and half go to server 2. After completing service at server 2, $\frac{3}{4}$ of the customers complete service, and $\frac{1}{4}$ return to server 1.

1 What fraction of the time is server 1 idle?

2 Find the expected number of customers at each server.

3 Find the average time a customer spends in the system.

4 How would the answers to parts (1)–(3) change if server 2 could serve only an average of 20 customers per hour?

Solution We have an open queuing network with $r_1 = 8$ customers/hour and $r_2 = 17$ customers/hour. Also, $p_{12} = .5$, $p_{21} = .25$, and $p_{11} = p_{22} = 0$. We can find λ_1 and λ_2 by solving $\lambda_1 = 8 + .25\lambda_2$ and $\lambda_2 = 17 + .5\lambda_1$. This yields $\lambda_1 = 14$ customers/hour and $\lambda_2 = 24$ customers/hour.

1 Server 1 may be treated as an $M/M/1/GD/\infty/\infty$ system with $\lambda = 14$ customers/hour and $\mu = 20$ customers/hour. Then $\pi_0 = 1 - \rho = 1 - .7 = .3$. Thus, server 1 is idle 30% of the time.

2 From (26), we find L at server $1 = \frac{14}{20-14} = \frac{7}{3}$ and L at server $2 = \frac{24}{30-24} = 4$. Thus, an average of $4 + \frac{7}{3} = \frac{19}{3}$ customers will be present in the system.

3 $W = \frac{L}{\lambda}$, where $\lambda = 8 + 17 = 25$ customers/hour. Thus,

$$W = \frac{\left(\dfrac{19}{3}\right)}{25} = \frac{19}{75} \text{ hour}$$

4 In this case, $s_2\mu_2 = 20 < \lambda_2$, so no steady state exists.

Network Models of Data Communication Networks

Queuing networks are commonly used to model data communication networks. The queuing models enable us to determine the typical delay faced by transmitted data and also to design the network. Our discussion is based on Tannenbaum (1981). See file Compnetwork.xls.

Compnetwork.xls

Consider a data communication network with 5 nodes (A, B, C, D, E). Suppose each data packet transmitted consists of 800 bits, and the number of packets per second that must be transmitted between each pair of nodes is as shown in Figure 26.

For example, an average of 5 packets per second must be sent from node A to node B. Packets are not always transmitted over the most direct route. Suppose the routings used to transmit each type of message are as shown in Figure 27.

For example, all messages that must go from A to D are transmitted via the route A–B–D. Each arc or route connecting two nodes has a capacity measured in thousands of bits per second. For example, an arc with 16,000 bits/second of capacity can "serve" $16,000/800 = 20$ packets/second. Each arc's capacity in thousands of bits per second is given in Figure 28. We are interested, of course, in the expected delay for a packet. Also, if total network capacity is limited, it is important to determine the capacity on each arc that will minimize the expected delay for a packet. The usual way to approach this problem is to treat each arc as if it were an independent $M/M/1$ queue and determine the expected time spent by each packet transmitted through that arc by the formula

	A	B	C	D	E	F	G
23	Packets/second		A	B	C	D	E
24		A	0	5	4	1	7
25		B	5	0	6	3	2
26		C	4	6	0	3	3
27		D	1	3	3	0	3
28		E	7	2	3	3	0

FIGURE 26

	A	B	C	D	E	F	G
30			A	B	C	D	E
31	Route used	A	-	AB	ABC	ABD	AE
32		B	BA	-	BC	BD	BDE
33		C	CBA	CB	-	CD	CDE
34		D	DBA	DB	DC	-	DCE
35		E	EA	EDB	EDC	ECD	-

FIGURE 27

	B	C	D	E	F
		Packets per second	Capacity (000) bits per second	Service Rate in Packets per second	W in seconds
4	Line				
5	AB	10	20	25	0.066667
6	AE	7	20	25	0.055556
7	BC	10	15	18.75	0.114286
8	BD	6	10	12.5	0.153846
9	CD	9	10	12.5	0.285714
10	CE	3	10	12.5	0.105263
11	DE	5	10	12.5	0.133333
12	BA	10	20	25	0.066667
13	EA	7	20	25	0.055556
14	CB	10	15	18.75	0.114286
15	DB	6	10	12.5	0.153846
16	DC	9	10	12.5	0.285714
17	EC	3	10	12.5	0.105263
18	ED	5	10	12.5	0.133333

FIGURE 28

$$W = \frac{1}{\mu - \lambda}$$

To illustrate, consider arc AB. Packets that are to be transmitted from A to B, A to C and A to D will use this arc. This is a total of $5 + 4 + 1 = 10$ packets per second. Suppose arc AB has a capacity of 20,000 bits per second. Then for arc AB, $\mu = 20,000/800 = 25$ packets per second, and $\lambda = 10$ packets per second. Then

$$W = \frac{1}{25 - 10} = .06667 \text{ second}$$

In rows 5–18 of Figure 28, we compute W for each arc in the communications network. Note that the network is assumed symmetric (that is, AB arrival rate and capacity equals BA arrival rate and capacity), so rows 12–18 are just copies of rows 5–11.

To determine the average delay faced by a packet, we use the following formula:

$$\text{Average delay per packet} = \frac{\sum_{\text{all arcs}} (\text{Arc arrival rate}) * (\text{expected time spent in arc})}{\text{Total number of arrivals}}$$

In Figure 29, we computed the average delay per packet in cell C20 with the formula

$$=\text{SUMPRODUCT(C5:C18,F5:F18)/SUM(C24:G28)}$$

Thus, average delay per packet is .18 second per packet.

	B	C
19		
20	Mean	0.180416
21	Time in system	
22	seconds	

FIGURE 29

	B	C	D	E	F	G
1		800				
2		bits/packet				
3						
4	Line	Packets per second	Capacity (000) bits per second	Service Rate in Packets per second	W in seconds	Diff
5	AB	10	18.31869	22.89836	0.077529	12.89836
6	AE	7	14.23287	17.79109	0.092669	10.79109
7	BC	10	18.31662	22.89577	0.077545	12.89577
8	BD	6	12.79577	15.99471	0.100053	9.994709
9	CD	9	16.98872	21.2359	0.081727	12.2359
10	CE	3	8.052237	10.0653	0.141537	7.065296
11	DE	5	11.2951	14.11888	0.109663	9.118877
12	BA	10	18.31869	22.89836	0.077529	12.89836
13	EA	7	14.23287	17.79109	0.092669	10.79109
14	CB	10	18.31662	22.89577	0.077545	12.89577
15	DB	6	12.79577	15.99471	0.100053	9.994709
16	DC	9	16.98872	21.2359	0.081727	12.2359
17	EC	3	8.052237	10.0653	0.141537	7.065296
18	ED	5	11.2951	14.11888	0.109663	9.118877
19		Total cap	200			
20	Mean	0.121843				

FIGURE 30

Suppose we had only 200,000 bits/second of total capacity to allocate to the network. How should we allocate capacity to minimize the expected delay per packet? See the sheet Optimization in file Compnetwork.xls (Figure 30). In G5:G18, we compute Service rate (in packets/second) − Arrival rate (in packets/second). We constrain this to be at least .01 so that a steady-state exists. Then our Solver window is as shown in Figure 31.

We choose capacities D5:D11 (remember that D12:D18 are just copies of D5:D11) to minimize expected system time (C20). We ensure that each arc's service rate exceeds its arrival rate (G5:G11≥.01), each capacity is nonnegative (D5:D11≥0), and total capacity is at most 200,000 (D19≤200). We find that we can reduce expected time in the system for a packet to .1218 second.

Of course, we are assuming a static routing, in which arrival rates to each node do not vary with the state of the network. In reality, many sophisticated dynamic routing schemes have been developed. A dynamic routing scheme would realize, for example, that if arc *AB* is congested and arc *AD* is relatively free, we should send messages directly from *A* to *D* instead of via route *A–B–D*.

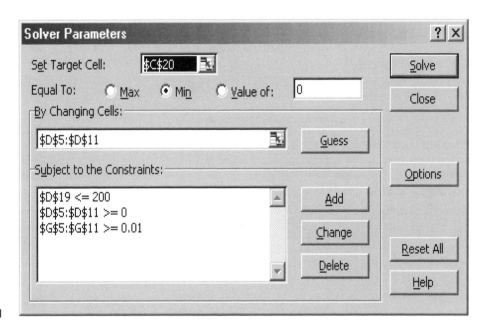

FIGURE 31

PROBLEMS

Group A

1 A Social Security Administration branch is considering the following two options for processing applications for social security cards:

Option 1 Three clerks process applications in parallel from a single queue. Each clerk fills out the form for the application in the presence of the applicant. Processing time is exponential with a mean of 15 minutes. Interarrival times are exponential.

Option 2 Each applicant first fills out an application without the clerk's help. The time to accomplish this is exponentially distributed, with a mean of 65 minutes. When the applicant has filled out the form, he or she joins a single line

to wait for one of the three clerks to check the form. It takes a clerk an average of 4 minutes (exponentially distributed) to review an application.

The interarrival time of applicants is exponential, and an average of 4.8 applicants arrive each hour. Which option will get applicants out of the office more quickly?

2 Consider an automobile assembly line in which each car undergoes two types of service: painting, then engine installation. Each hour, an average of 22.4 unpainted chassis arrive at the assembly line. It takes an average of 2.4 minutes to paint a car and an average of 3.75 minutes to install an engine. The assembly line has one painter and two engine installers. Assume that interarrival times and service times are exponential.

a On the average, how many painted cars without completely installed engines will be in the facility?

b On the average, how long will a painted car have to wait before installation of its engine begins?

3 Consider the following queuing systems:

System 1 An average of 40 customers arrive each hour; interarrival times are exponential. Customers must complete two types of service before leaving the system. The first server takes an average of 30 seconds (exponentially distributed) to perform type 1 service. After waiting in line, each customer obtains type 2 service (exponentially distributed with a mean of 1 minute) from a single server. After completing type 2 service, a customer leaves the system.

System 2 The arrival process for system 2 is identical to the interarrival process for system 1. In system 2, a customer must complete only one type of service. Service time averages 1.5 minutes and is exponentially distributed. Two servers are available.

In which system does a typical customer spend less time?

4 An average of 120 students arrive each hour (interarrival times are exponential) at State College's Registrar's Office to change their course registrations. To complete this process, a person must pass through three stations. Each station consists of a single server. Service times at each station are exponential, with the following mean times: station 1, 20 seconds; station 2, 15 seconds; station 3, 12 seconds. On the average, how many students will be present in the registrar's office for changing courses?

5 An average of 10 jobs per hour arrive at a job shop. Interarrival times of jobs are exponentially distributed. It takes an average of $\frac{10}{3}$ minutes (exponentially distributed) to complete a job. Unfortunately, $\frac{1}{3}$ of all completed jobs need to be reworked. Thus, with probability $\frac{1}{3}$, a completed job must wait in line to be reworked. In the steady state, how many jobs would one expect to find in the job shop? What would the answer be if it took an average of 5 minutes to finish a job?

6 Consider a queuing system consisting of three stations in series. Each station consists of a single server, who can process an average of 20 jobs per hour (processing times at each station are exponential). An average of 10 jobs per hour arrive (interarrival times are exponential) at station 1. When a job completes service at station 2, there is a .1 chance that it will return to station 1 and a .9 chance that it will move on to station 3. When a job completes service at station 3, there is a .2 chance that it will return to station 2 and a .8 chance that it will leave the system. All jobs completing service at station 1 immediately move on to station 2.

a Determine the fraction of time each server is busy.

b Determine the expected number of jobs in the system.

c Determine the average time a job spends in the system.

7 Before completing production, a product must pass through three stages of production. On the average, a new product begins at stage 1 every 6 minutes. The average time it takes to process the product at each stage is as follows: stage 1, 3 minutes; stage 2, 2 minutes; stage 3, 1 minute. After finishing at stage 3, the product is inspected (assume this takes no time). Ten percent of the final products are found to have a defective part and must return to stage 1 and go through the entire system again. After completing stage 3, 20% of the final products are found to be defective. They must return to stage 2 and pass through 2 and 3 again. On the average, how many jobs are in the system? Assume that all interarrival times and service times are exponential and that each stage consists of a single server.

8 A data communication network consists of three nodes, A, B, and C. Each packet transmitted contains 500 bits of information. The number of packets per second to be transmitted between each pair of nodes is as follows:

$$\begin{array}{c} & \begin{array}{ccc} A & B & C \end{array} \\ \begin{array}{c} A \\ B \\ C \end{array} & \begin{bmatrix} 0 & 4 & 3 \\ 4 & 0 & 6 \\ 3 & 6 & 0 \end{bmatrix} \end{array}$$

The routing used for each pair of nodes is as follows:

$$\begin{array}{c} & \begin{array}{ccc} A & B & C \end{array} \\ \begin{array}{c} A \\ B \\ C \end{array} & \begin{bmatrix} - & ACB & AC \\ BCA & - & BAC \\ CA & CAB & - \end{bmatrix} \end{array}$$

Assume that the capacities (in thousands of bits per second) for each arc are as follows:

Arc	Capacity
AB	12
AC	13
BC	15
BA	12
CA	13
CB	15

a Compute the expected delay for a packet.

b If a total of 75,000 bits/second of capacity is available, how should it be allocated?

9[†] Jobs arrive to a file server consisting of a CPU and two disks (disk 1 and disk 2). Currently there are six clients, and an average of three jobs per second arrive. Each visit to the CPU takes an average of .01 second, each visit to disk 1 takes an average of .02 second, and each visit to disk 2 takes an average of .03 second. An entering job first visits the CPU. After each visit to the CPU, with probability 7/16 the job next visits disk 1, with probability 8/16 the job next visits disk 2, and with probability 1/16 the job is completed. After visiting disk 1 or 2, the job immediately returns to the CPU.

a On the average, how many times does a job visit the CPU? How about disk 1? How about disk 2?

b On the average, how long does a job spend in the CPU? How long in disk 1? How long in disk 2? How long in the system?

10 Suppose the file server in Problem 9 now has 8 clients. Answer the questions in Problem 9.

[†]Based on Jain (1991).

11 Suppose we install a cache for disk 2. This will increase the mean time taken for a CPU visit by 30% and the mean time for a visit to disk 2 by 10%. On the other hand, the cache for disk 2 ensures that half the time the job was going to go to disk 2 the job will actually stay at the CPU and be processed there. Does the cache improve the operation of the system?

12 Suppose we eliminate disk 2. What will happen to the system response time? In this problem, you may assume that all requests that leave the CPU go to disk 1.

8.11 The $M/G/s/GD/s/\infty$ System (Blocked Customers Cleared)

In many queuing systems, an arrival who finds all servers occupied is, for all practical purposes, lost to the system. For example, a person who calls an airline for a reservation and gets a busy signal will probably call another airline. Or suppose that someone calls in a fire alarm and no engines are available; the fire will then burn out of control. Thus, in some sense, a request for a fire engine that occurs when no engines are available may be considered lost to the system. If arrivals who find all servers occupied leave the system, we call the system a **blocked customers cleared,** or BCC, system. Assuming that interarrival times are exponential, such a system may be modeled as an $M/G/s/GD/s/\infty$ system.

For an $M/G/s/GD/s/\infty$ system, L, W, L_q, and W_q are of limited interest. For example, since a queue can never occur, $L_q = W_q = 0$. If we let $\frac{1}{\mu}$ be the mean service time and λ be the arrival rate, then $W = W_s = \frac{1}{\mu}$.

In most BCC systems, primary interest is focused on the fraction of all arrivals who are turned away. Since arrivals are turned away only when s customers are present, a fraction π_s of all arrivals will be turned away. Hence, an average of $\lambda\pi_s$ arrivals per unit time will be lost to the system. Since an average of $\lambda(1 - \pi_s)$ arrivals per unit time will actually enter the system, we may conclude that

$$L = L_s = \frac{\lambda(1 - \pi_s)}{\mu}$$

For an $M/G/s/GD/s/\infty$ system, it can be shown that π_s depends on the service time distribution only through its mean ($\frac{1}{\mu}$). This fact is known as **Erlang's loss formula.** In other words, any $M/G/s/GD/s/\infty$ system with an arrival rate λ and mean service time of $\frac{1}{\mu}$ will have the same value of π_s. If we define $\rho = \frac{\lambda}{\mu}$, then for a given value of s, the value of π_s can be found from Figure 32. Simply read the value of ρ on the x-axis. Then the y-value on the s-server curve that corresponds to ρ will equal π_s. The following example illustrates the use of Figure 32.

EXAMPLE 15   **Ambulance Calls**

An average of 20 ambulance calls per hour are received by Gotham City Hospital. An ambulance requires an average of 20 minutes to pick up a patient and take the patient to the hospital. The ambulance is then available to pick up another patient. How many ambulances should the hospital have to ensure that there is at most a 1% probability of not being able to respond immediately to an ambulance call? Assume that interarrival times are exponentially distributed.

Solution   We are given that $\lambda = 20$ calls per hour, and $\frac{1}{\mu} = \frac{1}{3}$ hour. Thus, $\rho = \frac{\lambda}{\mu} = \frac{20}{3} = 6.67$. For $\rho = 6.67$, we seek the smallest value of s for which π_s is .01 or smaller. From Figure 32, we see that for $s = 13$, $\pi_s = .011$; and for $s = 14$, $\pi_s = .005$. Thus, the hospital needs 14 ambulances to meet its desired service standards.

FIGURE 32
Loss Probabilities for *M*/*GI*/*s*/*GD*/*s*/∞ Queuing System

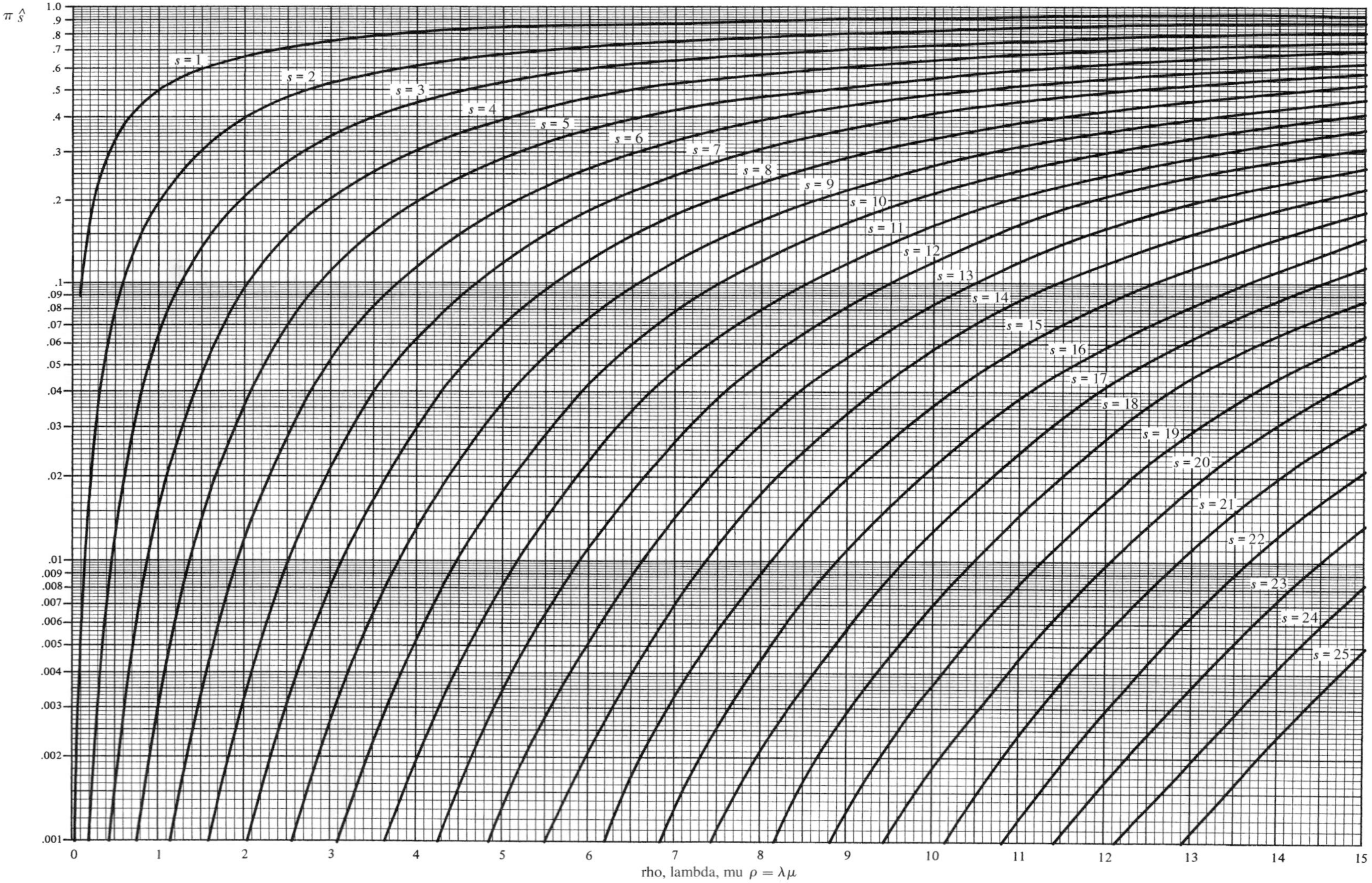

Source: Reprinted by permission of the publisher from *Introduction to Queuing Theory* by Robert B. Cooper. p. 316. Copyright©1980 by Elsevier Science Publishing Co., Inc.

A Spreadsheet for the BCC MODEL

Bcc.xls

In Figure 33 (file Bcc.xls) we give a spreadsheet template for the $M/G/s/GD/s/\infty$ queuing system. In cell B2, we input λ; in cell C2, μ; and in cell D2, the number of servers. In B4, we compute the expected number (in the steady state) of busy servers. In cell C4, we compute the value of π_s tabulated in Figure 32. Column E gives the steady-state probabilities for this model. We are assuming $s \le 1,000$. In Figure 33, we have input the values of λ, μ, and s for Example 15.

Using LINGO for BCC Computations

The LINGO function $@\mathbf{PEL}(\lambda/\mu,s)$ will yield π_s. For Example 15, the function $@\mathbf{PEL}(20/3,13)$ yields .010627, as in Figure 32. The $@\mathbf{PEL}$ function may be used to solve a problem (such as Problem 6) where we seek the number of servers minimizing expected cost per-unit time when cost is the sum of service cost and cost due to lost business.

	A	B	C	D	E	F	G
1	BCC MODEL	LAMBDA?	MU?	s?			
2		20	3	14			
3		L OR LS	PI(s)				
4		6.63320534	0.0050192				
5							
6							
7							
8							
9							
10							
11							
12	STATE	LAMBDA(J)	MU(J)	CJ	PROB	#IN QUEUE	COLA*COLE
13	0	20	0	1	0.00127738	0	0
14	1	20	3	6.66666667	0.00851587	0	0.008515872
15	2	20	6	22.2222222	0.02838624	0	0.056772479
16	3	20	9	49.382716	0.06308053	0	0.189241596
17	4	20	12	82.3045267	0.10513422	0	0.420536879
18	5	20	15	109.739369	0.14017896	0	0.700894799
19	6	20	18	121.932632	0.1557544	0	0.934526398
20	7	20	21	116.126316	0.14833752	0	1.038362665
21	8	20	24	96.7719303	0.1236146	0	0.988916824
22	9	20	27	71.6829114	0.09156637	0	0.824097353
23	10	20	30	47.7886076	0.06104425	0	0.610442484
24	11	20	33	28.9627925	0.03699651	0	0.406961656
25	12	20	36	16.0904403	0.02055362	0	0.246643428
26	13	20	39	8.25150783	0.01054032	0	0.137024127
27	14	0	42	3.92928944	0.0050192	0	0.070268783
28	15	0	42	0	0	1	0
29	16	0	42	0	0	2	0
30	17	0	42	0	0	3	0
31	18	0	42	0	0	4	0
32	19	0	42	0	0	5	0

FIGURE 33

PROBLEMS

Group A

1 Suppose that a fire department receives an average of 24 requests for fire engines each hour. Each request causes a fire engine to be unavailable for an average of 20 minutes. To have at most a 1% chance of being unable to respond to a request, how many fire engines should the fire department have?

2 A telephone order sales company must determine how many telephone operators are needed to staff the phones during the 9-to-5 shift. It is estimated that an average of 480 calls are received during this period and that the average call lasts for 6 minutes. If the company wants to have at most 1 chance in 100 of a caller receiving a busy signal, how many operators should be hired for the 9-to-5 shift? What assumption does the answer require?

3 In Example 15, suppose the hospital had 10 ambulances. On the average, how many ambulances would be en route or returning from a call?

4 A phone system is said to receive 1 Erlang of usage per hour if callers keep lines busy for an average of 3,600 seconds per hour. Suppose a phone system receives 2 Erlangs of usage per hour. If you want only 1% of all calls blocked, how many phone lines do you need?[†]

5 (Requires the use of a spreadsheet or LINGO) At the peak usage time, an average of 200 people per hour attempt to log on the Jade Vax. The average length of time somebody spends on the Vax is 20 minutes. If the Indiana University Computing Service wants to ensure that during peak usage only 1% of all users receive an "All ports busy" message, how many ports should the Jade Vax have?

6 (Requires the use of a spreadsheet or LINGO) US Airlines receives an average of 500 calls per hour from customers who want to make a reservation (time between calls follows an exponential distribution). It takes an average of 3 minutes to handle each call. Each customer who buys a ticket contributes $100 to US Airlines profit. It costs $15 per hour to staff a telephone line. Any customer who receives a busy signal will purchase a ticket on another airline. How many telephone lines should US Airlines have?

Group B

7 On the average, 26 patrons per year come to the I.U. library to borrow the *I Ching* (assume that interarrival times are exponential). Borrowers who find the book unavailable leave and never return. A borrower keeps a copy of the *I Ching* for an average of 4 weeks.

    a If the library has only one copy, what is the expected number of borrowers who will come to borrow the *I Ching* each year and find that the book is not available?

    b Suppose that each person who comes to borrow the *I Ching* and is unable to is considered to cost the library $1 in goodwill. A copy of the *I Ching* lasts two years and costs $11. A thief has just stolen the library's only copy. To minimize the sum of purchasing and goodwill costs over the next two years, how many copies of the *I Ching* should be purchased?

8[‡] A company's warehouse can store up to 4 units of a good. Each month, an average of 10 orders for the good are received. The times between the receipt of successive orders are exponentially distributed. When an item is used to fill an order, a replacement item is immediately ordered, and it takes an average of one month for a replacement item to arrive. If no items are on hand when an order is received, the order is lost. What fraction of all orders will be lost due to shortages? (*Hint:* Let the storage space for each item be a server and think about what it means for a server to be busy. Then come up with an appropriate definition of "service" time.)

[†]Based on Green (1987).

[‡]Based on Karush (1957).

8.12 How to Tell Whether Interarrival Times and Service Times Are Exponential[§]

How can we determine whether the actual data are consistent with the assumption of exponential interarrival times and service times? Suppose, for example, that interarrival times of $t_1, t_2, \ldots, t_n$ have been observed. It can be shown that a reasonable estimate of the arrival rate λ is given by

$$\hat{\lambda} = \frac{n}{\sum_{i=1}^{n} t_i}$$

[§]This section covers topics that may be omitted with no loss of continuity.

For example, if $t_1 = 20$, $t_2 = 30$, $t_3 = 40$, and $t_4 = 50$, we have seen 4 arrivals in 140 time units, or an average of 1 arrival per 35 time units. In this case, our estimate of the arrival rate $\hat{\lambda}$ is given by

$$\hat{\lambda} = \frac{4}{20 + 30 + 40 + 50} = \frac{1}{35}$$

customer per unit time. Given $\hat{\lambda}$, we can try to determine whether $t_1, t_2, \ldots, t_n$ are consistent with the assumption that interarrival times are governed by an exponential distribution with rate $\hat{\lambda}$ and density $\hat{\lambda}e^{-\hat{\lambda}t}$. The easiest way to test this conjecture is by using a chi-square goodness-of-fit test to determine whether it is reasonable to conclude that t_1, $t_2, \ldots, t_n$ represent a random sample from a random variable with a given density function $f(t)$. A Kolmogorov–Smirnov test may also be used (see Law and Kelton (1990)).

To begin, we break up the set of possible interarrival times into k categories. Under the assumption that $f(t)$ does govern interarrival times, we determine the number of the t_i's that we would expect to fall into category i. We call this number e_i. Then we count up how many of the observed t_i's actually were in category i. We call this number o_i. Next, we use the following formula to compute the observed value of the chi-square statistic, written $\chi^2(\text{obs})$:

$$\chi^2(\text{obs}) = \sum_{i=1}^{i=k} \frac{(o_i - e_i)^2}{e_i}$$

The value of $\chi^2(\text{obs})$ follows a chi-square distribution, with $k - 2$ degrees of freedom. Important percentile points of the chi-square distribution are tabulated in Table 9.

If $\chi^2(\text{obs})$ is small, it is reasonable to assume that the t_i's are samples from a random variable with density function $f(t)$. (After all, a perfect fit would have $o_i = e_i$ for $i = 1$, $2, \ldots, k$, resulting in a χ^2 value of zero.) If $\chi^2(\text{obs})$ is large, it is reasonable to assume that the t_i's do not represent a random sample from a random variable with density $f(t)$.

More formally, we are interested in testing the following hypotheses:

H_0: $t_1, t_2, \ldots, t_n$ is a random sample from a random variable
       with density $f(t)$

H_a: $t_1, t_2, \ldots, t_n$ is not a random sample from a random variable
       with density function $f(t)$

Given a value of α (the desired Type I error), we accept H_0 if $\chi^2(\text{obs}) \leq \chi^2_{k-r-1}(\alpha)$ and accept H_a if $\chi^2(\text{obs}) > \chi^2_{k-r-1}(\alpha)$. From Table 9, we obtain $\chi^2_{k-r-1}(\alpha)$ which represents the point in the χ^2_{k-r-1} table that has an area α to the right of it. Here, r is the number of parameters that must be estimated to specify the interarrival time distribution. To find $\chi^2_{k-r-1}(\alpha)$ in Excel, we simply enter the formula CHINV(Alpha, $k-r-1$). Thus, if interarrival times are exponential, $r = 1$, and if interarrival times follow a normal distribution or an Erlang distribution, $r = 2$. When choosing the boundaries for the k categories, it is desirable to ensure that each e_i is at least 5, $k \leq 30$, and the e_i's be kept as equal as possible. Example 16 illustrates the use of the chi-square test.

EXAMPLE 16   **Interarrival Times: Exponential or Not Exponential?**

The following interarrival times (in minutes) have been observed: 0.01, 0.07, 0.03, 0.08, 0.04, 0.10, 0.05, 0.10, 0.11, 1.17, 1.50, 0.93, 0.54, 0.19, 0.22, 0.36, 0.27, 0.46, 0.51, 0.11, 0.56, 0.72, 0.29, 0.04, 0.73. Does it seem reasonable to conclude that these observations come from an exponential distribution?

Solution   There are 25 observations with $\sum_{i=1}^{i=25} t_i = 9.19$. Thus, $\bar{\lambda} = \frac{25}{9.19} = 2.72$ arrivals per minute. We now test whether or not our data are consistent with an exponential random variable

TABLE 9
Percentiles of Chi-Square Distribution

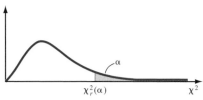

d.f. v	.990	.950	.900	.500	α .100	.050	.025	.010	.005
1	.0002	.004	.02	.45	2.71	3.84	5.02	6.63	7.88
2	.02	.10	.21	1.39	4.61	5.99	7.38	9.21	10.60
3	.11	.35	.58	2.37	6.25	7.81	9.35	11.34	12.84
4	.30	.71	1.06	3.36	7.78†	9.49	11.14	13.28	14.86
5	.55	1.15	1.61	4.35	9.24	11.07	12.83	15.09	16.75
6	.87	1.64	2.20	5.35	10.64	12.59	14.45	16.81	18.55
7	1.24	2.17	2.83	6.35	12.02	14.07	16.01	18.48	20.28
8	1.65	2.73	3.49	7.34	13.36	15.51	17.53	20.09	21.95
9	2.09	3.33	4.17	8.34	14.68	16.92	19.02	21.67	23.59
10	2.56	3.94	4.87	9.34	15.99	18.31	20.48	23.21	25.19
11	3.05	4.57	5.58	10.34	17.28	19.68	21.92	24.72	26.76
12	3.57	5.23	6.30	11.34	18.55	21.03	23.34	26.22	28.30
13	4.11	5.89	7.04	12.34	19.81	22.36	24.74	27.69	29.82
14	4.66	6.57	7.79	13.34	21.06	23.68	26.12	29.14	31.32
15	5.23	7.26	8.55	14.34	22.31	25.00	27.49	30.58	32.80
16	5.81	7.96	9.31	15.34	23.54	26.30	28.85	32.00	34.27
17	6.41	8.67	10.09	16.34	24.77	27.59	30.19	33.41	35.72
18	7.01	9.39	10.86	17.34	25.99	28.87	31.53	34.81	37.16
19	7.63	10.12	11.65	18.34	27.20	30.14	32.85	36.19	38.58
20	8.26	10.85	12.44	19.34	28.41	31.41	34.17	37.57	40.00
21	8.90	11.59	13.24	20.34	29.62	32.67	35.48	38.93	41.40
22	9.54	12.34	14.04	21.34	30.81	33.92	36.78	40.29	42.80
23	10.20	13.09	14.85	22.34	32.01	35.17	38.08	41.64	44.18
24	10.86	13.85	15.66	23.34	33.20	36.42	39.36	42.98	45.56
25	11.52	14.61	16.47	24.34	34.38	37.65	40.65	44.31	46.93
26	12.20	15.38	17.29	25.34	35.56	38.89	41.92	45.64	48.29
27	12.88	16.15	18.11	26.34	36.74	40.11	43.19	46.96	49.64
28	13.56	16.93	18.94	27.34	37.92	41.34	44.46	48.28	50.99
29	14.26	17.71	19.77	28.34	39.09	42.56	45.72	49.59	52.34
30	14.95	18.49	20.60	29.34	40.26	43.77	46.98	50.89	53.67
40	22.16	26.51	29.05	39.34	51.81	55.76	59.34	63.69	66.77
50	29.71	34.76	37.69	49.33	63.17	67.50	71.42	76.15	79.49
60	37.48	43.19	46.46	59.33	74.40	79.08	83.30	88.38	91.95
70	45.44	51.74	55.33	69.33	85.53	90.53	95.02	100.43	104.21
80	53.54	60.39	64.28	79.33	96.58	101.88	106.63	112.33	116.32
90	61.75	69.13	73.29	89.33	107.57	113.15	118.14	124.12	128.30
100	70.06	77.93	82.36	99.33	118.50	124.34	129.56	135.81	140.17

Source: Richard A. Johnson and Dean W. Wichern, *Applied Multivariate Statistical Analysis,* © 1982, p. 583. Reprinted by permission of Prentice Hall, Inc., Englewood Cliffs, New Jersey.
†*Note:* For example, $P(\chi_4^2 > 7.78) = .10$.

(call it **A**) having a density $f(t) = 2.72e^{-2.72t}$. We choose five categories so as to ensure that the probability that an observation from **A** falls into each of the five categories is .20. This yields $e_i = 25(.20) = 5$ for each category. To set the category boundaries, we need to determine the cumulative distribution function, $F(t)$, for **A**:

$$F(t) = P(\mathbf{A} \leq t) = \int_0^t 2.72e^{-2.72s} \, ds = 1 - e^{-2.72t}$$

Then we choose the categories to be as follows:

Category 1 $0 \leq t < m_1$ minutes

Category 2 $m_1 \leq t < m_2$ minutes

Category 3 $m_2 \leq t < m_3$ minutes

Category 4 $m_3 \leq t < m_4$ minutes

Category 5 $m_4 \leq t$ minutes

where $F(m_1) = .20$, $F(m_2) = .40$, $F(m_3) = .60$, and $F(m_4) = .80$.

Since $F(t) = 1 - e^{-2.72t}$, we see that for any number p, the value of t satisfying $F(t) = p$ may be found as follows:

$$1 - e^{-2.72t} = p$$
$$1 - p = e^{-2.72t}$$

Taking logarithms (to base e) of both sides yields

$$t = \frac{\ln(1 - p)}{-2.72}$$

$$m_1 = \frac{\ln .80}{-2.72} = 0.08$$

$$m_2 = \frac{\ln .60}{-2.72} = 0.19$$

$$m_3 = \frac{\ln .40}{-2.72} = 0.34$$

$$m_4 = \frac{\ln .20}{-2.72} = 0.59$$

Hence, our categories are as follows:

Category 1 $0 \leq t < 0.08$ minute

Category 2 $0.08 \leq t < 0.19$ minute

Category 3 $0.19 \leq t < 0.34$ minute

Category 4 $0.34 \leq t < 0.59$ minute

Category 5 $0.59 \leq t$

After classifying the data into these categories, we find that $o_1 = 6$, $o_2 = 5$, $o_3 = 4$, $o_4 = 5$, and $o_5 = 5$. By the construction of our categories, $e_1 = e_2 = e_3 = e_4 = e_5 = .20(25) = 5$. We now compute $\chi^2(\text{obs})$:

$$\chi^2(\text{obs}) = \frac{(6 - 5)^2}{5} + \frac{(5 - 5)^2}{5} + \frac{(4 - 5)^2}{5} + \frac{(5 - 5)^2}{5} + \frac{(5 - 5)^2}{5}$$

$$= .20 + 0 + .20 + 0 + 0 = .40$$

We arbitrarily choose $\alpha = .05$. Since we are trying to fit an exponential distribution to interarrival times, $r = 1$. Then $\chi_3^2(.05) = 7.81$, and we see that for $\alpha = .05$, we can accept the hypothesis that the observed interarrival times come from an exponential distribution with $\lambda = 2.72$ arrivals per minute.

Alternatively, we could have found the cutoff point for the chi-square test with the formula

$$=CHINV(.05,3)$$

This formula yields the value 7.81.

To test whether service times are exponentially distributed, simply apply the preceding approach to observed service times $s_1, s_2, \ldots, s_n$. Begin by obtaining an estimate (call it $\hat{\mu}$) of the actual service rate μ from

$$\hat{\mu} = \frac{n}{s_1 + s_2 + \cdots + s_n}$$

Then use the chi-square test to test whether or not it is reasonable to assume that the observed service times are observations from an exponential distribution with density $\hat{\mu}e^{-\hat{\mu}t}$.

PROBLEM

Group A

1 A travel agency wants to determine if the length of customers' phone calls can be adequately modeled by an exponential distribution. Last week, the agency recorded the length of all phone calls and obtained the following results (in seconds): 4, 6, 5, 8, 9, 10, 12, 8, 16, 20, 24, 27, 33, 37, 43, 50, 58, 68, 70, 78, 88, 100, 120, 130. Do these data indicate that the length of phone calls to the travel agency is governed by an exponential distribution?

8.13 Closed Queuing Networks

For manufacturing units attempting to implement just-in-time manufacturing, it makes sense to maintain a constant level of work in process. For a busy computer network, it may be convenient to assume that as soon as a job leaves the system, another job arrives to replace it. Such manufacturing and computer systems, where there is a constant number of jobs present, may be modeled as **closed queuing networks.** Recall that in an open queuing network, the numbers of jobs at each server were independent random variables. Since the number of jobs in the system is always constant, the distribution of jobs at different servers cannot be independent. We now discuss **Buzen's algorithm,** which can be used to determine steady-state probabilities for closed queuing networks.

We let P_{ij} be the probability that a job will go to server j after completing service at station i. Let P be the matrix whose $(i-j)$th entry is P_{ij}. We assume that service times at server j follow an exponential distribution with parameter μ_j. The system has s servers, and at all times, exactly N jobs are present. We let n_i be the number of jobs present at server i. Then the state of the system at any given time can be defined by an n-dimensional vector $\mathbf{n} = (n_1, n_2, \ldots, n_s)$. The set of possible states is given by $S_N = \{\mathbf{n}$ such that all $n_i \geq 0$ and $n_1 + n_2 + \cdots + n_s = N\}$.

Let λ_j equal the arrival rate to server j. Since there are no external arrivals, we may set all $r_j = 0$ and obtain the values of the λ_j's from the equation used in the open network situation. That is,

$$\lambda_j = \sum_{i=1}^{i=s} \lambda_i P_{ij} \qquad j = 1, 2, \ldots, s \tag{59}$$

Since jobs never leave the system, for each i, $\sum_{j=1}^{j=s} P_{ij} = 1$. This fact causes equation (59) to have no unique solution. Fortunately, it turns out that we can use any solution to (59) to help us get steady-state probabilities. If we define

$$\rho_i = \frac{\lambda_i}{\mu_i}$$

then we determine, for any state n, its steady-state probability $\Pi_N(\mathbf{n})$ from the following equation:

$$\Pi_N(\mathbf{n}) = \frac{\rho_1^{n_1} \rho_2^{n_2} \cdots \rho_n^{n_s}}{G(N)} \tag{60}$$

Here, $G(N) = \sum_{\mathbf{n} \in S_N} \rho_1^{n_1} \rho_2^{n_2} \cdots \rho_s^{n_s}$.

Buzen's algorithm gives us an efficient way to determine (in a spreadsheet) $G(N)$. Once we have the steady-state probability distribution, we can easily determine other measures of effectiveness, such as expected queue length at each server and expected time a job spends during each visit to a server, fraction of time a server is busy, and the throughput for each server (jobs per second processed by each server).

To obtain $G(N)$, we recursively compute the quantities $C_i(k)$, for $i = 1, 2, \ldots, s$ and $k = 0, 1, \ldots, N$. We initialize the recursion with $C_1(k) = \rho_1^k$, $k = 0, 1, \ldots, N$ and $C_i(0) = 1$, $i = 1, 2, \ldots, s$. For other values of k and i, we build up the values of $C_i(k)$ recursively via the following relationship:

$$C_i(k) = C_{i-1}(k) + \rho_i C_i(k - 1)$$

Then it can be shown that $G(N) = C_s(N)$. We illustrate the use of Buzen's algorithm with the following example.[†]

EXAMPLE 17 **Flexible Manufacturing System**

Consider a flexible manufacturing system in which 10 parts are always in process. Each part requires two operations. Each part begins by having operation 1 done at machine 1. Then, with probability .75 the part has operation 2 processed on machine 2, and with probability .25 the part has operation 2 processed on machine 3. Once a part completes operation 2, the part leaves the system and is immediately replaced by another part. We are given the following machine rates (the time for each operation is exponentially distributed): $\mu_1 = .25$ minute, $\mu_2 = .48$ minute, and $\mu_3 = .08$ minute.

a Find the probability distribution of the number of parts at each machine.

b Find the expected number of parts present at each machine.

c What fraction of the time is each machine busy?

d How many parts per minute are completed by each machine?

Buzen.xls **Solution** Our work is in file Buzen.xls. To begin, we need to compute one solution to the equations (59) defining λ_1, λ_2, and λ_3. We must solve

$$\lambda_1 = \lambda_2 + \lambda_3$$
$$\lambda_2 = .75\lambda_1$$
$$\lambda_3 = .25\lambda_1$$

[†]From Kao (1996).

There are an infinite number of solutions to this system. Arbitrarily choosing $\lambda_1 = 1$ yields the solution $\lambda_2 = .75$ and $\lambda_3 = .25$. In cells G8:I8, we compute $\rho_i = \frac{\lambda_i}{\mu_i}$. In G10:G20, we compute $C_1(k) = \rho_1^k$, $k = 0, 1, \ldots, 10$, and in G10:I10, we enter $C_i(0) = 1$, $i = 1, 2, 3$. Copying from H11 to H11:I20 the formula

$$=G11+H\$8*H10$$

implements the recursion $C_i(k) = C_{i-1}(k) + \rho_i C_i(k - 1)$. Then we can find G(10) = 7,231,883 from the value of $C_3(10)$ in cell H20. See Figure 34.

We can now generate all possible system states efficiently by starting with $n_1 = 0$ and listing those states in order of increasing values of n_2. Then we increase n_1 to 1 and list all states in increasing values of n_2, etc. Once we have $n_1 = 10$, we will have listed all states. (See Figure 35.) To efficiently generate all possible states, we copy down from C25 the formula

$$=IF(D25=0,B25+1,B25)$$

This formula increments n_1 by 1 if $n_3 = 0$ (which is the same as having $n_2 = 10 - n_1$). Otherwise, the formula keeps n_1 constant.

Then we copy down from D25 the formula

$$=IF(B25-B24=1,0,C24+1)$$

This formula makes $n_2 = 0$ if we have just increased the value of n_1; otherwise, the formula increments the value of n_2 by 1.

Finally, from E25, we copy down the formula

$$=10-B24-C24$$

This ensures that $n_3 = 10 - n_1 - n_2$.

In E24:E89 we use (60) to compute the steady-state probability for each state by copying from E24 to E25:E89 the formula

$$=(\$G\$8^B24)*(\$H\$8^C24)*(\$I\$8^D24)/\$I\$20$$

Part (a) Next, we answer part (a) by determining the probability distribution of the number of parts at each machine. We use the SUMIF function and a one-way data table to accomplish this goal. To begin, compute in H24 the probability of 0 parts at machine 1 with the formula

$$=SUMIF(\$B\$24:\$B\$89,I23,E24:E89)$$

This formula adds up every number in column D (which contains state probabilities) for the rows in which column B (which is parts at machine 1) has a 0 entry. See Figure 36.

	F	G	H	I
7	Mui	0.25	0.48	0.08
8	phoi	4	1.5625	3.125
9		1	2	3
10	0	1	1	1
11	1	4	5.5625	8.6875
12	2	16	24.6914063	51.83984
13	3	64	102.580322	264.5798
14	4	256	416.281754	1243.094
15	5	1024	1674.44024	5559.108
16	6	4096	6712.31287	24084.53
17	7	16384	26871.9889	102136.1
18	8	65536	107523.483	426698.9
19	9	262144	430149.442	1763583
20	10	1048576	1720684.5	7231883

FIGURE 34

	B	C	D	E
23	Parts at 1	Parts at 2	Parts at 3	Probability
24	0	0	10	0.01228143
25	0	1	9	0.00614071
26	0	2	8	0.00307036
27	0	3	7	0.00153518
28	0	4	6	0.00076759
29	0	5	5	0.00038379
30	0	6	4	0.0001919
31	0	7	3	9.5949E-05
32	0	8	2	4.7974E-05
33	0	9	1	2.3987E-05
34	0	10	0	1.1994E-05
35	1	0	9	0.01572023
36	1	1	8	0.00786011
37	1	2	7	0.00393006
38	1	3	6	0.00196503
39	1	4	5	0.00098251
40	1	5	4	0.00049126
41	1	6	3	0.00024563
42	1	7	2	0.00012281
43	1	8	1	6.1407E-05
44	1	9	0	3.0704E-05
45	2	0	8	0.02012189
46	2	1	7	0.01006094
47	2	2	6	0.00503047
48	2	3	5	0.00251524
49	2	4	4	0.00125762
50	2	5	3	0.00062881
51	2	6	2	0.0003144
52	2	7	1	0.0001572
53	2	8	0	7.8601E-05
54	3	0	7	0.02575602
55	3	1	6	0.01287801
56	3	2	5	0.006439
57	3	3	4	0.0032195
58	3	4	3	0.00160975
59	3	5	2	0.00080488
60	3	6	1	0.00040244
61	3	7	0	0.00020122
62	4	0	6	0.0329677
63	4	1	5	0.01648385
64	4	2	4	0.00824193
65	4	3	3	0.00412096
66	4	4	2	0.00206048
67	4	5	1	0.00103024
68	4	6	0	0.00051512
69	5	0	5	0.04219866
70	5	1	4	0.02109933
71	5	2	3	0.01054967
72	5	3	2	0.00527483
73	5	4	1	0.00263742
74	5	5	0	0.00131871
75	6	0	4	0.05401429
76	6	1	3	0.02700714
77	6	2	2	0.01350357
78	6	3	1	0.00675179
79	6	4	0	0.00337589
80	7	0	3	0.06913829
81	7	1	2	0.03456914
82	7	2	1	0.01728457
83	7	3	0	0.00864229
84	8	0	2	0.08849701
85	8	1	1	0.0442485
86	8	2	0	0.02212425
87	9	0	1	0.11327617
88	9	1	0	0.05663809
89	10	0	0	0.1449935

FIGURE 35

Selecting the table range G24:H35 and column input cell I23 enables us to loop through and compute the steady-state probabilities for each number of parts at machine 1. In a similar fashion, we obtain the following steady-state probability distributions for machines 2 and 3. See Figure 37.

Part (b) The mean number of parts present at machine 1 may be computed as $\sum_{i=0}^{i=10} i *$ (Probability of i parts at machine 1). In cell K31, we compute the mean number of parts at machine 1 with the formula

$$=\text{SUMPRODUCT(G25:G35,H25:H35)}$$

In a similar fashion, we compute the mean number of parts at machines 2 and 3 in cells K32 and K33. See Figure 38. Note that machine 1 is clearly the bottleneck.

Part (c) To compute the probability that each machine is busy, we just subtract from 1 the probability that each machine has 0 parts. These computations are done in L31:L33. We find that machine 1 is busy 97% of the time, machine 2 38% of the time, and machine 3 76% of the time.

	G	H	I
21			
22			Parts
23		Prob	0
24	Machine 1 parts	0.02455086	
25	0	0.02455086	
26	1	0.03140975	
27	2	0.04016518	
28	3	0.05131082	
29	4	0.06542029	
30	5	0.08307862	
31	6	0.10465268	
32	7	0.12963429	
33	8	0.15486976	
34	9	0.16991426	
35	10	0.1449935	

FIGURE 36

	G	H
	Machine 2	
37	Parts	0.61896518
38	0	0.61896518
39	1	0.23698584
40	2	0.09017388
41	3	0.03402481
42	4	0.01269126
43	5	0.00465769
44	6	0.00166949
45	7	0.00057718
46	8	0.00018798
47	9	5.4691E-05
48	10	1.1994E-05

	G	H
	Machine 3	
50	Parts	0.23793036
51	0	0.23793036
52	1	0.18587372
53	2	0.14519511
54	3	0.1133962
55	4	0.08851582
56	5	0.06900306
57	6	0.0536088
58	7	0.0412822
59	8	0.03105236
60	9	0.02186094
61	10	0.01228143

FIGURE 37

	J	K	L	M
29				
30	Mean Number	Mean Number	Prob busy	Completions per second
31	Machine 1	6.696224299	0.97544914	0.243862285
32	Machine 2	0.609634749	0.38103482	0.182896714
33	Machine 3	2.694140952	0.76206964	0.060965571

FIGURE 38

Part (d)   To compute the mean number of service completions per minute by each machine, we simply multiply the probability that a machine is busy by the machine's service rate. These computations are done in M31:M33. We find that machine 1 on average completes .24 part/minute, machine 2 .18 part/minute, and machine 3 .06 part/minute.

PROBLEMS

Group A

1 Jobs arrive to a file server consisting of a CPU and two disks (disk 1 and disk 2). With probability 13/20, a job goes from CPU to disk 1, and with probability 6/20, a job goes from CPU to disk 2. With probability 1/20, a job is finished after its CPU operation and is immediately replaced by another job. There are always 3 jobs in the system. The mean time to complete the CPU operation is .039 second. The mean time to complete the disk 1 operation is .18 second, and the mean time to complete the disk 2 operation is .26 second.

a Determine the steady-state distribution of the number of jobs at each part of the system.

b What is the average number of jobs at CPU? Disk 1? Disk 2?

c What is the probability that CPU is busy? Disk 1? Disk 2?

d What is the average number of jobs completed per second by CPU? Disk 1? Disk 2?

2 A manufacturing process always has 8 parts in process. A part must successfully complete two steps (step 1 and step 2) to be completed. A single machine performs step 1 and can process an average of 8 parts per minute. A single machine performs step 2 and can process 11 parts per minute. Unfortunately, step 2 is not totally reliable. (Step 1 is totally reliable, however.) Each time a part is sent through step 2, there is a 10% chance that step 2 must be repeated.

a Find the steady-state distribution of parts at each machine.

b Find the average number of parts at each machine.

c Find the probability that each machine is busy.

d Find the number of parts per minute successfully completing service at each machine.

8.14 An Approximation for the *G/G/m* Queuing System

In most situations, interarrival times follow an exponential random variable. (See Denardo (1982) for an explanation of this fact.) Often, however, service times do not follow an exponential distribution. When interarrival times and service times each follow a nonexponential random variable, we call the queuing system a *G/G/m* system. The first *G* indicates that interarrival times always follow the same (but not necessarily exponential) random variable, while the second *G* indicates that service times always follow the same (but not necessarily exponential) random variable. For these situations, the templates discussed in the previous sections of this chapter are not valid. Fortunately, the Allen–Cunneen approximation (see Tanner (1995)) often gives a good approximation to *L*, *W*,

L_q, and W_q for *G/G/m* systems. The file ggm.xls contains a spreadsheet implementation of the Allen–Cunneen approximation. The user need only input the following information:

- The average number of arrivals per unit time (λ) in cell B3.
- The average rate at which customers can be serviced (μ) in cell B4.
- The number of servers (*s*) in cell B5.
- The squared coefficient of variation—(variance of interarrival times)/(mean interarrival time)^2—of interarrival times in cell B6.
- The squared coefficient of variation—(variance of service times)/(mean service time)^2—of service times in cell B7.

The squared coefficient of variation for interarrival or service times can easily be estimated with the Excel functions =AVERAGE and =VARP. Recall that the exponential random variable has the property that variance = mean^2. Thus, the squared coefficient of variation for exponential interarrival or service times will equal 1, and the amount by which the squared coefficient of variation for interarrival or service times differs from 1 indicates the degree of departure from exponentiality. The Allen–Cunneen approximation is exact if interarrival times and service times are exponential. Extensive testing by Tanner indicates that in a wide variety of situations, the values of L, W, L_q, and W_q obtained by the approximation are within 10% of their true values. Here is an illustration of the Allen–Cunneen approximation.

EXAMPLE 18   NBD Bank

The NBD Bank branch in Bloomington, Indiana has 6 tellers. At peak times, an average of 4.8 customers per minute arrive at the bank. It takes a teller an average of 1 minute to serve a customer. The squared coefficient of variation for both interarrival times and service times is .5. Estimate the average time a customer will have to wait before seeing a teller. On average, how many customers will be present?

Solution   After inputting the relevant information in cells B3 through B7 (see Figure 39), we find that on average, a customer will wait .216 minute for a teller. On average, 5.83 customers will be present in the bank. Congestion seems to be well under control. This favorable outcome is largely due to the low squared coefficient of variation for both interarrival and service times. For example, if both squared coefficients of variation were 4, then W_q would be 1.73 minutes, an 800% increase.

	A	B	C	D
1		**G/G/m Template**		
2		Allen-Cunneen Approximation		
3	Lambda	**4.8**		
4	Mu	**1**		
5	s	**6**		
6	CV arrive	**0.5**		
7	CV service	**0.5**		
8	u	4.8		
9	ro	0.8		
10	R(s,mu)	0.82322		
11	E_C(s,mu)	0.517772		
12	W_q	0.215738		
13	L_q	1.035544		
14	W	1.215738		
15	L	5.835544		

FIGURE 39

PROBLEMS

Group A

Problems 1–4 refer to Example 18.

1 NBD believes the congestion level is satisfactory if the average number of customers in line equals the number of servers. For the information given in the example, what is the maximum arrival rate that can be satisfactorily handled with 6 servers?

2 Show how the average time a customer must wait for a teller depends on the number of servers.

3 Using a two-way data table, determine how changes in the squared coefficient of variation for interarrival and service times affect the average number of customers in the NBD branch.

4 Suppose a teller costs $30 per hour. Suppose the bank values a customer's time at NBD at c per hour. Show how variations in c affect the number of tellers that NBD should use.

5 Southbest Airlines has an average of 230 customers per hour arriving at a ticket counter where 8 agents are working. Each agent can serve an average of 30 customers per hour. The squared coefficient of variation for the interarrival times is 1.5 and 2 for the service times.

    a On average, how many customers will be present at the ticket counter?

    b On average, how long will a customer have to wait for an agent?

8.15 Priority Queuing Models[†]

There are many situations in which customers are not served on a first come, first served (FCFS) basis. In Section 8.1, we also discussed the service in random order (SIRO) and last come, first served (LCFS) queue disciplines. Let $\mathbf{W}_{FCFS}$, $\mathbf{W}_{SIRO}$, and $\mathbf{W}_{LCFS}$ be the random variables representing a customer's waiting time in queuing systems under the disciplines FCFS, SIRO, and LCFS, respectively. It can be shown that

$$E(\mathbf{W}_{FCFS}) = E(\mathbf{W}_{SIRO}) = E(\mathbf{W}_{LCFS})$$

Thus, the average time (steady-state) that a customer spends in the system does not depend on which of these three queue disciplines is chosen. It can also be shown that

$$\text{var } \mathbf{W}_{FCFS} < \text{var } \mathbf{W}_{SIRO} < \text{var } \mathbf{W}_{LCFS} \tag{61}$$

Since a large variance is usually associated with a random variable that has a relatively large chance of assuming extreme values, (61) indicates that relatively large waiting times are most likely to occur with an LCFS discipline and least likely to occur with an FCFS discipline. This is reasonable, because in an LCFS system, a customer can get lucky and immediately enter service but can also be bumped to the end of a long line. In FCFS, however, the customer cannot be bumped to the end of a long line, so a very long wait is relatively unlikely.

    In many organizations, the order in which customers are served depends on the customer's "type." For example, hospital emergency rooms usually serve seriously ill patients before they serve nonemergency patients. Also, in many computer systems, longer jobs do not enter service until all shorter jobs in the queue have been completed. Models in which a customer's type determines the order in which customers undergo service are called **priority queuing models.**

    The following scenario encompasses many priority queuing models (including all the models discussed in this section). Assume there are n types of customers (labeled type 1, type 2, . . . , type n). The interarrival times of type i customers are exponentially distributed with rate λ_i. Interarrival times of different customer types are assumed to be independent. The service time of a type i customer is described by a random variable $\mathbf{S}_i$ (not necessarily exponential). We assume that lower-numbered customer types have priority over higher-numbered customer types.

[†]This section covers topics that may be omitted with no loss of continuity.

Nonpreemptive Priority Models

We begin by considering nonpreemptive priority models. In a nonpreemptive model, a customer's service cannot be interrupted. After each service completion, the next customer to enter service is chosen by giving priority to lower-numbered customer types (with ties broken on an FCFS basis). For example, if $n = 3$ and three type 2 and four type 3 customers are present, the next customer to enter service would be the type 2 customer who was the first of that type to arrive.

In the Kendall–Lee notation, a nonpreemptive priority model is indicated by labeling the fourth characteristic as NPRP. To indicate multiple customer types, we subscript the first two characteristics with i's. Thus, $M_i/G_i/\cdots$ would represent a situation in which the interarrival times for the ith customer type are exponential and the service times for the ith customer type have a general distribution. In what follows, we let

$$W_{qk} = \text{expected steady-state waiting time in line spent by a type } k \text{ customer}$$
$$W_k = \text{expected steady-state time in the system spent by a type } k \text{ customer}$$
$$L_{qk} = \text{expected steady-state number of type } k \text{ customers waiting in line}$$
$$L_k = \text{expected steady-state number of type } k \text{ customers in the system}$$

The $M_i/G_i/1/\text{NPRP}/\infty/\infty$ Model

Our first results concern the single-server, nonpreemptive $M_i/G_i/1/\text{NPRP}/\infty/\infty$ system. Define $\rho_i = \frac{\lambda_i}{\mu_i}$, $a_0 = 0$, and $a_k = \sum_{i=1}^{i=k}\rho_i$. We assume[†] that

$$\sum_{i=1}^{i=n} \frac{\lambda_i}{\mu_i} < 1$$

Then

$$W_{qk} = \frac{\sum_{k=1}^{k=n} \lambda_k E(\mathbf{S}_k^2)/2}{(1 - a_{k-1})(1 - a_k)}$$

$$L_{qk} = \lambda_k W_{qk}$$

$$W_k = W_{qk} + \frac{1}{\mu_k} \tag{62}$$

$$L_k = \lambda_k W_k$$

The following example illustrates the use of (62).

EXAMPLE 19   Copying Priority

A copying facility gives shorter jobs priority over long jobs. Interarrival times for each type of job are exponential, and an average of 12 short jobs and 6 long jobs arrive each hour. Let type 1 job = short job and type 2 job = long job. Then we are given that

$$E(\mathbf{S}_1) = 2 \text{ minutes} \qquad E(\mathbf{S}_1^2) = 6 \text{ minutes}^2 = \frac{1}{600} \text{ hour}^2$$

$$E(\mathbf{S}_2) = 4 \text{ minutes} \qquad E(\mathbf{S}_2^2) = 18 \text{ minutes}^2 = \frac{1}{200} \text{ hour}^2$$

Determine the average length of time each type of job spends in the copying facility.

[†]If this condition does not hold, then for one or more customer types, no steady-state waiting time will exist.

Solution We are given that $\lambda_1 = 12$ jobs per hour, $\lambda_2 = 6$ jobs per hour, and $\mu_1 = 30$ jobs per hour, and $\mu_2 = 15$ jobs per hour. Then $\rho_1 = \frac{12}{30} = .4$ and $\rho_2 = \frac{6}{15} = .4$. Since $\rho_1 + \rho_2 < 1$, a steady state will exist. Now $a_0 = 0$, $a_1 = .4$, and $a_2 = .4 + .4 = .8$. Equations (62) now yield

$$W_{q1} = \frac{\dfrac{12\left(\dfrac{1}{600}\right)}{2} + \dfrac{6\left(\dfrac{1}{200}\right)}{2}}{(1-0)(1-.4)} = \frac{\dfrac{30}{1,200}}{.6} = 0.042 \text{ hour}$$

$$W_{q2} = \frac{\dfrac{12\left(\dfrac{1}{600}\right)}{2} + \dfrac{6\left(\dfrac{1}{200}\right)}{2}}{(1-.4)(1-.8)} = \frac{\dfrac{30}{1,200}}{.12} = 0.208 \text{ hour}$$

Also,

$$W_1 = W_{q1} + \frac{1}{\mu_1} = 0.042 + 0.033 = 0.075 \text{ hour}$$

$$W_2 = W_{q2} + \frac{1}{\mu_2} = 0.208 + 0.067 = 0.275 \text{ hour}$$

Thus, as expected, the long jobs spend much more time in the copying facility than the short jobs do.

The $M_i/G_i/1/NPRP/\infty/\infty$ Model with Customer-Dependent Waiting Costs

Consider a single-server, nonpreemptive priority system in which a cost c_k is charged for each unit of time that a type k customer spends in the system. If we want to minimize the expected cost incurred per unit time (in the steady state), what priority ordering should be placed on the customer types? Suppose the n customer types are numbered such that

$$c_1\mu_1 \geq c_2\mu_2 \geq \cdots \geq c_n\mu_n \tag{63}$$

Then expected cost is minimized by giving the highest priority to type 1 customers, the second-highest priority to type 2 customers, and so forth, and the lowest priority to type n customers. To see why this priority ordering is reasonable, observe that when a type k customer is being served, cost leaves the system at a rate $c_k\mu_k$. Thus, cost can be minimized by giving the highest priority to customer types with the largest values of $c_k\mu_k$.

As a special case of this result, suppose we want to minimize L, the expected number of jobs in the system. Let $c_1 = c_2 = \cdots = c_n = 1$. Then at any time, the cost per unit time is equal to the number of customers in the system. Thus, the expected cost per unit time will equal L. Now (63) becomes

$$\mu_1 \geq \mu_2 \geq \cdots \geq \mu_n \qquad \text{or} \qquad \frac{1}{\mu_1} \leq \frac{1}{\mu_2} \leq \cdots \leq \frac{1}{\mu_n}$$

Thus, we may conclude that the expected number of jobs in the system will be minimized if the highest priority is given to the customer types with the shortest mean service time. This priority discipline is known as the *shortest processing time* (SPT) discipline.

The $M_i/M/s/\text{NPRP}/\infty/\infty$ Model

To obtain tractable analytic results for multiserver priority systems, we must assume that each customer type has exponentially distributed service times with a mean of $\frac{1}{\mu}$, and that type i customers have interarrival times that are exponentially distributed with rate λ_i. Such a system with s servers is denoted by the notation $M_i/M/s/\text{NPRP}/\infty/\infty$. For this model,

$$W_{qk} = \frac{P(j \geq s)}{s\mu(1 - a_{k-1})(1 - a_k)} \tag{64}$$

In (64),

$$a_k = \sum_{i=1}^{i=k} \frac{\lambda_i}{s\mu} \qquad (k \geq 1)$$

$a_0 = 0$, and $P(j \geq s)$ is obtained from Table 6 for an s-server system having

$$\rho = \frac{\lambda_1 + \lambda_2 + \cdots + \lambda_n}{s\mu}$$

Example 20 illustrates the use of (64).

EXAMPLE 20   Police Response

Gotham Township has 5 police cars. The police department receives two types of calls: emergency (type 1) and nonemergency (type 2) calls. Interarrival times for each type of call are exponentially distributed, with an average of 10 emergency and 20 nonemergency calls being received each hour. Each type of call has an exponential service time, with a mean of 8 minutes (assume that, on the average, 6 of the 8 minutes is the travel time from the police station to the call and back to the station). Emergency calls are given priority over nonemergency calls. On the average, how much time will elapse between the placement of a nonemergency call and the arrival of a police car?

Solution   We are given that $s = 5$, $\lambda_1 = 10$ calls per hour, $\lambda_2 = 20$ calls per hour, $\mu = 7.5$ calls per hour, $\rho = \frac{10+20}{5(7.5)} = .80$, $a_0 = 0$, $a_1 = \frac{10}{37.5} = .267$, and $a_2 = \frac{10+20}{37.5} = .80$. From Table 6, with $s = 5$ and $\rho = .80$, $P(j \geq 5) = .55$. Then (64) yields

$$W_{q2} = \frac{.55}{5(7.5)(1 - .267)(1 - .80)} = \frac{.55}{5.50} = 0.10 \text{ hour} = 6 \text{ minutes}$$

The average time between the placement of a nonemergency call and the arrival of the car is $W_{q2} + (\frac{1}{2})$ (total travel time per call) $= 6 + 3 = 9$ minutes.

Preemptive Priorities

We close our discussion of priority queuing systems by discussing a single-server **preemptive queuing system.** In a preemptive queuing system, a lower-priority customer (say, a type i customer) can be bumped from service whenever a higher-priority customer arrives. Once no higher-priority customers are present, the bumped type i customer reenters service. In a **preemptive resume model,** a customer's service continues from the point at which it was interrupted. In a **preemptive repeat model,** a customer begins service anew each time he or she reenters service. Of course, if service times are exponentially distributed, the resume and repeat disciplines are identical. (Why?) In the Kendall–Lee notation, we denote a preemptive queuing system by labeling the fourth characteristic PRP. We now consider a single-server $M_i/M/1/\text{PRP}/\infty/\infty$ system in which the service time of each customer is exponential with mean $\frac{1}{\mu}$ and the interarrival times for the ith customer type are exponentially distributed with rate λ_i. Then

$$W_k = \frac{\frac{1}{\mu}}{(1 - a_{k-1})(1 - a_k)} \tag{65}$$

where $a_0 = 0$ and

$$a_k = \sum_{i=1}^{i=k} \frac{\lambda_i}{\mu}$$

For obvious reasons, preemptive disciplines are rarely used if the customers are people. Preemptive disciplines are sometimes used, however, for "customers" like computer jobs. The following example illustrates the use of (65).

EXAMPLE 21   University Computer System

On the Podunk U computer system, faculty jobs (type 1) always preempt student jobs (type 2). The length of each type of job follows an exponential distribution, with mean 30 seconds. Each hour, an average of 10 faculty and 50 student jobs are submitted. What is the average length of time between the submission and completion of a student's computer job? Assume that interarrival times are exponential.

Solution We are given that $\mu = 2$ jobs per minute, $\lambda_1 = \frac{1}{6}$ job per minute, and $\lambda_2 = \frac{5}{6}$ job per minute. Then

$$a_0 = 0, \qquad a_1 = \frac{\frac{1}{6}}{2} = \frac{1}{12}, \qquad a_2 = \frac{1}{12} + \frac{\frac{5}{6}}{2} = \frac{1}{2}$$

Equation (65) yields

$$W_2 = \frac{\frac{1}{2}}{\left(1 - \frac{1}{12}\right)\left(1 - \frac{1}{2}\right)} = \frac{12}{11} \text{ minutes} = 1.09 \text{ minutes}$$

An average of 1.09 minutes will elapse between the time a student submits a job and the time the job is completed.

PROBLEMS

Group A

1 English professor Jacob Bright has one typist, who types for 8 hours per day. He submits three types of jobs to the typist: tests, research papers, and class handouts. The information in Table 10 is available. Professor Bright has told the typist that tests have priority over research papers, and research papers have priority over class handouts. Assuming a nonpreemptive system, determine the expected time that Professor Bright will have to wait before each type of job is completed.

TABLE **10**

Type of Job	Frequency (number per day)	$E(S_i)$ (hours)	$E(S_i^2)$ (hours)^2
Test	2	1	2
Research paper	0.5	4	20
Class handout	5	0.5	0.50

2 Suppose a supermarket uses a system in which all customers wait in a single line for the first available cashier. Assume that the service time for a customer who purchases k items is exponentially distributed, with mean k seconds. Also, a customer who purchases k items feels that the cost of waiting in line for 1 minute is $\frac{\$1}{k}$. If customers can be assigned priorities, what priority assignment will minimize the expected waiting cost incurred by the supermarket's customers? Why would a customer's waiting cost per minute be a decreasing function of k?

3 Four doctors work in a hospital emergency room that handles three types of patients. The time a doctor spends with each type of patient is exponentially distributed, with a mean of 15 minutes. Interarrival times for each customer type are exponential, with the average number of arrivals per hour for each patient type being as follows: type 1, 3 patients; type 2, 5 patients; type 3, 3 patients. Assume that type 1 patients have the highest priority, and type 3 patients have the lowest priority (no preemption is allowed). What is the average length of time that each type of patient must wait before seeing a doctor?

4 Consider a computer system to which two types of computer jobs are submitted. The mean time to run each type of job is $\frac{1}{\mu}$. The interarrival times for each type of job are exponential, with an average of λ_i type i jobs arriving each hour. Consider the following three situations.

a Type 1 jobs have priority over type 2 jobs, and preemption is allowed.

b Type 1 jobs have priority over type 2 jobs, and no preemption is allowed.

c All jobs are serviced on a FCFS basis.

Under which system are type 1 jobs best off? Worst off? Answer the same questions for type 2 jobs.

8.16 Transient Behavior of Queuing Systems

Throughout the chapter, we have assumed that the arrival rate, service rate, and number of servers have stayed constant over time. This allows us to talk reasonably about the existence of a steady state. In many situations, the arrival rate, service rate, and number of servers may vary. Here are some examples.

- A fast-food restaurant is likely to experience a much larger arrival rate during the time from noon to 1:30 P.M. than during other hours of the day. Also, the number of servers (in a restaurant with parallel servers) will vary during the day, with more servers available during the busier periods.

- Since most heart attacks occur during the morning, a coronary care unit will experience more arrivals during the morning.

- Most voters vote either before or after work, so a polling place will be less busy during the middle of the day.

When the parameters defining the queuing system vary over time, we say that the system is **nonstationary.** Consider, for example, a fast-food restaurant that opens at 10 A.M. and closes at 6 P.M. We are interested in the probability distribution of the number of customers present at all times between 10 A.M. and closing. We call these probability distributions **transient probabilities.** For example, if we want to determine the probability that at least 10 customers are present, this probability will surely be larger at 12:30 P.M. than at 3 P.M.

We now assume that at time t, interarrival times are exponential with rate $\lambda(t)$. Also, $s(t)$ servers are available at time t, with service times exponential with rate $\mu(t)$. We assume that the maximum number of customers present at any time is given by N. To determine transient probabilities, we choose a small length of time Δt and assume at most one event (an arrival or service completion) can occur during an interval of length Δt. We assume that k customers are currently present at time t, and that

- The probability of an arrival during an interval of length Δt is $\lambda(t)*(\Delta t)$.
- The probability of more than one arrival during a time interval of length Δt is $o(\Delta t)$.
- Arrivals during different intervals are independent.
- The probability of a service completion during an interval of length Δt is given by $\min(s(t), k)*\mu t \Delta t$.
- The probability of more than one service completion during a time interval of length Δt is $o(\Delta t)$.

When arrivals are governed by the first three assumptions, we say that arrivals follow a **nonhomogeneous Poisson process.** Our assumptions imply that, given the arrival rate and service rate, the expected number of arrivals and/or service completions during the next Δt will match what we expect. The source of the error in our approximation is the fact that at least two events can occur during a length of time Δt. The probability of this occurring is $o(\Delta t)$, so if we make Δt small enough, our approximation should not cause large errors in computing transient probabilities.

We now define $P_i(t)$ to be the probability that i customers are present at time t. We will assume (although this is not necessary) that the system is initially empty, so $P_0(0) = 1$ and for $i > 0$, $P_0(i) = 0$. Then, given knowledge of $P_i(t)$, we may compute $P_i(t + \Delta t)$ as follows:

$$P_0(t + \Delta t) = (1 - \lambda(t)\Delta t)P_0(t) + \mu(t)\Delta t P_1(t)$$
$$P_i(t + \Delta t) = \lambda(t)\Delta t P_{i-1}(t) + (1 - \lambda(t)\Delta t - \min(s(t), i)\mu(t)\Delta t)P_i(t) + \min(s(t), i+1)\mu(t)\Delta t P_{i+1}(t), \quad N - 1 \geq i \geq 1$$
$$P_N(t + \Delta t) = \lambda(t)\Delta t P_{N-1}(t) + (1 - \min(s(t), N)\mu(t))\Delta t P_N(t)$$

As previously stated, these equations are based on the assumption that if the state at time t is i, then during the next Δt, the probability of an arrival is $\lambda(t)\Delta t$, and the probability of a service completion is $\min(s(t), i)\mu(t)\Delta t$. The first equation then follows after observing that being in state 0 at time $t + \Delta t$ can only happen if we were in state 1 at time t and had a service completion during the next Δt or were in state 0 at time t and had no arrival during the next Δt. The second equation follows after observing that for $N - 1 \geq i \geq 1$, we can only be in state i at time $t + \Delta t$ if one of the following occurs.

- We were in state $i - 1$ at time t and had an arrival during the next Δt.
- We were in state $i + 1$ at time t and had a service completion during the next Δt.
- We were in state i at time t, and no arrival or service completion occurred during the next Δt.

The final equation follows after observing that to be in state N at time $t + \Delta t$, one of the following must occur:

- We were in state $N - 1$ at time t, and an arrival occurs during the next Δt.
- We were in state N at time t, and no service completion occurs during the next Δt.

The following example shows how we can use our approximations to determine transient probabilities for a nonstationary queuing system.

EXAMPLE 22   **Lunchtime Rush**

A small fast-food restaurant is trying to model the lunchtime rush. The restaurant opens at 11 A.M., and all customers wait in one line to have their orders filled. The arrival rate per hour at different times is as shown in Table 11. Arrivals follow a nonhomogeneous

TABLE 11

Time	Hourly Arrival Rate
11–11.30 A.M.	30
11:30 A.M.–noon	40
Noon–12:30 P.M.	50
12:30 P.M.–1 P.M.	60
1 P.M.–1:30 P.M.	35
1:30 P.M.–2 P.M.	25

Poisson process. The restaurant can serve an average of 50 people per hour. Service times are exponential. Management wants to model the probability distribution of customers from 11 A.M. through 2 P.M.

a At 12:30 P.M., estimate the average number of people in line or in service.

b At 11:30 A.M., estimate the average number of people in line or in service.

restaurant.xls **Solution** Our work is in the file restaurant.xls. (See Figure 40.) We use 5-second time increments and proceed as follows:

Step 1 In E4, we compute the probability of a service completion in 5 seconds by multiplying the hourly service rate by $\Delta t = 1/720$.

Step 2 In column A, we use the Excel DATA FILL command to generate times ranging in 5-second increments from 0 to 10,800 (2 P.M.).

Step 3 By copying from B11 to B11:B2171 the formula

$$A11/3600$$

we convert the time in seconds to hours.

Step 4 By copying from C11 to C12:C2171 the formula

$$=VLOOKUP(B11,\$G\$2:\$H\$7,2)/720$$

we look up the hourly arrival rate for the current time and convert it to a 5-second arrival rate by multiplying the hourly arrival rate by $\Delta t = 1/720$. Note that the arrival rate is highly nonstationary. This fact will greatly affect the system's level of congestion.

Step 5 We assume that a maximum of $N = 30$ customers will be present. Therefore, we need 31 columns to compute the probability of 0, 1, . . . 30 people being present at each time. At time 0, we assume that the restaurant is empty, so the probability that 0 people are present equals 1. For i at least 1, there is a 0 probability of i people being present. These probabilities are entered in row 11 of columns D–AH.

Step 6 In cell D12, we compute the probability that nobody is in the system at time 5 seconds with the formula

$$=(1-C11)*D11+sprob*E11$$

This formula implements the first of our approximating equations.

Step 7 By copying from cell E12 to E12:AG12 the formula

$$=\$C11*D11+(1-\$C11-sprob)*E11+sprob*F11$$

we compute the probability that 1, 2, . . . , 29 people are present after 5 seconds. This formula implements our second approximating equation.

FIGURE 40

	A	B	C	D	E	F	G	H
1								
2							0	30
3				srate	50		0.5	40
4				sprob	0.069444		1	50
5							1.5	60
6							2	35
7							2.5	25
8								
9				0	1	2	3	4
10	Time	Hour	Arrival Prob	Prob 0	Prob 1	Prob 2	Prob 3	Prob 4
11	0	0	0.04166667	1	0	0	0	0
12	5	0.001389	0.04166667	0.958333	0.041667	0	0	0
13	10	0.002778	0.04166667	0.921296	0.076968	0.001736	0	0
14	15	0.004167	0.04166667	0.888254	0.106924	0.00475	7.23E-05	0
15	20	0.005556	0.04166667	0.858669	0.132384	0.008683	0.000262	3.01E-06
16	25	0.006944	0.04166667	0.832084	0.154055	0.013252	0.000595	1.36E-05
17	30	0.008333	0.04166667	0.808112	0.172528	0.01824	0.001082	3.69E-05
18	35	0.009722	0.04166667	0.786422	0.188297	0.023477	0.001724	7.79E-05
19	40	0.011111	0.04166667	0.766731	0.201773	0.028834	0.002516	0.000141
20	45	0.0125	0.04166667	0.748795	0.213303	0.034212	0.003448	0.000231

Step 8 In cell AH12, we compute the probability that 30 people are present after 5 seconds with the formula

$$=(1\text{-sprob})*AH11+C11*AG11$$

This implements the third approximating equation.

Step 9 Select the cell range D12:AH12 and position the cursor over the crosshair in the lower right-hand corner of cell AH12. Now double-clicking the left mouse button will copy the formulas in D12:AH12 down to match the number of rows in column C. Thus, we have now completed our computation of the probability distribution of customers from 11 A.M. to 2 P.M. See Figure 41.

Part (a) In cell K5, we compute the expected number of customers present at 12:30 P.M. (note that row 1091 has time 1.5 hours or 5,400 seconds) with the formula

$$=SUMPRODUCT(\$D\$9:\$AH\$9,D1091:AH1091)$$

We find that an average of 6.25 customers will be present at 12:30 P.M.

FIGURE 41

	D	E	F	G	H	I	J	K	L	M	AH
2				0	30						
3	srate	50		0.5	40		minutes				
4	sprob	0.069444		1	50						
5				1.5	60		Mean #	6.253338	Mean#	1.430961	
6				2	35		12:30		11:30		
7				2.5	25						
8											
9	0	1	2	3	4	5	6	7	8	9	30
10	Prob 0	Prob 1	Prob 2	Prob 3	Prob 4	Prob 5	Prob 6	Prob 7	Prob 8	Prob 9	Prob 30
11	1	0	0	0	0	0	0	0	0	0	0
12	0.958333	0.041667	0	0	0	0	0	0	0	0	0
13	0.921296	0.076968	0.001736	0	0	0	0	0	0	0	0
14	0.888254	0.106924	0.00475	7.23E-05	0	0	0	0	0	0	0

Note that row 371 is time 11:30 A.M. In cell M5, the formula

$$=\text{SUMPRODUCT(D9:AH9,D371:AH371)}$$

shows that an average of only 1.43 customers are expected to be present at 11:30.

PROBLEMS

Group A

1[†] A single machine is used between 8 A.M. and 4 P.M. to perform EKGs (electrocardiograms). There are 3 waiting spaces, and any arrival finding no available waiting space is lost to the system. The arrival rate per hour at time t ($t = 0$ is 8 A.M., and $t = 8$ is 4 P.M.) is given by

$$\lambda(t) = 9.24 - 1.584 \cos\left(\frac{\pi t}{1.51}\right) + 7.897 \sin\left(\frac{\pi t}{3.02}\right)$$
$$- 10.434 \cos\left(\frac{\pi t}{4.53}\right) + 4.293 \cos\left(\frac{\pi t}{6.04}\right)$$

Assume that service times are exponential and an average of 7 EKGs can be completed per hour. Also assume that arrivals follow a nonhomogeneous Poisson process. Determine how the probability that an arriving patient is lost to the system varies during the day.

2 The polls are open in Gotham City from 11 A.M. to 6 P.M. The city has 3 voting machines. It takes an average

of 1.5 minutes (exponentially distributed) for a voter to complete voting. The arrival rate of voters throughout the day is as shown in Table 12. What is the probability that all voting will be completed by 6:30 P.M.?

TABLE 12

Time	Hourly Arrival Rate
11 A.M.–noon	80
Noon–1 P.M.	125
1 P.M.–2 P.M.	110
2 P.M.–3 P.M.	90
3 P.M.–4 P.M.	80
4 P.M.–5 P.M.	70
5 P.M.–6 P.M.	100

[†]Based on Kao (1996).

SUMMARY Exponential Distribution

A random variable $\mathbf{X}$ has an exponential distribution with parameter λ if the density of $\mathbf{X}$ is given by

$$f(t) = \lambda e^{-\lambda t} \qquad (t \geq 0)$$

Then

$$E(\mathbf{X}) = \frac{1}{\lambda} \qquad \text{and} \qquad \text{var } \mathbf{X} = \frac{1}{\lambda^2}$$

The exponential distribution has the no-memory property. This means, for instance, that if interarrival times are exponentially distributed with rate or parameter λ, then no matter how long it has been since the last arrival, there is a probability $\lambda \Delta t$ that an arrival will occur during the next Δt time units.

Interarrival times are exponential with parameter λ if and only if the number of arrivals to occur in an interval of length t follows a Poisson distribution with parameter λt. The mass function for a Poisson distribution with parameter λ is given by

$$P(\mathbf{N} = n) = \frac{e^{-\lambda} \lambda^n}{n!} \qquad (n = 0, 1, 2, \ldots)$$

Erlang Distribution

If interarrival or service times are not exponential, an Erlang random variable can often be used to model them. If $\mathbf{T}$ is an Erlang random variable with rate parameter R and shape parameter k, the density of $\mathbf{T}$ is given by

$$f(t) = \frac{R(Rt)^{k-1}e^{-Rt}}{(k-1)!} \qquad (t \geq 0)$$

and

$$E(\mathbf{T}) = \frac{k}{R} \qquad \text{and} \qquad \text{var } \mathbf{T} = \frac{k}{R^2}$$

Birth–Death Processes

For a birth-death process, the steady-state probability (π_j) or fraction of the time that the process spends in state j can be found from the following flow balance equations:

$$
\begin{array}{ll}
(j = 0) & \pi_0 \lambda_0 = \pi_1 \mu_1 \\
(j = 1) & (\lambda_1 + \mu_1)\pi_1 = \lambda_0 \pi_0 + \mu_2 \pi_2 \\
(j = 2) & (\lambda_2 + \mu_2)\pi_2 = \lambda_1 \pi_1 + \mu_3 \pi_3 \\
\vdots & \\
(j\text{th equation}) & (\lambda_j + \mu_j)\pi_j = \lambda_{j-1}\pi_{j-1} + \mu_{j+1}\pi_{j+1}
\end{array}
$$

The jth flow balance equation states that the expected number of transitions per unit time out of state j = (expected number of transitions per unit time into state j). The solution to the balance equations is found from

$$\pi_j = \pi_0 \frac{\lambda_0 \lambda_1 \cdots \lambda_{j-1}}{\mu_1 \mu_2 \cdots \mu_j} \qquad (j = 1, 2, \ldots)$$

and the fact that $\pi_0 + \pi_1 + \cdots = 1$.

Notation for Characteristics of Queuing Systems

π_j = steady-state probability that j customers are in system

L = expected number of customers in system

L_q = expected number of customers in line (queue)

L_s = expected number of customers in service

W = expected time a customer spends in system

W_q = expected time a customer spends waiting in line

W_s = expected time a customer spends in service

λ = average number of customers per unit time

μ = average number of service completions per unit time (service rate)

$\rho = \dfrac{\lambda}{s\mu}$ = traffic intensity

The *M/M/1/GD/∞/∞* Model

If $\rho \geq 1$, no steady state exists. For $\rho < 1$,

$$\pi_j = \rho^j (1 - \rho) \qquad (j = 0, 1, 2, \ldots)$$

$$L = \frac{\lambda}{\mu - \lambda}$$

$$L_q = \frac{\lambda^2}{\mu(\mu - \lambda)}$$

$$L_s = \rho$$

$$W = \frac{1}{\mu - \lambda}$$

$$W_q = \frac{\lambda}{\mu(\mu - \lambda)}$$

$$W_s = \frac{1}{\mu}$$

(The last three formulas were obtained from the L, L_q, and L_s formulas via the relation $L = \lambda W$.)

The *M/M/1/GD/c/∞* Model

If $\lambda \neq \mu$,

$$\pi_0 = \frac{1 - \rho}{1 - \rho^{c+1}}$$

$$\pi_j = \rho^j \pi_0 \qquad (j = 1, 2, \ldots, c)$$

$$\pi_j = 0 \qquad (j = c + 1, c + 2, \ldots)$$

$$L = \frac{\rho[1 - (c + 1)\rho^c + c\rho^{c+1}]}{(1 - \rho^{c+1})(1 - \rho)}$$

If $\lambda = \mu$,

$$\pi_j = \frac{1}{c + 1} \qquad (j = 0, 1, \ldots, c)$$

$$L = \frac{c}{2}$$

For all values of λ and μ,

$$L_s = 1 - \pi_0$$

$$L_q = L - L_s$$

$$W = \frac{L}{\lambda(1 - \pi_c)}$$

$$W_q = \frac{L_q}{\lambda(1 - \pi_c)}$$

$$W_s = \frac{1}{\mu}$$

The *M/M/s/GD/∞/∞* Model

For $\rho \geq 1$, no steady state exists. For $\rho < 1$,

$$\pi_0 = \frac{1}{\displaystyle\sum_{i=0}^{i=s-1} \frac{(s\rho)^i}{i!} + \frac{(s\rho)^s}{s!(1-\rho)}}$$

$$\pi_j = \frac{(s\rho)^j \pi_0}{j!} \qquad (j = 1, 2, \ldots, s)$$

$$\pi_j = \frac{(s\rho)^j \pi_0}{s!s^{j-s}} \qquad (j = s, s+1, s+2, \ldots)$$

$$P(j \geq s) = \frac{(s\rho)^s \pi_0}{s!(1-\rho)} \qquad \text{(tabulated in Table 6)}$$

$$L_q = \frac{P(j \geq s)\rho}{1-\rho}$$

$$W_q = \frac{P(j \geq s)}{s\mu - \lambda}$$

$$L_s = \frac{\lambda}{\mu}$$

$$W_s = \frac{1}{\mu}$$

$$L = L_q + \frac{\lambda}{\mu}$$

$$W = \frac{L}{\lambda}$$

The *M/G/∞/GD/∞/∞* Model

$$L = L_s = \frac{\lambda}{\mu}$$

$$W = W_s = \frac{1}{\mu}$$

$$W_q = L_q = 0$$

The *M/G/1/GD/∞/∞* Model

$$\sigma^2 = \text{variance of service time distribution}$$

$$L_q = \frac{\lambda^2\sigma^2 + \rho^2}{2(1-\rho)}$$

$$L = L_q + \rho$$

$$L_s = \lambda\left(\frac{1}{\mu}\right)$$

$$W_q = \frac{L_q}{\lambda}$$

$$W = W_q + \frac{1}{\mu}$$

$$W_s = \frac{1}{\mu}$$

$$\pi_0 = 1 - \rho$$

Machine Repair (*M/M/R/GD/K/K*) Model

$$\rho = \frac{\lambda}{\mu}$$

L = expected number of broken machines

L_q = expected number of machines waiting for service

W = average time a machine spends broken

W_q = average time a machine spends waiting for service

π_j = steady-state probability that j machines are broken

λ = rate at which machine breaks down

μ = rate at which machine is repaired

Also,

$$\pi_j = \binom{K}{j} \rho^j \pi_0 \qquad (j = 0, 1, \ldots, R)$$

$$= \frac{\binom{K}{j} \rho^j j! \pi_0}{R! R^{j-R}} \qquad (j = R + 1, R + 2, \ldots, K)$$

$$L = \sum_{j=0}^{j=K} j \pi_j$$

$$L_q = \sum_{j=R}^{j=K} (j - R) \pi_j$$

$$\bar{\lambda} = \sum_{j=0}^{j=K} \pi_j \lambda_j = \sum_{j=0}^{j=K} \lambda(K - j)\pi_j = \lambda(K - L)$$

$$W = \frac{L}{\bar{\lambda}}$$

$$W_q = \frac{L_q}{\bar{\lambda}}$$

Exponential Queues in Series

If a steady state exists and if (1) interarrival times for a series queuing system are exponential with rate λ; (2) service times for each stage i server are exponential; and (3) each stage has an infinite-capacity waiting room, then interarrival times for arrivals to each stage of the queuing system are exponential with rate λ.

The *M/G/s/GD/s/∞* Model

A fraction π_s of all customers are lost to the system, and π_s depends only on the arrival rate λ and on the mean $\frac{1}{\mu}$ of the service time. Figure 21 can be used to find π_s.

What to Do If Interarrival or Service Times Are Not Exponential

A chi-square test may be used to determine if the actual data indicate that interarrival or service times are exponential. If interarrival and/or service times are not exponential, then L, L_q, W, and W_q may be approximated by Allen–Cunneen formula.

For many queuing systems, there is no formula or table that can be used to compute the system's operating characteristics. In this case, we must resort to simulation (see Chapters 9 and 10).

Closed Queuing Network

Manufacturing and computer systems in which there is a constant number of jobs present may be modeled as **closed queuing networks.**

We let P_{ij} be the probability that a job will go to server j after completing service at station i. Let P be the matrix whose $(i - j)$th entry is P_{ij}. We assume that service times at server j follow an exponential distribution with parameter μ_j. The system has s servers, and at all times, exactly N jobs are present. We let n_i be the number of jobs present at server i. Then the state of the system at any given time can be defined by an n-dimensional vector $\mathbf{n} = (n_1, n_2, \ldots, n_s)$. The set of possible states is given by $S_N = \{\mathbf{n}$ such that all $n_i \geq 0$ and $n_1 + n_2 + \cdots + n_s = N\}$.

Let λ_j equal the arrival rate to server j. Since there are no external arrivals, we may set all $r_j = 0$ and obtain the values of the λ_j's from the equation used in the open network situation. That is,

$$\lambda_j = \sum_{i=1}^{i=s} \lambda_i P_{ij} \qquad j = 1, 2, \ldots, s$$

Since jobs never leave the system, for each i, $\sum_{j=1}^{j=s} P_{ij} = 1$. This fact causes the above equation to have no unique solution. Fortunately, it turns out that we can use any solution to help us get steady-state probabilities. If we define

$$\rho_i = \frac{\lambda_i}{\mu_i}$$

then we determine, for any state n, its steady-state probability $\Pi_N(\mathbf{n})$ from the following equation:

$$\Pi_N(\mathbf{n}) = \frac{\rho_1^{n_1} \rho_2^{n_2} \cdots \rho_n^{n_s}}{G(N)}$$

Here, $G(N) = \sum_{\mathbf{n} \in S_N} \rho_1^{n_1} \rho_2^{n_2} \cdots \rho_s^{n_s}$.

Buzen's algorithm gives us an efficient way to determine (in a spreadsheet) $G(N)$. Once we have the steady-state probability distribution, we can easily determine other measures of effectiveness, such as expected queue length at each server and expected time a job spends during each visit to a server, fraction of time a server is busy, and the throughput for each server (jobs per second processed by each server).

To obtain $G(N)$, we recursively compute the quantities $C_i(k)$ for $i = 1, 2, \ldots, s$ and $k = 0, 1, \ldots, N$. We initialize the recursion with $C_1(k) = \rho_1^k$, $k = 0, 1, \ldots, N$ and $C_i(0) = 1$, $i = 1, 2, \ldots, s$. For other values of k and i, we build up the values of $C_i(k)$ recursively via the following relationship:

$$C_i(k) = C_{i-1}(k) + \rho_i C_i(k - 1)$$

Then it can be shown that $G(N) = C_s(N)$.

An Approximation for the *G/G/m* Queuing System

In most situations, interarrival times follow an exponential random variable. Often, however, service times do not follow an exponential distribution. When interarrival times and service times each follow a nonexponential random variable, we call the queuing system a *G/G/m* system. For these situations, the templates discussed in the previous sections of this chapter are not valid. Fortunately, the Allen–Cunneen approximation often gives a good approximation to L, W, L_q, and W_q for *G/G/m* systems. The file ggm.xls contains a spreadsheet implementation of the Allen–Cunneen approximation. The user need only input the following information:

ggm.xls

- The average number of arrivals per unit time (λ) in cell B3.
- The average rate at which customers can be serviced (μ) in cell B4.
- The number of servers (s) in cell B5.
- The squared coefficient of variation—(variance of interarrival times)/(mean interarrival time)^2—of interarrival times in cell B6.
- The squared coefficient of variation—(variance of service times)/(mean service time)^2—of service times in cell B7.

The Allen–Cunneen approximation is exact if interarrival times and service times are exponential. Extensive testing by Tanner indicates that in a wide variety of situations, the values of L, W, L_q, and W_q obtained by the approximation are within 10% of their true values.

Transient Behavior of Queuing Systems

We define $P_i(t)$ to be the probability that i customers are present at time t. We then assume (although this is not necessary) that the system is initially empty, so $P_0(0) = 1$ and, for $i > 0$, $P_0(i) = 0$. Then, given knowledge of $P_i(t)$, we may compute $P_i(t + \Delta t)$ as follows:

$$P_0(t + \Delta t) = (1 - \lambda(t)\Delta t)P_0(t) + \mu(t)\Delta t P_1(t)$$
$$P_i(t + \Delta t) = \lambda(t)\Delta t P_{i-1}(t) + (1 - \lambda(t)\Delta t - \min(s(t), i)\mu(t)\Delta t)P_i(t) + \min(s(t),$$
$$i + 1)\mu(t)\Delta t P_{i+1}(t), \quad N - 1 \geq i \geq 1$$
$$P_N(t + \Delta t) = \lambda(t)\Delta t P_{N-1}(t) + (1 - \min(s(t), N)\mu(t))\Delta t P_N(t)$$

These equations are based on the assumption that, if the state at time t is i, then during the next Δt, the probability of an arrival is $\lambda(t)\,\Delta t$, and the probability of a service completion is $\min(s(t), i)\mu(t)\Delta t$.

REVIEW PROBLEMS

Group A

1 Buses arrive at the downtown bus stop and leave for the mall stop. Past experience indicates that 20% of the time, the interval between buses is 20 minutes; 40% of the time, the interval is 40 minutes; and 40% of the time, the interval is 2 hours. If I have just arrived at the downtown bus stop, how long, on the average, should I expect to wait for a bus?

2 Registration at State University proceeds as follows: Upon entering the registration hall, the students first wait in line to register for classes. A single clerk handles registration for classes, and it takes the clerk an average of 2 minutes to handle a student's registration. Next, the student must wait in line to pay fees. A single clerk handles the payment of fees. The clerk takes an average of 2 minutes to process a student's fees. Then the student leaves the registration building. An average of 15 students per hour arrive at the registration hall.

    a If interarrival and service times are exponential, what is the expected time a student spends in the registration hall?

    b What is the probability that during the next 5 minutes, exactly 2 students will enter the registration hall?

    c Without any further information, what is the probability that during the next 3 minutes, no student will arrive at the fee clerk's desk?

    d Suppose the registration system is changed so that a student can register for classes and pay fees at the same station. If the service time at this single station follows an Erlang distribution with rate parameter 1.5 per minute and shape parameter 2, what is the expected time a student spends waiting in line?

3 At the Smalltown post office, patrons wait in a single line for the first open window. An average of 100 patrons per hour enter the post office, and each window can serve an average of 45 patrons per hour. The post office estimates a cost of 10¢ for each minute a patron waits in line and believes that it costs $20 per hour to keep a window open. Interarrival times and service times are exponential.

    a To minimize the total expected hourly cost, how many windows should be open?

    b If the post office's goal is to ensure that at most 5% of all patrons will spend more than 5 minutes in line, how many windows should be open?

4 Each year, an average of 500 people pass the New York state bar exam and enter the legal profession. On the average, a lawyer practices law in New York State for 35 years. Twenty years from now, how many lawyers would you expect there to be in New York State?

5 There are 5 students and one keg of beer at a wild and crazy campus party. The time to draw a glass of beer follows an exponential distribution, with an average time of 2 minutes. The time to drink a beer also follows an exponential distribution, with a mean of 18 minutes. After finishing a beer, each student immediately goes back to get another beer.

    a On the average, how long does a student wait in line for a beer?

    b What fraction of the time is the keg not in use?

    c If the keg holds 500 glasses of beer, how long, on the average, will it take to finish the keg?

6 The manager of a large group of employees must decide if she needs another photocopying machine. The cost of a machine is $40 per 8-hour day whether or not the machine is in use. An average of 4 people per hour need to use the copying machine. Each person uses the copier for an average of 10 minutes. Interarrival times and copying times are exponentially distributed. Employees are paid $8 per hour, and we assume that a waiting cost is incurred when a worker is waiting in line or is using the copying machine. How many copying machines should be rented?

7 An automated car wash will wash a car in 10 minutes. Arrivals occur an average of 15 minutes apart (exponentially distributed).

    a On the average, how many cars are waiting in line for a wash?

    b If the car wash could be speeded up, what wash time would reduce the average waiting time to 5 minutes?

8 The Newcoat Painting Company has for some time been experiencing high demand for its automobile repainting service. Since it has had to turn away business, management is concerned that the limited space available to store cars awaiting painting has cost lost revenue. A small vacant lot next to the painting facility has recently been made available for lease on a long-term basis at a cost of $10 per day. Management believes that each lost customer costs $20 in profit. Current demand is estimated to be 21 cars per day with exponential interarrival times (including those turned away), and the facility can service at an exponential rate of 24 cars per day. Cars are processed on an FCFS basis. Waiting space is now limited to 9 cars but can be increased to 20 cars with the lease of the vacant lot. Newcoat wants to determine whether the vacant lot should be leased. Management also wants to know the expected daily lost profit due to turning away customers if the lot is leased. Only one car can be painted at a time.

9 At an exclusive restaurant, there is only one table and waiting space for only one other group; others that arrive when the waiting space is filled are turned away. The arrival rate follows an exponential distribution with a rate of one group per hour. It takes the average group 1 hour (exponentially distributed) to be served and eat the meal. What is the average time that a group spends waiting for a table?

10 The owner of an exclusive restaurant has two tables but only one waiter. If the second table is occupied, the owner waits on that table himself. Service times are exponentially distributed with mean 1 hour, and the time between arrivals is exponentially distributed with mean 1.5 hours. When the restaurant is full, people must wait outside in line.

a What percentage of the time is the owner waiting on a table?

b If the owner wants to spend at most 10% of his time waiting on tables, what is the maximum arrival rate that can be tolerated?

11 Ships arrive at a port facility at an average rate of 2 ships every 3 days. On the average, it takes a single crew 1 day to unload a ship. Assume that interarrival and service times are exponential. The shipping company owns the port facility as well as the ships using that facility. It is estimated to cost the company $1,000 per day that each ship spends in port. The crew servicing the ships consists of 100 workers, who are each paid an average of $30 per day. A consultant has recommended that the shipping company hire an additional 40 workers and split the employees into two equal-sized crews of 70 each. This would give each crew an unloading or loading time averaging $\frac{3}{2}$ days. Which crew arrangement would you recommend to the company?

12 An average of 40 jobs per day arrive at a factory. The time between arrivals of jobs is exponentially distributed. The factory can process an average of 42 jobs per day, and the time to process a job is exponentially distributed.

a What is the probability that exactly 180 jobs arrive at the factory during a 5-day period?

b On the average, how long does it take before a job is completed (measured from the time the job arrives at the factory)?

c What fraction of the time is the factory idle?

d What is the probability that work on a job will begin within 2 days of its arrival at the factory?

13 A printing shop receives an average of 1 order per day. The average length of time required to complete an order is .5 day. At any time, the print shop can work on at most one job.

a On the average, how many jobs are present in the print shop?

b On the average, how long will a person who places an order have to wait until it is finished?

c What is the probability that an order will be finished within 2 days of its arrival?

Group B

14 The mail order firm of L. L. Pea receives an average of 200 calls per hour (times between calls are exponentially distributed). It takes an L. L. Pea operator an average of 3 minutes to handle a call. If a caller gets a busy signal, L. L. Pea assumes that he or she will call Seas Beginning (a competing mail order house), and L. L. Pea will lose an average of $30 in profit. The cost of keeping a phone line open is $9 per hour. How many operators should L. L. Pea have on duty?

15 Each hour, an average of 3 type 1 and 3 type 2 customers arrive at a single-server station. Interarrival times for each customer type are exponential and independent. The average service time for a type 1 customer is 6 minutes, and the average service time for a type 2 customer is 3 minutes (all service times are exponentially distributed). Consider the following three service arrangements:

Arrangement 1 All customers wait in a single line and are served on an FCFS basis.

Arrangement 2 Type 1 customers are given nonpreemptive priority over type 2 customers.

Arrangement 3 Type 2 customers are given nonpreemptive priority over type 1 customers.

Which arrangement will result in the smallest average per-customer waiting time? Which arrangement will result in the largest average per-customer waiting time?

16 Podunk University Operations Research Department has two phone lines. An average of 30 people per hour try to call the OR Department, and the average length of a phone call is 1 minute. If a person attempts to call when both lines are busy, he or she hangs up and is lost to the system. Assume that the time between people attempting to call and service times is exponential.

a What fraction of the time will both lines be free? What fraction of the time will both lines be busy? What fraction of the time will exactly one line be free?

b On the average, how many lines will be busy?

c On the average, how many callers will hang up each hour?

17[†] Smalltown has two ambulances. Ambulance 1 is based at the local college, and ambulance 2 is based downtown. If a request for an ambulance comes from the college, the college-based ambulance is sent if it is available. Otherwise, the downtown-based ambulance is sent (if available). If no ambulance is available, the call is assumed to be lost to the system. If a request for an ambulance comes from anywhere else in the town, the downtown-based ambulance is sent if it is available. Otherwise, the college-based ambulance is sent if available. If no ambulance is available, the call is considered lost to the system. The time between calls is exponentially distributed. An average of 3 calls per hour are received from the college, and an average of 4 calls per hour are received from the rest of the town. The average time (exponentially distributed) it takes an ambulance to respond to a call and be ready to respond to another call is shown in Table 13.

a What fraction of the time is the downtown ambulance busy?

b What fraction of the time is the college ambulance busy?

c What fraction of all calls will be lost to the system?

d On the average, who waits longer for an ambulance, a college student or a town person?

18 An average of 10 people per hour arrive (interarrival times are exponential) intending to swim laps at the local

TABLE 13

Ambulance Comes From	Ambulance Goes to	
	College	Noncollege
College	4 minutes	7 minutes
Downtown	5 minutes	4 minutes

[†]Based on Carter (1972).

YMCA. Each intends to swim an average of 30 minutes. The YMCA has three lanes open for lap swimming. If one swimmer is in a lane, he or she swims up and down the right side of the lane. If two swimmers are in a lane, each swims up and down one side of the lane. Swimmers always join the lane with the fewest number of swimmers. If all three lanes are occupied by two swimmers, a prospective swimmer becomes disgusted and goes running.

a What fraction of the time will 3 people be swimming laps?

b On the average, how many people are swimming laps in the pool?

c How many lanes does the YMCA need to allot to lap swimming to ensure that at most 5% of all prospective swimmers will become disgusted and go running?

19[†] (Requires use of a spreadsheet) An average of 140 people per year apply for public housing in Boston. An average of 20 housing units per year become available. During a given year, there is a 10% chance that a family on the waiting list will find private housing and remove themselves from the list. Assume that all relevant random variables are exponentially distributed.

a On the average, how many families will be on the waiting list?

b On the average, how much time will a family spend on the list before obtaining housing (either public or private)? For the last question, remember that $L = \lambda W$!

[†]Based on Kaplan (1986)

REFERENCES

The following books contain excellent discussions of queuing theory at an intermediate level:

Cooper, R. *Introduction to Queuing Theory,* 2d ed. New York: North-Holland, 1981.

Gross, D., and C. Harris. *Fundamentals of Queuing Theory,* 3d ed. New York: Wiley, 1997.

Karlin, S., and H. Taylor. *A First Course in Stochastic Processes,* 2d ed. Orlando, Fla.: Academic Press, 1975.

Lee, A. *Applied Queuing Theory.* New York: St. Martin's Press, 1966.

Ross, S. *Applied Probability Models with Optimization Applications.* San Francisco, Calif.: Holden-Day, 1970.

Saaty, T. *Elements of Queuing Theory with Applications.* New York: Dover, 1983.

The following three books contain excellent discussions of queuing theory at a more advanced level:

Heyman, D., and M. Sobel. *Stochastic Models in Operations Research,* vol. 1. New York: McGraw-Hill, 1984.

Kao, E. *An Introduction to Stochastic Processes.* Belmont, Cal.: Duxbury, 1997.

Kleinrock, L. *Queuing Systems,* vols. 1 and 2. New York: Wiley, 1975.

For an excellent applications-oriented study of queuing theory we recommend:

Hall, R. *Queuing Methods for Service and Manufacturing.* Englewood Cliffs, N.J.: Prentice Hall, 1991.

Tanner, M. *Practical Queuing Analysis.* New York: McGraw-Hill, 1995.

Brigham, G. "On a Congestion Problem in an Aircraft Factory," *Operations Research* 3(1955):412–428.

Buzen, J. P. "Computational Algorithms for Closed Queuing Networks with Exponential Servers," *Communications of the ACM* 16:9(1973):527–531.

Carter, G., J. Chaiken, and E. Ignall. "Response Areas for Two Emergency Units," *Operations Research* 20(1972):571–594.

Denardo, E. *Dynamic Programming: Theory and Applications.* Englewood Cliffs, N.J.: Prentice Hall, 1982. Explains why interarrival times are often exponential.

Erickson, W. "Management Science and the Gas Shortage," *Interfaces* 4(1973):47–51.

Feller, W. *An Introduction to Probability Theory and Its Applications,* 2d ed., vol. 1. New York: Wiley, 1957. Proves that the exponential distribution is the only continuous random variable with the no-memory property.

Gilliam, R. "An Application of Queuing Theory to Airport Passenger Security Screening," *Interfaces* 9(1979): 117–123.

Green, J. "Managing a Telephone System Demands Skill," *The Office* (November 1987):144–145.

Hillier, F., and O. Yu. *Queuing Tables and Graphs.* New York: North-Holland, 1981.

Jackson, J. "Networks of Waiting Lines," *Operations Research* 5(1957):518–521.

Jain, R. *The Art of Computer System Performance Analysis.* New York: Wiley, 1991.

Kaplan, E. "Tenant Assignment Models," *Operations Research* 34(no. 6, 1986):833–843.

Karmarker, U. "Lot-Sizing and Lead-Time Performance in a Manufacturing Cell," *Interfaces* 15(no.2, 1985):1–9.

Karush, W. "A Queuing Model for an Inventory Problem," *Operations Research* 5(1957):693–703.

Kendall, D. "Some Problems in the Theory of Queues," *Journal of the Royal Statistical Society,* Series B, 13(1951):151–185.

Kolesar, P. "A Quick and Dirty Response to the Quick and Dirty Crowd: Particularly to Jack Byrd's 'The Value of Queuing Theory,' " *Interfaces* 9(1979):77–82.

Law, A., and W. Kelton. *Simulation Modeling and Analysis.* New York: McGraw-Hill, 1990. Discusses simulation of queuing systems and fitting random variables to actual interarrival and service time data.

Tannenbaum, A. *Computer Networks.* Englewood Cliffs, N.J.: Prentice Hall, 1981.

Vogel, M. "Queuing Theory Applied to Machine Manning," *Interfaces* 9(1979):1–8.

9

Simulation

Simulation is a very powerful and widely used management science technique for the analysis and study of complex systems. In previous chapters, we were concerned with the formulation of models that could be solved analytically. In almost all of those models, our goal was to determine optimal solutions. However, because of complexity, stochastic relations, and so on, not all real-world problems can be represented adequately in the model forms of the previous chapters. Attempts to use analytical models for such systems usually require so many simplifying assumptions that the solutions are likely to be inferior or inadequate for implementation. Often, in such instances, the only alternative form of modeling and analysis available to the decision maker is simulation.

Simulation may be defined as a technique that imitates the operation of a real-world system as it evolves over time. This is normally done by developing a simulation model. A *simulation model* usually takes the form of a set of assumptions about the operation of the system, expressed as mathematical or logical relations between the objects of interest in the system. In contrast to the exact mathematical solutions available with most analytical models, the simulation process involves executing or running the model through time, usually on a computer, to generate representative samples of the measures of performance. In this respect, simulation may be seen as a sampling experiment on the real system, with the results being sample points. For example, to obtain the best estimate of the mean of the measure of performance, we average the sample results. Clearly, the more sample points we generate, the better our estimate will be. However, other factors, such as the starting conditions of the simulation, the length of the period being simulated, and the accuracy of the model itself, all have a bearing on how good our final estimate will be. We discuss such issues later in the chapter.

As with most other techniques, simulation has its advantages and disadvantages. The major advantage of simulation is that simulation theory is relatively straightforward. In general, simulation methods are easier to apply than analytical methods. Whereas analytical models may require us to make many simplifying assumptions, simulation models have few such restrictions, thereby allowing much greater flexibility in representing the real system. Once a model is built, it can be used repeatedly to analyze different policies, parameters, or designs. For example, if a business firm has a simulation model of its inventory system, various inventory policies can be tried on the model rather than taking the chance of experimenting on the real-world system. However, it must be emphasized that simulation is not an optimizing technique. It is most often used to analyze "what if" types of questions. Optimization with simulation is possible, but it is usually a slow process. Simulation can also be costly. However, with the development of special-purpose simulation languages, decreasing computational cost, and advances in simulation methodologies, the problem of cost is becoming less important.

In this chapter, we focus our attention on simulation models and the simulation technique. We present several examples of simulation models and explore such concepts as random numbers, time flow mechanisms, Monte Carlo sampling, simulation languages, and statistical issues in simulation.

9.1 Basic Terminology

We begin our discussion by presenting some of the terminology used in simulation. In most simulation studies, we are concerned with the simulation of some system. Thus, in order to model a system, we must understand the concept of the system. Among the many different ways of defining a system, the most appropriate definition for simulation problems is the one proposed by Schmidt and Taylor (1970).

DEFINITION ■ A **system** is a collection of entities that act and interact toward the accomplishment of some logical end. ■

In practice, however, this definition generally tends to be more flexible. The exact description of the system usually depends on the objectives of the simulation study. For example, what may be a system for a particular study may be only a subset of the overall system for another.

Systems generally tend to be dynamic—their status changes over time. To describe this status, we use the concept of the state of a system.

DEFINITION ■ The **state** of a system is the collection of variables necessary to describe the status of the system at any given time. ■

As an example of a system, let us consider a bank. Here, the system consists of the servers and the customers waiting in line or being served. As customers arrive or depart, the status of the system changes. To describe these changes in status, we require a set of variables called the **state variables.** For example, the number of busy servers, the number of customers in the bank, the arrival time of the next customer, and the departure time of the customers in service together describe every possible change in the status of the bank. Thus, these variables could be used as the state variables for this system. In a system, an object of interest is called an **entity,** and any properties of an entity are called **attributes.** For example, the bank's customers may be described as the entities, and the characteristics of the customers (such as the occupation of a customer) may be defined as the attributes.

Systems may be classified as discrete or continuous.

DEFINITION ■ A **discrete system** is one in which the state variables change only at discrete or countable points in time. ■

A bank is an example of a discrete system, since the state variables change only when a customer arrives or when a customer finishes being served and departs. These changes take place at discrete points in time.

DEFINITION ■ A **continuous system** is one in which the state variables change continuously over time. ■

A chemical process is an example of a continuous system. Here, the status of the system is changing continuously over time. Such systems are usually modeled using differential equations. We do not discuss any continuous systems in this chapter.

There are two types of simulation models: static and dynamic.

DEFINITION ■   A **static simulation model** is a representation of a system at a particular point in time. ■

We usually refer to a static simulation as a **Monte Carlo simulation.**

DEFINITION ■   A **dynamic simulation** is a representation of a system as it evolves over time. ■

Within these two classifications, a simulation may be deterministic or stochastic. A **deterministic simulation model** is one that contains no random variables; a **stochastic simulation model** contains one or more random variables. Discrete and continuous simulation models are similar to discrete and continuous systems. In this chapter, we concentrate mainly on discrete stochastic models. Such models are called *discrete-event* simulation models. Discrete-event simulation concerns the modeling of a stochastic system as it evolves over time by a representation in which state variables change only at discrete points in time.

9.2 An Example of a Discrete-Event Simulation

Before we proceed to the details of simulation modeling, it will be useful to work through a simple simulation example to illustrate some of the basic concepts in discrete-event simulation. The model we have chosen as our initial example is a single-server queuing system. Customers arrive into this system from some population and either go into service immediately if the server is idle or join a waiting line (queue) if the server is busy. Examples of this kind of a system are a one-person barber shop, a small grocery store with only one checkout counter, and a single ticket counter at an airline terminal.

The same model was studied in Chapter 8 in connection with queuing theory. In that chapter, we used an analytical model to determine the various operating characteristics of the system. However, we had to make several restrictive assumptions to use queuing theory. In particular, when we studied an $M/M/1$ system, we had to assume that both interarrival times and service times were exponentially distributed. In many situations, these assumptions may not be appropriate. For example, arrivals at an airline counter generally tend to occur in bunches, because of such factors as the arrivals of shuttle buses and connecting flights. For such a system, an empirical distribution of arrival times must be used, which implies that the analytical model from queuing theory is no longer feasible. With simulation, any distribution of interarrival times and service times may be used, thereby giving much more flexibility to the solution process.

To simulate a queuing system, we first have to describe it. For this single-server system, we assume that arrivals are drawn from an infinite calling population. There is unlimited waiting room capacity, and customers will be served in the order of their arrival—that is, on a first come, first served (FCFS) basis. We further assume that arrivals occur one at a time in a random fashion, with the distribution of interarrival times as specified in Table 1. All arrivals are eventually served, with the distribution of service times shown in Table 2. Service times are also assumed to be random. After service, all customers return to the calling population. This queuing system can be represented as shown in Figure 1.

Before dealing with the details of the simulation itself, we must define the state of this system and understand the concepts of events and clock time within a simulation. For this

TABLE 1
Interarrival Time Distribution

Interarrival Time (minutes)	Probability
1	.20
2	.30
3	.35
4	.15

TABLE 2
Service Time Distribution

Service Time (minutes)	Probability
1	.35
2	.40
3	.25

example, we use the following variables to define the state of the system: (1) the number of customers in the system; (2) the status of the server—that is, whether the server is busy or idle; and (3) the time of the next arrival.

Closely associated with the state of the system is the concept of an event. An **event** is defined as a situation that causes the state of the system to change instantaneously. In the single-server queuing model, there are only two possible events that can change the state of the system: an arrival into the system and a departure from the system at the completion of service. In the simulation, these events will be scheduled to take place at certain points in time. All the information about them is maintained in a list called the **event list.** Within this list, we keep track of the type of events scheduled and, more important, the time at which these events are scheduled to take place. Time in a simulation is maintained using a variable called the **clock time.** The concept of clock time will become clearer as we work through the example.

We begin this simulation with an empty system and arbitrarily assume that our first event, an arrival, takes place at clock time 0. This arrival finds the server idle and enters service immediately. Arrivals at other points in time may find the server either idle or busy. If the server is idle, the customer enters service. If the server is busy, the customer joins the waiting line. These actions can be summarized as shown in Figure 2.

Next, we schedule the departure time of the first customer. This is done by randomly generating a service time from the service time distribution (described later in the chapter) and setting the departure time as

$$\text{Departure time} = \text{clock time now} + \text{generated service time} \qquad (1)$$

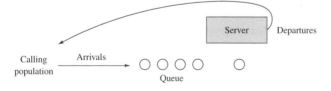

FIGURE 1
Single-Server Queuing System

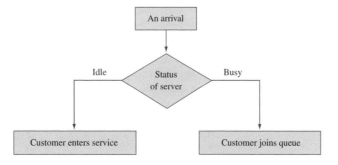

FIGURE 2
Flowchart for an Arrival

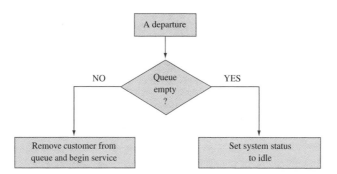

FIGURE **3**
Flowchart for a
Departure

Also, we now schedule the next arrival into the system by randomly generating an interarrival time from the interarrival time distribution and setting the arrival time as

$$\text{Arrival time} = \text{clock time now} + \text{generated interarrival time} \qquad (2)$$

If, for example, we have generated a service time of 2 minutes, then the departure time for the first customer will be set at clock time 2. Similarly, if we have generated an interarrival time of 1 minute, the next arrival will be scheduled for clock time 1.

Both these events and their scheduled times are maintained on the event list. Once we have completed all the necessary actions for the first arrival, we scan the event list to determine the next scheduled event and its time. If the next event is determined to be an arrival, we move the clock time to the scheduled time of the arrival and go through the preceding sequence of actions for an arrival. If the next event is a departure, we move the clock time to the time of the departure and process a departure. For a departure, we check whether the length of the waiting line is greater than zero. If it is, we remove the first customer from the queue and begin service on this customer by setting a departure time using Equation (1). If no one is waiting, we set the status of the system to idle. These departure actions are summarized in Figure 3.

This approach of simulation is called the **next-event time-advance mechanism,** because of the way the clock time is updated. We advance the simulation clock to the time of the most imminent event—that is, the first event in the event list. Since the state variables change only at event times, we skip over the periods of inactivity between the events by jumping from event to event. As we move from event to event, we carry out the appropriate actions for each event, including any scheduling of future events. We continue in this manner until some prespecified stopping condition is satisfied. However, the procedure requires that at any point in the simulation, we have an arrival and a departure scheduled for the future. Thus, a future arrival is always scheduled when processing a new arrival into the system. A departure time, on the other hand, can only be scheduled when a customer is brought into service. Thus, if the system is idle, no departures can be scheduled. In such instances, the usual practice is to schedule a dummy departure by setting the departure time equal to a very large number—say, 9,999 (or larger if the clock time is likely to exceed 9,999). This way, our two events will consist of a real arrival and a dummy departure.

The jump to the next event in the next-event mechanism may be a large one or a small one; that is, the jumps in this method are variable in size. We contrast this approach with the **fixed-increment time-advance method.** With this method, we advance the simulation clock in increments of Δt time units, where Δt is some appropriate time unit, usually 1 time unit. After each update of the clock, we check to determine whether any event is scheduled to take place at the current clock time. If an event is scheduled, we carry out the appropriate actions for the event. If none is scheduled, or if we have completed all the required actions for the current time, we update the simulation clock by Δt units and repeat the process. As with the next-event approach, we continue in this manner until the prespecified stopping condition is reached. The fixed-increment time-advance mechanism is often simpler to comprehend, because of its fixed steps in time. For most models, however, the next-event mechanism tends to be more efficient computationally. Consequently, we use only the next-event approach in developing the models for the rest of the chapter.

We now illustrate the mechanics of the single-server queuing system simulation, using a numerical example. In particular, we want to show how the simulation model is represented in the computer as the simulation progresses through time. The entire simulation process for the single-server queuing model is presented in the flowchart in Figure 4. All the blocks in this flowchart are numbered for easy reference. For simplicity, we assume that both the interarrival times (ITs) and the service times (STs) have already been generated for the first few customers from the given probability distributions in Tables 1 and 2. These times are shown in Table 3, from which we can see that the time between the first and the second arrival is 2 time units, the time between the second and the third arrival is also 2 time units, and so on. Similarly, the service time for the first customer is 3 time units, ST for the second customer is also 3 time units, and so on.

To demonstrate the simulation model, we need to define several variables:

$$TM = \text{clock time of the simulation}$$
$$AT = \text{scheduled time of the next arrival}$$
$$DT = \text{scheduled time of the next departure}$$
$$SS = \text{status of the server } (1 = \text{busy}, 0 = \text{idle})$$
$$WL = \text{length of the waiting line}$$
$$MX = \text{length (in time units) of a simulation run}$$

Having taken care of these preliminaries, we now begin the simulation by initializing all the variables (block 1 in Figure 4). Since the first arrival is assumed to take place at time 0, we set $AT = 0$. We also assume that the system is empty at time 0, so we set $SS = 0$, $WL = 0$, and $DT = 9,999$. (Note that DT must be greater than MX). This implies that our list of events now consists of two scheduled events: an arrival at time 0 and a dummy departure at time 9,999. This completes the initialization process and gives us the computer representation of the simulation shown in Table 4.

We are now ready for our first action in the simulation: searching through the event list to determine the first event (block 2). Since our simulation consists of only two events, we simply determine the next event by comparing AT and DT. (In other simulations, we might have more than two events, so we would have to have an efficient system of searching through the event list.) An arrival is indicated by $AT < DT$, a departure by $DT < AT$. At this point, $AT = 0$ is less than $DT = 9,999$, indicating that an arrival will take place next. We label this event 1 and update the clock time, TM, to the time of event 1 (block 3). That is, we set $TM = 0$.

The arrival at time 0 finds the system empty, indicated by the fact that $SS = 0$ (block 4). Consequently, the customer enters service immediately. For this part of the simulation, we first set $SS = 1$ to signify that the server is now busy (block 6). We next generate a service time (block 7) and set the departure time for this customer (block 8). From Table 3, we see that ST for customer 1 is 3. Since $TM = 0$ at this point, we set $DT = 3$ for the first customer. In other words, customer 1 will depart from the system at clock time 3. Finally, to complete all the actions of processing an arrival, we schedule the next arrival into the system by generating an interarrival time, IT (block 9), and setting the time of this arrival using the equation $AT = TM + IT$ (block 10). Since $IT = 2$, we set $AT = 2$. That is, the second arrival will take place at clock time 2. At the end of event 1, our computer representation of the simulation will be as shown in Table 4.

FIGURE 4

Flowchart for Simulation Model for Single-Server Queuing System

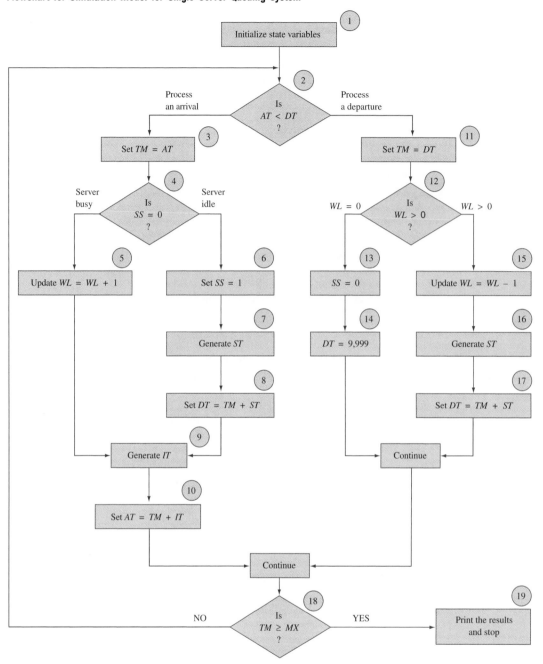

At this stage of the simulation, we proceed to block 18 to determine whether the clock time, *TM*, has exceeded the specified time length of simulation, *MX*. If it has, we print out the results (block 19) and stop the execution of the simulation model. If it has not, we continue with the simulation. We call this the termination process. We execute this process at the end of each event. However, for this example, we assume that *MX* is a large number. Consequently, from here on, we will not discuss the termination process.

TABLE 3

Generated Interarrival and Service Times

Customer Number	Interarrival Time (*IT*)	Service Time (*ST*)
1	—	3
2	2	3
3	2	2
4	3	1
5	4	1
6	2	2
7	1	1
8	3	2
9	3	—

TABLE 4

Computer Representation of the Simulation

End of Event	Type of Event	Customer Number	System Variables			Event List	
			TM	SS	WL	AT	DT
0	Initialization	—	0	0	0	0	9,999
1	Arrival	1	0	1	0	2	3
2	Arrival	2	2	1	1	4	3
3	Departure	1	3	1	0	4	6
4	Arrival	3	4	1	1	7	6
5	Departure	2	6	1	0	7	8
6	Arrival	4	7	1	1	11	8
7	Departure	3	8	1	0	11	9
8	Departure	4	9	0	0	11	9,999
9	Arrival	5	11	1	0	13	12
10	Departure	5	12	0	0	13	9,999
11	Arrival	6	13	1	0	14	15
12	Arrival	7	14	1	1	17	15
13	Departure	6	15	1	0	17	16
14	Departure	7	16	0	0	17	9,999
15	Arrival	8	17	1	0	20	19

At this point, we loop back to block 2 to determine the next event. Since $AT = 2$ and $DT = 3$, the next event, event 2, will be an arrival at time 2. Having determined the next event, we now advance the simulation to the time of this arrival by updating TM to 2.

The arrival at time 2 finds the server busy, so we put this customer in the waiting line by updating WL from 0 to 1 (block 5). Since the present event is an arrival, we now schedule the next arrival into the system. Given that $IT = 2$ for arrival 3, the next arrival takes place at clock time 4. This completes all the necessary actions for event 2. We again loop back to block 2 to determine the next event. From the computer representation of the system in Table 4, we see that at this point (end of event 2), $DT = 3$ is less than $AT = 4$. This implies that the next event, event 3, will be a departure at clock time 3. We advance the clock to the time of this departure; that is, we update TM to 3 (block 11).

FIGURE 5
Time Continuum
Representation of
Single-Server
Simulation

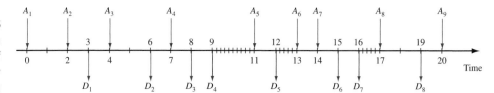

At time 3, we process the first departure from the system. With the departure, the server now becomes idle. We check the status of the waiting line to see whether there are any customers waiting for service (block 12). Since $WL = 1$, we have one customer waiting. We remove this customer from the waiting line, set $WL = 0$ (block 15), and bring this customer into service by generating a service time, ST (block 16), and setting the departure time using the relation $DT = TM + ST$ (block 17). From Table 3, we see that for customer 2, $ST = 3$. Since $TM = 3$, we set $DT = 6$. We have now completed all the actions for event 3, giving us the computer representation shown in Table 4.

From here on, we leave it to the reader to work through the logic of the simulation for the rest of the events in this example. Table 4 shows the status of the simulation at the end of each of these events. Note that at the end of events 8, 10, and 14 (all departures), the system becomes idle. During the sequence of actions for these events, we set $SS = 0$ (block 13) and $DT = 9,999$ (block 14). In each case, the system stays idle until an arrival takes place. This simulation is summarized in the time continuum diagram in Figure 5. Here, the A's represent the arrivals and the D's the departures. Note that the hatched areas, such as the one between times 9 and 11, signify that the system is idle.

This simple example illustrates some of the basic concepts in simulation and the way in which simulation can be used to analyze a particular problem. Although this model is not likely to be used to evaluate many situations of importance, it has provided us with a specific example and, more important, has introduced a variety of key simulation concepts. In the rest of the chapter, we analyze some of these simulation concepts in more detail. No mention was made in the example of the collection of statistics, but procedures can be easily incorporated into the model to determine the measures of performance of this system. For example, we could expand the flowchart to calculate and print the mean waiting time, the mean number in the waiting line, and the proportion of idle time. We discuss statistical issues in detail later in the chapter.

9.3 Random Numbers and Monte Carlo Simulation

In our queuing simulation example, we saw that the underlying movement through time is achieved in the simulation by generating the interarrival and the service times from the specified probability distributions. In fact, all event times are determined either directly or indirectly by these generated service and interarrival times. The procedure of generating these times from the given probability distributions is known as sampling from probability distributions, or random variate generation, or **Monte Carlo sampling.** In this section, we present and discuss several different methods of sampling from discrete distributions. We initially demonstrate the technique using a roulette wheel and then expand it by carrying out the sampling using random numbers.

The principle of sampling from discrete distributions is based on the frequency interpretation of probability. That is, in the long run, we would like the outcomes to occur with the frequencies specified by the probabilities in the distribution. For example, if we consider the service time distribution in Table 2, we would like, in the long run, to generate

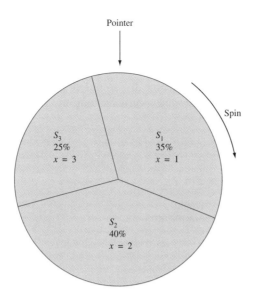

Pointer

Spin

S_3
25%
$x = 3$

S_1
35%
$x = 1$

S_2
40%
$x = 2$

FIGURE 6
Segmentation of
Roulette Wheel

a service time of 1 minute 35% of the time, a service time of 2 minutes 40% of the time, and a service time of 3 minutes 25% of the time. In addition to obtaining the right frequencies, the sampling procedure should be independent; that is, each generated service time should be independent of the service times that precede it and follow it.

To achieve these two properties using a roulette wheel, we first partition the wheel into three segments, each proportional in area to a probability in the distribution (see Figure 6). For example, the first segment (say, S_1) is allocated 35% of the area of the roulette wheel. This area corresponds to the probability of .35 and the service time of 1 minute. The second segment, S_2, covers 40% of the area and corresponds to the probability of .40 and the service time of 2 minutes. Finally, the third segment, S_3, is allocated the remaining 25% of the area, corresponding to the probability .25 and the service time of 3 minutes. If we now spin the roulette wheel and the pointer falls in segment S_1, it means that we have generated a service time of 1 minute; in segment S_2, 2 minutes; and in segment S_3, 3 minutes. If the roulette wheel is fair, as we assume, then in the long run, (1) we will generate the service times with approximately the same frequency as specified in the distribution, and (2) the results of each spin will be independent of the results that precede and follow it.

We now expand on this technique by using numbers for segmentation instead of areas. We assume that the roulette wheel has 100 numbers on it, ranging from 00 to 99, inclusive. We further assume that the segmentation is such that each number has the same probability, .01, of showing up. Using this method of segmentation, we allocate 35 numbers (say, from 00 to 34) to the service time of 1 minute. Since each number has a probability .01 of showing up, the 35 numbers together are equivalent to a probability of .35. Similarly, if we allocate the numbers from 35 to 74 to the service time of 2 minutes, and the numbers from 75 to 99 to the service time of 3 minutes, we achieve the desired probabilities. As before, we spin the roulette wheel to generate the service times, but with this method, the numbers directly determine the service times. In other words, if we generate a number between 00 and 34, we set the service time equal to 1 minute; between 35 and 74, to 2 minutes; and between 75 and 99, to 3 minutes.

This procedure of segmentation and using a roulette wheel is equivalent to generating integer random numbers between 00 and 99. This follows from the fact that each random number in a sequence (in this case from 00 to 99) has an equal probability (in this case, .01) of showing up, and each random number is independent of the numbers that precede and follow it. If we now had a procedure for generating the 100 random numbers between 00 and 99, then instead of spinning a roulette wheel to obtain a service time, we could use a generated random number. Technically, a random number, R_i, is defined as an independent random sample drawn from a continuous uniform distribution whose probability density function (pdf) is given by

$$f(x) = \begin{cases} 1 & 0 \leq x \leq 1 \\ 0 & \text{otherwise} \end{cases}$$

Thus, each random number will be uniformly distributed over the range between 0 and 1. Because of this, these random numbers are usually referred to as $U(0, 1)$ random numbers, or simply as uniform random numbers.

Random Number Generators

Uniform random numbers can be generated in many different ways. Since our interest in random numbers is for use within simulations, we need to be able to generate them on a computer. This is done using mathematical functions called **random number generators.**

Most random number generators use some form of a congruential relationship. Examples of such generators include the linear congruential generator, the multiplicative generator, and the mixed generator. The linear congruential generator is by far the most widely used. In fact, most built-in random number functions on computer systems use this generator. With this method, we produce a sequence of integers $x_1, x_2, x_3, \ldots$ between 0 and $m - 1$ according to the following recursive relation:

$$x_{i+1} = (ax_i + c) \text{ modulo } m \qquad (i = 0, 1, 2, \ldots)$$

The initial value of x_0 is called the seed, a is the constant multiplier, c is the increment, and m is the modulus. These four variables are called the parameters of the generator. Using this relation, the value of x_{i+1} equals the remainder from the division of $ax_i + c$ by m. The random number between 0 and 1 is then generated using the equation

$$R_i = \frac{x_i}{m} \qquad (i = 1, 2, 3, \ldots)$$

For example, if $x_0 = 35$, $a = 13$, $c = 65$, and $m = 100$, the algorithm works as follows:

Iteration 0 Set $x_0 = 35$, $a = 13$, $c = 65$, and $m = 100$.

Iteration 1 Compute

$$x_1 = (ax_0 + c) \text{ modulo } m$$
$$= [13(35) + 65] \text{ modulo } 100$$
$$= 20$$

Deliver

$$R_1 = \frac{x_1}{m}$$
$$= \frac{20}{100}$$
$$= 0.20$$

Iteration 2   Compute

$$x_2 = (ax_1 + c) \text{ modulo } m$$
$$= [13(20) + 65] \text{ modulo } 100$$
$$= 25$$

Deliver

$$R_2 = \frac{x_2}{m}$$
$$= \frac{25}{100}$$
$$= 0.25$$

Iteration 3   Compute

$$x_3 = (ax_2 + c) \text{ modulo } m$$
$$\vdots$$

and so on.

Each random number generated using this method will be a decimal number between 0 and 1. Note that although it is possible to generate a 0, a random number cannot equal 1. Random numbers generated using congruential methods are called **pseudorandom numbers.** They are not true random numbers in the technical sense, because they are completely determined once the recurrence relation is defined and the parameters of the generator are specified. However, by carefully selecting the values of a, c, m, and x_0, the pseudorandom numbers can be made to meet all the statistical properties of true random numbers. In addition to the statistical properties, random number generators must have several other important characteristics if they are to be used efficiently within computer simulations. (1) The routine must be fast; (2) the routine should not require a lot of core storage; (3) the random numbers should be replicable; and (4) the routine should have a sufficiently long cycle—that is, we should be able to generate a long sequence without repetition of the random numbers.

There is one important point worth mentioning at this stage: Most programming languages have built-in library functions that provide random (or pseudorandom) numbers directly. Therefore, most users need only know the library function for a particular system. In some systems, a user may have to specify a value for the seed, x_0, but it is unlikely that a user would have to develop or design a random number generator. However, for more information on random numbers and random number generators, the interested reader may consult Banks and Carson (1984), Knuth (1998), or Law and Kelton (1991).

Computer Generation of Random Numbers

We now take the method of Monte Carlo sampling a stage further and develop a procedure using random numbers generated on a computer. The idea is to transform the $U(0, 1)$ random numbers into integer random numbers between 00 and 99 and then to use these integer random numbers to achieve the segmentation by numbers. The transformation is a relatively straightforward procedure. If the $(0, 1)$ random numbers are multiplied by 100, they will be uniformly distributed over the range from 0 to 100. Then, if the fractional part of the number is dropped, the result will be integers from 00 to 99, all equally likely. For example, if we had generated the random number 0.72365, multiplying it by 100 gives

TABLE 5
Two-Digit Integer Random Numbers

69	56	30	32	66	79	55	24	80	35	10	98
92	92	88	82	13	04	86	31	13	23	44	93
13	42	51	16	17	29	62	08	59	41	47	72
25	96	58	14	68	15	18	99	13	05	03	83
34	78	50	89	98	93	70	11	49	01	9	35
64	43	71	48	36	78	53	67	37	57	25	17
84	59	68	45	12	53	68	38	18	60	02	82
31	28	52	89	27	35	34	74	96	93	45	63
21	17	71	55	32	74	20	68	44	34	53	68
91	84	39	25	20	83	60	62	99	61	32	98
55	86	18	93	51	77	68	37	69	02	85	60
43	16	20	42	82	17	41	50	54	21	25	43
40	98	71	03	68	05	37	02	86	17	38	99
42	37	72	33	72	43	51	60	17	94	51	39
18	06	28	75	69	80	33	69	12	25	53	36
13	20	42	92	57	08	24	06	41	12	89	95
58	18	98	89	08	60	89	93	58	13	29	34
63	68	69	62	07	49	95	48	20	03	71	90
92	54	29	31	80	28	48	45	92	71	31	33
84	11	57	64	93	69	86	22	23	84	38	60
33	24	65	76	87	95	98	47	00	71	31	97
53	08	80	85	73	13	25	35	22	82	26	43
02	19	61	38	00	21	42	79	31	70	00	17
22	81	43	44	78	88	30	31	15	63	09	99
38	25	32	92	11	55	18	52	47	30	43	87
04	61	82	18	82	75	12	19	44	87	77	93
06	54	51	64	81	98	63	47	57	52	74	56
51	51	00	41	78	84	42	79	06	82	58	53
99	93	87	86	83	79	16	33	53	34	40	32
29	12	64	73	38	08	49	32	53	33	91	90
31	78	93	25	37	51	68	40	34	47	83	76
81	69	27	35	71	12	69	78	96	93	35	96
26	73	28	81	38	09	55	10	27	29	52	46
92	29	08	15	73	26	33	05	89	08	26	99
00	86	32	46	80	22	97	19	99	95	53	20
39	25	07	41	74	71	01	64	23	69	74	95
38	86	41	38	71	91	75	54	65	73	47	86
41	74	68	21	74	89	43	19	98	74	09	50
63	53	45	07	47	15	58	75	88	51	88	99
00	54	86	59	77	09	54	55	99	15	67	63
01	38	88	03	71	88	72	39	76	45	11	07
38	05	53	31	18	11	26	65	61	77	19	03
34	43	19	12	35	02	09	86	69	90	53	50
23	41	56	34	77	30	50	02	34	68	49	16
57	24	80	69	51	81	83	05	19	45	30	20
93	86	08	08	99	62	75	97	29	51	68	96
16	10	38	33	32	25	34	66	72	17	51	97
75	28	35	14	01	00	98	51	74	10	79	30
53	38	65	32	78	77	64	11	31	06	73	47
91	90	95	95	66	80	10	90	51	24	81	06

Interarrival Time (minutes)	Probability	Cumulative Probability	Random Number Ranges
1	.20	.20	00–19
2	.30	.50	20–49
3	.35	.85	50–84
4	.15	1.00	85–99

us 72.365. Truncating the decimal portion of the number will leave us with the integer random number 72. On the computer, we achieve this transformation by first generating a $U(0, 1)$ random number. Next, we multiply it by 100. Finally, we store the product using an integer variable; this final stage will truncate the decimal portion of the number. This procedure will give us integer random numbers between 00 and 99. Table 5 lists some integer pseudorandom numbers obtained using this procedure. (These random numbers will be used in several examples later in the chapter.)

We now formalize this procedure and use it to generate random variates for a discrete random variable. The procedure consists of two steps: (1) We develop the cumulative probability distribution (cdf) for the given random variable, and (2) we use the cdf to allocate the integer random numbers directly to the various values of the random variable. To illustrate this procedure, we use the distribution of interarrival times from the queuing example of Section 9.2 (see Table 1). If we develop the cdf for this distribution, we get the probabilities shown in Table 6. The first interarrival time of 1 minute occurs with a probability of .20. Thus, we need to allocate 20 random numbers to this outcome. If we assign the 20 numbers from 00 to 19, we utilize the decimal random number range from 0 to 0.19999. Note that the upper end of this range lies just below the cumulative probability of .20. For the interarrival time of 2 minutes, we allocate 30 random numbers. If we assign the integer numbers from 20 to 49, we notice that this covers the decimal random number range from 0.20 to 0.49999. As before, the upper end of this range lies just below the cumulative probability of .50, but the lower end coincides with the previous cumulative probability of .20. If we now allocate the integer random numbers from 50 to 84 to the interarrival time of 3 minutes, we notice that these numbers are obtained from the decimal random number range from 0.50 (the same as the cumulative probability associated with an interarrival time of 2 minutes) to 0.84999, which is a fraction smaller than .85. Finally, the same analyses apply to the interarrival time of 4 minutes. In other words, the cumulative probability distribution enables us to allocate the integer random number ranges directly. Once these ranges have been specified for a given distribution, all we must do to obtain the value of a random variable is generate an integer random number and match it against the random number allocations. For example, if the random number had turned out to be 35, this would translate to an interarrival time of 2 minutes. Similarly, the random number 67 would translate to an interarrival time of 3 minutes, and so on. We now demonstrate these concepts in an example of a Monte Carlo simulation.

9.4 An Example of Monte Carlo Simulation

In this section, we use a Monte Carlo simulation to simulate a news vendor problem (see Chapter 4).

EXAMPLE 1   Pierre's Bakery

Pierre's Bakery bakes and sells french bread. Each morning, the bakery satisfies the demand for the day using freshly baked bread. Pierre's can bake the bread only in batches of a dozen loaves each. Each loaf costs 25¢ to make. For simplicity, we assume that the total daily demand for bread also occurs in multiples of 12. Past data have shown that this demand ranges from 36 to 96 loaves per day. A loaf sells for 40¢, and any bread left over at the end of the day is sold to a charitable kitchen for a salvage price of 10¢/loaf. If demand exceeds supply, we assume that there is a lost-profit cost of 15¢/loaf (because of loss of goodwill, loss of customers to competitors, and so on). The bakery records show that the daily demand can be categorized into three types: high, average, and low. These demands occur with probabilities of .30, .45, and .25, respectively. The distribution of the demand by categories is given in Table 7. Pierre's would like to determine the optimal number of loaves to bake each day to maximize profit (revenues + salvage revenues − cost of bread − cost of lost profits).

Solution   To solve this problem by simulation, we require a number of different policies to evaluate. Here, we define a policy as the number of loaves to bake each day. Each given policy is then evaluated over a fixed period of time to determine its profit margin. The policy that gives the highest profit is selected as the best policy.

In the simulation process, we first develop a procedure for generating the demand for the day:

Step 1   Determine the type of demand—that is, whether the demand for the day is high, average, or low. To do this, calculate the cdf for this distribution and set up the random number assignments (see Table 8). Then, to determine the type of demand, all we have to do is to generate a two-digit random number and match it against the random number allocations in this table.

Step 2   Generate the actual demand for the day from the appropriate demand distribution. The cdf and the random number allocations for the distribution of each of the three demand types are presented in Table 9. Then, to generate a demand, we simply generate an integer random number and match it against the appropriate random number assignments. For example, if our demand type was "average" in step 1, the random number 80 would translate into a demand of 72. Similarly, if the type of demand was "high" in step 1, the random number 9 would translate into a demand of 48.

The simulation process for this problem is relatively simple. For each day, we generate a demand for the day. Then we evaluate the various costs for a given policy. Suppose, for example, that the policy is to bake 60 loaves each day. If the demand for a particular

TABLE 7
Demand Distribution by Demand Categories

Demand	Demand Probability Distribution		
	High	Average	Low
36	.05	.10	.15
48	.10	.20	.25
60	.25	.30	.35
72	.30	.25	.15
84	.20	.10	.05
96	.10	.05	.05

TABLE 8
Distribution of Demand Type

Type of Demand	Probability	Cumulative Distribution	Random Number Ranges
High	.30	.30	00–29
Average	.45	.75	30–74
Low	.25	1.00	75–99

TABLE 9
Distribution by Demand Type

Demand	Cumulative Distribution			Random Number Ranges		
	High	Average	Low	High	Average	Low
36	.05	.10	.15	00–04	00–09	00–14
48	.15	.30	.40	05–14	10–29	15–39
60	.40	.60	.75	15–39	30–59	40–74
72	.70	.85	.90	40–69	60–84	75–89
84	.90	.95	.95	70–89	85–94	90–94
96	1.00	1.00	1.00	90–99	95–99	95–99

day turns out to be 72, we have 60(0.40) = $24.00 in revenues, 60(0.25) = $15.00 in production costs, and 12(0.15) = $1.80 in lost-profit costs (because of the shortfall of 12 loaves). This gives us a net profit of 24.00 − 15.00 − 1.80 = $7.20 for that day.

Using this procedure, we calculate a profit margin for each day in the simulation. To evaluate a policy, we run the simulation for a fixed number of days for the given policy. At the end of the simulation, we average the profit margins over the set number of days to obtain the expected profit margin per day for the policy. Note that the procedure in this simulation is different from the queuing simulation, in that the present simulation does not evolve over time in the same way. Here, each day is an independent simulation. Such simulations are commonly referred to as **Monte Carlo simulations.**

To illustrate this procedure, we present in Table 10 a manual simulation for the first 15 days for a policy where we bake 60 loaves per day. From this table, the demand for both day 1 and day 2 turns out to be 60 loaves. (Random numbers used in this example were obtained from Table 5.) This demand generates a revenue of $24.00 for each of these days. Since the 60 loaves cost $15.00 to bake, our profit margin for each of the first 2 days is $9.00. On day 3, the demand is 72, giving us a shortfall of 12 loaves. As shown in the table, the profit margin for day 3 is $7.20 (24.00 − 15.00 − 1.80). On day 4, we generate a demand of 48. Since our policy is to bake 60 loaves, we will have 12 loaves left over. The 48 loaves sold give us revenues of only $19.20. However, the 12 loaves left over provide an additional $1.20 in salvage revenue, yielding a profit of $5.40 (19.20 + 1.20 − 15.00) for day 4.

If we now complete the manual simulation for the period of 15 days, the total profit earned during this time comes to $97.20. This gives us an average daily profit figure of $\frac{97.20}{15}$ = $6.48. However, this cannot be accepted as the final profit margin for this policy. The simulation results over this short a period are likely to be highly dependent on the sequence of random numbers generated, so they cannot be accepted as statistically valid. The simulation would have to be carried out over a long period of time before the profit margin could be accepted as truly representative. These statistical issues are discussed later. In the meantime, we have evaluated several different policies for this problem using

TABLE **10**
Simulation Table for Baking 60 Loaves per Day

Day	Random No. for Demand Type	Type of Demand	Random No. for Demand	Demand	Revenue	Lost Profit	Salvage Revenue	Profit
1	69	Average	56	60	$24.00	—	—	$9.00
2	30	Average	32	60	$24.00	—	—	$9.00
3	66	Average	79	72	$24.00	$1.80	—	$7.20
4	55	Average	24	48	$19.20	—	$1.20	$5.40
5	80	Low	35	48	$19.20	—	$1.20	$5.40
6	10	High	98	96	$24.00	$5.40	—	$3.60
7	92	Low	88	72	$24.00	$1.80	—	$7.20
8	82	Low	17	48	$19.20	—	$1.20	$5.40
9	04	High	86	84	$24.00	$3.60	—	$5.40
10	31	Average	13	48	$19.20	—	$1.20	$5.40
11	23	High	44	72	$24.00	$1.80	—	$7.20
12	93	Low	13	36	$14.40	—	$2.40	$1.80
13	42	Average	51	60	$24.00	—	—	$9.00
14	16	High	17	60	$24.00	—	—	$9.00
15	29	High	62	72	$24.00	$1.80	—	$7.20

TABLE **11**
Evaluation of Policies

Policy	No. of Loaves Baked Daily	Average Daily Profit	
		Exact	Simulation
A	36	$1.273	$1.273
B	48	$4.347	$4.349
C	60	$6.435	$6.436
D	72	$6.917	$6.915
E	84	$6.102	$6.104
F	96	$4.653	$4.642

a simulation model on a computer. The results of these policies are presented in Table 11. We see that the best policy for Pierre's Bakery is to bake 72 loaves each day. This table also compares the results from the simulation with the exact solution for each policy. We can see that simulation does a remarkable job of converging to the right solution. The closeness of the two solutions is not totally unexpected, since we ran the simulation model for 10,000 days for each policy.

PROBLEMS

Group A

Use the random numbers in Table 5 to solve the following problems.

1 Simulate the single-server queuing system described in Section 9.2 for the first 25 departures from the system to

develop an estimate for the expected time in the waiting line. Is this a reasonable estimate? Explain.

2 Perform the simulation for Pierre's Bakery for 25 more days (days 16 through 40) for policy C in Table 11. Compare the answer with the results in the table.

Group B

3 Consider the simplest form of craps. In this game, we roll a pair of dice. If we roll a 7 or an 11 on the first throw, we win right away. If we roll a 2 or a 3 or a 12, we lose right away. Any other total (that is, 4, 5, 6, 8, 9, or 10) gives us a second chance. In this part of the game, we keep rolling the dice until we get either a 7 or the total rolled on the first throw. If we get a 7, we lose. If we roll the same total as on the first throw, we win. Assuming that the dice are fair, develop a simulation experiment to determine what percentage of the time we win.

4 Tankers arrive at an oil port with the distribution of interarrival times shown in Table 12. The port has two terminals, A and B. Terminal B is newer and therefore more efficient than terminal A. The time it takes to unload a tanker depends on the tanker's size. A supertanker takes 4 days to unload at terminal A and 3 days at terminal B. A midsize tanker takes 3 days at terminal A and 2 days at terminal B. The small tankers take 2 days at terminal A and 1 day at terminal B. Arriving tankers form a single waiting line in the port area until a terminal becomes available for service. Service is given on an FCFS basis. The type of tankers and the frequency with which they visit this port is given by the distribution in Table 13. Develop a simulation model for this port. Compute such statistics as the average number of

TABLE 12

Interarrival Times (days)	Probability
1	.20
2	.25
3	.35
4	.15
5	.05

TABLE 13

Type of Tanker	Probability
Supertanker	.40
Midsize tanker	.35
Small tanker	.25

tankers in port, the average number of days in port for a tanker, and the percentage of idle time for each of the terminals. (*Hint:* Use the flowchart in Figure 4 and modify it for a multiserver queuing system.)

9.5 Simulations with Continuous Random Variables

The simulation examples presented thus far used only discrete probability distributions for the random variables. However, in many simulations, it is more realistic and practical to use continuous random variables. In this section, we present and discuss several procedures for generating random variates from continuous distributions. The basic principle is very similar to the discrete case. As in the discrete method, we first generate a $U(0, 1)$ random number and then transform it into a random variate from the specified distribution. The process for carrying out the transformation, however, is quite different from the discrete case.

There are many different methods for generating continuous random variates. The selection of a particular algorithm will depend on the distribution from which we want to generate, taking into account such factors as the exactness of the random variables, the computational and storage efficiencies, and the complexity of the algorithm. The two most commonly used algorithms are the inverse transformation method (ITM) and the acceptance–rejection method (ARM). Between these two methods, it is possible to generate random variables from almost all of the most frequently used distributions. We present a detailed description of both these algorithms, along with several examples for each method. In addition to this, we present two methods for generating random variables from the normal distribution.

Inverse Transformation Method

The inverse transformation method is generally used for distributions whose cumulative distribution function can be obtained in closed form. Examples include the exponential, the uniform, the triangular, and the Weibull distributions. For distributions whose cdf does not exist in closed form, it may be possible to use some numerical method, such as a power-series expansion, within the algorithm to evaluate the cdf. However, this is likely to complicate the procedure to such an extent that it may be more efficient to use a different algorithm to generate the random variates. The ITM is relatively easy to describe and execute. It consists of the following three steps:

Step 1 Given a probability density function $f(x)$ for a random variable **X**, obtain the cumulative distribution function $F(x)$ as

$$F(x) = \int_{-\infty}^{x} f(t)dt$$

Step 2 Generate a random number r.

Step 3 Set $F(x) = r$ and solve for x. The variable x is then a random variate from the distribution whose pdf is given by $f(x)$.

We now describe the mechanics of the algorithm using an example. For this, we consider the distribution given by the function

$$f(x) = \begin{cases} \dfrac{x}{2} & 0 \le x \le 2 \\ 0 & \text{otherwise} \end{cases}$$

A function of this type is called a **ramp function.** It can be represented graphically as shown in Figure 7. The area under the curve, $f(x) = \frac{x}{2}$, represents the probability of the occurrence of the random variable **X**. We assume that in this case, **X** represents the service times of a bank teller. To obtain random variates from this distribution using the inverse transformation method, we first compute the cdf as

$$F(x) = \int_{0}^{x} \frac{t}{2} \, dt$$

$$= \frac{x^2}{4}$$

This cdf is represented formally by the function

$$F(x) = \begin{cases} 0 & x < 0 \\ \dfrac{x^2}{4} & 0 \le x \le 2 \\ 1 & x \ge 2 \end{cases}$$

Next, in step 2, we generate a random number r. Finally, in step 3, we set $F(x) = r$ and solve for x.

$$\frac{x^2}{4} = r$$

$$x = \pm 2\sqrt{r}$$

Since the service times are defined only for positive values of x, a service time of $x = -2\sqrt{r}$ is not feasible. This leaves us with $x = 2\sqrt{r}$ as the solution for x. This equation is called a **random variate generator** or a **process generator.** Thus, to obtain a service time, we first generate a random number and then transform it using the preceding equation. Each execution of the equation will give us one service time from the given distribution. For instance, if a random number $r = 0.64$ is obtained, a service time of $x = 2\sqrt{0.64} = 1.6$ will be generated.

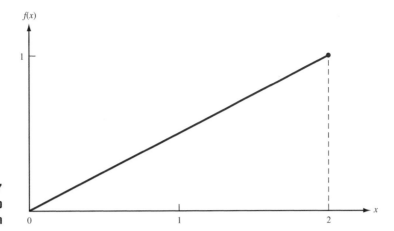

FIGURE 7
The pdf of a Ramp
Function

Graphically, the inverse transformation method can be represented as shown in Figure 8. We see from this graph that the range of values for the random variable (that is, $0 \leq x \leq 2$) coincides with the cumulative probabilities, $0 \leq F(x) \leq 1.0$. In other words, for any value of $F(x)$ over the interval $[0, 1]$, there exists a corresponding value of the random variable, given by x. Since a random number is also defined in the range between 0 and 1, this implies that a random number can be translated directly into a corresponding value of x using the relation $r = F(x)$. The solution for x in terms of r is known as taking the inverse of $F(x)$, denoted by $x = F^{-1}(r)$—hence the name *inverse transformation*. Note that if r is equal to 0, we will generate a random variate equal to 0, the smallest possible value of x. Similarly, if we generate a random number equal to 1, it will be transformed to 2, the largest possible value of x.

To show that the ITM generates numbers with the same distribution as x, consider the fact that for any two numbers x_1 and x_2, the probability $P(x_1 \leq \mathbf{X} \leq x_2) = F(x_2) - F(x_1)$. Then what we have to show is that the probability that the generated value of $\mathbf{X}$ lies between x_1 and x_2 is also the same. From Figure 8, we see that the generated value of $\mathbf{X}$ will be between x_1 and x_2 if and only if the chosen random number is between $r_1 = F(x_1)$ and $r_2 = F(x_2)$. Thus, the probability that the generated value of $\mathbf{X}$ is between x_1 and x_2 is also $F(x_2) - F(x_1)$. This shows that the ITM does indeed generate numbers with the same distribution as $\mathbf{X}$.

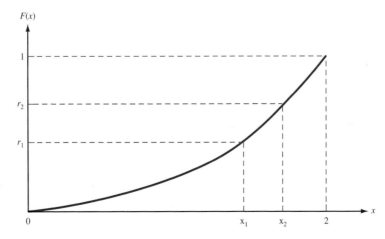

FIGURE 8
Graphical
Representation of
Inverse Transformation
Method

As this example shows, the major advantage of the inverse transformation method is its simplicity and ease of application. However, as mentioned earlier, we must be able to determine $F(x)$ in closed form for the desired distribution before we can use the method efficiently. Also, in this example, we see that we need exactly one random number to produce one random variable. Other methods, such as the acceptance–rejection method, may require several random numbers to generate a single value of $\mathbf{X}$. The following three examples illustrate the application of the ITM.

EXAMPLE 2   The Exponential Distribution

As mentioned in Chapter 8, the exponential distribution has important applications in the mathematical representation of queuing systems. The pdf of the exponential distribution is given by

$$f(x) = \begin{cases} \lambda e^{-\lambda x} & x \geq 0, \lambda > 0 \\ 0 & \text{otherwise} \end{cases}$$

Use the inverse transformation method to generate observations from an exponential distribution.

Solution   In step 1, we compute the cdf. This is given by

$$F(x) = \begin{cases} 0 & x < 0 \\ 1 - e^{-\lambda x} & x \geq 0 \end{cases}$$

Next, we generate a random number r and set $F(x) = r$ to solve for x. This gives us

$$1 - e^{-\lambda x} = r$$

Rearranging to

$$e^{-\lambda x} = 1 - r$$

and taking the natural logarithm of both sides, we have

$$-\lambda x = \ln(1 - r)$$

Finally, solving for x gives the solution

$$x = -\frac{1}{\lambda} \ln(1 - r)$$

To simplify our computations, we can replace $(1 - r)$ with r. Since r is a random number, $(1 - r)$ will also be a random number. This means that we have not changed anything except the way we are writing the $U(0,1)$ random number. Thus, our process generator for the exponential distribution will now be

$$x = -\frac{1}{\lambda} \ln r$$

For instance, $r = \frac{1}{e}$ yields $x = \frac{1}{\lambda}$, and $r = 1$ yields $x = 0$.

EXAMPLE 3   The Uniform Distribution

Consider a random variable $\mathbf{X}$ that is uniformly distributed on the interval $[a, b]$. The pdf of this distribution is given by the function

$$f(x) = \begin{cases} \dfrac{1}{b - a} & a \leq x \leq b \\ 0 & \text{otherwise} \end{cases}$$

Use the ITM to generate observations from this random variable.

Solution The cdf of this distribution is given by

$$F(x) = \begin{cases} 0 & x < a \\ \dfrac{x-a}{b-a} & a \le x \le b \\ 1 & x > b \end{cases}$$

To use the ITM to generate observations from a uniform distribution, we first generate a random number r and then set $F(x) = r$ to solve for x. This gives

$$\frac{x-a}{b-a} = r$$

Solving for x yields

$$x = a + (b-a)r$$

as the process generator for the uniform distribution. For example, $r = \frac{1}{2}$ yields $x = \frac{a+b}{2}$, $r = 1$ yields $x = b$, $r = 0$ yields $x = a$, and so on.

EXAMPLE 4 | **The Triangular Distribution**

Consider a random variable **X** whose pdf is given by

$$f(x) = \begin{cases} \frac{1}{2}(x-2) & 2 \le x \le 3 \\ \frac{1}{2}(2 - \frac{x}{3}) & 3 \le x \le 6 \\ 0 & \text{otherwise} \end{cases}$$

Use the ITM to generate observations from the distribution. This distribution, called a *triangular* distribution, is represented graphically in Figure 9. It has the endpoints [2, 6], and its mode is at 3. We can see that 25% of the area under the curve lies in the range of x from 2 to 3, and the other 75% lies in the range from 3 to 6. In other words, 25% of the values of the random variable **X** lie between 2 and 3, and the other 75% fall between 3 and 6. The triangular distribution has important applications in simulation. It is often used to represent activities for which there are few or no data. (For a detailed account of this distribution, see Banks and Carson (1984) or Law and Kelton (1991).)

Solution The cdf of this triangular distribution is given by the function

$$F(x) = \begin{cases} 0 & x < 2 \\ \frac{1}{4}(x-2)^2 & 2 \le x \le 3 \\ -\frac{1}{12}(x^2 - 12x + 24) & 3 \le x \le 6 \\ 1 & \text{otherwise} \end{cases}$$

For simplicity, we redefine $F(x) = (\frac{1}{4})(x-2)^2$, for $2 \le x \le 3$, as $F_1(x)$, and $F(x) = (-\frac{1}{12})(x^2 - 12x + 24)$, for $3 \le x \le 6$, as $F_2(x)$.

This cdf can be represented graphically as shown in Figure 10. Note that at $x = 3$, $F(3) = 0.25$. This implies that the function $F_1(x)$ covers the first 25% of the range of the cdf, and $F_2(x)$ applies over the remaining 75% of the range. Since we now have two separate functions representing the cdf, the ITM has to be modified to account for these two functions, their ranges, and the distribution of the ranges. As far as the ITM goes, the distribution of the ranges is the most important. This distribution is achieved by using the random number from step 2. In other words, if $r < 0.25$, we use the function $F_1(x) = (\frac{1}{4})(x-2)^2$ in step 3. Otherwise, we use $F_2(x) = (-\frac{1}{12})(x^2 - 12x + 24)$. Since $r < 0.25$ for 25% of the time and $r \ge 0.25$ for the other 75%, we achieve the desired distribution. In

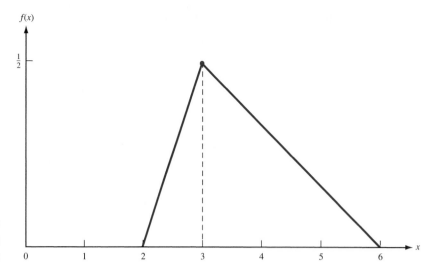

FIGURE 9
Density Function for a
Triangular Distribution

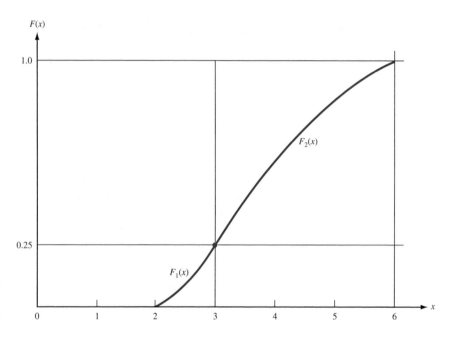

FIGURE 10
The cdf of a Triangular
Distribution

either case, we set the function $F_1(x)$ or $F_2(x)$ equal to r and solve for x. That is, we solve one of the following equations:

$$(\tfrac{1}{4})(x - 2)^2 = r \qquad \text{for } 0 \le r < 0.25$$

$$(-\tfrac{1}{12})(x^2 - 12x + 24) = r \qquad \text{for } 0.25 \le r \le 1.0$$

x will then be our random variable of interest.

As the graph in Figure 10 shows, a random number between 0 and 0.25 will be transformed into a value of x between 2 and 3. Similarly, if $r \ge 0.25$, it will be transformed into a value of x between 3 and 6.

To solve the first equation, $(\frac{1}{4})(x - 2)^2 = r$, we multiply the equation by 4 and then take the square root of both sides. This gives us

$$x - 2 = \pm \sqrt{4r}$$
$$x = 2 \pm 2\sqrt{r}$$

Since x is defined only for values greater than 2, $x = 2 - 2\sqrt{r}$ is infeasible, leaving

$$x = 2 + 2\sqrt{r}$$

as the process generator for a random number in the range from 0 to 0.25. Note that when $r = 0$, $x = 2$, the smallest possible value for this range. Similarly, when we generate $r = 0.25$, it will be transformed to $x = 3$.

To solve the second equation, $(-\frac{1}{12})(x^2 - 12x + 24) = r$, we can use one of two methods: (1) employing the quadratic formula or (2) completing the square. (See Banks and Carson (1984) for details of the quadratic formula method.) Here, we use the method of completing the square. We multiply the equation by -12 and rearrange the terms to get

$$x^2 - 12x = -24 - 12r$$

To complete the square, we first divide the x term's coefficient by 2. This gives us -6. Next, we square this value to get 36. Finally, we add this resultant to both sides of the equation. This leaves us with the equation

$$x^2 - 12x + 36 = 12 - 12r$$
$$(x - 6)^2 = 12 - 12r$$

Writing the equation in this form enables us to take the square root of both sides. That is,

$$x - 6 = \pm\sqrt{12 - 12r}$$
$$x = 6 \pm 2\sqrt{3 - 3r}$$

As before, part of the solution is infeasible. In this case, x is feasible only for values less than 6. Thus, we use only the equation $x = 6 - 2\sqrt{3 - 3r}$ as our process generator. Note that when $r = 0.25$, our random variate is equal to 3. Similarly, when $r = 1$, we generate a random variate equal to 6.

Acceptance–Rejection Method

There are several important distributions, including the Erlang (used in queuing models) and the beta (used in PERT), whose cumulative distribution functions do not exist in closed form. For these distributions, we must resort to other methods of generating random variates, one of which is the acceptance–rejection method (ARM). This method is generally used for distributions whose domains are defined over finite intervals. Thus, given a distribution whose pdf, $f(x)$, is defined over the interval $a \leq x \leq b$, the algorithm consists of the following steps.

Step 1 Select a constant M such that M is the largest value of $f(x)$ over the interval $[a, b]$.

Step 2 Generate two random numbers, r_1 and r_2.

Step 3 Compute $x^* = a + (b - a)r_1$. (This ensures that each member of $[a, b]$ has an equal chance to be chosen as x^*.)

Step 4 Evaluate the function $f(x)$ at the point x^*. Let this be $f(x^*)$.

Step 5 If

$$r_2 \leq \frac{f(x^*)}{M}$$

deliver x^* as a random variate from the distribution whose pdf is $f(x)$. Otherwise, reject x^* and go back to step 2.

Note that the algorithm continues looping back to step 2 until a random variate is accepted. This may take several iterations. For this reason, the algorithm can be relatively inefficient. The efficiency, however, is highly dependent on the shape of the distribution. There are several ways by which the method can be made more efficient. One of these is to use a function in step 1 instead of a constant. See Fishman (1978) or Law and Kelton (1991) for details of the algorithm.

We now illustrate the details of the algorithm using a ramp function. Consider a random variable **X** whose pdf is given by the function

$$f(x) = \begin{cases} 2x & 0 \leq x \leq 1 \\ 0 & \text{otherwise} \end{cases}$$

In step 1 of the ARM, it is generally useful to graph the pdf. Since our objective is to obtain the largest value of $f(x)$ over the domain of the function, graphing will enable us to determine the value of M simply by inspection. The graph of the pdf is shown in Figure 11. We see that the largest value of $f(x)$ occurs at $x = 1$ and is equal to 2. In other words, we set $M = 2$ in step 1. Next, we generate two random numbers, r_1 and r_2. In step 3, we transform the first random number, r_1, into an **X** value, x^*, using the relationship $x^* = a + (b - a)r_1$. This step is simply a procedure for randomly generating a value of the random variable **X**. Given that we are using a random number r_1 to determine x^*, every value over the interval $[a, b]$ has an equal probability of showing up. Note that if $r_1 = 0$, x^* will be equal to a, the left endpoint of the domain. Similarly, if $r_1 = 1$, x^* will be equal to b, the right endpoint of the domain. Since $a = 0$ and $b = 1$ for this distribution, it follows that $x^* = r_1$. This value of x^* now becomes our potential random variate for the current iteration. In steps 4 and 5, we have to determine whether to accept or reject x^*. We first evaluate the function $f(x)$ at $x = x^*$ to obtain $f(x^*)$ and then compute $\frac{f(x^*)}{M}$. If $r_2 \leq \frac{f(x^*)}{M}$, we accept x^*. Otherwise, we reject x^*. Substituting $x^* = r_1$ in $f(x)$ gives us $f(x^*) = 2r_1$. Since $M = 2$, the term $\frac{f(x^*)}{M}$ reduces to r_1. Given this, our decision rule (in step 5) for accepting x^* simplifies to a comparison of r_2 and r_1. If $r_2 \leq r_1$, we accept x^* as our random variate. Otherwise, we go back to step 2 and repeat the process until we obtain an acceptance.

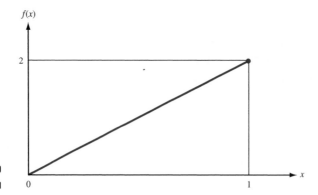

FIGURE 11
The pdf of a Ramp Function

For example, if $r_1 = 0.7$ and $r_2 = 0.6$, we choose $x = 0.7$, and if $r_1 = 0.7$ and $r_2 = 0.8$, no value of the random variable is generated. For this problem, exactly half the random variates generated in step 3 will be rejected in step 5.

EXAMPLE 5 **The Acceptance–Rejection Method**

Use the acceptance–rejection method to generate random variates from a triangular distribution whose pdf is given by

$$f(x) = \begin{cases} -\frac{1}{6} + \frac{x}{12} & 2 \leq x \leq 6 \\ \frac{4}{3} - \frac{x}{6} & 6 \leq x \leq 8 \end{cases}$$

Solution   For simplicity, we redefine $f(x) = -\frac{1}{6} + \frac{x}{12}$ as $f_1(x)$, and $f(x) = \frac{4}{3} - \frac{x}{6}$ as $f_2(x)$. This distribution is represented graphically in Figure 12.

Since this distribution is defined over two intervals, we must modify steps 4 and 5 of the acceptance–rejection method to account for these ranges. The first three steps of the algorithm, however, stay the same as before. That is, step 1 determines M, step 2 generates r_1 and r_2, and step 3 transforms r_1 into a value x^* of $\mathbf{X}$.

From the graph of the pdf in Figure 12, it is clear that $M = \frac{1}{3}$. This distribution has the endpoints [2, 8], which implies that $a = 2$ and $b = 8$. If we now substitute these endpoints in step 3, the x^* values are generated by the equation $x^* = 2 + 6r_1$. Then we see that if r_1 is between 0 and $\frac{2}{3}$, x^* will lie in the range from 2 to 6. If $r_1 > \frac{2}{3}$, x^* will lie in the interval [6, 8]. To account for this, we make our first modification in step 4. If x^* lies between 2 and 6, then in step 4, we use the function $f_1(x)$ to evaluate $f(x^*)$. Otherwise, we use $f_2(x)$ to compute $f(x^*)$. Step 4 now can be summarized as follows: If $2 \leq x^* \leq 6$,

$$f(x^*) = f_1(x^*)$$
$$= -\frac{1}{6} + \frac{x^*}{12}$$
$$= \frac{r_1}{2}$$

If $6 \leq x^* \leq 8$,

$$f(x^*) = f_2(x^*)$$
$$= \frac{4}{3} - \frac{x^*}{6}$$
$$= 1 - r_1$$

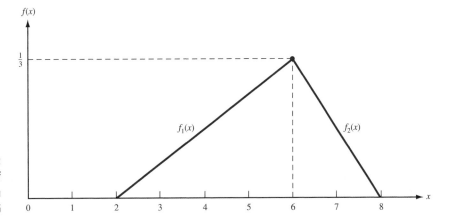

FIGURE 12
The pdf of
Triangular Distribution
of Example 5

The next step in the algorithm is either to accept or to reject the current value of x^*. We accept x^* if the condition $r_2 \leq \frac{f(x^*)}{M}$ is satisfied. However, following step 4, we need to evaluate this condition over the two intervals by substituting the appropriate function, $f(x^*)$, into the relation. In other words, step 5 for this distribution will now be as follows: For $2 \leq x^* \leq 6$, we accept x^* if $r_2 \leq \frac{f_1(x^*)}{M}$—that is, if $r_2 \leq \frac{3r_1}{2}$. For $6 \leq x^* \leq 8$, we accept x^* if $r_2 \leq \frac{f_2(x^*)}{M}$—that is, if $r_2 \leq 3(1 - r_1)$. If x^* is rejected, we go back to step 2 and repeat the process.

As before, some of the x^* values will be rejected. In this case, also, the probability of accepting a random variate is .5. That is, one half of all random variates generated in step 3 will, in the long run, be rejected in step 5.

We now give an intuitive justification of the validity of the ARM. In particular, we want to show that the ARM does generate observations from the given random variable $\mathbf{X}$. For any number x the ARM should yield $P(x \leq \mathbf{X} \leq x + \Delta) = f(x)\Delta$. Now the probability that the ARM generates an observation between x and $x + \Delta$ is given by

$$\sum_{i=1}^{i=\infty} \quad \text{(probability first } i - 1 \text{ iterations yield no value and } i\text{th iteration yields a value between } x \text{ and } x + \Delta)$$

$$= \sum_{i=1}^{i=\infty} \left(1 - \frac{1}{M(b - a)}\right)^{i-1} \frac{f(x)\Delta}{M(b - a)}$$

$$= \frac{f(x)\Delta}{M(b - a)} \left(\frac{1}{1 - (1 - 1/(M(b - a)))}\right) = f(x)\Delta$$

where we have used the fact that on any ARM iteration, there is a probability $1/M(b - a)$ that a value of the random variable will be generated (see Problem 6), and that for $c < 1$,

$$\sum_{i=1}^{i=\infty} c^{i-1} = \frac{1}{1 - c}$$

Direct and Convolution Methods for the Normal Distribution

Because of the importance of the normal distribution, considerable attention has been paid to generating normal random variates. This has resulted in many different algorithms for the normal distribution. Both the inverse transformation method and the acceptance–rejection method are inappropriate for the normal distribution, because (1) the cdf does not exist in closed form and (2) the distribution is not defined over a finite interval. Although it is possible to use numerical methods within the ITM and to truncate the distribution for the acceptance–rejection method, other methods tend to be much more efficient. In this section, we describe two such methods—first, an algorithm based on convolution techniques, and then a direct transformation algorithm that produces two standard normal variates with mean 0 and variance 1.

The Convolution Algorithm

In the convolution algorithm, we make direct use of the Central Limit Theorem. The Central Limit Theorem states that the sum $\mathbf{Y}$ of n independent and identically distributed random variables (say, $\mathbf{Y}_1, \mathbf{Y}_2, \ldots, \mathbf{Y}_n$, each with mean μ and finite variance σ^2) is approximately normally distributed with mean $n\mu$ and variance $n\sigma^2$. If we now apply this to $U(0, 1)$ random variables, $\mathbf{R}_1, \mathbf{R}_2, \ldots, \mathbf{R}_n$, with mean $\mu = 0.5$ and $\sigma^2 = \frac{1}{12}$, it follows that

$$\mathbf{Z} = \frac{\sum_{i=1}^{n} \mathbf{R}_i - 0.5n}{\left(\frac{n}{12}\right)^{1/2}}$$

is approximately normal with mean 0 and variance 1. We would expect this approximation to work better as the value of n increases. However, most simulation literature suggests using a value of $n = 12$. Using 12 not only seems adequate but, more important, has the advantage that it simplifies the computational procedure. If we now substitute $n = 12$ into the preceding equation, the process generator simplifies to

$$\mathbf{Z} = \sum_{i=1}^{12} \mathbf{R}_i - 6$$

This equation avoids a square root and a division, both of which are relatively time-consuming routines on a computer.

If we want to generate a normal variate $\mathbf{X}$ with mean μ and variance σ^2, we first generate $\mathbf{Z}$ using this process generator and then transform it using the relation $\mathbf{X} = \mu + \sigma\mathbf{Z}$. Note that this convolution is unique to the normal distribution and cannot be extended to other distributions. Several other distributions do, of course, lend themselves to convolution methods. For example, we can generate random variates from an Erlang distribution with shape parameter k and rate parameter $k\lambda$, using the fact that an Erlang random variable can be obtained by the sum of k iid exponential random variables, each with parameter $k\lambda$.

The Direct Method

The direct method for the normal distribution was developed by Box and Muller (1958). Although it is not as efficient as some of the newer techniques, it is easy to apply and execute. The algorithm generates two $U(0, 1)$ random numbers, r_1 and r_2, and then transforms them into two normal random variates, each with mean 0 and variance 1, using the direct transformations

$$\mathbf{Z}_1 = (-2 \ln r_1)^{1/2} \sin 2\pi r_2$$
$$\mathbf{Z}_2 = (-2 \ln r_1)^{1/2} \cos 2\pi r_2$$

As in the convolution method, it is easy to transform these standardized normal variates into normal variates $\mathbf{X}_1$ and $\mathbf{X}_2$ from the distribution with mean μ and variance σ^2, using the equations

$$\mathbf{X}_1 = \mu + \sigma\mathbf{Z}_1$$
$$\mathbf{X}_2 = \mu + \sigma\mathbf{Z}_2$$

The direct method produces exact normal random variates, whereas the convolution method gives us only approximate normal random variates. For this reason, the direct method is much more commonly used. For details of these and other normal algorithms, see Fishman (1978) or Law and Kelton (1991).

PROBLEMS

Group A

1 Consider a continuous random variable with the following pdf:

$$f(x) = \begin{cases} \frac{1}{2} & 0 \le x \le 1 \\ \frac{3}{4} - \frac{x}{4} & 1 \le x \le 3 \end{cases}$$

Develop a process generator for these breakdown times using the inverse transformation method and the acceptance–rejection method.

2 A job shop manager wants to develop a simulation model to help schedule jobs through the shop. He has evaluated the completion times for all the different types of jobs. For one particular job, the times to completion can be represented by the following triangular distribution:

$$f(x) = \begin{cases} \frac{x}{8} - \frac{1}{4} & 2 \le x \le 4 \\ \frac{10}{24} - \frac{x}{24} & 4 \le x \le 10 \end{cases}$$

Develop a process generator for this distribution using the inverse transformation method.

3 Given the continuous triangular distribution in Figure 13, develop a process generator using the inverse transformation method.

Group B

4 For Problem 2, develop a computer program for the process generator. Generate 100 random variates and compare the mean and variance of this sample against the theoretical mean and variance of this distribution. Now repeat the experiment for the following numbers of random variates: 250; 500; 1,000; and 5,000. From these experiments, what can be said about the process generator?

5 A machine operator processes two types of jobs, A and B, during the course of the day. Analyses of past data show that 40% of all jobs are type A jobs, and 60% are type B jobs. Type A jobs have completion times that can be represented by an Erlang distribution with rate parameter 5 and shape parameter 2. Completion times of type B jobs can be represented by the following triangular distribution:

$$f(x) = \begin{cases} \dfrac{x-1}{12} & 1 \le x \le 4 \\ \dfrac{9-x}{20} & 4 \le x \le 9 \end{cases}$$

If jobs arrive at an exponential rate of 10 per hour, develop a simulation model to calculate the percentage of idle time of the operator and the average number of jobs in the line waiting to be processed.

6 Show that on any iteration of the acceptance–rejection method, there is a probability $\frac{1}{M(b-a)}$ that a value of the random variable is generated.

7 We all hate to bring small change to the store. Using random numbers, we can eliminate the need for change and give the store and the customer a fair shake.

    a Suppose you buy something that costs $.20. How could you use random numbers (built into the cash register system) to decide whether you should pay $1.00 or nothing? This eliminates the need for change!

    b If you bought something for $9.60, how would you use random numbers to eliminate the need for change?

    c In the long run, why is this method fair to both the store and the customer?

FIGURE 13

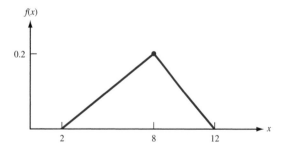

9.6 An Example of a Stochastic Simulation

We now present an example of a simulation using some of the concepts covered in Section 9.5. We consider the case of Cabot, Inc., a large mail order firm in Chicago. Orders arrive into the warehouse via telephones. At present, Cabot maintains 10 operators at work 24 hours a day. The operators take the orders and feed them directly into a central computer, using terminals. Each operator has one terminal. At present, the company has a total of 11 terminals. That is, if all terminals are working, there will be 1 spare terminal.

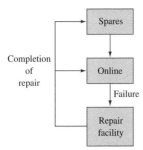

FIGURE 14
Flow of Terminals

A terminal that is online may break down. If that happens, the terminal is removed from the workstation and is replaced with a spare, if one is available. If none is available, the operator must wait until one becomes available. During this time, this operator does not take any orders. The broken terminal is sent to the workshop, where the company has one repair channel allocated to repairing terminals. At the completion of a repair, the terminal either acts as a spare or goes directly into service if an operator is waiting for a terminal. The flow of terminals in the system is shown in Figure 14.

The Cabot managers believe that the terminal system needs evaluation, because the downtime of operators due to broken terminals has been excessive. They feel that the problem can be solved by the purchase of some additional terminals for the spares pool. Accountants have determined that a new terminal will cost a total of $75 per week in such costs as investment cost, capital cost, maintenance, and insurance. It has also been estimated that the cost of terminal downtime, in terms of delays, lost orders, and so on, is $1,000 per week. Given this information, the Cabot managers would like to determine how many additional terminals they should purchase.

This model is a version of the machine repair problem (see Section 8.9). In such models (if both the breakdown and the repair times can be represented by the exponential distribution), it is easy to find an analytical solution to the problem using birth–death processes. However, in analyzing the historical data for the terminals, it has been determined that although the breakdown times can be represented by the exponential distribution, the repair times can be adequately represented only by the triangular distribution. This implies that analytical methods cannot be used; we must use simulation.

To simulate this system, we first require the parameters of both the distributions. For the breakdown time distribution, the data show that the breakdown rate is exponential and equal to 1 per week per terminal. In other words, the time between breakdowns for a terminal is exponential with a mean equal to 1 week. Analysis for the repair times (measured in weeks) shows that this distribution can be represented by the triangular distribution

$$f(x) = \begin{cases} -10 + 400x & 0.025 \leq x \leq 0.075 \\ 50 - 400x & 0.075 \leq x \leq 0.125 \end{cases}$$

which has a mean of 0.075 week. That is, the repair staff can, on the average, repair 13.33 terminals per week. We represent the repair time distribution graphically in Figure 15.

To find the optimum number of terminals for the system, we must balance the cost of the additional terminals against the increased revenues (because of reduced downtime costs) generated as a result of the increase in the number of terminals. In the simulation, we increase the number of terminals in the system, n, from the present total of 11 in increments of 1. For this fixed value of n, we then run our simulation model to estimate the net revenue. Net revenue here is defined as the difference between the increase in revenues due to the additional terminals and the cost of these additional terminals. We keep

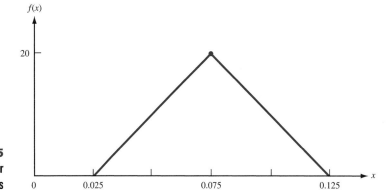

FIGURE **15**
The pdf of the Repair
Times

on adding terminals until the net revenue position reaches a peak. Thus, our primary objective in the simulation is to determine the net revenue for a fixed number of terminals.

To calculate the net revenue, we first compute the average number of online terminals, EL_n (or equivalently, the average number of downtime terminals, ED_n), for a fixed number of terminals in the system, n. In fact, ED_n is simply equal to $n - EL_n$. Once we have a value for EL_n, we can compute the expected weekly downtime costs, given by $1{,}000(10 - EL_n)$. Then the increase in revenues as a result of increasing the number of terminals from 11 to n is $1{,}000(EL_n - EL_{11})$.

Mathematically, we compute EL_n as

$$EL_n = \frac{\int_0^T N(t)\,dt}{T} = \frac{\sum_{i=1}^m A_i}{T}$$

where

$$T = \text{length of the simulation}$$

$$N(t) = \text{number of terminals online at time } t \qquad (0 \le t \le T)$$

$$A_i = \text{area of rectangle under } N(t) \text{ between } e_{i-1} \text{ and } e_i$$
$$(\text{where } e_i \text{ is the time of the } i\text{th event})$$

$$m = \text{number of events that occur in the interval } [0, T]$$

This computation is illustrated in Figure 16 for $n = 10$. In this example, we start with 10 terminals online at time 0. Between time 0 and time e_1, the time of the first event, the total online time for all the terminals is given by $10e_1$, since each terminal is online for a period of e_1 time units. Similarly, the total online time between events 1 and 2 is 9 $(e_2 - e_1)$, given that the breakdown at time e_1 leaves us with only 9 working terminals between time e_1 and time e_2. If we now run this simulation over T time units and sum up the areas $A_1, A_2, A_3, \ldots$, we can get an estimate for EL_{10} by dividing this sum by T. This statistic is called a **time-average statistic.** As long as the simulation is run for a sufficiently long period of time, our estimate for EL_{10} should be fairly close to the actual.

In this simulation, we would like to set up the process in such a way that it will be possible to collect the statistics to compute the areas $A_1, A_2, A_3, \ldots$. That is, as we move from event to event, we would like to keep track of at least the number of terminals online between the events and the time between events. To do this, we first define the state of the system as the number of terminals in the repair facility. From this definition, it follows that the only time the state of the system will change is when there is either a breakdown or a completion of a repair. This implies that there are two events in this simulation: breakdown and completion of repairs.

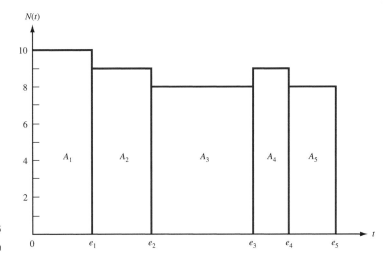

FIGURE 16
Computation of EL_{10}

To set up the simulation, our first task is to determine the process generators for both the breakdown and the repair times. Since both these distributions have cdf's in closed form, we use the ITM to develop the process generators. For the exponential distribution, the process generator is simply

$$x = -\log r$$

In the case of the repair times, applying the ITM gives us

$$x = 0.025 + \sqrt{0.005r} \qquad (0 \leq r \leq 0.5)$$

and

$$x = 0.125 - \sqrt{0.005(1 - r)} \qquad (0.5 \leq r \leq 1.0)$$

as the process generators.

Within this experiment, we run several different simulations, one for each different value of n. Since n at present equals 11, we begin the experiment with this number and increase n until the net revenues reach a peak. For each n, we start the simulation in the state where there are no terminals in the repair facility. In this state, all 10 operators are online and any remaining terminals are in the spares pool.

Our first action in the simulation is to schedule the first series of events, the breakdown times for the terminals presently online. We do this in the usual way, by generating an exponential random variate for each online terminal from the breakdown distribution and setting the time of breakdown by adding this generated time to the current clock time, which is zero. Having scheduled these events, we next determine the first event, the first breakdown, by searching through the current event list. We then move the simulation clock to the time of this event and process this breakdown.

To process a breakdown, we take two separate series of actions: (1) Determine whether a spare is available. If one is available, bring the spare into service and schedule the breakdown time for this terminal. If none is available, update the back-order position. (2) Determine whether the repair staff is idle. If so, start the repair on the broken terminal by generating a random variate from the service times distribution and scheduling the completion time of the repair. If the repair staff is busy, place the broken terminal in the repair queue. Having completed these two series of actions, we now update all the statistical counters. These actions are summarized in the system flow diagram in Figure 17. We

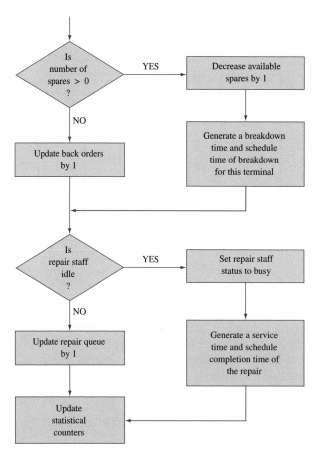

FIGURE 17
Flowchart for a
Breakdown

proceed with the simulation by determining the next event and moving the clock to the time of this event. If the next event is another breakdown, we repeat the preceding series of actions. Otherwise, we process a completion of a repair.

To process the completion of a repair, we also undertake two series of actions. (1) At the completion of a repair, we have an additional working terminal, so we determine whether the terminal goes directly to an operator or to the spares pool. If a back order exists, we bring the terminal directly into service and schedule the time of the breakdown for this terminal in the usual manner. If no operator is waiting for a terminal, the terminal goes into the spares pool. (2) We check the repair queue to see whether any terminals are waiting to be repaired. If the queue is greater than zero, we bring the first terminal from the queue into repair and schedule the time of the completion of this repair. Otherwise, we set the status of the repair staff to idle. Finally, at the completion of these actions, we update all the statistical counters. This part of the simulation is summarized in Figure 18.

We proceed with the simulation (for a given n) by moving from event to event until the termination time T. At this time, we calculate all the relevant measures of performance from the statistical counters. Our key measure is the net revenue for the current value of n. If this revenue is greater than the revenue for a system with $n - 1$ terminals, we increase the value of n by 1 and repeat the simulation with $n + 1$ terminals in the system.

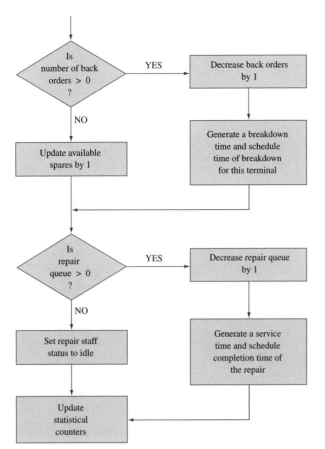

FIGURE 18
Flowchart for
Completion of a Repair

Otherwise, the net revenue has reached a peak. If this is the case, we stop the experiment and accept $n - 1$ terminals as the optimal number of terminals to have in the system. This simulation experiment is summarized in Figure 19. For this experiment, we assume that the maximum number of terminals we can have is 25. (Note that the variable $REVO$ in the flowchart is revenue for the system with $n - 1$ terminals.)

For this problem, we ran a complete experiment, whose results are summarized in Table 14. In this table, we show the overall effect on the net revenues as we increase the number of terminals from 11. For example, when we increase n from 11 to 12, the expected number of online terminals increases from 9.362 to 9.641, for a net increase of 0.279. This results in an increase of $279 in revenues per week at a cost of $75, giving us a net revenue increase of $204 per week. Similarly, if we increase the number of terminals from 11 to 13, we have a net increase of $289. The net increase peaks with 14 terminals in the system. This is further highlighted by the graph in Figure 20.

The simulation outlined in this example can be used to analyze other policy options that management may have. For example, instead of purchasing additional terminals, Cabot could hire a second repair worker or choose a preventive maintenance program for the terminals. Alternatively, the company might prefer a combination of these policies. The simulation model provides a very flexible mechanism for evaluating alternative policies.

FIGURE 19
Flowchart for Terminal Simulation

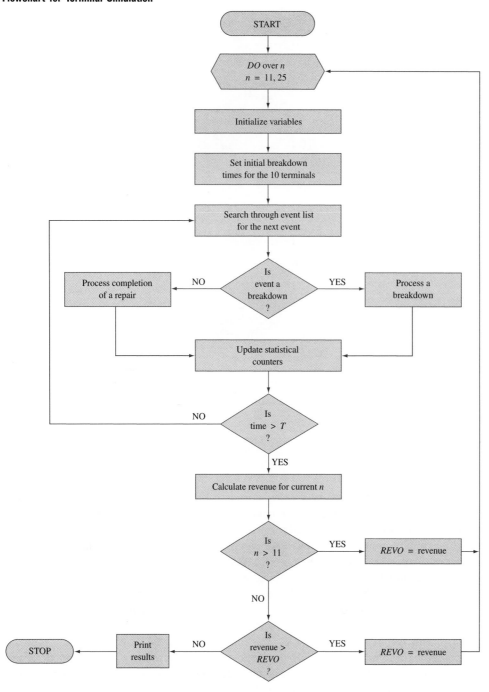

TABLE 14
Simulation Results for the Terminal System

	No. of Terminals (n)				
	11	12	13	14	15
EL_n	9.362	9.641	9.801	9.878	9.931
ED_n	0.638	0.359	0.199	0.122	0.069
$EL_n - EL_{11}$	—	0.279	0.439	0.516	0.569
Increase in revenue	—	$279	$439	$516	$569
Cost of terminals	—	$75	$150	$225	$300
Net revenue	—	$204	$289	$291	$269

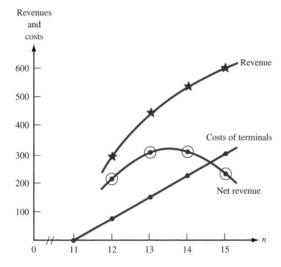

FIGURE 20
Revenue and Costs for
Terminal System

9.7 Statistical Analysis in Simulations

As previously mentioned, output data from simulation always exhibit random variability, since random variables are input to the simulation model. For example, if we run the same simulation twice, each time with a different sequence of random numbers, the statistics generated for the two simulations will almost certainly have different values. Because of this, we must utilize statistical methods to analyze output from simulations. If the performance of the system is measured by a parameter (say, θ), then our objective in the simulation will be to develop an estimate $\hat{\theta}$ of θ, and determine the accuracy of the estimator $\hat{\theta}$. We measure this accuracy by the standard deviation (also called the *standard error*) of $\hat{\theta}$. The overall measure of variability is generally stated in the form of a confidence interval at a given level of confidence. Thus, the purpose of the statistical analysis is to estimate this confidence interval.

Determination of the confidence intervals in simulation is complicated by the fact that output data are rarely, if ever, independent. That is, the data are autocorrelated. For example, in a queuing simulation, the waiting time of a customer often depends on the prior customers. Similarly, in an inventory simulation, the models are usually set up such that the beginning inventory on a given day is the ending inventory from the previous day, thereby creating a correlation. This means that the classical methods of statistics, which assume independence, are not directly applicable to the analysis of simulation output data. Thus, we must modify the statistical methods to make proper inferences from simulation data.

In addition to the problem of autocorrelation, we may have a second problem, in that the specification of the initial conditions of the system at time 0 may influence the output data. For example, suppose that in the queuing simulation from Section 9.2, the arrival and the service distributions are such that the average waiting time per customer exceeds 15 minutes. In other words, the system is heavily congested. If we were to begin this simulation with no one in the system, the first few customers would have either zero or very small waiting times. These initial waiting times are highly dependent on the starting conditions and may therefore not be representative of the steady-state behavior of the system. This initial period of time before a simulation reaches steady state is called the **transient period** or **warmup period.**

There are two ways of overcoming the problems associated with the transient period. The first approach is to use a set of initial conditions that are representative of the system in steady state. However, in many simulations, it may be difficult to set such initial conditions. This is particularly true of queuing simulations. The alternative approach is to let the simulation run for a while and discard the initial part of the simulation. With this approach, we are assuming that the initial part of the simulation warms the model up to an equilibrium state. Since we do not collect any statistics during the warmup stage, we can reduce much of the initialization bias. Unfortunately, there are no easy ways to assess how much initial data to delete to reduce the initialization bias to negligible levels. Since each simulation model is different, it is up to the analyst to determine when the transient period ends. Although this is a difficult process, there are some general guidelines one can use. For these and other details of this topic, see Law and Kelton (1991).

Simulation Types

For the purpose of analyzing output data, we generally categorize simulations into one of two types: terminating simulations and steady-state simulations. A **terminating simulation** is one that runs for a duration of time T_E, where E is a specified event (or events) that stops the simulation. The event E may be a specified time, in which case the simulation runs for a fixed amount of time. Or, if it is a specified condition, the length of the simulation will be a random variable. A **steady-state simulation** is one that runs over a long period of time; that is, the length of the simulation goes to "infinity."

Often, the type of model determines which type of output analysis is appropriate for a particular simulation. For example, in the simulation of a bank, we would most likely use a terminating simulation, since the bank physically closes every evening, giving us an appropriate terminating event. When simulating a computer system, a steady-state simulation may be more appropriate, since most large computer systems do not shut down except in cases of breakdowns or maintenance. However, the system or model may not always be the best indicator of which simulation would be the most appropriate. It is quite possible to use the terminating simulation approach for systems more suited to steady-state simulations, and vice versa. In this section, we provide a detailed description of the statistical analysis associated with terminating simulations. The analysis for steady-state simulations is much more involved. For details of the latter, see Banks and Carson (1984) or Law and Kelton (1991).

Suppose we make n independent replications using a terminating simulation approach. If each of the n simulations is started with the same initial conditions and is executed using a different sequence of random numbers, then each simulation can be treated as an independent replication. For simplicity, we assume that there is only a single measure of performance, represented by the variable X. Thus, X_j is the estimator of the measure of performance from the jth replication. Then, given the conditions of the replications, the sequence $X_1, X_2, \ldots, X_n$ will be iid random variables. With these iid random variables, we can use classical statistical analysis to construct a $100(1 - \alpha)\%$ confidence interval for $\theta = E(X)$ as follows:

$$\bar{X} \pm t_{(\alpha/2, n-1)} \sqrt{\frac{S^2}{n}}$$

where

$$\bar{X} = \sum_{i=1}^{n} \frac{X_i}{n}$$

$$S^2 = \sum_{i=1}^{n} \frac{(X_i - \bar{X})^2}{n - 1}$$

and $t_{(\alpha, n-1)}$ is the number such that for a t-distribution with $n - 1$ degrees of freedom,

$$P(t_{n-1} \geq t_{(\alpha, n-1)}) = \alpha$$

(see Table 13 in Chapter 15). This probability can also be computed in Excel with the formula

$$=TINV(2*alpha, degrees\ of\ freedom)$$

The overall mean $\bar{X}$ is simply the average of the X-values computed over the n samples and can be used as the best estimate of the measure of performance. The quantity S^2 is the sample variance.

To illustrate the terminating simulation approach, we use an example from the Cabot, Inc. case. For this illustration, we assume there are 11 terminals in the system, and we perform only 10 independent terminating runs of the simulation model. The terminating event, E, is a fixed time. That is, all 10 simulations are run for the same length of time. The results from these runs are shown in Table 15. The overall average for these 10 runs

TABLE 15

Sample Averages of Number of OnLine Terminals from 10 Replications

Run Number	x_j
1	9.252
2	9.273
3	9.413
4	9.198
5	9.532
6	9.355
7	9.155
8	9.558
9	9.310
10	9.269

for the expected number of terminals online turns out to be 9.331. (Compare this average with the result in Table 14 of 9.362, which was obtained using the steady-state approach.) If we now calculate the sample variance, we find $S^2 = 0.018$. Since $t_{(.025,9)} = 2.26$, we obtain

$$9.331 \pm 2.26 \sqrt{\frac{0.0180}{10}} = 9.331 \pm 0.096$$

as the 95% confidence interval for this sample.

We could have also computed $t_{(.025,9)}$ in Excel with the formula

$$=\text{TINV}(.05,9)$$

This yields 2.26, which is consistent with the table in Chapter 15.

The length of the confidence interval will, of course, depend on how good our sample results are. If this confidence interval is unacceptable, we can reduce its length by either increasing the number of terminating replications or the length of each simulation. For example, if we increase the number of runs from 10 to 20, we improve the results on two fronts. First, the overall average (9.359) approaches the result from the steady-state simulation; second, the confidence interval length decreases from 0.192 to 0.058. As we saw in this example, the terminating simulation approach offers a relatively easy method for analyzing output data. However, it must be emphasized that other methods for analyzing simulation data may be more efficient for a given problem. For a detailed treatment of this topic, see Banks and Carson (1984) or Law and Kelton (1991).

9.8 Simulation Languages

One of the most important aspects of a simulation study is the computer programming. Writing the computer code for a complex simulation model is often a difficult and arduous task. Because of this, several special-purpose computer simulation languages have been developed to simplify the programming. In this section, we describe several of the best-known and most readily available simulation languages, including GPSS, GASP IV, and SLAM.

Most simulation languages use one of two different modeling approaches or orientations: event scheduling or process interaction. As we have seen, in the event-scheduling approach, we model the system by identifying its characteristic events and writing routines to describe the state changes that take place at the time of each event. The simulation evolves over time by updating the clock to the next scheduled event and making whatever changes are necessary to the system and the statistics by executing the routines. In the process-interaction approach, we model the system as a series of activities that an entity (or a customer) must undertake as it passes through the system. For example, in a queuing simulation, the activities for an entity consist of arriving, waiting in line, getting service, and departing from the system. Thus, using the process-interaction approach, we model these activities instead of events. When programming in a general-purpose language such as FORTRAN or BASIC, we generally use the event-scheduling approach. GPSS uses the process-interaction approach. SLAM allows the modeler to use either approach or even a mixture of the two, whichever is the most appropriate for the model being analyzed.

Of the general-purpose languages, FORTRAN is the most commonly used in simulation. In fact, several simulation languages, including GASP IV and SLAM, use a FORTRAN base. Generally, simulation programs in FORTRAN are written as a series of subroutines, one for each major function of the simulation process. This is particularly true of the FORTRAN-based simulation languages. For example, in GASP IV, there are approximately 30 FORTRAN subroutines and functions. These include a time-advance routine, random variate generation routines, routines to manage the future events list, routines to collect statistics, and so on. To use GASP IV, we must provide a main program, an initialization routine, and the event routines. For the rest of the program, we use the GASP routines. Because of these prewritten routines, GASP IV provides a great deal of programming flexibility. For more details of this language, see Pritsker (1974).

GPSS, in contrast to GASP, is a highly structured special-purpose language. It was developed by IBM. GPSS does not require writing a program in the usual sense. The language is made up of about 40 standard statements or blocks. Building a GPSS model then consists of combining these sets of blocks into a flow diagram so that it represents the path an entity takes as it passes through the system. For example, for a single-server queuing system, the statements are of the form GENERATE (arrive in the system), QUEUE (join the waiting line), DEPART (leave the queue to enter service), ADVANCE (advance the clock to account for the service time), RELEASE (release the service facility at the end of service), and TERMINATE (leave the system). The simulation program is then compiled from these statements of the flow diagram. GPSS was designed for relatively easy simulation of queuing systems. However, because of its structure, it is not as flexible as GASP IV, especially for the nonqueuing type of simulations. For a more detailed description of GPSS, see Schriber (1974).

SLAM was developed by Pritsker and Pegden (1979). It allows us to develop simulation models as network models, discrete-event models, continuous models, or any combination of these. The discrete-event orientation is an extension of GASP IV. The network representation can be thought of as a pictorial representation of a system through which entities flow. In this respect, the structure of SLAM is similar to that of GPSS. Once the network model of the system has been developed, it is translated into a set of SLAM program statements for execution on the computer.

In Chapter 10, we will show how to use the powerful, user-friendly Process Model package that is included on this book's CD-ROM.

The decision of which language to use is one of the most important that a modeler or an analyst must make in performing a simulation study. The simulation languages offer several advantages. The most important of these is that the special-purpose languages provide a natural framework for simulation modeling and most of the features needed in programming a simulation model. However, this must be balanced against the fact that the general-purpose languages allow greater programming flexibility, and that languages like FORTRAN and BASIC are much more widely used and available.

9.9 The Simulation Process

In this chapter, we have considered several simulation models and presented a number of key simulation concepts. We now discuss the process for a complete simulation study and present a systematic approach of carrying out a simulation. A simulation study normally consists of several distinct stages. These are presented in Figure 21. However, not all simulation studies consist of all these stages or follow the order stated here. On the other hand, there may even be considerable overlap between some of these stages.

The initial stage of any scientific study, including a simulation project, requires an explicit *statement of the objectives* of the study. This should include the questions to be answered, the hypothesis to be tested, and the alternatives to be considered. Without a clear understanding and description of the problem, the chances of successful completion and implementation are greatly diminished. Also in this step, we address issues such as the

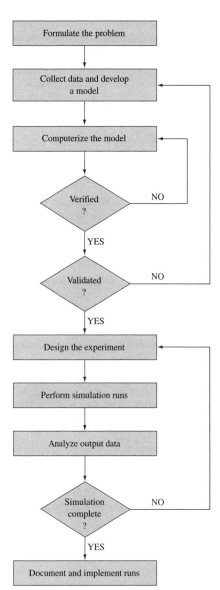

FIGURE 21
Steps in a Simulation Study

performance criteria, the model parameters, and the identification and definition of the state variables. It is, of course, very likely that the initial formulation of the problem will undergo many modifications as the study proceeds and as we learn more about the situation being studied. Nevertheless, a clear initial statement of the objectives is essential.

The next stage is the *development* of the model and the *collection of data*. The development of the model is probably the most difficult and critical part of a simulation study. Here, we try to represent the essential features of the systems under study by mathematical or logical relations. There are few firm rules to guide an analyst on how to go about this process. In many ways, this is as much an art as a science. However, most experts agree that the best approach is to start with a simple model and make it more detailed and complex as one learns more about the system.

Having developed the model, we next put it into a form in which it can be analyzed on the computer. This usually involves *developing a computer program* for the model. One of the key decisions here is the choice of the language. As noted earlier, the special-purpose languages require less programming than the general-purpose languages but are less flexible and tend to require longer computer running times. In either case, the programming part of the study is likely to be a time-consuming process, since simulation programs tend to be long and complex. Once the program has been developed and debugged, we determine whether the program is working properly. In other words, is the program doing what it is supposed to do? This process is called the *verification* step and is usually difficult, since for most simulations, we will not have any results with which to compare the computer output.

If we are satisfied with the program, we now move to the *validation* stage. This is another critical part of a simulation study. In this step, we validate the model to determine whether it realistically represents the system being analyzed and whether the results from the model will be reliable. As with the verification stage, this is generally a difficult process. Each model presents a different challenge. However, there are some general guidelines that one can follow. For more on these procedures, see Law and Kelton (1991) or Shannon (1979). If we are satisfied at this stage with the performance of the model, we can use the model to conduct the experiments to answer the questions at hand. The data generated by the simulation experiments must be collected, processed, and analyzed. The results are analyzed not only as the solution to the model but also in terms of statistical reliability and validity. Finally, after the results are processed and analyzed, a decision must be made whether to perform any additional experiments.

The primary emphasis in this chapter has been on sampling procedures and model construction. As a result, many topics of the simulation process are either not covered or treated only briefly. However, these are important issues in simulation, and the reader interested in using simulation should consult Law and Kelton (1991), Shannon (1979), Banks and Carson (1984), or Ross (1996).

SUMMARY Introduction to Simulation

Simulation may be defined as a technique that imitates the operation of a real-world system as it evolves over a period of time. There are two types of simulation models: static and dynamic. A **static simulation model** represents a system at a particular point in time. A **dynamic simulation model** represents a system as it evolves over time. Simulations can be deterministic or stochastic. A **deterministic simulation** contains no random variables, whereas a **stochastic simulation** contains one or more random variables. Finally, simulations may be represented by either discrete or continuous models. A **discrete simulation** is one in which the state variables change only at discrete points in time. In a **continuous simulation,** the state variables change continuously over time. In this chapter, we have dealt only with discrete stochastic models. Such models are called **discrete-event simulation models.**

The Simulation Process

The simulation process consists of several distinct stages. Each study may be somewhat different, but in general, we use the following framework:

1 Formulate the problem.

2 Collect data and develop a model.

3 Computerize the model.

4 Verify the computer model.

5 Validate the simulation model.

6 Design the experiment.

7 Perform the simulation runs.

8 Document and implement.

Generating Random Variables

Random variables are represented using probability distributions. The procedure for generating random variables from given probability distributions is known as random variate generation or **Monte Carlo sampling.** The principle of sampling is based on the frequency interpretation of probability and requires a steady stream of random numbers. We generate random numbers for this procedure using congruential methods. The most commonly used of the congruential methods is the **linear congruential method.** Random numbers generated from a linear congruential generator use the following relation:

$$x_{i+1} = (ax_i + c) \text{ modulo } m \qquad (i = 0, 1, 2, \ldots)$$

This gives us the remainder from the division of $(ax_i + c)$ by m. The random numbers are delivered using the relation

$$R_i = \frac{x_i}{m} \qquad (i = 1, 2, 3, \ldots)$$

For discrete distributions, Monte Carlo sampling is achieved by allocating ranges of random numbers according to the probabilities in the distribution. For continuous distributions, we generate random variates using one of several algorithms, including the **inverse transformation method** and the **acceptance–rejection method.** The inverse transformation method requires a cdf in closed form and consists of the following steps.

Step 1 Given a probability density function $f(x)$, develop the cumulative distribution function as

$$F(x) = \int_{-\infty}^{x} f(t)dt$$

Step 2 Generate a random number r.

Step 3 Set $F(x) = r$ and solve for x. The variable x is then a random variate from the distribution whose pdf is given by $f(x)$.

The acceptance–rejection method requires the pdf to be defined over a finite interval. Thus, given a probability density function $f(x)$ over the interval $a \leq x \leq b$, we execute the acceptance–rejection algorithm as follows.

Step 1 Select a constant M such that M is the largest value of $f(x)$ over the interval $[a, b]$.

Step 2 Generate two random numbers, r_1 and r_2.

Step 3 Compute $x^* = a + (b - a)r_1$.

Step 4 Evaluate the function $f(x)$ at the point x^*. Let this be $f(x^*)$.

Step 5 If $r_2 < \frac{f(x^*)}{M}$, deliver x^* as a random variate. Otherwise, reject x^* and go back to step 2.

Between these two methods, it is possible to generate random variates from almost all of the commonly used distributions. The one exception is the normal distribution. For the normal distribution, we generate random variates directly by transforming the random numbers r_1 and r_2 into standardized normal variates, $\mathbf{Z}_1$ and $\mathbf{Z}_2$, using the relations

$$\mathbf{Z}_1 = (-2 \ln r_1)^{1/2} \sin 2\pi r_2$$
$$\mathbf{Z}_2 = (-2 \ln r_1)^{1/2} \cos 2\pi r_2$$

Types of Simulations

In discrete-event simulations, we generally simulate using the next-event time-advance approach. In this procedure, the simulation evolves over time by updating the clock to the next scheduled event and taking whatever actions are necessary for each event. The events are scheduled by generating random variates from probability distributions. Data from a simulation can be analyzed using either a terminating simulation approach or a steady-state simulation approach. In terminating simulations, we make n independent replications of the model, using the same initial conditions but running each replication with a different sequence of random numbers. If the measure of performance is represented by the variable X, this approach gives us the estimators $X_1, X_2, \ldots, X_n$ from the n replications. These estimators are used to develop a $100(1 - \alpha)\%$ confidence interval as

$$\bar{X} \pm t_{(\alpha/2, n-1)} \sqrt{\frac{S^2}{n}}$$

for a fixed value of n.

Simulation gives us the flexibility to study systems that are too complex for analytical methods. However, it must be put into proper perspective. Simulation models are time consuming and costly to construct and run. Additionally, the results may not be very precise and are often hard to validate. Simulation can be a powerful tool, but only if it is used properly.

REVIEW PROBLEMS

Group A

1 Use the linear congruential generator to obtain a sequence of 10 random numbers, given that $a = 17$, $c = 43$, $m = 100$, and $x_0 = 31$.

2 A news vendor sells newspapers and tries to maximize profits. The number of papers sold each day is a random variable. However, analysis of the past month's data shows the distribution of daily demand in Table 16. A paper costs the vendor 20¢. The vendor sells the paper for 30¢. Any unsold papers are returned to the publisher for a credit of 10¢. Any unsatisfied demand is estimated to cost 10¢ in goodwill and lost profit. If the policy is to order a quantity equal to the preceding day's demand, determine the average daily profit of the news vendor by simulating this system. Assume that the demand for day 0 is equal to 32.

3 An airport hotel has 100 rooms. On any given night, it takes up to 105 reservations, because of the possibility of no-shows. Past records indicate that the number of daily reservations is uniformly distributed over the integer range [96, 105]. That is, each integer number in this range has an

TABLE 16

Demand per Day	Probability
30	.05
31	.15
32	.22
33	.38
34	.14
35	.06

equal probability, .1, of showing up. The no-shows are represented by the distribution in Table 17. Develop a simulation model to find the following measures of performance of this booking system: the expected number of rooms used per night and the percentage of nights when more than 100 rooms are claimed.

TABLE 17	
Number of No-Shows	Probability
0	.10
1	.20
2	.25
3	.30
4	.10
5	.05

TABLE 18	
Interarrival Time (minutes)	Probability
1	.20
2	.25
3	.40
4	.10
5	.05

TABLE 21	
Length of Appointment (minutes)	Probability
24	.10
27	.20
30	.40
33	.15
36	.10
39	.05

TABLE 22	
Daily Demand (units)	Probability
12	.05
13	.15
14	.25
15	.35
16	.15
17	.05

4 The university library has one copying machine for the students to use. Students arrive at the machine with the distribution of interarrival times shown in Table 18. The time to make a copy is uniformly distributed over the range [16, 25] seconds. Analysis of past data has shown that the number of copies a student makes during a visit has the distribution in Table 19. The librarian feels that under the present system, the lines in front of the copying machine are too long and that the time a student spends in the system (waiting time + service time) is excessive. Develop a simulation model to estimate the average length of the waiting line and the expected waiting time in the system.

5 A salesperson in a large bicycle shop is paid a bonus if he sells more than 4 bicycles a day. The probability of selling more than 4 bicycles a day is only .40. If the number of bicycles sold is greater than 4, the distribution of sales is as shown in Table 20. The shop has four different models of bicycles. The amount of the bonus paid out varies by type. The bonus for model A is $10; 40% of the bicycles sold are of this type. Model B accounts for 35% of the sales and pays a bonus of $15. Model C has a bonus rating of $20 and makes up 20% of the sales. Finally, model D pays a bonus of $25 for each sale but accounts for only 5% of the sales. Develop a simulation model to calculate the bonus a salesperson can expect in a day.

6 A heart specialist schedules 16 patients each day, 1 every 30 minutes, starting at 9 A.M. Patients are expected to arrive for their appointments at the scheduled times. However, past experience shows that 10% of all patients arrive 15 minutes early, 25% arrive 5 minutes early, 50% arrive exactly on time, 10% arrive 10 minutes late, and 5% arrive 15 minutes late. The time the specialist spends with a patient varies, depending on the type of problem. Analysis of past data shows that the length of an appointment has the

distribution in Table 21. Develop a simulation model to calculate the average length of the doctor's day.

Group B

7 Suppose we are considering the selection of the reorder point, R, of a (Q, R) inventory policy. With this policy, we order up to Q when the inventory level falls to R or less. The probability distribution of daily demand is given in Table 22. The lead time is also a random variable and has the distribution in Table 23. We assume that the "order up to" quantity for each order stays the same at 100. Our interest here is to determine the value of the reorder point, R, that minimizes the total variable inventory cost. This variable cost is the sum of the expected inventory carrying cost, the expected ordering cost, and the expected stockout cost. All stockouts are backlogged. That is, a customer waits until an item is available. Inventory carrying cost is estimated to be 20¢/unit/day and is charged on the units in inventory at the end of a day. A stockout costs $1 for every unit short. The cost of ordering is $10 per order. Orders arrive at the beginning of a day. Develop a simulation model to simulate this inventory system to find the best value of R.

8 A large car dealership in Bloomington, Indiana, employs five salespeople. All salespeople work on commission; they are paid a percentage of the profits from the cars they sell. The dealership has three types of cars: luxury, midsize, and subcompact. Data from the past few years show that the car sales per week per salesperson have the distribution in Table 24. If the car sold is a subcompact, a salesperson is given a commission of $250. For a midsize car, the commission is either $400 or $500, depending on the model sold. On the

TABLE 19	
Number of Copies	Probability
6	.20
7	.25
8	.35
9	.15
10	.05

TABLE 20	
No. of Bicycles Sold	Probability
5	.35
6	.45
7	.15
8	.05

TABLE 23	
Lead Time (days)	Probability
1	.20
2	.30
3	.35
4	.15

TABLE 24	
No. of Cars Sold	Probability
0	.10
1	.15
2	.20
3	.25
4	.20
5	.10

TABLE 25

Type of Car Sold	Probability
Subcompact	.40
Midsize	.35
Luxury	.25

TABLE 26

Length of Queue (q)	Probability of Reneging
$6 \leq q \leq 8$	.20
$9 \leq q \leq 10$	.40
$11 \leq q \leq 14$	.60
$q > 14$	.80

midsize cars, a commission of $400 is paid out 40% of the time, and $500 is paid out the other 60% of the time. For a luxury car, commission is paid out according to three separate rates: $1,000 with a probability of 35%, $1,500 with a probability of 40%, and $2,000 with a probability of 25%. If the distribution of type of cars sold is as shown in Table 25, what is the average commission for a salesperson in a week?

9 Consider a bank with 4 tellers. Customers arrive at an exponential rate of 60 per hour. A customer goes directly into service if a teller is idle. Otherwise, the arrival joins a waiting line. There is only one waiting line for all the tellers. If an arrival finds the line too long, he or she may decide to leave immediately (reneging). The probability of a customer reneging is shown in Table 26. If a customer joins the waiting line, we assume that he or she will stay in the system until served. Each teller serves at the same service rate. Service times are uniformly distributed over the range [3, 5]. Develop a simulation model to find the following measures of performance for this system: (1) the expected time a customer spends in the system, (2) the percentage of customers who renege, and (3) the percentage of idle time for each teller.

10 Jobs arrive at a workshop, which has two work centers (A and B) in series, at an exponential rate of 5 per hour. Each job requires processing at both these work centers, first on A and then on B. Jobs waiting to be processed at each center can wait in line; the line in front of work center A has unlimited space, and the line in front of center B has space for only 4 jobs at a time. If this space reaches its capacity, jobs cannot leave center A. In other words, center A stops processing until space becomes available in front of B. The processing time for a job at center A is uniformly distributed over the range [6, 10]. The processing time for a job at center B is represented by the following triangular distribution:

$$f(x) = \begin{cases} \frac{1}{4}(x - 1) & 1 \leq x \leq 3 \\ \frac{1}{4}(5 - x) & 3 \leq x \leq 5 \end{cases}$$

Develop a simulation model of this system to determine the following measures of performance: (1) the expected number of jobs in the workshop at any given time, (2) the percentage of time center A is shut down because of shortage of queuing space in front of center B, and (3) the expected completion time of a job.

REFERENCES

There are several outstanding books on simulation. For a beginning book, we recommend Watson (1981); for an intermediate approach, Banks and Carson (1984); and for a more advanced treatment, Law and Kelton (1991).

Banks, J., and J. Carson. *Discrete-Event System Simulation.* Englewood Cliffs, N.J.: Prentice Hall, 1984.

Box, G., and M. Muller. "A Note on the Generation of Random Normal Deviates," *Annals of Mathematical Statistics* 29(1958):610–611.

Fishman, G. *Principles of Discrete Event Simulation.* New York: Wiley, 1978.

Fishman, G. *Monte Carlo: Concepts, Algorithms and Applications.* Berlin: Springer-Verlag, 1996.

Kelton, D., Sadowski, R., and Sadowski, S. *Simulation with Arena.* New York: McGraw-Hill, 2001.

Knuth, D.W. *The Art of Computer Programming: II. Seminumerical Algorithms.* Reading, Mass.: Addison-Wesley, 1998.

Law, A.M., and W. Kelton. *Simulation Modeling and Analysis.* New York: McGraw-Hill, 1991.

Pritsker, A. *The GASP IV Simulation Language.* New York: Wiley, 1974.

Pritsker, A., and C. Pegden. *Introduction to Simulation and SLAM.* New York: Wiley, 1979.

Ross, S. *Simulation.* San Francisco: Academic Press, 1996.

Schmidt, J.W., and R.E. Taylor. *Simulation and Analysis of Industrial Systems.* Homewood, Ill.: Irwin, 1970.

Schriber, T. *Simulation Using GPSS.* New York: Wiley, 1974.

Shannon, R. E. *Systems Simulation: The Art and Science.* Englewood Cliffs, N.J.: Prentice Hall, 1979.

Watson, H. *Computer Simulation in Business.* New York: Wiley, 1981.

Kelly, J. "A New Interpretation of Information Rate," *Bell System Technical Journal* 35(1956):917–926.

Marcus, A. "The Magellan Fund and Market Efficiency," *Journal of Portfolio Management* Fall (1990):85–88.

Morrison, D., and R. Wheat. "Pulling the Goalie Revisited," *Interfaces* 16(no. 6, 1984):28–34.

10

Simulation with Process Model

In Chapter 9, we learned how to build simulation models of many different situations. In this chapter, we will explain how the powerful, user-friendly simulation package Process Model can be used to simulate queuing systems.

10.1 Simulating an *M/M/*1 Queuing System

After installing Process Model from the book's CD-ROM, you can start Process Model by selecting Start Programs Process Model 4. You will see the screen shown in Figure 1, where some key icons have been labeled.

MM1.igx

It is simple to simulate an *M/M/*1 queuing system having $\lambda = 10$ arrivals/hour and $\mu = 15$ customers/hour. See file MM1.igx. Assume that these are calls for directory assistance.

Step 1 Click on one of the arrival icons (a person or a phone) and drag the icon to the blank part of the screen (called the Layout portion). We have chosen to use the phone icon. Your screen should look like Figure 2.

Step 2 Select the Process rectangle and drag it right over the arrival icon. Click on it and drag it to the right. You will now have a double-arrowed connection between the arrival icon and the Process rectangle. The double-arrowed icon indicates the arrival of entities into the system. Later we will tell Process Model that interarrival times are exponential with mean 6 minutes. After Taking Calls is typed within the Process rectangle, the Layout window looks as shown in Figure 3.

Step 3 Choose one of the server icons to represent a telephone operator (say, the person with the computer) and drag this icon to the Layout window above the Take Calls Process rectangle. Then type the word "operator" to indicate a phone operator. Next, click on the Connector Line tool in the Toolbox and place the cursor over the operator. We then click once and drag a connection down to the Take Calls activity. This indicates that the operator can take calls. The Layout window should now look as shown in Figure 4.

Step 4 Next, tell Process Model to make interarrival times exponential. Process Model works off the mean interarrival time or service time, not the arrival or service rates. Process Model supports many distributions, including the triangular, normal, and Erlang random variables. For now, we will use the exponential distribution. Since the average time between arrivals is 6 minutes, we will model the interarrival times as E(6). (E stands for exponential.) To enter the interarrival time distribution, click on the double arrow connecting Call to Take Calls and fill in the dialog box as shown in Figure 5. Entering Periodic and E(6) ensures that interarrival times will be generated over and over as independent exponential random variables with mean 6.

FIGURE 1

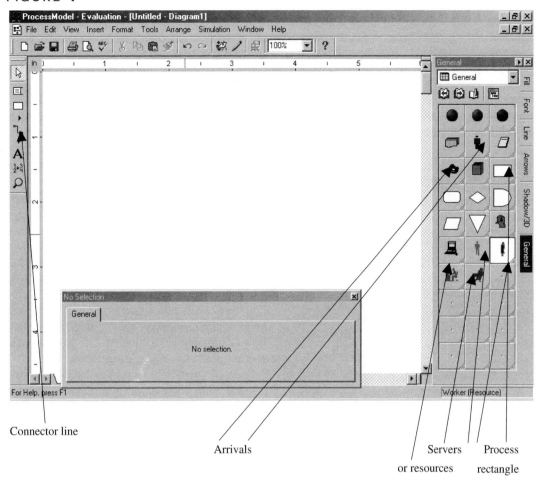

Connector line

Arrivals

Servers
or resources

Process
rectangle

Step 5 We now need to tell Process Model that service times are exponential with mean 4. To do this, click on the Take Calls Process rectangle and fill in the dialog box as shown in Figure 6.

Step 6 We have now completed the model setup. Select File Save As and save the model. (All models have the suffix .igx.)

Step 7 To run the simulation, select Simulation and then Options and fill in the dialog box as shown in Figure 7.

We have chosen to run the system for 4,000 hours. Choosing a Warmup length of 1 hour, the first hour of running the simulation will not be used in the collection of statistics. To start the simulation, choose Simulation Save and Simulation. As the simulation progresses, telephone calls moving through the flowchart illustrate the flow of calls through the process. Resources or servers will show a green light when the resource is being utilized and a blue light when idle. Counters above and to the left of each activity represent the number of calls waiting to be processed. The speed of the simulation can be

FIGURE 2

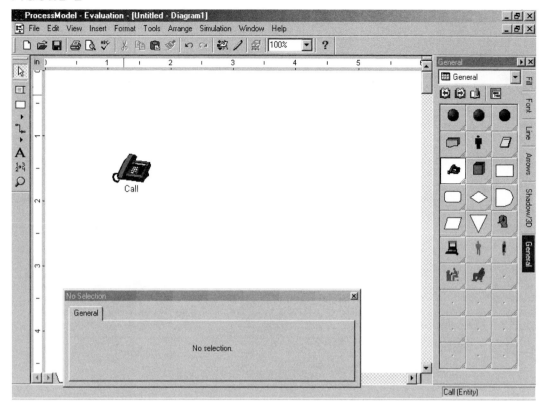

FIGURE 3

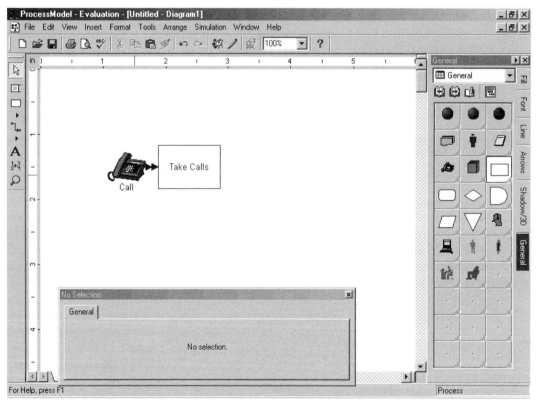

FIGURE 4

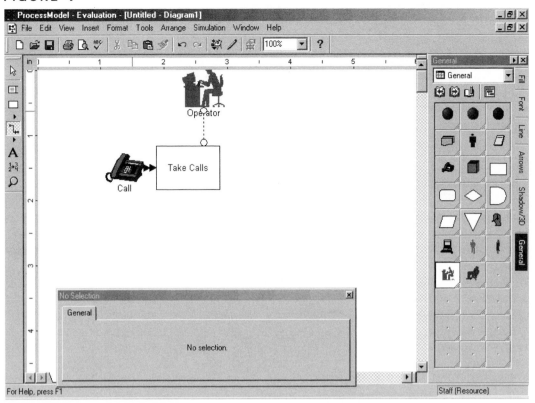

FIGURE 5

FIGURE 6

controlled by moving the Speed Control bar, left for slower and right for faster. By choosing Simulation End Simulation, you may stop the simulation at any time. During the simulation, an on-screen scoreboard tracks the following quantities:

■ Quantity Processed (total number of units to leave the system)

■ Cycle Time (average time a unit spends in the system)

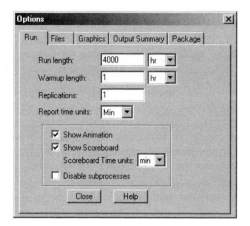

FIGURE 7

- Value Added Time (time a unit spends in service)
- Cost Per Unit (if costs are associated with the resources, the cost incurred per unit serviced is computed)

After completing the simulation, you are asked if you want to view the output. If so, you will see an output similar to Figure 8, and the output may be saved as a text file. Figure 8 includes comments (in **boldface**) to explain the key portions of the output.

If we treat this output as representative of the system's steady state, we have the following parameter estimates:

- $\Pi_0 = .3363$
- $\Pi_0 + \Pi_1 = .5634$
- $W_s = 4.01$ minutes
- $W_q = 7.71$ minutes
- $W = 11.72$ minutes

For an $M/M/1$ system, we can compute the steady-state values of these quantities exactly. Since $\rho = 10/15 = .667$, we find from Equation (24) of Chapter 8 that $\Pi_0 = 1 - .667 = .333$. From (25) of Chapter 8, we find that $\Pi_1 = .667(1 - .667) = .222$. Thus, $\Pi_0 + \Pi_1 = .555$. Clearly $W_s = 4$ minutes. From Formula (31) of Chapter 8, $W = \frac{1}{15 - 10} = .2$ hour $= 12$ minutes. Then $W_q = 12 - 4 = 8$ minutes. Note that the simulation yields very close agreement with the steady-state estimates.

10.2 Simulating an *M/M/2* System

MM2.igx

Let us modify the previous example by changing the number of operators to 2 and ensuring that up to 2 operators can be working on calls at the same time. See file MM2.igx. To change the number of operators to 2, click on the resource and fill in the dialog box as shown in Figure 9.

To ensure that two operators can work on calls at the same time click on the Process rectangle Take Calls and modify it as in Figure 10.

After saving this file as MM2.igx and running it for 1,000 hours, we obtain the output shown in Figure 11. (Boldface comments explain key portions.)

```
-------------------------------------------------------------------------
---------
General Report
Output from C:\Program Files\ProcessModel 4\mm1.mod
Date: Aug/13/2002   Time: 07:50:42 AM
-------------------------------------------------------------------------
---------
Scenario       : Normal Run
Replication    : 1 of 1
Warmup Time    : 1 hr
Simulation Time : 3986.90
-------------------------------------------------------------------------
---------


ACTIVITIES

                                            Average
Activity        Scheduled        Total    Minutes   Average
Maximum    Current
Name              Hours  Capacity  Entries  Per Entry  Contents
Contents  Contents  % Util
------------  ---------  --------  -------  ---------  --------  -----
---  --------  ------
Take Call in Q   3985.90      999    39581       7.71      1.27
21         1     0.13
Take Call        3985.90        1    39581       4.01      0.66
1         1    66.37
```

In the nearly 3,986 hours for which data was collected, almost exactly 10
arrivals per hour occurred. The total time spent by a call in queue
(waiting) averaged 7.71 minutes, and total service time averaged 4.01
minutes.

ACTIVITY STATES BY PERCENTAGE (Multiple Capacity)

```
                                      %
Activity        Scheduled     %  Partially    %
Name              Hours   Empty  Occupied  Full
------------  ---------  -----  ---------  ----
Take Call inQ    3985.90  56.34     43.66  0.00
```

The queue for calls was empty 56.34% of the time.

ACTIVITY STATES BY PERCENTAGE (Single Capacity)

```
Activity   Scheduled      %      %       %        %
Name         Hours   Operation  Idle  Waiting  Blocked
---------  ---------  ---------  -----  -------  -------
Take Call   3985.90      66.37  33.63    0.00     0.00
```

FIGURE 8   The operator was idle 33.63% of the time.

RESOURCES

Resource Name	Units	Scheduled Hours	Number Of Times Used	Average Minutes Per Usage	% Util
Staff	1	3985.90	39581	4.01	66.37

Staff was busy 66.37% of the time, and average staff usage per call processed was 4.01 minutes.

RESOURCE STATES BY PERCENTAGE

Resource Name	Scheduled Hours	% In Use	% Idle	% Down
Staff	3985.90	66.37	33.63	0.00

Staff was busy 66.37% of the time.

ENTITY SUMMARY (Times in Scoreboard time units)

Entity Name	Qty Processed	Average Cycle Time (Minutes)	Average VA Time (Minutes)	Average Cost
Call	39580	11.72	4.01	1.33

FIGURE 8
(Continued) **Each call spent an average of 11.72 minutes in the system.**

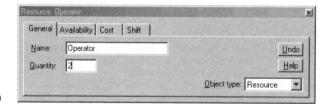

FIGURE 9

FIGURE 10

```
----------------------------------------------------------------------
---------
General Report
Output from C:\Program Files\ProcessModel 4\second.mod
Date: Aug/13/2002   Time: 03:08:49 PM
----------------------------------------------------------------------
---------
Scenario       : Normal Run
Replication    : 1 of 1
Warmup Time    : 1 hr
Simulation Time : 4001 hr
----------------------------------------------------------------------
---------


ACTIVITIES

                                          Average
Activity         Scheduled       Total    Minutes   Average
Maximum    Current
Name              Hours  Capacity Entries Per Entry Contents
Contents   Contents  % Util
--------------  --------- -------- ------- --------- -------- ----
----   -------- ------
Take Calls inQ    4000      999    39968     0.51     0.08
7        0   0.01
```

We see that an average of almost exactly 10 arrivals per hour have been observed.

```
Take Calls        4000       2     39970     4.03     0.67
2        0  33.60


ACTIVITY STATES BY PERCENTAGE (Multiple Capacity)

                                    %
Activity         Scheduled     %  Partially    %
Name              Hours    Empty  Occupied   Full
--------------  --------- ----- --------- -----
Take Calls inQ    4000    94.34     5.66    0.00
Take Calls        4000    49.58    33.64   16.78
```

We see that 5.66% of the time, people are waiting. Note from the *M/M/s* template below that probability of people waiting = probability >2 people present = 1 - .5 - .333 - .111 = .0566.

```
RESOURCES

                                Average
                        Number  Minutes
Resource        Scheduled Of Times  Per
Name      Units   Hours   Used   Usage   % Util
--------- ----- --------- -------- ------- ------
Operator.1   1     4000    19971    4.03   33.60
Operator.2   1     4000    19999    4.03   33.60
Operator     2     8000    39970    4.03   33.60
```

Each operator's mean service time is 4.03 minutes (compared to the 4 minutes we input).

FIGURE 11

```
RESOURCE STATES BY PERCENTAGE

Resource    Scheduled      %       %       %
Name          Hours     In Use   Idle    Down
----------  ---------   ------   -----   ----
Operator.1     4000     33.60    66.40   0.00
Operator.2     4000     33.60    66.40   0.00
Operator       8000     33.60    66.40   0.00
```

**Each operator was busy 33.60% of the time. Note from the steady-state
probabilities in the *M/M/s* template that the probability that a server is busy is
.5*(Prob. 1 person present) + Prob(>=2 people present) = .5*(.333) + .167 = .333.**
ENTITY SUMMARY (Times in Scoreboard time units)

```
                         Average    Average
                          Cycle        VA
Entity           Qty      Time        Time    Average
Name        Processed  (Minutes)   (Minutes)   Cost
------      ---------  ---------   ---------  -------
Call           39970      4.55        4.03      0.00
```

**The average time a call spends in the system is 4.55 minutes. From the *M/M/s*
spreadsheet, we find that *W* = 4.5 minutes.**

```
VARIABLES
                                     Average
Variable                  Total      Minutes   Minimum   Maximum   Current
Average
Name                      Changes   Per Change   Value     Value     Value
Value
------------------------  -------   ----------  -------   -------   ------- --
-----
Avg BVA Time Entity           1       0.00         0         0         0
0
Avg BVA Time Call         39971       6.00         0         0         0
0
```

	A	B	C	D	E	F	G
1	M/M/s/GD	LAMBDA?	MU?	s?	RO		
2		10	15	2	0.33333333		
3		L	LS	LQ	W	WS	WQ
4		0.75	0.66666667	0.08333333	0.075	0.066666667	0.008333333
5	STATE	P(j>=s)					
6		1	0.16666667				
7	P(Wq>t)	t?	P(W>t)				
8	9.57313E-60	6.70521931	3.1296E-44				
9							
10							
11							
12	STATE	LAMBDA(J)	MU(J)	CJ	PROB	#IN QUEUE	COLA*COLE
13	0	10	0	1	0.5	0	0
14	1	10	15	0.66666667	0.33333333	0	0.333333333
15	2	10	30	0.22222222	0.11111111	0	0.222222222

FIGURE 11
(Continued)

10.3 Simulating a Series System

In this section, we use Process Model to simulate a series queuing system. We will take
the case of the auto assembly line (Example 13 of Section 8.10).

EXAMPLE 1   Auto Assembly

The last two things that are done to a car before its manufacture is complete are installing
the engine and putting on the tires. An average of 54 cars per hour arrive requiring these

FIGURE **12**

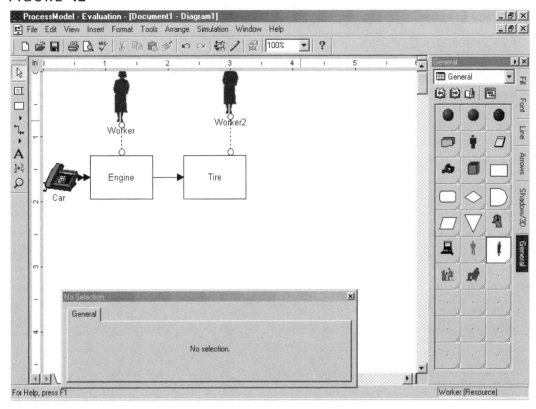

two tasks. One worker is available to install the engine and can service an average of 60 cars per hour. After the engine is installed, the car goes to the tire station and waits for its tires to be attached. Three workers serve at the tire station. Each works on one car at a time and can put tires on a car in an average of 3 minutes. Assume that interarrival times and service times are exponential. Simulate this system for 400 hours.

Solution

Carassembly.igx

See file Carassembly.igx. The key to creating a queuing network with Process Model is to build the diagram one service center at a time. We begin by creating the arrivals as in Section 10.1. Then we create the engine production center as in Section 10.1. Then we drag the Process rectangle over the Engine production center and pull it to the right to create the Tire production center. Of course, we must change the number of servers at the Tire production center to 3 (and also change the capacity of the tire operation to 3). We must enter service times of E(1) for the Engine production center and E(3) for the Tire production center. For the interarrival times, we must enter E(1.11), since there is an arrival an average of every 60/54 = 1.1 minutes. The flowchart looks like Figure 12.

Note that clicking on the arrow connecting the Engine and Tire rectangles results in the dialog box shown in Figure 13. This indicates that 100% of all cars completing Engine installation are sent to the Tire station. We also adjusted the Move time from 1 minute to 0 minute. The Move time indicates how many minutes are needed to move from the Engine to the Tire station. Suppose, for example 70% of the jobs completing Engine installation are sent to other stations (such as Final inspection), and 30% are sent on to the Tire station. We would model this by changing the percentage on the arrow joining Engine and Final inspection to 70%. The percentage going from Engine to Tire would automatically adjust to 30%. We will see how this works in the next example.

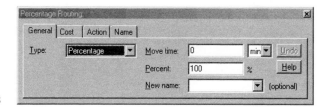

FIGURE 13

After running the simulation for 400 hours (with a 1-hour warmup period), we obtained the results shown in Figure 14 (explanatory comments in boldface).

```
-----------------------------------------------------------------------
---------
General Report
Output from C:\Program Files\ProcessModel 4\carassembly.mod
Date: Aug/15/2002   Time: 10:14:13 AM
-----------------------------------------------------------------------
---------
Scenario      : Normal Run
Replication   : 1 of 1
Warmup Time   : 1 hr
Simulation Time : 404.88
-----------------------------------------------------------------------
---------


ACTIVITIES

                                         Average
Activity    Scheduled          Total    Minutes   Average    Maximum
Current
Name              Hours  Capacity  Entries  Per Entry  Contents   Contents
Contents  % Util
----------  ---------  --------  -------  ---------  --------  --------
--------  ------
Engine inQ     403.88       999    21741      8.32      7.46        54
1    0.75
Engine         403.88         1    21741      1.00      0.89         1
1   89.82
Tire inQ       403.88      1000    21743      8.00      7.18        46
0    0.72
Tire           403.88         3    21746      3.02      2.71         3
3   90.45
```

An average of 21,741/403.88 = 53.83 arrivals per hour were observed.

```
ACTIVITY STATES BY PERCENTAGE (Multiple Capacity)

                            %
Activity    Scheduled    %   Partially    %
Name            Hours  Empty  Occupied   Full
----------  ---------  -----  ---------  -----
Engine inQ     403.88  19.24    80.76    0.00
Tire inQ       403.88  25.74    74.26    0.00
Tire           403.88   2.37    15.20   82.42
```

19% of the time, no cars are waiting for engine installation. 26% of the time, no cars are waiting for tire installation. 82.4% of the time, all tire installers are busy; 2.37% of the time, no tire installers are busy; and 15.2% of the time, some (but not all) tire installers are busy.

FIGURE 14

ACTIVITY STATES BY PERCENTAGE (Single Capacity)

Activity Name	Scheduled Hours	% Operation	% Idle	% Waiting	% Blocked
Engine	403.88	89.82	10.18	0.00	0.00

The engine installer is busy 89.8% of the time. (In the steady state, she should be busy 90% of the time according to $P_0 = r$.

RESOURCES

Resource Name	Units	Scheduled Hours	Number Of Times Used	Average Minutes Per Usage	% Util
Worker	1	403.88	21741	1.00	89.82
Worker2.1	1	403.88	7250	3.02	90.45
Worker2.2	1	403.88	7258	3.02	90.46
Worker2.3	1	403.88	7238	3.02	90.45
Worker2	3	1211.64	21746	3.02	90.45

Mean service time at the Engine station is 1 minute. At the Tire station, mean service time is 3.02 minutes.

RESOURCE STATES BY PERCENTAGE

Resource Name	Scheduled Hours	% In Use	% Idle	% Down
Worker	403.88	89.82	10.18	0.00
Worker2.1	403.88	90.45	9.55	0.00
Worker2.2	403.88	90.46	9.54	0.00
Worker2.3	403.88	90.45	9.55	0.00
Worker2	1211.64	90.45	9.55	0.00

Tire workers appear to be busy around 90.5% of the time, and the engine installer is busy 89.9% of the time.

ENTITY SUMMARY   (Times in Scoreboard time units)

Entity Name	Qty Processed	Average Cycle Time (Minutes)	Average VA Time (Minutes)	Average Cost
Car	21743	20.36	4.02	0.00

Average time in the system is 20.4 minutes. In our discussion of Example 20.13, we found total time (in steady state) in the system to equal mean engine service time + mean tire installation time + mean time waiting for engine + mean time waiting for tires = 1 + 3 + 60(.15) + 60(.138) = 21.4 minutes.

VARIABLES

Variable Name Average Value	Total Changes	Average Minutes Per Change	Minimum Value	Maximum Value	Current Value
Avg BVA Time Entity 0	1	0.00	0	0	0
Avg BVA Time Car 0	21744	1.11	0	0	0

FIGURE 14 (Continued)

The Effect of a Finite Buffer

Suppose we have only enough space for two cars to wait for tire installation. This is called a *buffer* of size 2. We assume that if there are two cars waiting for tire installation, the engine installation center must shut down until there is room to "store" a car waiting for tire installation. To model this, change the Input Capacity in the Tire Activity dialog box to 2. (See Figure 15.) Rerunning the simulation, we now find the average time for a car in the system is nearly 6 hours! Clearly, we need more storage space. Even with a buffer of size 10, total time in the system is increased by around 50%.

10.4 Simulating Open Queuing Networks

In this section, we show how to use Process Model to simulate open queuing networks. To illustrate, we simulate Example 14 from Section 8.10.

EXAMPLE 2 **Open Queuing Network**

An open queuing network consists of two servers: server 1 and server 2. An average of 8 customers per hour arrive from outside at server 1. An average of 17 customers per hour arrive from outside at server 2. Interarrival times are exponential. Server 1 can serve at an exponential rate of 20 customers per hour, and server 2 can serve at an exponential rate of 30 customers per hour. After completing service at server 1, half the customers leave the system and half go to server 2. After completing service at server 2, 75% of the customers complete service and 25% return to server 1. Simulate this system for 400 hours.

Open.igx　　**Solution**　　See file Open.igx. To begin, we need to create two arrival entities: one representing external arrivals to server 1 and one representing external arrivals to server 2. After creating Process rectangles for server 1 and server 2, we use the Connector tool to create a link from server 1 to server 2, a link from server 2 to server 1, and a link from Server 1 and 2 to exit the system. For server 1, arrivals are E(7.5), and for server 2, arrivals are E(3.53). Service time for server 1 is E(3), while service time for server 2 is E(2). After clicking on the link from server 1 to server 2, we make sure that the routing percentage is 50% (this is the default). After clicking on the link from server 2 to server 1, we change the routing percentage to 25%. Note that the routing percentages on the exit links automatically adjust so that the total routing percentage leaving a service center is 100%. As an example, the dialog box on the link leaving server 2 and going to server 1 should be filled in

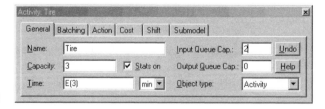

FIGURE 15

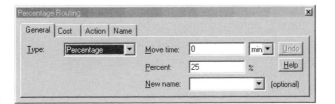

FIGURE 16

FIGURE 17

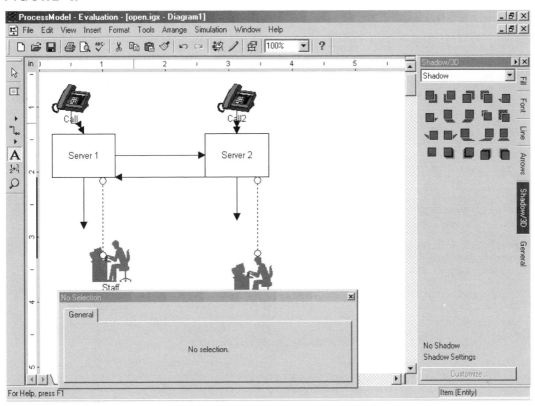

as shown in Figure 16. There, the Move time from server 2 to server 1 is made equal to 0, and we ensured that 25% of all customers completing server 2 go instantly to server 1. This implies that 75% of all customers completing server 2 instantly leave the system. (Click on the arc leaving server 2 if you do not believe this.) Note that the default Move times must always be adjusted from 1 minute unless you want a move time of 1 minute.

The flowchart is shown in Figure 17. Note that as the simulation runs, some calls move between the servers, and some exit the system. Seeing this movement really makes the concept of an open queuing network come alive. The simulation output is shown in Figure 18 (with explanatory comments in boldface).

```
--------------------------------------------------------------------
---------
General Report
Output from C:\Program Files\ProcessModel 4\open.mod
Date: Aug/15/2002   Time: 10:17:11 AM
--------------------------------------------------------------------
---------
Scenario         : Normal Run
Replication      : 1 of 1
Warmup Time      : 4 hr
Simulation Time  : 404 hr
--------------------------------------------------------------------
---------


ACTIVITIES

                                          Average
Activity         Scheduled        Total   Minutes    Average
Maximum    Current
Name             Hours  Capacity  Entries Per Entry  Contents
Contents   Contents  % Util
-----------      --------- -------- ------- --------- -------- ------
--  -------- ------
Server 1 inQ       400     999     5485     6.69      1.52
14         3    0.15
Server 1           400       1     5482     3.02      0.69
1          1   69.07
Server 2 inQ       400     999     9627     7.69      3.08
32         0    0.31
Server 2           400       1     9628     2.00      0.80
1          0   80.28
```

Note: Server 1 processes around 14 calls per hour, while server 2
processes around 24 calls per hour. This agrees with the results we
obtained in our analysis of Example 8.14.

```
ACTIVITY STATES BY PERCENTAGE (Multiple Capacity)

                                  %
Activity      Scheduled    %  Partially    %
Name          Hours    Empty  Occupied   Full
------------  --------- ----- --------- ----
Server 1 inQ    400    45.19    54.81   0.00
Server 2 inQ    400    35.44    64.56   0.00
```

45% of the time, no jobs are waiting at server 1. 35% of the time, no jobs are
waiting at server 2.

```
ACTIVITY STATES BY PERCENTAGE (Single Capacity)

Activity  Scheduled     %       %       %        %
Name      Hours   Operation  Idle   Waiting  Blocked
--------  --------- --------- ----- ------- -------
Server 1    400     69.07    30.93   0.00    0.00
Server 2    400     80.28    19.72   0.00    0.00
```

Server 1 is busy 69.1% of the time, while server 2 is busy 80.3% of the
time.

FIGURE 18

```
RESOURCES

                                        Average
                                Number  Minutes
Resource          Scheduled   Of Times     Per
Name      Units      Hours        Used    Usage  % Util
--------  -----   ---------   ---------  -------  ------
Staff       1         400        6787     2.83   80.04
Staff2      1         400        8323     1.99   69.32


RESOURCE STATES BY PERCENTAGE

Resource  Scheduled      %       %     %
Name        Hours     In Use   Idle  Down
--------  ---------   ------   -----  ----
Staff         400     80.04   19.96  0.00
Staff2        400     69.32   30.68  0.00


ENTITY SUMMARY   (Times in Scoreboard time units)

                        Average    Average
                          Cycle        VA
Entity        Qty         Time       Time    Average
Name      Processed    (Minutes)  (Minutes)    Cost
------    ---------    ---------  ---------   -------
Call         3114        16.30       4.57      0.00
Call2        6885        13.92       3.13      0.00
```

Calls that first arrive from outside to server 1 spend an average of 16.3 minutes in the system and 4.57 minutes in service. Calls that first arrive from outside to server 2 spend an average of 13.92 minutes in the system and 3.13 minutes in service.

```
VARIABLES

                             Average
Variable           Total     Minutes   Minimum  Maximum  Current
Average
Name              Changes   Per Change   Value    Value    Value
Value
-----------------  -------  ----------  -------  -------  -------  --
-----
Avg BVA Time Entity     1      0.00        0        0        0
0
Avg BVA Time Call    3115      7.70        0        0        0
0
Avg BVA Time Call2   6886      3.48        0        0        0
0
```

FIGURE 18
(Continued)

10.5 Simulating Erlang Service Times

As we saw in Chapter 8, service times often do not follow an exponential distribution. The Erlang distribution is often used to model nonexponential service times. An Erlang distribution can be defined by a mean and a shape parameter k. The shape parameter must be an integer. It can be shown that

$$\text{Standard deviation of Erlang} = \frac{\text{mean}}{\sqrt{k}}$$

Therefore, if we know the mean and standard deviation of the service times, we may determine an appropriate value of k. The syntax for generating Erlang service times in Process Model is ER(Mean, k). We now illustrate how to simulate a queuing system with Erlang service times.

EXAMPLE 3   **Walk-in Clinic**

A walk-in hospital clinic has four doctors. An average of 12 patients per hour arrive at the clinic (interarrival times are assumed to be exponential). A doctor can see an average of 4 patients per hour, with the standard deviation of service times being 8.66 minutes. Simulate the operation of this clinic for 1,000 hours.

Doctor.igx   **Solution**   See file Doctor.igx. The flowchart is shown in Figure 19. We must remember to adjust the number of doctors to 4 and the See Doctor capacity to 4. By clicking on the arrow connecting Customer and See Doctor, we can input the interarrival times as E(5). We know that the mean service time is 15 minutes. To estimate the shape parameter k, we solve $\frac{15}{\sqrt{k}} = 8.66$ minutes. This yields $k = 3$. We now fill in the See Doctor dialog box as shown in Figure 20. This ensures that up to 4 patients can be seen at once and that service times will follow an Erlang random variable with mean 15 minutes and shape parameter 3. (This implies a standard deviation of 8.66 minutes.)

The output is shown in Figure 21, with explanatory comments in boldface.

FIGURE 19

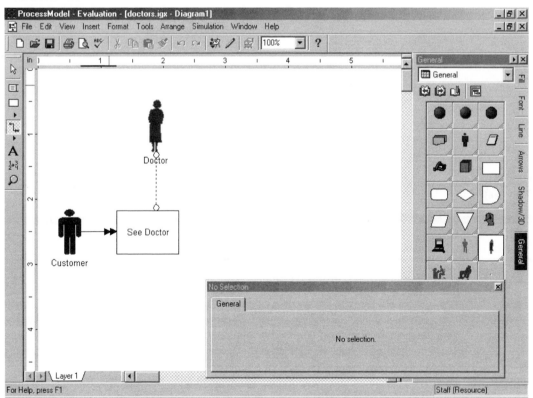

FIGURE 20

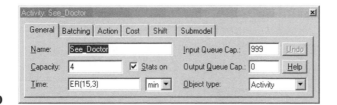

```
General Report
Output from C:\Program Files\ProcessModel 4\doctors.mod
Date: Aug/14/2002   Time: 10:29:27 AM
------------------------------------------------------------------------
---------
Scenario       : Normal Run
Replication    : 1 of 1
Simulation Time : 1000 hr
------------------------------------------------------------------------
---------


ACTIVITIES

                                            Average
Activity         Scheduled          Total   Minutes   Average
Maximum    Current
Name              Hours  Capacity  Entries  Per Entry  Contents
Contents  Contents  % Util
-------------    ----------  --------  -------  ---------  --------  ----
----    --------   ------
See Doctor inQ     1000      999     11964     5.37       1.07
23          0    0.11
See Doctor         1000        4     11964    15.07       3.00
4           2   75.12
```

This implies that average time waiting for doctor is 5.37 minutes and average service time is 15.07 minutes.

```
ACTIVITY STATES BY PERCENTAGE (Multiple Capacity)

                                      %
Activity        Scheduled      %  Partially     %
Name              Hours    Empty  Occupied    Full
--------------  ---------  -----  ---------   -----
See Doctor inQ     1000    64.88     35.12     0.00
See Doctor         1000     3.65     45.95    50.40
```
35% of all patients will have to wait for a doctor. All doctors are busy 50% of the time, and all doctors are idle 4% of the time. Between 1 and 3 doctors are busy 46% of the time.

```
RESOURCES

                                Average
                       Number   Minutes
Resource      Scheduled Of Times   Per
Name     Units  Hours     Used    Usage   % Util
-------- ----- --------- -------- ------- ------
Doctor.1   1     1000      2983    15.11   75.13
Doctor.2   1     1000      3025    14.90   75.12
Doctor.3   1     1000      2976    15.14   75.12
Doctor.4   1     1000      2980    15.12   75.12
Doctor     4     4000     11964    15.07   75.12
```
The average service time varies from a low of 14.9 minutes for Doctor 1 to a high of 15.14 minutes for Doctor 3.

FIGURE 21

RESOURCE STATES BY PERCENTAGE

Resource Name	Scheduled Hours	% In Use	% Idle	% Down
Doctor.1	1000	75.13	24.87	0.00
Doctor.2	1000	75.12	24.88	0.00
Doctor.3	1000	75.12	24.88	0.00
Doctor.4	1000	75.12	24.88	0.00
Doctor	4000	75.12	24.88	0.00

Each doctor is busy about 75% of the time. This is reasonable because (12 patients/hour)*(15 minutes/patient)= 180 minutes/hour of work arrives for doctors, and doctors have 240 minutes per hour to work, so they should be busy 75% of the time.

ENTITY SUMMARY (Times in Scoreboard time units)

Entity Name	Qty Processed	Average Cycle Time (Minutes)	Average VA Time (Minutes)	Average Cost
Customer	11962	20.45	15.07	0.00

On average, a patient spends a total of 20.45 minutes in the system. (Send me to this clinic!!)

VARIABLES

Variable Name	Total Changes	Average Minutes Per Change	Minimum Value	Maximum Value	Current Value	Average Value
Avg BVA Time Entity	1	0.00	0	0	0	0
Avg BVA Time Customer	11963	5.01	0	0	0	0

FIGURE 21
(Continued)

10.6 What Else Can Process Model Do?

Our discussion of Process Model has only scratched the surface of its capabilities. Other modeling features include the following.

- **Bulk arrivals and services.** At a restaurant, people often arrive in groups. This arrival pattern is called *bulk arrivals.* Consider an amusement park ride seating 40 people. The attendant waits until 40 people are present and then runs the ride. This service mechanism is known as *bulk service.*

- **Reneging.** Perhaps people hang up when calling an 800 number if they are put on hold more than 5 minutes. Process Model can accommodate such balking or reneging behavior.

- **Variation in arrival pattern.** At a restaurant or bank, the arrival rate varies substantially over the course of a day (or a whole week). Variable arrival rate patterns can easily be simulated with Process Model.

- **Variation in number of servers.** During the day, workers take breaks and go to lunch. Also, many companies vary the number of servers during the day. Process Model can easily accommodate variation in service capacity.

- **Priorities.** In an emergency room, more seriously ill patients are given priority over earlier arriving, less ill patients. Process Model can handle complex priority mechanisms.

For more details on these and other features of Process Model, consult the online manual.

REVIEW PROBLEMS

Group A

1 At a manufacturing assembly line, 30 jobs arrive per hour. Each job must pass through two production stages: stage 1 and stage 2. Stage 1 takes an average of 1 minute to complete, and 1 worker is available to perform stage 1. After completing stage 1, the job immediately passes to stage 2. Stage 2 takes an average of 2 minutes to complete, and 2 workers are available to work on stage 2. After completing stage 2, each job is inspected. Inspection takes an average of 3 minutes, and 3 workers are available to perform inspection. After inspection, 10% of the jobs must be returned to stage 1, and they then repeat both stages 1 and 2. After inspection, 20% of all jobs return to stage 2 and repeat stage 2. Assume that interarrival times and service times are exponential.

a What is the average time a job spends in the system from arrival to completion?

b What percentage of the time is each worker busy?

2 The United Airlines security station for Terminal C in Indianapolis has 3 X-ray machines. During the busy early morning hours, an average of 400 passengers per hour arrive at Terminal C (with exponential interarrival times). Each X-ray machine can handle an average of 150 passengers per hour (with exponential service times for X-ray machines).

After going through security, 90% of the customers are free to go to their flight, but 10% must be "wanded." Three people are available to do the wanding. Wanding requires a mean of 4 minutes, with a standard deviation of 2 minutes.

a How long does it take the average passenger to pass through security?

b If there were no wanding, how long would it take the average passenger to pass through security?

c Which would improve the situation more: adding an X-ray machine or adding an additional person to perform wanding?

3 Consider an emergency room. An average of 10 patients arrive per hour (interarrival times are exponential). Upon entering, the patient fills out a form. Assume that this always takes 5 minutes. Then each patient is processed by one of two registration clerks. This takes an average of 7 minutes (exponentially distributed). Then each patient walks 2 minutes to a waiting room and waits for one of 4 doctors. The time a doctor takes to see a patient averages 20 minutes, with a standard deviation of 10 minutes.

a On the average, how long does a patient spend in the emergency room?

b On the average, how much of this time is spent waiting for a doctor?

c What percentage of the time is each doctor busy?

4 The Indiana University Credit Union has 4 tellers working. It takes an average of 3 minutes (exponentially distributed) to serve a customer. Assume that an average of 60 customers per hour arrive at the Credit Union (interarrival times are exponential).

a How long do customers have to wait for a teller?

b What percentage of the time is a teller busy?

5 A pharmacist has to fill an average of 15 orders per hour (interarrival times are exponentially distributed). 80% of the orders are relatively simple and take 2 minutes to fill. 20% of the orders take 10 minutes to fill.

a What percentage of the time is the pharmacist busy?

b On average, how long does it take to get a prescription filled?

6 Solve Problem 5 if the service times followed a normal distribution with mean 3 minutes and standard deviation .5 minute. Use the syntax N(3,.5) to generate service times.

7 At Indiana Pacer games, 10,000 fans must enter through 10 checkpoints in the hour before each game (interarrival times are exponential). It takes exactly 3 seconds to have a ticket processed. How long does an average ticketholder spend from arrival to passing through the checkpoint?

8 Since September 11, 2001, each Pacer ticketholder's clothing and handbags are searched. Assume that this takes exactly 10 seconds and occurs right after the ticketholder passes through the checkpoint. Four people are available at each checkpoint to do the searching. How long does the average ticketholder spend from arrival to passing through the checkpoint?

Simulation with the Excel Add-in @Risk

Many simulations, particularly those involving financial applications can easily be performed with the Excel add-in @Risk. @Risk makes it easy to generate random variables. For example, to generate a standard normal random variable in a cell, just enter the formula =RISKNORMAL(0,1). If you want to run 10,000 iterations of a spreadsheet, just tell @Risk to run 10,000 iterations. Then @Risk provides a complete statistical or graphical summary of the results. In this chapter, we will see how @Risk can be used to simulate a wide variety of situations, ranging from the NPV of a new project to the probability of winning at craps.

11.1 Introduction to @Risk: The News Vendor Problem

@Risk is used to model situations where decisions are to be made under uncertainty. Here is an easy example. See the @Risk crib sheet in Appendix 1.

EXAMPLE 1 Ordering Calendars

Our bookstore must determine how many 2005 nature calendars to order in August 2004. It costs $2.00 to order each calendar, and we sell each calendar for $4.50. After January 1, 2005, leftover calendars are returned for $.75. Our best guess is that the number of calendars demanded is governed by the following probabilities.

Demand	Probability
100	.3
150	.2
200	.3
250	.15
300	.05

How many calendars should we order?

Solution The final result is in file Newsdiscrete.xls. See Figure 1.

Newsdiscrete.xls

Step 1 Enter parameter values in C3:C5.

Step 2 It can be shown that ordering an amount equal to one of the possible demands for calendars always maximizes expected profit. For now, we enter a trial order quantity of 200 calendars in cell C1.

	A	B	C	D	E	F	G
1	Order quantity		100				
2	Quantity demanded		100				
3	Sales price		$4.50				
4	Salvage value		$0.75			demand	prob
5	Purchase price		$2.00			100	0.3
6						150	0.2
7	Full price revenue	$450.00				200	0.3
8	Salvage revenue	$0.00				250	0.15
9	Costs	$200.00				300	0.05
10	Profit	$250.00					

FIGURE 1

Step 3 To tell @Risk to generate demand according to the above probabilities, type in C2 the formula

$$=RISKDISCRETE(F5:F9,G5:G9)$$

This generates a demand for calendars of 100 30% of the time, 150 20% of the time, etc. Essentially, for each iteration, @Risk generates a random number between 0 and 1. Then random numbers $<.3$ yield a demand of 100, random numbers $\geq.3$ and $<.5$ yield a demand of 150, random numbers $\geq.5$ and $<.8$ yield a demand of 200, random numbers $\geq.8$ and $<.95$ yield a demand of 250, and random numbers $\geq.95$ yield a demand of 300. Of course, successive random numbers generated by @Risk are independent of each other.

This demand could also have been generated with the formula

$$=RISKDISCRETE(\{100,150,200,250,300\},\{.3,.2,.3,.15,.05\})$$

In either format, the demands are listed first, followed by the probabilities. To see the spreadsheet recalculate when you hit F9, select Simulation Settings (the third icon from left) and choose from the Sampling tab Recalculation, and then choose Monte Carlo. Approximately 30% of the time a demand of 100 will occur, around 20% of the time a demand of 150 will occur, etc. If you change Sampling Recalculation to True EV, the mean of the random variable (172.5) will appear. If you change Sampling Recalculation to Expected Value, the value of the random variable nearest to the mean (in this case 150) will occur. We recommend always leaving Sampling Type on Latin Hypercube, because it is much more accurate than Monte Carlo. To illustrate how Latin Hypercube sampling works, suppose we told @Risk to sample from a normal distribution with mean 100 and standard deviation 15. The 5th, 10th, . . . , 95th percentile of a standard normal distribution can be found (using the NORMSINV function) to equal the values shown in Figure 2.

Suppose we want to simulate 100 values of a normal random variable with mean 100 and standard deviation 15. Then @Risk will ensure that 5 are less than or equal to 75.33, 5 are between 75.33 and 80.78, etc. Thus, the simulation will yield a very accurate representation of the random variable's distribution. In particular, our simulated means, variances, and other statistics will be much more accurate than if we used the Monte Carlo simulation. With Monte Carlo, 8 of 100 generated values could be <75.33, 3 out of 100 generated values between 75.33 and 80.78, etc.

Step 4 In cell B7, compute full-price revenue with the formula

$$=C3*MIN(C1,C2)$$

This ensures that we sell at full price the minimum of quantity ordered and quantity demanded.

Step 5 In B8, compute salvage revenue with the formula

$$=C4*IF(C1>C2,(C1-C2),0)$$

FIGURE 2

	D	E
5	Percentile	Value
6	0.05	75.3272
7	0.1	80.77672
8	0.15	84.4535
9	0.2	87.37568
10	0.25	89.88266
11	0.3	92.13399
12	0.35	94.22019
13	0.4	96.19979
14	0.45	98.11508
15	0.5	100
16	0.55	101.8849
17	0.6	103.8002
18	0.65	105.7798
19	0.7	107.866
20	0.75	110.1173
21	0.8	112.6243
22	0.85	115.5465
23	0.9	119.2233
24	0.95	124.6728

FIGURE 2

This ensures that the number left over is (number ordered) − (number demanded)—as long as that is >0.

Step 6 In B9, compute ordering costs with the formula

$$=C1*C5$$

Step 7 In cell B10, compute profit with the formula

$$=B7+B8-B9$$

We now want to compute profit for each possible order quantity (100, 150, 200, 250, or 300). The RISKSIMTABLE function makes this easy to do.

FIGURE 3

	D	E	F	G	H	I	J	K	L
13									
14		Name	Workbook	Worksheet	Cell	Sim#	Minimum	Mean	Maximum
15	Output 1	Profit	newsdiscrete.xl	Sheet1	B10	1	250	250	250
16	Output 1	Profit	newsdiscrete.xl	Sheet1	B10	2	187.5	318.75	375
17	Output 1	Profit	newsdiscrete.xl	Sheet1	B10	3	125	350	500
18	Output 1	Profit	newsdiscrete.xl	Sheet1	B10	4	62.5	325	625
19	Output 1	Profit	newsdiscrete.xl	Sheet1	B10	5	0	271.875	750
20	Input 1	Order quant	newsdiscrete.xl	Sheet1	C1	1	100	100	100
21	Input 1	Order quant	newsdiscrete.xl	Sheet1	C1	2	150	150	150
22	Input 1	Order quant	newsdiscrete.xl	Sheet1	C1	3	200	200	200
23	Input 1	Order quant	newsdiscrete.xl	Sheet1	C1	4	250	250	250
24	Input 1	Order quant	newsdiscrete.xl	Sheet1	C1	5	300	300	300
25	Input 2	Quantity de	newsdiscrete.xl	Sheet1	C2	1	100	172.5	300
26	Input 2	Quantity de	newsdiscrete.xl	Sheet1	C2	2	100	172.5	300
27	Input 2	Quantity de	newsdiscrete.xl	Sheet1	C2	3	100	172.5	300
28	Input 2	Quantity de	newsdiscrete.xl	Sheet1	C2	4	100	172.5	300
29	Input 2	Quantity de	newsdiscrete.xl	Sheet1	C2	5	100	172.5	300

	D	E	F	G	H	I
37						
38	Name	Profit	Profit	Profit	Profit	Profit
39	Description	Output (Sim	Output (Sim#2)	Output (Sim	Output (Sim	Output (Sim
40	Cell	B10	B10	B10	B10	B10
41	Minimum	250	187.5	125	62.5	0
42	Maximum	250	375	500	625	750
43	Mean	250	318.75	350	325	271.875
44	Std Deviatic	0	85.96629	163.5405	208.8956	225.6981
45	Variance	0	7390.203	26745.5	43637.39	50939.61
46	Skewness	Error!	-0.8715626	-0.397861	3.47E-02	0.2893988
47	Kurtosis	Error!	1.758383	1.42927	1.627334	2.06803
48	Errors Calcu	0	0	0	0	0
49	Mode	250	375	500	62.5	0
50	5% Perc	250	187.5	125	62.5	0
51	10% Perc	250	187.5	125	62.5	0
52	15% Perc	250	187.5	125	62.5	0
53	20% Perc	250	187.5	125	62.5	0
54	25% Perc	250	187.5	125	62.5	0
55	30% Perc	250	187.5	125	62.5	0
56	35% Perc	250	375	312.5	250	187.5
57	40% Perc	250	375	312.5	250	187.5
58	45% Perc	250	375	312.5	250	187.5
59	50% Perc	250	375	312.5	250	187.5
60	55% Perc	250	375	500	437.5	375
61	60% Perc	250	375	500	437.5	375
62	65% Perc	250	375	500	437.5	375
63	70% Perc	250	375	500	437.5	375
64	75% Perc	250	375	500	437.5	375
65	80% Perc	250	375	500	437.5	375
66	85% Perc	250	375	500	625	562.5
67	90% Perc	250	375	500	625	562.5
68	95% Perc	250	375	500	625	562.5
69	Filter Minimum					
70	Filter Maximum					
71	Type (1 or 2)					
72	# Values Fil	0	0	0	0	0
73	Scenario #1	>75%	>75%	>75%	>75%	>75%
74	Scenario #2	<25%	<25%	<25%	<25%	<25%
75	Scenario #3	>90%	>90%	>90%	>90%	>90%
76	Target #1 (V	400	400	400	400	400
77	Target #1 (F	100%	100%	50%	50%	80%
78	Target #2 (V	250	375	500	625	750
79	Target #2 (F	99%	99%	99%	99%	99%
80	Target #3 (V	360	360	360	360	360
81	Target #3 (F	100%	30%	50%	50%	50%

FIGURE 4

Step 8   In cell C1, enter the possible order quantities (100, 150, 200, 250, 300) with the formula

$$=RISKSIMTABLE(\{100,150,200,250,300\})$$

We could also have entered this RISKSIMTABLE function with the formula

$$=RISKSIMTABLE(F5:F9)$$

Note that if we obtain the arguments of an @Risk function such as RISKSIMTABLE or RISKDISCRETE by pointing to a different cell, we need to omit the { and } brackets.

On the first simulation, @Risk will put 100 in this cell and run the desired number of iterations. On the second simulation, @Risk will put 150 in this cell and run the desired number of iterations. Finally, on the fifth simulation, @Risk will put 300 in this cell and run the desired number of iterations.

Step 9 With the cursor in B10, select B10 as an output cell by selecting the single arrow icon. Note that the phrase RiskOutput() + appears before our Profit formula, indicating that Profit is an output cell. We could have entered this phrase instead of using the icon.

Step 10 Select the Simulations Settings icon. From the Iteration tab, select 1,000 iterations and 5 simulations. From Sampling tab, choose Latin Hypercube from the Sampling option. This will cause @Risk to recalculate demand and profit 1,000 times for each of the five order quantities. In general, if you have a RISKSIMTABLE in your spreadsheet, the number of simulations should equal the number of values in the RISKSIMTABLE. If you do not use a RISKSIMTABLE, leave Simulations at 1.

Step 11 Select the Run Simulation icon shown here. After running the simulation, you will see the summary statistics shown in Figure 3. The first simulation is for 100 calendars ordered, the second for 150 calendars ordered, etc.

To obtain detailed statistics, select Insert Detailed Statistics and obtain Figure 4. To paste the statistics into the spreadsheet, right click on Results and then select Copy. Click on the X icon and choose Paste to insert the results into the original spreadsheet.

Interpretation of Statistical Output Figures 3 and 4 show that average profit for 1,000 trials when 200 calendars are ordered (for example) is $350.00. From Figure 4, the standard deviation for 1,000 trials is $163.54. It appears that ordering 200 calendars maximizes expected profit, but a case can be made for ordering 150 calendars. For 10% less expected profit, we can cut risk in half. The decision depends on the store's degree of aversion to risk.

REMARKS **1** The RISKSIMTABLE function uses the same set of random numbers to generate demand for each simulation. Thus, for each order quantity, the profit keys off the same set of demands.
2 You can return to the Results at any time by selecting the Results icon.
3 You can return to your worksheet from Results by selecting Window Show Excel Window. You may also click on the X icon (for Excel).

Finding a Confidence Interval for Expected Profit

If we ran 1,000 more trials in Example 1, @Risk would generate a different set of profits[†], and we would get a different estimate of average profit. So no simulation gives average profit exactly. How accurate is the estimate of average profit @Risk gives?

From Section 9.9, we can be 95% sure that average or expected profit for 200 calendars is between

$$\text{(Mean profit)} \pm t_{(.025, 199)} \text{ mean standard error}$$

Using the Excel formula TINV(.05,199) = 1.97, we find $t_{(.025, 199)} = 1.97$. Here,

$$\text{Mean standard error} = \frac{\text{standard deviation}}{\sqrt{\text{iterations}}} = \frac{163.54}{\sqrt{1,000}} = 5.17$$

[†]When the seed is set (from Simulation Settings Sampling) to 0, each time you run a simulation you will obtain different results. Other possible seed values are integers between 1 and 32,767. Whenever a nonzero seed is chosen, the same values for the input cells and output cells will occur. For example, if we choose a seed value of 10, then each time we run the simulation we will obtain exactly the same results. We often choose a seed of 1. If you also choose a seed of 1, your statistical output should exactly match ours.

Thus, we are 95% sure that expected profit is between $350 \pm 1.97(163.54)/\sqrt{1{,}000}$, or $339.81 and $360.19.

To be 95% confident of estimating the mean within $1, how many iterations are needed? The required number of iterations must satisfy

$$\frac{1.97(163.54)}{\sqrt{\text{iterations}}} = 1 \quad \text{or} \quad \text{iterations} = 322.17^2 = 103{,}796$$

To achieve a precise estimate of expected profit requires many iterations!

Modeling Normal Demand with the RISKNORMAL Function

In Example 1, the assumption of discrete demand is unrealistic. Let's suppose demand is normally distributed, with a mean of 200 and a standard deviation of 30. Then we are 68% sure that demand is between 170 and 230, 95% sure between 140 and 270, etc. To model normal demand, simply change cell C2's formula to

$$=\text{RISKNORMAL}(200{,}30)$$

Normalsim.xls

(See file Normalsim.xls.) This implies (for example), by the well-known rule of thumb, that 68% of the time demand will be between 170 and 230, 95% of the time between 140 and 260, and 99.7% of the time between 110 and 290.

@Risk generates a normal random variable by the inverse transformation method. First, we generate a random number that is equally likely to be any value between 0 and 1. Suppose we generate .6. Then the generated value of the normal random variable will be the 60th percentile of the random variable ($=$NORMINV(.6,200,30)).

With normal demand, any order quantity is reasonable, because demand may assume any value. We will still try the same set of order quantities, however. After running the simulation and selecting Insert Detailed Statistics, we obtain the output shown in Figure 5.

Figure 5 shows that ordering 200 calendars yields a higher mean profit than ordering 100, 150, 250, or 300 calendars. Plotting the expected profit for each order quantity yields the graph shown in Figure 6.

Under the assumption that profit is a unimodal function of order quantity (which is indeed correct), Figure 6 shows that expected profit is maximized by ordering between 150 and 250 calendars. Another RISKSIMTABLE (with values 160, 170, 180, 190, 200, 210, 220, 230, 240, 250) would help zero in on the actual best order quantity (which turns out to be 213 calendars).

REMARK To preclude the demand for calendars being a fraction, you could change the formula in cell C2 to

$$=\text{ROUND(RISKNORMAL}(200{,}30){,}0)$$

Then each demand generated by the RISKNORMAL function will be rounded to the nearest integer.

Finding Targets and Percentiles

At the bottom of the Detailed Statistics output, we may enter targets as values or percentages. Enter a value and @Risk tells you for what fraction of iterations the output cell was less than or equal to target. For example, we entered 400 under value and found that the profit for ordering 200 calendars was less than or equal to $400 18.7% of the time. We entered 34% under percentage and found that 34% of the time, profit was less than or equal to $453.44. We entered 99% under percentage and found that 99% of the time,

	D	E	F	G	H	I
76						
77	Name	Profit	Profit	Profit	Profit	Profit
78	Description	Output (Sim	Output (Sim#2)	Output (Sim	Output (Sim	Output (Sim
79	Cell	B10	B10	B10	B10	B10
80	Minimum	238.7325	176.2325	113.7325	51.23248	-11.26752
81	Maximum	250	375	500	625	722.9146
82	Mean	249.9887	372.7469	455.0987	435.2473	374.9326
83	Std Deviatic	0.3563102	13.68735	65.79792	107.9455	112.4699
84	Variance	0.1269569	187.3436	4329.366	11652.22	12649.48
85	Skewness	-31.52797	-8.285047	-1.647313	-0.230145	-1.48E-02
86	Kurtosis	996.006	85.60149	5.455792	2.672569	2.967134
87	Errors Calcu	0	0	0	0	0
88	Mode	250	375	500	625	390.5545
89	5% Perc	250	375	314.0754	251.5754	189.0754
90	10% Perc	250	375	355.4924	292.9924	230.4924
91	15% Perc	250	375	382.9496	320.4496	257.9496
92	20% Perc	250	375	405.072	342.572	280.072
93	25% Perc	250	375	423.7888	361.2888	298.7888
94	30% Perc	250	375	440.8098	378.3098	315.8098
95	35% Perc	250	375	456.6077	394.1077	331.6077
96	40% Perc	250	375	471.358	408.858	346.358
97	45% Perc	250	375	485.665	423.165	360.665
98	50% Perc	250	375	499.9558	437.4558	374.9558
99	55% Perc	250	375	500	451.3856	388.8856
100	60% Perc	250	375	500	465.8172	403.3172
101	65% Perc	250	375	500	480.6487	418.1487
102	70% Perc	250	375	500	496.4514	433.9514
103	75% Perc	250	375	500	513.1733	450.6732
104	80% Perc	250	375	500	532.158	469.658
105	85% Perc	250	375	500	553.6349	491.135
106	90% Perc	250	375	500	581.4592	518.9592
107	95% Perc	250	375	500	621.5056	559.0056
108	Filter Minimum					
109	Filter Maximum					
110	Type (1 or 2)					
111	# Values Fil	0	0	0	0	0
112	Scenario #1	>75%	>75%	>75%	>75%	>75%
113	Scenario #2	<25%	<25%	<25%	<25%	<25%
114	Scenario #3	>90%	>90%	>90%	>90%	>90%
115	Target #1 (V	400	400	400	400	400
116	Target #1 (F	100%	100%	18.70%	36.90%	58.82%
117	Target #2 (V	250	375	500	625	634.03864
118	Target #2 (F	99%	99%	99%	99%	99%
119	Target #3 (V	360	360	360	360	360
120	Target #3 (F	100%	3.60%	10.70%	24.60%	44.72%
121	Target #4 (Value)			453.44424		
122	Target #4 (Perc%)			34%		

FIGURE 5

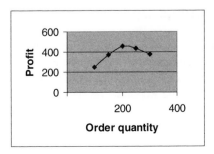

FIGURE 6

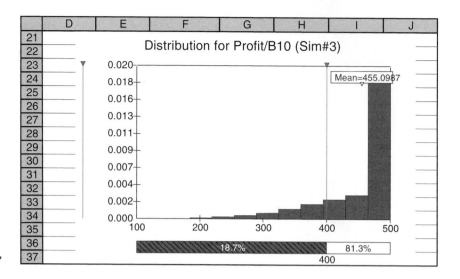

FIGURE 7

profit was less than or equal to $500. We entered $360 under value and found that profit was less than or equal to $360 10.7% of the time.

Creating Graphs with @Risk

To create a histogram of possible profits in Example 1, go to the Results menu and right click on the output cell Profit from the Explorer style list. Then choose Histogram and the third simulation (for 200 calendars ordered) to obtain a histogram similar to Figure 7. By moving the sliders at the bottom of the graph, you may zero in on the probability of any range of values. For example, there is an 18.7% chance that profit will be $400 or less. To paste any graph into Excel, right click on the graph and select Copy. You may also copy a graph into Excel by selecting Graph in Excel Option.

If we right click on a selected graph, we may change it to a **cumulative ascending graph.** See Figure 8.

Figure 8 gives the probability that profit is less than or equal to the x-value. Thus, there is around a 19% chance that profit is ≤$400.

By right clicking on a histogram or cumulative ascending graph and selecting Format, we can obtain a **cumulative descending graph.** (See Figure 9.) In a cumulative descending graph, the y-coordinate is the probability that profit exceeds the x-coordinate. For example, there is approximately an 81% probability that profit will exceed $400.

Using the Report Settings Option

You may also create graphs and statistical reports directly with the Report Settings option. By choosing the Report Settings icon, any output may be sent directly to the current workbook or a new workbook. For example, see Figure 10. Checking the dialog box as shown there, and choosing Generate Reports Now, would place the output that has been generated in a new workbook.

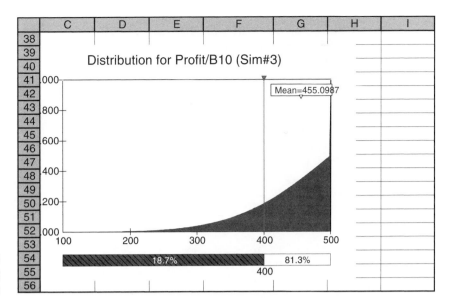

FIGURE 8
Cumulative Ascending Graph

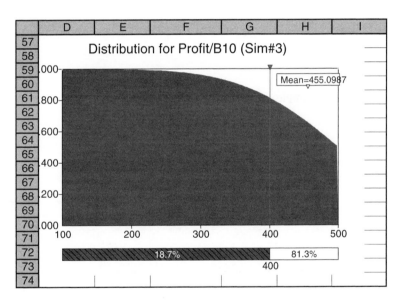

FIGURE 9
Cumulative Descending Graph

Using @Risk Statistics

Instead of generating large reports, you may just want your spreadsheet to show the mean and standard deviation (and possibly other statistics) of your output cells. @Risk 4.5 contains statistical functions that accomplish this goal. To see how this works, enter in cell F12 the formula

$$=RISKMEAN(\$B\$10,F11)$$

and copy this formula to G12:J12. In F12, this keeps track of the mean of the first simulation. In G12, it keeps track of the mean for the second simulation, etc.

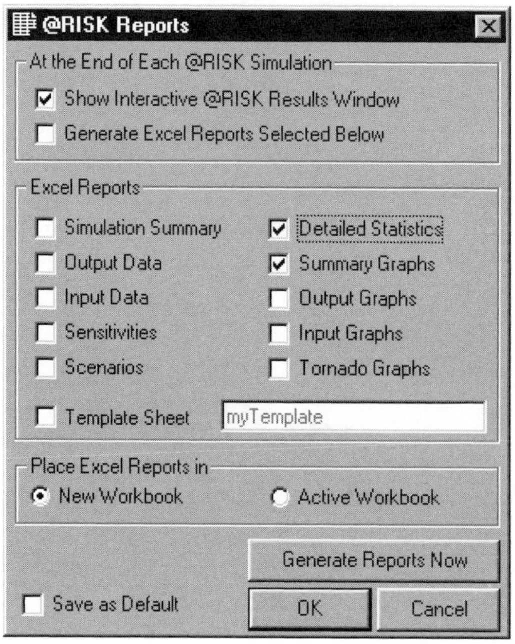

FIGURE 10

For example, in cell F12 we entered the formula

$$=RISKSTDDEV(\$B\$10,F11)$$

Then we copied this formula from F12 to G12:J12. This formula keeps track of the standard deviation from each order quantity. The results are as follows:

	E	F	G	H	I	J
10						
11		1	2	3	4	5
12	mean	$249.99	$372.75	$455.10	$435.25	$374.93
13	sigma	$0.36	$13.69	$65.80	$107.95	$112.47

For example, in the third simulation, for which we ordered 200 calendars, the mean profit for 1,000 iterations was $455.10, with a standard deviation of $65.80.

PROBLEMS

Group A

1 Explain why expected profit must be maximized by ordering a quantity equal to some possible demand for calendars. (*Hint:* If this is not the case, then some order quantity, such as 190 calendars, must maximize expected profit. If ordering 190 calendars maximizes expected profit, then it must yield a higher expected profit than an order size

of 150. But then an order of 200 calendars must also yield a larger expected profit than 190 calendars. This contradicts the assumed optimality of ordering 190 calendars!)

2 In August 2004, a car dealer is trying to determine how many 2005 cars should be ordered. Each car ordered in

TABLE 1

No. of Cars Demanded	Probability
20	.30
25	.15
30	.15
35	.20
40	.20

August 2004 costs $10,000. The demand for the dealer's 2005 models has the probability distribution shown in Table 1. Each car sells for $15,000. If demand for 2005 cars exceeds the number of cars ordered in August, the dealer must reorder at a cost of $12,000 per car. Excess cars may be disposed of at $9,000 per car. Use simulation to determine how many cars should be ordered in August. For your optimal order quantity, find a 95% confidence interval for expected profit.

3 Suppose that the bookstore in Example 1 receives no money for the first 50 excess calendars returned, but still receives $.75 for each subsequent calendar returned. Does this change the optimal order quantity?

4 A TSB (Tax Saver Benefit plan) allows you to put money into an account at the beginning of the calendar year to use for medical expenses. This amount is not subject to federal tax (hence the phrase TSB). As you pay medical expenses during the year, you are reimbursed by the administrator of the TSB, until the TSB account is exhausted. The catch is, however, that any money left in the TSB at the end of the year is lost to you. You estimate that it is equally likely that your medical expenses for next year will be $3,000, $4,000, $5,000, $6,000, or $7,000. Your federal income tax rate is 40%. Assume your annual salary is $50,000.

a How much should you put in a TSB? Consider both expected disposable income and the standard deviation of disposable income in your answer. (*Hint:* Your simu-

lation will indicate that two options have nearly the same expected disposable income.)

b Does your annual salary influence the correct decision?

Group B

5 For Problem 2, suppose that the demand for cars is normally distributed with $\mu = 40$ and $\sigma = 7$. Use simulation to determine an optimal order quantity. For your optimal order quantity, determine a 95% confidence interval for expected profit.

6 Six months before its annual convention, the American Medical Association must determine how many rooms to reserve. At this time, the AMA can reserve rooms at a cost of $50 per room. The AMA must pay the $50 room cost even if the room is not occupied. The AMA believes that the number of doctors attending the convention will be normally distributed, with a mean of 5,000 and a standard deviation of 1,000. If the number of people attending the convention exceeds the number of rooms reserved, extra rooms must be reserved at a cost of $80 per room. Use simulation to determine the number of rooms that should be reserved to minimize the expected cost to the AMA.

7 A ticket from Indianapolis to Orlando on Deleast Airlines sells for $150. The plane can hold 100 people. It costs $8,000 to fly an empty plane. The airline incurs variable costs of $30 (food and fuel) for each person on the plane. If the flight is overbooked, anyone who cannot get a seat receives $300 in compensation. On the average, 95% of all people who have a reservation show up for the flight. To maximize expected profit, how many reservations for the flight should be taken by Deleast? (*Hint:* The @Risk function RISKBINOMIAL can be used to simulate the number of passengers who show up. If the number of reservations taken is in cell A2, then the formula

$$=\text{RISKBINOMIAL}(A2,.95)$$

will generate the number of customers who actually show up for a flight!)

11.2 Modeling Cash Flows from a New Product

In this section, we will show how GM and Eli Lilly model the cash flows from new products. We begin by discussing the important triangular random variable.

The Triangular Random Variable

Managers often analyze in terms of best case, worst case, and most likely outcome. They often fail to realize that any value between the best and worst cases may occur. The triangular random variable can help.

Suppose we want to model first-year market share for a new product. We feel that the worst case is 20%, the most likely share is 40%, and the best case is 70%. We will model year 1 market share with a **triangular random variable.** See Figure 11. Basically, @Risk generates year 1 market share by making the likelihood of a given share proportional to

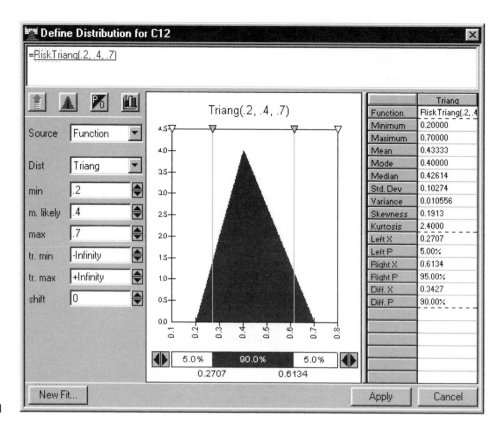

FIGURE 11

the height of the triangle in Figure 11. Thus, a 40% year 1 market share is most likely; all simulated market shares will be between 20% and 70%. A 30% market share occurs half as often as a year 1 40% market share, etc. The maximum height of the triangle is 4, because that makes the total area under the triangle equal to 1. The probability of market share being in a given range is equal to the area in that range under the triangle. For example, the chance of market share being at most 40% is .5*(4)*(.4 − .2) = .4 or 40%. To display this distribution, choose the Define Distributions icon and select the triangular random variable. Enter min = 0.2, m. likely = 0.4, and max = 0.7. You will see the picture in Figure 11.

EXAMPLE 2 General Motors

GM is trying to estimate the cash flows from a new car that will sell for 5 years. During the current year (year 0), a fixed development cost of $1.4 billion is incurred. This cost is depreciated on a straight-line basis over the next 5 years. Year 1 unit sales of the new model are assumed to follow a triangular random variable with worst case of 100,000 units, most likely case of 150,000 units, and best case of 170,000 units. Sales during years 2–5 are assumed to "decay" at the same rate each year. This annual decay rate is assumed to follow a triangular random variable with best case of 5%, most likely case of 8%, and worst case of 10%. Each year, a car sells for $15,000. During year 1, each car sold incurs a variable cost of $10,000. Due to increased labor costs, the variable cost of producing the car increases 4% a year. The tax rate is 40%, and cash flows are discounted at 15% a year. (Assume all cash flows occur at the end of the year.)

a Estimate the mean NPV of the cash flows from the new car.

b What fraction of the time will the new model add value to GM?

FIGURE **12**

	C	D	E	F	G	H	I
2							
3							
4	tax rate	0.4					
5	cost growth	0.04					
6	discount rate	0.15					
7	decay rate	0.072255268					
8							
9		Time					
10		0	1	2	3	4	5
11	Cost	1.40E+09					
12	Unit Sales		144227.4769	133806.2819	124138.0731	115168.4433	106846.9166
13	Price		$ 15,000.00	$ 15,000.00	$ 15,000.00	$ 15,000.00	$ 15,000.00
14	Unit cost		$ 10,000.00	$ 10,400.00	$ 10,816.00	$ 11,248.64	$ 11,698.59
15	Revenues		$ 2,163,412,154.19	$2,007,094,228.55	$ 1,862,071,096.57	$ 1,727,526,649.90	$ 1,602,703,748.32
16	Variable Cost		$ 1,442,274,769.46	$1,391,585,331.79	$ 1,342,677,398.70	$ 1,295,488,358.34	$ 1,249,957,799.41
17	Depreciation		$ 280,000,000.00	$ 280,000,000.00	$ 280,000,000.00	$ 280,000,000.00	$ 280,000,000.00
18	Before tax profit		$ 441,137,384.73	$ 335,508,896.76	$ 239,393,697.87	$ 152,038,291.56	$ 72,745,948.91
19	After tax profit		$ 264,682,430.84	$ 201,305,338.05	$ 143,636,218.72	$ 91,222,974.93	$ 43,647,569.34
20	Cash flow	-1400000000	$ 544,682,430.84	$ 481,305,338.05	$ 423,636,218.72	$ 371,222,974.93	$ 323,647,569.34
21							
22	npv cash flows	$77,633,524.27					
23							

Solution

Gmcashflow.xls

Our work is in file Gmcashflow.xls. (See Figure 12.)

Recall that in years 1–5, cash flow = after-tax profit + depreciation.

Step 1   In cell B11, enter the fixed cost of 1.4e9. In cell B20, enter the year 0 cash flow with the formula

$$=-\text{B}11$$

Step 2   In cell E12, compute year 1 unit sales with the formula

$$=\text{RISKTRIANG}(100000,150000,170000)$$

The syntax of the RISKTRIANG function requires that the lowest value of the random variable be entered first, followed by the most likely value, followed by the largest value.

Step 3   In cell D7, simulate the decay rate with the formula

$$=\text{RISKTRIANG}(0.05,0.08,0.1)$$

Step 4   In cells F12:I12, compute unit sales for years 2–5 by copying from F12 to G12:I12 the formula

$$=(1\text{-decay\_rate})*\text{E}12$$

The cell D7 has been named decay_rate.

Step 5   Enter in E13:I13 the unit price of $15,000.

Step 6   In cell E14, enter the year 1 variable cost of $10,000. Then in cells F14:I14, compute the variable cost for years 1–5 by copying from F14 to G14:I14 the formula

$$=\text{E}14*(1+\$\text{D}\$5)$$

Step 7   Copying from E15 to F15:I15 the formula

$$=\text{E}13*\text{E}12$$

computes the sales revenue for each year.

	H	I
29		gmcashflowdecay.xls
30	Name	npv cash flows / Time
31	Description	Output
32	Cell	Sheet1!D22
33	Minimum	-2.19E+08
34	Maximum	2.55E+08
35	Mean	4.31E+07
36	Std Deviation	9.92E+07
37	Variance	9.84E+15
38	Skewness	-0.3451601
39	Kurtosis	2.396719
40	Errors Calculated	0
41	Mode	6.10E+07
42	5% Perc	-1.35E+08
43	10% Perc	-1.03E+08
44	15% Perc	-7.06E+07
45	20% Perc	-4.68E+07
46	25% Perc	-3.02E+07
47	30% Perc	-8040124
48	35% Perc	9848326
49	40% Perc	2.64E+07
50	45% Perc	4.13E+07
51	50% Perc	5.76E+07
52	55% Perc	6.90E+07
53	60% Perc	8.02E+07
54	65% Perc	9.26E+07
55	70% Perc	1.04E+08
56	75% Perc	1.18E+08
57	80% Perc	1.32E+08
58	85% Perc	1.49E+08
59	90% Perc	1.66E+08
60	95% Perc	1.90E+08
61	Filter Minimum	

	L	M	N
34			
35	95% CI		
36	for Mean		
37	NPV		
38	Lower	3.69E+07	43-2(99)/sqrt(1000)
39	Upper	4.94E+07	43+2(99)/sqrt(1000)

FIGURE 13

Step 8 Copying from E16 to F16:I16 the formula

$$=E14*E12$$

computes the variable cost for each year.

Step 9 In cells E17:I17, compute the depreciation for each of years 1–5 by copying from E17 to F17:I17 the formula

$$=\$D\$11/5$$

Step 10 By copying from E18 to F18:I18 the formula

$$=E15-E16-E17$$

we determine before-tax profit for years 1–5.

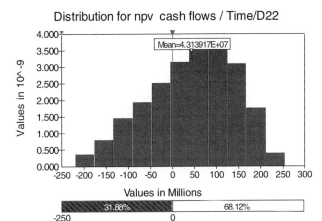

Distribution for npv cash flows / Time/D22

FIGURE 14

Step 11 By copying from E19 to F19:I19 the formula

$$=(1\text{-tax\_rate})*E18$$

we determine after-tax profit for years 1–5.

Step 12 By copying from E20 to F20:I20 the formula

$$=E19+E17$$

we add each year's depreciation to its after-tax profit to compute the year's cash flow.

Step 13 Assuming end-of-year cash flows, the formula

$$=NPV(0.15,D20:I20)$$

in cell D22 computes the NPV of all cash flows.

Step 14 After making cell D22 an output cell and running 1,000 iterations, we obtain the statistical output shown in Figure 13 and the graphical output in Figure 14.

From Figure 13, the mean NPV of cash flows (or risk-adjusted NPV) is $43 million. We are 95% certain that mean NPV is between $37 million and $49 million. Figure 14 shows that there is a 32% chance the project will have cash flows with a negative NPV (thereby reducing the company's value) and a 68% chance that cash flows will have a positive NPV.

The Lilly Model

In the car business, a new model virtually always has reduced sales every year. A new drug, however, sees increased sales in the first few years, followed by reduced sales. To model this form of the product life cycle, we must incorporate the following sources of uncertainty. (Note that we assume that total number of years for which the drug is sold is known).

- Number of years for which unit sales increase
- Average annual percentage increase in sales during the sales-increase portion of the sales period
- Average annual percentage decrease in sales during the sales-decrease portion of the sales period

FIGURE 15

	B	C	D	E	F	G	H	I	J	K	L	M	N
1													
2			Growth then decay										
3		length of growth	5										
4		tax rate	0.4										
5		cost growth	0.04										
6		discount rate	0.15										
7		growth rate	0.055313219										
8		decay rate	0.117781276										
9			Time										
10			0	1	2	3	4	5	6	7	8	9	10
11		Cost	1.60E+09										
12		Unit Sales		1.12E+05	1.18E+05	1.25E+05	1.32E+05	1.39E+05	1.47E+05	1.29E+05	1.14E+05	1.01E+05	8.88E+04
13		Price		1.50E+04	1.50E+04	1.50E+04	1.50E+04	1.50E+04	1.50E+04	1.50E+04	1.50E+04	1.50E+04	1.50E+04
14		Unit cost		1.00E+04	1.04E+04	1.08E+04	1.12E+04	1.17E+04	1.22E+04	1.27E+04	1.32E+04	1.37E+04	1.42E+04
15		Revenues		1.68E+09	1.77E+09	1.87E+09	1.98E+09	2.08E+09	2.20E+09	1.94E+09	1.71E+09	1.51E+09	1.33E+09
16		Variable Cost		1.12E+09	1.23E+09	1.35E+09	1.48E+09	1.63E+09	1.78E+09	1.64E+09	1.50E+09	1.38E+09	1.26E+09
17		Depreciation		1.60E+08	1.60E+08	1.60E+08	1.60E+08	1.60E+08	1.60E+08	1.60E+08	1.60E+08	1.60E+08	1.60E+08
18		Before tax profit		4.00E+08	3.84E+08	3.62E+08	3.34E+08	2.99E+08	2.56E+08	1.44E+08	5.01E+07	-2.77E+07	-9.19E+07
19		After tax profit		2.40E+08	2.30E+08	2.17E+08	2.00E+08	1.79E+08	1.53E+08	8.62E+07	3.01E+07	-1.66E+07	-5.51E+07
20		Cash flow	-1600000000	4.00E+08	3.90E+08	3.77E+08	3.60E+08	3.39E+08	3.13E+08	2.46E+08	1.90E+08	1.43E+08	1.05E+08
21													
22		npv cash flows	($290,597,621.28)										

Lillygrowth.xls

Example 3 shows how to model this type of product life cycle. See file Lillygrowth.xls and Figure 15.

EXAMPLE 3 Eli Lilly

Lilly is producing a new drug that will be sold for 10 years. Year 1 unit sales are assumed to follow a triangular random variable with worst case 100,000 units, most likely case 150,000, and best case 170,000. The year 0 fixed cost of developing the drug is $1.6 billion, to be depreciated on a 10-year straight-line basis. Sales are equally likely to increase for 3, 4, 5, or 6 years, with the average percentage increase during those years following a triangular random variable with worst case 5%, most likely case 8%, and best case 10%. During the remainder of the 10-year sales life of the drug, unit sales will decrease at a rate governed by a triangular random variable having best case 8%, most likely case 12%, and worst case 18%. During each year, a unit of the drug sells for $15,000. Year 1 variable cost of producing a unit of the drug is $10,000. The unit variable cost of producing the drug increases at 4% a year.

a Estimate the mean NPV of the drug's cash flows.

b What is the probability that the drug will add value to Lilly?

c What source of uncertainty is the most important driver of the drug's NPV?

Solution After dragging our formulas to create years 6–10 and changing the depreciation in row 17 to be over a 10-year period, we simulate random variables in D3 (length of sales increase), D7 (annual percentage rate of sales increase), and D8 (annual percentage rate of sales decrease) with the following formulas

Cell D3: =RISKDUNIFORM({3,4,5,6})

The RISKDUNIFORM variable is a discrete random variable that assigns equal probability to each listed value.

Cell D7: =RISKTRIANG(0.05,0.08,0.1)

Cell D8: =RISKTRIANG(0.08,0.12,0.18)

In cell E12, we generate year 1 units sales with the formula

$$=\text{RISKTRIANG}(100000,150000,170000)$$

Copying from F12 to G12:N12 the formula

$$=\text{IF}(\text{F10}\leq\text{length\_of\_growth}+1,\text{E12}*(1+\text{growth\_rate}),\text{E12}*(1-\text{decay\_rate}))$$

generates unit sales for years 2–10. Note that our formula increases annual sales by the growth rate for length-of-growth years and decreases annual sales by decay rate during later years. (D3 is named length_of_growth, D7 is named growth_rate, and D8 is named decay_rate.)

We used Autoconvergence to determine the number of iterations for @Risk to run. Under Simulation Settings, selecting Iterations Auto and a change of 1% ensures that @Risk will keep running iterations until, during the last 100 iterations, the mean, standard deviation, and selected other statistics change by 1% or less. In this example, @Risk ran 1,800 iterations, yielding the results in Figure 16. There was an estimated mean of −$29 million and a 54% chance of negative NPV. Right clicking on NPV from the Explorer interface yields the histogram in Figure 17. The histogram shows a 53% chance that the drug will decrease Lilly's NPV.

For part (c), use a **tornado graph** to determine the key drivers of NPV. To obtain a tornado graph, you must have selected the Collect All Outputs box from the Simulation Settings Sampling dialog box. (Unless you want a tornado graph, it is probably best to uncheck that box. Checking that box adds a column to your output for each @Risk function in the model, and this can clutter up the output.) Right click on NPV in the Explorer interface and select Tornado Graph. We can obtain a correlation and/or regression tornado graph as shown in Figures 18 and 19.

Each bar of the correlation tornado graph (Figure 18) gives the correlation of the @Risk random variable with NPV. For example,

- Year 1 unit sales has a .98 correlation with NPV.
- Annual growth rate has a .14 correlation with NPV.

In short, the uncertainty about year 1 unit sales is very important for determining NPV, but other random variables could probably be replaced by their mean without changing the distribution of NPV by much.

For each @Risk random variable, the regression tornado graph (Figure 19) computes the *standardized regression coefficient* for the @Risk random variable when we try to predict NPV from all @Risk random variables in the spreadsheet. A standardized regression coefficient tells us (after adjusting for other variables in the equation) the number of standard deviations by which NPV changes when the given @Risk random variable changes by one standard deviation. For example,

- A one standard deviation change in year 1 unit sales will (ceteris paribus) change NPV by .98 standard deviation.
- A one standard deviation change in annual growth rate will increase NPV by .15 standard deviation (ceteris paribus).

Again it is clear that the uncertainty for year 1 sales is really all that matters here; other random variables may as well be replaced by their means.

	C	D	E
24			
25			
26		Name	npv cash fl(
27		Description	Output
28		Cell	D22
29		Minimum	-3.54E+08
30		Maximum	2.37E+08
31		Mean	-2.86E+07
32		Std Deviation	1.23E+08
33		Variance	1.52E+16
34		Skewness	-0.34653
35		Kurtosis	2.440396
36		Errors Calculated	0.00E+00
37		Mode	2.33E+07
38		5% Perc	-2.52E+08
39		10% Perc	-2.06E+08
40		15% Perc	-1.71E+08
41		20% Perc	-1.41E+08
42		25% Perc	-1.14E+08
43		30% Perc	-9.07E+07
44		35% Perc	-6.99E+07
45		40% Perc	-5.01E+07
46		45% Perc	-3.21E+07
47		50% Perc	-1.31E+07
48		55% Perc	4.01E+06
49		60% Perc	1.94E+07
50		65% Perc	3.26E+07
51		70% Perc	4.98E+07
52		75% Perc	6.48E+07
53		80% Perc	8.02E+07
54		85% Perc	9.80E+07
55		90% Perc	1.23E+08
56		95% Perc	1.56E+08
57		Filter Minimum	
58		Filter Maximum	
59		Type (1 or 2)	
60		# Values Filtered	0
61		Scenario #1	>75%
62		Scenario #2	<25%
63		Scenario #3	>90%
64		Target #1 (Value)	0
65		Target #1 (Perc%)	53.73%

FIGURE **16**

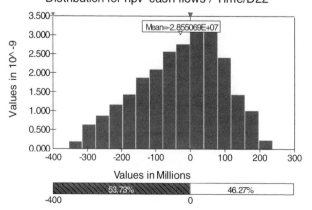

FIGURE **17**

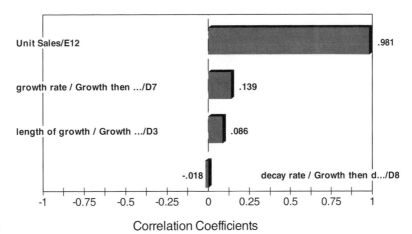

FIGURE 18

Correlations for npv cash flows / Time/D22

Correlation Coefficients

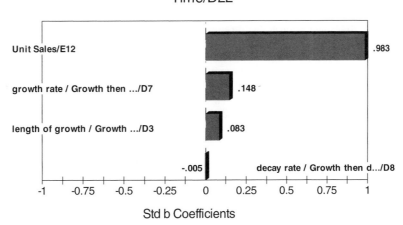

FIGURE 19

Regression Sensitivity for npv cash flows/ Time/D22

Std b Coefficients

PROBLEMS

Group A

1 Dord Motors is considering whether to introduce a new model: the Racer. The profitability of the Racer will depend on the following factors:

- Fixed cost of developing Racer: Equally likely to be $3 billion or $5 billion.
- Sales: Year 1 sales will be normally distributed with $\mu = 200,000$ and $\sigma = 50,000$.
  Year 2 sales will be normally distributed with $\mu =$ year 1 sales and $\sigma = 50,000$.
  Year 3 sales will be normally distributed with $\mu =$ year 2 sales and $\sigma = 50,000$.
  For example, if year 1 sales = 180,000, then the mean for year 2 sales will be 180,000.

- Price: Year 1 price = $13,000
  Year 2 price = 1.05*{(year 1 price) + $30*(% by which year 1 sales exceed expected year 1 sales)}
  The 1.05 is the result of inflation!
  Year 3 price = 1.05*{(year 2 price) + $30*(% by which year 2 sales exceed expected year 2 sales)}
  For example, if year 1 sales = 180,000, then year 2 price = 1.05*{13,000 + 30(−10)} = $13,335
- Variable cost per car: During year 1, the variable cost per car is equally likely to be $5,000, $6,000, $7,000, or $8,000.
  Variable cost for year 2 = 1.05*(year 1 variable cost)
  Variable cost for year 3 = 1.05*(year 2 variable cost)

TABLE 2

Year	1	2	3
GNP	3%	5%	4%
INF	4%	7%	3%

TABLE 3

Number of Competitors	Probability
0	.50
1	.30
2	.10
3	.10

TABLE 4

	Year 1	Year 2	Year 3
Sales price	$15,000	$16,000	$17,000
Variable cost	$12,000	$13,000	$14,000

TABLE 5

Time Abandoned	Value Received
End of year 1	$3,000
End of year 2	$2,600
End of year 3	$1,900
End of year 4	$900

Your goal is to estimate the NPV of the new car during its first three years. Assume that cash flows are discounted at 10%; that is, $1 received now is equivalent to $1.10 received a year from now.

a Simulate 400 iterations and estimate the mean and standard deviation of the NPV the first three years of sales.

b I am 95% sure that the expected NPV of this project is between _____ and _____.

c Use the Target option to determine a 95% confidence interval for the actual NPV of the Racer during its first three years of production.

d Use a tornado graph to analyze which factors are most influential in determining the NPV of the Racer.

2 Trucko produces the Goatco truck. The company wants information about the discounted profits earned during the next three years. During a given year, the total number of trucks sold in the United States is 500,000 + 50,000*GNP − 40,000*INF, where

GNP = % increase in GNP during year

INF = % increase in Consumer Price Index during year

Value Line has made the predictions given in Table 2 for the increase in GNP and INF during the next three years.

In the past, 95% of Value Line's GNP predictions have been accurate within 6% of the actual GNP increase, and 95% of Value Line's INF predictions have been accurate within 5% of the actual inflation increase.

At the beginning of each year, a number of competitors may enter the trucking business. At the beginning of a year, the probability that a certain number of competitors will enter the trucking business is given in Table 3.

Before competitors join the industry at the beginning of year 1, there are two competitors. During a year that begins (after competitors have entered the business, but before any have left) with c competitors, Goatco will have a market share given by $.5*(.9)^c$. At the end of each year, there is a 20% chance that each competitor will leave the industry.

The sales price of the truck and production cost per truck are given in Table 4.

a Simulate 500 times the next three years of Truckco's profit. Estimate the mean and variance of the discounted three-year profits (use a discount rate of 10%).

b Do the same if during each year there is a 50% chance that each competitor leaves the industry.

(*Hint:* You can model the number of firms leaving the industry in a given period with the RISKBINOMIAL function. For example, if the number of competitors in the industry is in cell A8, then the number of firms leaving the industry during a period can be modeled with the statement =RISKBINOMIAL(A8,.20). Just remember that the RISKBINOMIAL function is not defined if its first argument equals 0.)

Group B

3 You have the opportunity to buy a project that yields at the end of years 1–5 the following (random) cash flows:

End of year 1 cash flow is normal with mean 1,000 and standard deviation 200.

For $t > 1$, end of year t cash flow is normal with Mean = actual end of year $(t − 1)$ cash flow and Standard deviation = .2*(mean of year t cash flow).

a Assuming cash flows are discounted at 10%, determine the expected NPV (in time 0 dollars) of the cash flows of this project.

b Suppose we are given the following option: At the end of year 1, 2, 3, or 4, we may give up our right to future cash flows. In return for doing this, we receive the *abandonment value* given in Table 5.

Assume that we make the abandonment decision as follows: We abandon if and only if the expected NPV of the cash flows from the remaining years is smaller than the abandonment value. For example, suppose end of year 1 cash flow is $900. At this point in time, our best guess is that cash flows from years 2–5 will also be $900. Thus, we would abandon the project at the end of year 1 if $3,000 exceeded the NPV of receiving $900 for four straight years. Otherwise, we would continue. What is the expected value of the abandonment option?

4 Mattel is developing a new Madonna doll. Managers have made the following assumptions.

It is equally likely that the doll will sell for two, four, six, eight, or ten years.

At the beginning of year 1, the potential market for the doll is 1 million. The potential market grows by an average of 5% per year. They are 95% sure that the growth in the potential market during any year will be between 3% and 7%.

They believe their share of the potential market during year 1 will be at worst 20%, most likely 40%, and at best 50%. All values between 20% and 50% are possible.

The variable cost of producing a doll during year 1 is equally likely to be $4 or $6.

The sales price of the doll during year 1 will be $10.

Each year, the sales price and variable cost of producing the doll will increase by 5%.

The fixed cost of developing the doll (incurred in year 0) is equally likely to be $4, $8, or $12 million.

At time 0, there is one competitor in the market. During each year that begins with four or fewer competitors, there is a 20% chance that a new competitor will enter the market.

To determine year t unit sales (for $t > 1$), proceed as follows. Suppose that at the end of year $t - 1$, x competitors were present. Then assume that during year t, a fraction $.9 - .1*x$ of loyal customers (last year's purchasers) will buy a doll during the next year and a fraction $.2 - .04*x$ of people currently in the market who did not purchase a doll last year will purchase a doll from the company this year. We now generate a prediction for year t unit sales. Of course, this prediction will not be precise. We assume that it is sure to be accurate within 15%, however.

Cash flows are discounted at 10% per year.

a Estimate the expected NPV (in time 0 dollars) of this project.

b You are 95% sure the expected NPV of this project is between _____ and _____.

c You are 95% sure that the actual NPV of the project is between _____ and _____.

d What two factors does the tornado diagram indicate are key drivers of the project's profitability?

5 GM is thinking of marketing a new car, the Batmobile. It is equally likely that the car will take 1, 2, or 3 years to develop. This may be modeled by a RISKDUNIFORM random variable. A RISKDUNIFORM function is equally likely to assume any of the values listed in the cell.

Development cost is assumed equally split over development time. The best case is development cost of

TABLE 6

Years	Probability
4	.1
5	.3
6	.4
7	.2

$300 million, the most likely case is $800 million, and the worst case is $1.7 billion.

The product will begin sales during the year after development concludes. The number of years the car will be sold is assumed to be governed by the probability distribution in Table 6.

The size of the market during the first year of sales is unknown, but the worst case is a market size of 100,000, the most likely case is 145,000, and the best case is 165,000. Annual growth in market size is unknown, but is assumed to have a worst case of 1% per year, a most likely case of 6% a year, and a best case of 8% per year.

First-year market share is unknown, but the worst case is a 30% market share, the most likely case is 45%, and the best case is 50%. After the first year of sales, market share will fluctuate. On average, next year's share will equal this year's share. We are 95% sure that next year's market share will be within 40% of this year's market share.

During the first year of sales, price is unknown, with a worst-case price of $16,000, a most likely price of $17,500, and a best-case price of $18,000. Each year, price increases by 5%.

During the first year of sales, the best-case estimate for the cost of producing a car is $11,000, the most likely cost is $13,000, and the worst-case cost is $14,500. Each year, variable cost increases by 5%.

The discount rate for this project is 15%.

a You are 95% sure that mean NPV for this project is between _____ and _____.

b What is the probability that the project will add value to the company?

c What are the key drivers of the project's success?

d Construct a graph that illustrates the range of possible NPVs that might be generated by this project.

11.3 Project Scheduling Models

In Chapter 7 of Volume 1, we used linear programming to determine the length of time needed to complete a project. We also learned how to identify critical activities, where an activity is critical if increasing its activity time by a small amount increases the length of time needed to complete the project by the same amount. Our discussion there required the assumption that all activity times are known with certainty. In reality, these times are usually uncertain. Of course, this implies that the length of time needed to complete the project is also uncertain. It also implies that for each activity, there is a *probability* (not necessarily equal to 0 or 1) that the activity is critical.

To illustrate, suppose that activities *A* and *B* can begin immediately. Activity *C* can then begin as soon as activities *A* and *B* are both completed, and the project is completed as soon as activity *C* is completed. Activity *C* is clearly on the critical path, but what about *A* and *B*? Let's say that the *expected* activity times of *A* and *B* are 10 and 12. If we use these expected times and ignore any uncertainty about the actual times—that is, if we proceed as we did in Chapter 7 of Volume 1—then activity *B* is definitely a critical activity. However, suppose there is some positive probability that *A* can have duration 12 and *B* can have duration 11. Under this scenario, *A* is a critical activity. Therefore, we cannot say in advance which of the activities, *A* or *B*, will be critical. However, by using simulation we can see how *likely* it is that each of these activities is critical. We can also see how long the entire project is likely to take. We illustrate with the following example.

EXAMPLE 4	Construction Project with Uncertain Activity Times

Tom Lingley, an independent contractor, has agreed to build a new room on an existing house. He plans to begin work on Monday morning, June 1. The main question is when he will complete his work, given that he works only on weekdays. The owner of the house is particularly hopeful that the room will be ready by Saturday, June 27, that is, in 20 or fewer working days. The work proceeds in stages, labeled A through J, as summarized in Table 7. Three of these activities, E, F, and G, will be done by separate independent subcontractors. The *expected* durations of the activities (in days) are shown in the table. However, these are only best guesses. Lingley knows that the *actual* activities times can vary because of unexpected delays, worker illnesses, and so on. He would like to use computer simulation to see (1) how long the project is likely to take, (2) how likely it is that the project will be completed by the deadline, and (3) which activities are likely to be critical.

Solution   We first need to choose distributions for the uncertain activity times. Then, given any randomly generated activity times, we will illustrate a method for calculating the length of the project and identifying the activities on the critical path.

The Pert Distribution   As always, there are several reasonable candidate probability distributions we could use for the random activity times. Here we illustrate a distribution that

TABLE 7
Activity Time Data

Description	Index	Predecessors	Expected Duration
Prepare foundation	A	None	4
Put up frame	B	A	4
Order custom windows	C	None	11
Erect outside walls	D	B	3
Do electrical wiring	E	D	4
Do plumbing	F	D	3
Put in ductwork	G	D	4
Hang drywall	H	E, F, G	3
Install windows	I	B, C	1
Paint and clean up	J	H	2

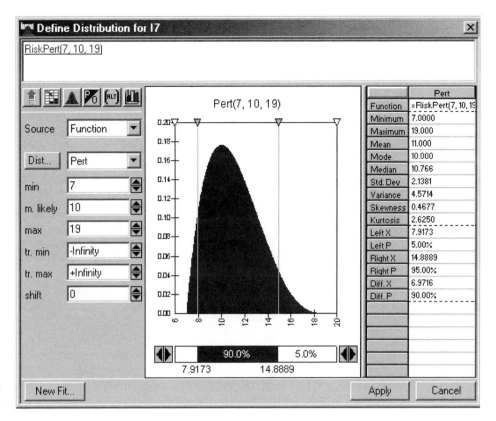

FIGURE 20
Pert Distribution

has become popular in project scheduling, called the *Pert distribution*.[†] As shown in Figure 20, it is a "rounded" version of the triangular distribution that is specified by three parameters: a minimum value, a most likely value, and a maximum value. The distribution in the figure uses the values 7, 10, and 19 for these three values, which implies a mean of 11. We will use this distribution for activity C. Similarly, for the other activities, we choose parameters for the Pert distribution that lead to the means in Table 7. In reality, it would be done the other way around. The contractor would estimate the minimum, most likely, and maximum parameters for the various activities, and the means would follow from these.

Developing the Simulation Model   The key to the model is representing the project network in activity-on-arc form, as in Figure 21, and then finding E_j for each j, where E_j is the earliest time we can get to node j. When the nodes are numbered so that all arcs go from lower-numbered nodes to higher-numbered nodes, we can calculate the E_j's iteratively, starting with $E_1 = 0$, with the equation

$$E_j = \max(E_i + t_{ij}) \tag{1}$$

Here, the maximum is taken over all arcs leading into node j, and t_{ij} is the activity time on such an arc. Then E_n is the time to complete the project, where n is the index of the finish node. This will make it very easy to calculate the project length.

[†]It is named after the acronym PERT (Program Review and Evaluation Technique) that is synonymous with project scheduling in an uncertain environment.

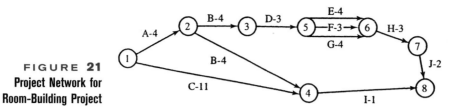

FIGURE 21
Project Network for Room-Building Project

FIGURE 22
Project Scheduling Simulation Model

	A	B	C	D	E	F	G	H	I	J
1	Room construction project									
2										
3	Data on activity network					Parameters of PERT distributions				
4	Activity	Code	Numeric index	Predecessors	Min	Most likely	Max	Implied mean	Duration	Duration+
5	Prepare foundation	A	1	None	1.5	3.5	8.5	4	2.158	2.159
6	Put up frame	B	2	A	3	4	5	4	4.513	4.513
7	Order custom windows	C	3	None	7	10	19	11	9.572	9.572
8	Erect outside walls	D	4	B	2	2.5	6	3	3.322	3.322
9	Do electrical wiring	E	5	D	3	3.5	7	4	3.282	3.282
10	Do plumbing	F	6	D	2	2.5	6	3	2.377	2.377
11	Put in duct work	G	7	D	2	4	6	4	4.668	4.668
12	Hang dry wall	H	8	E,F,G	2.5	3	3.5	3	3.197	3.197
13	Install windows	I	9	B,C	0.5	1	1.5	1	1.384	1.384
14	Paint and clean up	J	10	H	1.5	2	2.5	2	1.677	1.677
15										
16	Index of activity to increase	1								
17										
18	Event times									
19		Node	Event time	Event time+						
20		1	0	0						
21		2	2.158	2.159						
22		3	6.671	6.672						
23		4	9.572	9.572						
24		5	9.993	9.994						
25		6	14.661	14.662						
26		7	17.858	17.859						
27		8	19.536	19.537						
28										
29	Increase in project time?	1								
30										

We also need a method for identifying the critical activities for any given activity times. By definition, an activity is critical if a small increase in its activity time causes the project time to increase. Therefore, we will keep track of two sets of activity times and associated project times. The first uses the simulated activity times. The second adds a small amount, such as 0.001 day, to a "selected" activity's time. By using the RISKSIMTABLE function with a list as long as the number of activities, we can make each activity the "selected" activity in this method. The spreadsheet model appears in Figure 22, and the details are as follows. (See the Projectsim.xls file.)

Projectsim.xls

Inputs Enter the parameters of the Pert activity time distributions in the shaded cells and the implied means next to them. As discussed above, we actually chose the minimum, most likely, and maximum values while in @Risk's Model window to achieve the means in Table 7. Note that some of these distributions are symmetric about the most likely value, whereas others are skewed.

Activity Times Generate random activity times in column I by entering the formula

$$=RISKPERT(E5,F5,G5)$$

in cell I5 and copying it down.

Augmented Activity Times   We want to successively add a small amount to each activity's time to determine whether it is on the critical path. To do this, enter the formula

$$=RISKSIMTABLE(\{1, 2, 3, 4, 5, 6, 7, 8, 9, 10\})$$

in cell B16. (We use a list of length 10 because there are 10 activities.) Then enter the formula

$$=I5+IF(Index=C5,0.001,0)$$

in cell J5 and copy it down. (Here, Index is the range name of cell B16.) For example, if we are checking whether activity D (the 4th activity) is critical, the Index cell will be 4, and we will run a simulation where activity D's time is augmented by 0.001 and the other activity times are unchanged.

Event Times   We want to use Equation (1) to calculate the node event times in the range B20:B27. There is no quick way to enter the required formulas. (We see no way of using Copy and Paste.) We need to use the project network as a guide for each node. Begin by entering 0 in cell B20. Then enter the appropriate formulas in the other cells. For example, the formulas in cells B22, B23, and B27 are

$$=B21+I6$$
$$=MAX(B20+I7,B21+I6)$$

and

$$=RISKOUTPUT()+MAX(B23+I13,B26+I14)$$

To understand these, note that node 3 has only one arc leading into it, and this arc originates at node 2. No MAX is required for this node's equation. In contrast, node 4 has two arcs leading into it, from nodes 1 and 2, so a MAX is required. Similarly, node 8 requires a MAX, because it has two arcs leading into it. Also, it is the finish node, so we designate its event time cell as an @Risk output cell—it contains the time to complete the project.

Augmented Event Times   Copy the formulas in the range B20:B27 to the range C20:C27 to calculate the event times when the selected activity's time is augmented by 0.001.

Project Time Increases?   To check whether the selected activity's increased activity time increases the project time, enter the formula

$$=RISKOUTPUT()+IF(C27>B17,1,0)$$

If this calculates to 1, then the selected activity is critical for these particular activity times. Otherwise, it is not. Note that this cell is also designated as an @Risk output cell.

Using @Risk   We set the number of iterations to 1,000 and the number of simulations to 10 (one for each activity that we want to check for being critical). After running @Risk, we request the histogram of project times in Figure 23. In Volume 1, Chapter 7, when the activity times were not considered random, the project time was 20 days. Now it varies from a low of 15.89 days to a high of 25.50 days, with an average of 20.42 days.[†] Although the 5th and 95th percentiles appear in the figure, it might be more interesting (and depressing) to Tom Lingley to see the probabilities of various project times being exceeded. For example, we entered 20 in the Left X box next to the histogram. The Left P value implies that there is about a 57% chance that the project will not be completed

---

[†]It can be shown mathematically that the expected project time is *always* greater than when the expected activity times are used to calculate the project time, as we did in Volume 1. In other words, an assumption of certainty always leads to an underestimation of the true expected project time.

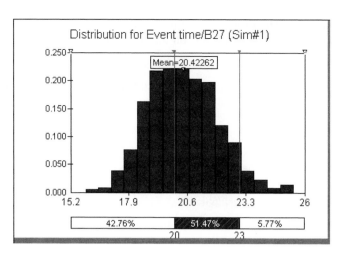

FIGURE 23
Histogram of Project Completion Time

	Name	Event time
	Cell	B27 Output (Sim#1)
	Minimum	15.89254
	Mean	20.42262
	Maximum	25.50042
	Std Dev	1.602013
	Variance	2.566447
	Skewness	0.292087
	Kurtosis	2.924874
	Mode	20.30279
	Left X	20
	Left P	42.75847%
	Right X	23
	Right P	94.2309%
	Diff. X	3
	Diff. P	51.47242%
	5th Perc.	17.9489
	95th Perc.	23.11927
	#Errors	0
	Filter Min	
	Filter Max	
	#Filtered	0

FIGURE 24
Probabilities of Activities Being Critical

	Name	Cell	Sim#	Minimum	Mean	Maximum
Output 2	Increase in project time? / Event time	B29	1	0	0.998	1
Output 2	Increase in project time? / Event time	B29	2	0	0.998	1
Output 2	Increase in project time? / Event time	B29	3	0	0.002	1
Output 2	Increase in project time? / Event time	B29	4	0	0.998	1
Output 2	Increase in project time? / Event time	B29	5	0	0.446	1
Output 2	Increase in project time? / Event time	B29	6	0	0.063	1
Output 2	Increase in project time? / Event time	B29	7	0	0.491	1
Output 2	Increase in project time? / Event time	B29	8	0	0.998	1
Output 2	Increase in project time? / Event time	B29	9	0	0.002	1
Output 2	Increase in project time? / Event time	B29	10	0	0.998	1

within 20 days. Similarly, the values in the Right X and Right P boxes imply that the chance of the project lasting longer than 23 days is slightly greater than 5%. This is certainly not good news for Lingley, and he might have to resort to the crashing we discussed in Volume 1.

The summary measures for the B29 output cell appear in Figure 24. Each "simulation" in this output represents one selected activity being increased slightly. The Mean column indicates the fraction of iterations where the project time increases as a result of the selected activity's time increase. Hence, it represents the probability that this activity is critical. For example, the first activity (A) is always critical, the third activity (C) is never critical, and the fifth activity (E) is critical about 45% of the time. More specifically, we see that the critical path always includes activities A, B, D, H, J, and one of the three "parallel" activities E, F, and G.

PROBLEMS

Group A

1 The city of Bloomington is about to build a new water treatment plant. Once the plant is designed (D), we can select the site (S), the building contractor (C), and the operating personnel (P). Once the site is selected, we can erect the building (B). We can order the water treatment machine (W) and prepare the operations manual (M) only after the contractor is selected. We can begin training (T) the operators when both the operations manual and operating personnel selection are completed. When the treatment plant and the building are finished, we can install the treatment machine (I). Once the treatment machine is installed and operators are trained, we can obtain an operating license (L). The estimated mean and standard deviation of the time

TABLE 8

	Mean	Standard Deviation
Activity D	6	1.5
Activity S	2	3.0
Activity C	4	1.0
Activity P	3	1.0
Activity B	24	6.0
Activity W	14	4.0
Activity M	3	0.4
Activity T	4	1.0
Activity I	6	1.0
Activity L	3	6.0

(in months) needed to complete each activity are given in Table 8. Use simulation to estimate the probability that the project will be completed in (a) under 50 days and (b) more than 55 days. Also estimate the probabilities that B, I, and T are critical activities.

2 To complete an addition to the Business Building, the activities in Table 9 need to be completed (all times are in months). The project is completed once Room 111 has been destroyed and the main structure has been built.

a Estimate the probability that it will take at least 3 years to complete the addition.

b For each activity, estimate the probability that it will be a critical activity.

3 To build Indiana University's new law building, the activities in Table 10 must be completed (all times are in months).

a Estimate the probability that the project will take less than 30 months to complete.

b Estimate the probability that the project will take more than 3 years to complete.

c For each of the activities A, B, C, and G, estimate the probability that it is a critical activity.

TABLE 9

	Predecessors	Mean Time	Standard Deviation
Activity A: Hire workers	—	4	0.6
Activity B: Dig big hole	A	9	2.5
Activity C: Pour foundation	B	5	1.0
Activity D: Destroy room	A	7	2.0
Activity E: Build main structure	C	10	1.5

TABLE 10

	Predecessors	Mean Time	Standard Deviation
Activity A: Obtain funding	—	6	0.6
Activity B: Design building	A	8	1.3
Activity C: Prepare site	A	2	0.2
Activity D: Lay foundation	B, C	2	0.3
Activity E: Erect walls and roof	D	3	1.0
Activity F: Finish exterior	E	3	0.6
Activity G: Finish interior	D	7	1.5
Activity H: Landscape grounds	F, G	5	1.2

11.4 Reliability and Warranty Modeling

In today's high-tech world, it is very important to be able to compute the probability that a system made up of machines will work for a desired amount of time. The subject of estimating the distribution of machine failure times and the distribution of time to failure of a system is known as **reliability theory.**

Distribution of Machine Life

We assume the length of time (call it **X**) until failure of a machine is a continuous random variable having a distribution function $F(t) = P(\mathbf{X} \le t)$ and a density function $f(t)$. Thus, for small Δt, the probability that a machine will fail between time t and $t + \Delta t$ is approximately $f(t)\Delta t$. The **failure rate** of a machine at time t [call it $r(t)$] is defined to be $(1/\Delta t)$ times the probability that the machine will fail between time t and time $t + \Delta t$, given that the machine has not failed by time t. Thus,

$$r(t) = \left(\frac{1}{\Delta t}\right) \text{Prob}(\mathbf{X} \text{ is between } t \text{ and } t + \Delta t | \mathbf{X} > t) = \frac{\Delta t f(t)}{\Delta t (1 - F(t))} = \frac{f(t)}{(1 - F(t))}$$

If $r(t)$ is an increasing function of t, the machine is said to have an **increasing failure rate (IFR).** If $r(t)$ is a decreasing function of t, the machine is said to have a **decreasing failure rate (DFR).**

Consider an exponential distribution which has $f(t) = \lambda e^{-\lambda t}$ and $F(t) = 1 - e^{-\lambda t}$. Then we find that

$$r(t) = \frac{\lambda e^{-\lambda t}}{e^{-\lambda t}} = \lambda$$

Thus, a machine whose lifetime follows an exponential random variable has **constant failure rate.** This is analogous to the no-memory property of the exponential distribution discussed in Chapter 8.

The random variable that is most frequently used to model the time till failure of a machine is the **Weibull random variable.** The Weibull random variable has the following density and distribution functions:

$$f(t) = \frac{\alpha x^{\alpha-1}}{\beta^{\epsilon}} e^{-(t/\beta)^{\epsilon}}$$
$$F(t) = 1 - e^{(-t/\beta)^{\alpha}}$$

It can be shown that if $\beta < 1$, the Weibull random variable exhibits DFR, and if $\beta > 1$, the Weibull random variable exhibits IFR. The @Risk function RISKWEIBULL(alpha, beta) will generate an observation for a Weibull random variable having parameters α and β. If you input the mean and variance of observed machine times to failure into cells D4 and D5, respectively, of workbook Weibest.xls, the workbook computes the unique values of α and β that yield the observed mean and variance of times to failure. For example, we see in Figure 25 that if the mean time to machine failure were 12 months and the standard deviation were 6 months, then a Weibull with $\alpha = 2.2$ and $\beta = 13.55$ would yield the desired mean and variance.

Weibest.xls

Common Types of Machine Combinations

Three common types of machine combinations are as follows:

- **A series system.** A series system functions only as long as each machine functions. See Figure 26(a).
- **A parallel system.** A parallel system functions as long as at least one machine functions. See Figure 26(b).

	A	B	C	D	E	F	G
1		**Estimating Weibull**					
2		**Distribution Parameters**					
3							
4		Mean time to failure		12			
5		Variance of time to Failure		36			
6		Second Moment of failure time		180			
7		Second moment/(mean)^2		1.25		Beta	13.54976
8		Alpha				Alpha	2.2

FIGURE 25

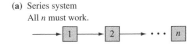
(a) Series system
All *n* must work.

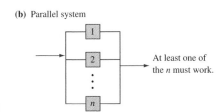
(b) Parallel system

At least one of the *n* must work.

FIGURE 26

■ **A *k* out of *n* system.** A *k* out of *n* system consists of *n* machines and is considered working as long as *k* machines are working.

Of course, by combining these types, a very complex system may be modeled. We now show how to use @Risk to model the probability that a machine system will last a desired amount of time.

EXAMPLE 5 **Hubble Telescope**

Assume that the Hubble telescope contains four large mirrors. The time (in months) until a mirror fails follows a Weibull random variable with $\alpha = 25$ and $\beta = 50$.

a For certain types of pictures to be useful, all mirrors must be working. What is the probability that the telescope can produce these types of pictures for at least 5 years?

b Certain types of pictures can be taken as long as at least one mirror is working. What is the probability that these pictures can be taken for at least 7 years?

c Certain types of pictures can be taken as long as at least two mirrors are working. What is the probability that these pictures can be taken for at least 6 years?

Solution See file Reliability.xls.

Reliability.xls

Step 1 We begin by generating the length of time until each mirror fails in C3:C6 by copying from C3 to C4:C6 the formula

$$=\text{RISKWEIBULL}(25,50)$$

FIGURE 27

	A	B	C
1			
2	**Hubble Telescope**		
3		Mirror 1	49.30487
4		Mirror 2	30.19602
5		Mirror 3	38.99237
6		Mirror 4	37.64995
7			
8		Time all 4 work	30.19602
9		Time till last one fails	49.30487
10		Last time 2 are working	38.99237

FIGURE 27

	F	G	H	I
10	Name	Time all 4 v	Time till las	Last time 2
11	Description	Output	Output	Output
12	Cell	C8	C9	C10
13	Minimum	3.733206	31.46502	27.71785
14	Maximum	66.0223	101.8234	80.49436
15	Mean	35.63382	64.18716	54.32821
16	Std Deviati	10.08293	9.306231	8.747266
17	Variance	101.6655	86.60596	76.51466
18	Skewness	-0.104045	0.121514	-2.73E-02
19	Kurtosis	2.737432	3.290648	2.910271
20	Errors Calc	0	0	0
21	Mode	34.15707	62.93507	58.45681
22	5% Perc	18.52796	49.15086	39.91564
23	10% Perc	22.40516	52.22929	42.77759
24	15% Perc	24.85496	54.84017	44.97655
25	20% Perc	26.80984	56.5674	46.99073
26	25% Perc	28.67021	57.84864	48.3152
27	30% Perc	30.25738	59.51218	49.89228
28	35% Perc	31.90257	60.52841	50.94405
29	40% Perc	33.26531	61.73524	52.26505
30	45% Perc	34.4916	62.83329	53.25808
31	50% Perc	35.7727	63.89499	54.54087
32	55% Perc	37.04685	65.06183	55.4865
33	60% Perc	38.58305	66.16101	56.72353
34	65% Perc	39.88355	67.578	58.00496
35	70% Perc	41.17931	68.97778	58.9762
36	75% Perc	42.88946	70.39309	60.089
37	80% Perc	44.47398	71.75684	61.61526
38	85% Perc	46.13106	73.66335	63.45758
39	90% Perc	48.46651	76.06507	65.64239
40	95% Perc	51.94818	79.72974	68.19598
41	Filter Minimum			
42	Filter Maximum			
43	Filter Type			
44	# Values F	0	0	0
45	Scenario #	>75%	>75%	>75%
46	Scenario #	<25%	<25%	<25%
47	Scenario #	>90%	>90%	>90%
48	Target #1 (	60	84	72
49	Target #1 (	99.54%	98.29%	98.00%

FIGURE 28

Step 2 Part (a) is a series system. We can take the desired pictures until the first mirror fails. The first mirror fails at the smallest of the four mirror failure times. Thus, the length of time for which the first type of picture can be taken is computed in cell C8 with the formula

$$=MIN(C3:C6)$$

Step 3 Part (b) is a parallel system. We can take the desired pictures until the time the last mirror fails. We compute the time the last mirror fails in cell C9 with the formula

$$=MAX(C3:C6)$$

Step 4 Part (c) is a 2 out of 4 system. We can take the desired pictures until the time of the third mirror failure. The time of the third mirror failure is the second largest of the failure times. We compute the time of the third mirror failing in cell C10 with the formula

$$=LARGE(C3:C6,2)$$

This formula computes the second largest of the mirror failure times. Of course, this is the time the third mirror fails. See Figure 27.

Step 5 We now select cells C8:C10 as output cells and run 1,000 iterations. After using targets with the Detailed Statistics output, we obtain the results in Figure 28.

We find in part (a) that there is a 99.54% chance that all four mirrors will fail in 60 months or less, and only a .46% chance that all four mirrors will work for at least 60 months. In part (b), we find that there is a 98.29% chance that all four mirrors will fail within 7 years, and only a 1.71% chance that all four mirrors will be working for at least 7 years. In part (c), we find that there is a 98% chance that two or more mirrors will be working for 72 months or less, and only a 2% chance that two or more mirrors will be working for at least 72 months.

Estimating Warranty Expenses

If we know the distribution of the time till failure of a purchased product, @Risk makes it a simple matter to estimate the distribution of warranty costs associated with a product. The idea is illustrated in the following example.

EXAMPLE 6 **Refrigerator Failure**

The time until first failure of a refrigerator (in years) follows a Weibull random variable with $\alpha = 6.7$ and $\beta = 8.57$. If a refrigerator fails within 5 years, we must replace it with a new refrigerator costing $500. If the replacement refrigerator fails within 5 years, we must also replace that refrigerator with a new one costing $500. Thus, the warranty stays in force until a refrigerator lasts at least 5 years. Estimate the average warranty cost incurred with the sale of a new refrigerator. (Do not worry about discounting costs.)

Solution
Refrigerator.xls

See file Refrigerator.xls. We enter the length of time a refrigerator lasts in cell C6 with the formula

$$=RISKWEIBULL(6.7,8.57)$$

	A	B	C	D	E	F	G
1							
2		**Refrigerator**					
3		**Warranty**					
4							
5		Number	Lasts	Cost			
6		1	8.113087	0			
7		2	6.91762	0			.027^5
8		3	7.233594	0			1.43489E-08
9		4	8.776642	0			
10		5	7.120917	0			
11			Total cost	0			
12							

FIGURE 29

We are not sure how many replacement refrigerators we might have to provide for the customer. By selecting the Define Distributions icon when we are in cell C6, we can move the sliders on the Weibull density function and determine the probability that we will have to replace a given refrigerator. We find that there is only a 2.7% chance that a refrigerator will have to be replaced. Then the chance that at least 5 refrigerators will have to be replaced is $(.027)^5 = .000014$. Thus, generating only 5 refrigerator lifetimes should give us an accurate estimate of total cost. We therefore copy the RISKWEIBULL formula from C6 to C7:C10. See Figure 29.

In cell D6, we compute the cost associated with a sold refrigerator with the formula

$$=IF(C6<5,500,0)$$

In cells D7:D10, we compute the cost (if any) associated with any replacement refrigerators by copying from D7 to D8:D10 the formula

$$=IF(AND(D6>0,C7<5),500,0)$$

This formula picks up the cost of a replacement if and only if the previous refrigerator failed and the current refrigerator lasts less than 5 years.

In cell D11, we compute total cost with the formula

$$=SUM(D6:D10)$$

After running 1,000 iterations and making cell D11 an output cell (see below), we find the mean warranty cost per refrigerator to be $14.50. Note that maximum cost was $1,000, so on at least one iteration, two refrigerators needed to be replaced.

	F	G	H	I	J	K	L	M
11								
12		Name	Workbook	Worksheet	Cell		Minimum	Mean	Maximum
13	Output 1	Total cost / Co⦚	refrigerator	Sheet1	D11		0	**14.5**	1000

PROBLEMS

Group A

Assume that the lifetimes of all machines described follow a Weibull random variable.

1 Suppose an auto engine consists of 12 components in series. The mean lifetime of each component is 5 years, with a standard deviation of 2 years.

   a What is the probability that the engine will work for at least 2 years?

   b If the engine were a parallel system, what is the probability that the engine would work for at least 10 years?

   c If at least 8 engine components need to work for the engine to work, what is the probability that the engine will work for at least 7 years?

2 An aircraft engine lasts an average of 5 years, with a standard deviation of 3 years before it needs to be replaced. Consider a plane with 4 new engines. On the average, how long will it be until an engine needs to be replaced?

3 A one-mile length of street has 5 street lights, equally spaced. The mean lifetime of a street light is 3 years, with a standard deviation of 1 year. Assume that all 5 lights have just been replaced. The street is considered too dark if at least one part of the street has no light working within .5 mile. On the average, how long will it be until the street is considered too dark?

4 In the refrigerator example, suppose the warranty works as follows. If a refrigerator fails at any time within 5 years of purchase, we give the consumer a prorated refund on the $500 purchase price. For example, if the refrigerator fails after 4 years, we pay the customer $100. If the refrigerator fails after 3 years, we pay the customer $200. Estimate our expected warranty expense per refrigerator sold.

5 The time to failure of a TV picture tube averages 5 years, with a standard deviation of 3 years. It costs an average of $250 to repair or replace a TV picture tube. Determine fair prices for a 3-year, 4-year, or 5-year warranty.

11.5 The RISKGENERAL Function

What if a continuous random variable (such as market share) does not appear to follow a normal or triangular distribution? We can model it with the **RISKGENERAL** function.

EXAMPLE 7   RISKGENERAL Distribution

Suppose that market shares between 0% and 60% are possible. A 45% share is most likely. There are five market-share levels for which we feel comfortable about comparing the relative likelihoods (see Table 11).

From the table, a market share of 45% is 8 times as likely as 10%; 20% and 55% are equally likely, etc. This distribution cannot be triangular, because then 20% would be (20/45) as likely as the peak of 45%. In fact, 20% is .75 as likely as 45%. See Figure 30 and file Riskgeneral.xls for our analysis.

Riskgeneral.xls

To model market share, enter the formula

$$=\text{RISKGENERAL}(0,60,\{10,20,45,50,55\},\{1,6,8,7,6\})$$

TABLE 11

Market Share	Relative Likelihood
10%	1
20%	6
45%	8
50%	7
55%	6

	B	C	D	E	F	G
1	EXAMPLE OF					
2	RISKGENERAL					
3	DISTRIBUTION					
4						
5			Minimum	0		
6			Maximum	60		
7			**Specified Points**			
8			10	1		
9			20	6		
10			45	8		
11			50	7		
12			55	6		
13	35.75	=RISKGENERAL(0,60,{10,20,45,50,55},{1,6,8,7,6})				

FIGURE 30

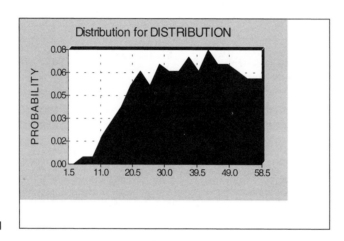

FIGURE 31

	C	D
29	Share	Likelihood
30	0	0
31	10	1
32	20	6
33	45	8
34	50	7
35	55	6
36	60	0

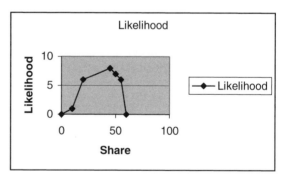

FIGURE 32

The syntax of RISKGENERAL is as follows.

- Begin with the smallest and largest possible values.
- Then enclose in {} the numbers for which you feel you can compare relative likelihoods.
- Finally, enclose in {} the relative likelihoods of the numbers you have previously listed.

Running this in @Risk yields the output in Figure 31. Note that 20 is 6/8 as likely as 45; 10 is 1/8 as likely as 45; 50 is 7/8 as likely as 45; 55 is 6/8 as likely as 45, etc. In between the given points, the density function changes at a linear rate. Thus, 30 would have a likelihood of

$$6 + \frac{(30 - 20)*(8 - 6)}{(45 - 20)} = 6.8$$

Basically what @Risk has done is to take the curve constructed by connecting (with straight lines) the points (0, 0), (10,1), . . . , (55,6), (60,0). @Risk rescales the height of this curve so that the area under it equals 1, and then randomly selects points based on the height of the curve. Thus, a share around 45 is 8/6 as likely as a share around 20, etc. Figure 32 illustrates this idea.

REMARK For the spreadsheet in Figure 30, the syntax

$$=RISKGENERAL(0,60,D8:D12,E8:E12)$$

is also acceptable.

Suppose we select the Define Distributions icon. Then we choose the RISKGENERAL random variable and select Apply. Now we can directly insert the RISKGENERAL (or any other) random variable into a cell.

After entering the appropriate parameters for the RISKGENERAL random variable, we will see the histogram shown in Figure 33. We are also given statistical information, such as the mean and variance, for the random variable. If we select Apply, the formula defining the desired RISKGENERAL random variable will be entered into the cell.

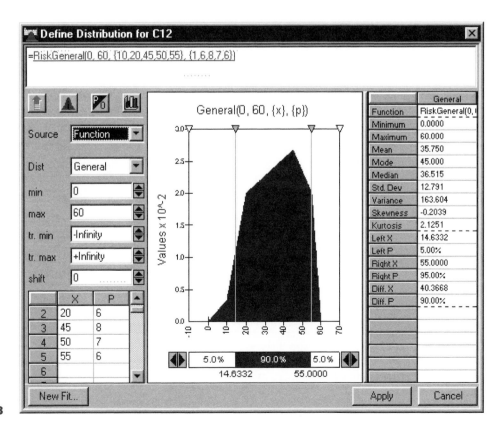

FIGURE 33

11.6 The RISKCUMULATIVE Random Variable

With the RISKGENERAL function, we estimated the relative likelihood of a random variable taking on various values. With the RISKCUMULATIVE function, we estimate the cumulative probability that the random variable is less than or equal to several given values. The RISKCUMULATIVE function can be used to approximate the cumulative distribution function for any continuous random variable.

EXAMPLE 8 **RISKCUMULATIVE**

A large auto company's net income for North American operations (NAO) for the next year may be between 0 and $10 billion. The auto company estimates there is a 10% chance that net income will be less than or equal to $1 billion, a 70% chance that net income will be less than or equal to $5 billion, and a 90% chance that net income will be less than or equal to $9 billion. Use @Risk to simulate NAO's net income for the next year.

	A	B	C	D	E	F	G	H
1	Cumulative distribution							
2								
3	Min	0						
4	Max	10		4.2				
5	x	P(X<=x)	Slope	4.2	RiskCumul(B3,B4,A6:A8,B6:B8)			
6	1	0.1	0.1					
7	5	0.7	0.15					
8	9	0.9	0.05		Name	P(X<=x)		
9	>9		0.1		Description	Output		
10					Cell	D5		
11					Minimum =	4.89E-03		
12					Maximum	9.999967		
13					Mean =	4.199986		
14					Std Deviat	2.773699		
15					Variance =	7.693407		
16					Skewness	0.589373		
17					Kurtosis =	2.285831		
18					Errors Calc	0		
19					Mode =	3.43314		
20					5% Perc =	0.497997		
21					10% Perc	0.999338	10%ile is 1!	
22					15% Perc	1.333212		
23					20% Perc	1.665637		
24					25% Perc	1.996866		
25					30% Perc	2.332803		
26					35% Perc	2.664376		
27					40% Perc	2.996635		
28					45% Perc	3.330816		
29					50% Perc	3.663554		
30					55% Perc	3.995894		
31					60% Perc	4.33135		
32					65% Perc	4.664128		
33					70% Perc	4.997442	70%ile is 5!	
34					75% Perc	5.995409		
35					80% Perc	6.993743		
36					85% Perc	7.99109		
37					90% Perc	8.989162	90%ile is near 9	
38					95% Perc	9.499336		

FIGURE 34

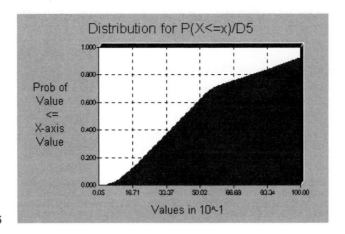

Distribution for P{X<=x}/D5

FIGURE **35**

Solution

Cumulative.xls

Our work is in the file Cumulative.xls. See Figure 34. The RISKCUMULATIVE function takes as inputs (in order) the following quantities:

- The smallest value assumed by the random variable
- The largest value assumed by the random variable
- Intermediate values assumed by the random variable
- For each intermediate value, the cumulative probability that the random variable is less than or equal to the intermediate value

In cell D5, we enter the following formula to simulate NAO's annual net income:

$$=RISKCUMUL(B3,B4,A6:A8,B6:B8)$$

We could have also used the following formula in cell D4:

$$=RISKCUMUL(0,10,\{1,5,9\},\{0.1,0.7,0.9\})$$

@Risk will now ensure that

- For net income x between 0 and \$1 billion, the cumulative probability that net income is less than or equal to x rises with a slope equal to $\frac{.1 - 0}{1 - 0} = .1$.
- For net income x between \$1 billion and \$5 billion, the cumulative probability that net income is less than or equal to x rises with a slope equal to $\frac{.7 - .1}{5 - 1} = .15$.
- For net income x between \$5 billion and \$9 billion, the cumulative probability that net income is less than or equal to x rises with a slope equal to $\frac{.9 - .7}{9 - 5} = .05$.
- For net income x greater than \$9 billion, the cumulative probability that net income is less than or equal to x rises with a slope equal to $\frac{1 - .9}{10 - 9} = .10$.

After running 1,600 iterations we found the output in Figure 34. Note that the 10th percentile of the random variable is near 1, the 70th percentile is near 5, and the 90th percentile is near 9. Figure 35 displays a cumulative ascending graph of net income. Note that (as described previously) the slope of the graph is relatively constant between 0 and 1, between 1 and 5, between 5 and 9, and between 9 and 10.

11.7 The RISKTRIGEN Random Variable

When we use the RISKTRIANG function, we are assuming we know the absolute worst and absolute best case that can occur. Many companies, such as Eli Lilly, prefer to use a triangular random variable in which the worst case and best case are defined by a percentile of the random variable. For example, at Eli Lilly the 10th percentile of demand, most likely demand, and 90th percentile of demand often define forecasts. The following example shows how to use the RISKTRIGEN function to model uncertainty.

EXAMPLE 9   RISKTRIGEN

Eli Lilly believes there is a 10% chance that its new drug Niagara's market share will be 25% or less, a 10% chance that market share will be 70% or more, and the most likely market share is 40%. Use @Risk to model the market share for Niagara.

Solution   Our work is in the file Risktrigen.xls. See Figure 36. In B7, we just entered the formula

Risktrigen.xls

$$=RISKTRIGEN(B3,B4,B5,10,90)$$

	A	B
1	trigen function	
2		
3	10%ile	0.25
4	Most likely	0.4
5	90 %ile	0.7
6		
7	share	0.464537

FIGURE 36

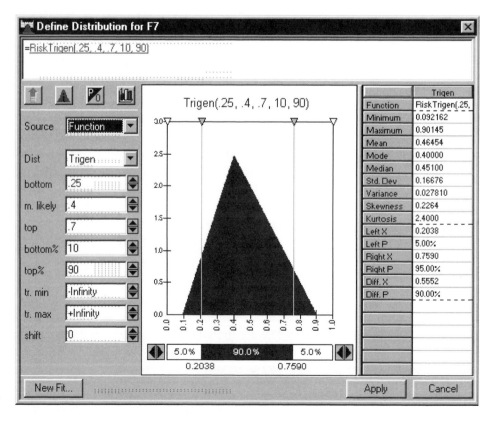

FIGURE 37

	C	D
34	Name	
35	Description	Output
36	Cell	[trigen.xls]
37	Minimum =	9.73E-02
38	Maximum	0.886495
39	Mean =	0.464533
40	Std Deviat	0.166746
41	Variance =	2.78E-02
42	Skewness	0.22598
43	Kurtosis =	2.398804
44	Errors Calc	0
45	Mode =	0.401881
46	5% Perc =	0.203626
47	10% Perc	0.249634
48	15% Perc	0.285171
49	20% Perc	0.315337
50	25% Perc	0.341713
51	30% Perc	0.365485
52	35% Perc	0.387192
53	40% Perc	0.407964
54	45% Perc	0.428942
55	50% Perc	0.450952
56	55% Perc	0.473918
57	60% Perc	0.498488
58	65% Perc	0.524349
59	70% Perc	0.552427
60	75% Perc	0.58265
61	80% Perc	0.61619
62	85% Perc	0.654338
63	90% Perc	0.699825
64	95% Perc	0.758373

FIGURE 38

The syntax of the RISKTRIGEN function is as follows:

=RISKTRIGEN(lower value, most likely value, higher value, percentile for lower value, percentile for higher value)

In Figure 37, we show the density function for the market share. Note that @Risk picks the worst case for RISKTRIGEN (around 10%), so the chance of a market share below 25% is .10. @Risk picks the best case for RISKTRIGEN (around 89%), so the probability of a share exceeding 70% is .10. When we ran 1,600 iterations, with cell B7 being the output cell, we obtained the output in Figure 38.

Note that the 10th percentile is almost exactly 25%, and the 90th percentile is almost exactly 70%.

11.8 Creating a Distribution Based on a Point Forecast

We are constantly inundated by forecasts:

- The government predicts the GDP will grow by 4% during the next year.
- The Eli Lilly marketing department predicts that demand for a given drug will be 400,000,000 d.o.t. (days of therapy) during the next year.

- A Wall Street guru predicts that the Dow will go up 20% during the next 12 months.
- The bookmakers forecast that the Pacers will beat the Rockets by 6 in the opening game of the 2005 NBA season.

Although the forecasts may be the best available, they are almost sure to be incorrect. For example, the bookmakers' prediction that the Pacers will win by 6 points is incorrect unless the Pacers win by exactly 6 points. In short, any single-valued (or *point*) forecast implies a distribution for the quantity being forecasted. How can we find a random variable that correctly models the uncertainty inherent in the point forecast? The key to putting a distribution around a point forecast is to have some historical data about the accuracy of past forecasts of the quantity of interest. For example, with regard to our forecast for the Dow, we might have the forecast made in January of each of the past 10 years for the percentage change in the Dow and the actual change in the Dow for each of those years. We begin by seeing if past forecasts exhibit any bias. For each past forecast, we determine (actual value)/(forecast value). Then we average these ratios. If our forecasts are unbiased, this average should be around 1. Any significant deviation from 1 would indicate a significant bias.[†] For example, if the average of actual/forecast is 2, the actual results tend to be around twice our forecast. To correct for this bias, we should automatically double our forecast. If the average of actual/forecast is .5, the actual results tend to be around half our forecast; to eliminate bias, we should automatically halve our forecast. Once we have eliminated forecast bias, we look at the standard deviation of the percentage errors of the unbiased forecast. We use the following @Risk random variable to model the quantity being forecast.

> RISKNORMAL(unbiased forecast, (percentage standard deviation of unbiased forecasts)*(unbiased forecast))

EXAMPLE 10 **Drug Forecast**

Drugforecast.xls

The file Drugforecast.xls contains actual and forecast sales (in millions of d.o.t.) for the years 1995–2002. See Figure 39. The forecast for 2003 is that 60 million d.o.t. will be sold. How would you model actual sales of the drug for 2003?

Solution **Step 1** In cells F5:F12, check for bias by computing actual sales/forecast sales for each year. To do this, copy from F5 to F6:F12 the formula

$$=D5/E5$$

Step 2 In cell F2, compute the bias of the original forecasts by averaging each year's actual/forecast sales.

$$=AVERAGE(F5:F12)$$

We find that actual sales tend to come in 8% under forecast.

Step 3 In G5:G12, correct past biased forecasts by multiplying them by .92. Simply copy from G5 to G6:G12 the formula

$$=\$F\$2*E5$$

[†]To see if the bias is significantly different from 1, compute

$$\frac{\text{Average of (actual)/(forecast)} - 1}{\text{Standard deviation of actual/forecast}}$$

If this exceeds $t_{(\alpha/2, n-1)}$ then there is significant bias. We usually choose $\alpha = .05$.

	C	D	E	F	G	H	I
1						mean	std dev
2			mean	0.918031		1	0.113753
3							
4	Year	Actual Sales	Forecast	A/F	Unbiased forecast	%age error	
5	1995	17	22	0.772727	20.19668	84%	
6	1996	59	61	0.967213	55.9999	105%	
7	1997	46	51	0.901961	46.81959	98%	
8	1998	85	86	0.988372	78.95067	108%	
9	1999	98	103	0.951456	94.5572	104%	
10	2000	94	118	0.79661	108.3277	87%	
11	2001	24	22	1.090909	20.19668	119%	
12	2002	14	16	0.875	14.6885	95%	

FIGURE 39

	E	F
14		
15	Mean 2003	55.08187
16	Sigma 2003	6.2657

FIGURE 40

Step 4 In H5:H12, compute each year's percentage error for the unbiased forecast. Copy from H5 to H6:H12 the formula

$$=D5/G5$$

Step 5 In cell I2, compute the standard deviation of the percentage errors with the formula

$$=STDEV(H5:H12)$$

We find that the standard deviation of past unbiased forecasts has been around 11% of the unbiased forecast. We now model the 2003 sales of the drug (in millions of d.o.t.) with the formula

$$=RISKNORMAL(60*(.918), (60*.918)*.114) \quad or \quad RISKNORMAL(55.08, 6.27)$$

See Figure 40.

11.9 Forecasting the Income of a Major Corporation

In many large corporations, different parts of a company make forecasts for quarterly net income. An analyst in the CEO's office pulls together the individual predictions to forecast the entire company's net income. In this section, we show an easy way to pool forecasts from different portions of a company and create a probabilistic forecast for the entire company.

So far, we have usually assumed that @Risk functions in different cells are independent. For example, the value of a RISKNORMAL(0,1) in cell A6 has no effect on the value of a RISKNORMAL(0,1) in any other cell. In many situations, however, variables of interest might be correlated. For example, a weak yen will lower the price of a Japanese car in the United States and hurt GM market share. Since higher price incentives increase market share, GM market share may also be negatively correlated with car price. Also, net income of NAO (North American operations) is often correlated with net income in Europe. The following example shows how to model correlations with @Risk. Recall that the correlation between two random variables must lie between −1 and +1.

- Correlation near +1 implies a strong positive linear relationship.
- Correlation near −1 implies a strong negative linear relationship.
- Correlation near +.5 implies a moderate positive linear relationship.
- Correlation near −.5 implies a moderate negative linear relationship.
- Correlation near 0 implies a weak linear relationship.

EXAMPLE 11   Forecasting GM Net Income

Corrinc.xls

Suppose GM CEO Rick Waggoner has received the following forecast for quarterly net income (in billions of dollars) for Europe, NAO, Latin America, and Asia. See Figure 41 and file Corrinc.xls.

For example, we believe Latin American income will be on average $.4 billion. Based on past forecast records, the standard deviation of forecast errors is 25%, so the standard deviation of net income is $.1 billion. We assume that actual income will follow a normal distribution. Historically, net income in different parts of the world has been correlated. Suppose the correlations are as given in B10:F13. Latin America and Europe are most correlated, and Asia and NAO are least correlated. What is the probability that total net income will exceed $4 billion?

Solution   To correlate the net incomes of the different regions, we use the RISKCORRMAT function. The syntax is as follows:

= Actual @Risk formula, RISKCORRMAT(correlation matrix, relevant column of matrix)

where

Correlation matrix: cells where correlations between variables are located

Relevant column: column of correlation matrix that gives correlations for this cell

Actual @Risk formula: distribution of the random variable

	A	B	C	D	E	F	G
1	Net Income Consolidation						
2	with correlation					Goal is 4 billion!	
3			Mean	Std. Dev	Actual		
4	1	LA	0.4	0.1	0.449011	0.521472	
5	2	NAO	2	0.4	1.256578	1.264837	
6	3	Europe	1.1	0.3	1.14203	0.994558	
7	4	Asia	0.8	0.3	0.685143	0.707549	
8				Total!!	3.532761	3.488417	
9							
10		Correlations	LA	NAO	Europe	Asia	
11		LA	1	0.6	0.7	0.5	
12		NAO	0.6	1	0.6	0.4	
13		Europe	0.7	0.6	1	0.5	
14		Asia	0.5	0.4	0.5	1	
15							
16							

FIGURE 41

	B	C	D	E	F
54	Scenario #3 =	>90%		36% chance we fail	
55	**Target #1 (Value)**	4		to meet target	
56	**Target #1 (Perc%)**	35.72%	▲		
57					

	B	C	D
17	Name	Total!! / Actual	
18	Description	Output	
19	Cell	E8	
20	Minimum =	1.858541	
21	Maximum =	6.71191	
22	Mean =	4.300031	
23	Std Deviation =	0.895158	
24	Variance =	0.801308	
25	Skewness =	-5.82E-02	
26	Kurtosis =	2.894021	
27	Errors Calculated	0	
28	Mode =	4.470891	
29	5% Perc =	2.756473	
30	10% Perc =	3.186955	
31	15% Perc =	3.364678	
32	20% Perc =	3.554199	
33	25% Perc =	3.715597	
34	30% Perc =	3.854618	
35	35% Perc =	3.96633	
36	40% Perc =	4.080534	
37	45% Perc =	4.173182	
38	50% Perc =	4.306374	
39	55% Perc =	4.413318	
40	60% Perc =	4.530555	
41	65% Perc =	4.632649	
42	70% Perc =	4.7776	
43	75% Perc =	4.907873	
44	80% Perc =	5.04496	
45	85% Perc =	5.216321	
46	90% Perc =	5.456462	
47	95% Perc =	5.758535	

FIGURE 42

Step 1 Generate actual Latin American income in cell E4 with the formula

$$=RISKNORMAL(C4,D4,RISKCORRMAT(\$C\$11:\$F\$14,A4))$$

This ensures that the correlation of Latin American income with other incomes is created according to the first column of C11:F14. Also, Latin American income will be normally distributed, with a mean of $.4 billion and standard deviation of $.1 billion.

Step 2 Copying the formula in E4 to E5:E7 (respectively) generates the net income in each region and tells @Risk to use the correlations in C11:F14.

Step 3 In cell E8, compute total income with the formula

$$=SUM(E4:E7)$$

Step 4 Cell E8 has been made the output cell. We find from Targets (value of 4) that there is a 36% chance of not meeting the $4 billion target. Also, the standard deviation of net income is $895 million. See Figure 42.

	B	C	D
15	Name	Total!! / Actual	
16	Description	Output	
17	Cell	E8	
18	Minimum =	2.174825	
19	Maximum =	6.290998	
20	Mean =	4.299921	
21	Std Deviation =	0.605397	
22	Variance =	0.366506	

	B	C	D	E	F
53	Target #1 (Value)=	4			
54	Target #1 (Perc%)=	30.76%		31% chance we fail to meet target	
55					
56					
57					
58					

FIGURE 43

FIGURE 44

	B	C	D	E	F	G	H	I	J	K	L
5											
6	Name	Total!! / Ac	LA / Actual	NAO / Actu	Europe / A	Asia / Actual					
7	Descriptior	Output	Normal(C4	Normal(C5	Normal(C6	Normal(C7,D7)					
8	Iteration#	E8	E4	E5	E6	E7		LA	NAO	Europe	Asia
9	1	4.804644	0.478546	2.196594	1.351783	0.777721	LA	1			
10	2	4.132098	0.441263	1.699526	1.184871	0.806438	NAO	0.591262	1		
11	3	6.129157	0.496915	2.453791	1.91255	1.265901	Europe	0.702735	0.587704	1	
12	4	6.54744	0.57896	2.424948	1.968532	1.574999	Asia	0.498132	0.399115	0.496651	1
13	5	3.057065	0.319965	1.517732	0.968105	0.251263					
14	6	5.324339	0.488499	2.292126	1.084479	1.459235					
907	899	4.735623	0.469691	2.19903	1.466369	0.600534					
908	900	4.901974	0.507751	2.242637	1.004801	1.146786					

What If Net Incomes Are Not Correlated?

Nocorrinc.xls

In workbook Nocorrinc.xls, we ran the simulation of Example 10, assuming that the net incomes in different regions were independent (that is, had 0 correlation). The results appear in Figure 43. Note that the absence of correlation has reduced the standard deviation to $600 million and our chance of not meeting our $4 billion income target. This is because if the incomes of all the regions are independent, then it is likely that a high income in one region will be cancelled out by a low income in another region. If the incomes of the regions are positively correlated, these correlations reduce the diversification or hedging effect.

Checking the Correlations

We can check that @Risk actually did correctly correlate net incomes. Make sure to check Collect Distribution Samples when you run the simulation. Once you have run the simulation, select the Data option from the Results menu. The results of each iteration will appear in the bottom half of the screen. You can Edit Copy Paste this data to a blank worksheet. See Figure 44. Now check the correlations between each region's net income with Data Analysis Tools Correlation. Select Data Analysis Tools Correlations and fill in the

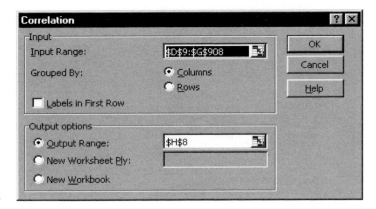

FIGURE 45

dialog box as in Figure 45. Note that the correlations between the net incomes are virtually identical to what we entered in the spreadsheet.

11.10 Using Data to Obtain Inputs for New Product Simulations

Many companies use subjective estimates to obtain inputs for new product simulations. For example, market size may be subjectively modeled as a triangular random variable, with the marketing department coming to a consensus on best-case, worst-case, and most likely scenarios. In many situations, however, past data may be used to obtain estimates of key variables. We now discuss how past data on similar products or projects can be used to model share, price, volume, and cost uncertainty. The utility of any model will depend on the type of data available.

The Scenario Approach to Modeling Volume Uncertainty

When trying to model volume of sales for a new product in the auto and drug industries, it is common to look for similar products sold in the past. We often have knowledge of the following:

- Accuracy of forecasts for year 1 sales volume
- Data on how sales change after the first year

Consider Figure 46—data on actual and forecast year 1 sales for seven similar products. See file Volume.xls. For example, for product 1, actual year 1 sales were 80,000; the forecast for year 1 was 44,396. The percentage change in sales from year to year for the seven products is given in Figure 47.

For example, product 1 sales went up 43% during the second year, 33% during the third year, etc.

Suppose we forecast year 1 sales to be 90,000 units. How can we model the uncertain volume in product sales?

Step 1 From cell D11 (formula =AVERAGE(D4:D10)) of Figure 46, we see that past forecasts for year 1 sales of similar products have overforecast the actual sales by 36.3%.

Volume.xls

FIGURE 46

	B	C	D	E	F
3	Actual	Forecast	Actual/Forecast	Unbiased forecast	%age error
4	80000	44396	1.8019641	60516.733	1.3219484
5	100000	99209	1.0079731	135233.01	0.7394644
6	120000	94808	1.265716	129233.95	0.9285486
7	150000	96813	1.5493787	131966.99	1.1366479
8	180000	172862	1.0412931	235630.31	0.7639085
9	200000	108770	1.8387423	148265.72	1.3489295
10	55000	53052	1.0367187	72315.832	0.7605527
11		mean	1.3631123	stdev	0.2677479

FIGURE **47**

	A	B	C	D	E	F	G	H	I	J
13	Scenario	Year 2	Year 3	Year 4	Year 5	Year 6	Year 7	Year 8	Year 9	Year 10
14	1	1.43	1.33	0.93	0.75	0.57	0.40	0.37	0.38	0.24
15	2	1.39	1.13	0.96	0.59	0.49	0.45	0.46	0.40	0.24
16	3	1.30	1.38	0.98	0.84	0.80	0.65	0.57	0.48	0.35
17	4	1.47	1.49	1.36	1.15	1.20	1.15	0.93	0.99	0.71
18	5	1.23	1.06	0.73	0.45	0.39	0.31	0.28	0.23	0.15
19	6	1.26	1.22	1.08	0.79	0.77	0.70	0.60	0.60	0.49
20	7	1.30	1.02	0.84	0.62	0.45	0.32	0.27	0.24	0.22

Step 2 Therefore, we can create unbiased forecasts in column E by copying the formula

$$=\$D\$11*C4$$

from E4 to E5:E10.

Step 3 In column F, we compute the percentage error of our unbiased forecasts. In cell F4, we compute the percentage error for product 1 with the formula

$$=B4/E4$$

Copying this formula from F4 to F5:F10 generates percentage errors for the other products.

Step 4 In cell F11, we compute the standard deviation (26.7%) of these percentage errors with the formula

$$=STDEV(F4:F10)$$

We are now ready to model 10 years of sales for the new product. To generate year 1 sales, we model year 1 sales to be normally distributed, with a mean of 1.36*90,000 and a standard deviation of .267*(90,000*1.267). To model sales for years 2–10, we use @Risk to randomly choose one of the seven volume-change patterns (or **scenarios**) from Figure 47. Then we use the chosen scenario to generate sales growth for years 2–10.

Step 5 In cell G4, we choose a scenario with the formula

$$=RISKDUNIFORM(A14:A20)$$

This formula gives a 1/7 chance of choosing each scenario.

FIGURE **48**

	G	H	I	J	K	L	M	N	O	P	Q
1	Year 1 Forecast		90000								
2		Year									
3	Scenario	1	2	3	4	5	6	7	8	9	10
4	4	102588.9	151164	225922	306360.9	351610	420801.1	484511.5	451618.5	445300.1	314821.9

Step 6 In H4, we generate year 1 sales with the formula

$$=\text{RISKNORMAL}(I1*D11,(I1*D11)*F11)$$

This implies that

> Mean year 1 sales = (biased forecast)(factor to correct for bias)

> (Standard deviation year 1 sales) = (unbiased forecast for year 1 sales)*(standard deviation of errors as percentage of unbiased forecast)

Step 7 In cell I4, we generate year 2 sales with the formula

$$=\text{H4*VLOOKUP}(\$G\$4,\$A\$14{:}\$J\$20,I3)$$

This formula takes year 1 generated sales and multiplies it by the year 2 growth factor for the chosen scenario. Copying this formula to I4:Q4 generates sales for years 2–10. See Figure 48.

Modeling Statistical Relationships with One Independent Variable

Suppose we want to model the dependence of a variable Y on a single independent variable X. We proceed as follows.

Step 1 Try to find the straight line, power curve, and exponential curve that best fit the data. The easiest way to do this is to plot the points with Excel and use the Trend Curve feature.

- The straight line is of the form $Y = a + bX$.
- The power function is of the form $Y = ax^b$.
- The exponential function is of the form $Y = ae^{bX}$.

Step 2 For each curve and each data point, compute the percentage error

$$\frac{\text{Actual value of } Y - \text{predicted value of } Y}{\text{Predicted value of } Y}$$

Step 3 For each curve, compute mean absolute percentage error (MAPE) by averaging the absolute percentage errors.

Step 4 Choose the curve that yields the lowest MAPE as the best fit.

Step 5 Does at least one of the three curves appear to have some predictive value? Check the plot for this, or look at the *p*-value from the regression; it should be $\leq .15$. If so, model the uncertainty associated with the relationship between X and Y as follows:

- If the straight line is the best fit, then model Y as

    =RISKNORMAL(prediction, standard deviation of actual (not percentage) errors)

- If the power curve or the exponential curve is the best fit, then model Y as

    =RISKNORMAL(prediction, prediction*(standard deviation of percentage errors))

EXAMPLE 12   **Modeling the Cost of Building Capacity**

We are not sure of the cost of building capacity for a new drug, but we believe that costs will run around 50% more (in real terms) than for the drug Zozac. Table 12 gives data on the costs incurred when capacity was built for Zozac.

For example, when 110,000 units of capacity for Zozac were built, the cost was $654,000 (in today's dollars). How would you model the uncertain cost of building capacity for the new product?

Capacity.xls   **Solution**   See the file Capacity.xls.

Step 1 To begin, we plot the best-fitting straight line, power curve, and exponential curve. To do this, use Chart Wizard (X-Y option 1) and click on points till they turn gold. Next, choose the desired curve and select R-SQ and the Equation option. We obtain the graphs in Figures 49–51.

Step 2 In C3:E8 (see Figure 52), we compute the predictions for each curve. In C3:C8, we compute the straight-line predictions by copying from C3 to C3:C8 the formula

$$=5.0623*A3+77.516$$

In D3:D8, we compute the power curve prediction by copying from D3 to D3:D8 the formula

$$=13.483*A3\wedge 0.8229$$

In E3:E8, we compute the exponential curve predictions by copying from E3 to E3:E8 the formula

$$=164.52*EXP(0.0114*A3)$$

Step 3 In F3:H8, we use

$$\frac{\text{Actual value of } Y - \text{predicted value of } Y}{\text{Predicted value of } Y}$$

TABLE 12

Capacity (thousands)	Cost ($ thousands)
20	156
50	350
80	490
110	654
140	760
160	890

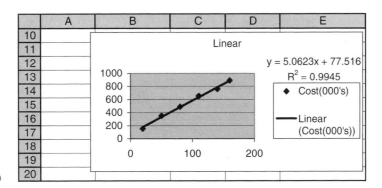

FIGURE 49

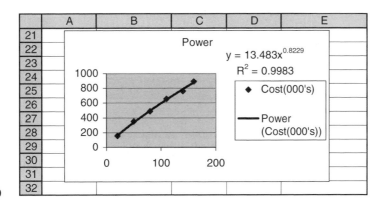

FIGURE 50

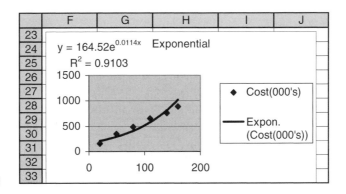

FIGURE 51

	A	B	C	D	E
1	Capacity Cost Modeling				
2	Capacity(0	Cost(000's)	Linear Prediction	Power Prediction	Exponential Prediction
3	20	156	178.762	158.6369	206.6511577
4	50	350	330.631	337.1855	290.9152953
5	80	490	482.5	496.4086	409.5390027
6	110	654	634.369	645.132	576.5327482
7	140	760	786.238	786.7474	811.6199132
8	160	890	887.484	878.1261	1019.463863

FIGURE 52

	F	G	H	I	J	K
1						
2	%age Error Linear	%age Error Power	%age Error Exponential	APE Linear	APE Power	APE Exponential
3	-0.127331	-0.016622	-0.2451046	0.127331	0.016622	0.24510464
4	0.058582	0.038004	0.20309934	0.058582	0.038004	0.20309934
5	0.015544	-0.01291	0.19646724	0.015544	0.01291	0.19646724
6	0.030946	0.013746	0.13436748	0.030946	0.013746	0.13436748
7	-0.033372	-0.033997	-0.0636011	0.033372	0.033997	0.06360109
8	0.002835	0.013522	-0.1269921	0.002835	0.013522	0.12699211
9	St dev	0.026132		0.044768	0.021467	0.16160532
10				MAPE		

FIGURE 53

to compute the percentage error for each model. (See Figure 53.) To do this, simply copy the formula

$$=(\$B3-C3)/C3$$

from F3 to F3:H8.

Step 4   In I3:K9, we compute the MAPE for each equation. We begin by computing the absolute percentage error for each point and each curve by copying the formula

$$=ABS(F3)$$

from I3:K8.

Next we compute the MAPE for each equation by copying the formula

$$=AVERAGE(I3:I8)$$

from I9:K9.

Step 5   We find that the power curve (see J9) has the lowest MAPE. Therefore, we model the cost of adding capacity with a power curve. By entering in G9 the formula

$$=STDEV(G3:G8)$$

we find 2.6% to be the standard deviation of the percentage errors for the power curve. We now model the cost of adding capacity for the new product with the formula

$$=1.5*RISKNORMAL(13.483*(Capacity)^{.8229},.026*13.483*(Capacity)^{.8229})$$

That is, our best guess for the cost of adding capacity has a mean equal to the power curve forecast and a standard deviation equal to 2.6% of our forecast.

EXAMPLE 13   Bidding on a Construction Project

We are bidding against a competitor for a construction project and want to model her bid. In the past, her bid has been closely related to our (estimated) cost of completing the project. See file Biddata.xls and Figure 54.

Biddata.xls

Figures 55–57 give the best fitting linear, power, and exponential curves.

As in Example 12, we compute predictions and MAPEs for each curve (see Figure 58). The linear curve has the smallest MAPE. Computing the actual errors for the linear curve's predictions (in column F) and their standard deviation, we find a standard deviation of .94. Therefore, we model our competitor's bid as

$$=RISKNORMAL(1.489*(Our cost) - 1.7893, .94)$$

	A	B	C	D	E	F
1	(All numbers in 000's)					
2	Our cost	Comp1 bid	Linear prediction	Power prediction	Exponential prediction	Actual Linear Error
3	10	13	13.1027	13.35697	16.3795084	-0.1027
4	14	20	19.0587	19.07213	19.5315493	0.9413
5	16	22	22.0367	21.96795	21.3282198	-0.0367
6	18	25	25.0147	24.88511	23.2901627	-0.0147
7	30	44	42.8827	42.73548	39.4893521	1.1173
8	25	34	35.4377	35.23444	31.6909474	-1.4377
9	38	56	54.7947	54.88668	56.1502464	1.2053
10	44	63	63.7287	64.10133	73.114819	-0.7287
11	24	33	33.9487	33.74424	30.3267775	-0.9487
12					stdev	0.94189151

FIGURE 54

FIGURE 55

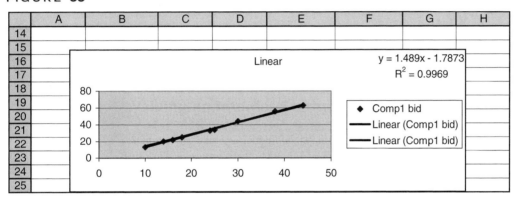

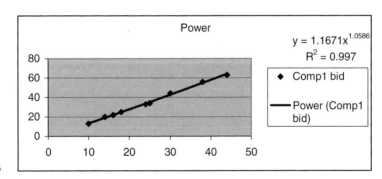

FIGURE 56

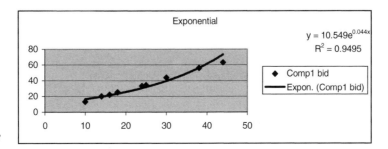

FIGURE 57

	G	H	I	J	K	L
2	Linear %age error	Power %age error	Exponential %age error	Linear abs %age error	Power abs %age error	Exponential %age error
3	-0.00784	-0.02673	-0.20633	0.007838	0.026726	0.206325
4	0.04939	0.048651	0.023984	0.04939	0.048651	0.023984
5	-0.00167	0.001459	0.031497	0.001665	0.001459	0.031497
6	-0.00059	0.004617	0.073415	0.000588	0.004617	0.073415
7	0.026055	0.029589	0.114224	0.026055	0.029589	0.114224
8	-0.04057	-0.03504	0.072862	0.04057	0.035035	0.072862
9	0.021997	0.020284	-0.00268	0.021997	0.020284	0.002676
10	-0.01143	-0.01718	-0.13834	0.011434	0.017181	0.138342
11	-0.02795	-0.02206	0.088147	0.027945	0.022055	0.088147
12				0.020831	0.022844	0.083497
13				MAPE		

FIGURE 58

EXAMPLE 14   The Effects of New Competition on Price

For similar products, the year after the first competitor comes in has historically shown a significant price drop. Figure 59 contains data on this situation.

For example, for the first product, a competitor entered in year 1. During year 2, a 22% price drop was observed, after allowing for a normal inflationary increase of 5% during the second year. Model the effect on price the year after the first competitor enters the market. See file Pricedata.xls.

Pricedata.xls

Solution   Figures 60–62 give the best-fitting linear, power, and exponential curves. The extremely low R^2 values imply that the year of entry has little or no effect on the price drop the year after the first competitor comes in. Therefore, we model price drop as a RISKNORMAL function, using the mean and standard deviation found in D14 and D15. If a competitor enters during year t, we would model the year $t + 1$ price with the formula

$$=1.05*(\text{year } t \text{ price})*\text{RISKNORMAL}(.803,.0366)$$

Note: $.803 = 1 - .197$.

	B	C	D
3	Year competitor enters	Share drop next year	Price drop next year
4	1	35	22
5	1	33	21
6	2	20	17
7	3	15	15
8	3	13	19
9	4	14	24
10	5	10	15
11	6	9	22
12	5	11	25
13	4	13	17
14		Mean	19.7
15		Std Dev	3.622461

FIGURE 59

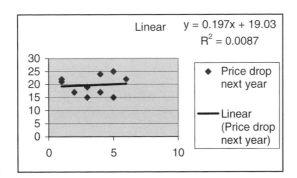

FIGURE 60

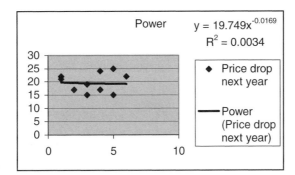

FIGURE 61

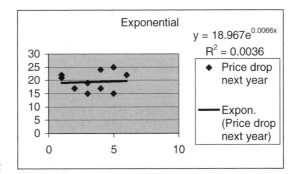

FIGURE 62

Here, the assumption is that the market drop during a year is normally distributed. To check this, we could compute the skewness (with the SKEW function) and kurtosis (with the KURT function) of the data. If both the skewness and kurtosis are near 0, the market drop is probably normally distributed. An alternate approach to modeling the drop in price is to use the formula RISKDUNIFORM(D4:D13). This ensures that the drop in price is equally likely to assume one of the observed values. This approach has the advantage of not automatically assuming normality. The disadvantage, however, is that using the RISKDUNIFORM function implies that only 10 values of price drop are possible.

PROBLEMS

Group A

1 You are considering developing a new product. Forecast year 1 sales are for 80,000 units, and the year 1 price is $4.00 per unit. In file Simidata.xls you are given data on seven similar products from the past. (See Figure 63.)

For example,

For product 1, actual sales were 92.26% of forecast sales.

Year 2 price (in real dollars) was 76.7% of year 1 price.

Year 2 demand was 30.7% more than year 1 demand.

Product 1 only sold for 6 years.

The risk-adjusted discount rate is 11% per year. We assume that the price index will climb 5% per year.

We are unsure about the fixed cost of developing the product. It is equally likely to be $50,000 or $150,000. We are also unsure about the year 1 variable cost of producing it. It is equally likely to be $1, $1.50, or $2. After year 1, variable cost will climb by 5% per year. It costs $3 to build one unit of annual capacity.

a Assuming 80,000 units of annual capacity, estimate the 10-year risk-adjusted NPV of this product.

b What capacity level do you recommend?

2 You are trying to estimate NPV of profit for a new computer product, which you are confident will sell for ten years. You are given the following information.

The hurdle rate is 15%. Assume end of year for profits.

The total cost of developing the product will be spread equally over the product's life. Total development cost will be between $2 billion and $11 billion. There is a 25% chance that total fixed cost is $3 billion or less, a 50% chance of $6 billion or less, and a 75% chance of $9 billion or less.

The total year 1 market size (in terms of annual unit sales) is unknown but is believed to be between 0 and 600 million units. Unit sales of 100 million and 500 million are equally likely. Unit sales of 200 million and 400 million are equally likely, and are 4 times as likely as sales of 100

FIGURE 63

	A B	C	D	E	F	G	H
6	**Other Products**	1	2	3	4	5	6	7
7	Forecast Year 1	10000	15000	20000	25000	30000	18000	12000
8	Actual Year 1	9226	18544	20147	24093	27517	21670	12345
9	Year 1 Price	$ 10.00	$ 11.00	$ 12.00	$ 9.00	$ 8.00	$ 7.00	$ 9.00
10	Year 2 Price %age change	0 .7671376	1.01958	1.157148	0.799233	0.66222	0.96338	1.108995
11	Year 3 Price %age change	0.77301544	0.916731	0.629211	0.763033	0.785409	1.12807	0.770459
12	Year 4 Price %age change	0.93641094	1.07326	0.704279	1.032535	1.214266	0.607159	0.635232
13	Year 5 Price %age change	0.62486148	0.838744	0.730323	1.128628	1.222691	0.915762	0.709054
14	Year 6 Price %age change	0 .9909713	0.836317	1.178683	0.835511	1.186193	0.98035	0.870983
15	Year 7 Price %age change	0	1.154532	0.778286	1.008012	1.155539	0.83953	1.044165
16	Year 8 Price %age change	0	1.002691	0.991726	0.70686	0.871703	0	0.789561
17	Year 9 Price %age change	0	0.866046	0.933498	0	0.748535	0	0.800709
18	Year 10 Price %age change	0	0	1.137376	0	0.804221	0	0.963354
19	Year 2 % change in demand	1.30771172	1.257895	1.30467	1.326283	1.371715	1.203553	1.246827
20	Year 3 %age change change in demand	1.35463816	1.176022	1.439216	1.681935	1.186842	1.105765	0.705081
21	Year 4 % change change in demand	0.89116031	1.156565	0.940743	1.060037	0.953028	1.280444	0.927083
22	Year 5 %age change change in demand	0.62	0.728722	0.956427	0.744345	0.711915	0.710494	0.536207
23	Year 6 % change change in demand	0.53	0.529757	0.571999	0.6862	0.613999	0.572403	0.393907
24	Year 7 %age change change in demand	0	0.298447	0.193459	0.375018	0.432	0.269806	0.561965
25	Year 8 %age change change in demand	0	0.305065	0.314531	0.298697	0.294049	0	0.288539
26	Year 9 %age change change in demand	0	0.2	0.25	0	0.2	0	0.22
27	Year 10 %age change change in demand	0	0	0.15	0	0.16	0	0.12

TABLE 13

Year Competitor Entered	Drop in Share
2	21
3	17
4	15
5	13
6	12
2	20
4	16
5	12
6	11
7	10
8	9
8	10
10	9
12	8

TABLE 14

Predicted	Actual
40,000	37,000
50,000	42,000
60,000	56,000
70,000	67,000
80,000	75,000

million. Sales of 300 million are 5 times as likely as sales of 100 million. Each year, market growth is expected to average 5%, and during each year we are 95% sure that market growth will be between 3% and 7%.

Our most likely year 1 market share is 30%. There is a 5% chance that our market share will be less than or equal to 10% and a 5% chance that our market share will be more than 40%. A triangular distribution appears to be reasonable for market share. In later years, we expect market share, on average, to equal the previous year's share, but there is a 95% chance that market share could change by up to 20% of its current value.

The year 1 price charged for each unit follows a triangular random variable, with the most likely value $50, worst case $45, and best case $60. Each year, unit price will increase 5%.

The year 1 unit variable cost of production follows a triangular random variable with worst case $30, best case $20, and most likely case $24. Each year, variable costs will increase 5%.

a You are 95% sure that the mean NPV of the project is between _____ and _____. Run 1,600 iterations.

b What is the chance that this project will meet its hurdle rate?

c What are key drivers of the project's profitability?

3 You are trying to model what fraction of market share a new drug will lose the year a competitor comes in. Table 13 gives information for similar drugs. For example, competition for one drug entered the market 2 years after our drug, and we then lost 21% of our market share. How would you model the effect of competition on our product sales?

Group B

4 You own a small biotech firm. Eli Daisy wants to buy the rights to a potential cancer drug you are developing.

There is no way you could sell the product yourself. It will cost you $350,000 (payable at end of year 0) to develop the drug. Here's what Daisy has offered. At the end of years 1–8, Daisy will pay you 10% of the sales revenue for the drug, up to a maximum of $700,000. You discount cash flows at 20% per year. Each year, the drug sells for $20 per unit. You believe that the drug will sell 50,000 units during year 1. Table 14 shows your forecasts and actual year 1 sales for similar products in the past.

The pattern of sales for similar products is as follows. For a certain number of years, sales increase by a given percentage. Then, for all remaining years, sales decrease by a given percentage. You believe there is a 20% chance that sales will increase for 2 years, a 50% chance for 3 years, and a 30% chance for 4 years.

The percentage increase during the first path of the product life cycle will be between 2% and 20% per year. There is one chance in four that the annual percentage increase during this part of the product life cycle will be 5% or less; one chance in two of 15% or less, and three chances in 4 of 18% or less.

The annual percentage decrease during the remaining portion of the product life cycle will be between 2% and 10%. A 6% annual decrease is four times as likely as an 8% annual decrease. A 4% annual decrease is twice as likely as an 8% annual decrease.

Based on this information, would you take the deal? Explain your answer. What is the single most important driver of the deal's NPV?

5 You are trying to evaluate the profitability of a new drug produced by Eli Lilly. The drug will be sold during the years 2005–2010.

Development cost will be charged on September 10, 2004. The development cost will be between $.5 million and $5 million. A development cost of $2 million is four times as likely as a development cost of $1 million. A development cost of $4 million is twice as likely as a development cost of $1 million.

Unit sales during 2005 will be between 80,000 and 240,000. There is a 25% chance that 2005 unit sales will be less than or equal to 100,000 units, a 50% chance that they will be less than or equal to 140,000 units, and a 75% chance that they will be less than or equal to 200,000 units.

After year 1, sales will decay at a constant annual rate. For similar products, the decay rates have been 5%, 6%, 8%, 9%, 10%, 4%, 3%, and 8%.

Each year, you will charge $45 for the product.

Each year's variable production cost will depend on the number of units sold. For a drug with similar cost structure,

FIGURE **64**

	F	G
6	000's	000's
7	Units produced	Cost
8	40	813.323
9	50	999.459
10	60	1230.911
11	70	1399.077
12	80	1592.645
13	90	1812.399
14	100	2013.139
15	110	2709.943
16	120	3405.542
17	130	4096.212
18	140	4815.177
19	150	5516.294
20	160	6200.432
21	170	6914.829
22	180	7613.689
23	190	8320.617
24	200	9012.181

FIGURE **65**

	K	L
11	Actual	Forecast
12	20000	26000
13	30000	35000
14	10000	14000
15	40000	48000
16	50000	62000

(*Hint:* In modeling annual variable cost, you need not do a MAPE. From the proper plot, the appropriate model should be clear.)

6 GM is considering producing a new car. GM's current net income for each of the next 6 years is assumed to be $100 million, which we assume to be received on June 30 of the years 2004–2009. The tax rate is 40%. Assume that no other GM projects involve depreciation. The fixed cost of developing the new car will be between $20 million and $40 million, with a most likely value of $25 million. The entire fixed cost is incurred on June 30, 2004 and is depreciated on a straight-line basis during the years 2005–2009. All future cash flows are received midyear. The car is assumed to be sold during the years 2005–2009. Forecast for 2005 unit sales is 15,000. Past forecasts and actual sales during the first year of similar models are as shown in Figure 65.

During the years 2006–2009, sales are assumed to decay at the same rate each year. This rate will be between 5% and 20%, with a 12% decay rate twice as likely as an 8% decay rate and a 16% decay rate three times as likely as an 8% decay rate. During 2005, the car will sell for $13,000. The price will increase by the same percentage each year, with 1%, 2%, and 3% price increases being equally likely. During 2005, variable costs are $11,000. During 2006–2009, variable costs will increase by the same percentage, with increases of 2%, 4%, and 6% being equally likely. Discount cash flows at 15%. Should GM produce the car?

the variable cost as a function of units produced was as shown in Figure 64. For example, during year 1, 40,000 units were produced, and the cost was $813,323. (See file Sim3data.xls.)

Assume that cash flows are discounted at 10% and cash flows for years 2005–2010 may be considered to be received midyear (June 30).

a After running 900 iterations, you are 95% sure that actual NPV earned by the drug (in 09/10/04 dollars) is between _____ and _____.

b What is the key driver of the drug's NPV?

11.11 Simulation and Bidding

In situations in which you must bid against competitors, simulation can often be used to determine an appropriate bid. Usually you do not know what a competitor will bid, but you may have an idea about the range of bids a competitor may choose. In this section, we show how to use simulation to determine a bid that maximizes your expected profit. First, we briefly discuss generating observations from a uniformly distributed random variable.

Uniform Random Variables

A random variable is said to be uniformly distributed on the closed interval [a, b] (written U(a, b)) if the random variable is equally likely to assume any value between a and b inclusive. To generate samples from a U(a, b) random variable, enter the formula

$$=RISKUNIFORM(a,b)$$

into a cell.

We now show how to use simulation to determine a bid that maximizes expected profit.

EXAMPLE 15   **Bidding**

You are going to make a bid on a construction project. You believe it will cost you $10,000 to complete the project. Four competitors are going to bid against you. Based on past history, you believe that each competitor's bid is equally likely to be any value between your cost of completing the project and triple your cost of completing the project. You also believe that each competitor's bid is independent of the other competitors' bids. What bid maximizes your expected profit?

Solution   In our solution, all amounts will be in thousands of dollars. The statement of the problem implies that each competitor's bid is U(10, 30), and the bids of the competitors are independent. Our simulation is shown in Figure 66 (file Bid.xls). We proceed as follows.

Bid.xls

Step 1   In cell C3, we enter the cost of the project.

Step 2   In cell C4, we enter ten possible bids (11, 12, 13, 14, 15, 16, 17, 18, 19, and 20) with the formula

$$=RISKSIMTABLE(\{11,12,13,14,15,16,17,18,19,20\})$$

	A	B	C	D	E	F
1	Bidding Example					
2						
3	My cost(thousands)		10			
4	My bid(thousands)		11			
5	Competitor 1 Bid		15.23686647			
6	Competitor 2 Bid		24.37239289			
7	Competitor 3 Bid		23.26008201			
8	Competitor 4 Bid		25.59600472			
9	Profit(thousands)		1			
10						
11		Cell	Name	Minimum	Mean	Maximum
12						
13		C9	(Sim#1) Profit(thousa...	0	0.81	1
14		C9	(Sim#2) Profit(thousa...	0	1.3	2
15		C9	(Sim#3) Profit(thousa...	0	1.575	3
16		C9	(Sim#4) Profit(thousa...	0	1.66	4
17		C9	(Sim#5) Profit(thousa...	0	1.5625	5
18		C9	(Sim#6) Profit(thousa...	0	1.41	6
19		C9	(Sim#7) Profit(thousa...	0	1.155	7
20		C9	(Sim#8) Profit(thousa...	0	0.96	8
21		C9	(Sim#9) Profit(thousa...	0	0.7425	9
22		C9	(Sim#10) Profit(thous...	0	0.55	10

FIGURE 66
Bidding Simulation

Step 3   In C5, we generate the bid of the first competitor by entering the formula

$$=RISKUNIFORM(C\$3,3*C\$3)$$

Copying this formula to the range C6:C8 generates the bids of the other three competitors. (Why does this ensure that their bids are independent?)

Step 4   In cell C9, we compute the actual profit for this trial by entering the formula

$$=IF(C4<=MIN(C5:C8),C4-C3,0)$$

This ensures that if we win the bid (C4≤MIN(C5:C8)), then our profit equals our bid less the project cost of $10,000; if we don't win the bid (C4>MIN(C5:C8)), then we earn no profit. This statement assumes that we win all ties, but the chance of a tie bid is negligible (why?), so this really does not matter. To see how things work, hit the recalculation (F9) button and see how the cells of the spreadsheet change.

Step 5   To determine the bid that maximizes expected profit, we ran 400 iterations of this spreadsheet for each bid with @Risk. From Figure 66, it appears that a bid between $13,000 and $15,000 will maximize expected profit (with an expected profit of $1,660).

Step 6   To zero in on the bid that maximizes expected profit, we replaced the formula in cell C4 with

$$=RISKSIMTABLE(\{13.2,13.4,13.6,13.8,14,14.2,14.4,14.6,14.8\})$$

One hundred iterations of this spreadsheet indicate that a bid of around $14,200 maximizes expected profit (an expected profit of around $1,800 is earned).

PROBLEMS

Group A

1   If the number of competitors in Example 15 were to double, how would the optimal bid change?

2   If the average bid for each competitor stayed the same, but their bids exhibited less variability, would the optimal bid increase or decrease? To study this question, assume that each competitor's bid follows each of the following random variables:

   a $U(15, 25)$

   b $U(18, 22)$

3   Warren Millken is attempting to take over Biotech Corporation. The worth of Biotech depends on the success or failure of several drugs under development. Warren does not know the actual (per share) worth of Biotech, but the current owners of Biotech do know the actual worth of the company. Warren assumes that Biotech's actual worth is equally likely to be between $0 and $100 per share. Biotech will accept Warren's offer if it exceeds the true worth of the company. For example, if the current owners think Biotech is worth $40 per share and Warren bids $50 per share, they will accept the bid. If the current owners accept Warren's bid, then Warren's corporate strengths immediately increase Biotech's market value by 50%. How much should Warren bid?

11.12 Playing Craps with @Risk

Craps is a very complex game. With @Risk, it is easy to estimate the probability of winning at craps.

EXAMPLE 16   Craps

In the game of craps, a player tosses two dice. If the first toss yields a 2, 3, or 12, the player loses. If the player rolls a 7 or 11 on the first toss, he or she wins. Otherwise, the player continues tossing the dice until he or she either matches the number thrown on the first roll (called the *point*) or tosses a 7. Rolling the point before rolling a 7 wins. Rolling a 7 before the point loses. By complex calculations, it can be shown that a player wins at craps 49.3% of the time. Use @Risk to verify this.

Solution   The key observation is that we do not know how many rolls the game will take. Suppose the game does not end on the first toss. The least likely points to be made are 4 and 10 which have probability $3/36 = 1/12$ of being made. Therefore, after the first toss, there is at least a $(1/12)$ + probability of 7 = $(1/12) + (1/6) = (1/4)$ chance that the game will end on each toss. Thus, the chance of the game continuing on each toss is at most $(3/4)$. After (say) 50 tosses, the probability that the game is still going on is at most $.75^{49} = 7$ in 10,000,000. Therefore, we can cut off the game after 50 tosses and not worry about the (fewer than 1 in a million) games that go on beyond 50 tosses. After each dice roll, we keep track of the game status:

$$0 = \text{game lost}$$
$$1 = \text{game won}$$
$$2 = \text{game still going}$$

Craps.xls

The output cell will keep track of the status of the game after the 50th toss. A 1 will indicate a win, and a 0 will indicate a loss. The work is in the file Craps.xls. See Figure 67.

Step 1   In B2, we use the RISKDUNIFORM function (discrete uniform random variable) to generate the roll of the dice on the first toss with the formula

$$= \text{RISKDUNIFORM}(\$AD\$9:\$AD\$14)$$

The RISKDUNIFORM function ensures that each of its arguments is equally likely. Therefore, each die has an equal $(1/6)$ chance of yielding a 1, 2, 3, 4, 5, or 6.

FIGURE 67

	A	B	C	D	E	F	G	H	AX	AY
1	TOSS#	1	2	3	4	5	6	7	49	50
2	Die Toss 1	3.5	3.5	3.5	3.5	3.5	3.5	3.5	3.5	3.5
3	Die Toss 2	3.5	3.5	3.5	3.5	3.5	3.5	3.5	3.5	3.5
4	Total	7	7	7	7	7	7	7	7	7
5	GAME STATUS	1	1	1	1	1	1	1	1	1
6	0=LOSS	WIN??	1							
7	1=WIN									
8	2=STILL GOING	95% CI								
9	LOWER									
10	UPPER									

Copying this formula to the range B2:AY3 generates both dice rolls for 30 tosses. Note that we have hidden rolls 8–28.

Step 2 In B4:AY4, we compute the total dice roll on all 30 rolls by copying from B4 to C4:AY4 the formula

$$=SUM(B4:C4)$$

Step 3 In cell B5, we determine the game status after the first roll with the formula

$$=IF(OR(B4=2,B4=3,B4=12),0,IF(OR(B4=7,B4=11),1,2))$$

Note that a 2, 3, or 12 will result in a loss, a 7 or 11 will result in a win, and any other roll will result in the game continuing.

Step 4 In cell C5, we compute the status of the game after the second roll with the formula

$$=IF(OR(B5=0,B5=1),B5,IF(C4=\$B4,1,IF(C4=7,0,2)))$$

Note that if the game ended on the first roll, we maintain the status of the game. If we make our point, we record a win with a 1. If we roll a 7, we record a loss. Otherwise, the game is still going.

Copying this formula from C5 to D5:AY5 records the game status after rolls 2–50. The game result is in AY5, which we copy to C6 so that we can easily see it. After running 4,000 iterations with output cell C6, we obtain a 48.3% chance of winning. With 10,000 iterations, we usually obtain a probability very close to 49.3%.

11.13 Simulating the NBA Finals

The Indiana Pacers came within two plays (one questionable foul call on Dale Davis in game 6 and Travis Best missing a shot in game 4) of winning the 2000 NBA championship. Before the series, what was the probability that the Lakers would win the series? From the Sagarin ratings (found at http://www.kiva.net/~jsagarin/), we found that the Lakers are around 4 points better than the Pacers. The home team has a 3-point edge, and games play out according to a normal distribution, with mean equal to our prediction and a standard deviation of 12 points. Past history shows that the Sagarin forecasts exhibit no bias. In the file Finals.xls and Figure 68, we simulate the 2000 NBA Finals. Recall that the Lakers were at home during games 1, 2, 6, and 7, while the Pacers were at home during games 3–5. (Note: We always make a series go 7 games, because we do not know when it will actually end.) If the Lakers win at least 4 of the 7 games, they win the series, which is indicated by a 1 in cell I14. We have named the cells in D2:D5 with the range names given in C2:C5.

Step 1 In G5:G11, we generate our forecast for each game by copying the formula

$$=IF(F5="LA",HE+LA-IND,-HE+LA-IND)$$

from G5 to G6:G11.

Step 2 In H5:H11, we generate the Lakers' margin of victory in each game as normally distributed with a standard deviation of 12 and mean given in column G. Just copy from H5 to H6:H11 the formula

$$=RISKNORMAL(G5,STDEV)$$

Finals.xls

FIGURE 68

	B	C	D	E	F	G	H	I
1	NBA Finals 2000							
2		IND	5					
3		LA	9					
4		HE	3	Game	Home	LA Forecast	LA Margin	LA Win
5		STDEV	12	1	LA	7	1.312640391	1
6				2	LA	7	0.911581437	1
7				3	IND	1	-0.423239113	0
8				4	IND	1	22.02026177	1
9				5	IND	1	-13.825966	0
10				6	LA	7	-5.963644407	0
11				7	LA	7	-2.505744158	0
12							LA total wins	3
13								
14							LA Wins series?	0
15								
16								
17								
18		Name	NPV	LA Wins series? / LA Win				
19		Description	Output	Output				
20		Cell	[]Sheet1!F{	[finals.xls]Sheet1!I14				
21		Minimum =	0	0				
22		Maximum =	0	1				
23		Mean =	0	**0.796875**				

Step 3 In I5:I11, we determine if the Lakers won the game by copying from I5 to I6:I11 the formula

$$=IF(H6>0,1,0)$$

Step 4 In cell I12, we compute the total number of Lakers wins in the series with the formula

$$=SUM(I5:I11)$$

Step 5 Note that if the Lakers win at least 4 games, they win the series. In cell I14, we determine if the Lakers win the series with the formula

$$=IF(I12>=4,1,0)$$

From the @Risk output, we find that the Lakers had an 80% chance to win the series. The bookmakers had L.A. as a 7-1 favorite, which means (after taking out a 10% profit) they believed that the Lakers had around a 90% chance to win.

REVIEW PROBLEMS

Group A

1 The New York Knicks and the Chicago Bulls are ready for the best-of-seven NBA Eastern finals. The two teams are evenly matched, but the home team wins 60% of the games between the two teams. The sequence of home and away games is to be chosen by the Knicks. The Knicks have the home edge and will be the home team for four of the seven scheduled games. They have the following choices (home team is listed for each game):

Sequence 1: NY, NY, CHIC, CHIC, NY, CHIC, NY
Sequence 2: NY, NY, CHIC, CHIC, CHIC, NY, NY

Use simulation to show that either sequence gives the Knicks the same chance of winning the series.

2[†] You currently have $100. Each week, you can invest any amount of money you currently have in a risky investment. With probability .4, the amount you invest is tripled (e.g., if you invest $100, you increase your asset

[†]Based on Kelly (1956).

position by \$300), and with probability .6, the amount you invest is lost. Consider the following investment strategies:

(1) Each week, invest 10% of your money.

(2) Each week, invest 30% of your money.

(3) Each week, invest 50% of your money.

Simulate 100 weeks of each strategy 50 times. Which strategy appears to be best? In general, if you can multiply your investment by M with probability p and lose your investment with probability q, you should invest a fraction $\frac{p(M-1)-q}{M-1}$ of your money each week. This strategy maximizes (for a favorable game) the expected growth rate of your fortune and is known as the **Kelly criterion.**

3[†]   The Magellan mutual fund has beaten the Standard and Poor's 500 during 11 of the last 13 years. People use this as an argument that you can "beat the market." Here's another way to look at it that shows that Magellan's beating the market 11 out of 13 times is not unusual. Consider 50 mutual funds, each of which has a 50% chance of beating the market during a given year. Use simulation to estimate the probability that over a 13-year period the "best" of the 50 mutual funds will beat the market for at least 11 out of 13 years. This probability turns out to exceed 40%, which means that the best mutual fund's beating the market 11 out of 13 years is not an unusual occurrence!

4   You have made it to the final round of "Let's Make a Deal." You know that there is \$1 million behind either door 1, door 2, or door 3. It is equally likely that the prize is behind any of the three doors. The two doors without a prize have nothing behind them. You randomly choose door 2. Before you see whether the prize is behind door 2, Monty chooses to open a door that has no prize behind it. For the sake of definiteness, suppose that before door 2 is opened, Monty reveals that there is no prize behind door 3. You now have the opportunity to switch and choose door 1. Should you switch?

Use a spreadsheet to simulate this situation 400 times. For each "trial" use an @Risk function to generate the door behind which the prize lies. Then use another @Risk function to generate the door that Monty will open. Assume that Monty plays as follows: Monty knows where the prize is and will open an empty door, but he cannot open door 2. If the prize is really behind door 2, Monty is equally likely to open door 1 or door 3. If the prize is really behind door 1, Monty must open door 3. If the prize is really behind door 3, Monty must open door 1.

5   Star-crossed soap-opera lovers Noah and Julia have had a big argument. Julia's sister Maria wants Noah and Julia to make up, so she has told them both to go to the romantic gazebo at 1 P.M. Unfortunately, Noah and Julia are not punctual. Each is equally likely to show up at the gazebo any time between 1 and 2 P.M. Assuming that each will stay for 20 minutes, what is the probability that they will meet? You can model the arrival of each person using a RISKUNIFORM random variable. For example, RISKUNIFORM(1,2) is equally likely to choose any number between 1 and 2 (including the endpoints 1 and 2).

6   The game of Chuck-a-Luck is played as follows: You pick a number between 1 and 6 and toss three dice. If your number does not appear, you lose \$1. If your number appears x times, you win \$$x$. On the average, how much money will you win or lose on each play of the game?

7   I toss a die several times until the total number of spots I have seen is at least 13. What is the most likely total that will occur?

Group B

8[‡]   When the team is behind late in the game, a hockey coach usually waits until there is one minute left before pulling the goalie. Actually, coaches should pull their goalies much sooner. Suppose that if both teams are at full strength, each team scores an average of .05 goal per minute. Also suppose that if you pull your goalie, you score an average of .08 goal per minute, while your opponent scores an average of .12 goal per minute. Suppose you are one goal behind with five minutes left in the game. Consider the following two strategies:

Strategy 1:   Pull your goalie if you are behind at any point in the last five minutes of the game; put him back in if you tie the score or go ahead.

Strategy 2:   Pull your goalie if you are behind at any point in the last minute of the game; put him back in if you tie the score or go ahead.

Which strategy maximizes your chance of winning or tying the game? Simulate the game using ten-second increments of time. Use the RISKBINOMIAL function to determine whether a team scores a goal in a given ten-second segment. It is acceptable to do this because the probability of scoring two or more goals in a ten-second period is near 0.

9   Suppose we toss an ordinary die 5 times. A 4-straight occurs if exactly 4 (not 5) of our rolls are consecutive integers. For example, if we roll 1, 2, 3, 4, 6 we have a 4-straight. Also 3, 4, 5, 6, 1, 1 is a 4-straight. However 2, 3, 4, 5, 6 is not a 4-straight. After running 4,000 iterations, you are 95% sure that the chance of tossing a 4-straight is between _____ and _____.

10   Buffie the Vampire Slayer is going to Las Vegas to relax. She is going to play the following game of blackjack. She throws a pair of dice until the cumulative total of her tosses is at least 4. If her total is 8 or more, she loses. Assuming that Buffie has not yet lost, the croupier (Spike) tosses the dice until his total is at least 4. If Spike's total is 8 or more, then Buffie (assuming she did not total 8 or more) wins. Otherwise, we compare Spike's and Buffie's totals. The high total wins, with a tie going to Spike. After running 900 iterations, you are 95% sure Buffie's chance of winning the game is between _____ and _____.

11   Wheaties is producing cereals with five different sets of trading cards:

- Rock stars
- NBA stars
- Baseball stars
- Hockey stars
- Football stars

[†]Based on Marcus (1990).

[‡]Based on Morrison and Wheat (1984).

Each box contains one set of trading cards, and you do not know which set is in a box until you open it.

a On the average, how many boxes are needed to obtain all five sets of trading cards?

b You are 95% sure that between _____ and _____ boxes of Wheaties must be purchased to obtain all five sets of trading cards.

REFERENCES

Kelly, J. "A New Interpretation of Information Rate," *Bell System Technical Journal* 35(1956):917–926.

Marcus, A. "The Magellan Fund and Market Efficiency," *Journal of Portfolio Management* (1990, Fall):85–88.

Morrison, D., and R. Wheat. "Pulling the Goalie Revisited," *Interfaces* 16(no. 6, 1984):28–34.

Optimization under Uncertainty with Riskoptimizer

In Volume 1, we studied optimization and learned how to maximize or minimize an objective, subject to certain constraints. In Chapters 9 and 11 of this book, we studied simulation. In many simulation problems, optimization is an important issue. For example, in Section 11.1, how many calendars should be ordered? We now show how the Excel add-in Riskoptimizer can be used to optimize simulations (in the sense of maximizing expected profit or minimizing risk).

12.1 Introduction to Riskoptimizer: The News Vendor Problem

The Excel add-in @Risk is used to obtain descriptive statistics for situations in which we make decisions under uncertainty. With the add-in Riskoptimizer, we can actually find the *best* decisions to make under uncertainty. The following easy example will introduce the power of Riskoptimizer.

We need to determine how many year 2005 nature calendars to order in August 2004. It costs $2.00 to order each calendar, which we sell for $4.50. After January 1, 2005, left-over calendars are returned for $0.75. Our best guess is that the number of calendars demanded is governed by the following probabilities:

Demand	Probability
100	.3
150	.2
200	.3
250	.15
300	.05

How many calendars should we order?

Newsdis.xls

Our final result is in file Newsdis.xls. (See Figure 1.) We proceed as follows.

Step 1 Enter parameter values in C3:C5.

Step 2 In cell C1, enter a trial value for the number of calendars to order. Later, Riskoptimizer will be used to determine the best order quantity for calendars.

Step 3 Use @Risk to generate demand according to the above probabilities. Type in C2 the formula

$$=RISKDISCRETE(F5:F9,G5:G9)$$

	A	B	C	D	E	F	G
1	Order quantity		200				
2	Quantity demanded		172.5				
3	Sales price		$4.50			Order Quant	Mean Profit
4	Salvage value		$0.75			100	0.3
5	Purchase price		$2.00			150	0.2
6						200	0.3
7	Full price revenue	$776.25				250	0.15
8	Salvage revenue	$20.63				300	0.05
9	Costs	$400.00					
10	Profit	$396.88					
11							
12							
13			Mean = 350				
14							
15							

FIGURE 1

This generates a demand for 100 calendars 30% of the time, 150 calendars 20% of the time, etc. This demand could also have been generated with the formula

$$=RISKDISCRETE(\{100,150,200,250,300\},\{.3,.2,.3,.15,.05\})$$

Note that in either format, the demands are listed first, followed by the probabilities. Approximately 30% of the time a demand of 100 will occur, around 20% of the time a demand of 150 will occur, etc.

Step 4 In cell B7, compute full-price revenue with the formula

$$=C3*MIN(C1,C2)$$

This ensures that we will sell at full price the minimum of quantity ordered and quantity demanded.

Step 5 In B8, compute salvage revenue with the formula

$$=C4*IF(C1>C2,(C1-C2),0)$$

This ensures that the number left over is (number ordered) − (number demanded), as long as that is greater than 0.

Step 6 In B9, compute ordering costs with the formula

$$=C1*C5$$

Step 7 In cell B10, compute profit with the formula

$$= B7+B8-B9$$

Our goal is to find the order quantity that maximizes expected profit. For the time being, we ignore the risk associated with our decision. We are now ready to use Riskoptimizer. Basically, Riskoptimizer tries to optimize a (possibly random) function of a target cell by changing adjustable cells. If desired, the adjustable cells or other spreadsheet cells may have to satisfy desired constraints. Riskoptimizer uses genetic algorithms (Goldberg (1989); Davis (1991)) to find the best values for adjustable cells. To explain how Riskoptimizer works, we focus on the calendar example. We want to maximize the mean profit (this is the target cell) by adjusting the number of calendars ordered (this is the adjustable cell). It is unreasonable to order more than 300 or less than 0 calendars, so we constrain the number of ordered calendars to be an integer between 0 and 300. Riskoptimizer now

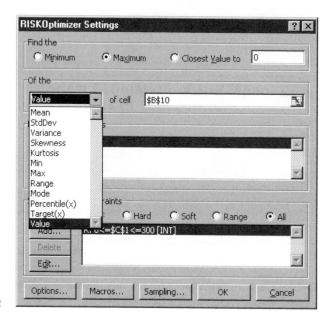

FIGURE 2

proceeds as follows. For the trial value of the adjustable cell in the spreadsheet, Riskoptimizer runs enough iterations for the mean profit to satisfactorily converge. (We may limit the number of iterations, if desired.) Then the mean profit for this order quantity is recorded. Riskoptimizer next selects another order quantity and runs enough iterations to obtain a satisfactory estimate of the mean profit for the second order quantity. Riskoptimizer continues in this fashion, determining the mean profit associated with different order quantities. Genetic algorithm technology is used to operationalize the idea of survival of the fittest. Riskoptimizer is much more likely to try order quantities with large mean profit than with small mean profit. After a while, Riskoptimizer will zero in on the order quantity that has the largest mean profit.

We begin with a brief discussion of the Riskoptimizer icons.

The Settings Icon

This icon is used to define the problem Riskoptimizer is to solve.

Target Cells

With the Settings icon, you may choose a target cell in which to maximize or minimize a random function of the target cell. As shown in Figure 2, you may maximize the mean of a cell or minimize the variance or standard deviation of a cell. To maximize the 5th percentile of a cell, select Percentile (.05). To minimize the probability that a company's profit is less than $1,000,000, select Target(1000000).

Adjustable Cells

You can select the cells that Riskoptimizer is allowed to adjust. (See Figure 3.) These are called adjustable cells. They can be assigned lower and upper bounds. Adjustable cells are created by clicking on the Add button under By Adjusting the Cells. The Edit button lets you modify adjustable cells, and the Delete button lets you delete them. Riskoptimizer

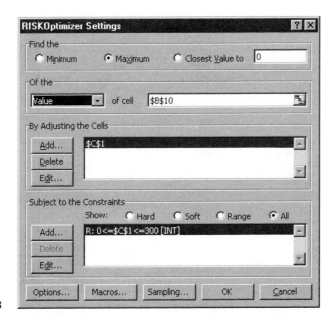

FIGURE 3

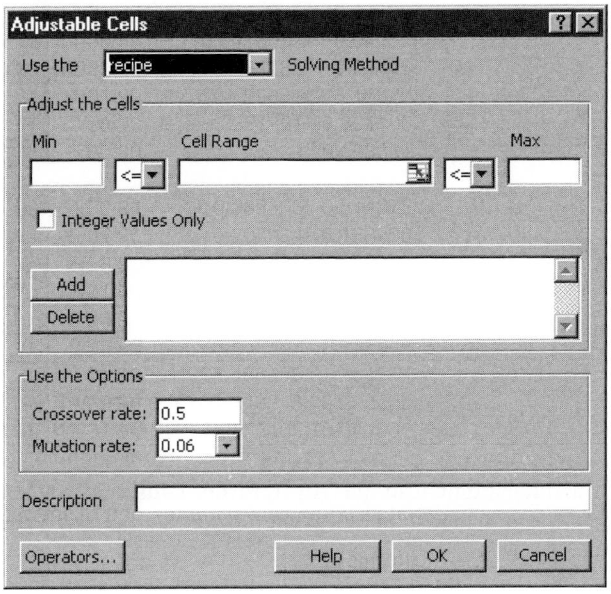

FIGURE 4

will attempt to change the adjustable cells to optimize the desired random function of the target cell. Clicking on the Add button in Figure 3 brings up the Adjustable Cells dialog box in Figure 4.

From the Adjustable Cells dialog box, select the type of solution method. For most applications, this will be Recipe. The Recipe method works like Excel Solver, but it does not take advantage of any special structure your problem may have. We will later give examples of the Budget, Grouping, and Order methods.

From the Adjustable Cells dialog box, you can also select (if desired) lower and upper bounds on adjustable cells and require adjustable cells to have (if desired) integer values.

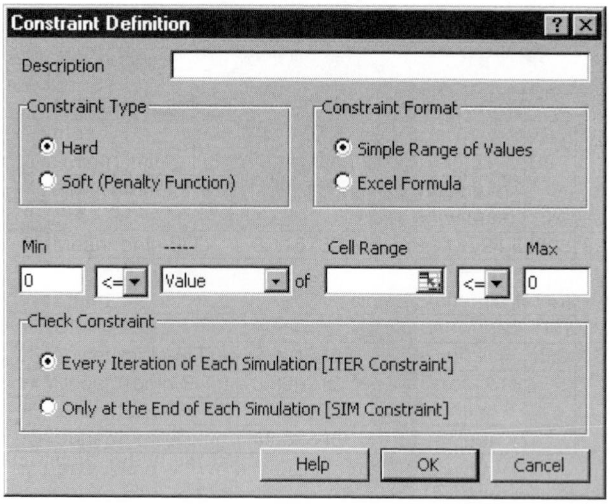

FIGURE 5

From the Adjustable Cells dialog box, you can also modify the Crossover and Mutation rates. It is difficult to understand the functions of these settings without a full discussion of genetic algorithms; see Goldberg (1989) and Davis (1991). Suffice it to say that we have obtained the best results with Mutations set to Auto and Crossover set to .50.

From the Adjustable Cells menu, you may also select operators, which are described in detail in the Riskoptimizer manual. Checking every available operator cannot hurt you; it simply increases the arsenal of weapons Riskoptimizer uses to try to find improved solutions.

Constraints

By clicking on the Add button under Subject to the Constraints in the Riskoptimizer Settings dialog box, you may add constraints on spreadsheet cells. The Constraint Definition box is displayed in Figure 5. The Excel Formula option lets you input a constraint with a formula, such as $B10 \geq D10$. We will focus only on hard constraints—ones that *must* be satisfied. Constraints may depend on random functions (such as mean standard deviation, or percentile) of spreadsheet cells. A constraint may be defined as either an iteration constraint or a simulation constraint. An **iteration constraint** must be satisfied during each iteration of all simulations run by Riskoptimizer. A **simulation constraint** is checked only at the end of each simulation. Constraints may be entered using simple values or Excel formulas.

Options

If you select the Options button (see Figure 3), you are sent to the dialog box displayed in Figure 6. General Options and Random Number Seed need not concern us. Optimization Stopping Conditions tells Riskoptimizer when to terminate its search for an optimal solution. We usually select Minutes. This tells Riskoptimizer to run for a stated amount of time. If you select Trials equal to 1,000, Riskoptimizer will examine 1,000 combinations of adjustable cells. For Simulation Stopping Conditions, we usually select Actual Convergence: For each combination of adjustable cells examined, Riskoptimizer will run enough simulation iterations to ensure that the random function defined in the target cell and all constraints converge. If this takes too long, you may limit the number of iterations run for each adjustable cell combination to 1,000, for example.

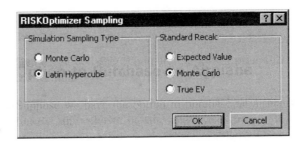

FIGURE 6

Sampling

For sampling, select the Latin Hypercube sampling type, which is much more accurate than Monte Carlo sampling, and the Monte Carlo standard recalculation. See Figure 7. The Monte Carlo recalculation option ensures that when you hit F9, all @Risk functions (and cells depending on them) will recalculate according to each @Risk function's probability distribution. For instance, there will be a 30% chance that demand for calendars is 100, a 20% chance that calendar demand is 150, etc.

Macros

With this option, you may run macros at various times during the use of Riskoptimizer. See Figure 8. See Winston (1999) for a discussion of how to link macros to Riskoptimizer.

The Start Optimization Icon

 This icon enables you to begin an optimization.

FIGURE 7

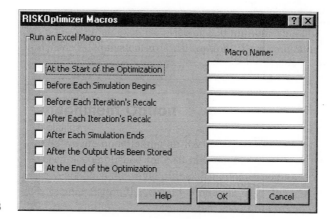

FIGURE 8

The Pause Optimization Icon

This icon enables you to pause an optimization.

The Stop Optimization Icon

This icon enables you to stop an optimization.

The Display Watcher Icon

This icon enables you to watch the progress of the optimization or to change the Mutation and Crossover parameters during a simulation. For details on the Display Watcher icon, see the Riskoptimizer manual.

Using Riskoptimizer for the Calendar Example

We are now ready to use Riskoptimizer to determine the calendar order quantity that maximizes expected profit. The relevant Riskoptimizer settings are given in Figure 9. We want to maximize expected profit (the mean of cell B10) by varying the order quantity (cell C1) among the integers between 0 and 300 (inclusive). In Figure 1, Riskoptimizer selected 200 calendars to maximize expected profit. The expected profit ($350) associated with the optimal order quantity is logged in a comment in Figure 1.

Normal Demand

The assumption of discrete demand is unrealistic. Let's suppose demand is normal, with a mean of 200 and a standard deviation of 30. To model normal demand, simply change cell C2's formula to

$$=RISKNORMAL(200,30)$$

Then, by the well-known rule of thumb, 68% of the time demand will be between 170 and 230, 95% of the time between 140 and 260, and 99.7% of the time between 110 and 290. See file Newsnorm.xls.

Newsnorm.xls

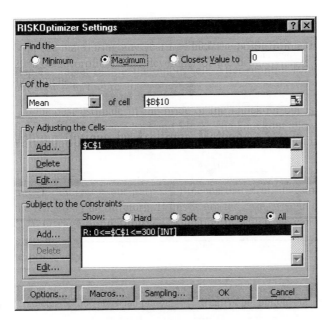

FIGURE 9

Let's use Riskoptimizer to determine the order quantity that maximizes expected profit. We will change the allowable range for order quantity to be between 0 and 400 calendars. After a while, Riskoptimizer obtains the results in Figure 10. Riskoptimizer indicates that ordering 219 calendars optimizes expected profit. The maximum expected profit is $459.39.[†] In reality, expected profit will be maximized by ordering 213 calendars. Should we be disappointed that Riskoptimizer did not find the actual profit-maximizing order quantity? Not really; the expected profit for ordering 219 calendars is within $1 of the expected profit for ordering 213 calendars.

Managing Risk

With normal demand, the standard deviation of profit when ordering 219 calendars is $88. Suppose we decide that the strategy of ordering 219 calendars involves too much risk. Can we use Riskoptimizer to maximize expected profit, subject to a constraint that the standard deviation of profit does not exceed $50? To accomplish this goal, we enter in cell E6 the formula

$$=RISKSTDDEV(B10)$$

Newsnorm2.xls

See Figures 11–14 and file Newsnorm2.xls for the work. This formula keeps track of the standard deviation of profit on each simulation. We may now incorporate this cell in a constraint, as shown in Figure 11 or 12. Our Settings window now looks as displayed in Figure 13. We chose a Simulation constraint because we want the profit standard deviation to be ≤50 after running all iterations for a given order quantity. If we had chosen an Iteration constraint, we would be requiring our standard deviation of profit for a given order quantity to be ≤50 after each iteration; this is too stringent a constraint.

After running Riskoptimizer, we find the solution given in Figure 14. Thus, constraining the standard deviation of profit to at most $50 causes us to reduce our order quantity

---

[†]Actually, the expected profit indicated by Riskoptimizer is only an estimate of the true mean profit. We can use the mean profit found by Riskoptimizer to determine a confidence interval for the true mean profit. See Chapter 26 of Winston (1998) for details.

	A	B	C	D	E
1	Order quantity		219		
2	Quantity demanded		208.44176		
3	Sales price		$4.50		
4	Salvage value		$0.75		
5	Purchase price		$2.00		
6					
7	Full price revenue	$937.99			
8	Salvage revenue	$7.92			
9	Costs	$438.00			
10	Profit	$507.91			
11					
12					
13				Mean = 459.3899	
14					
15					

FIGURE 10

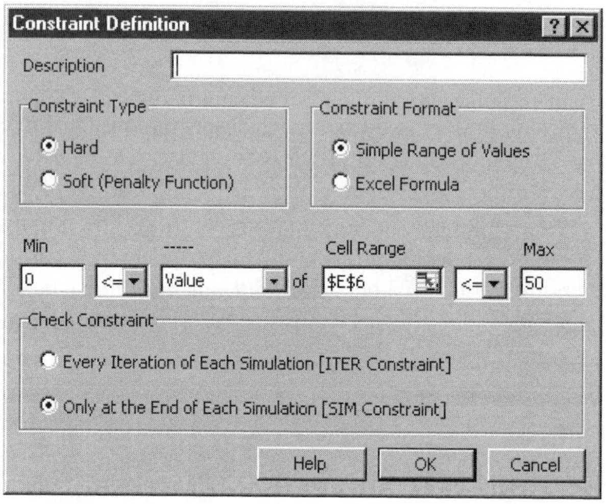

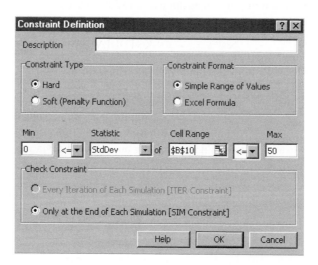

FIGURE 11

FIGURE 12

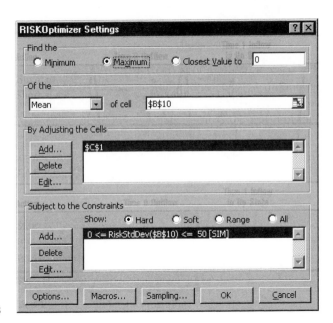

FIGURE 13

	A	B	C	D	E
1	Order quantity		187		
2	Quantity demanded		200		
3	Sales price		$4.50		
4	Salvage value		$0.75		
5	Purchase price		$2.00		
6				Std dev	48.73
7	Full price revenue	$841.50			
8	Salvage revenue	$0.00			
9	Costs	$374.00			
10	Profit	$467.50			
11					
12					
13				Mean = 442.622	
14					

FIGURE 14

from 219 calendars to 187. Our expected profit drops from $459 to $443. This is the price we pay for our constraint on risk.

12.2 The News Vendor Problem with Historical Data

Next, we show how historical data can be used to model full-price and clearance-price demand in a news vendor problem.

EXAMPLE 1   Levi's Order

Sears is trying to determine how many pairs of Levi's slacks to order for the 1997 fashion season. Past demand for Levi's at full price and at clearance price is given in Figure 15. For most of the season, Levi's sell for $30 a pair. Sears pays $15 per pair of Levi's.

	A	B	C	D	E
3	Year	Full-price Demand	Growth	Clearance demand	Clearance /FP
4	1990	5.00E+05			
5	1991	6.00E+05	1.20E+00	2.00E+05	3.33E-01
6	1992	6.60E+05	1.10E+00	2.80E+05	4.24E-01
7	1993	8.50E+05	1.29E+00	3.00E+05	3.53E-01
8	1994	9.50E+05	1.12E+00	3.00E+05	3.16E-01
9	1995	1.00E+06	1.05E+00	3.60E+05	3.60E-01
10	1996	1.20E+06	1.20E+00	5.00E+05	4.17E-01
11		mean	1.16E+00		3.67E-01
12		sigma	0.08541		0.04415

FIGURE 15

	A	B
15	Levi's Ordered	1.73E+06
16	Full-price demand	1.39E+06
17	Clearance demand	5.11E+05
18	Unit cost	$15.00
19	Full price	$30.00
20	Clearance price	$18.00
21	Full-price units sold	1.39E+06
22	Available for clearance	3.36E+05
23	Clearance units sold	3.36E+05
24		
25		
26		
27	Full price revenue	$41,748,944.55
28	clearance revenue	$6,042,573.27
29	order cost	$25,909,950.00
30	profit	$21,881,567.82
31		
32	Mean profit	21713775.34

FIGURE 16

At the end of the season, leftover Levi's are put up for sale at a clearance price of $18. How many Levi's should Sears order at the beginning of the season?

Levis.xls   **Solution**   The work is in file Levis.xls. See Figures 15 and 16. The key to this problem is modeling the full-price and clearance-price demand for Levi's. From Figure 15, it seems reasonable to model the year-to-year growth of full-price demand for Levi's as a normal random variable, with mean 1.16 and standard deviation .085. It also seems reasonable to model the clearance-price demand to be a fraction of full-price demand that is normally distributed, with a mean of .367 and a standard deviation of .044. We now proceed as follows.

Step 1   Enter in cell B15 a trial value for the number of Levi's ordered.

Step 2   In cell B16, generate full-price demand for Levi's with the formula

$$=B10*RISKNORMAL(C11,C12)$$

This formula embodies the assumption that full-price demand for 1997 will grow by an average of 16% over 1996, with a standard deviation of 8.5%.

Step 3 In cell B17, generate clearance demand with the formula

$$=B16*RISKNORMAL(E11,E12)$$

This formula ensures that actual clearance-price demand (as a fraction of full-price demand) is normally distributed, with a mean of 36.7% and a standard deviation of 4.4%.

Step 4 In cells B18–B20, enter the unit cost, full price, and clearance price.

Step 5 In cell B21, compute the number of units sold at full price with the formula

$$=MIN(B15,B16)$$

Step 6 In cell B22, compute the number of units available to be sold at clearance with the formula

$$=MAX(0,B15-B21)$$

Step 7 In cell B23, compute the number of units actually sold at clearance with the formula

$$=MIN(B22,B17)$$

Step 8 In cell B27, compute full-price revenue with the formula

$$=B19*B21$$

Step 9 In cell B28, compute clearance revenue with the formula

$$=B23*B20$$

Step 10 In cell B29, compute ordering cost with the formula

$$=B18*B15$$

Step 11 In cell B30, compute profit with the formula

$$=B27+B28-B29$$

Step 12 In cell B32, we enter the formula

$$=RISKMEAN(B30)$$

This cell will record the best mean profit found by Riskoptimizer. This formula is used because in some instances (as in this example), the box containing the mean yields erroneous results.

We now use Riskoptimizer to find an order quantity that maximizes expected profit. Our Settings window looks as shown in Figure 17. We choose to maximize the mean of cell B30 (profit) by adjusting the number of Levi's ordered (cell B15). We constrain the number ordered to be an integer between 500,000 and 3,000,000. From Figure 16, Riskoptimizer indicates that ordering 1,730,000 Levi's will maximize mean profit, with mean profit being $21.7 million.

REMARK It is important to note that if we run out of Levi's at full price or at clearance price, sales will not equal actual demand. This causes a lot of difficulty in using historical data to forecast future demand. For example, if in 1991 we ran out of Levi's at the clearance price and believed we could have sold 50,000 more at the clearance price (if they were available), we should have bumped clearance demand for 1991 up to 250,000.

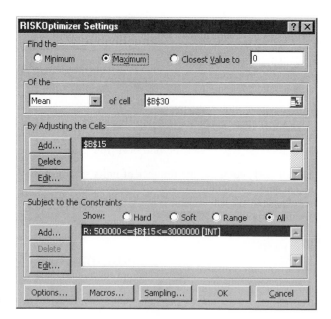

FIGURE 17

12.3 Manpower Scheduling under Uncertainty

Banks, retailers, phone companies, fast-food restaurants, and many other organizations must schedule a work force in the face of uncertain manpower requirements. Riskoptimizer makes it easy to determine a schedule that can handle uncertain manpower demand.

EXAMPLE 2   Post Office Staffing

Post.xls

The Smallville Post Office does not know for certain how many employees will be needed each day of the week, but the number is believed to follow a Poisson random variable with the means given in row 13 of Figure 18 (see file Post.xls). Employees work five consecutive days. Determine the minimum number of employees needed to ensure that there is less than a 10% chance that a shortage of employees will occur during the week.

Solution The adjustable cells will be the number of employees starting work on each day of the week. We proceed as follows.

Step 1 In cells E4:E10, enter trial values for the number of employees starting work each day (E4 for Sunday starters, E5 for Monday starters, . . . , E10 for Saturday starters).

Step 2 In cells F4:L10, enter a 1 if an employee is working on that day and a 0 if not. For example, row 4 refers to workers starting on Sunday, so we enter a 1 for Sunday–Thursday and a 0 for Friday and Saturday.

Step 3 In F11:L11, compute the number of people working each day of the week by copying from F11 to G11:L11 the formula

$$=\text{SUMPRODUCT}(\$E\$4{:}\$E\$10,F4{:}F10)$$

FIGURE 18

	E	F	G	H	I	J	K	L
2	32	Work Matrix						
3	Started	Sunday	Monday	Tuesday	Wednesda	Thursday	Friday	Saturday
4	13	1	1	1	1	1	0	0
5	3		1	1	1	1	1	0
6	3	32	0	1	1	1	1	1
7	0	1	0	0	1	1	1	1
8	7	1	1	0	0	1	1	1
9	4	1	1	1	0	0	1	1
10	2	1	1	1	1	0	0	1
11	Working	26	29	25	21	26	17	16
12		>=	>=	>=	>=	>=	>=	>=
13	Mean needed	16	10	14	12	16	9	11
14	Actual needed	20	7	19	16	15	12	9
15	Shortage	0	0	0	0	0	0	0
16								
17	shortage?	0						
18				Standard				
19	Mean shortage	0.079412	<=	0.1				

Step 4   In F14:L14, compute the actual number of employees needed each day by copying from F14 to G14:L14 the formula

$$=RISKPOISSON(F13)$$

Step 5   In F15:L15, compute the daily shortage of employees by copying from F15 to G15:L15 the formula

$$=MAX(F14-F11,0)$$

Step 6   In F17, determine if there is any shortage at all, using the formula

$$=IF(MAX(F15:L15)>0,1,0)$$

Step 7   In cell F19, compute the fraction of iterations for each simulation that result in a shortage, with the formula

$$=RISKMEAN(F17)$$

Step 8   In cell E2, compute the total number of workers:

$$=SUM(E4:E10)$$

We can now use Riskoptimizer to determine the minimum number of employees needed to provide adequate service. The Settings window is shown in Figure 19. Our goal is to minimize the total number of employees (cell E2). The adjustable cells are the number of employees starting each day (E4:E10). We constrain the number of employees starting each day to be an integer between 1 and 30. The second constraint ensures that each employee schedule will cause a shortage during at most 10% of all weeks.

From Figure 18, Riskoptimizer utilizes 32 workers according to the schedule shown in Figure 20. This schedule causes a shortage of employees during around 8% of all weeks.

FIGURE 19

	D	E
3	Day	Started
4	Sunday	13
5	Monday	3
6	Tuesday	3
7	Wednesda	0
8	Thursday	7
9	Friday	4
10	Saturday	2

FIGURE 20

Using a Cost Structure to Deal with Uncertainty

Another way to approach this problem is to deal with shortages by paying employees to work overtime. Suppose that each postal employee is paid $700 per week for the regular five-day stint, but manpower shortages are made up at a cost of $280 per day short. We now determine a schedule for this situation. See file Post2.xls and Figure 21. Few changes are needed from the previous model.

Post2.xls

Step 1 In cell F17, compute the total shortage for the week with the formula

$$=SUM(F15:L15)$$

Step 2 In cell C5, compute the regular-time wage cost for the week with the formula

$$=D2*E2$$

Step 3 In cell C6, compute the overtime-wage cost for the week with the formula

$$=F17*C2$$

Step 4 In cell E7, compute the total weekly cost with the formula

$$=SUM(E5:E6)$$

We are now ready to use Riskoptimizer to find a work schedule that minimizes expected weekly costs. See Figure 22. We choose to minimize mean weekly cost (cell E2) by adjusting the number of workers starting each day (E4:E10). We assume that no more than 30 workers will start each day. Riskoptimizer obtains a mean weekly cost of $16,995, with the schedule shown in Figure 23. Note that a total of 17 employees are needed. This

FIGURE 21

	B	C	D	E	F	G	H	I	J	K	L
1		Ot Wage	RT Wage								
2		280	700	17	Work Matrix						
3				Started	Sunday	Monday	Tuesday	Wednesda	Thursday	Friday	Saturday
4	Cost			0	1	1	1	1	1	0	0
5	RT	11900		6	0	1	1	1	1	1	0
6	OT	7000		3	0	0	1	1	1	1	1
7	total	18900		0	1	0	0	1	1	1	1
8				3	1	1	0	0	1	1	1
9				3	1	1	1	0	0	1	1
10				2	1	1	1	1	0	0	1
11			Mean = 16695		8	14	14	11	12	15	11
12					>=	>=	>=	>=	>=	>=	>=
13				Mean needed	16	10	14	12	16	9	11
14				Actual needed	27	15	13	11	16	8	12
15				Shortage	19	1	0	0	4	0	1
16											
17				Total shortage	25						

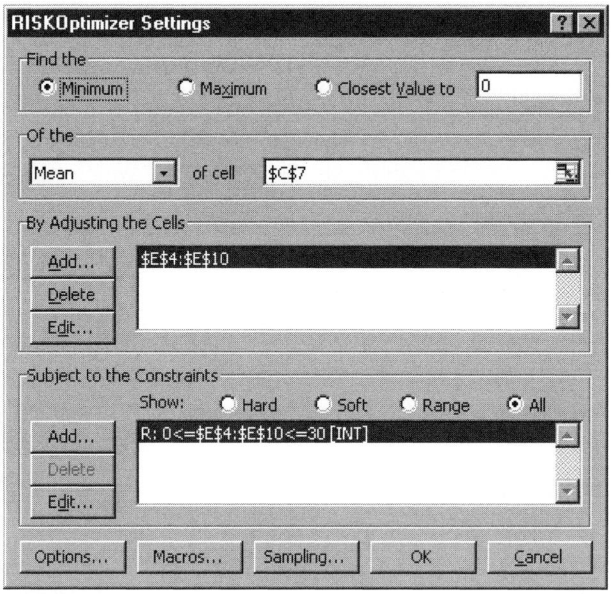

FIGURE 22

	D	E
3	Day	Started
4	Sunday	0
5	Monday	6
6	Tuesday	3
7	Wednesda	0
8	Thursday	3
9	Friday	3
10	Saturday	2

FIGURE 23

accounts for $11,900 in weekly salary. Therefore, an average of around $5,000 in overtime is paid each week. There are around 7 days of overtime each week, or .4 day of overtime work per employee—a very reasonable answer!

12.4 The Product Mix Problem

Virtually every management science book begins the study of linear programming with the classical product mix problem. The object is to determine a mix of products to produce that maximizes profit, subject to limited resources and known demand for each product. The problem with this setup is, of course, that many parameters of the problem are not known with certainty. Demand is always unknown; the amount of each resource used by a product may be unknown; the price of each product may even be unknown. Despite this uncertainty, companies must determine what to produce. Riskoptimizer lets us determine a production schedule that is "best"—maximizes expected profit in the presence of multiple sources of uncertainty.

EXAMPLE 3 **The Drugco Product Mix**

Drugco produces four drugs. Each drug uses the amount of raw material, labor, and machine time given in Figure 24. For example, drug 1 uses 4 units of raw material, 3 units of labor, and 2 units of machine time. The sales price of each drug is also given. Demand for each drug is unknown, but best-case, worst-case, and most likely demand for each product are also given in Figure 24. During the current month, 8,000 units of raw material, 10,000 units of labor, and 14,000 units of machine time are available. How can we schedule this month's production to maximize expected profit? The drugs will spoil at the end of the month, so leftover units of each drug have no value.

Prodmix.xls   **Solution**   The work is in the file Prodmix.xls. See Figure 24. We proceed as follows.

Step 1   In cells B4:E4, enter trial values for the number of units of each drug to be produced.

FIGURE 24

	A	B	C	D	E	F	G	H	I
1	**Product Mix**		**R**			**evenues**			
2							$ 23,125.67		
3		Product 1	Product 2	Product 3	Product 4				
4	Produced	303	1541	774	205				
5	Units Sold	303	943.0077	715.322	205		Mean = 24664.7222		
6	Sales price	$ 14.00	$ 12.00	$ 8.00	$ 9.00				
7	RM	4	3	2	3		Available		
8	Labor	3	3	2	1	RM	8000		
9	Machine Time	2	3	1	2	Labor	10000		
10	Actual Demand	1259.804	943.0077	715.322	481.9714	Machine Time	14000		
11	Worst	1200	900	300	0				
12	Most Likely	1400	1000	800	600		Used		Available
13	Best	1800	1400	1000	1600	RM	7998	<=	8000
14						Labor	7285	<=	10000
15						Machine Time	6413	<=	14000

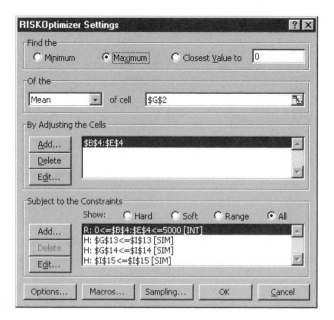

FIGURE 25

Step 2 In cells B10:E10, compute the actual demand (assuming a triangular distribution) for each drug. We copy the demand for drug 1, which is computed in cell B10, with the formula

$$=RISKTRIANG(B11,B12,B13)$$

from cell B10 to C10:E10.

Step 3 In cells B5:E5, compute the number of each product sold by copying from B5 to C5:E5 the formula

$$=MIN(B4,B10)$$

Step 4 In cell G2, compute the total revenue with the formula

$$=SUMPRODUCT(B5:E5,B6:E6)$$

Step 5 In cells G13:G15, compute the usage of each resource. In G13, compute the usage of raw material with the formula

$$=SUMPRODUCT(\$B\$4:\$E\$4,B7:E7)$$

Copying this formula to G14:G15 computes the usage of labor and machine time.

We are now ready to use Riskoptimizer to determine a product mix that maximizes expected profit. Figure 25 displays the Riskoptimizer settings. Our goal is to maximize expected profit (G2). The adjustable cells are the number of units of each drug produced (cells B4:E4). We add the constraints G13 ≤ I13, G14 ≤ I14, and G15 ≤ I15 to be satisfied at the end of each simulation. These constraints must be added individually as Excel formulas; Riskoptimizer does not let you create groups of constraints by dragging and dropping. For example, to enter the raw material usage constraint, select Add and then choose the Excel formula and enter G13 ≤ I13 as a simulation constraint. Then Riskoptimizer will ensure that at the end of each simulation, no more raw material is used than is available.

	I	J
11		
12	Available	
13	8000	TRUE
14	10000	TRUE
15	14000	TRUE
16		
17	all ok?	TRUE

FIGURE 26

There is a more efficient way to enter multiple constraints. Enter in J13 the formula

$$=G13<I13$$

and copy this formula down to J14:J15. This is an Excel logical formula, so True will be placed in J13 if we are using less raw material than we have available, and False will be placed in J13 if we are using more raw material than is available. Now, in cell J17, enter the formula

$$=AND(J13:J15)$$

This formula is true if and only if we do not violate any of the resource usage constraints. Now, instead of entering the resource usage constraints individually, we can simply enter

$$=J17$$

as an Excel constraint (simulation, not iteration). Since J13, J14, and J15 must be satisfied for J17 to be true, this will ensure that all three of the resource usage constraints are satisfied. See Figure 26.

From Figure 24, Riskoptimizer indicates that the maximum expected profit we can obtain is $24,665. We should plan on producing 303 units of drug 1, 1,541 units of drug 2, 774 units of drug 3, and 205 units of drug 4.

What If Resource Usage Is Uncertain?

In the Drugco example, let's suppose that we do not know exactly how much of each resource will be needed to produce the drugs. Assume that the usage per unit of each drug produced is normally distributed and has a standard deviation of .3. For example, this would imply that the amount of labor used per unit of drug 1 produced would be normally distributed with a mean of 3 and standard deviation of .3. Now suppose we want to have only a 5% chance of running short of any resource during the month. How can we maximize expected profit? The work is in file Prodmix2.xls. See Figure 27.

Prodmix2.xls

We modify our previous formulation as follows.

Step 1 In B7:E9, insert Risknormal formulas to ensure that resource usage is random. For example, the formula

$$=RISKNORMAL(4,0.3)$$

in cell B7 ensures that raw material usage per unit of drug 1 produced is normally distributed, with mean 4 and standard deviation .3.

Step 2 In H13:H15, determine if a shortage occurs for each resource by copying from H13 to H14:H15 the formula

$$=IF(G13>I13,1,0)$$

A 1 indicates that we have used more of the resource than is available.

FIGURE 27

	A	B	C	D	E	F	G	H	I
1	**Product Mix**		R			evenues			
2							$ 26,041.00		
3		Product 1	Product 2	Product 3	Product 4		Mean = 24477.2109		
4	Produced	646	529	412	817				
5	Units Sold	646	529	412	817				
6	Sales price	$ 14.00	$ 12.00	$ 8.00	$ 9.00				
7	RM	4.205199	3.102702	1.997795	3.381408		Available		
8	Labor	2.908318	3.293896	3.380615	0.835023	RM	8000		
9	Machine Time	1.820009	2.372704	0.981585	2.392784	Labor	10000		
10	Actual Demand	1347.803	1200.599	434.4585	1204.438	Machine Time	14000		
11	Worst	1200	900	300	0				
12	Most Likely	1400	1000	800	600		Used	Shortage?	Available
13	Best	1800	1400	1000	1600	RM	7943.58884	0	8000
14						Labor	5696.27183	0	10000
15						Machine Time	4790.20371	0	14000
16							Shortage	0	
17							Mean Short	0.05	
18								<=	
19								0.05	

Step 3 In cell H16, determine if any shortage has occurred with the formula

$$=MAX(H13:H15)$$

A 1 in this cell indicates that a shortage has occurred.

Step 4 In cell H17, compute the fraction of all iterations of a simulation that yield a shortage with the formula

$$=RISKMEAN(H16)$$

We are now ready to use Riskoptimizer. See Figure 28 for the settings. We choose to maximize mean profit (G2) by adjusting the production of each drug (B4:E4). We constrain production of each drug to be an integer between 0 and 5,000. To ensure that the

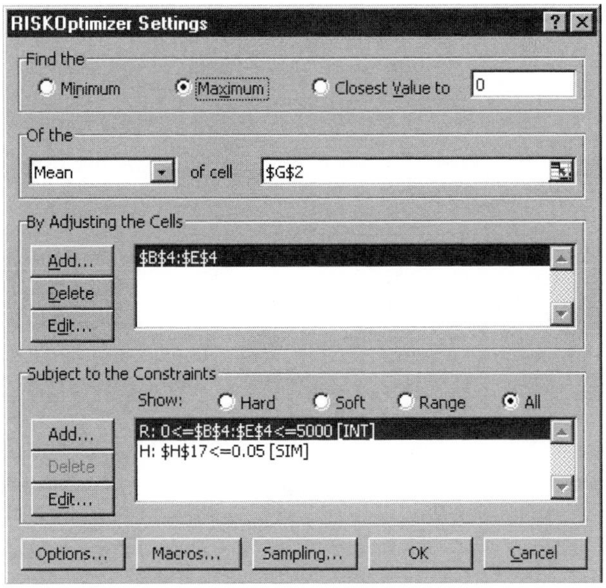

FIGURE 28

probability of a shortage for any resource does not exceed 5% we add the (simulation) constraint H17≤.05.

As shown in Figure 27, Riskoptimizer finds that the maximum expected profit of $24,477 is obtained by producing 646 units of drug 1, 529 units of drug 2, 412 units of drug 3, and 817 units of drug 4. Note that introducing the uncertainty in resource usage has slightly reduced our profit. If we wanted to be (for example) 99.9% sure that we would not run out of any resource, we would suffer a much greater profit reduction.

12.5 Agricultural Planning under Uncertainty

At the beginning of each season, farmers must determine how to utilize their land. The weather is highly uncertain. One strategy might be best for a dry season, and another for a wet season. In the presence of uncertainty about climate, how can a farmer determine a land utilization strategy that maximizes expected profit?

EXAMPLE 4   Farmer Jones

Farmer.xls

Farmer Jones owns a 100-acre farm on which he plants corn, wheat, and soybeans and raises cattle. The per-acre yield of each crop depends on rainfall—dry, medium, or wet. The yield per acre (in bushels per acre) is given in Figure 29 (see file Farmer.xls).

The figure also gives the per-acre labor requirement for each crop. Each cow requires 1/10 acre of land, 30 bushels of corn, and 50 hours of labor. Each cow contributes $400 of profit. It is estimated that there is a 20% chance of a dry season, a 50% chance of a medium season, and a 30% chance of a wet season. Farmer Jones has 10,000 hours of labor available. How can he maximize his expected profit?

Solution   **Step 1**   In G5:G7, enter trial values for the number of acres assigned to each crop. In G9, enter a trial value for the number of acres assigned to cows. For reasons that will become apparent later, make sure that a total of 100 acres is used by your trial solution.

FIGURE 29

	A	B	C	D	E	F	G	H	I
1	**Farmer Jones**			Weather	2				
2	Code	1	2	3					
3	Probability	0.2	0.5	0.3					
4	Bushels/acre	Dry	Medium	Wet	Labor	Profit/bushel	Acres	Bushels made	Bushels sold
5	Corn	70	170	80	50	$ 4.00	43.508	7396.3432	4351.197
6	Wheat	80	160	100	60	$ 5.00	45.096	7215.3914	7215.391
7	Soybeans	100	120	90	35	$ 3.00	1.2454	149.44991	149.4499
8		Corn/cow	Profit/cow	Labor/cov	Cows/acre				
9	Cows	30	$ 400.00	50	10		10.15		
10									
11			Profit						
12			$ 94,532.04						
13									
14		Used		Available					
15	Labor	10000	<=	10000		Mean = 78637.6315			
16		Made		Corn fed to cows					
17	Corn	7396.343	>=	3045.15					
18									
19				TRUE					

Step 2   In cell E1, generate the type of weather (1 = dry, 2 = medium, 3 = wet) with the formula

$$=RISKDISCRETE(B2:D2,B3:D3)$$

Step 3   In cells H5:H7, compute the number of bushels of each crop produced by copying from H5 to H6:H7 (and changing the 4 to a 5 in H6 and the 4 to a 6 in H7) the formula

$$=G5*HLOOKUP(\$E\$1,\$B\$2:\$E\$7,4)$$

Step 4   In cell D17, compute the corn fed to cows with the formula

$$=E9*B9*G9$$

Step 5   In cells I5:I7, compute the number of bushels sold of each product. After subtracting corn fed to the cows, compute the number of bushels of corn sold in cell I5 with the formula

$$=H5-D17$$

In cells I6:I7, compute the bushels of wheat and soybeans sold by copying the formula

$$=H6$$

from I6 to I7.

Step 6   In cell C12, compute profit with the formula

$$=C9*E9*G9+SUMPRODUCT(F5:F7,I5:I7)$$

Step 7   In cell B15, compute total labor usage with the formula

$$=SUMPRODUCT(E5:E7,G5:G7)+E9*D9*G9$$

Step 8   In cell B17, recompute the amount of corn produced. This will allow for a constraint on corn produced so that it will exceed the corn needed for the cows.

Step 9   In cell D19, enter the logical formula

$$=B17>=D17$$

This cell will contain a True if enough corn to feed the cows is produced, and a False otherwise. By making D19 an iteration constraint, we ensure that we will always have enough corn to feed the cows.

We are now ready to invoke Riskoptimizer. Figure 30 contains the settings. The goal is to maximize mean revenue (C12) by choosing the number of acres devoted to cows and each crop. Constraints (G9 and G5:G7) ensure that each adjustable cell will be between 0 and 100. We select the Budget solution method, which ensures that on each simulation, the adjustable cells will always add to their initial sum (100). This guarantees that all simulations will use all available land. The constraint B15≤10,000 is a simulation constraint to ensure that each land allocation uses at most 10,000 labor hours. The constraint D19 is an iteration constraint, ensuring that on each iteration of any simulation, we will have enough corn to feed the cows. We enter this constraint by choosing Add and typing =D19 after selecting the Excel Formula option. See Figure 31.

Riskoptimizer reports a maximum expected profit of $78,637. To obtain this maximum, we plant 43.51 acres of corn, 45.10 acres of wheat, and 1.25 acres of soybeans, and

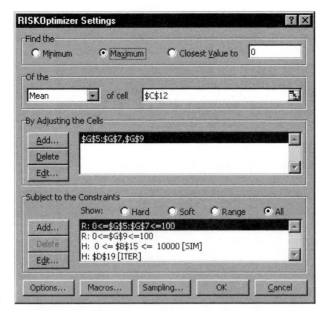

FIGURE 30

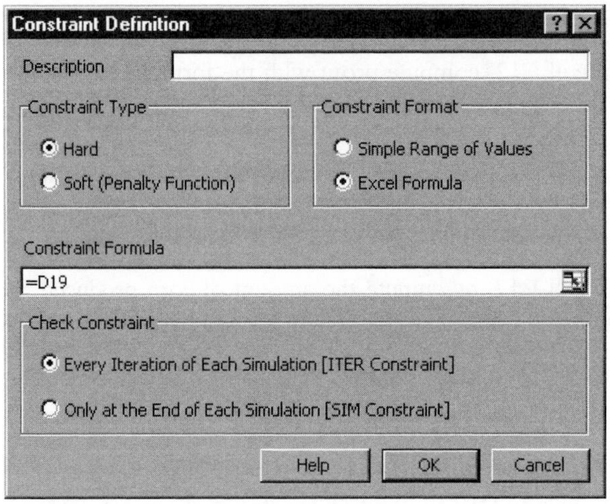

FIGURE 31

we devote 10.15 acres to cows. Note that even during a dry season, we have enough corn to feed the cows. Soybeans, although not very profitable, are utilized because of their low labor requirement.

12.6 Job Shop Scheduling

Consider a job shop with a single machine. The length of time needed to complete each job is uncertain. Each job has a known due date. In what order should the jobs be processed? Three reasonable objectives spring to mind.

- Maximize the expected number of jobs completed on time.
- Minimize the mean total of the tardiness of the jobs.
- Minimize the maximum lateness of any job.

The following example shows how to solve job shop scheduling problems with Riskoptimizer.

EXAMPLE 5   **Print Shop Scheduling**

Jobshop1.xls

A small print shop has 10 jobs scheduled. For each job, the best-case duration, the most likely duration, and the worst-case duration are given in Figure 32. The due date for each job is also given (see file Jobshop1.xls). For example, for job 1, the best case is that the job will take 5 days, the most likely case 8 days, and the worst case 15 days. Job 1 is due at the end of day 6.

If the goal is to maximize the expected number of jobs completed on time, in what order should the jobs be processed?

Solution   To solve this problem, we select the Order solution method. For a given set of adjustable cells, the Order method will try different permutations of the numbers originally placed in the adjustable cells. For example, if we put the numbers 1, 2, 3, . . . , 10 in 10 adjustable cells and select the Order method, Riskoptimizer will try different orderings such as 3, 4, 5, 6, 8, 9, 10, 1, 2, 7, etc. These different orderings correspond to different job schedule sequences. We proceed as follows:

Step 1   In cells G19:G28, use the triangular random variable to generate the duration of each job. Copy from G19 to G20:G28 the formula

$$=RISKTRIANG(D19,E19,F19)$$

Step 2   In cells C5:C14, enter any permutation of the integers 1, 2, . . . , 10. This permutation represents a potential job schedule. See Figure 33. The figure indicates that job 2 is done first, then job 5, . . . , and finally job 9.

Step 3   Name the cell range C19:H28 Lookup, and create the duration of each job in D5:D14 by copying from D5 to D14 the formula

$$=VLOOKUP(\$C5,LOOKUP,5)$$

Step 4   In E5:E14, compute the due date of each job by copying from E5 to E6:E14 the formula

$$=VLOOKUP(C5,LOOKUP,6)$$

Step 5   In F5, compute the time the first job scheduled is completed with the formula

$$=D5$$

In F6:F14, compute the time all remaining jobs are completed by copying from F6 to F7:F14 the formula

$$=F5+D6$$

	C	D	E	F	G	H
18	Job	Lowest	Most likely	Worst	Time	Due Date
19	1	5	8	15	7.752926	6
20	2	3	7	12	7.208382	21
21	3	1	3	9	5.15733	46
22	4	2	4	8	4.693512	47
23	5	10	12	18	11.12812	35
24	6	6	8	17	14.58061	66
25	7	2	5	9	5.452041	50
26	8	4	6	11	6.09866	38
27	9	10	11	13	10.30965	14
28	10	6	7	9	7.772322	55

FIGURE 32

	B	C	D	E	F	G	H	I
4	Order	Actual job	Actual time	Due Date	Time completed	On time	Time Late	
5	1	2	10.29628014	21	10.29628	1	0	
6	2	5	11.42196385	35	21.718244	1	0	
7	3	8	6.708545465	38	28.426789	1	0	
8	4	4	5.422226365	47	33.849016	1	0	
9	5	3	3.21222167	46	37.061237	1	0	
10	6	10	6.669641845	55	43.730879	1	0	
11	7	7	5.061215362	50	48.792095	1	0	
12	8	1	5.324259711	6	54.116354	0	48.11635	
13	9	6	11.46979883	66	65.586153	1	0	
14	10	9	11.20713324	14	76.793286	0	62.79329	
15					Total	8	110.9096	
16								Mean = 6.86
17								

FIGURE 33

Step 6 In G5:G14, determine if each job is completed on time by copying from G5 to G6:G14 the formula

$$=IF(F5<=E5,1,0)$$

Step 7 In H5:H14, compute the lateness (if any) of each job by copying from H5 to H6:H14 the formula

$$=IF(F5>E5,F5-E5,0)$$

Step 8 In cell G15, compute the total number of jobs completed on time with the formula

$$=SUM(G5:G14)$$

Copying this formula to H15 computes the total tardiness of the jobs.

Step 9 We can now use Riskoptimizer to determine a job sequence that maximizes the expected number of jobs completed on time. The settings are as shown in Figure 34.

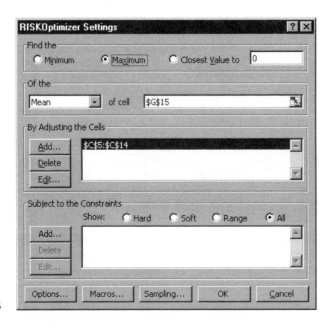

FIGURE 34

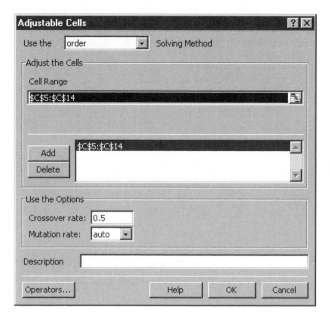

FIGURE 35

In Figure 35, note the selection of the Order method in the Adjustable Cells dialog box. We maximize the mean number of jobs completed on time (G15) by trying different permutations of C5:C14. From Figure 33, the best schedule averages completing 6.65 jobs on time. The optimal sequence found is 2, 5, 8, 4, 3, 10, 7, 1, 6, and 9.

12.7 The Traveling Salesperson Problem

Willie Lowman is a salesman who lives in Boston. He needs to visit certain cities and then return to Boston. In what order should Willie visit the cities in order to minimize the total distance traveled? The distances between each pair of cities are given in Figure 36.

For obvious reasons, this type of problem is called a **traveling salesperson problem** (TSP). Here are two other situations where the TSP is relevant.

- A UPS driver needs to make 20 stops today. In which order should she deliver the packages to minimize her total time on the road?
- A robot must drill 10 holes to produce a single printed circuit board. Which order of drilling the holes will minimize the total time needed to produce a circuit board?

To model Willie Lowman's problem in a spreadsheet, note that any ordering or permutation of the numbers 1–11 represents an order of visiting the cities. For example, the ordering

$$2\text{-}4\text{-}6\text{-}8\text{-}10\text{-}1\text{-}3\text{-}5\text{-}7\text{-}9\text{-}11$$

FIGURE 36

		Boston	Chicago	Dallas	Denver	LA	Miami	NY	Phoenix	Pittsburgh	SF	Seattle
1	Boston	0	983	1815	1991	3036	1539	213	2664	792	2385	2612
2	Chicago	983	0	1205	1050	2112	1390	840	1729	457	2212	2052
3	Dallas	1815	1205	0	801	1425	1332	1604	1027	1237	1765	2404
4	Denver	1991	1050	801	0	1174	1332	1780	836	1411	1765	1373
5	LA	3036	2112	1425	1174	0	2757	2825	398	2456	403	1909
6	Miami	1539	1390	1332	1332	2757	0	1258	2359	1250	3097	3389
7	NY	213	840	1604	1780	2825	1258	0	2442	386	3036	2900
8	Phoenix	2664	1729	1027	836	398	2359	2442	0	2073	800	1482
9	Pittsburgh	792	457	1237	1411	2456	1250	386	2073	0	2653	2517
10	SF	2385	2212	1765	1765	403	3097	3036	800	2653	0	817
11	Seattle	2612	2052	2404	1373	1909	3389	2900	1482	2517	817	0

could be viewed as going from Boston (city 1) to Dallas (city 3), then to L.A. (city 5), ..., and finally visiting S.F. (city 10) before returning to Boston. Since we are viewing the ordering from the location of city 1, there are

$$10 \times 9 \times 8 \times 7 \times 6 \times \cdots \times 2 \times 1 = 10! = 3,628,800$$

possible orderings for Willie to consider. In Riskoptimizer, the Order method will find the ordering of the numbers $1, 2, \ldots, n$ (in this case $1, \ldots, 11$) that maximizes or minimizes a spreadsheet cell. This is just what is needed to solve a TSP. The question becomes how to find the total distance that corresponds to a given ordering. The Excel function INDEX is perfect for this situation. The syntax of the INDEX function is

$$=\text{INDEX(Range, row\#, column\#)}$$

In this case, Excel will look in the range of cells named Range and pick out the entry in row row# and column# of Range. To use the INDEX function to find the total distance traveled in visiting all cities, proceed as follows.

Step 1 In F16:F26, enter any ordering of the integers 1–11. See Figure 37.

Step 2 Name the range G4:Q14 Distances. Then enter in cell G16 the formula

$$=\text{INDEX(Distances,F26,F16)}$$

This determines the distance between the last city listed (in F26) and the first city listed (in F16).

Step 3 Enter the formula

$$=\text{INDEX(Distances,F16,F17)}$$

in cell G17 and copy it to the range G18:G26. In G17, the formula computes the distance between the first and second cities listed, etc.

Step 4 In cell G27, compute total distance traveled with the formula

$$=\text{SUM(G16:G26)}$$

Step 5 We can now invoke Riskoptimizer. Choose to minimize cell G27. Then select the Order method and select F16:F26 as adjustable cells. Set the number of iterations to 1 (this is OK because we have no uncertainty!). Riskoptimizer will try to find the permutation of the numbers currently in F16:F26 that minimizes total distance. The Settings window looks as shown in Figure 38.

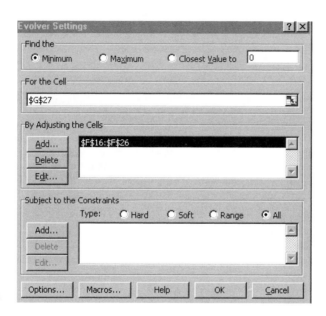

	E	F	G	H
15		Order	Distance	City
16	1	8	398	Phoenix
17	2	3	1027	Dallas
18	3	6	1332	Miami
19	4	1	1539	Boston
20	5	7	213	NY
21	6	9	386	Pittsburgh
22	7	2	457	Chicago
23	8	4	1050	Denver
24	9	11	1373	Seattle
25	10	10	817	SF
26	11	5	403	LA
27		Total	8995	

FIGURE 37

FIGURE 38

We find the minimum possible distance to be 8,995 miles, with the cities to be visited in the following order:

Boston–N.Y.–Pittsburgh–Chicago–Denver–Seattle–S.F.–L.A.–Phoenix–Dallas–Miami–Boston

As you will see in the problems, the Order method is useful in a variety of situations.

REVIEW PROBLEMS

Group A

1 Fly-by-Day airlines is flying a 145-seat plane from Indianapolis to Detroit. Tickets for the flight are nonrefundable and cost $250. On average, 10% of all ticket holders fail to show up (but the airline still keeps their money!). If more than 145 people show up for the flight, each customer who does not get a seat receives $420 in compensation. The variable cost (meals, fuel consumption, etc.) incurred for each person on the flight is $60.00. How many reservations should Fly-by-Day accept?

2 A new shoe store is trying to determine how many large pairs and small pairs of its most popular shoe to order. The following information is available. (See file Shoe.xls.)

Large	Small
337	194
165	88
105	74
223	239
437	391
401	362
140	96
405	350
424	329
135	115
450	328
403	277
395	305
117	139
311	215
196	151
190	160
193	99
424	362
217	159
342	280
321	324
285	185
127	147
115	144
438	294

It costs $35.00 to order a pair of shoes, and each pair sells for $75.00. At the end of the season, the store can sell back leftover shoes for $15.00 a pair. The display above lists last year's sales of large and small shoes at stores similar to ours. For example, one store sold 337 pairs of large shoes and 194 pairs of small shoes last year. The store has enough warehouse space to order up to 500 pairs of shoes. How many pairs of each size should be ordered?

3 At HMO Hell Hospital, nurses work for four consecutive days and are paid $300 per day. Each nurse can handle 10 patients per day in regular time. Each extra patient a nurse handles results in $50 of overtime pay. The average number of patients in the hospital each day is: Monday 123, Tuesday 97, Wednesday 104, Thursday 80, Friday 55, Saturday 91, and Sunday 131. How would you schedule the nurses?

4 Solve Example 5 assuming that jobs 6 and 9 must be among of the first three jobs completed.

5 Drugco must determine how many units of drugs 1 and 2 to produce this month. Demand for drug 1 is unknown but is thought to follow a truncated normal random variable with $\mu = 500$ and $\sigma = 140$. Demand for drug 2 is unknown but is thought to follow a triangular random variable with worst case 200, most likely case 400, and best case 500. This month, 600 hours of machine time are available. The mean time required to produce a unit of a drug during the month is unknown. We believe that the mean time needed to produce a unit of drug 1 follows a triangular random variable with best case .8 hour, most likely case .9 hour, and worst case 1.1 hours. The mean time needed to produce a unit of drug 2 follows a triangular random variable with best case .5 hour, most likely case .52 hour, and worst case .78 hour. There are 900 units of raw material currently available, and extra units of raw material can be purchased at $2 per unit. The mean raw material needed to produce a unit of drug 1 is unknown but follows a triangular random variable with best case 0.8 unit, most likely case 0.9 unit, and worst case 1.1 units. The mean raw material needed to produce a unit of drug 2 is unknown but follows a triangular random variable with best case 1.15 units, most likely case 1.2 units, and worst case 1.23 units. Drug 1 sells for $10 a unit, while drug 2 sells for $8 per unit. It costs $1.50 (in addition to raw material cost) to produce a unit of a drug. At the end of the month, all leftover raw material and drugs spoil and are worthless. We want to have at most a 5% chance of using more machine time than is available. How many units of each drug should we plan to produce this month?

6 For Example 5, find a schedule that will minimize the expected total lateness of all the jobs.

7 For Example 5, find a schedule that will minimize the maximum amount by which a job is late.

8 The Limited 3 store is trying to determine how many of a new T-shirt to order. On January 1, an order can be placed at a cost of $4 per shirt. Those shirts arrive in stores on February 1. By April 1, Limited 3 will know if the T-shirt has high, low, or medium demand. There is .5 chance of low demand, .3 chance of medium demand, and .2 chance of high demand. Total demand under each scenario follows a triangular random variable, as listed in Table 1. The shirt is sold for $13. On April 1, more shirts may be ordered at a cost of $8 per shirt. Find an ordering policy that will maximize Limited 3's expected profit.

9 We have $10,000 to invest during the next year. Four investment vehicles are available: cash, bonds, stocks, and gold. During the next year, the economy can be in one of the following states: boom, bust, or moderate growth. The probability of each scenario and the annual percentage return for each investment under each scenario are as shown in Table 2. What investment allocation minimizes the standard deviation of our final asset position subject to earning an expected return of at least 4%?

10 A factory has six machines. Each product must be processed on two machines. The number of products per week that need to be processed on each combination of two machines is as shown in Table 3. For example, 100 products must be processed first on machine 1 and then on machine 5. Each machine can be assigned to one of six locations. The distances (in feet) between each combination of

TABLE 1

Demand	Worst Case	Most Likely	Best Case
Low	5,000	8,500	9,500
Medium	7,000	11,000	13,000
High	10,000	14,000	17,000

TABLE 2

Scenario	Probability	Cash	Bonds	Stocks	Gold
Boom	.2	0%	−2%	45%	−8%
Bust	.25	0%	5%	−35%	15%
Moderate growth	.55	0%	3%	10%	2%

TABLE 3

From	To	Products
1	2	80
1	3	150
1	4	170
1	5	100
1	6	120
2	3	160
2	4	200
2	5	240
2	6	360
3	4	120
3	5	240
3	6	160
4	5	90
4	6	230
5	6	190

locations are as shown in Table 4. For example, locations 1 and 2 are 5,000 feet apart. Where should each machine be located?

11 How could the method of Problem 10 be used to determine where to locate stores in a shopping mall?

12 UPS driver Amanda Woodward begins her day at the UPS station. Then she must deliver packages to the airport, the Uptown Café, the convention center, the mall, the business school, the hospital, and the high school. In which order should she deliver the packages? The time (in minutes) needed to travel between these locations is as shown in Table 5.

13 A robot must drill seven holes into a printed circuit board. The x and y coordinates of these holes are as follows:

Hole	x	y
1	2	4
2	3	6
3	0	7
4	5	2
5	4	3
6	5	1
7	9	6

In what order should the holes be drilled?

Group B

14 You are operating a web site to match up sellers and buyers of a product. As Table 6 shows, 35 sellers and 35 buyers have input their reservation prices. For example, buyer 1 is willing to pay up to $8 for an item. Seller 1 is willing to accept $9 or more for an item. This means that buyer 1 and seller 1 cannot be matched. The goal of Splitthedifference.com (an actual web site run by Yale professor Barry Nalebuff) is to pair up buyers and sellers in order to maximize the sum of the seller's and the buyer's surplus. For example, if buyer 31 and seller 31 are matched, a deal can be made by splitting the difference with a price of $57. Then buyer 31 earns a surplus of $60 − 57 = \$3$, while seller 31 earns a surplus of $57 − 54 = \$3$. What is the maximum sum of seller's and buyer's surplus that can be obtained? The data are in file Splitdata.xls.

15 Eli Lilly is trying to determine an optimal production schedule for the drug Electionitis. The mean demand per week for the drug is 700 units. Weekly demand follows a Poisson random variable. At the beginning of each week, Lilly must determine whether any batches of the drug should be produced. The company will produce only during weeks in which beginning inventory is less than or equal to R.

TABLE 4

	1	2	3	4	5	6
1	0	5,000	2,000	7,000	4,000	8,000
2	5,000	0	3,000	8,000	5,000	7,000
3	2,000	3,000	0	3,000	6,000	4,000
4	7,000	8,000	3,000	0	4,500	3,500
5	4,000	5,000	6,000	4,500	0	4,000
6	8,000	7,000	4,000	3,500	4,000	0

TABLE 5

	UPS	Airport	Uptown Café	Convention Center	Mall	B-School	Hospital	High School
UPS	0	4	8	9	18	13	7	12
Airport	4	0	6	7	14	11	6	11
Uptown Café	8	6	0	2	10	6	4	10
Convention Center	9	7	2	0	8	7	2	9
Mall	18	14	10	8	0	7	9	15
B-School	13	11	6	7	7	0	8	9
Hospital	7	6	4	2	9	8	0	11
High School	12	11	10	9	15	9	11	0

TABLE 6

Buyer	Buyer's Reservation Price	Seller	Seller's Reservation Price
1	8	1	9
2	58	2	60
3	19	3	20
4	89	4	95
5	54	5	60
6	61	6	64
7	26	7	31
8	65	8	56
9	54	9	64
10	92	10	78
11	14	11	17
12	50	12	46
13	19	13	24
14	18	14	16
15	53	15	65
16	59	16	60
17	69	17	65
18	84	18	80
19	36	19	39
20	20	20	24
21	20	21	18
22	26	22	27
23	94	23	87
24	35	24	31
25	57	25	49
26	69	26	72
27	27	27	27
28	10	28	13
29	31	29	39
30	69	30	58
31	60	31	54
32	26	32	32
33	54	33	64
34	36	34	37
35	19	35	17

During such a week, the company will produce B batches. If any batches are produced, a setup cost of $5,000 is incurred. Each batch consists of 1,000 units. At the end of any week, a shortage cost of $10 per unit is incurred, and a holding cost of $1 per unit is incurred. Lilly begins week 1 with 2,000 units in stock. How can Lilly minimize its expected cost per week (over, say, a 100-week period)?

16 Eli Lilly plans to sell a drug for the next ten years. The company must decide today how much capacity to build. After two years, the company can, if desired, add capacity. It costs $6 per unit to build a unit of annual capacity now (time 0) and $8 per unit to build a unit of annual capacity in two years (at time 2). The drug sells for $7.00 per unit and has a variable cost of $2 per unit. Lilly believes annual demand will be governed by one of the following random variables:

- With probability .2, each year's demand will be normally distributed with a mean of 1,000 and a standard deviation of 300. Each year's demand will be a different value, however.
- With probability .3, each year's demand will be normally distributed with a mean of 3,000 and a standard deviation of 500. Each year's demand will be a different value, however.
- With probability .3, each year's demand will be normally distributed with a mean of 4,000 and a standard deviation of 800. Each year's demand will be a different value, however.
- With probability .2, each year's demand will be normally distributed with a mean of 9,000 and a standard deviation of 200. Each year's demand will be a different value, however.

A reasonable strategy for Lilly is to build C units of capacity now and in two years, review average demand to date. Then add enough capacity (if needed) to bring total capacity to a multiple M of average demand to date. For example, suppose $C = 3,000$ and $M = 1.5$. Then if the first two years' average demand were 4,600, Lilly would add $6,900 - 3,000 = 3,900$ units of capacity. If the first two years' average demand were 1,500, capacity would remain unchanged. Lilly discounts profits at 10%. How can Lilly maximize its expected discounted profit?

17 Camp Farwell will enroll 20 campers this summer. The camp has four five-person cabins. The camp's goal is to ensure that the campers are assigned to cabins with as many

	E	F	G	H	I
29		**Pals**			
30	Kid				
31	1	19	11	20	10
32	2	8	9	10	12
33	3	17	10	19	15
34	4	14	20	6	5
35	5	17	13	3	2
36	6	7	13	16	14
37	7	2	8	5	16
38	8	14	11	13	3
39	9	1	13	14	15
40	10	1	3	17	8
41	11	18	4	7	10
42	12	19	7	5	6
43	13	8	10	20	16
44	14	12	8	11	8
45	15	6	9	7	16
46	16	7	8	11	12
47	17	12	16	8	5
48	18	2	19	17	11
49	19	6	13	3	10
50	20	2	9	3	15

FIGURE 39

of their friends as possible. Figure 39 lists each camper's friends. (See file Campdata.xls.) For example, if camper 1 were to bunk with campers 4, 8, 10, and 11, she would be with two friends, etc. Use the Order method to assign the campers to cabins so as to maximize the total number of friends all campers have in their cabins. Run one iteration, for 1 hour.

18 Microsoft is coming out with a new .net service to compete against a Real Player service. The market for these products consists of three market segments: low-volume, medium-volume, and high-volume users. As shown in Table 7, 60% of all potential customers are low-volume users. It is equally likely that a low-volume user will have between 5,000 and 20,000 transactions per year. Real Player is charging a $12,000 fixed fee plus $0.40 per transaction as the annual fee for using its product. Each customer can develop his or her own product for an annual cost of $30,000 + $0.15*(number of transactions). Microsoft is considering

TABLE 7

User Type	Probability	Lower Size	Upper Size
Low-volume	.6	5,000	20,000
Medium-volume	.3	20,000	100,000
High-volume	.1	100,000	1,000,000

a quantity discount price structure. If a customer is involved in t transactions, the following price is charged:

$$\text{If } t \leq T, p_1(t)$$
$$\text{If } t > T, p_2(t)$$

Of course $p_2 \leq p_1$ must hold.

For example, if $T = 100,000$, $p_1 = \$0.30$, and $p_2 = \$0.20$, then Microsoft would charge $24,000 for 80,000 transactions and $24,000 for 120,000 transactions. Assume each customer chooses the cheapest product. What pricing strategy maximizes Microsoft's expected revenue per customer? Run 200 iterations and run for 1 hour. To get a forecast for Microsoft's expected annual revenue, multiply the total number of customers in the market by Microsoft's expected revenue per customer.

19 Eli Lilly has asked doctors to compare (pairwise) ten drugs. The results are in file Drugcomp.xls. For example, (we have hidden many rows), 2 doctors preferred drug 1 over drug 2; 3 doctors preferred drug 1 over drug 4; etc. Based on this data, come up with a ranking of the drugs from best (number 1) to worst (number 10). One way to do this is to choose the set of rankings that minimizes the number of "preference reversals." For example, if we rank drug 1 second and drug 3 first, we would have incurred 2 preference reversals. Find a ranking of the drugs from best to worst that minimizes the number of preference reversals.

20 In 1980, the United States attempted to use helicopters to rescue the hostages then being held by Iran. Here are the essential facts:

- The mission would succeed if at least 6 helicopters lasted throughout the mission.
- The mission might take 9 hours or 14 hours.
- If too many helicopters were sent, the Russians might spot them on radar and tip off Iran, ensuring failure of the mission.

How many helicopters should have been sent on the mission in order to maximize the chance of success? In order to make this decision, we need to model the following sources of uncertainty:

- The probability a helicopter would work for both 9 and 14 hours.
- The probability the mission would take 9 hours or 14 hours.
- The probability (as a function of number of helicopters sent) that Russian radar would spot our helicopters.

Here are our modeling assumptions:

- Based on past testing of helicopters, we believe a helicopter had a .5 chance of making it through a 14-hour mission and a .66 chance of making it through a 9-hour mission.
- Our best estimate is that there was a 60% chance of a 9-hour mission and a 40% chance of a 14-hour mission.

Sending 8 or fewer helicopters, there was no chance of being spotted by Russian radar. Each helicopter beyond 8 increased the chances of being spotted by 3%.

How many helicopters should have sent been on the mission?

REFERENCES

Davis, M. *Handbook of Genetic Algorithms.* Princeton, N.J.: Van Nostrand-Reinhold, 1991.

Goldberg, D. *Genetic Algorithms in Search Optimization and Machine Learning.* Reading, Mass.: Addison-Wesley, 1989.

Winston, W. *Financial Models Using Simulation and Optimization.* Ithaca, N.Y.: Palisades Publishing, 1998.

Winston, W. *Decision-Making Using Riskoptimizer.* Ithaca, N.Y.: Palisades Publishing, 1999.

Riskoptimizer Manual. Ithaca, N.Y.: Palisades Publishing, 1999.

13

Option Pricing and Real Options

Fisher Black, Robert Merton, and Myron Scholes developed the Black–Scholes option pricing formula (Black and Scholes, 1973). More recently, the theory of options has been used to value capital investment projects. This is referred to as "the real options revolution." This chapter presents the basics of options and real options analysis. Our discussion will use many tools from earlier chapters, such as simulation (Chapters 9–12) and probabilistic dynamic programming (Chapter 7).

13.1 The Lognormal Model of Stock Prices

The lognormal model for asset value (or stock price) assumes that in a small time Δt, the stock price changes by an amount that is normally distributed, with

$$\text{Standard deviation} = \sigma S \sqrt{\Delta t}$$

$$\text{Mean} = \mu S \Delta t$$

where S = current stock price. μ may be thought of as the instantaneous rate of return on the stock. This model leads to "jumpy" changes in stock prices, just as in real life. During a small period of time, the standard deviation of the stock's movement will greatly exceed the mean of the stock's movement. This follows because for small Δt, $\sqrt{\Delta t}$ will be much larger than Δt.

In a small time Δt, the natural logarithm $\ln(S)$ of the current stock price will change (by Ito's Lemma, see Hull (1997)) by an amount that is normally distributed, with

$$\text{Mean} = (\mu - .5\sigma^2)\Delta t$$

$$\text{Standard deviation} = \sigma \sqrt{\Delta t}$$

See Chapter 16 for a more rigorous discussion of the lognormal random variable.

Let S_t = stock price at time t. In Hull (1997) and Chapter 16 of this book, it is shown that at time t, $\ln S_t$ is normally distributed, with

$$\text{Mean} = \ln S_0 + (\mu - .5\sigma^2)\, t$$

$$\text{Standard deviation} = \sigma \sqrt{t}$$

We refer to $(\mu - .5\sigma^2)$ as the **continuously compounded rate of return** on the stock. Note that the continuously compounded rate of return on S is less than instantaneous return. Since $\ln S_t$ follows a normal random variable, we say that S_t is a **lognormal random variable.**

To simulate S_t, we get $\ln(S_t)$ by entering in @Risk the formula

$$= \text{LN}(S_0) + (\mu - .5\sigma^2)t + \sigma\sqrt{t}\ \text{RISKNORMAL}(0,1)$$

Therefore, to get S_t, we must take the antilog of this equation and get

$$S_t = S_0 e^{(\mu - .5\sigma^2)t + \sigma\sqrt{t}\text{RISKNORMAL}(0,1)} \qquad (1)$$

Historical Estimation of the Mean and Volatility of Stock Return

If we average the values of

$$\ln \frac{S_t}{S_{t-1}}$$

we obtain an estimate of $(\mu - .5\sigma^2)$. If we take the standard deviation of

$$\ln \frac{S_t}{S_{t-1}}$$

we obtain an estimate of σ.

Dell.xls

Using the monthly returns of Dell Computer for 1988–1996, we may estimate μ and σ. See the file Dell.xls and Figure 1. We begin by estimating σ and μ for a monthly lognormal process. Then our estimate of σ for an annual lognormal process is just $\sqrt{12}$ (monthly estimate of σ), and our estimate of μ for an annual lognormal process is just 12*(monthly estimate of μ).

Step 1 Note that for any month, $\frac{S_{t+1}}{S_t} = (1 + \text{month } t \text{ return})$. Therefore, in C6:C107, we compute for each month $\ln \frac{S_{t+1}}{S_t}$ by copying from C6 to C7:C107 the formula

$$=\text{LN}(1+\text{B6})$$

Step 2 In C2, we estimate $\mu - .5\sigma^2$ with the formula

$$= \text{AVERAGE}(\text{C6:C107})$$

Step 3 In C3, we estimate σ with the formula

$$=\text{STDEV}(\text{C6:C107})$$

Step 4 In cell F7, we find our estimate of μ for a monthly lognormal process with the formula

$$=\text{C2}+0.5*\text{F6}^\wedge 2$$

Thus, for the monthly lognormal, we estimate that $\mu = .0475$ and $\sigma = .161$.

Step 5 In cells F9 and F10, we find annualized estimates of μ and σ with the formulas

$$=12*\text{F7} \qquad \text{for } \mu$$
$$=\text{SQRT}(12)*\text{F6} \qquad \text{for } \sigma$$

The annualized estimates are $\mu = 57.0\%$ and $\sigma = 55.7\%$.

Finding the Mean and Variance of a Lognormal Random Variable

It is important to point out that μ is not actually the mean of a lognormal random variable, and σ is not really the standard deviation. Assume that a stock follows a lognormal random variable with parameters μ and σ. Let S = current price of the stock (which is known) and S_t = price of the stock at time t (unknown). Then (Luenberger (1997), p. 310)

FIGURE 1

	A	B	C	D	E	F
1		estimate of	monthly			
2		mu-.5*sigma^2	0.034572			
3		sigma	0.160929			
4	DATE	Dell	Ln(1+Dell)			
5	6/30/88				monthly estimate	
6	7/29/88	0.106667	0.101353		sigma	0.160929
7	8/31/88	-0.228916	-0.25996		mu	0.047521
8	9/30/88	0.28125	0.247836		annual estimate	
9	10/31/88	0.158537	0.147158		sigma	0.557473
10	11/30/88	-0.0842105	-0.08797		mu	0.570254
11	12/30/88	-0.0804598	-0.08388			
12	1/31/89	-0.05	-0.05129			
13	2/28/89	-0.171053	-0.1876			
14	3/31/89	-0.0952381	-0.10008			
15	4/28/89	0.105263	0.100083			
16	5/31/89	0.0793651	0.076373			
17	6/30/89	-0.0735294	-0.07637			
18	7/31/89	-0.142857	-0.15415			
19	8/31/89	0.037037	0.036368			
20	9/29/89	0.0178571	0.0177			
21	10/31/89	-0.157895	-0.17185			
22	11/30/89	-0.0416667	-0.04256			
23	12/29/89	-0.0434783	-0.04445			
24	1/31/90	-0.159091	-0.17327			
25	2/28/90	0.351351	0.301105			
26	3/30/90	0.22	0.198851			
27	4/30/90	0.114754	0.108634			
99	4/30/96	0.369403	0.314375			
100	5/31/96	0.207084	0.188208			
101	6/28/96	-0.0812641	-0.08476			
102	7/31/96	0.0909091	0.087011			
103	8/30/96	0.209459	0.190173			
104	9/30/96	0.158287	0.146942			
105	10/31/96	0.0466238	0.04557			
106	11/29/96	0.248848	0.222222			
107	12/31/96	0.0455105	0.044505			

the mean and variance of S_t are as follows:

$$\text{Mean of } S_t = Se^{\mu t}$$
$$\text{Variance of } S_t = S^2 e^{2\mu t}(e^{\sigma^2 t} - 1)$$

Lognormal.xls

The file Lognormal.xls contains a template to determine the mean and variance of a stock price at any future time. See Figure 2. For example, consider a stock currently selling for $20 and following a lognormal random variable with $\mu = .20$ and $\sigma = .40$. The mean stock price one year from now is $24.43, with a standard deviation of $10.18.

Confidence Intervals for a Lognormal Random Variable

The file Lognormal.xls computes a confidence interval for a future stock price. For a 95% confidence interval, enter .95 for alpha, etc. From Figure 2, we find that for a stock currently selling for $20 with $\mu = .20$ and $\sigma = .40$, we are 95% sure that the price of the stock one year from now will be between $10.30 and $49.39.

	A	B	C	D
2				
3	S=current price	20		
4	t=time	1		
5	mu	0.2		
6	sigma	0.4		
7	alpha	0.95		
8				
9	Mean for ln S(T)	3.115732	Mean S(T)	24.42806
10	Sigma for ln S(T)	0.4	var S(T)	103.5391
11			sigma S(T)	10.17542
12	CI			
13	Lower(for ln S(T))	2.331748		
14	Upper (for ln S(T))	3.899717		
15	Lower for S(T)	10.29592		
16	Upper for S(T)	49.38846		

FIGURE 2

PROBLEMS

Group A

1 Use the monthly stock returns in file Volatility.xls to determine estimates of μ and σ for Intel, Microsoft, and GE.

2 Consider a stock currently selling for $50. Assume that the stock's future price follows a lognormal random variable with $\mu = .20$ and $\sigma = .30$.

   a What is the probability that the stock will be selling for at least $60 in three years?

   b Assume that the stock will sell for at least $55 in two years. What is the probability that the stock will be selling for at least $60 three years from now?

3 For the stock in Problem 2, I am 95% sure that in two years, the stock will be selling for between _____ and _____.

13.2 Option Definitions

We begin our study of options with some definitions. A **call option** gives the owner the right to buy a share of stock for a price called the **exercise price**. A **put option** gives the owner the right to sell a share of stock for the exercise price. An **American option** can be exercised on or before a time known as the **exercise date**. A **European option** can be exercised only at the exercise date.

For example, let's look at cash flows from a six-month European call option on Microsoft with an exercise price of $110. Let P = price of Microsoft in six months. Then the payoff from the call option is $0 if $P \le 110$ and $P - 110$ if $P \ge 110$. This is because, for P below $110, we would not exercise the option; if P is larger than $110, we would exercise our option to buy stock for $110 and immediately sell stock for P, thereby earning a profit of $P - 110$. Figure 3 displays the payoff from this put option. The call payoff may be written as $\max(0, P - 110)$. Note that the call option graph has slope 0 for P smaller than the exercise price and has slope 1 for P greater than the exercise price.

It can be shown that if a stock pays no dividends, it is never optimal to exercise an American call option early. See Section 13.7 for a proof of this result. Therefore, for a non-dividend-paying stock, an American and a European option have the same value.

Now let's look at cash flows from a six-month European put option on Microsoft with an exercise price of $110. Let P = price of Microsoft. Then the payoff from the put option is $0 if $P \ge 110$ and $110 - P$ if $P \le 110$. This is because for P below $110, we

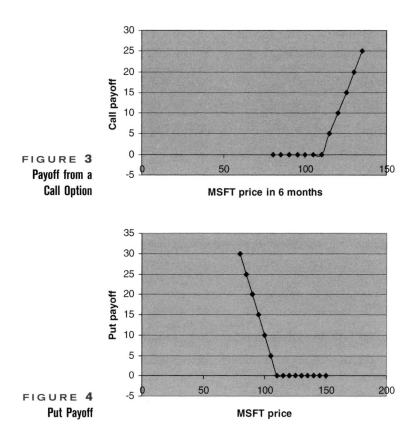

FIGURE 3
Payoff from a
Call Option

FIGURE 4
Put Payoff

would buy a share of stock for $P and immediately sell the stock for $110. This yields a profit of $110 - P$. If P is larger than $110, it would not pay to buy the stock for $P and sell it for $110, so we would not exercise our option to sell the stock for $110. Figure 4 displays the payoff from this put option. The put payoff may be written as $\max(0, 110 - P)$. Note that the slope of the put payoff is -1 for P less than the exercise price, and 0 for P greater than the exercise price.

An American put option may be exercised early, so the cash flows from an American put cannot be determined without knowledge of the stock price at times before the expiration date.

13.3 Types of Real Options

The key observation that led to the field of real options is generally attributed to Judy Lewent, the CFO of Merck. She noted that many actual investment opportunities (not just those involving stocks) may be viewed as combinations of puts and calls. Therefore, if we know how to value puts and calls, we can value many actual investment opportunities. Here are some examples of real options.

Option to Purchase an Airplane

Suppose we have the option to purchase an airplane 3 years from now for $20 million. Let P = the value of the airplane 3 years from now. P is uncertain, depending on the economic cycle, fuel prices, etc. Then cash flows in 3 years from the option to purchase would equal max($P-20$, 0). This is the same equation that would define the cash flows from a call option with an exercise price of $20 million. This implies that the option to purchase an airplane is equivalent to a call option. If we can value a call option, we can value an option to purchase an airplane.

Abandonment Option

Suppose we are undertaking an R&D project. Five years from now, we can sell what has then been accomplished for $80 million. If we let P = the value of the ideas developed after 5 years, then the value of the abandonment option would equal max($80 - P$, 0). This is the same equation that would define the cash flows from a put option with an exercise price of $80 million. This implies that the option to abandon a project is equivalent to a put option. Thus, if we can value a put option, we can value an abandonment option.

Other Real Option Opportunities

Many other real investment opportunities can (with some effort) be represented as combinations of puts and calls. As we proceed, we will learn how to value such options.

Expansion Option

Suppose that three years from now, we have an option to double the size of a project. What is this option worth?

Contraction Option

Suppose that three years from now, we have the option to cut the scale of a project in half. What is this option worth?

Postponement Option

We are thinking of developing a new SUV-minivan hybrid. In two years, we will know more about the size of the market. We have the option to wait two years before deciding to develop the car. What is this option worth?

Pioneer Option

Microsoft decided to buy Web TV even though the deal had a negative NPV. Perhaps this was because Web TV was a *pioneer option* that gave Microsoft the opportunity to enter new markets that might or might not be profitable. Without having purchased Web TV, Microsoft would not have been able to enter these new markets. If the value of the option to enter new markets later exceeds the negative NPV of Web TV, then the purchase would be a good idea.

Flexibility Option

An auto company is thinking of building a plant that can produce three types of cars. The cost per unit of capacity for such a flexible plant far exceeds the unit capacity cost for a plant that can produce only one type of car. Is the increased flexibility worthwhile?

Licensing Option

Suppose that during any year in which the profit from a drug exceeds $50 million, we pay 20% of all profits to the developer of the drug. What is a fair price for such a licensing agreement?

13.4 Valuing Options by Arbitrage Methods

The Black–Scholes option pricing formula is based on arbitrage pricing methods. Arbitrage pricing (for our purposes) means that if an investment has no risk, it should yield the risk-free rate of return. If this is not the case, we can create a money-making machine or **arbitrage opportunity**—a situation in which we can spend $0 today with no chance of losing money and a positive chance of making money. We can use this simple insight to price very complex financial derivatives.

EXAMPLE 1   Arbitrage

A stock is currently selling for $40. One period from now, the stock will either increase in price to $50 or decrease to $32. The risk-free rate of interest is $11\frac{1}{9}\%$. What is a fair price for a European call option with an exercise price of $40?

Solution   The key is the realization that if we create a portfolio consisting of x shares of stock and short one call option, then for some value of x, the portfolio will have no risk. This is because an increase in the stock price benefits the stock owned but hurts the value of our shorted call, while a decrease in the stock price hurts the stock owned but helps the value of our shorted call. Then this portfolio must earn the risk-free rate. The work is in file Arbvalue.xls. See Figure 5. Any model in which a stock price can only increase or decrease by a certain amount during a period is called a **binomial model.**

Arbvalue.xls

FIGURE 5

	A	B	C	D	E	F	G	H	I	J	K
1					Step 1						
2	Arbitrage value				x shares of stock						
3	of call option				1 call						
4					risk-free if						
5					50x-10=32x						
6					x=5/9						
7					Step 2						
8					risk-free portfolio			c = call price today			
9					must earn risk-free rate			value in one period=32(5/9)=160/9			
10					so value today= (1/(1+r)*value in one period						
11					(5/9)*40-c=(1/(1+(1/9))*(160/9)						
12					200/9-c =16						
13					c = 56/9						
14								Using Arbitrage method			
15								to Value a			
16								Call with exercise price of $40			
17								11 1/9% risk free rate			
18										1 period later	
19										Stock value	Call Value
20							today		up	$50	$10
21											
22							$40		down	$32	$0
23											
24											

We begin by computing the value of our portfolio if the stock increases or decreases in value:

Stock price	Portfolio value
$50	$50x - 10$
$32	$32x - 0$

To create a portfolio with no risk, we simply set the portfolio value to be the same for both stock prices. That is,

$$50x - 10 = 32x$$
$$18x = 10$$
$$x = \frac{5}{9}$$

Thus, a portfolio consisting of $\frac{5}{9}$ shares of stock and short one call has no risk. This portfolio must yield the risk-free rate. This means that

$$\frac{\text{Value of portfolio in one period}}{1 + r} = \text{initial value of portfolio}$$

where r = the risk-free rate. Therefore,

$$\frac{5}{9}(40) - c = \frac{1}{1 + \frac{1}{9}}\left(\frac{160}{9}\right)$$

$$\frac{200}{9} - c = 16$$

$$c = \frac{56}{9}$$

Creating a Money-Making Machine if a Call Option Is Incorrectly Priced

To gain greater insight on why the call price in Example 1 must equal $\frac{56}{9}$, let's show that if the call price is any other value, we can make guaranteed positive profits at time 1 with an investment today of $0. In the real world, such an arbitrage opportunity would never exist. If the call price today (call it c) is less than $\frac{56}{9}$, the call is "underpriced," and shorting our risk-free portfolio can yield an arbitrage opportunity. We will show that if we buy one call, short $\frac{5}{9}$ share of stock, and lend out $\frac{200}{9} - c$ dollars, we have a $0 cash outlay today and a guaranteed positive cash position at time 1. These investments yield the cash flows shown in Table 1.

The time 0 cash outflow is $0. We chose the size of the loan to make this work. If the stock goes up to $50, time 1 cash inflow is positive if and only if

$$\frac{-250}{9} + 10 + \frac{2000}{81} - \frac{10c}{9} > 0$$

$$\frac{560}{81} - \frac{10c}{9} > 0$$

$$c < \frac{56}{9}$$

TABLE 1

Action	Time 0 Outflow	Time 1 Inflow in Up State	Time 1 Inflow in Down State
Buy call	c	10	0
Short stock	$\frac{-200}{9} = -(\frac{5}{9})(40)$	$-(\frac{5}{9})(50) = \frac{-250}{9}$	$(\frac{-5}{9})(32) = \frac{-160}{9}$
Lend money	$\frac{200}{9} - c$	$(\frac{10}{9})(\frac{200}{9} - c)$	$\frac{10}{9}(\frac{200}{9} - c)$

TABLE 2

Action	Time 0 Outflow	Time 1 Inflow in Up State	Time 1 Inflow in Down State
Short call	$-c$	-10	-0
Buy stock	$\frac{200}{9} = (\frac{5}{9})(40)$	$(\frac{5}{9})(50) = \frac{250}{9}$	$(\frac{5}{9})(32) = \frac{160}{9}$
Borrow money	$c - \frac{200}{9}$	$-(\frac{10}{9})(\frac{200}{9} - c)$	$-(\frac{10}{9})(\frac{200}{9} - c)$

If the stock price drops, time 1 cash inflow is positive if and only if

$$\frac{2,000}{81} - \frac{10c}{9} - \frac{160}{9} > 0$$

$$\frac{560}{9} - \frac{10c}{9} > 0$$

$$c < \frac{56}{9}$$

Thus, if the call is underpriced, we have created a money-making machine!

Now suppose that the call is overpriced ($c > \frac{56}{9}$). At time 0, we short the call, buy $\frac{5}{9}$ share of stock, and borrow $\frac{200}{9} - c$ dollars. This investment strategy yields the cash flows in Table 2.

Time 0 cash outflow is $0. We chose the amount of the borrowing to make this work. If the stock goes up to $50, time 1 cash inflow is positive if and only if

$$\frac{250}{9} - 10 - \frac{2,000}{81} + \frac{10c}{9} > 0$$

$$\frac{-560}{81} + \frac{10c}{9} > 0$$

$$c > \frac{56}{9}$$

If the stock price drops, time 1 cash inflow is positive if and only if

$$\frac{-2,000}{81} + \frac{10c}{9} + \frac{160}{9} > 0$$

$$\frac{-560}{9} + \frac{10c}{9} > 0$$

$$c > \frac{56}{9}$$

Thus, if the call is overpriced, we have created a money-making machine.

Why Doesn't the Stock's Growth Rate Influence the Call Price?

From our analysis, the key factor that influences the option price are the two values of the stock price. The further spread out the up and down prices are, the more value the call has. Note that the probability that the stock goes up or down does not influence the option value. In effect, the average growth rate of the stock does not affect the call's value. At first glance, this doesn't make sense. If the stock has more chance of increasing in value, shouldn't the call sell for more, since the call pays off for high stock prices? The answer to this conundrum is that *the current price of the stock incorporates information about the stock's growth rate.* In effect, the probabilities of the stock increasing or decreasing in value are included in today's price, so they do not affect the value of the call option.

PROBLEMS

Group A

1 A stock currently sells for $20. In three months, the price of the stock will be either $18 or $22. The risk-free rate is 1% per month.

a Use the arbitrage approach to value a three-month European call option with an exercise price of $20 on this stock.

b Use the arbitrage approach to value a three-month European put option with an exercise price of $20 on this stock.

2 A stock currently sells for $40. In six months, the price of the stock will be either $50 or $15. The risk-free rate is 1% per month.

a Use the arbitrage approach to value a six-month European call option having an exercise price of $24.

b Use the arbitrage approach to value a six-month European put option having an exercise price of $24.

Group B

3 It is February 15, 2000. Three bonds are for sale, each having a face value of $100. The coupon rates (paid semiannually), expiration dates, and prices are given in Table 3. Every six months, starting six months from the current date and ending at the expiration date, each bond pays .5*(coupon rate)*(face value). At the expiration date, the face value is paid. For example, the second bond pays

TABLE 3

Current Price	Expiration Date	Coupon Rate
$101.625 (bond 1)	8-15-00	6.875
$101.5625 (bond 2)	2-15-01	5.5
$103.80 (bond 3)	2-15-01	7.75

$2.75 on 8-15-00 and $102.75 on 2-15-01. Given the current price structure, is there a way to make an infinite amount of money? To answer this, look for an arbitrage. An arbitrage exists if there is a combination of bond sales and purchases today that yields a positive cash flow today and nonnegative cash flows at all future dates. If such a strategy exists, it is indeed possible to make an infinite amount of money. For instance, if buying 10 of bond 1 today and selling 5 of bond 2 today yielded $1 today and nothing at all on future dates, then we could make k by purchasing $10k$ of bond 1 today and selling $5k$ of bond 2 today. We would also be able to cover all payments at future dates from money received on those dates. Clearly, bond prices at any point in time should be set so that an arbitrage does not exist. Show that for bonds 1–3, an arbitrage opportunity exists. (*Hint:* Set up an LP that maximizes today's cash flow subject to constraints that the cash flow at each future date is nonnegative.)

13.5 The Black–Scholes Option Pricing Formula

In 1973, the brilliant trio of Fisher Black, Robert Merton, and Myron Scholes developed the Black–Scholes option pricing formula, which earned Scholes and Merton the Nobel Prize in economics in 1997. Assuming that the price of the underlying stock follows a lognormal random variable, they found a closed-form solution for the price of a European call and a European put. Essentially, their method was to extend the arbitrage pricing approach developed in Section 13.4 by letting the length of a period in the binomial model go to zero. See Chapter 16 for more details on the Black–Scholes derivation. They were able to find formulas for pricing European puts and calls. Let

$$S = \text{today's stock price}$$
$$t = \text{duration of the option}$$
$$X = \text{exercise or strike price}$$
$$r = \text{risk-free rate}$$
$$\sigma = \text{annual volatility of stock}$$
$$y = \text{percentage of stock value paid annually in dividends}$$

The risk-free rate is assumed to be continuously compounded. Thus, if $r = .05$, in one year, \$1 will grow to $e^{.05}$ dollars. For a European call option, the price is computed as follows. Define

$$d_1 = \frac{\ln\left(\dfrac{S}{X}\right) + \left(r - y + \dfrac{\sigma^2}{2}\right)t}{\sigma\sqrt{t}}$$
$$d_2 = d_1 - \sigma\sqrt{t}$$

Then the call price C is given by

$$C = Se^{-yt} N(d_1) - Xe^{-rt} N(d_2)$$

where $N(x)$ is the cumulative normal probability for a normal random variable having mean 0 and $\sigma = 1$. For example, $N(-1) = .16$, $N(0) = .5$, $N(1) = .84$, $N(1.96) = .975$. The cumulative normal probability may be computed in Excel with the =NORMSDIST( ) function.

The price of a European put P may be written as

$$P = Se^{-yt}(N(d_1) - 1) - Xe^{-rt}(N(d_2) - 1)$$

American options are usually modeled using binomial trees.

Bstemp.xls

In file Bstemp.xls, we have set up a template that lets you enter S, X, r, t, σ, and y and read off the European call and put prices. For example, suppose the current stock price of Microsoft is \$100, and you own a 7-year European call option with an exercise price of \$95. Assume $r = .05$ and $\sigma = 47\%$. (We will see where this comes from in the next section.) From Figure 6, we see that this call option is worth \$57.15. If Microsoft gave you 1,000 of these options, their worth would be around \$57,150. Note that an American call on a stock that does not pay dividends should be exercised early, so this is also the value of an American call option.

Figure 6 also shows that a European put with an exercise price of \$95 would be worth \$24.10. An American put might be exercised early, so it could be worth more than \$24.10.

Comparative Statics Results

It is natural to ask how changes in key inputs will change the value of an option. An Excel one-way data table makes it easy. For example, Figure 7 shows how making the duration of a European call option from 1 to 10 years changes the value of the call option.

To obtain this data table, proceed as follows.

Step 1   Enter possible durations in J10:J19.

	A	B	C	D	E
1	**Call with dividends**				
2					
3					
4	**Input data**				
5	Stock price	$100			
6	Exercise price	$95			
7	Duration	7			
8	Interest rate	5.00%			
9	dividend rate	0			
10	volatility	47.00%			
11					
12				Predicted	
13	Call price	=		$57.15	
14	put			$24.10	
15					
16					
17	Other quantities for option price				
18	d1	0.944463		N(d1)	0.827534
19	d2	-0.29904		N(d2)	0.382455
20					

FIGURE 6

	J	K
7		
8	Duration	
9		$57.15
10	1	22.90206
11	2	31.82951
12	3	38.65993
13	4	44.31017
14	5	49.15378
15	6	53.39188
16	7	57.14977
17	8	60.51304
18	9	63.5439
19	10	66.28968

FIGURE 7

Step 2 In cell K9, enter the formula you want calculated (the call price) for different durations:

$$=D13$$

Step 3 Select the table range J9:K19.

Step 4 From the Data menu, choose the Table option. Then select cell B7 as the Column Input Cell. This will ensure that the possible durations in J10:J19 are input to cell B7.

We see that lengthening the duration of the call option increases its value. If the option paid dividends, this might not be the case. (Why?)

In Figure 8 we construct another data table to show how changing stock volatility between 10% and 100% affects the value of a call option. Note that an increase in volatility always increases the value of any kind of option.

In general, the comparative statics results for puts and calls (ceteris paribus) are as given in Table 4.

	N	O
8	Volatility	
9		$57.15
10	0.1	33.66137
11	0.2	38.53958
12	0.3	45.28839
13	0.4	52.33248
14	0.47	57.14977
15	0.5	59.15782
16	0.6	65.53342
17	0.7	71.33781
18	0.8	76.50927
19	0.9	81.02667
20	1	84.89906

FIGURE 8

TABLE 4

Parameter	European Call	European Put	American Call	American Put
Stock price	+	−	+	−
Exercise price	−	+	−	+
Time to expiration	?	?	+	+
Volatility	+	+	+	+
Risk-free rate	+	−	+	−
Dividends	−	+	−	+

- An increase in today's stock price always increases the value of a call and decreases the value of a put.

- An increase in the exercise price always increases the value of a put and decreases the value of a call.

- An increase in the duration of an option always increases the value of an American option. In the presence of dividends, an increase in the duration of an option can either increase or decrease the value of a European option.

- An increase in volatility always increases option value.

- An increase in the risk-free rate increases the value of a call. This is because higher rates tend to increase the growth rate of the stock price, which is good for the call. This more than cancels out the fact that the option payoff is worth less due to the higher interest rate. An increase in the risk-free rate always decreases the value of a put, because the higher growth rate tends to hurt the put, as does the fact that future payoffs from the put are worth less. Again, this assumes that interest rates do not affect current stock prices.

- Dividends tend to reduce the growth rate of a stock price, so increased dividends reduce the value of a call and increase the value of a put.

PROBLEMS

Group A

1 A stock is selling for $42. The stock has an annual volatility of 40%, and the annual risk-free rate is 10%.

a What is a fair price for a six-month European call option with an exercise price of $40?

b How much does the current stock price have to increase for the purchaser of the call option to break even in six months?

c What is a fair price for a six-month European put option with an exercise price of $40?

d How much does the current stock price have to decrease for the purchaser of the put option to break even in six months?

e What level of volatility would make the option in part (a) sell for $6? (*Hint:* Use Excel Goal Seek or Solver.)

2 A stock is selling for $52. The stock has an annual volatility of 50%, and the annual risk-free rate is 10%.

    a What is a fair price for a six-month European call option with an exercise price of $50?

b How much does the current stock price have to increase for the purchaser of the call option to break even in six months?

c What is a fair price for a six-month European put option with an exercise price of $50?

d How much does the current stock price have to decrease for the purchaser of the put option to break even in six months?

e What level of volatility would make the option in part (a) sell for $6? (*Hint:* Use Excel Goal Seek or Solver.)

13.6   Estimating Volatility

In a real options analysis or the valuation of an ordinary financial option, the value of the option depends crucially on the volatility of the underlying project or asset. Under the assumption that a stock follows a lognormal random variable, volatility is the σ characterizing the stock's lognormal price random variable. Most experts believe that the best way to estimate the volatility of, say, an Internet startup or a biotech drug is to look at the volatility of companies in a similar line of business. It is important to understand how to estimate the volatility of a stock. Basically, there are two approaches to estimating volatility.

- Estimate volatility based on historical data, as described in Section 13.1.
- Look at a traded option and estimate volatility as the value of σ that makes the actual option price match the predicted Black–Scholes price. This approach is called **implied volatility estimation.**

Most experts prefer to use the implied volatility approach, because it is forward-looking, while the historical approach is based on the past.

The Implied Volatility Approach

Volest.xls

At the end of trading on February 8, 2000, Microsoft stock sold for $106.61. A March put option (expiring on March 25, 2000) with exercise price $100 sold for $3.75. The risk-free rate on 90-day T-bills was 5.5%. How can we estimate Microsoft's implied volatility at that point in time? See File Volest.xls and Figure 9.

Step 1 The work is in the Sheet option of file Volest.xls. Enter all parameters (stock price, duration, risk-free rate, actual put price, and exercise price) into the Black–Scholes template. Note: Enter ln (1 + .055) as the risk-free rate, because the Black–Scholes assumed risk-free rate is compounded continuously. The duration of the option is 45/366 days.

Step 2 See Figure 10. Use Goal Seek to determine the value of σ that makes the predicted Black–Scholes price (in D14) equal the actual price (in B14). Fill out the Goal Seek window to set cell D14 to the value B14 by changing the volatility (B10). We find the implied volatility of Microsoft to be 47.07%.

    A similar analysis has been done for Amazon.com in Figure 11. Amazon.com's implied volatility is around 75%.

	A	B	C	D	E
1					
2	**Microsoft Implied Volatility**			February 9 ,2000	
3				March $100 put	
4	**Input data**			expires	
5	Stock price	$106.61		25-Mar	
6	Exercise price	$100			
7	Duration	0.1229508			
8	Interest rate	5.33%			
9	dividend rate	0			
10	volatility	47.07%			
11					
12		Actual		Predicted	
13	Call price	=		$11.01	
14	put	$3.75		$3.75	
15					
16					
17	Other quantities for option price				
18	d1	0.5099518		N(d1)	0.694957
19	d2	0.3448948		N(d2)	0.634913

FIGURE 9

Goal Seek	? X
Set cell:	D14
To value:	3.75
By changing cell:	B10
	OK Cancel

FIGURE 10

	A	B	C	D	E	F
1						
2	**Amazon Implied Volatility**			February 9 ,2000		
3				March $75 put		
4	**Input data**			expires		
5	Stock price	$74.94		25-Mar		
6	Exercise price	$75				
7	Duration	0.1229508				
8	Interest rate	5.33%				
9	dividend rate	0				
10	volatility	75.24%				
11						
12		Actual		Predicted		
13	Call price	=		$8.06		
14	put	$7.63		$7.63		
15						
16						
17	Other quantities for option price					
18	d1	0.1535665		N(d1)	0.561024	
19	d2	-0.1102461		N(d2)	0.456107	

FIGURE 11

REMARKS

1   Of course, different puts (or calls) will yield slightly different implied volatilities. You can use Solver to combine several options into a single volatility estimate. See Problem 4 for an example of this approach.

2   In a real options model, you should estimate implied volatility for a time horizon that is as close as possible to the duration of the real option.

PROBLEMS

Group A

1 On September 25, 2000, JDS Uniphase stock sold for $106.81. A $100 European put expiring on January 20, 2001 sold on September 25, 2000 for $11.875. Compute an implied volatility for JDS Uniphase based on this information. Use a risk-free rate of 5%.

2 On August 9, 2002, Microsoft stock was selling for $48.58. A $35 European call option expiring on January 17, 2003 was selling for $13.85. Use this information to estimate Microsoft's implied volatility. Use a risk-free rate of 4%.

3 On August 9, 2002, Microsoft stock was selling for $48.58. A $45 European put option expiring on January 17, 2003 was selling for $4.65. Use this information to estimate Microsoft's implied volatility. Use a risk-free rate of 4%.

Group B

4 The current price of IBM stock is $145\frac{1}{8}$. The risk-free rate is 6.36%. The current prices of three European call options expiring in 64 days are as follows:

Exercise Price	Call Price
$140	$12
$150	$6.25
$160	$3.125

Find a single volatility that best predicts these prices. Use the following approach to estimate the IBM 64-day volatility. Use Solver to choose a single volatility that minimizes the sum over the three options of (actual option price − Black–Scholes option price)^2. This is called the *kappa approach* to estimating implied volatility. Explain why it should provide a reasonable estimate of a stock's volatility.

13.7 The Risk-Neutral Approach to Option Pricing

The arbitrage pricing approach can be difficult to implement, since it is often hard to determine what risk-free portfolio yields an arbitrage opportunity. One alternative method of pricing options is the **risk-neutral pricing approach** (Cox, Ross, and Rubenstein, 1979). This approach often greatly simplifies the process of pricing options.

The Logic behind the Risk-Neutral Approach

Recall that the arbitrage argument works regardless of a person's risk preferences. A risk-averse, risk-seeking, or risk-neutral decision maker would accept the option price obtained by the arbitrage pricing argument.

This observation leads to the following conclusions.

1 In a world where everyone is risk-neutral, arbitrage pricing is valid.

2 In a risk-neutral world, all assets yield a return equal to the risk-free rate.

3 In a risk-neutral world, any asset (including an option) is worth the expected value of its discounted cash flows.

4 We can set up a risk-neutral world in which all stocks grow at a risk-free rate and use Monte Carlo simulation (or a binomial tree) to determine the expected discounted cash flows from an option.

5 By applying steps 1–4, we may easily find the correct price of an option in a risk-neutral world. Since arbitrage pricing is valid in a risk-neutral world (as well as our own world), the option price we find by applying steps 1–4 must match the option price found using arbitrage pricing. Finally, the arbitrage price is the price of the option in our world, so whatever option price we find by applying steps 1–4 must be the correct price for the option in our world.

Essentially we use step 2 to adjust the probabilities of various states of the world. Then we use step 3 to value the derivative.

Example of Risk-Neutral Pricing

Let's price the call option of Example 1 via risk-neutral pricing. The situation was as follows. A stock is currently selling for $40. One period from now, the stock will either increase to $50 or decrease to $32. The risk-free rate of interest is $11\frac{1}{9}\%$. What is a fair price for a European call option with an exercise price of $40?

We now value our call option in a risk-neutral world. Let p = risk-neutral probability that the stock increases to $50. Then $1 - p$ = risk-neutral probability that the stock decreases to $32. From step 2, we know that in a risk-neutral world, the stock (and any other asset) must yield on average the risk-free rate of return. This requires

$$50p + 32(1 - p) = 40\left(1 + \frac{1}{9}\right)$$

$$18p + 32 = \frac{400}{9}$$

$$18p = \frac{112}{9}$$

$$p = \frac{56}{81}$$

Thus, in a risk-neutral world, the stock has a $\frac{56}{81}$ chance of increasing to $50 and a $\frac{25}{81}$ chance of decreasing to $32.

In the $50 state, the option is worth $10. In the $32 state, the option is worth $0. In a risk-neutral world, any asset is worth the expected discounted value of its cash flows. Therefore, in a risk-neutral world, the call option is worth

$$\frac{1}{1 + \frac{1}{9}}\left(\left(\frac{56}{81}\right)(10) + \frac{25}{81}(0)\right) = \frac{56}{9}$$

Since, in the risk-neutral world, the arbitrage-pricing price must be correct, we know that the real-world price for this call equals the risk-neutral price $\frac{56}{9}$.

The beauty of the risk-neutral approach is that it is usually easy to find the risk-neutral probabilities. Then we may value the derivative as the expected discounted value of its cash flows. This is very intuitive and much more satisfying than the original Black–Scholes derivation.

Proof That an American Call Should Never Be Exercised Early

As an example of the power of the risk-neutral approach, we will prove that early exercise is never optimal for a non-dividend-paying American option. Assume we are using the risk-neutral approach to value an American call with a binomial model.

The exercise price is c, and the per-period risk-free rate is r. The up move is by a factor u, and the down move is by a factor d. Let p be the probability of an up move. Then we know that for a non-dividend-paying stock,

$$pu + (1 - p)d = 1 + r \tag{2}$$

Assume that there are N total periods in the binomial tree approach.

Let $V_n(x)$ be the value of American call at time n when the price is x. Then

$$V_n(x) = \max(x - c) \frac{1}{1 + r} (pV_{n+1}(ux) + (1 - p)V_{n+1}(dx))$$

We need to show that this maximum is always attained by

$$\frac{1}{1 + r} (pV_{n+1}(ux) + (1 - p)V_{n+1}(dx))$$

We will have shown that it is always optimal to continue if

$$\frac{1}{1 + r} (pV_{n+1}(ux) + (1 - p)V_{n+1}(dx)) \geq x - c$$

Now, $V_{n+1}(ux) \geq ux - c$ and $V_{n+1}(dx) \geq dx - c$. This implies that

$$\frac{1}{1 + r} (pV_{n+1}(ux) + (1 - p)V_{n+1}(dx)) \geq \frac{1}{1 + r} (p(ux - c) + (1 - p)(dx - c))$$

$$= \frac{1}{1 + r} ((1 + r)x - c) = x - \frac{c}{1 + r} > x - c$$

The first equality follows from (2). The last inequality follows from the fact that for $r > 0$ and $c > 0$, $-\frac{c}{1+r} > -c$. This shows that continuing is at least as good as exercising the option. Therefore an American call (on a non-dividend-paying stock) should never be exercised early.

PROBLEMS

Group A

1 A stock currently sells for $20. In three months, the price of the stock will be either $18 or $22. The risk-free rate is 1% per month.

a Use the risk-neutral approach to value a three-month European call option with an exercise price of $20 on this stock.

b Use the risk-neutral approach to value a three-month European put option with an exercise price of $20 on this stock.

2 A stock currently sells for $40. In six months, the price of the stock will be either $50 or $15. The risk-free rate is 1% per month.

a Use the risk-neutral approach to value a six-month European call option having an exercise price of $24.

b Use the risk-neutral approach to value a six-month European put option having an exercise price of $24.

Group B

3 Explain why it might be optimal to exercise an American put early.

4 Explain why it might be optimal to exercise early an American call on a dividend-paying stock.

13.8 Valuing an Internet Startup and Web TV with the Black–Scholes Formula

Traditionally, projects have been evaluated on the basis of the net present value of their discounted cash flow. This is often referred to as the **DCF approach** for evaluating investments. In this section, we will show that the DCF approach can often lead a company to make an incorrect investment decision.

We now show how the real options approach can be applied to make correct business decisions in two situations:

- Valuing an investment in an Internet startup
- Valuing the opportunity to purchase a company with negative NPV projections

FIGURE 12

	B	C	D	E	F	G	H	I	J	K	L	M
2	Internet Startup											
3	Classic NPV					Cost through Q7		Option value				
4	Analysis					3.834325		$5.02				
5												
6						NPV with option		$1.19				
7												
8												
9												
10			rf rate	0.05								
11			int rate	0.23								
12			Static NPV									
13												
14			time	0	0.25	0.5	0.75	1	1.25	1.5	1.75	2
15			investment	0.5	0.5	0.5	0.5	0.5	0.5	0.5	0.5	12
16			revenue									22
17			dfcosts	1	0.987877	0.9759	0.964069	0.952381	0.940835	0.929429	0.918161	0.907029
18			PV costs	0.5	0.493938	0.48795	0.482034	0.47619	0.470417	0.464714	0.45908	10.88435
19			NPV rev	14.54161								0.660982
20			NPV costs	14.71868								
21			NPV inv	-0.17707								

In both situations, we will find that using real options leads to a different (and more satisfying) decision than the traditional DCF approach.

Valuing an Internet Startup

Netstart.xls

See file Netstart.xls and Figure 12. We are thinking of investing in an Internet startup. We are quite sure that today and every three months for the next 21 months, we will have to invest $.5 million to keep the project going. Our present estimate of the NPV of revenues (less future operating costs) to be received from this project, beginning two years from now, is $22 million. We will have to spend $12 million two years from now to complete the startup. Assume a risk-free rate of 5%. Should we go ahead with this project? We will assume that the revenue stream has an annual volatility of 40%.

Figure 12 contains a traditional DCF analysis of the project. We are sure about the project costs, so we discount them at 5%. This yields an NPV (cell E20) of $14.72 million. The costs, excluding the $12 million expense, have an NPV of $3.83 million. The revenue stream is, of course, risky, so we discount the revenue stream at the appropriate risk-adjusted rate. (Let's assume that it equals 23%.) Then the NPV of the revenue is $14.54 million. Thus, on a traditional DCF basis, we should not do the project.

What are we missing here? The traditional DCF argument has at least two flaws:

- The riskiness of the project changes over time. Do we really know the appropriate discount rate for all cash flows?

- The revenue stream is highly uncertain. In two years, we will have a much better idea of the value of the revenue stream. If things look bad, we will probably not invest $12 million. If things look good, we will. In short, we have an option to invest in this project, which we may or may not exercise.

How can we apply Black–Scholes to this situation? The uncertain quantity here is the NPV of future operating profits from the startup. Our best current estimate is $14.54 million. In two years, we have the option to spend $12 million to obtain the operating profits from the startup. The operating profits (at time 2) from the startup are the analog of the stock price at expiration. To model the operating profits at time 2, we need to estimate volatility. Suppose similar projects have a volatility of 40%. Then our option to invest in

	A	B	C	D	E	F	G	H
1	**Call with dividends**							
2								
3								
4	**Input data**							
5	Stock price	$14.54						
6	Exercise price	$12					Cost through quarter 7	$ 3.83
7	Duration	2						
8	Interest rate	5.00%						
9	dividend rate	0					Volatility	$ 5.02
10	volatility	40.00%					0.1	3.694589
11							0.11	3.704829
12				Predicted			0.12	3.718931
13	Call price		=	$5.02			0.13	3.736975
14	put	$		1.34			0.14	3.758878
15							0.15	3.784453
16							0.16	3.813459
17	Other quantities for option price						0.17	3.845628
18	d1	0.799025		N(d1)	0.787862		0.18	3.880691
19	d2	0.23334		N(d2)	0.592251		0.19	3.918384
20							0.2	3.958461
21							0.21	4.000694
22							0.22	4.044874
23							0.23	4.090811
24							0.24	4.138334
25							0.25	4.187288
26							0.26	4.237536
27							0.27	4.288952
28							0.28	4.341425
29							0.29	4.394852
30							0.3	4.449144
31							0.31	4.504218
32							0.32	4.56
33							0.33	4.616423
34							0.34	4.673426
35							0.35	4.730952
36							0.36	4.788953
37							0.37	4.847381
38							0.38	4.906194
39							0.39	4.965353
40							0.4	5.024824

FIGURE 13

two years is equivalent to a two-year call option with an exercise price of $12 million. Plugging the relevant parameters

> Current stock price: $14.54 million
>
> Duration = 2 years
>
> Volatility = 40%
>
> Dividend rate = 0 (no profits are earned before year 2)
>
> Exercise price = $12 million
>
> Risk-free rate = 5%

into our Black–Scholes template yields the result in Figure 13.

We find that the option opportunity is worth $5.02 million. Since the NPV of costs prior to time 2 is only $3.83 million, our opportunity to pursue the Internet startup is worth $5.02 − $3.83 = $1.19 million. The use of option thinking has turned our no-go decision into a go! Later, we will learn how to value this opportunity if we are allowed to abandon the project at any time in the next two years.

Sensitivity Analysis

Of course, we are not really sure about the volatility value. Therefore, we used a one-way data table to determine how the value of the option varies as volatility changes between

10% and 40%. We see that the option value outweighs the costs for the next 1.75 years as long as the volatility is at least 17%. Of course, any high-tech project will have volatility at least this large, so we feel confident about going ahead with the project.

Valuing a "Pioneer Option": Web TV

In April 1997, Microsoft purchased Web TV for $425 million. Let's assume that Microsoft felt that the true value of Web TV at this point in time was $300 million. Could the purchase of Web TV still be worthwhile? If the purchase of Web TV gave Microsoft the "option" to enter another business in the future, then the value of this option might outweigh the −$125 million NPV of the Web TV deal. More concretely, let's suppose that buying Web TV gives the opportunity (for $2 billion) to enter (in three years) another Internet-related business that has a current value of $1 billion. Assume a risk-free rate of 5%. Assume an annual volatility of 50% for the value of the Internet-related business. Does the "pioneer" option created by Web TV exceed Web TV's NPV of −$125 million?

Again, the pioneer option is a European call option with the following parameters:

Current stock price = $1 billion
Exercise price = $2 billion
Risk-free rate = 5%
Duration = 3 years
Volatility = 50%
Dividend rate = 0% (assuming none of the value of the Internet-related business is paid out prior to the exercise date)

From Figure 14, we find that the value of our option to enter the Internet-related business is $.17 billion = $170 million. This more than compensates for the negative NPV of the Web TV deal.

Sensitivity Analysis

Of course, the actual volatility of the Internet-related business is unknown. In Figure 14, we used a one-way data table to see how the value of the pioneer option varies as the volatility changes between 30% and 60%. As long as the volatility exceeds 43%, the pioneer option is worth more than Web TV's negative NPV of $125 million.

PROBLEMS

Group A

1 You have an option to buy a new plane in three years for $25 million. Your current estimate of the value of the plane is $21 million. The annual volatility for change in plane value is 25%, and the risk-free rate is 5%. What is the worth of the option to buy the plane?

2 The current price of copper is 95 cents. The annual volatility for copper prices is 20%, and the risk-free rate is 5%. In one year, we have the option to spend $1.25 million to mine 8 million pounds of copper. The copper can be sold at the copper price in one year. It costs 85 cents to extract a pound of copper from the ground. What is the value of this situation to us?

3 We own the rights to a biotech drug. Our best estimate of the current value of these rights is $50 million. Assuming that the annual volatility of biotech companies is 90% and the risk-free rate is 5%, value an option to sell the rights to the drug 5 years from now for $40 million.

FIGURE 14

	A	B	C	D	E	F	G	H	I	J	K	L	M
1	**Call with dividends**												
2													
3													
4	**Input data**												
5	Stock price	$1											
6	Exercise price	$2											
7	Duration	3											
8	Interest rate	5.00%											
9	dividend rate	0											
10	volatility	50.00%											
11													
12				Predicted									
13	Call price		=	$0.17									
14	put			$0.90									
15												Volatility	
16													$0.17
17	Other quantities for option price											0.3	0.05095
18	d1	-0.19416		N(d1)	0.423025							0.31	0.05611
19	d2	-1.060185		N(d2)	0.14453							0.32	0.061432
20												0.33	0.066902
21												0.34	0.07251
22												0.35	0.078246
23												0.36	0.0841
24												0.37	0.090062
25												0.38	0.096124
26												0.39	0.102279
27												0.4	0.108518
28												0.41	0.114835
29												0.42	0.121223
30												0.43	0.127676
31												0.44	0.134188
32												0.45	0.140754
33												0.46	0.147369
34												0.47	0.154029
35												0.48	0.160728
36												0.49	0.167463
37												0.5	0.174229
38												0.51	0.181023
39												0.52	0.187842
40												0.53	0.194681
41												0.54	0.201539
42												0.55	0.208412
43												0.56	0.215297
44												0.57	0.222192
45												0.58	0.229094
46												0.59	0.236001
47												0.6	0.24291

4 Merck is debating whether to invest in a pioneer biotech project. The estimate of the worth of this project is −$56 million. Investing in the pioneer project gives Merck the option to own a much bigger technology that will be available in four years. If Merck does not participate in the pioneer project, the company cannot own the bigger project. The bigger project will require $1.5 billion in cash four years from now. Currently, Merck estimates the NPV of the cash flows from the bigger project to be $597 million. What should Merck do? The risk-free rate is 10%, and the annual volatility of the big project is 35%. (This was the case that started the whole field of real options.)

13.9 The Relationship between the Binomial Model and the Lognormal Model

The Black–Scholes option pricing formula for pricing European puts and calls was derived under the assumption that future stock prices follow a lognormal random variable. To price American options, however, we often need to approximate stock prices in discrete time and work backward from the option's expiration date. It therefore becomes crucial to know how to approximate a lognormal random variable using a discrete random variable. The usual method is a **binomial tree.** In a binomial tree, we assume that in each period of length Δt, one of the following occurs:

- With probability p, the stock price is multiplied by a factor $u > 1$.
- With probability $q = 1 - p$, the stock price is multiplied by a factor $d = 1/u < 1$.

Binlognorm.xls

In short, the stock either increases or decreases each period. Assume a lognormal random variable having a mean return of $\mu\Delta t$ and a standard deviation of $\sigma\sqrt{\Delta t}$ in a length of time Δt. The file Binlognorm.xls computes values of p, u, and d that closely approximate the lognormal. We use Solver to choose values of p, u, and d for which the binomial model yields a mean growth by a factor of $e^{\mu\Delta t}$ and a standard deviation of $\sigma\sqrt{\Delta t}$ in one time period. Figure 15 illustrates the idea. Consider a stock having $\mu = .30$ and $\sigma = .38$. If we choose a period of 1 month $= \frac{1}{12}$ year with $\mu = .30$ and $\sigma = .38$, we insert those values into the spreadsheet. Then we run Solver and obtain the results in Figure 15.

We find that the best binomial approximation to model the stock's growth is that in each period the stock price increases (with probability .586) by 11.8% or decreases by 10.5% (with probability .414).

Using Simulation to Show How the Binomial Approximation Works

Binomsim.xls

In the spreadsheet Binomsim.xls (see Figure 16), we show how the approximation works. Our goal is to simulate the price of the stock in 60 months given $\mu = .30$, $\sigma = .38$, and a current price of $100. We use time intervals of 1 month ($\frac{1}{12}$ year) and apply the values of u, d, and p given in Figure 16.

Step 1 In cells F8:F67, enter $=$ RISKUNIFORM(0,1) to generate random numbers between 0 and 1.

Step 2 In cells E8:E67, use the binomial model to generate monthly changes in stock prices by copying from E8 to E9:E67

$$=IF(F8<p,u*E7,d*E7)$$

	A	B	C	D	E
1	Given mu and sigma				
2	find p and u				
3	**Exact**				
4				mu	0.3
5	p	0.585919		sigma	0.38
6	u	1.117534			
7	d	0.894828			
8	time	0.083333			
9		Prob	Value	Sq Dev	
10	up	0.585919	1.117534	0.008504	
11	down	0.414081	0.894828	0.017027	
12		binomial		lognormal	
13	mean growth	1.025315	=	1.025315	
14	var growth	0.012033	=	0.012033	
15					
16	**Approximate**				
17	a	1.025315			
18	up move	1.115939			
19	down move	0.896106			
20	prob up	0.587759			
21	prob down	0.412241			

FIGURE 15

FIGURE 16

	A	B	C	D	E	F	G	H	I	J	K
1					p	0.585919		76.67	5th %ile		80.07337
2	Microsoft simulation				u	1.117534		1275	95th %ile		1152.769
3	discrete time				d	0.894828		449.16	Mean		445.769
4								464.8	Sigma		437.712
5								Lognormal			Binomial
6				Date	Price	RN					
7				Today	100						
8				1	89.4828	0.653927					
9				2	100.0001	0.476678					
10				3	111.7535	0.500347					
11				4	124.8883	0.203208					
12				5	111.7536	0.655102					
13				6	124.8884	0.154175					
14				7	111.7536	0.936949					

With probability p, this formula increases stock price by u, and with probability $1 - p$, it increases stock price by d.

Step 3 To track the results of the simulation, use @Risk's statistical functions. Enter

=RISKPERCENTILE(E67,.05)   (cell K2 to compute 5th percentile)

=RISKPERCENTILE(E67,.95)   (cell K3 to compute 95th percentile)

=RISKMEAN(E67)   (cell K4 to compute mean)

=RISKSTDDEV(E67)   (cell K5 to compute standard deviation)

From Figure 17, we find that the simulated results from the binomial approximation are close to the lognormal simulation results. If we had let the length of a time period be smaller and run more iterations, our approximation would have been even better.

An Approximation to the Approximation

Most aficionados of real options do not know about Solver, so they use an approximation to the values of u, d, and p that we found with Solver. For Δt small, these approximations are extremely accurate. The formulas used in most books are as follows:

$$u = \frac{1}{d}$$

$$p = \frac{a - d}{u - d}$$

$$u = e^{\sigma \sqrt{\Delta t}}$$

$$a = e^{\mu \Delta t}$$

	H	I	J	K
1	76.67	5th %ile		80.07337
2	1275	95th %ile		1152.769
3	449.16	Mean		445.769
4	464.8	Sigma		437.712
5	Lognormal			Binomial
6				

FIGURE 17

The formulas for these approximations have been entered in the spreadsheet in cells B17:B21. As Figure 15 shows, the differences between the exact values of p, u, and d and the approximate values are very small. To be consistent with other authors, we always use these approximations in our binomial tree analyses.

13.10 Pricing an American Option with Binomial Trees

We cannot use the Black–Scholes model to price many American options. To determine the cash flows from the option, we need to account for the possibility of early exercise. American options are usually priced using **binomial trees.** We divide the duration of the option into smaller time periods (usually weeks or months). During each period, the stock price either increases by a factor u or decreases by a factor d. We assume that $d = 1/u$. Let Δt equal the length of a period in the tree. The probability of an increase each period (p) is chosen in conjunction with u and d such that the stock price grows on average at the risk-free rate r and has an annual volatility of σ. We let $q = 1 - p$ equal the probability of a decrease during each period. To perform risk-neutral valuation, we assume (given that the stock pays no dividends[†]) that the underlying asset grows according to a lognormal random variable having μ = the risk-free rate (r). If σ is the volatility parameter for the underlying stock's lognormal random variable, then we know from Section 13.9 that the lognormal movement of prices can be approximated by a binomial model with parameters given by

$$u = \frac{1}{d}$$

$$p = \frac{a - d}{u - d}$$

$$u = e^{\sigma\sqrt{\Delta t}}$$

$$a = e^{\mu\Delta t}$$

$$q = 1 - p$$

In one part of the spreadsheet, we model the evolution of the stock price. Then we value the option by working backward from the last time period. At each node (a combination of stock price and time), we recursively determine the value of the option from that point on. By the time we have reached today's node, we have well approximated the value of the option today. Note that our approximation will improve as Δt is reduced. To begin, we value an American put option. Recall that an American call option on a non-dividend-paying stock will never be exercised early. As we will soon see, an American put option may be exercised early. The work is in file American.xls.

American.xls

Let's price a 5-month American put having

$$\text{Current stock price} = \$50$$
$$\text{Exercise price} = \$50$$
$$\text{Risk-free rate} = 10\%$$
$$\text{Annual volatility} = 40\%$$
$$\Delta t = 1 \text{ month} = .083 \text{ year}$$

[†]If a stock pays dividends at rate q (or a project pays out a percentage q of its value each year), then we assume the asset grows in the risk-neutral world at rate $r - q$.

All this data, plus the previous definitions of u, d, a, q, and p, are input in the range A1:B13 of the spreadsheet. We have used range names to make the tree easier to explain.

The Stock Price Tree

We begin by determining the possible stock prices during the next 5 months. Column B has today's (month 0) price, column C month 1, etc.

Step 1 Enter today's stock price ($50) in B16 with the formula

$$=B3$$

Step 2 By copying the formula

$$=u*B16$$

from C16 to D16:G16, obtain the price each month when there have been no down moves.

Step 3 To compute all other prices, note that in each column as we move down a row, the price is multiplied by a factor (d/u). Also note that during month i, there are $i + 1$ possible prices. This allows us to compute all prices by copying from C17 to C17:G21 the formula

$$=IF(\$A17<=C\$15,(d/u)*C16,\text{"-"})$$

As we move down each column, the prices are successively multiplied by d/u. Also, the formula places a "-" where a price does not exist.

The Optimal Decision Strategy

We now work backward to find the value of the American put. Remember that at each node, the value of the put equals the expected discounted value of future cash flows from the put.

Step 1 At month 5, the option is just worth maximum (0, exercise price − current stock price). Thus, we enter

$$=MAX(0,\$B\$4\text{-}G16)$$

in cell G24 and copy this formula down to G29.

Step 2 In month 4 (and all previous months), the value of the option at any node is

$$\text{Maximum (value from exercising now, } \left(\frac{1}{1 + \frac{.1}{12}}\right) *(p*\text{(value of option for up move)}$$
$$+ \ q*\text{(value of option for down move)))}$$

For example, in F28, the value of the option is

$$\max(50 - 31.50, (1/1.0083)*(.507*(\$14.64) + .493*(\$21.93))) = \$18.50$$

Since this maximum is attained by exercising now, if this node occurs, we would exercise the option now. At the node in F26, the maximum is attained by not exercising. To implement this decision-making procedure, we enter in cell F24 the formula

$$=IF(\$A24<=F\$23,MAX(\$B\$4\text{-}F16,(1/(1+r\_*deltat))*(p*G24+(1\text{-}p)*G25)),\text{"-"})$$

Copying this formula to B24:F29 generates the value of the option for all possible prices during months 1–4 and places a "-" in any cell where there is no actual stock price. In cell B24, we find the estimated value of the put to equal $4.49. Of course, $4.49 is an approximation to the put value. As Δt grows small, however, our price will converge to the actual price of the put if the stock follows a lognormal random variable.

Using Conditional Formatting to Describe the Optimal Exercise Policy

Assuming the stock grows at the risk-free rate, how would we react to actual price changes? Suppose the first three months have down moves. We do not exercise during the first two months, but we exercise after the third down move. Suppose the first four months are down–up–down–down. Then we exercise after the fourth month. To make this clearer, we use Excel's Conditional Format Option to format the spreadsheet so that cells for which the option would be exercised are in boldface. We begin by noting that it is optimal to exercise the option at a month and price if and only if the value of the cell corresponding to the month and price equals (exercise price) − (stock price). To indicate the cells where exercising the option is optimal, begin in cell G24 and select the cell range B24:G29, where you want the format to be valid. Then select Format Conditional Format and make the entries shown in Figure 18. Then click on Format Bold. This dialog box ensures that if the option is exercised in the state with 5 up moves in period 5, then a bold font is used. The interpretation of

$$=(\$B\$4-G16)=G24$$

is that the format takes hold only if (B4-G16)=G24, which is equivalent to the option being exercised. We have ensured that any period and number of up moves for which exercise of the option is optimal will be indicated in the bold font. See Figure 19.

Sensitivity Analysis

Using one-way data tables, it is easy to see how changes in various input parameters change the price of the put. We have varied the annual volatility of the stock from 10% to 70%. Figure 20 (a one-way data table with input cell sigma) shows how an increase in volatility greatly increases the value of the put. Increased volatility gives a larger chance of a big price drop, which increases the value of the put.

Figure 21 shows how a change in the exercise price of the stock changes the value of the put. We used a one-way data table with input cell B4. As the exercise price increases, the value of the put increases, because an increased exercise price increases the number of values for which the put is "in the money."

FIGURE 18

	A	B	C	D	E	F	G
1	**American Option**						
2	**Put**						
3	Current price	$ 50.00					
4	Exercise price	$ 50.00					
5	r	0.1					
6	sigma	0.4					
7	t	0.416667					
8	deltat	0.083333					
9	u	1.122401					
10	d	0.890947					
11	a	1.008368					
12	p	0.507319					
13	q	0.492681					
14		**Time**					
15	**Stock Prices**	0	1	2	3	4	5
16	0	$ 50.00	$ 56.12	$ 62.99	$ 70.70	$ 79.35	$ 89.07
17	1		$ 44.55	$ 50.00	$ 56.12	$ 62.99	$ 70.70
18	2		-	$ 39.69	$ 44.55	$ 50.00	$ 56.12
19	3		-	-	$ 35.36	$ 39.69	$ 44.55
20	4		-	-	-	$ 31.50	$ 35.36
21	5		-	-	-	-	$ 28.07
22	**Down Moves**						
23	**Put Value**	0	1	2	3	4	5
24	0	$ 4.49	$ 2.16	$ 0.64	$ -	$ -	$ -
25	1	-	$ 6.96	$ 3.77	$ 1.30	$ -	$ -
26	2	-	-	$ 10.36	$ 6.38	$ 2.66	$ -
27	3	-	-	-	**$ 14.64**	**$ 10.31**	**$ 5.45**
28	4	-	-	-	-	**$ 18.50**	**$ 14.64**
29	5	-	-	-	-	-	**$ 21.93**

FIGURE 19

	H	I
7	**Volatility**	4.489053
8	0.1	0.680146
9	0.2	1.91777
10	0.3	3.197364
11	0.4	4.489053
12	0.5	5.799995
13	0.6	7.104648
14	0.7	8.400682

FIGURE 20

	K	L
7	**Ex Price**	4.489053
8	$ 45.00	2.139609
9	$ 46.00	2.609498
10	$ 47.00	3.079387
11	$ 48.00	3.549275
12	$ 49.00	4.019164
13	$ 50.00	4.489053
14	$ 51.00	4.980604
15	$ 52.00	5.484364
16	$ 53.00	5.988124
17	$ 54.00	6.52469
18	$ 55.00	7.092245
19	$ 56.00	7.733071

FIGURE 21

	N	O
7	r	4.489053
8	0.01	5.300677
9	0.02	5.207424
10	0.03	5.114923
11	0.04	5.023183
12	0.05	4.932208
13	0.06	4.842007
14	0.07	4.752584
15	0.08	4.663947
16	0.09	4.576102
17	0.1	4.489053
18	0.11	4.405002
19	0.12	4.333566

FIGURE **22**

Finally, Figure 22 shows that an increase in the risk-free rate decreases the value of the put. This is because an increase in the risk-free rate makes the payoff from the put (which occurs in the future) less valuable.

Relationship to an Abandonment Option

Often, an option to abandon a project may be thought of as a put. To see this in the current context, suppose the current value of a project is $50 million, the risk-free rate is 10%, and the project has a 40% annual volatility. Any time in the next five months, we may abandon the project and receive $50 million. To determine the value of this abandonment option, we would proceed exactly as we proceeded to value the put. We would have found the value of the abandonment option to be $4.49 million.

Computing the Early Exercise Boundary

If the price of the stock today is $50, it would be nice to know in advance what we would do (exercise or not exercise) for any given price during a future period. For example, would we exercise if the month 1 price were $42? If we have not exercised during the first three months, would we exercise during month 4 if the price were $40? Answering this question requires that we compute the **early exercise boundary** for each period. It turns out that for each month, there exists a boundary price $p(t)$ such that we will exercise during month t (assuming the option has not been exercised) if and only if the month t price is less than or equal to $p(t)$. Together, $p(1)$, $p(2)$, $p(3)$, $p(4)$, and $p(5)$ define the early exercise boundary for the put. To find the early exercise boundary, it is convenient to make four copies of the original sheet. To copy a sheet, put the cursor on the sheet name, hold down the left mouse button and Control key, and drag the sheet to another tab. We have renamed our four copies Ex Bound 1, Ex Bound 2, etc. In sheet Ex Bound 1, we determine $p(1)$ as follows. The value $p(1)$ for which we exercise during month 1 if and only if $p \leq p(1)$ can be found by observing that $p(1)$ is the largest month 1 price for which exercise price $- p(1)$ equals the month 1 value of the option. To find $p(1)$ (see Figure 23), we proceed as follows:

Step 1 In cell C16, insert a trial value for $p(1)$. Note that the way we have set up the price tree ensures that the prices in B16 and C17 have no effect on the value of the put computed in C24.

	A	B	C	D	E	F	G
1	**American Option**						
2	**Put**						
3	Current price	$ 50.00					
4	Exercise price	$ 50.00					
5	r	0.1					
6	sigma	0.4					
7	t	0.416667					
8	deltat	0.083333					
9	u	1.122401					
10	d	0.890947					
11	a	1.008368					
12	p	0.507319					
13	q	0.492681					
14		Time					
15	**down moves**	0	1	2	3	4	5
16	0	$ 50.00	$ 39.19	$ 43.98	$ 49.37	$ 55.41	$ 62.19
17	1		$ 31.11	$ 34.91	$ 39.19	$ 43.98	$ 49.37
18	2		-	$ 27.71	$ 31.11	$ 34.91	$ 39.19
19	3		-	-	$ 24.69	$ 27.71	$ 31.11
20	4		-	-	-	$ 22.00	$ 24.69
21	5		-	-	-	-	$ 19.60
22		ex-price	$ 10.81				
23	**Put Value**	0	1	2	3	4	5
24	0	14.6728	**10.81387**	6.84145	3.096029	0.309755	0.00
25	1		**18.89456**	**15.08723**	**10.81387**	**6.017455**	0.63
26	2		-	**22.28669**	**18.89456**	**15.08723**	10.81
27	3		-	-	**25.30891**	**22.28669**	$ 18.89
28	4		-	-	-	**28.00154**	$ 25.31
29	5		-	-	-	-	$ 30.40

FIGURE 23

Step 2 Assuming that the month 1 price equals the value in C16, compute in C22 the value if the put is exercised in month 1 with the formula

$$=B4-C16$$

Step 3 Now use Solver to determine $p(1)$. Note that $p(1)$ is the largest price (entered in C16) for which the value of exercising now (in cell C22) equals the month 1 value of the option (computed in C24). Therefore, the Solver settings in Figure 24 enable us to compute $p(1)$.

We find that $p(1) = \$39.19$. Thus if the month 1 price is below \$39.19, we should exercise. Otherwise, we go on. Try a variety of month 1 prices in B16 to convince yourself that it is optimal to exercise for any price below \$39.19 and continue for any price above \$39.19. Of course, we are assuming that exercise can only occur at time $= 1, 2, \ldots, 5$. But it turns out that even if many more points of exercise were allowed, $p(1)$ would be fairly close to \$39.19. In a similar fashion, we find the rest of the early exercise boundary to be the following:

Time	Exercise if price $\leq$
1	$39.19
2	$39.41
3	$41.64
4	$45.28
5	$50.00

Let's reconsider a previous example.

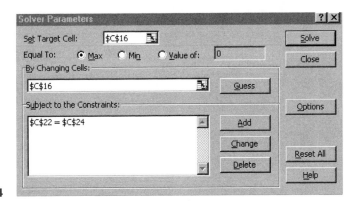

FIGURE 24

EXAMPLE 2 Internet Startup

Startupabandon.xls

We are thinking of investing in an Internet startup. (See file Startupabandon.xls.) We are quite sure that today and every three months for the next 21 months, we will have to invest $.5 million to keep the project going. Our present estimate of the NPV of revenues (less future operating costs) to be received from this project, beginning two years from now, is $22 million. We will have to spend $12 million two years from now to complete the startup. Assume a risk-free rate of 5%. Should we go ahead with this project? Recall that we believe the NPV of project revenues to be $14.54 million. We also assume a 40% annual volatility.

By looking at this situation as a European call option, we found its value to be $1.19 million. Now let's suppose that during each of the next seven quarters we may, if things look bad, abandon the project. For example, if during quarter 3 our estimate of the value of revenues is $3 million, it hardly seems worthwhile to pump in $.5 million during each quarter. By how much does the abandonment option increase the project's value?

Solution Our work is in the sheet Abandon of file Startupabandon.xls. See Figure 25. In A1:B7, we compute the relevant parameters for our binomial tree. We use $\Delta t = .25$ year. In cell B8, we compute our discount factor per period as $\frac{1}{1 + .25(.05)} = .99$. We now proceed as follows:

Step 1 Begin by generating the set of possible revenue values for each quarter. Enter the current estimate of revenue value ($14.54 million) in cell B19. Then generate the possible prices for each quarter by copying from J11 to C11:J19 the formula

$$=IF(\$A11>J\$9,``\ ",IF(\$A11=J\$9,u*I12,(d/u)*J10))$$

If a price is not possible, a blank is entered in the cell. For example, there cannot be 7 up moves in 6 periods, so cell H12 is blank. The highest possible price during each period is computed as u times the highest possible price during the previous period. In any other cell, the price is computed as d/u times the price in the cell directly above.

Step 2 We now value the situation without the abandonment option. This will, of course, approximate the Black–Scholes answer of $1.19 million. We begin by finding the value of each period 8 node. This is simply max (revenue value − 12, 0). By copying from J22 to J23:J30 the formula

$$=MAX(J11-12,0)$$

we compute the value of the situation at time 2 (period 8) for each possible stock price.

FIGURE 25

	A	B	C	D	E	F	G	H	I	J
1	r	0.05		Abandonment						
2	sigma	0.40								
3	deltat	0.25								
4	a	1.01								
5	u	1.22								
6	d	0.82								
7	p	0.48		1.01						
8	df	0.99								
9	**Firm Value**	0.00	1.00	2.00	3.00	4.00	5.00	6.00	7.00	8.00
10	up moves	0.00	0.25	0.50	0.75	1.00	1.25	1.50	1.75	2.00
11	8.00									72.02
12	7.00								58.96	48.27
13	6.00							48.27	39.52	32.36
14	5.00						39.52	32.36	26.49	21.69
15	4.00					32.36	26.49	21.69	17.76	14.54
16	3.00				26.49	21.69	17.76	14.54	11.90	9.75
17	2.00			21.69	17.76	14.54	11.90	9.75	7.98	6.53
18	1.00		17.76	14.54	11.90	9.75	7.98	6.53	5.35	4.38
19	0.00	14.54	11.90	9.75	7.98	6.53	5.35	4.38	3.59	2.94
20	**No abandonment**	0.00	1.00	2.00	3.00	4.00	5.00	6.00	7.00	8.00
21	Value from now on	0.00	0.25	0.50	0.75	1.00	1.25	1.50	1.75	2.00
22	8.00									60.02
23	7.00								46.62	36.27
24	6.00							35.58	27.18	20.36
25	5.00						26.49	19.67	14.14	9.69
26	4.00					18.99	13.46	9.00	5.41	2.54
27	3.00				12.87	8.47	5.02	2.43	0.71	0.00
28	2.00			7.98	4.61	2.12	0.44	-0.42	-0.50	0.00
29	1.00		4.17	1.71	0.04	-0.91	-1.21	-0.99	-0.50	0.00
30	0.00	1.27	-0.42	-1.44	-1.87	-1.83	-1.48	-0.99	-0.50	0.00
31	**Abandonment**	0.00	1.00	2.00	3.00	4.00	5.00	6.00	7.00	8.00
32		0.00	0.25	0.50	0.75	1.00	1.25	1.50	1.75	2.00
33	8.00									60.02
34	7.00								46.62	36.27
35	6.00							35.58	27.18	20.36
36	5.00						26.49	19.67	14.14	9.69
37	4.00					18.99	13.46	9.00	5.41	2.54
38	3.00				12.87	8.47	5.02	2.43	0.71	0.00
39	2.00			8.01	4.67	2.23	0.66	0.00	0.00	0.00
40	1.00		4.33	2.01	0.56	0.00	0.00	0.00	0.00	0.00
41	0.00	1.79	0.45	0.00	0.00	0.00	0.00	0.00	0.00	0.00

Step 3 We now compute that value of each node in periods 0–7. Let node (t, j) be the period t node with j up moves of the stock. Let the value of the situation from the node (t, j) onward equal $V(t, j)$. Then

$$V(t, j) = -.5 + \frac{1}{df}(pV(t + 1, j + 1) + (1 - p)V(t + 1, j))$$

where df = discount factor and p = risk-neutral probability of an up move. This result follows, because at any node, we obtain with probability p a value $V(t + 1, j + 1)$, and with probability $1 - p$ we obtain a value $V(t + 1, j)$. Of course, we must also pay \$.5 million during the current period. Copying from I23 to B30:I23 the following formula computes all node values:

$$=IF(\$A23>I\$20,``\ ",-0.5+df*(p*J22+(1-p)*J23))$$

This formula also puts a blank space in any cells for which there is no defined value.

We find that the value of the situation is $1.27 million. Choosing a smaller value of Δt would have resulted in a value closer to our Black–Scholes value of $1.19 million.

Step 4 We can now determine the amount by which the abandonment option increases the value of the situation. In J33:J42, compute the period 8 value by copying from J33 to J34:J42 the formula

$$=MAX(J19-12,0)$$

If we define $V(t, j)$ as before, then

$$V(t, j) = \max\left(0, -.5 + \frac{1}{df}(pV(t+1, j+1) + (1-p)V(t+1, j))\right)$$

This follows, because at any node we can abandon (and earn 0 from that point onward) or pay $.5 million and continue as before. By copying from I34 to B41:I34 the formula

$$=IF(\$A34>I\$31,``\ ",MAX(0,-0.5+df*(p*J33+(1-p)*J34)))$$

we compute the value of each (time, stock price) combination.

We find that the value today (with the abandonment option) is $1.79 million. Thus, the abandonment option is worth around $600,000.

When Should We Abandon?

To obtain the value of the abandonment option in Example 2, we must know at each time the range of revenue values for which abandonment is optimal. For example, during quarter 4, what ranges of revenue values lead to abandonment? To answer this question (see the sheet Value), we go to the quarter 4 stock price with the highest value (cell F15) and put any number in that cell. The largest number in F15 (call it r^*) that will make the project value (found in F37) equal to 0 represents the boundary between abandoning and not abandoning the project. If the quarter 4 revenue value exceeds r^*, we should keep going. If the quarter 4 revenue estimate is less than r^*, we should abandon the project. To find r^*, we use the Solver model shown in Figure 26.

We find the largest quarter 4 revenue value that makes the project value from quarter 4 onward equal to 0. We find that if the quarter 4 revenue estimate is less than $10.37 million, we should abandon the project. Otherwise, we should continue.

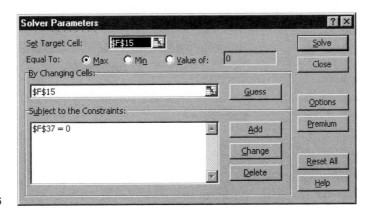

FIGURE 26

PROBLEMS

Group A

1 A 9-month American put option on a non-dividend-paying stock has a strike price of $49. The stock price is $50 today; the risk-free rate is 5% per year (continuously compounded); and the annual volatility is 30%. Use a three-step binomial tree to value this put.

2 Corpco is currently worth $50 million. The value of Corpco has an annual volatility of 40%, and the risk-free rate is 10%. During each of the next five years, we will have the option to buy Corpco at the following prices (in millions):

Year	Purchase Price
Now	$40
1 year later	$41
2 years later	$42
3 years later	$43
4 years later	$50
5 years later	$70

Assume that Corpco earns roughly 12% of its value during each year. Value the option to purchase Corpco. How sensitive is the value of this option to changes in volatility and interest rates? Determine a range of values for Corpco during each year for which you would purchase Corpco.

3 You are thinking of buying Walco. You believe the current value of this business to be $100 million, with a 45% annual volatility. The risk-free rate is 5%, and Walco pays out 20% of its value annually in cash flow. If you purchase Walco and things go sour at any time during the next five years, you can move Walco's assets to Europe, where they are worth $75 million. Walco is willing to sell for $115 million. Should you take the offer?

Group B

4 A drug will be sold for 20 years; the rights to the drug have a current value of $50 million. Assume that an equal percentage of this value is derived from each year of sales. At any time during the next five years, I can sell the rights to the drug for $40 million. If the annual volatility of changes in the value of the rights is 40% and the risk-free rate is 6%, what is the value of this option?

5 Georgia Woodward is trying to determine whether or not she should develop vacant land into a 6-unit or a 9-unit apartment complex. It will cost $480,000 to build a 6-unit complex and $810,000 to build a 9-unit complex. (The cost will remain unchanged if she builds next year.) The current market value of an apartment unit is $100,000. The risk-free rate is 10%, and the annual volatility of the value of an apartment unit is 30%. Also, the apartment pays 8% of its value in profits each year. Georgia has the option to build now or at the beginning of each of the next ten years. What is the worth of this option?

6 You are developing a new drug. You believe the value of the drug to be $100 million, with a 40% annual volatility. During the next ten years, you have two options available. In four years, you may sell the project for $80 million. If you have not sold the project in year 4, then 10 years from now you may, at a cost of $40 million, expand the size of the project by 50%. The risk-free rate is 5%. By how much do these options increase the value of the project?

7 Drug company managers are undertaking development of a new drug. Today they will make a $25 million investment to screen many potential compounds. After three years, the screening process will be complete, and they can conduct clinical trials, if desired, at a cost of $90 million. Finally, 10 years from now, they can, if desired, invest $500 million and bring the drug to market. Their current estimate is that the cash inflows from the drug will have a value of $800 million (in time 10 dollars). This estimate would, of course, have some adjustment for risk contained in its computation. The annual volatility of revenues is estimated to be 60%. The company's cost of capital is 14%, and the risk-free rate for a ten-year bond is 4%. How would you value this opportunity?

Group C

8 The current price of gold is $400. The risk-free rate is 10%, and the annual volatility for changes in gold prices is 30%. Gold prices are assumed to follow a lognormal random variable. Each year our gold mine is open, a fixed cost of $1 million is incurred. This cost is incurred even if no gold is mined during the year. If we open the mine, a cost of $2 million is incurred. If we shut the mine, a cost of $1.5 million is incurred. During the current year and each of the next ten years, we can, if the mine is open, mine up to 10,000 ounces of gold at a variable cost of $250 per ounce. What is the worth of this situation? (*Hint:* Use two lattices; one for the value given that the mine is shut down at the beginning of the year and one for the value given that the mine is open at the beginning of the year.)

13.11 Pricing European Puts and Calls by Simulation

Even though European puts and calls can easily be priced by the Black–Scholes (BS) formula, it is instructive to use Monte Carlo simulation to price European options.

EXAMPLE **3**   **Dell Computer Stock**

On June 30, 1998, Dell Computer stock sold for $94. A European put with an exercise price of $80, expiring on November 22, 1998, was selling for $5.25. The 90-day T-bill rate is 5.5%. Assume that the annual volatility of this option is 53.27%. Use simulation to value this put option.

Solution

Option.xls

We will use Monte Carlo simulation to price the put. We begin by entering relevant parameters in cells B2:B4 of sheet Dell Sim in file Option.xls. See Figure 27. Recall from Section 13.7 that a fair price for the put is the expected discounted value of the put's cash flows in a risk-neutral world. In a risk-neutral world, the stock will grow at the risk-free rate. Therefore, we will use equation (1) with μ = risk-free rate to price the put. We have used the following range names:

$$r\_ = \text{the risk-free rate}$$
$$p = \text{the current stock price}$$
$$v = \text{volatility}$$
$$d = \text{duration}$$
$$x = \text{exercise price}$$

We proceed as follows.

Step 1   In cell B7, use Equation (1) to generate Dell's price at the expiration date with the formula

$$=p*EXP((r\_-0.5*v^2)*d+RISKNORMAL(0,1)*v*SQRT(d))$$

Step 2   In cell B8, compute the cash flows from the put. Recall that the put pays nothing if Dell's price on the expiration date exceeds $80; otherwise the put pays $80 − (Dell's price at expiration).

$$=IF(B7>x,0,x-B7)$$

	A	B	C	D	E
1	**Pricing Put by Simulation**				
2	Current stock price	$ 94.00		95% CI	
3	Risk free rate	0.053541		Lower Limit	5.090718
4	Duration	0.39726		Upper Limit	5.469088
5	volatility	53.265%			
6	Exercise price	$ 80.00			
7	Stock price at expiration	91.93506			Name
8	Put cash flows at expiration	0			Description
9	Discounted value of put cash flows	0			Cell

FIGURE **27**

	D	E
2	95% CI	
3	Lower Limit	5.090718
4	Upper Limit	5.469088

	E	F
9	Cell	B9
10	Minimum =	0
11	Maximum	57.99311
12	Mean =	5.279903
13	Std Deviat	9.459261
14	Variance =	89.47761
15	Skewness	1.900912
16	Kurtosis =	5.931631
17	Errors Calc	0
18	Mode =	0
19	5% Perc =	0
20	10% Perc	0
21	15% Perc	0
22	20% Perc	0
23	25% Perc	0
24	30% Perc	0
25	35% Perc	0
26	40% Perc	0
27	45% Perc	0
28	50% Perc	0
29	55% Perc	0
30	60% Perc	0
31	65% Perc	0.246574
32	70% Perc	3.815132
33	75% Perc	7.46743
34	80% Perc	11.33537
35	85% Perc	15.56934
36	90% Perc	20.52942
37	95% Perc	27.15265

FIGURE 28

Step 3 In cell B9, compute the expected discounted value of the put's cash flows with the formula

$$=EXP(-r\_*d)*B8$$

Step 4 We select B9 as our output cell. After 10,000 iterations, Figure 28 gives our best estimate of the put price as $5.28. We are 95% sure that the put price is between $5.09 and $5.47 (see cells E3 and E4). After 10,000 iterations, why have we not come closer to the price of the put? The reason is that this put pays off only on extreme results (Dell's price dropping a lot). It takes many iterations to accurately represent extreme values of a lognormal random variable.

REMARK If a stock pays dividends at a rate q% per year, then in a risk-neutral world, the stock price must grow at a rate $r - q$. Therefore, for stocks that pay dividends at a rate of q% per year, we should price their options by using Equation (1) with $r - q$ replacing r.

The Black–Scholes Formula is a fairly simple closed-form formula to price European puts and calls. However, it is much more difficult to come up with a closed-form formula to price so-called *exotic* options:

- As you like it options: The owner decides if the option is a put or a call.

- Knockout options: The option terminates if the stock price reaches a certain value. Otherwise, a knockout option is treated as a normal put or call.

- Lookback option: The exercise price is determined (for a call) by the minimum price observed before expiration or (for a put) by the maximum price observed before expiration.

- Bermuda option: Exercise is restricted to specific dates.

- Asian option: The payoff from the option depends on the average value of the stock price during the option's duration. The average may be used as either a strike price or be substituted for the stock's final price. The average is sometimes computed as an arithmetic mean and sometimes as a geometric average (the nth root of the product of the stock's price at the end of each of n weeks).

There are many other exotic options, but they all share the property that the payoff from the option depends on the entire path of the stock price during the option's duration, not just the stock's price at expiration. To illustrate how Monte Carlo simulation and the risk-neutral approach can be used to price exotic options, we show how to price a knockout put.

EXAMPLE 4   **Knockout Put**

Let's reconsider the put on Dell discussed in Example 3. Suppose you are confident that Dell will not fall below $60 during the next 21 weeks, but you are worried about it falling to a price of between $60 and $80. The put priced in Example 3 is pretty expensive, so you opt to buy a put that gets knocked out if Dell's price drops below $60 during the next 21 weeks. What is a fair price for this put? Of course, if Dell drops below $60, you are in trouble. But as we will see, the knockout put will be much cheaper. For simplicity, we are assuming a 21-week duration rather than a 145-day put.

Knockout.xls   **Solution**   The work is in the file Knockout.xls. See Figure 29. We begin by defining the following range names:

$$S = \text{current price}$$
$$X = \text{exercise price}$$
$$K = \text{knockout value for the put}$$
$$R = \text{risk-free rate}$$
$$V = \text{volatility}$$

We will approximate the path of Dell's price by looking at the price at the end of each week.

Step 1   Enter today's price in cell B10 and generate Dell's price in a week in cell B11 with the formula (based on (1))

$$=B10*EXP((r\_-0.5*v^2)*(1/52)+RISKNORMAL(0,1)*SQRT(1/52)*v)$$

Again, we assume that the stock grows at the risk-free rate.

Step 2   Copying this formula to B11:B31 generates Dell's price at the end of each week.

Step 3   In cell B32, compute the lowest price of Dell during the option's duration with the formula

$$=MIN(B10:B31)$$

	A	B
1		
2	Knockout put	
3	Today's price	$ 94.00
4	Duration(weeks)	21
5	Exercise price	$ 80.00
6	Knockout value	$ 60.00
7	riskfree rate	0.053541
8	volatility	0.53265
9	Week	Price
10	0	$ 94.00
11	1	99.8283
12	2	83.70867
13	3	75.21299
14	4	63.84219
15	5	59.77915
16	6	56.7612
17	7	55.70837
28	18	71.05029
29	19	67.65725
30	20	67.65375
31	21	69.29394
32	Min price	$ 50.41
33	Option payoff	0
34	Discounted value of option payoff	0

FIGURE 29

Step 4 In cell B33, compute the payoff from the option with the formula

$$=IF(B32<k,0,IF(B31>x,0,x-B31))$$

This ensures that the put is knocked out if Dell's price ever drops below $60. If the price of Dell at expiration exceeds $80, there is no value to the put. Otherwise, the put pays (exercise price) − (stock price at expiration).

	C	D	E	F
3	95% CI			
4	Lower	1.429149	Name	Discounted
5	Upper	1.583197	Description	Output
6			Cell	B34
7			Minimum =	0
8			Maximum	19.57092
9			Mean =	1.506173
10			Std Deviat	3.851216
11			Variance =	14.83186
12			Skewness	2.754046
13			Kurtosis =	9.835486
14			Errors Calc	0
15			Mode =	0
16			5% Perc =	0
17			10% Perc	0
28			65% Perc	0
29			70% Perc	0
30			75% Perc	0
31			80% Perc	0
32			85% Perc	2.951085
33			90% Perc	6.718378
34			95% Perc	11.73514

FIGURE 30

Step 5 Compute the discounted value of the option's payoffs in B34 with the formula

$$=EXP(-r\_*(B4/52))*B33$$

Step 6 Select B34 as the output cell. After running 10,000 iterations, we obtain the output shown in Figure 30.

Our best estimate of the price of the knockout put is $1.51. We are 95% sure that a fair price is between $1.43 and $1.58. The knockout allows us to protect ourselves against a moderate drop in Dell's price much more cheaply than an ordinary put, but a catastrophic drop in Dell's price will be a real wipeout!

PROBLEMS

Group A

1 Consider the Dell knockout option in Example 4. For the same time period and parameters, value an Asian put on Dell with an exercise price of $75. The payoff on the put is keyed off the average price of Dell for weeks 1–21.

2 On January 1, 1996, a stock sells for $44. On April 1, 1996, you observe the price of the stock and have the choice of obtaining a put or a call option with an expiration date of January 1, 1997 and an exercise price of $42. Suppose the stock grows at 15% per year, with an annual volatility of 20%. The risk-free rate is currently 10%. Run 400 iterations. From your simulation, find a 95% confidence interval for the value of the option.

3 Consider a 52-week Asian call option. Today's price is $100, and the exercise price is $110. The annual volatility for the stock is 30%, and the risk-free rate is 9%. The payoff to the option depends on the average price of the stock at weekly intervals (beginning with today). What is a fair price for this option?

4 A knockout call option loses all value at the instant the price of the stock drops below a given "knockout level." Determine a fair price for a knockout call option in the following situation:

Current stock price: $20   Knockout price: $19.50
Exercise price: $21   Annual volatility: 40%
Risk-free rate: 10%   Mean growth rate of stock: 12%
Duration of option: 1 month = 21 days;
                    assume 250 days = 1 year

5 An increase in volatility will surely increase the price of a call or put option, but an increase in volatility may decrease the value of a knockout option. Why?

6 If you wanted to hedge holdings in a stock, what advantage would a knockout put have over a regular put?

7 Consider the stock described in Problem 2.

a Price a lookback call option with a duration of 6 months.

b Price a lookback put option with a duration of 6 months.

8 For the Asian option of Problem 2, determine how a change in volatility affects the option's value.

9 Suppose that 20 weeks before the averaging period for the Asian option of Problem 2 begins, the stock sells for $100. At this time, what would be a fair price for the option?

10 Suppose that 20 weeks into the averaging period (i.e., in row 33) for the Asian option of Problem 2, the stock is selling for $100, and the average price of the stock during the first 21 averaging points (initial price plus first 20 weeks) has been $100. At this point in time, what is a fair price for the option?

13.12 Using Simulation to Value Real Options

In this section, we will learn how to use simulation and risk-neutral valuation to value European real options. The key idea is to let the value of the underlying asset grow at the risk-free rate (less the percentage of asset value, if any, paid out each year). Then the option is worth the expected discounted value of the cash flows generated in this risk-neutral world.

EXAMPLE 5   Abandonment Option

An asset is currently worth $553,000 and has an annual volatility of 28%. The risk-free rate is 5%. A year from now, the asset may be sold for a salvage value of $500,000. How much is this abandonment option worth?

Solution   Let V = value of the asset one year from now. Then a year from now, the option returns

$$\max(0, \$500,000 - V) \tag{3}$$

If we let the asset grow for one year at the risk-free rate and with the given volatility and take the discounted value of (3) as the output cell, then the mean of the output cell is the value of the abandonment option. The work is in Figure 31 and the file Abandonment.xls. We proceed as follows.

Abandonment.xls

Step 1   In D4, enter a standard normal (μ 0, σ of 1) random variable with the formula

$$=\text{RISKNORMAL}(0,1)$$

Step 2   In E4, generate the value of the asset a year from now using the lognormal random variable. The formula is

$$=\text{C4*EXP}((\text{A4-0.5*B4\^2})+\text{B4*D4})$$

Step 3   In F4, exploit (3) to generate the value of the option's cash flows one year from now with the formula

$$=\text{IF(E4<C2,C2-E4,0)}$$

Step 4   In cell G4, compute the discounted value of the option's cash flows with the formula

$$=\text{EXP(-A4)*F4}$$

Step 5   After choosing G4 as the output cell, we find that the option is worth $34,093. See Figure 32.

	A	B	C	D	E	F	G
1	Abandonment						
2	Abandonment value	$500,000.00					
3	r		Volatility	Current Price	Normal(0,1	Price 1 year from now	Value of Option	Discounted value
4	0.05	0.28	$553,000.00	0.387636	$623,093.16	0	0

FIGURE 31

	D	E	F
6			
7	Name	Discounted value	
8	Description	Output	
9	Cell	G4	
10	Minimum =	0	
11	Maximum =	280536.9	
12	Mean =	34,092.91	
13	Std Deviati	54,870.25	
14	Variance =	3.01E+09	

FIGURE 32

EXAMPLE 6　An Option to Postpone[†]

The current risk-free rate is 8%. We can build a plant now costing $104 million and gain revenues worth (risk-adjusted) $100 million. Revenues begin one year from now. Therefore, the current value of the project is $-\$4$ million, and the project does not appear worthwhile. Suppose, however, the project's value has a 60% annual volatility, and we can wait one year before investing. What is the worth of this option? Assume that construction costs grow at the risk-free rate.

Postpone.xls　**Solution**　The work is in Figure 33 and the file Postpone.xls. Let V = the value of the project one year from now. Then the value of the option to postpone is

$$\max(0, V - e^{.08}(104))$$

This is because we will invest one year from now only if the value of the project exceeds the cost (which, one year from now, will be $e^{.08}(104)$).

We proceed as follows.

Step 1　Enter needed parameters in A5–E5.

Step 2　In F5, enter the formula

$$=\text{RISKNORMAL}(0,1)$$

Step 3　In G5, use the lognormal random variable to compute the value of the asset one year from now with the formula

$$=\text{B5*EXP}((\text{A5-0.5*E5}^\wedge2)+\text{F5*E5})$$

Step 4　In H5, use (3) to determine the cash flows obtained in one year with the formula

$$=\text{MAX}(0,\text{G5-C5*EXP}(\text{A5}))$$

Step 5　Discount the value of the cash flows back to time 0 in cell I5 with the formula

$$=\text{EXP}(-\text{A5})\text{*H5}$$

Step 6　Make I5 an output cell. With the option, the situation is worth $22.14 million. See Figure 34.

FIGURE **33**

	A	B	C	D	E	F	G	H	I
1	Postpone Investment								
2									
3									
4	r	Current value	Investment cost	Value with no option	Volatility	Normal (0, 1)	Value One year from now	Cash flows in one year from waiting	Discounted value of cash flows
5	0.08	100	104	-4	0.6	1.6431	242.5075	129.8456	119.86261
6									

[†]Based on Trigeorgis (1995).

	D	E
8		
9	Name	Discounte
10	Description	Output
11	Cell	I5
12	Minimum =	0
13	Maximum =	664.869
14	Mean =	22.1453
15	Std Deviation =	50.32
16	Variance =	2532.11
17	Skewness =	4.17166
18	Kurtosis =	31.0005

FIGURE 34

Thus, the option to postpone improves our position by $22.14 - (-4) = \$26.14$ million relative to our position if we did the project without having the option to postpone.

We now modify Example 6 to show how we evaluate an option to expand a project.

EXAMPLE 7 — An Option to Expand

Assume you have the option to spend $40 million one year from now on a plant expansion that will increase the project's value by 50%. Value this expansion option.

Expand.xls **Solution** The work is in the file Expand.xls. See Figure 35. We proceed as follows.

Step 1 In A5:E5 and J5:K5, enter relevant parameters for the problem.

Step 2 In cell F5, enter

$$=\text{RISKNORMAL}(0,1)$$

This will help generate the value of the project one year from now.

Step 3 In cell G5, generate the (random) value of the project one year from now with the formula

$$=\text{B5*EXP}((\text{A5-0.5*E5}\char`\^2)+\text{F5*E5})$$

	A	B	C	D	E	F
1	Expand for 40 million by 50%?					
2						
3						
4	r	Current value	Investment cost	Value with no option	Volatility	Normal (0,1)
5	0.08	100	104	-4	0.6	-0.16196

	G	H	I	J	K
3					
4	Value One year from now	Cash Flows in One Year	Discounted value of cash flows	Expansion cost	Expansion factor
5	82.10484	83.15726	-27.236177	40	0.5

FIGURE 35

	D	E
11	Cell	I5
12	Minimum =	-91.8832
13	Maximum =	987.4983
14	Mean =	14.08345
15	Std Deviati	95.02
16	Variance =	9029.213
17	Skewness	2.609704

FIGURE 36

Step 4 Note that if we choose to expand, our cash flows in one year will equal

$$1.5*(\text{value in one year}) - 40$$

If we do not expand, our cash flows in one year will simply equal the value of the project. Therefore, in H5, compute cash flows in one year with the formula

$$=MAX((1+K5)*G5-J5,G5)$$

Step 5 In cell I5, compute the discounted value of cash flows with the formula

$$=EXP(-A5)*H5-C5$$

Step 6 Now select cell I5 as the output cell. After 4,000 iterations, with the option to expand, the situation is worth an average of $14.08 million. Thus, the option to expand improves our position over doing the project right away by $18.08 million. See Figure 36.

PROBLEMS

Note: The problems in this section should be solved using Monte Carlo simulation, not the Black–Scholes formula or binomial trees.

Group A

1 Merck is debating whether to invest in a pioneer biotech project. The company estimates the worth of this project to be −$56 million. Investing in the pioneer project gives Merck the option to own a much bigger technology that will be available in four years. If Merck does not participate in the pioneer project, it cannot own the bigger project. The bigger project will require $1.5 billion in cash four years from now. Currently, Merck estimates the NPV of the cash flows from the bigger project to be $597 million. What should Merck do? The risk-free rate is 10%, and the annual volatility of the bigger project is 35%.

2 Let's reconsider Example 6. Instead of paying the entire $104 million plant cost now, we must only pay $50 million now. A year from now, we may pay the remaining $54 million cost (with interest) and obtain the full project value, or we may reduce the scale of the project by paying only $25 million. If we reduce the scale of the project, it will be worth only 50% of what it would have been worth. Value this reduction option.

3 We can invest in a project today costing $104 million and gain profits worth (risk-adjusted) $100 million. Therefore, the current value of the project is −$4 million, and the project does not appear worthwhile. Suppose, however, that the project's future profits have a 60% annual volatility, and we have the following option. After paying $44 million today to begin construction of the plant, an installment payment of $64.8 million is planned for one year from now. $24.8 million of this amount is a required fixed cost to complete construction, but $40 million is for discretionary variable costs such as advertising and maintenance. If the plant is operated in a given year, cash profits from the project equal 30% of project value. The project will not begin operation until time 1 (a year from now). After seeing the value of the project one year from now, we may choose one of the following alternatives:

- Operate at time 1 and obtain (at time 1) the complete project value, less variable costs and $24.8 million construction costs.
- Shut down (temporarily) at time 1 and obtain (at time 1) the project value, less the year's cash profits, less $24.8 million construction costs.

The current risk-free rate is 8%. By how much does this option improve the value of the project?

4 Georgia Woodward is trying to determine whether or not she should develop vacant land into a 6-unit or a 9-unit apartment complex. It will cost $480,000 to build a 6-unit complex and $810,000 to build a 9-unit complex. (The cost will remain unchanged if she builds next year.) The current market value of an apartment unit is $100,000. The risk-free rate is 10%, and the annual volatility of the value of an apartment unit is 30%. Also, the apartment pays 8% of its value in profits each year. Georgia has the option to build now or wait a year. What value would you place on the option to wait a year?

5 The current price of IBM stock is $145\frac{1}{8}$. In 64 days, Lou will be paid as follows. For every $1 increase in stock price up to $10, he receives $1 million; for every $1 increase in stock price over $10, he receives $500,000. What is a fair market value for Lou's option? Assume annual volatility of 32.7% and an annual risk-free rate of 6.36%.

6 We are thinking of purchasing the rights to sell a software package. Our current estimate of the value (to us) of future sales is $400 million. We are offering to compensate the developer by paying him each year (beginning in year 1 and through the end of year 20) 40% of all profits in excess of $50 million. What is this payment worth? Assume a risk-free rate of 6%, annual volatility of 50%, and an average of 12% return on the project's value each year.

Group B

7 You are developing a new drug. You believe the value of the drug to be $100 million, with a 40% annual volatility. During the next ten years, you have two options available. In four years, you may sell the project for $80 million. If you have not sold the project in year 4, then 10 years from now you may, at a cost of $40 million, expand the size of the project by 50%. The risk-free rate is 5%. By how much do these options increase the value of the project?

8 Drug company managers are undertaking development of a new drug. Today they will make a $25 million investment to screen many potential compounds. After three years, the screening process will be complete, and they can conduct clinical trials, if desired, at a cost of $90 million. Finally, 10 years from now, they can, if desired, invest $500 million and bring the drug to market. Their current estimate is that the cash inflows from the drug will have a value of $800 million (in time 10 dollars). This estimate would, of course, have some adjustment for risk contained in its computation. The annual volatility of revenues is estimated to be 60%. The company's cost of capital is 14%, and the risk-free rate for a 10-year bond is 4%. How would you value this opportunity?

REFERENCES

Black, F., and Scholes, M. "The Pricing of Options and Corporate Liabilities," *Journal of Political Economy* 7(1973):637–654.

Cox, J., Ross, S., and Rubenstein, M. "Option Pricing: A Simplified Approach," *Journal of Financial Economics* 7(1979):229–264.

Hull, J. *Options Futures and Derivative Securities.* Englewood Cliffs, N.J.: Prentice Hall, 1997.

Hull J. *Options and Other Derivatives,* 4th ed. Englewood Cliffs, N.J.: Prentice Hall, 2000.

Luenberger, D. *Investment Science.* 1997, Oxford, U.K.: Oxford University Press.

Trigeorgis, L. *Real Options.* Cambridge, Mass.: MIT Press, 1996.

14

Portfolio Risk, Optimization, and Hedging

In recent years, there has been considerable uncertainty in the world's stock markets. This chapter describes scientific techniques for measuring and managing this uncertainty.

14.1 Measuring Value at Risk

No matter where money is invested, the value of the investments at a given date in the future is uncertain. The concept of **value at risk (VAR)** is useful in describing a portfolio's uncertainty. Simply stated, the value at risk of a portfolio at a future time is considered to be the loss associated with the fifth percentile of the portfolio's value at that point in time. In short, there is considered to be only one chance in 20 that the portfolio's loss will exceed the VAR. To illustrate, suppose a portfolio today is worth $100. We simulate the portfolio's value one year from now and find that there is a 5% chance that the portfolio's value will be $80 or less. Then the portfolio's VAR for a one-year horizon is $20 or 20%. The following example shows how the Excel add-in @Risk can be used to measure VAR. The example also demonstrates how buying puts can hedge the risk of a long position in a stock.

EXAMPLE 1 | **Dell Stock**

On June 30, 1998, Dell Computer stock sold for $94.125. Assume that future prices of Dell stock are governed by a lognormal random variable. The method of implied volatility yields an annual volatility estimate (for the next four months) of 53.27%. We own one share of Dell Computer. The mean annual return on Dell during the next four months is equally likely to be 0%, 10%, 20%, or 30%. To hedge the risk involved in owning Dell, we are considering buying (for $5.25) a European put on Dell with exercise price $80 and expiration date November 22, 1998. Our work is in file Varduniform.xls.

Varduniform.xls

a Compute the VAR on November 22, 1998, if we own Dell Computer and do not buy the put.

b Compute the VAR on November 22, 1998, if we own Dell Computer and buy the put.

Solution The work is in file Varduniform.xls. We have created range names as indicated in Figure 1.

Step 1 In cell C7, create the (random) mean growth rate (value of μ) for the next four months with the formula

$$=RISKDUNIFORM(\{0,0.1,0.2,0.3\})$$

Step 2 In cell B11, generate Dell's price on November 22, 1998 with the formula

$$=S*EXP((g-0.5*v^2)*d+RISKNORMAL(0,1)*v*SQRT(d))$$

	A	B	C	D
1				
2	**Stress Testing Dell**	**Range Name**		
3	Current price	S	$ 94.13	
4	Put exercise price	x	$ 80.00	
5	put duration	d	0.39726	
6	risk free rate	r_	0.053541	
7	actual growth rate	g	0.1	
8	volatility	v	0.5327	
9	put price	p	$ 5.25	
10				
11	Dell price at expiration	94.73401045		
12	put value at expiration	0		
13				
14	% age Gain without put	0.6%		
15	%age Gain with put	-4.7%		
16				

FIGURE 1

Step 3 In cell B12, compute the payments from the put at expiration with the formula

$$=IF(B11>x,0,x-B11)$$

Step 4 If we just own Dell, the percentage gain on our portfolio is given by

$$\frac{\text{Ending Dell price} - \text{beginning Dell price}}{\text{Beginning Dell price}}$$

In B14, compute the percentage gain on our portfolio if we do not buy a put with the formula

$$=(B11-S)/S$$

Step 5 The percentage gain on our portfolio if we own Dell and a put is

$$\frac{\text{Ending Dell price} + \text{cash flows from put} - \text{beginning Dell price} - \text{put price}}{\text{Beginning Dell price} + \text{put price}}.$$

In cell B15, compute the percentage gain on our portfolio if we buy the put with the formula

$$=((B12+B11)-(S+p))/(S+ p)$$

Step 6 After selecting B14 and B15 as output cells and running 1,600 iterations, we obtain the @Risk output in Figure 2.

The simulation output in Figure 2 provides three ways to show how the purchase of a put has hedged our risk.

- We find the VAR by looking at the fifth percentile of the @Risk output. If we do not buy the put, the VAR is 42% of our invested cash. If we buy the put, the VAR drops to 19% of the invested cash. This is, of course, because if Dell stock drops below $80, every one-dollar decrease in the value of the stock is countered by a one-dollar increase in the value of the put.

- Also note that if we do not buy the put, the stock (despite its high growth rate) might lose up to 68% of its value. If we buy the put, the worst we can do is lose 19%.

- We also find that the standard deviation of our percentage return with the put is 30%. Without the put, the standard deviation is 37%.

Name	% age Gain without put / p	%age Gain with put / p
Description	Output	Output
Cell	B14	B15
Minimum	-0.67688	-0.19497
Maximum	2.284537	2.111014
Mean	6.20E-02	5.25E-02
Std Deviation	0.368249	0.300672
Variance	0.135607	9.04E-02
Skewness	1.050841	1.712664
Kurtosis	4.961527	6.904467
Errors Calculated	0	0
Mode	-1.60E-02	-0.19497
5% Perc	-0.42484	-0.19497
10% Perc	-0.34509	-0.19497
15% Perc	-0.29408	-0.19497
20% Perc	-0.24843	-0.19497
25% Perc	-0.20574	-0.19497
30% Perc	-0.16278	-0.19497
35% Perc	-0.11705	-0.1637
40% Perc	-7.99E-02	-0.12847
45% Perc	-3.39E-02	-8.50E-02
50% Perc	6.97E-03	-0.04623
55% Perc	5.19E-02	-3.69E-03
60% Perc	9.57E-02	3.78E-02
65% Perc	0.145802	0.085269
70% Perc	0.200943	0.137497
75% Perc	0.256346	0.189973
80% Perc	0.331439	0.261099
85% Perc	0.428579	0.353107
90% Perc	0.538941	0.457639
95% Perc	0.752852	0.660248

FIGURE 2

However, when we buy the put, the mean return on our portfolio is reduced from 6.2% to 5.3%. This leads people to refer to the cost of the put as **portfolio insurance.**

Creating Histograms of Percentage Profit with and without a Put We used @Risk to create histograms showing the distribution of the percentage gain on our portfolio: Figure 3 without the put, and Figure 4 with the put. The huge spike in Figure 4 corresponds to a minimum loss of 19% when hedged. Without the put, we can easily go below a 50% loss.

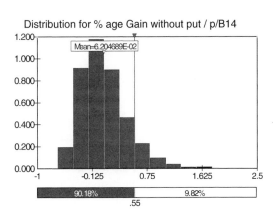

FIGURE 3
Percentage Return
without Put

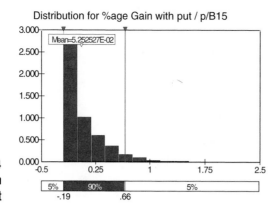

FIGURE 4
Percentage Return with Put

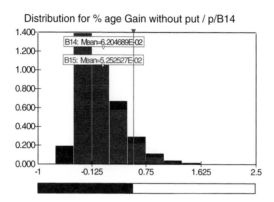

FIGURE 5
Overlaying Return with Put and without Put

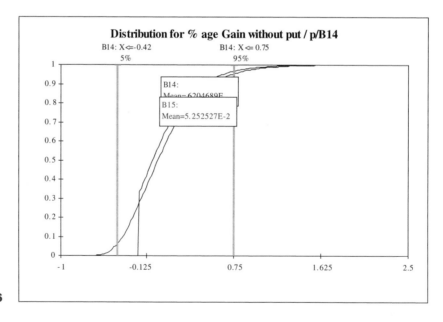

FIGURE 6

Overlaying Graphs We can easily report the histograms for our return with and without the put on the same graph. Simply right click on a histogram for either output cell and choose Format Graph and then Variables to Graph. Select the desired variables (return with put and without put). Figure 5 shows the result.

The gray bars (for cell B14) represent the histogram for the percentage return with the put, and the black bars (for cell B15) represent the percentage return without the put. The

TABLE 1

Date	Cash Flows to Purchaser	Cash Flows to Short Seller
Today	−$50	$50
Six months	$70	−$70

graph shows that percentage return without the put can go much lower than percentage return with the put.

By changing the graph type to Cumulative Ascending Line, we can also overlay the cumulative distributions of returns for the portfolio with and without the put. (See Figure 6.) Note that the dotted line (for gain without the put) starts accumulating probability before the solid line graph (for gain with the put).

VAR for Transactions Involving Short Sales

In recent years, many investors have followed strategies involving shorting stocks or options. When a stock or option is shorted, the investor receives cash flows that are the negative of the cash flows received by an investor who purchases the instrument. For example, if Microsoft stock sells for $50 today and for $70 in six months, the cash flows received by a purchaser today and an investor who sells the stock short today may be viewed as indicated in Table 1.

When investment strategies involve shorting stocks or options, it does not make sense to look at the portfolio's returns. This is because the short seller (unlike the purchaser of the stock) does not really have an outlay of cash at the time of the short sales. When investment strategies involve short selling, the best way to evaluate the risk is to evaluate the NPV of the cash flows associated with the portfolio. To do this, a discount rate is needed. An investor who desires a return of 10% per year should discount cash flows at 10%. Then a negative NPV would indicate that the strategy fails to meet the target rate of return.

PROBLEMS

Group A

1 If you own a stock, buying a put option on the stock will greatly reduce your risk. This is the idea behind portfolio insurance. To illustrate, consider a stock (Trumpco) that currently sells for $56 and has $\sigma = 30\%$. The risk-free rate is 8%, and you estimate $\mu = 12\%$.

a You own one share of Trumpco. Use simulation to estimate the probability distribution of the percentage return earned on this stock during a one-year period.

b Now suppose you also buy a put option (for $2.38) on Trumpco. The option has an exercise price of $50 and a one-year expiration date. Use simulation to estimate the probability distribution of the percentage return on your portfolio over a one-year period. Can you see why this is called a portfolio insurance strategy?

c Use simulation to show that the put option should, indeed, sell for $2.38.

2 For the data in Problem 1, consider the following strategy (called a **butterfly spread**). Buy two calls with an exercise price of $50. Short one call with an exercise price of $40, and short one call with an exercise price of $60. Assuming a 10% desired rate of return, determine the VAR for this strategy.

3 Cryco stock currently sells for $69. The annual growth rate of the stock is 15%, and the stock's annual volatility is 35%. The risk-free rate is currently 5%. You have bought a six-month European put option on this stock, with an exercise price of $70.

a Use @Risk to value this option.

b Use @Risk to analyze the distribution of percentage returns (for a six-month horizon) for the following portfolios:

Portfolio 1: Own one share of Cryco.
Portfolio 2: Own one share of Cryco and buy the put described in part (a).

Which portfolio has the larger expected return? Explain why portfolio 2 is known as portfolio insurance.

Group B

4 On December 18, 1998, AOL stock sold for $104.25 a share. An April 1999 $120 put sold for $24.50, and an April 1999 $90 put sold for $8.75. These puts were to expire in 125 days.

a Estimate the annual volatility of AOL. You may use

either put to get implied volatility, or average your two estimates.

Investor's Business Daily recommended the following strategy (called a **bull spread**). Short the April 1999 $120 put and buy the April 1999 $90 put. Compare the merits of the bull spread to the merits of buying one share of AOL under the following seven values for μ $\{-.2, -.1, 0, .1, .2, .3, .4\}$. For the seven values of μ listed, analyze the bull spread and purchase of one share of AOL. Answer the following questions.

b How do the mean NPV (using a 10% discount rate) and risk of the bull spread depend on μ?

c Why is the strategy called a bull spread?

d Overall, how would you compare the merits of the bull spread with simply purchasing one share of AOL?

14.2 Portfolio Optimization: The Markowitz Approach

Given a set of investments, how do we find the portfolio that has the minimum variance and yields an acceptable expected return? This question was answered by Harry Markowitz in the 1950s. For his work on this and other investment topics, he received the Nobel Prize in economics in 1991. The ideas discussed in this section are the basis for most methods of **asset allocation** used by Wall Street firms. **Asset allocation models** are used, for example, to determine the percentage of assets to invest in stocks, gold, and Treasury bills. To begin, we review some important formulas involving the mean and variance of sums of random variables.

Sums of Random Variables: Mean and Variance

Let S_i be the (random) return earned during a year on a dollar invested in investment i. Thus, if $S_i = .10$, a dollar invested at the beginning of the year grows to $1.10 at the end of the year, whereas if $S_i = -.20$, a dollar invested at the beginning of the year decreases in value to $0.80. We assume that n investments are available. Let x_i be the fraction of our money invested in investment i. We assume that $x_1 + x_2 + \cdots + x_n = 1$, so that all of our money is invested. To rule out shorting a stock, that is, selling shares we don't own, we assume that $x_i \geq 0$. Then the annual return on our investments is given by the random variable R, where

$$R = x_1 S_1 + x_2 S_2 + \cdots + x_n S_n$$

Let μ_i be the expected value of S_i, let σ_i^2 be the variance of S_i (so that σ_i is the standard deviation of S_i), and let ρ_{ij} be the correlation between S_i and S_j. To do any work with investments, you must understand how to use the following formulas, which relate the data for the individual investments to the expected return and the variance of return for a *portfolio* of investments.

$$\text{Expected value of } R = x_1 \mu_1 + x_2 \mu_2 + \cdots + x_n \mu_n \tag{1}$$

$$\text{Variance of } R = x_1^2 \sigma_1^2 + x_2^2 \sigma_2^2 + \cdots + x_n^2 \sigma_n^2 + \sum_{i \neq j} x_i x_j \rho_{ij} \sigma_i \sigma_j \tag{2}$$

Since we never actually know the true expected values (μ_i's), variances (σ_i^2's), and correlations (ρ_{ij}'s), we must estimate them. If historical data are used, we proceed as follows.

1 Estimate μ_i with $\bar{X}_i$, the sample average of returns on investment i over several previous years. You can use the Excel AVERAGE function to obtain $\bar{X}_i$.

2 Estimate σ_i^2 with s_i^2, the sample variance of returns on investment i over several previous years. You can use the Excel VAR function to obtain s_i^2.

3 Estimate σ_i with s_i, the sample standard deviation of returns on investment i. You can obtain s_i with the Excel STDEV function. (Alternatively, of course, you can obtain s_i as the square root of s_i^2.)

4 Estimate ρ_{ij} with r_{ij}, the sample correlation between past returns on investments i and j. You can find all of the r_{ij}'s by using Excel's Correlation option from the Tools/Data Analysis menu item. The input range should consist of n columns of historical returns.

We now estimate the mean and variance of the return on a portfolio by replacing each parameter in Equations (1) and (2) with its sample estimate. This yields

$$\text{Estimated expected value of } R = x_1\bar{X}_1 + x_2\bar{X}_2 + \cdots + x_n\bar{X}_n \tag{3}$$

and

$$\text{Estimated variance of } R = x_1^2 s_1^2 + x_2^2 s_2^2 + \cdots + x_n^2 s_n^2 + \sum_{i \neq j} x_i x_j r_{ij} s_i s_j \tag{4}$$

In keeping with common practice, we express the annual return on investments as decimals. Thus a return of .10 on a stock means that the stock has increased in value by 10%.

Most investors have two objectives in forming portfolios: to obtain a large expected return and to keep a small variance (in order to minimize risk). The most common way of handling this two-objective problem is to specify what minimal expected return we require and then minimize the variance subject to this expected return requirement. The following example shows how we can use a spreadsheet Solver to do this.

EXAMPLE 2 **Stockco Portfolio**

Stockco can invest in three stocks. From past data, the means and standard deviations of annual returns have been estimated as shown in Table 2 (where we express percentages as decimals). The correlation between the annual returns on the stocks are listed in Table 3. Stockco wants to find a minimum-variance portfolio that yields an expected annual return of at least .12.

Solution To model Stockco's problem, we must keep track of the following:

- The fraction of money invested in each stock
- The total fraction of Stockco's money invested
- The expected annual return of the portfolio
- The variance of the annual portfolio return

Developing the Spreadsheet Model The individual steps are now listed. (See Figure 7 and the file Portfol1.xls.)

Portfol1.xls

Step 1 Inputs. Enter the data from Tables 2 and 3 in the ranges B5:D6 and B10:D12. (Note that the correlation between any stock's return and itself is 1.)

Step 2 Fractions invested. Enter *any* trial values in the range B16:D16 for the fractions of Stockco's money placed in the three investments.

TABLE 2
Estimated Means and Standard Deviations for the Stockco Example

	Mean	Standard Deviation
Stock 1	.14	.20
Stock 2	.11	.15
Stock 3	.10	.08

TABLE 3
Estimated Correlations for the Stockco Example

Combination	Correlation
Stocks 1 and 2	.6
Stocks 1 and 3	.4
Stocks 2 and 3	.7

	A	B	C	D	E	F	G	H
1	**Stockco Portfolio Problem**							
2								
3	**Stock input data**							
4		Stock 1	Stock 2	Stock 3				
5	Mean return	0.14	0.11	0.1				
6	StDev of return	0.2	0.15	0.08				
7								
8	Correlations							
9		Stock 1	Stock 2	Stock 3				
10	Stock 1	1	0.6	0.4				
11	Stock 2	0.6	1	0.7				
12	Stock 3	0.4	0.7	1				
13								
14	**Investment decision**							
15		Stock 1	Stock 2	Stock 3	Total		Required	
16	Fractions to invest	0.5	0	0.5	1	=	1	
17								
18	Expected portfolio return							
19		Actual		Required				
20		0.12	>=	0.12				
21								
22	Standard deviations times fractions invested							
23		Stock 1	Stock 2	Stock 3				
24		0.1	0	0.04				
25								
26	Portfolio variance	0.0148						
27								
28	Portfolio stdev	0.1217						

FIGURE 7

Step 3 **Total fraction invested.** Calculate the total fraction of money invested in cell E16 with the formula

=SUM(B16:D16)

Then enter a 1 in cell G16 to indicate that 100% of the money will be invested in the available investments.

Step 4 Expected annual return. Use Equation (3) to compute the expected annual return. Specifically, enter the formula

$$=\text{SUMPRODUCT(B16:D16,B5:D5)}$$

in cell B20. Then enter the required annual mean return (.12) in cell D20.

Step 5 Variance of portfolio return. In preparation for computing the variance of the portfolio return, compute for each stock the quantity

(Standard deviation of return on stock i) * (Fraction of money invested in stock i)

Specifically, enter the formula

$$=\text{B6*B16}$$

in cell B24 for stock 1 and copy this formula to the range C24:D24 for the other two stocks. Now, using the entries in cells B24:D24, compute the estimated variance of the portfolio's annual return in cell B26 (from Equation (4)) with the formula

$$=\text{SUMPRODUCT(B24:D24,B24:D24)}+2\text{*B24*C24*C10}$$
$$+2\text{*B24*D24*D10}+2\text{*C24*D24*D11}$$

The term SUMPRODUCT(B24:D24,B24:D24) corresponds to the first n variance terms in Equation (4), while the remaining terms correspond to the summation expression in Equation (4).

Using Solver

Step 1 Objective. Select cell B26 (portfolio variance) as the objective to minimize.

Step 2 Changing cells. Choose the range B16:D16 as the changing cells and constrain these to be nonnegative.

Step 3 Constraint on expected return. Add the constraint B20≥D20. This ensures that the portfolio will have an expected return of at least .12.

Step 4 Investment constraint. Add the constraint E16=G16. This ensures that all of Stockco's money is invested.

Step 5 Optimize. Click on Solve.

The Solver solution is shown in Figure 7. This portfolio has a variance of .0148 (or a standard deviation of $\sqrt{.0148} = .1217$). Stockco's optimal portfolio places half of the money in stock 1 and half in stock 3.

Is the Solver Solution Optimal? It can be shown that the variance of a portfolio is a convex function of the fractions invested in the various stocks. All of the constraints are linear. Therefore, the assumptions for a minimization problem are satisfied, and we are assured that Solver will find the minimum-variance portfolio.

REMARKS **1** What does it mean to say that the standard deviation of the annual return on the portfolio is .1217? If portfolio returns are normally distributed (which is usually a reasonable assumption), then during the year there is a 68% chance that the portfolio return is between $-.0017\ (=.12-1217)$ and .2417 $(=.12+.1217)$, a 95% chance that the portfolio return is between $-.1234(=.12-.2434)$ and .3634$(=.12+.2434)$, and a 99.7% chance that the portfolio return is between $-.2451(=.12-.3651)$ and .4851 $(=.12+.3651)$

2 If Stockco is allowed to short a stock, we simply allow the fraction invested in that stock to be negative. That is, we simply eliminate the nonnegativity constraints on the changing cells.

3 An alternative objective might be to minimize the probability that the portfolio loses money. (See Problem 4.)

Matrix Multiplication and Portfolio Optimization

In Example 2, the formula for the estimated variance of Stockco's portfolio was very tedious to enter in the spreadsheet. If we had ten investments, typing this variance formula would take forever. Using matrix multiplication, however, it is easy to express the variance of a portfolio.

Suppose that n investments are available. Let x_i be the fraction of money invested in investment i; let s_i be the sample standard deviation of return on investment i; and let C be the $n \times n$ matrix whose ijth entry is r_{ij}, the sample correlation between the returns on investments i and j.

It is not hard to show (see Problem 3) that the estimated portfolio variance, computed from Equation (4), can be written as

$$\text{Portfolio variance} = [x_1 s_1 \quad x_2 s_2 \quad \cdots \quad x_n s_n] \, C \, [x_1 s_1 \quad x_2 s_2 \quad \cdots \quad x_n s_n]^T \qquad (5)$$

Here, $[x_1 s_1 \quad x_2 s_2 \quad \cdots \quad x_n s_n]^T$ is the $n \times 1$ matrix (or column vector)

$$\begin{bmatrix} x_1 s_1 \\ x_2 s_2 \\ \vdots \\ x_n s_n \end{bmatrix}$$

Referring to the Stockco spreadsheet in Figure 7, if we use the MMULT function to compute the estimated variance of the return on Stockco's portfolio, there is no change except in the portfolio variance cell B26. Note that we have already computed the row vector

$$[x_1 s_1 \quad x_2 s_2 \quad \cdots \quad x_n s_n]$$

in the range B24:D24. Also, the matrix C of correlations is already in the range B10:D12. So to compute the estimated portfolio variance, we enter the array formula

$$=\text{MMULT(B24:D24,MMULT(B10:D12,TRANSPOSE(B24:D24)))}$$

Portfol2.xls

in cell B26. (See the file Portfol2.xls.) Note that TRANSPOSE(B24:D24) creates the column vector

$$[x_1 s_1 \quad x_2 s_2 \quad \cdots \quad x_n s_n]^T$$

Therefore, the inner MMULT finds the product

$$C[x_1 s_1 \quad x_2 s_2 \quad \cdots \quad x_n s_n]^T$$

Then the outer MMULT multiplies this product on the left by the row vector

$$[x_1 s_1 \quad x_2 s_2 \quad \cdots \quad x_n s_n]$$

to obtain the variance.

We could now use Solver with the same settings as before to find the variance-minimizing portfolio that yields an expected return of at least 12%. Again, the main benefit from calculating the portfolio variance with the MMULT function is that it generalizes easily to *any* number of stocks, whereas the "brute force" formula used earlier becomes too tedious for a large number of stocks.

Note that in Equation (5), we first multiply a $1 \times n$ matrix by an $n \times n$ matrix. This yields a $1 \times n$ matrix, which is then multiplied by an $n \times 1$ matrix. The final result is a 1×1 matrix—a number.

PROBLEMS

Group A

1 The annual returns on three different types of assets (T-bonds, stocks, and gold) during the years 1968–88 are listed in Table 4. (See the file Inv68.xls.) For example, $1 invested in T-bonds at the beginning of 1978 grew to $1.07 by the end of 1978. You have $1,000 to invest in these three investments. Your goal is to minimize the variance of the annual dollar return of your portfolio, subject to the constraint that the expected return on the portfolio for a one-year period is to be at least .10.

    a Determine how much money you should invest in each investment.

    b Find an interval such that you are 95% sure that the change in the value of your assets during the next year will be within this interval (assuming normally distributed returns).

    c Find an interval such that you are 95% sure that the percentage annual return on your portfolio will be within this interval (assuming normally distributed returns).

2 Consider three investments. You are given the following means, standard deviations, and correlations for the annual return on these three investments. The means are .12, .15, and .20. The standard deviations are .20, .30, and .40. The correlation between stocks 1 and 2 is .65, between stocks 1 and 3 is .75, and between stocks 2 and 3 is .41. You have $10,000 to invest and can invest no more than half of your money in any single stock. Determine the minimum-variance portfolio that yields an expected annual return of at least .14.

TABLE **4**

	Stocks	Gold	T-Bonds
1968	.11	.11	.05
1969	−.09	.08	.07
1970	.04	−.14	.07
1971	.14	.14	.04
1972	.19	.44	.04
1973	−.15	.66	.07
1974	−.27	.64	.08
1975	.37	.00	.06
1976	.24	−.22	.05
1977	−.07	.18	.05
1978	.07	.31	.07
1979	.19	.59	.10
1980	.33	.99	.11
1981	−.05	−.25	.15
1982	.22	.04	.11
1983	.23	−.11	.09
1984	.06	−.15	.10
1985	.32	−.12	.08
1986	.19	.16	.06
1987	.05	.22	.05
1988	.17	−.02	.06

Group B

3 Using the notation from this section, show algebraically that the portfolio variance in Equation (4) can be written in matrix form as the product

$$[x_1s_1 \quad x_2s_2 \quad \cdots \quad x_ns_n]C[x_1s_1 \quad x_2s_2 \quad \cdots \quad x_ns_n]^T$$

4 Reconsider the Stockco portfolio example (Example 2). Suppose that your goal is to find (among all portfolios that invest all of your money in stocks 1, 2, and 3) the portfolio that minimizes the probability that you will lose money during the next year. Use the Excel Solver to solve this problem. (*Hint:* You will need to assume that the returns on any portfolio follow a normal distribution. Then the Excel

function NORMSDIST() will return the area under a standard normal curve to the left of a given number x. For example, NORMSDIST(1) returns .84, and NORMSDIST(−1) returns .16.

5 The file Stock83.xls contains the percentage return on the market for the years 1984–1991, as well as the closing stock price for Ford, Lilly, Kellogg, Merck, and Hewlett-Packard for the same years. Use the approach outlined in Example 2 to determine the minimum-variance portfolio (of the stocks listed) that yields an expected return of at least .22.

14.3 The Scenario Approach to Portfolio Optimization

Scenport97_02.xls

In this section, we will show how to find desirable portfolios based on the assumption that the distributions of future returns will be similar to those of past returns. For example, suppose we are given monthly returns on shares of Microsoft (MSFT), Intel (INT), and General Electric (GE) for the years 1997–2002. See file Scenport97_02.xls and Figure 8. How could we determine what fraction of our assets we should allocate to each investment during the next year? Assume that a 4% risk-free investment is available. We begin

	A	B	C	D	E
3	Code		MSFT	INTC	GE
4	1	8/2/2002	-0.08316	-0.153969	-0.121118
5	2	7/2/2002	-0.122852	0.028493	0.108434
6	3	6/2/2002	0.074445	-0.338528	-0.061389
7	4	5/2/2002	-0.025832	-0.03464	-0.012759
8	5	4/2/2002	-0.133477	-0.05894	-0.156578
9	6	3/2/2002	0.033768	0.064867	-0.028489
10	7	2/2/2002	-0.084288	-0.185143	0.041088
11	8	1/2/2002	-0.03834	0.114295	-0.07314
12	9	12/1/2002	0.031771	-0.037094	0.045898
13	10	11/1/2002	0.104213	0.337433	0.057167
14	11	10/1/2002	0.136408	0.194417	-0.021021
15	12	9/1/2002	-0.103067	-0.268887	-0.086534
16	13	8/1/2002	-0.138087	-0.061807	-0.059568
17	14	7/1/2002	-0.093288	0.019172	-0.10944
18	15	6/1/2002	0.055218	0.082654	0
19	16	5/1/2002	0.021107	-0.126012	0.009701
20	17	4/1/2002	0.238801	0.174658	0.159413
21	18	3/1/2002	-0.073051	-0.078864	-0.096532
22	19	2/1/2002	-0.033737	-0.228084	0.011168
23	20	1/1/2002	0.407561	0.231179	-0.040711
24	21	12/1/2000	-0.243987	-0.210208	-0.02973
25	22	11/1/2000	-0.166836	-0.154394	-0.095695
26	23	10/1/2000	0.141933	0.082872	-0.052041
27	24	9/1/2000	-0.136084	-0.444905	-0.011452
28	25	8/1/2000	0	0.121644	0.134066
29	26	7/1/2000	-0.127375	-0.001348	-0.024747
30	27	6/1/2000	0.278772	0.072105	0.007262
31	28	5/1/2000	-0.103082	-0.016738	0.003941
32	29	4/1/2000	-0.343529	-0.038853	0.010352
33	30	3/1/2000	0.188878	0.167641	0.175796
34	31	2/1/2000	-0.08685	0.142077	-0.012254
35	32	1/1/2000	-0.161713	0.201897	-0.133961
36	33	12/1/1999	0.282262	0.073368	0.189048
37	34	11/1/1999	-0.016314	-0.009568	-0.039561
38	35	10/1/1999	0.022085	0.042037	0.14267
39	36	9/1/1999	-0.021608	-0.095981	0.055724
40	37	8/1/1999	0.078662	0.191236	0.030415
41	38	7/1/1999	-0.048564	0.159879	-0.035372
42	39	6/1/1999	0.117735	0.10037	0.111213
43	40	5/1/1999	-0.007625	-0.116492	-0.03499
44	41	4/1/1999	-0.092725	0.02965	-0.047339
45	42	3/1/1999	0.193819	-0.009015	0.102873
46	43	2/1/1999	-0.142057	-0.148906	-0.043722
47	44	1/1/1999	0.261716	0.188851	0.02825
48	45	12/1/1998	0.136885	0.1016	0.128557
49	46	11/1/1998	0.152248	0.206556	0.032932
50	47	10/1/1998	-0.037979	0.040168	0.099688
51	48	9/1/1998	0.147175	0.204162	-0.005422
52	49	8/1/1998	-0.127342	-0.156546	-0.105336
53	50	7/1/1998	0.014394	0.138844	-0.016025
54	51	6/1/1998	0.278066	0.037556	0.089929
55	52	5/1/1998	-0.059032	-0.115956	-0.021099
56	53	4/1/1998	0.006927	0.035403	-0.011862
57	54	3/1/1998	0.055923	-0.129911	0.108808
58	55	2/1/1998	0.136193	0.107266	0.003199
59	56	1/1/1998	0.154441	0.153364	0.056166
60	57	12/1/1997	-0.086772	-0.094943	-0.006711
61	58	11/1/1997	0.088615	0.0078	0.142857
62	59	10/1/1997	-0.017533	-0.165727	-0.050523
63	60	9/1/1997	0.000908	0.002174	0.088162

FIGURE 8

by bootstrapping off our 1997–2002 returns to generate many (say, 1,000) possible distributions of returns for these three stocks. Then we will use Excel Premium Solver (see Chapter 15 of Volume 1) for an optimal asset allocation by each of the following criteria:

- Minimize the risk of the portfolio (standard deviation) subject, to a desired mean return of at least 10%.

- Minimize the probability of a loss, subject to a desired mean return of at least 8%.
- Maximize the portfolio's Sharpe ratio. The Sharpe ratio of a portfolio equals

$$\frac{\mu - r}{\sigma}$$

where μ = mean return on portfolio, r = risk-free rate and σ = portfolio standard deviation.

- Find the portfolio that yields at least 10% on average and yields the best worst-case return. This is called the minimax portfolio.
- Find the portfolio that yields at least an 8% expected return and gives the lowest VAR (that is, the portfolio that yields at least 8% on average and maximizes the fifth percentile of the portfolio return).
- Choose a desired expected annual return (say, 10%). Find the portfolio that minimizes the expected amount by which we fail to meet our desired target. This is portfolio optimization to minimize downside risk. This makes more sense than minimizing standard deviation, because standard deviation is increased by upside risk (variability that takes your return above a target), and this type of risk is not bad.

Bootstrapping to Future Annual Returns

To get 1,000 samples of annual returns for the next year, we simply generate 12 months of MSFT, INT, and GE returns, assuming that each of these 12 months is equally likely to be one of the last 60 months. We proceed as follows (see the sheet Data).

Step 1 Name the range A4:E63 (where the returns lie) Lookup.

Step 2 Current stock prices are listed in I4:K4 (actually, we could assume any value for the current stock prices).

Step 3 In cells H5:H16, generate the "scenario" or month of 1993–1997 that is chosen to generate each month's returns for the next year by copying from H5 to H6:H16 the formula

$$=\text{RISKDUNIFORM(\$A\$4:\$A\$63)}$$

The RISKDUNIFORM function makes it equally likely that any of the numbers 1, 2, ..., 60 are chosen. See Figure 9.

Step 4 Copying from I5 to I5:K16 the formula

$$=\text{I4*(1+VLOOKUP(\$H5,LOOKUP,I\$2))}$$

generates 12 months of MSFT, INT, and GE returns. For each month, the new price equals the old price times the return on the stock generated by the column H scenario for that month. For example, in month 1, we chose scenario 1, so returns for the stocks come from row 4. Note that this approach implicitly models the correlations between the returns on the assets.

Step 5 Select cells I16:K16 as output cells and run 1,000 iterations. Clicking on the Data button extracts the results of each iteration, which we then paste to a spreadsheet. We now have 1,000 scenarios (which we may assume to be equally likely) for the price of MSFT, INT, and GE stock in one year. Again, these scenarios are based on the assumption that

FIGURE 9

	G	H	I	J	K
2			3	4	5
3	Month	Scenario	MSFT	INTC	GE
4	0		33.05	23	20.19
5	1	1	30.30157	19.45871	17.74463
6	2	7	27.74751	15.85607	18.47373
7	3	46	31.97201	19.13123	19.0821
8	4	3	34.35217	12.65477	17.91067
9	5	28	30.81106	12.44296	17.98125
10	6	22	25.67066	10.52184	16.26054
11	7	30	30.51928	12.28574	19.11908
12	8	28	27.37328	12.0801	19.19442
13	9	56	31.60085	13.93275	20.27249
14	10	44	39.87129	16.56397	20.84519
15	11	18	36.95866	15.25766	18.83296
16	12	40	36.67684	13.48026	18.174

FIGURE 10

	B	C	D	E
1	Original Price	33.05	23	20.19
2	FINAL MEAN	39.15313	24.73775	22.40104
3	Weights	0.239309	0	0.365128
4	Name	MSFT	INTC	GE
5	Mean	1.184663	1.075554	1.109512

the next year will be similar to the last five years. What if we think the next year might be worse than the last five years? Note that in cells C5:E5 of sheet Scen(StDev), we computed the percentage mean return on each stock. See Figure 10.

For example, in our 1,000 scenarios, MSFT increased by an average of 18.4% per year. Suppose we thought that MSFT would only increase by an average of 5% per year during the next year. If we multiply each final price by 1.05/1.184 = .887, this will ensure that on average, MSFT price will increase by 5% in our scenarios. The scenarios will still maintain the volatility and correlation patterns of the recent past.

We are now ready to solve each of our portfolio problems.

Minimizing the Risk of the Portfolio's Standard Deviation

We now show how to minimize the portfolio's standard deviation while ensuring that we have an expected return of at least 10%. See sheet Scen(Min StDev) and Figure 11.

Step 1 In cells C3:F3, enter trial fractions for the percentages of our money invested in each stock and the risk-free investment.

FIGURE 11

	B	C	D	E	F	G	H	I	J	K	L	M	N
1	Original Price	33.05	23	20.19									
2	FINAL MEAN	39.15313	24.73775	22.40104									
3	Weights	0.239309	0	0.365128	0.39556243	1							
4	Name	MSFT	INTC	GE	riskfree								
5	Mean	1.184663	1.075554	1.109512							Mean		
6	Iteration# / Cell	I16	J16	K16	MSFT Return	INTC Return	GE Return	Risk free	Port return		0.1	>	0.1
7	1	49.77746	33.40337	18.10553	0.50612587	0.45232043	-0.103243	0.04	0.099246		Sigma		
8	2	42.14295	16.45524	16.32466	0.27512708	-0.28455478	-0.191448	0.04	0.01176		0.218492		

Step 2   In F7:H1006, compute the actual return on all three stocks for each scenario by copying from F7 to H7:F1006 the formula

$$=(C7-C\$1)/C\$1$$

Step 3   In I7:I1006, enter for each scenario a return of 4% for the risk-free investment. Then compute in J7:J1006 the mean return on our portfolio as the weighted average of the returns on the three stocks and the risk-free investment. To do this, copy from J7 to J8:J1006 the formula

$$=SUMPRODUCT(F7:I7,\$C\$3:\$F\$3)$$

Step 4   In cell L6, compute the average return on the portfolio over all scenarios with the formula

$$=AVERAGE(Scenarios)$$

Note: The range Scenarios refers to J7:J1006.

Step 5   In cell L8, compute the portfolio's standard deviation under the assumption that each scenario occurs with probability 1/1,000.

$$=STDEVP(Scenarios)$$

Note: If we had used the formula $=STDEV$, we would have obtained

$$\sum_{\text{all scenarios}} \frac{1}{999}(\text{scenario return} - \text{mean})^2$$

when we actually wanted to divide by 1,000 (not 999).

Step 6   In cell G3, compute the total fraction invested with the formula

$$=SUM(C3:F3)$$

We now are ready to use Solver. Even though the problem of minimizing standard deviation is nonlinear, Solver is guaranteed to solve it correctly. This is because the portfolio standard deviation is a convex function of our changing cells, and our constraints are linear. Our Solver window is shown in Figure 12.

We minimize the portfolio standard deviation (L8) by changing the fraction invested in each stock and the risk-free investment (C3:F3). We ensure that each fraction is non-negative (this rules out short sales) and that the total invested equals 1 (G3 = 1). To ensure at least a 10% mean return, we add constraint L6 ≥ .1. After selecting Solve, we find that we should use the investment strategy shown in Figure 13. A mean return of 10% and standard deviation of 21.8% is obtained. We invest 24% in MSFT, 0% in INT, 37% in GE, and 40% in the risk-free investment.

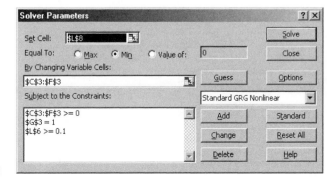

FIGURE 12

FIGURE 13

	B	C	D	E	F
1	Original Price	33.05	23	20.19	
2	FINAL MEAN	39.15313	24.73775	22.40104	
3	Weights	0.239309	0	0.365128	0.39556243
4	Name	MSFT	INTC	GE	riskfree
5	Mean	1.184663	1.075554	1.109512	

Minimizing the Probability of a Loss

Suppose we want to have an expected annual return of at least 8%, but our goal is to minimize the probability of losing money. We begin by copying our previous work to a blank sheet. Rename the sheet Min Prob Loss. (See Figure 14.) In cell L8 of the sheet, compute the probability of a loss with the formula

$$=COUNTIF(Scenarios,``<0")/1000$$

This formula gives the fraction of iterations that yield a nonpositive annual return. See Figure 15. The Solver window simply minimizes the probability of a loss (cell L8). We have all the previous constraints (except the desired expected mean return is changed to 8%), and we have added the constraint that each fraction invested is ≤ 1. We will soon see why this constraint was added.

The Target cell for minimizing probability of loss, unlike for portfolio standard deviation, is not a cell that the ordinary Solver handles well. Fortunately, we have access to the Evolutionary Solver, which does handle this type of problem well. To use Evolutionary Solver, we need to put a lower and upper bound on each changing cell; this is why we added the constraints that C3:F3 ≤ 1. We also go into Options and Limit Options to change the settings as shown in Figure 16.

Setting Max Time to 1000 lets Solver run for 1,000 seconds without asking you to continue. Population of 200 and Mutation of .25 seem to yield the best results. Max Subproblems of 50,000 and Feasible Solutions of 50,000 means that about 50,000 trial solutions will be tried before stopping. Tolerance of .0005 and Max Time without Improvement

FIGURE 14

	B	C	D	E	F	G	H	I	J	K	L	M	N
1	Original Price	33.05	23	20.19									
2													
3	Weights	0.228862	0.010504	0.376848	0.383785645	0.999999							
4	Name	MSFT	INTC	GE	riskfree								
5	Description	Output	Output	Output							Mean		
6	Iteration# / Cell	I16	J16	K16	MSFT Return	INTC Return	GE Return	RF Return	Port return		0.099677	>	0.08
7		1	49.77746	33.40337	18.10553	0.50612587	0.45232043	-0.103243	0.04	0.097029	Prob Loss		
8		2	42.14295	16.45524	16.32466	0.27512708	-0.28455478	-0.191448	0.04	0.003182	0.363		

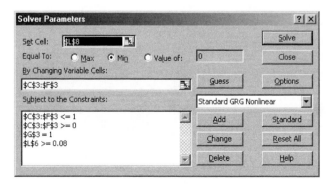

FIGURE 15

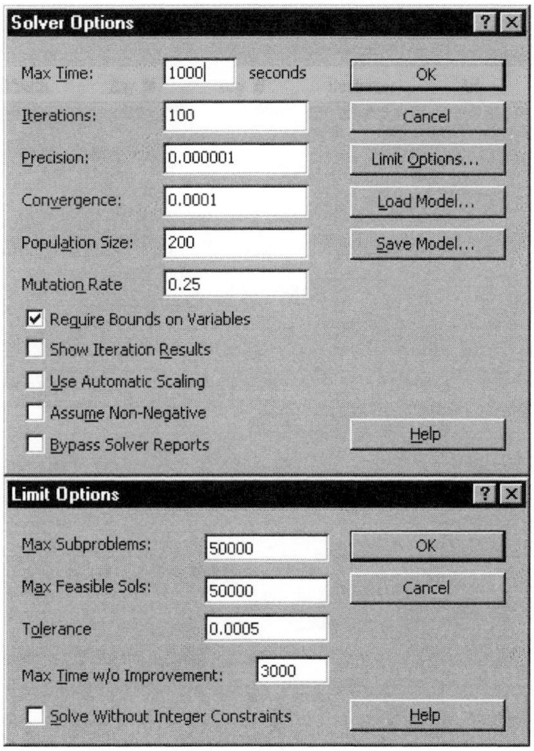

FIGURE 16

of 3000 mean that Solver will stop if, in 3,000 seconds, an improvement of less than .05% is found. The advantage of Evolutionary Solver is that it does a global search and is less likely to get stuck on a local hill and miss a mountain. The best strategy is to start with GRG2 and then run Evolutionary Solver. If you get a little better results, go back to GRG2, and continue in this fashion until no improvement is obtained. The GRG2 method is the generalized gradient method, used by the regular Excel Solver to solve nonlinear programming problems. Doing this, we obtained the optimal solution in Figure 17.

Thus, we can hold the probability of a loss to 36.3% and still obtain a mean return of at least 8%. We invest 23% of our money in MSFT, 1% in INT, 38% in GE, and 38% in the risk-free investment.

Maximizing the Sharpe Ratio

The Sharpe ratio of a portfolio equals

$$\frac{\mu - r}{\sigma}$$

FIGURE 17

	B	C	D	E	F	G	H	I	J	K	L
1	Original Price	33.05	23	20.19							
2											
3	Weights	0.228862	0.010504	0.376848	0.383785645	0.999999					
4	Name	MSFT	INTC	GE	riskfree						
5	Description	Output	Output	Output							Mean
6	Iteration# / Cell	I16	J16	K16	MSFT Return	INTC Return	GE Return	RF Return	Port return		0.099677
7	1	49.77746	33.40337	18.10553	0.50612587	0.45232043	-0.103243	0.04	0.097029		Prob Loss
8	2	42.14295	16.45524	16.32466	0.27512708	-0.28455478	-0.191448	0.04	0.003182		0.363

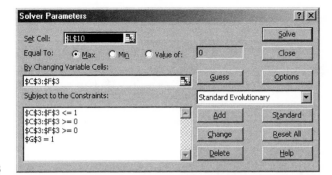

FIGURE 18

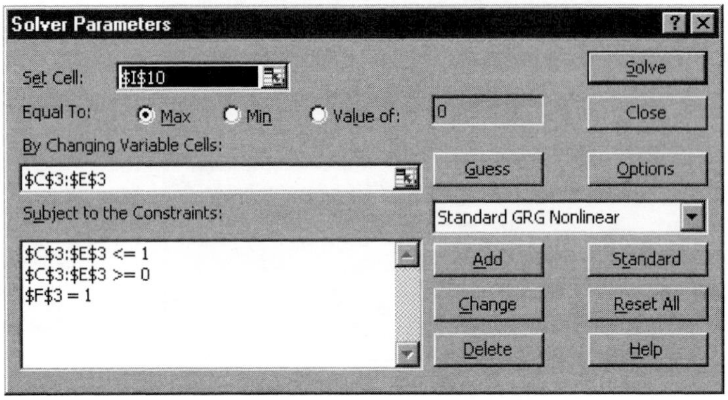

FIGURE 19

where μ = mean return on portfolio, r = risk-free rate, and σ = portfolio standard deviation. It turns out that whatever your risk-return preference, you should invest your money in the risky portfolio that maximizes the Sharpe ratio. We can easily use Solver to find the portfolio that maximizes the Sharpe ratio.

Intuitively, the act of maximizing the Sharpe ratio will maximize the excess return (relative to the risk-free rate) earned per unit of risk. Thus, maximizing the Sharpe ratio forces an investor to obtain an expected return commensurate with the assumed risk. For this reason, many Wall Street firms measure traders' success based on their Sharpe ratios.

Step 1 In cell L10, compute the Sharpe ratio with the formula

$$=(L6-0.04)/L8$$

Step 2 Our Solver window is as shown in Figure 18. We maximize (cell L10) the Sharpe ratio by changing the fraction invested in each stock and the risk-free investment (cells C3:F3). (See Figure 19.) We ensure that the fraction invested in each stock is between 0 and 1. Finally a total (cell F3) of 100% of our money is invested. Note that no constraint on expected return is needed. This is because the Sharpe ratio in some sense optimally trades off expected return against risk. After cycling between GRG2 and Evolutionary Solver, we obtain the solution in Figure 20: A maximum Sharpe ratio of .27 is obtained by placing 24% of our money in MSFT, 0% in INT, 37% in GE, and 38% in the risk-free investment. A mean return of 10.1% is obtained.

Minimizing Downside Risk

Although most finance professionals measure the riskiness of a portfolio by its variance or standard deviation, this is a flawed approach to measuring risk. For example, a portfo-

FIGURE 20

	B	C	D	E	F	G	H	I	J	K	L
1	Original Price	33.05	23	20.19							
2											
3	Weights	0.243208	1.51E-05	0.373787	0.384908928	1.00191866					
4	Name	MSFT	INTC	GE	riskfree						
5	Description	Output	Output	Output							Mean
6	Iteration# / Cell	I16	J16	K16	MSFT Return	INTC Return	GE Return	Risk free	Port return		0.10124315
7	1	49.77746	33.40337	18.10553	0.50612587	0.45232043	-0.103243	0.04	0.099906		Stdev
8	2	42.14295	16.45524	16.32466	0.27512708	-0.28455478	-0.191448	0.04	0.010744		0.22274249
9	3	85.65345	32.07594	30.80666	1.591632375	0.39460609	0.525838	0.04	0.599051		Sharpe Ratio
10	4	61.2311	29.30221	21.80993	0.852680787	0.27400913	0.080234	0.04	0.25277		0.27495046

lio with a .5 chance at yielding a 100% return and a .5 chance at yielding a 20% return has a standard deviation of 40%, while a portfolio which has a .5 chance at losing 40% and a .5 chance of breaking even has a 20% standard deviation. Clearly, the second portfolio is more risky, even though it has a lower standard deviation. The problem is that the risk we really care about is the chance of failing to meet a goal. The downside risk approach attempts to minimize the average amount by which we fail to meet a target. For example, we might set a target of a 10% return for our portfolio. A reasonable approach to portfolio selection is to set a desired expected return (say, 10%) and minimize the average amount by which the target is not met (the downside risk), subject to meeting the constraint on desired expected return. We implement this approach in the sheet Downside Risk. See Figure 21. By copying from K7 to K8:K1006 the formula

$$=IF(J7<0.1,0.1-J7,0)$$

we determine the downside risk for each scenario. Then, in cell N6, we compute the average downside risk with the formula

$$=AVERAGE(K7:K1006)$$

Our Solver Window is set as shown in Figure 22. We minimize the expected downside risk (in cell N6) subject to a desired expected return of at least 7% ($N9 \geq .07$).

The optimal solution is as shown in Figures 21 and 23. We invest 12% in MSFT, 0% in INT, 17% in GE, and 70% in the risk-free investment. On average, we fail to meet our 10% target by an average of 6%. Our expected return meets the target of 7%.

FIGURE 21

	B	C	D	E	F	G	H	I	J	K
1	Original Price	33.05	23	20.19						
2										
3	Weights	0.124391	0	0.172708	0.702901525		1			
4	Name	MSFT	INTC	GE	riskfree					
5	Description	Output	Output	Output						
6	Iteration# / Cell	I16	J16	K16	MSFT Return	INTC Return	GE Return	RF Return	Port return	Downside
7	1	49.77746	33.40337	18.10553	0.50612587	0.45232043	-0.103243	0.04	0.073243	0.026757
8	2	42.14295	16.45524	16.32466	0.27512708	-0.28455478	-0.191448	0.04	0.029275	0.070725
9	3	85.65345	32.07594	30.80666	1.591632375	0.39460609	0.525838	0.04	0.316917	0
10	4	61.2311	29.30221	21.80993	0.852680787	0.27400913	0.080234	0.04	0.148039	0
11	5	64.66952	50.53436	35.92397	0.9567177	1.19714609	0.779295	0.04	0.281713	0
12	6	50.67136	29.0577	25.48657	0.533172769	0.26337826	0.262336	0.04	0.139745	0
13	7	81.77549	28.48368	27.31935	1.474296218	0.23842087	0.353113	0.04	0.27249	0
14	8	18.50617	17.86171	16.24957	-0.44005537	-0.22340391	-0.195167	0.04	-0.06033	0.16033
15	9	15.19965	6.50584	18.05137	-0.54010136	-0.71713739	-0.105925	0.04	-0.057362	0.157362
16	10	15.72494	37.08724	23.72074	-0.52420756	0.6124887	0.174876	0.04	-0.006888	0.106888

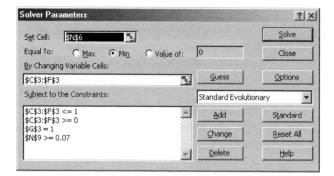

FIGURE 22

	M	N	O
2			
3			
4		Downside	
5		Risk	
6		0.060838	
7			
8		Mean return	
9		0.07	
10			
11			

FIGURE 23

The Minimax Approach

In the sheet Minimax, we find the portfolio that yields an expected return of at least 10% and maximizes the absolute worst-case scenario. In cell L8 (see Figure 24), we compute the worst-case scenario with the formula

$$=\text{MIN(Scenarios)}$$

Then we maximize the worst-case scenario (subject to a desired 10% expected return) with the Solver window shown in Figure 25.

	K	L	M	N	O
4					
5		Mean			
6		0.1	>	0.1	
7		Worst Case			
8		-0.290931			

FIGURE 24

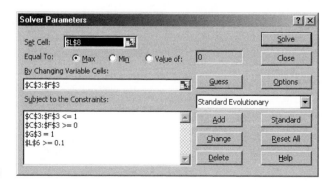

FIGURE 25

	A	B	C	D	E	F
1		Original Price	33.05	23	20.19	
2						
3		Weights	0.414756	0	0	0.585244109
4		Name	MSFT	INTC	GE	riskfree
5		Description	Output	Output	Output	
6		Iteration# / Cell	I16	J16	K16	MSFT Return

FIGURE 26

We find that the portfolio in Figure 26 yields a desired expected return of at least 10% and has a "best worst case" of losing 29%. Thus, the minimax approach suggests investing 42% in MSFT and 58% in the risk-free investment. We should entirely ignore investing in INT and GE.

Maximizing VAR

Finally, we find the portfolio yielding an expected return of at least 8% which has the best VAR. Recall that VAR is the loss associated with the fifth percentile of final portfolio return. Clearly, maximizing the fifth percentile of the portfolio's return will reduce the loss associated with the fifth percentile to the lowest possible level. See the sheet VAR.

In cell L8 (see Figure 27), we compute the VAR of our portfolio with the formula

$$=PERCENTILE(Scenarios,0.05)$$

After running the Solver model indicated in Figure 28, we find that the portfolio in Figure 29 has the most favorable VAR among all portfolios yielding an expected return of at least 8%. Thus, investing 27% in MSFT, .2% in INT, 2% in GE, and 71.4% in the risk-free investment yields an expected return of 8% and has a 5% chance of losing 11.5% or more.

	L	M	N
3			
4			
5	Mean		
6	0.080006	>	0.08
7	VAR		
8	-0.115457		

FIGURE 27

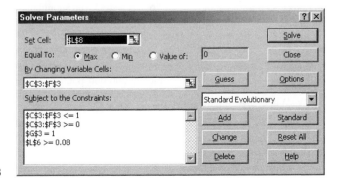

FIGURE 28

FIGURE 29

	B	C	D	E	F
1	Original Price	33.05	23	20.19	
2					
3	Weights	0.265863	0.001627	0.020332	0.714038173
4	Name	MSFT	INTC	GE	riskfree
5	Description	Output	Output	Output	
6	Iteration# / Cell	I16	J16	K16	MSFT Return

REMARKS **1** Suppose our portfolio contains equities and derivatives such as puts and calls. Then to find the return on the portfolio for each scenario, we use the fact that

$$\text{Portfolio return} = \frac{\text{final portfolio value} - \text{initial portfolio value}}{\text{initial portfolio value}}$$

2 Note that Proctor and Gamble used the scenario approach in its foreign exchange hedging and saved an estimated $25 million in annual hedging costs. See Problem 4.

PROBLEMS

Group A

1 Consider the Dell example in Section 14.1.

a Using the fact that the return on Dell follows a lognormal random variable with current price $94.125, μ equally likely to be 0%, 10%, 20%, or 30%, and $\sigma = 53.27\%$, generate 1,000 scenarios for the price of Dell in 4 months. Assume a 4-month planning horizon.

Recall that a 4-month $80 put on Dell sells for $5.25. Assume that we own 1,000 shares of Dell.

b How many puts should be purchased to minimize the portfolio standard deviation subject to an expected return of at least 10%?

c How many puts should be purchased to maximize the Sharpe ratio?

d How many puts should be purchased to minimize the probability of losing money, subject to a desired expected return of 10%?

e How many puts should be purchased to obtain the best VAR, subject to a desired expected return of at least 10%?

2 On January 4, 2000, Nortel stock was selling for $93.72. Suppose we buy 10,000 shares of Nortel. Perhaps we may want to trade in some 3-month $75, $80, and $85 puts to hedge our risk.

The historical volatility of Nortel is 48%. Plugging into Black–Scholes, we find the following prices for the puts.

Put Exercise Price	3-Month Put Price
$75	$1.43
$80	$2.50
$85	$4.02

We will evaluate our trades on NPV. When investment strategies involve shorting investments, this makes more sense than evaluating trades based on return. We assume that the investment firm desires an expected return of 20%, so a hurdle rate of 20% will be used. Use a lognormal random variable with $\mu = 20\%$ and $\sigma = 48\%$ to generate 400 scenarios for Nortel's price in three months.

a Minimize variance subject to a mean NPV that is nonnegative.

b Maximize VAR (the fifth percentile of NPV) subject to a 10% desired expected return.

c Maximize the worst-case scenario subject to a 10% desired expected return.

Group B

3 You now own 100 shares of GE, MSFT, and INT stock. Assume again that the risk-free rate is 4%. Assume that the current price of each stock is $1. You also have the option of buying today one-year European put options on GE, MSFT, and INT that have an exercise price of $0.90. The price of each put option on a single share is as follows: MSFT: 8.59 cents, INT: 11.98 cents, GE: 3.75 cents.

a You wish to obtain a mean return of 7% for the next year and minimize the standard deviation of your annual return. How many units of the risk-free investment and put options on each stock should you purchase? For the purposes of this problem, you may use the 1,000 scenarios for MSFT, GE and INT given in our example.

b With a desired mean return of 7%, determine how many units of the risk-free investment and put options on each stock you should purchase to minimize expected downside risk.

4 All major U.S. companies face significant exchange rate risk. As an example, consider a U.S. company that expected in 1998 to receive the following in six months: 5,000,000 francs, 2,000,000 pounds, and 100,000,000 yen. Suppose today's exchange rates are as follows:

Currency	Current Exchange Rate ($/unit of currency)
Franc	$0.1649
Pound	$1.6667
Yen	$0.0072

Given the current exchange rates, the value of our foreign accounts receivable would be

5,000,000(.1649) + 2,000,000(1.6667)

$$+ \ 100,000,000(.0072) = \$4.88 \ \text{million}$$

If the values of the franc, pound, and yen all drop, then the dollar value of our foreign accounts receivable will also drop. For example, if all currencies drop 10% against the dollar, the company has lost $488,000. Exchange rate risk makes it difficult for CFOs who want to ensure that the company meets analysts' forecasts for earnings per share, which are measured in dollars. To lessen the risk caused by foreign exchange fluctuations, many companies' treasury departments invest in puts or forward options. In our analysis, we will consider put options. A European put option on a currency having an exercise price x will pay out on the option's expiration date the greater of $0 or (exercise price) $-$ (currency value on expiration date). For example, a 6-month put with an exercise price of $0.15 on the franc would pay nothing if the franc were above $0.15 in value in six months and would pay $0.02 if the franc were worth $0.13 in six months. We see that put options pay more money when the value of a currency decreases. Since the dollar value of our foreign accounts receivable decreases as currency value decreases, we see that purchasing puts hedges our foreign exchange risk.

The natural question is how many puts to buy on each currency and exercise price to obtain the desired risk profile. Here is how to attack this problem.

TABLE 5

Currency	Exercise Price	Option Price (in cents)
Franc	$0.15	.0166
Franc	$0.13	.000058
Pound	$1.60	.7396
Pound	$1.50	.094
Yen	$0.006	.00163
Yen	$0.005	.000019

Consider the foreign exchange accounts receivable and spot rates (June 30, 1998) given above. The company can purchase on June 30, 1998, the six-month European put options shown in Table 5 (prices are given in cents).

a Use bootstrapping off the data in the file Fxdata.xls to generate 1,000 scenarios for the pound, yen, and franc exchange rates in six months. Assume 126 trading days = 6 months.

b How can we minimize the variability of our six-month profit in dollars, subject to the constraint that our average profit is at least $4.80 million?

c To meet analysts' targets, we need to have our six-month profit be at least $4.60 million. How can we minimize the probability of not meeting this target?

REFERENCE

Markowitz, H. *Portfolio Selection: Efficient Diversifications of Investments.* New Haven, Conn.: Yale University Press, 1971.

15

Forecasting Models

In previous chapters, we have often blindly substituted numbers into problems without considering where the numbers came from. For example, in the Giapetto LP (Example 1 in Chapter 3 of Volume 1), we assumed that the variable cost of producing a train was $21. In reality, we would have to estimate the cost of producing a train. This can be done using the method of *simple linear regression,* explained in Section 15.6.

In Chapters 3 and 4, we used inventory theory to determine production quantities and reorder points. To use the models of Chapters 3 and 4, we need to be able to forecast the demand for a product. In Sections 15.1–15.5, we discuss extrapolation and smoothing methods that can be used to forecast future demand for a product.

As another example of how we can use "good forecasts," suppose that we want to use the queuing models of Chapter 8 to determine how the number of tellers at a bank should vary with the day of the week and the time of day. To tackle this problem, we need to determine how the rate at which customers enter the bank depends on the time of day and the day of the week. For example, if we knew that over half the bank's customers arrived during the lunch hour (noon to 1 P.M.), that would have a significant effect on the optimal staffing policy.

In this chapter, we discuss two important types of forecasting methods: extrapolation methods and causal forecasting methods. In Sections 15.1–15.5, we discuss *extrapolation methods,* which are used to forecast future values of a time series from past values of a time series. To illustrate, consider Lowland Appliance Company's monthly sales of TVs, compact disc players (CDs), and air conditioners (ACs) for the last 24 months, given in Table 1. In an extrapolation forecasting method, it is assumed that past patterns and trends in sales will continue in future months. Thus, past data on appliance sales (and no other information) are used to generate forecasts for appliance sales during future months. Extrapolation methods (unlike the causal forecasting methods described in Sections 15.6–15.8) don't take into account what "caused" past data; they simply assume that past trends and patterns will continue in the future.

Causal forecasting methods attempt to forecast future values of a variable (called the dependent variable) by using past data to estimate the relationship between the dependent variable and one or more independent variables. For example, Lowland might try to forecast future monthly sales of air conditioners by using past data to determine how air conditioner sales are related to independent variables such as price, advertising, and the month of the year. Causal forecasting methods will be discussed in Sections 15.6–15.8.

15.1 Moving-Average Forecasting Methods

Let $x_1, x_2, \ldots, x_t, \ldots$ be observed values of a time series, where x_t is the value of the time series observed during period t. One of the most commonly used forecasting

TABLE 1
Lowland Appliance Sales

Month	TV Sales	CD Sales	AC Sales	Month	TV Sales	CD Sales	AC Sales
1	30	40	13	13	38	79	36
2	32	47	7	14	30	82	21
3	30	50	23	15	35	80	47
4	39	49	32	16	30	85	81
5	33	56	58	17	34	94	112
6	34	53	60	18	40	89	139
7	34	55	90	19	36	96	230
8	38	63	93	20	32	100	201
9	36	68	63	21	40	100	122
10	39	65	39	22	36	105	84
11	30	72	37	23	40	108	74
12	36	69	29	24	34	110	62

methods is the moving-average method. We define $f_{t,1}$ to be the forecast period for period $t + 1$ made after observing x_t. For the moving-average method,

$$f_{t,1} = \text{average of the last } N \text{ observations}$$
$$= \text{average of } x_t, x_{t-1}, x_{t-2}, \ldots, x_{t-N+1}$$

where N is a given parameter.

To illustrate the moving-average method, we choose $N = 3$ and use the moving-average method to forecast TV sales for the first six months of data in Table 1. The resulting computations are given in Table 2. For months 1–3, we have not yet observed three months of data, so (for $N = 3$) we cannot develop a moving-average forecast for sales for these months. For month 4, we find our forecast, $f_{3,1} = \frac{30+32+30}{3} = 30.67$. For month 5, our forecast is $f_{4,1} = \frac{32+30+39}{3} = 33.67$. For month 6, our forecast is $f_{5,1} = \frac{30+39+33}{3} = 34$.

Note that from one period to the next, our forecast "moves" by replacing the "oldest" observation in the average by the most recent observation.

Choice of *N*

How should we choose N, the number of periods used to compute the moving average? To answer this question, we need to define a measure of forecast accuracy. We will use the **mean absolute deviation** (MAD) as our measure of forecast accuracy. Before defining the MAD, we need to define the concept of a **forecast error.** Given a forecast for x_t, we define e_t to be the error in our forecast for x_t, to be given by

$$e_t = x_t - (\text{forecast for } x_t)$$

From Table 2, we find $e_4 = 39 - 30.67 = 8.33$, $e_5 = 33 - 33.67 = -0.67$, and $e_6 = 34 - 34 = 0$. The MAD is simply the average of the absolute values of all the e_t's. Thus, for periods 1–6, our moving-average forecast yields a MAD given by

$$\text{MAD} = \frac{|e_4| + |e_5| + |e_6|}{3} = \frac{8.33 + 0.67 + 0}{3} = 3$$

Thus, on the average, our forecasts for TV sales are off by 3 TVs per month.

TABLE 2
Moving-Average Forecasts ($N = 3$)
for TV Sales

Month	Actual Sales	Predicted Sales
1	30	—
2	32	—
3	30	—
4	39	$\frac{30 + 32 + 30}{3}$
5	33	$\frac{32 + 30 + 39}{3}$
6	34	$\frac{30 + 39 + 33}{3}$

We are trying to forecast next month's TV sales as an average of the last N months' actual sales. What value of N will minimize our mean absolute error (obtained by averaging the actual error incurred during each month)? We will try $N = 1, 2, \ldots, 12$.

We begin with an explanation of the Excel OFFSET function. This function lets you pick out a cell range relative to a given location in the spreadsheet. The syntax of the OFFSET function is as follows:

OFFSET(reference, rows, columns, height, width)

- **Reference** is the cell from which you base the row and column references.
- **Rows** helps locate the upper left-hand corner of the OFFSET range. Rows is measured by number of rows up or down (up is negative, and down is positive) from the cell reference.
- **Columns** helps locate the upper left-hand corner of the OFFSET range. Columns is measured by number of columns left or right (left is negative, and right is positive) from the cell reference.
- **Height** is the number of rows in the selected range.
- **Width** is the number of columns in the selected range.

Offsetexample.xls

File Offsetexample.xls contains some examples of how the OFFSET function works. See Figure 1. The nice thing about the OFFSET function is that it can be copied like any formula. The next section will show the true power of the OFFSET function.

Tvsales.xls

Our work is in file Tvsales.xls. We begin creating a forecast in month 13, because that is the first month in which 12 months of historical data are available. See Figures 2 and 3.

Step 1 By copying from C17 to C18:C28 the formula

=AVERAGE(OFFSET(B17,-D3,0,D3,1))

obtain the average of the last D3 months of data.

- B17 ensures that we define our range relative to the cell directly to the left of the cell where the formula is entered.
- -D3 ensures that our range begins D3 rows above the row where the formula is entered.
- The 0 ensures that the OFFSET range will always remain in column B.
- D3 ensures that we average the last D3 observations.
- 1 ensures that the OFFSET range includes a single column.

FIGURE 1

	A	B	C	D	E	F	G	H	I	J	K
1											
2											
3		Offset examples									
4											
5											
6		1	2	3	4			1	2	3	4
7		5	6	7	8			5	6	7	8
8		9	10	11	12			9	10	11	12
9											
10	=SUM(OFFSET(B7,-1,1,2,1))	8					=SUM(OFFSET(H6,0,1,3,2))	39			
11											
12											
13											
14		1	2	3	4						
15		5	6	7	8						
16		9	10	11	12						
17											
18	=SUM(OFFSET(E16,-2,-3,2,3))	24									
19											

	A	B
4	Month	TV Sales Actual
5	1.00	30
6	2.00	32
7	3.00	30
8	4.00	39
9	5.00	33
10	6.00	34
11	7.00	34
12	8.00	38
13	9.00	36
14	10.00	39
15	11.00	30
16	12.00	36
17	13.00	38
18	14.00	30
19	15.00	35
20	16.00	30
21	17.00	34
22	18.00	40
23	19.00	36
24	20.00	32
25	21.00	40
26	22.00	36
27	23.00	40
28	24.00	34

FIGURE 2

	A	B	C	D	E	F	G	H
2				# OF PERIODS				
3				1				
4	Month	TV Sales Actual	Moving average forecast	Abs error	MAD	5		
5	1.00	30						
6	2.00	32						
7	3.00	30						
8	4.00	39					# of periods	5
9	5.00	33					1	5
10	6.00	34					2	3.666666667
11	7.00	34					3	3.361111111
12	8.00	38					4	3.333333333
13	9.00	36					5	3.016666667
14	10.00	39					6	3.111111111
15	11.00	30					7	3.226190476
16	12.00	36					8	3.21875
17	13.00	38	36	2			9	3.055555556
18	14.00	30	38	8			10	3.083333333
19	15.00	35	30	5			11	3.045454545
20	16.00	30	35	5			12	3.111111111
21	17.00	34	30	4			Min	3.016666667
22	18.00	40	34	6			best #	5
23	19.00	36	40	4				
24	20.00	32	36	4				
25	21.00	40	32	8				
26	22.00	36	40	4				
27	23.00	40	36	4				
28	24.00	34	40	6				

FIGURE 3

Step 2 By copying from D17 to D18:D28 the formula

$$=ABS(B17-C17)$$

compute the absolute value of the error in each month's forecast (based on a D3-month moving average).

Step 3 In cell F4, compute the average of the absolute errors (often called the MAD) with the formula

$$=AVERAGE(D17:D28)$$

Step 4 Enter the trial number of periods for the moving average (1–12) in G9:G20, and in cell H8, enter the MAD with the formula

$$=F4$$

Step 5 After selecting the table range G8:H20 and choosing a one-way data table with the column input cell of D3, we find that a 5-period moving average yields the smallest MAD (3.02).

Step 6 We obtain the minimum MAD in cell H21 with the formula

$$=MIN(H9:H20)$$

Step 7 Entering in cell H22 the formula

$$=MATCH(H21,H9:H21,0)$$

gives the number of periods (5) yielding the smallest MAD.

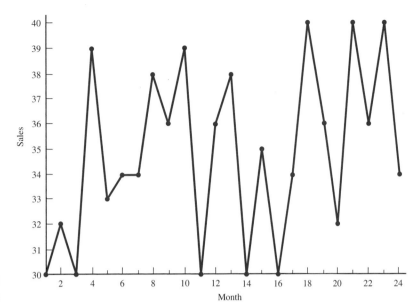

FIGURE 4
TV Sales

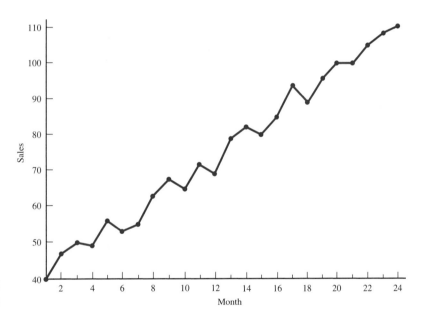

FIGURE 5
CD Player Sales

Moving-average forecasts perform well for a time series that fluctuates about a constant **base level.** From Figure 4, it appears that monthly TV sales fluctuate about a base level of 35. More formally, moving-average forecasts work well if

$$x_t = b + \varepsilon_t \tag{1}$$

where b is the base level for the series and ε_t is the random fluctuation in period t about the base level.

From Figures 5 and 6, we see that sales of CD players and air conditioners are not well described in Equation (1). From Figure 5, we see that there is an upward **trend** in CD player sales, so they do not fluctuate about a base level. From Figure 6, we find that air conditioner sales exhibit **seasonality:** The peaks and valleys of the series repeat at regular 12-month intervals. Figure 6 also shows that air conditioner sales exhibit an upward

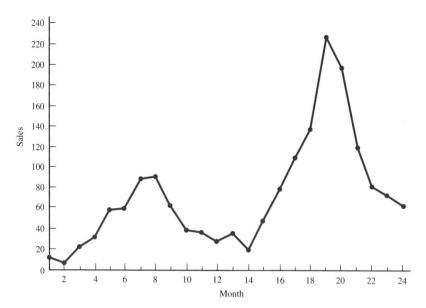

FIGURE 6
Air Conditioner Sales

trend. In situations where trend and/or seasonality are present, the moving-average method usually yields poor forecasts. To close this section, we note that in addition to trend and seasonality, a time series may exhibit **cyclic behavior.** For example, auto sales often follow the business cycle of the national economy. Cyclic behavior is much more irregular than a seasonal pattern and is often hard to detect.

15.2 Simple Exponential Smoothing

If a time series fluctuates about a base level, **simple exponential smoothing** may be used to obtain good forecasts for future values of the series. To describe simple exponential smoothing, let A_t = smoothed average of a time series after observing x_t. After observing x_t, A_t is the forecast for the value of the time series during any future period. The key equation in simple exponential smoothing is

$$A_t = \alpha x_t + (1 - \alpha)A_{t-1} \tag{2}$$

In (2), α is a **smoothing constant** that satisfies $0 < \alpha < 1$. To initialize the forecasting procedure, we must have (before observing x_1) a value for A_0. Usually, we let A_0 be the observed value for the period immediately preceding period 1. As with moving-average forecasts, we let $f_{t,k}$ be the forecast for x_{t+k} made at the end of period t. Then

$$A_t = f_{t,k} \tag{3}$$

Assuming that we are trying to forecast one period ahead, our error for predicting x_t (written again as e_t) is given by

$$e_t = x_t - f_{t-1,1} = x_t - A_{t-1} \tag{4}$$

To understand (2) better, we use (4) to rewrite (2) as

$$A_t = A_{t-1} + \alpha(x_t - A_{t-1}) = A_{t-1} + \alpha e_t$$

Thus, our new forecast $A_t = f_{t,1}$ is equal to our old forecast (A_{t-1}) plus a fraction of our period t error (e_t). This implies that if we "overpredict" x_t, we lower our forecast, and if we "underpredict" x_t, we raise our forecast. For larger values of the smoothing constant α, more weight is given to the most recent observation (see Remark 3 at the end of the section).

We illustrate simple exponential smoothing (with $\alpha = 0.1$) for the first six months of TV sales. The results are given in Table 3. We assume that 32 TVs were sold last month, so we initialize the procedure with $A_0 = 32$. Here are some illustrations of the computations:

$$A_t = 0.1x_1 + 0.9A_0 = 0.1(30) + 0.9(32) = 31.8$$
$$f_{0,1} = A_0 = 32$$
$$e_1 = x_1 - A_0 = 30 - 32 = -2$$
$$f_{1,1} = A_1 = 31.8$$
$$e_2 = x_2 - A_1 = 32 - 31.8 = 0.2$$
$$A_2 = 0.1x_2 + 0.9A_1 = 0.1(32) + 0.9(31.8) = 31.82$$

For months 1–6, the MAD of our forecast is given by

$$\text{MAD} = \frac{|-2| + |0.2| + |-1.82| + |7.36| + |0.63| + |1.56|}{6}$$
$$= 2.26$$

For the entire 24-month period, we can determine (using a one-way data table) the value of α yielding the lowest MAD. The results are given in Table 4. It appears that a value of α between 0.20 and 0.30 yields the lowest MAD.

REMARKS   **1**   Since $\alpha < 1$, exponential smoothing "smooths out" variations in a time series by not giving total weight to the last observation.
2   If $\alpha = \frac{2}{N+1}$, simple exponential smoothing (with smoothing parameter α) and an N-period moving-average forecast will both yield similar forecasts. For example, $\alpha = 0.33$ is roughly equivalent to a five-period moving average.
3   To see why we call the method *exponential* smoothing, consider (2) for $t - 1$:

$$A_{t-1} = \alpha x_{t-1} + (1 - \alpha)A_{t-2} \tag{5}$$

Substituting (5) into (2) yields

$$A_t = \alpha x_t + (1 - \alpha)[\alpha x_{t-1} + (1 - \alpha)A_{t-2}]$$
$$= \alpha x_t + \alpha(1 - \alpha)x_{t-1} + (1 - \alpha)^2 A_{t-2} \tag{6}$$

Note that

$$A_{t-2} = \alpha x_{t-2} + (1 - \alpha)A_{t-3} \tag{7}$$

TABLE 3
Simple Exponential Smoothing for TV Sales ($\alpha = .1$)

Month	Actual Sales	Forecast	A_t	e_t
1	30	32	31.8	−2.00
2	32	31.8	31.82	0.20
3	30	31.82	31.64	−1.82
4	39	31.64	32.37	7.36
5	33	32.37	32.44	0.63
6	34	32.44	32.60	1.56

TABLE 4
MAD for TV Sales

α	MAD
0.05	3.20
0.10	3.04
0.15	2.94
0.20	2.89
0.25	2.88
0.30	2.90
0.35	2.94
0.40	2.98
0.45	3.05
0.50	3.13

Substituting (7) into (6) yields

$$A_t = \alpha x_t + \alpha(1 - \alpha)x_{t-1} + \alpha(1 - \alpha)^2 x_{t-2} + (1 - \alpha)^3 A_{t-3}$$

Repeating this process yields

$$A_t = \alpha x_t + \alpha(1 - \alpha)x_{t-1} + \alpha(1 - \alpha)^2 x_{t-2} + \cdots + \alpha(1 - \alpha)^k x_{t-k} + \cdots \tag{8}$$

Since $\alpha + \alpha(1 - \alpha) + \alpha(1 - \alpha)^2 + \cdots = 1$, (8) shows that if we go back an "infinite" number of periods, our current smoothed average is a weighted average of all past observations. The weight given to the observation from k periods in the past declines exponentially (by a factor of $1 - \alpha$). The larger the value of α, the more weight is given to the most recent observations. For example, for $\alpha = 0.2$, the three most recent observations have 49% of the weight (20%, 16%, and 13%), whereas for $\alpha = 0.5$, the three most recent observations have 88% of the weight (50%, 25%, and 13%).

4 In practice, α is usually chosen to equal 0.10, 0.30, or 0.50. If the value of α that minimizes the MAD exceeds 0.5, then trend, seasonality, or cyclical variation is probably present, and simple exponential smoothing is not a recommended forecast technique. In such cases, better forecasts will probably be provided by either Holt's method (exponential smoothing with trend, discussed in Section 15.3) or Winter's method (exponential smoothing with trend and seasonality, discussed in Section 15.4).

5 Even if a time series is not fluctuating about a constant base level, simple exponential smoothing may still provide good forecasts. If $x_t = m_t + \varepsilon_t$ and $m_t = m_{t-1} + \delta_t$, where ε_t and δ_t are independent error terms each having mean 0, then simple exponential smoothing will provide good forecasts. This implies that if the mean demand (m_t) for a product is randomly shifting over time, simple exponential smoothing will still provide good forecasts of product demand.

15.3 Holt's Method: Exponential Smoothing with Trend

If we believe that a time series exhibits a linear trend (and no seasonality), **Holt's method** often yields good forecasts. At the end of the tth period, Holt's method yields an estimate of the base level (L_t) and the per-period trend (T_t) of the series. For example, suppose that $L_{20} = 20$ and $T_{20} = 2$. This means that after observing x_{20}, we believe that the base level of the series is 20 and that the base level is increasing by two units per period. Thus, five periods from now, we estimate that the base level of the series will equal 30.

After observing x_t, equations (9) and (10) are used to update the base and trend estimates. α and β are smoothing constants, each between 0 and 1.

$$L_t = \alpha x_t + (1 - \alpha)(L_{t-1} + T_{t-1}) \tag{9}$$

$$T_t = \beta(L_t - L_{t-1}) + (1 - \beta)T_{t-1} \tag{10}$$

To compute L_t, we take a weighted average of the following two quantities:

1 x_t, which is an estimate of the period t base level from the current period

2 $L_{t-1} + T_{t-1}$, which is an estimate of the period t base level based on previous data

To compute T_t, we take a weighted average of the following two quantities:

1 An estimate of trend from the current period given by the increase in the smoothed base from period $t-1$ to period t

2 T_{t-1}, which is our previous estimate of the trend

As before, we define $f_{t,k}$ to be the forecast for x_{t+k} made at the end of period t. Then

$$f_{t,k} = L_t + kT_t \tag{11}$$

To initialize Holt's method, we need an initial estimate (call it L_0) of the base and an initial estimate (call it T_0) of the trend. We might set T_0 equal to the average monthly increase in the time series during the previous year, and we might set L_0 equal to last month's observation.

From Figure 5, it is clear that CD player sales exhibit an upward trend, but no obvious seasonal pattern is present. Therefore, Holt's method should yield good forecasts. Let's assume that CD player sales during each of the last 12 months are given by 4, 6, 8, 10, 14, 18, 20, 22, 24, 28, 31, and 34. Then

$$T_0 = \frac{(6-4) + (8-6) + (10-8) + \cdots + (34-31)}{11}$$

$$= \frac{34-4}{11} = 2.73$$

We then estimate $L_0 = 34$.

Applying the Holt method to the first six months of sales (using $\alpha = 0.30$ and $\beta = 0.10$) we obtain the results shown in Table 5. Here are some illustrations of the calculations:

$$L_1 = 0.30x_1 + 0.70(L_0 + T_0) = 0.3(40) + 0.7(34 + 2.73) = 37.71$$
$$T_1 = 0.1(L_1 - L_0) + 0.9T_0 = 0.1(37.71 - 34) + 0.9(2.73) = 2.83$$
$$f_{1,1} = L_1 + T_1 = 37.71 + 2.83 = 40.54$$
$$e_2 = x_2 - f_{1,1} = 47 - 40.54 = 6.46$$

TABLE 5
Holt's Method for CD Player Sales ($\alpha = 0.30$, $\beta = 0.10$)

Month	Sales	L_t	T_t	$f_{t-1,1}$ $(L_{t-1} + T_{t-1})$	e_t $(x_t - f_{t-1,1})$
1	40	37.71	2.83	36.73	3.27
2	47	42.48	3.02	40.54	6.46
3	50	46.85	3.16	45.50	4.50
4	49	49.70	3.13	50.01	−1.01
5	56	53.78	3.22	52.83	3.17
6	53	55.80	3.10	57.00	−4.00

For the first six months of CD player sales, we find

$$\text{MAD} = \frac{3.27 + 6.46 + 4.5 + 1.01 + 3.17 + 4.00}{6} = 3.74$$

For the entire 24-month period, we find MAD = 2.85.

As one more illustration of Equation (11), suppose that we want to make a forecast at the end of month 6 for month 10 CD player sales. From (11), we find that $f_{6,4} = L_6 + 4T_6 = 55.80 + 4(3.10) = 68.2$. By trying various combinations of α and β, we could find the values of α and β that minimize the MAD. If these values are not both less than 0.5, then seasonality or cyclical behavior is probably present, and another forecasting method should be used.

In summary, Holt's method will provide good forecasts for a series with a linear trend. Such a series may be modeled as $x_t = a + bt + \varepsilon_t$, where

$$a = \text{base level at beginning of period 1}$$

$$b = \text{per-period trend}$$

$$\varepsilon_t = \text{error term for period } t$$

A multiplicative version of Holt's method (see Problem 15) can be used to generate good forecasts for a series of the form $x_t = ab^t\varepsilon_t$. Here, the value of b represents the percentage growth in the base level of the series during each period. Thus, $b = 1.1$ implies that the base level of the series is increasing by 10% per period. In this model, ε_t is a random error factor with a mean of 1.

A Spreadsheet Implementation of the Holt Method

Holt.xls

Figure 7 (obtained from the file Holt.xls) contains an implementation of the Holt method. In columns B and C, we have typed in the 24 months of CD player sales obtained from Table 1. In cells D4 and E4, we have input L_0 and T_0. Trial values of alpha and beta appear in cells E2 and F2. In cell D5, we compute L_1 by inputting the formula =E$2*C5+(1−E$2)*(D4+E4). In cell E5, we compute T_1 by inputting the formula =F$2*(D5−D4)+(1−F$2)*E4. In cell F5, we compute $f_{0,1}$ from the formula =D4+E4. In cell G5, we compute e_1 from the formula =C5−F5. In cell H5, we compute $|e_1|$ from the formula =ABS(G5). Copying the formulas from the range C5:H5 to the range C5:H28 completes the implementation of the Holt method. The formula =AVERAGE(H5:H28) in cell G2 computes the MAD (2.85) for the 24 months.

We can use an Excel two-way data table to determine values of α and β that yield a small MAD. We input possible values for α in a cell range B31:B39 and values for β in the cell range C30:K30. We input a formula to compute the MAD (=G2) into cell B30. Invoking the DATATABLE command, we choose the table range B30:K39. Then we select cell E2 as the column input cell and cell F2 as the row input cell. This causes the values in B1:B39 to be input into E2 and the values in C30:K30 to be input into cell F2. After selecting OK, for each combination of α and β in the table Excel computes the MAD. We see that of the combinations listed, $\alpha = .10$ and $\beta = .40$ yields the lowest MAD (2.70). If we wanted to obtain an even lower MAD, we could explore values of α and β near .10 and .40, respectively, by creating another data table. By the way, F9 will recalculate the last data table you have created in your spreadsheet.

FIGURE 7
Holt's Method

	B	C	D	E	F	G	H	I	J	K
1		HOLT	METHOD	ALPHA	BETA	MAD=				
2				0.3	0.1	2.84686937				
3	MONTH	CD SALES	Lt	Tt	f(t-1,1)	et	|et|			
4	0		34	2.73						
5	1	40	37.711	2.8281	36.73	3.27	3.27			
6	2	47	42.47737	3.021927	40.5391	6.4609	6.4609			
7	3	50	46.8495079	3.15694809	45.499297	4.500703	4.500703			
8	4	49	49.7045192	3.12675441	50.006456	-1.006456	1.00645599			
9	5	56	53.7818915	3.2218162	52.8312736	3.1687264	3.1687264			
10	6	53	55.8025954	3.10170497	57.0037077	-4.0037077	4.00370772			
11	7	55	57.7330103	2.98457596	58.9043004	-3.9043004	3.90430038			
12	8	63	61.4023104	3.05304837	60.7175862	2.28241378	2.28241378			
13	9	68	65.5187511	3.15938761	64.4553587	3.54464127	3.54464127			
14	10	65	67.5746971	3.04904345	68.6781387	-3.6781387	3.67813872			
15	11	72	71.0366184	3.09033123	70.6237406	1.37625945	1.37625945			
16	12	69	72.5888647	2.93652274	74.1269496	-5.1269496	5.12694962			
17	13	79	76.5677712	3.04076112	75.5253875	3.47461252	3.47461252			
18	14	82	80.3259726	3.11250515	79.6085324	2.39146765	2.39146765			
19	15	80	82.4069345	3.00935081	83.4384778	-3.4384778	3.4384778			
20	16	85	85.2913997	2.99686226	85.4162853	-0.4162853	0.41628527			
21	17	94	90.0017834	3.1682144	88.2882619	5.71173805	5.71173805			
22	18	89	91.9189984	3.04311447	93.1699978	-4.1699978	4.16999776			
23	19	96	95.273479	3.07425108	94.9621129	1.0378871	1.0378871			
24	20	100	98.8434111	3.12381918	98.3477301	1.65226989	1.65226989			
25	21	100	101.377061	3.06480227	101.96723	-1.9672303	1.96723025			
26	22	105	104.609304	3.08154636	104.441863	0.55813656	0.55813656			
27	23	108	107.783596	3.09082084	107.690851	0.30914923	0.30914923			
28	24	110	110.612091	3.06458835	110.874416	-0.8744164	0.87441638			
29					BETA					
30	2.84686937	0.1	0.2	0.3	0.4	0.5	0.6	0.7	0.8	0.9
31	0.1	2.85549229	2.79917857	2.73752333	2.70190155	2.74959079	2.79516165	2.83287197	2.86298631	2.92313769
32	0.2	2.76620734	2.73280037	2.76089108	2.78733151	2.84068253	2.92107922	2.97420106	2.98283649	2.99339222
33	0.3	2.84686937	2.87216928	2.90902939	2.95265285	2.98980036	3.04270174	3.13130108	3.21889032	3.2935572
34	0.4	2.96069374	3.0030147	3.05465632	3.110774	3.17038284	3.23467306	3.30196817	3.35141111	3.38544721
35	0.5	3.0707409	3.1289783	3.19166282	3.25002095	3.31027756	3.36404349	3.42489473	3.50405565	3.58298443
36	0.6	3.19398831	3.26198014	3.33253292	3.39877183	3.48348934	3.58574625	3.68496609	3.78162671	3.87646977
37	0.7	3.31493308	3.39345284	3.47449093	3.59818192	3.71792621	3.83504256	3.95075646	4.06889148	4.21031682
38	0.8	3.43163015	3.52755966	3.66950469	3.80654215	3.940503	4.07309251	4.26812614	4.4709749	4.67844324
39	0.9	3.53843635	3.69071551	3.844996	4.02099412	4.2282208	4.45607616	4.68692747	4.9520137	5.24579003

15.4 Winter's Method: Exponential Smoothing with Seasonality

The appropriately named **Winter's method** is used to forecast time series for which trend and seasonality are present. As previously mentioned, Figure 6 shows that air conditioner sales exhibit an upward trend and seasonality, so Winter's method is a logical candidate for forecasting these sales.

To describe Winter's method, we require two definitions. Let c = the number of periods in the length of the seasonal pattern ($c = 4$ for quarterly data, and $c = 12$ for monthly data). Let s_t be an estimate of a seasonal multiplicative factor for month t, obtained after observing x_t. For instance, suppose month 7 is July and $s_7 = 2$. Then after observing month 7's air conditioner sales, we believe that July air conditioner sales will (all other things being equal) be twice the sales expected during an average month. If month 24 is December, and $s_{24} = 0.4$, then after observing month 24 sales, we predict that December air conditioner sales will be 40% of the expected sales during an average month. In what follows, L_t and T_t have the same meaning as they did in Holt's method. Each period, L_t, T_t, and s_t are updated (in that order) by using Equations (12)–(14). Again, α, β, and γ are smoothing constants, each of which is between 0 and 1.

$$L_t = \alpha \frac{x_t}{s_{t-c}} + (1 - \alpha)(L_{t-1} + T_{t-1}) \tag{12}$$

$$T_t = \beta(L_t - L_{t-1}) + (1 - \beta)T_{t-1} \tag{13}$$

$$s_t = \gamma \frac{x_t}{L_t} + (1 - \gamma)s_{t-c} \tag{14}$$

Equation (12) updates the estimate of the series base by taking a weighted average of the following two quantities:

1 $L_{t-1} + T_{t-1}$, which is our base level estimate before observing x_t

2 The deseasonalized observation $\frac{x_t}{s_{t-c}}$, which is an estimate of the base obtained from the current period

Equation (13) is identical to the T_t equation (10) used to update trend in the Holt method.

Equation (14) updates the estimate of month t's seasonality by taking a weighted average of the following two quantities:

1 Our most recent estimate of month t's seasonality (s_{t-c})

2 $\frac{x_t}{L_t}$, which is an estimate of month t's seasonality, obtained from the current month

At the end of period t, the forecast ($f_{t,k}$) for month $t + k$ is given by

$$f_{t,k} = (L_t + kT_t)s_{t+k-c} \tag{15}$$

Thus, to forecast the value of the series during period $t + k$, we multiply our estimate of the period $t + k$ base ($L_t + kT_t$) by our most recent estimate of month ($t + k$)'s seasonality factor (s_{t+k-c}).

Initialization of Winter's Method

To obtain good forecasts with Winter's method, we must obtain good initial estimates of base, trend, and all seasonal factors. Let

$$L_0 = \text{estimate of base at beginning of month 1}$$
$$T_0 = \text{estimate of trend at beginning of month 1}$$
$$s_{-11} = \text{estimate of January seasonal factor at beginning of month 1} \tag{16}$$
$$s_{-10} = \text{estimate of February seasonal factor at beginning of month 1}$$
$$\vdots$$
$$s_0 = \text{estimate of December seasonal factor at beginning of month 1}$$

A variety of methods are available to estimate the parameters in (16). We choose a simple method that requires two years of data. Suppose that the last two years of sales (by month) were as follows:

$$\text{Year} -2: \quad 4, 3, 10, 14, 25, 26, 38, 40, 28, 17, 16, 13$$
$$\text{Year} -1: \quad 9, 6, 18, 27, 48, 50, 75, 77, 52, 33, 31, 24$$
$$\text{Total sales during year} - 2 = 234$$
$$\text{Total sales during year} - 1 = 450$$

We estimate T_0 by

$$T_0 = \frac{(\text{Avg. monthly sales during year} -1) - (\text{Avg. monthly sales during year} -2)}{12}$$

or

$$T_0 = \frac{\frac{450}{12} - \frac{234}{12}}{12} = 1.5$$

To estimate L_0, we first determine the average monthly demand during year $-1(\frac{450}{12})$. This estimates the base at the middle of year -1 (month 6.5 of year -1). To bring this estimate to the end of month 12 of year -1, we add $(12 - 6.5)T_0 = 5.5T_0$. Thus, our estimate of $L_0 = 37.5 + 5.5(1.5) = 45.75$.

To estimate the seasonality factor for a given month (say, January $= s_{-11}$), we take an estimate of January seasonality for year -2 and year -1 and average them. In year -2, average monthly demand was $\frac{234}{12} = 19.5$; in January of year -2, 4 air conditioners were sold. Therefore,

$$\text{Year} -2 \text{ estimate of January seasonality} = \frac{4}{19.5} = 0.205$$

Similarly,

$$\text{Year} -1 \text{ estimate of January seasonality} = \frac{9}{37.5} = 0.240$$

Finally, we obtain $s_{-11} = \frac{0.205 + 0.24}{2} = 0.22$. In similar fashion, we obtain

$$s_{-10} = 0.16, \quad s_{-9} = 0.50, \quad s_{-8} = 0.72, \quad s_{-7} = 1.28, \quad s_{-6} = 1.33,$$
$$s_{-5} = 1.97, \quad s_{-4} = 2.05, \quad s_{-3} = 1.41, \quad s_{-2} = 0.88, \quad s_{-1} = 0.82, \quad s_0 = 0.65$$

As a check, initial seasonal factor estimates should average to 1.

Before showing how (12)–(14) are used, we demonstrate how to use (15) for forecasting. At the beginning of month 1, our forecast for month 1 air conditioner sales is

$$f_{0,1} = (L_0 + T_0)s_{0+1-12} = (45.75 + 1.5)0.22 = 10.40$$

At the beginning of month 1, our forecast for month 7 air conditioner sales is

$$f_{0,7} = (L_0 + 7T_0)s_{0+7-12} = (45.75 + 7(1.5))1.97 = 110.81$$

For $\alpha = 0.5$, $\beta = 0.4$, $\gamma = 0.6$, applying Winter's method to the first 12 months of air conditioner sales data yields the results in Table 6.

We illustrate the computations by computing L_1, T_1, and s_1.

$$L_1 = 0.5\left(\frac{x_1}{s_{-11}}\right) + 0.5(L_0 + T_0) = 0.5\left(\frac{13}{0.22}\right) + 0.5(45.75 + 1.5) = 53.17$$

$$T_1 = 0.4(L_1 - L_0) + 0.6T_0 = 0.4(53.17 - 45.75) + 0.6(1.5) = 3.87$$

$$s_1 = 0.6\left(\frac{x_1}{L_1}\right) + 0.4s_{-11} = 0.6\left(\frac{13}{53.17}\right) + 0.4(0.22) = 0.23$$

TABLE 6
Winter's Method for Air Conditioners ($\alpha = 0.5$, $\beta = 0.4$, $\gamma = 0.6$)

Month	Sales	L_t	T_t	s_t	$f_{t-1,1}$	Error
1	13	53.17	3.87	0.23	10.40	2.60
2	7	50.39	1.21	0.15	9.13	−2.13
3	23	48.80	0.09	0.48	25.80	−2.80
4	32	46.67	−0.80	0.70	35.20	−3.20
5	58	45.59	−0.91	1.28	58.71	−0.71
6	60	44.90	−0.82	1.33	59.42	0.58
7	90	44.88	−0.50	1.99	86.82	3.18
8	93	44.87	−0.30	2.06	90.97	2.03
9	63	44.62	−0.28	1.41	62.84	0.16
10	39	44.33	−0.29	0.88	39.02	−0.02
11	37	44.58	−0.07	0.83	36.12	0.88
12	29	44.56	−0.05	0.65	28.93	0.07

Thus, at the end of month 1, our forecast for (say) month 7 air conditioner sales is $f_{1,6} = (L_1 + 6T_1)s_{1+6-12} = (53.17 + 6(3.87))1.97 = 150.49$. Our forecast for month 7 at the end of month 1 exceeds the forecast for month 7 made at the beginning of month 1, because month 1 sales were higher than predicted.

For all 24 months of data, spreadsheet calculations show that MAD = 10.48.

REMARKS **1** Since Winter's method uses three smoothing constants, it is quite a chore to find the combination of α, β, and γ values that yields the smallest MAD. The use of a spreadsheet to do Winter's method is discussed in Review Problem 3. The Excel Solver can aid in finding good values of α, β, and γ. Just use Solver to find parameter values that minimize MAD.

2 Although the values of α and β that minimize MAD should not exceed 0.5 (as in the Holt method), it is not uncommon for the best value of γ to exceed 0.5. This is because for monthly data, each monthly seasonal factor is updated during only $\frac{1}{12}$ of all periods. Since the seasonality factors are updated so infrequently, we may need to give more weight to each observation, so $\gamma > 0.5$ is not out of the question.

3 Figure 8 shows how well forecasts of air conditioner sales (for $\alpha = 0.5$, $\beta = 0.4$, and $\gamma = 0.6$) compare to actual air conditioner sales. The agreement between predicted and actual sales is quite good except during months 15 and 17. During these months, our forecasts are much too high. Perhaps new salespeople were hired during these two months, causing sales to be less than anticipated.

Forecasting Accuracy

For any forecasting model in which forecast errors are normally distributed, we may use MAD to estimate s_e = standard deviation of our forecast errors. The relationship between MAD and s_e is given in Formula (17).

$$s_e = 1.25 \text{ MAD} \tag{17}$$

Assuming that errors are normally distributed, we know that approximately 68% of our predictions should be within s_e of the actual value, and approximately 95% of our predictions should be within $2s_e$ of the actual value. Thus, for our air conditioner sales predictions, we find that $s_e = 1.25(10.48) = 13.10$. So we would expect that for about $0.68(24) = 16$ of 24 months, our predictions for sales would be off by at most 13.10 air conditioners, and for $0.95(24) = 23$ of 24 months, our predictions would be off by at most $2(13.10) = 26.2$ air conditioners. Actually, our predictions for air conditioner sales are accurate within 13.10 during 17 months and accurate within 26.2 during 22 months.

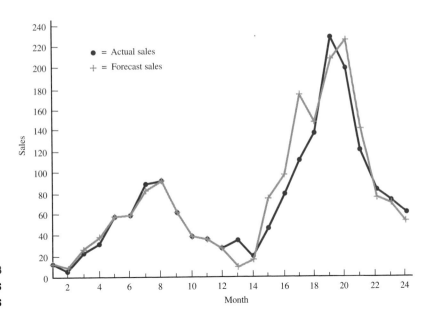

FIGURE 8
Air Conditioner Sales Predictions

We note that in most situations where a forecast is required, knowing something about the probable accuracy of the forecast is almost as important as the actual forecast. Thus, this short subsection is very important!

PROBLEMS

Group A

1 Simple exponential smoothing (with $\alpha = 0.2$) is being used to forecast monthly beer sales at Gordon's Liquor Store. After observing April's demand, the predicted demand for May is 4,000 cans of beer.

a At the beginning of May, what is the prediction for July's beer sales?

b Actual demand during May and June is as follows: May, 4,500 cans of beer; June, 3,500 cans of beer. After observing June's demand, what is the forecast for July's demand?

c The demand during May and June averages out to $\frac{4,500+3,500}{2} = 4,000$ cans per month. This is the same as the forecast for monthly sales before we observed the May and June data. Yet after observing the May and June demands for beer, our forecast for July demand has decreased from what it was at the end of April. Why?

2 We are predicting quarterly sales for soda at Gordon's Liquor Store using Winter's method. We are given the following information:

Seasonality factors: fall = 0.8   spring = 1.2

winter = 0.7   summer = 1.3

Current base estimate = 400 cases per quarter

Current trend estimate = 40 cases per quarter

$\alpha = 0.2$ $\beta = 0.3$ $\gamma = 0.5$

Now sales of 650 cases during the summer quarter are observed.

a Use this information to update the estimates of base, trend, and seasonality.

b After observing the summer demand, forecast demand for the fall quarter and the winter quarter.

3 We are using Winter's method and monthly data to forecast the GDP. (All numbers are in billions of dollars.) At the end of January 2005, $L_t = 600$ and $T_t = 5$. We are given the following seasonalities: January, 0.80; February, 0.85; December, 1.2. During February 2005, the GDP is at a level of 630. At the end of February what is the forecast for the December 2005 level of the GDP? Use $\alpha = \beta = \gamma = 0.5$.

4 We are using the Holt method to predict monthly VCR sales at Highland Appliance. At the end of October, 2005, $L_t = 200$ and $T_t = 10$. During November, 2005, 230 VCRs are sold. At the end of November, MAD = 25, and we are 95% sure that VCR sales for December, 2005 will be between _____ and _____. Use $\alpha = \beta = 0.5$.

5 We are using simple exponential smoothing to predict monthly electric shaver sales at Hook's Drug Store. At the end of October 2006, our forecast for December 2006 sales was 40. In November 50 shavers were sold, and during December 45 shavers were sold. Suppose $\alpha = 0.50$. At the end of December, 2006, what is our prediction for the total number of shavers that will be sold during March and April of 2007?

6 We are using simple exponential smoothing to predict monthly auto sales at Bloomington Ford. The company believes that sales do not exhibit trend or seasonality, so simple exponential smoothing has yielded satisfactory forecasts for the most part. Each March, however, Bloomington Ford has observed that sales tend to exceed the simple exponential smoothing forecast (A_{Feb}) by 200. Suppose that at the end of February 2004, $A_t = 600$. During March 2004, 900 cars are sold.

    a Using $\alpha = 0.3$, determine (at the end of March 2004) a forecast for April 2004 car sales.

    b Assume that at the end of March, MAD = 60. We are 95% sure that April sales will be between _____ and _____.

7 The University Credit Union is open Monday through Saturday. Winter's method is being used (with $\alpha = \beta = \gamma = 0.5$) to predict the number of customers entering the bank each day. After incorporating the arrivals of October 16, $L_t = 200$ customers, $T_t = 1$ customer, and the seasonalities are as follows: Monday, 0.90; Tuesday, 0.70; Wednesday, 0.80; Thursday, 1.1; Friday, 1.2; Saturday, 1.3. For example, this means that on a typical Monday, the number of customers is 90% of the number of customers entering the bank on an average day. On Tuesday, October 17, 182 customers enter the bank. At the close of business on October 17, make a prediction for the number of customers to enter the bank on October 25.

8 The Holt method (exponential smoothing with trend and without seasonality) is being used to forecast weekly car sales at TOD Ford. Currently, the base is estimated to be 50 cars per week, and the trend is estimated to be 6 cars per week. During the current week, 30 cars are sold. After observing the current week's sales, forecast the number of cars to be sold during the week that begins three weeks after the conclusion of the current week. Use $\alpha = \beta = 0.3$.

9 Winter's method (with $\alpha = 0.2$, $\beta = 0.1$, and $\gamma = 0.5$) is being used to forecast the number of customers served each day by Last National Bank. The bank is open Monday through Friday. At present, the following seasonalities have been estimated: Monday, 0.80; Tuesday, 0.90; Wednesday, 0.95; Thursday, 1.10; Friday, 1.25. A seasonality of 0.80 for Monday means that on a Monday, the number of customers served by the bank tends to be 80% of average. Currently, the base is estimated to be 20 customers, and the trend is estimated to equal 1 customer. After observing that on Monday 30 customers are served by the bank, predict the number of customers to be served by the bank on Wednesday.

10 We have been assigned to forecast the number of aircraft engines ordered each month by Engine Company. At the end of February, the forecast is that 100 engines will be ordered during April. During March, 120 engines are ordered.

    a Using $\alpha = 0.3$, determine (at the end of March) a forecast for the number of orders placed during April. Answer the same question for May.

    b Suppose at the end of March, MAD = 16. At the end of March, we are 68% sure that April orders will be between _____ and _____.

11 Winter's method is being used to forecast quarterly U.S. retail sales (in billions of dollars). At the end of the first

quarter, $L_t = 300$, $T_t = 30$, and the seasonal indexes are as follows: quarter 1, 0.90; quarter 2, 0.95; quarter 3, 0.95; quarter 4, 1.20. During the second quarter, retail sales are $360 billion. Assume $\alpha = 0.2$, $\beta = 0.4$, and $\gamma = 0.5$.

    a At the end of the second quarter, develop a forecast for retail sales during the fourth quarter of the year.

    b At the end of the second quarter, develop a forecast for the second quarter of the following year.

Group B

12 Simple exponential smoothing with $\alpha = 0.3$ is being used to predict sales of radios at Lowland Appliance. Predictions are made on a monthly basis. After observing August radio sales, the forecast for September is 100 radios.

    a During September, 120 radios are sold. After observing September sales, what is the prediction for October radio sales? For November radio sales?

    b It turns out that June sales were recorded as 10 radios. Actually, however, 100 radios were sold in June. After correcting for this error, what would be the prediction for October radio sales?

13 In our discussion of Winter's method, a monthly seasonality of (say) 0.80 for January means that during January, air conditioner sales are expected to be 80% of the sales during an average month. An alternative approach to modeling seasonality is to let the seasonality factor for each month represent how far above average air conditioner sales will be during the current month. For instance, if $s_{\text{Jan}} = -50$, then air conditioner sales during January are expected to be 50 less than air conditioner sales during an average month. If $s_{\text{July}} = 90$, then air conditioner sales during July are expected to be 90 more than air conditioner sales during an average month. Let

    s_t = the seasonality for month t after month t demand is observed

    L_t = the estimate of base after month t demand is observed

    T_t = the estimate of trend after month t demand is observed

Then the Winter's method equations given in the text are modified to be as follows (* indicates multiplication):

$$L_t = \alpha * (I) + (1 - \alpha) * (L_{t-1} + T_{t-1})$$
$$T_t = \beta * (L_t - L_{t-1}) + (1 - \beta) * T_{t-1}$$
$$s_t = \gamma * (II) + (1 - \gamma) * s_{t-12}$$

    a What should I and II be?

    b Suppose that month 13 is a January, $L_{12} = 30$, $T_{12} = -3$, $s_1 = -50$, and $s_2 = -20$. Let $\alpha = \gamma = \beta = 0.5$. Suppose 12 air conditioners are sold during month 13. At the end of month 13, what is the prediction for air conditioner sales during month 14?

14 Winter's method assumes a multiplicative seasonality but an additive trend. For example, a trend of 5 means that the base will increase by 5 units per period. Suppose there is actually a multiplicative trend. Then (ignoring seasonality) if the current estimate of the base is 50 and the current estimate of the trend is 1.2, we would predict demand to increase by 20% per period. Ignoring seasonality, we would thus forecast the next period's demand to be 50(1.2) and forecast the demand two periods in the future to be $50(1.2)^2$.

If we want to use a multiplicative trend in Winter's method, we should use the following equations:

$$L_t = \alpha * \left(\frac{x_t}{s_{t-c}}\right) + (1 - \alpha) * (I)$$

$$T_t = \beta * (II) + (1 - \beta) * T_{t-1}$$

$$s_t = \gamma * \left(\frac{x_t}{L_t}\right) + (1 - \gamma) * s_{t-12}$$

a Determine what I and II should be.

b Suppose we are working with monthly data and month 12 is a December, month 13 a January, and so on. Also suppose that $L_{12} = 100$, $T_{12} = 1.2$, $s_1 = 0.90$, $s_2 = 0.70$, and $s_3 = 0.95$. Suppose $x_{13} = 200$. At the end of month 13, what is the prediction for x_{15}? Assume $\alpha = \beta = \gamma = 0.5$.

15 Holt's method assumes an additive trend. For example, a trend of 5 means that the base will increase by 5 units per period. Suppose there is actually a multiplicative trend. Thus, if the current estimate of the base is 50 and the current estimate of the trend is 1.2, we would predict demand to increase by 20% per period. So we would forecast the next period's demand to be 50(1.2) and forecast the demand two periods in the future to be $50(1.2)^2$. If we want to use a multiplicative trend in Holt's method, we should use the following equations:

$$L_t = \alpha * (x_t) + (1 - \alpha) * (I)$$
$$T_t = \beta * (II) + (1 - \beta) * T_{t-1}$$

a Determine what I and II should be.

b Suppose we are working with monthly data and month 12 is a December, month 13 a January, and so on. Also suppose that $L_{12} = 100$ and $T_{12} = 1.2$. Suppose $x_{13} = 200$. At the end of month 13, what is the prediction for x_{15}? Assume $\alpha = \beta = 0.5$.

16 A version of simple exponential smoothing can be used to predict the outcome of sporting events. To illustrate, consider pro football. We first assume that all games are played on a neutral field. Before each day of play, we assume that each team has a rating. For example, if the Bears' rating is +10 and the Bengals' rating is +6, we would predict the Bears to beat the Bengals by $10 - 6 = 4$ points. Suppose the Bears play the Bengals and win by 20 points. For this observation, we "underpredicted" the Bears' performance by $20 - 4 = 16$ points. The best α for pro football is 0.10. After the game, we therefore increase the Bears' rating by $16(0.1) = 1.6$ and decrease the Bengals' rating by 1.6 points. In a rematch, the Bears would be favored by $(10 + 1.6) - (6 - 1.6) = 7.2$ points.

a How does this approach relate to the equation $A_t = A_{t-1} + \alpha(e_t)$?

b Suppose the home-field advantage in pro football is 3 points; that is, home teams tend to outscore visiting teams by an average of 3 points a game. How could the home-field advantage be incorporated into this system?

c How could we determine the best α for pro football?

d How might we determine ratings for each team at the beginning of the season?

e Suppose we tried to apply the above method to predict pro football (16-game schedule), college football (11-game schedule), college basketball (30-game schedule), and pro basketball (82-game schedule). Which sport would have the smallest optimal α? Which sport would have the largest optimal α?

f Why would this approach probably yield poor forecasts for major league baseball?

15.5 Ad Hoc Forecasting

Suppose we want to determine how many tellers a bank must have working each day to provide adequate service. In order to use the queuing models of Chapter 8 to answer this question, we need to be able to predict the number of customers who will enter the bank each day. The bank manager believes that the month of the year and the day of the week influence the number of customers entering the bank. (The bank is open Monday through Saturday, except for holidays.) Can we develop a simple forecasting model to help the bank predict the number of customers who will enter each day?

The number of customers entering the bank each day during the last year is given in Table 7. We have used 1 = Monday, 2 = Tuesday, ..., 6 = Saturday, and 7 = Sunday to denote the days of the week. A "Y" in the AH column means that the day is the day after the bank was closed for a holiday.

Let x_t = number of customers entering the bank on day t. We postulate that $x_t = B \times DW_t \times M_t \times \varepsilon_t$, where

B = base level of customer traffic corresponding to an average day

DW_t = day of the week factor corresponding to the day of the week on which day t falls

M_t = month factor corresponding to the month during which day t occurs

ε_t = random error term whose average value equals 1

TABLE 7
Arrivals to Bank

Month	Day M	Day W	Customer	AH	Forecast
1	1	1			
1	2	2	431	Y	399.13
1	3	3	271		415.88
1	4	4	362		416.51
1	5	5	696		560.10
1	6	6	315		356.32
1	7	7			
1	8	1	330		493.98
1	9	2	352		399.13
1	10	3	606		415.88
1	11	4	550		416.51
1	12	5	626		560.10
1	13	6	392		356.32
1	14	7			
1	15	1	540		493.98
1	16	2	474		399.13
1	17	3	457		415.88
1	18	4	401		416.51
1	19	5	691		560.10
1	20	6	388		356.32
1	21	7			
1	22	1	533		493.98
1	23	2	384		399.13
1	24	3	360		415.88
1	25	4	515		416.51
1	26	5	325		560.10
1	27	6	412		356.32
1	28	7			
1	29	1	592		493.98
1	30	2	366		399.13
1	31	3	512		415.88
2	1	4	476		425.33
2	2	5	531		571.97
2	3	6	303		363.87
2	4	7			
2	5	1	474		504.45
2	6	2	255		407.58
2	7	3	282		424.69
2	8	4	321		425.33
2	9	5	416		571.97
2	10	6	257		363.87
2	11	7			
2	12	1	638		504.45
2	13	2	506		407.58
2	14	3	420		424.69
2	15	4	459		425.33
2	16	5	515		571.97

(Continued)

TABLE 7

(Continued)

Month	Day M	Day W	Customer	AH	Forecast
2	17	6	501		363.87
2	18	7			
2	19	1	556		504.45
2	20	2	510		407.58
2	21	3	436		424.69
2	22	4	512		425.33
2	23	5	547		571.97
2	24	6	319		363.87
2	25	7			
2	26	1	637		504.45
2	27	2	474		407.58
2	28	3	487		424.69
2	29	4	402		425.33
3	1	5	778		574.26
3	2	6	374		365.32
3	3	7			
3	4	1	544		506.46
3	5	2	485		409.21
3	6	3	361		426.39
3	7	4	315		427.03
3	8	5	423		574.26
3	9	6	357		365.32
3	10	7			
3	11	1	649		506.46
3	12	2	351		409.21
3	13	3	405		426.39
3	14	4	404		427.03
3	15	5	483		574.26
3	16	6	411		365.32
3	17	7			
3	18	1	309		506.46
3	19	2	453		409.21
3	20	3	515		426.39
3	21	4	380		427.03
3	22	5	426		574.26
3	23	6	427		365.32
3	24	7			
3	25	1	489		506.46
3	26	2	341		409.21
3	27	3	471		426.39
3	28	4	517		427.03
3	29	5	647		574.26
3	30	6	415		365.32
3	31	7			
4	1	1	363		483.02
4	2	2	337		390.27
4	3	3	314		406.65

(Continued)

TABLE 7
(Continued)

Month	Day M	Day W	Customer	AH	Forecast
4	4	4	465		407.26
4	5	5	584		547.67
4	6	6	313		348.41
4	7	7			
4	8	1	376		483.02
4	9	2	292		390.27
4	10	3	484		406.65
4	11	4	227		407.26
4	12	5	496		547.67
4	13	6	395		348.41
4	14	7			
4	15	1	625		483.02
4	16	2	430		390.27
4	17	3	454		406.65
4	18	4	372		407.26
4	19	5	455		547.67
4	20	6	253		348.41
4	21	7			
4	22	1	432		483.02
4	23	2	469		390.27
4	24	3	392		406.65
4	25	4	467		407.26
4	26	5	684		547.67
4	27	6	349		348.41
4	28	7			
4	29	1	750		483.02
4	30	2	409		390.27
5	1	3	348		373.31
5	2	4	230		373.88
5	3	5	630		502.78
5	4	6	358		319.85
5	5	7			
5	6	1	269		443.43
5	7	2	107		358.27
5	8	3	360		373.31
5	9	4	208		373.88
5	10	5	547		502.78
5	11	6	325		319.85
5	12	7			
5	13	1	473		443.43
5	14	2	337		358.27
5	15	3	317		373.31
5	16	4	341		373.88
5	17	5	338		502.78
5	18	6	369		319.85
5	19	7			
5	20	1	618		443.43

(Continued)

TABLE **7**
(Continued)

Month	Day M	Day W	Customer	AH	Forecast
5	21	2	458		358.27
5	22	3	457		373.31
5	23	4	572		373.88
5	24	5	668		502.78
5	25	6	318		319.85
5	26	7			
5	27	1	300		443.43
5	28	2	469		358.27
5	29	3	434		373.31
5	30	4	419		373.88
5	31	5			
6	1	6	432	Y	354.08
6	2	7			
6	3	1	463		490.89
6	4	2	457		396.62
6	5	3	273		413.27
6	6	4	327		413.90
6	7	5	554		556.60
6	8	6	256		354.08
6	9	7			
6	10	1	465		490.89
6	11	2	479		396.62
6	12	3	437		413.27
6	13	4	585		413.90
6	14	5	616		556.60
6	15	6	318		354.08
6	16	7			
6	17	1	724		490.89
6	18	2	390		396.62
6	19	3	550		413.27
6	20	4	266		413.90
6	21	5	410		556.60
6	22	6	303		354.08
6	23	7			
6	24	1	514		490.89
6	25	2	353		396.62
6	26	3	397		413.27
6	27	4	539		413.90
6	28	5	411		556.60
6	29	6	413		354.08
6	30	7			
7	1	1	583		484.44
7	2	2	477		391.42
7	3	3	410		407.85
7	4	4			
7	5	5	615	Y	549.29
7	6	6	288		349.44

(Continued)

TABLE 7
(Continued)

Month	Day M	Day W	Customer	AH	Forecast
7	7	7			
7	8	1	478		484.44
7	9	2	298		391.42
7	10	3	253		407.85
7	11	4	366		408.46
7	12	5	410		549.29
7	13	6	270		349.44
7	14	7			
7	15	1	541		484.44
7	16	2	331		391.42
7	17	3	318		407.85
7	18	4	441		408.46
7	19	5	651		549.29
7	20	6	300		349.44
7	21	7			
7	22	1	608		484.44
7	23	2	401		391.42
7	24	3	390		407.85
7	25	4	391		408.46
7	26	5	619		549.29
7	27	6	391		349.44
7	28	7			
7	29	1	413		484.44
7	30	2	474		391.42
7	31	3	503		407.85
8	1	4	267		418.33
8	2	5	619		562.56
8	3	6	370		357.88
8	4	7			
8	5	1	406		496.15
8	6	2	432		400.87
8	7	3	333		417.70
8	8	4	327		418.33
8	9	5	647		562.56
8	10	6	407		357.88
8	11	7			
8	12	1	396		496.15
8	13	2	664		400.87
8	14	3	508		417.70
8	15	4	519		418.33
8	16	5	555		562.56
8	17	6	365		357.88
8	18	7			
8	19	1	492		496.15
8	20	2	420		400.87
8	21	3	360		417.70
8	22	4	469		418.33

(Continued)

TABLE 7

(Continued)

Month	Day M	Day W	Customer	AH	Forecast
8	23	5	488		562.56
8	24	6	326		357.88
8	25	7			
8	26	1	465		496.15
8	27	2	384		400.87
8	28	3	280		417.70
8	29	4	292		418.33
8	30	5	649		562.56
8	31	6	493		357.88
9	1	7			
9	2	1			
9	3	2	459	Y	391.76
9	4	3	353		408.21
9	5	4	287		408.82
9	6	5	471		549.77
9	7	6	266		349.74
9	8	7			
9	9	1	505		484.87
9	10	2	528		391.76
9	11	3	342		408.21
9	12	4	551		408.82
9	13	5	525		549.77
9	14	6	304		349.74
9	15	7			
9	16	1	479		484.87
9	17	2	258		391.76
9	18	3	263		408.21
9	19	4	450		408.82
9	20	5	540		549.77
9	21	6	297		349.74
9	22	7			
9	23	1	399		484.87
9	24	2	264		391.76
9	25	3	479		408.21
9	26	4	459		408.82
9	27	5	915		549.77
9	28	6	247		349.74
9	29	7			
9	30	1	725		484.87
10	1	2	197		390.39
10	2	3	326		406.78
10	3	4	374		407.39
10	4	5	477		547.85
10	5	6	367		348.52
10	6	7			
10	7	1	317		483.17
10	8	2	205		390.39

(Continued)

TABLE **7**
(Continued)

Month	Day M	Day W	Customer	AH	Forecast
10	9	3	519		406.78
10	10	4	483		407.39
10	11	5	489		547.85
10	12	6	345		348.52
10	13	7			
10	14	1	660		483.17
10	15	2	262		390.39
10	16	3	395		406.78
10	17	4	522		407.39
10	18	5	582		547.85
10	19	6	335		348.52
10	20	7			
10	21	1	503		483.17
10	22	2	396		390.39
10	23	3	548		406.78
10	24	4	471		407.39
10	25	5	528		547.85
10	26	6	344		348.52
10	27	7			
10	28	1	419		483.17
10	29	2	429		390.39
10	30	3	609		406.78
10	31	4	519		407.39
11	1	5	674		596.31
11	2	6	352		379.35
11	3	7			
11	4	1	360		525.91
11	5	2	500		424.92
11	6	3	339		442.76
11	7	4	326		443.43
11	8	5	459		596.31
11	9	6	255		379.35
11	10	7			
11	11	1	432		525.91
11	12	2	527		424.92
11	13	3	394		442.76
11	14	4	424		443.43
11	15	5	388		596.31
11	16	6	356		379.35
11	17	7			
11	18	1	635		525.91
11	19	2	309		424.92
11	20	3	613		442.76
11	21	4	580		443.43
11	22	5	627		596.31
11	23	6	514		379.35
11	24	7			

(Continued)

TABLE 7
(Continued)

Month	Day M	Day W	Customer	AH	Forecast
11	25	1	686		525.91
11	26	2	452		424.92
11	27	3	384		442.76
11	28	4			
11	29	5	701	Y	596.31
11	30	6	425		379.35
12	1	7			
12	2	1	291		510.06
12	3	2	407		412.12
12	4	3	458		429.42
12	5	4	243		430.06
12	6	5	449		578.34
12	7	6	315		367.91
12	8	7			
12	9	1	633		510.06
12	10	2	429		412.12
12	11	3	375		429.42
12	12	4	540		430.06
12	13	5	615		578.34
12	14	6	455		367.91
12	15	7			
12	16	1	385		510.06
12	17	2	472		412.12
12	18	3	576		429.42
12	19	4	321		430.06
12	20	5	679		578.34
12	22	7			
12	23	1	407		510.06
12	24	2	328		412.12
12	25	3			
12	26	4	491	Y	430.06
12	27	5	586		578.34
12	28	6	367		367.91
12	29	7			
12	30	1	707		510.06
12	31	2	400		412.12

To begin, we estimate B = average number of arrivals per day the bank is open = 438.33. We illustrate the estimation of the DW_t by

$$DW_t \text{ for Monday} = \frac{\text{average number of arrivals on Mondays bank is open}}{B}$$

$$= \frac{492.07}{438.33} = 1.122$$

Similarly, we find

$$DW_t \text{ for Tuesday} = 0.907$$
$$DW_t \text{ for Wednesday} = 0.945$$
$$DW_t \text{ for Thursday} = 0.947$$
$$DW_t \text{ for Friday} = 1.273$$
$$DW_t \text{ for Saturday} = 0.809$$

To estimate M_t (say, for May), we write

$$M_t \text{ for May} = \frac{\text{average number of arrivals on May day for which bank is open}}{B}$$

$$= \frac{395}{438.33} = 0.901$$

In a similar fashion, we find the M_t for the remaining months:

$$M_t \text{ for January} = 1.004$$
$$M_t \text{ for February} = 1.025$$
$$M_t \text{ for March} = 1.029$$
$$M_t \text{ for April} = 0.982$$
$$M_t \text{ for June} = 0.998$$
$$M_t \text{ for July} = 0.984$$
$$M_t \text{ for August} = 1.008$$
$$M_t \text{ for September} = 0.985$$
$$M_t \text{ for October} = 0.982$$
$$M_t \text{ for November} = 1.069$$
$$M_t \text{ for December} = 1.037$$

To illustrate how the forecasts in Table 7 were generated, consider how we would generate a forecast for the number of customers to enter the bank on Thursday, February 1, of the current year. Assuming ε_t equals its average value of 1, we would forecast $B \times (DW_t \text{ for Thursday}) \times (M_t \text{ for February}) = 438.33(0.947)(1.025) = 425.48$ customers would enter. (The difference from the printout value shown in the table is due to rounding of DW_t and M_t values.) To forecast customer arrivals for a future day (say, Saturday, February 8, of next year), we would obtain $B \times (DW_t \text{ for Saturday}) \times (M_t \text{ for February}) = 438.33 (0.809)(1.025) = 363.47$ customers.

For the data given in Table 7, our simple model yielded a MAD of 79.1. If this method were used to generate forecasts for the coming year, however, the MAD would probably exceed 79.1. This is because we have fit our parameters to past data; there is no guarantee that future data will "know" that they should follow the same pattern as past data. We have also neglected to consider whether or not an upward trend in the data is present (see Problem 3).

Suppose the bank manager observes that on the day after a holiday, bank traffic is much higher than the model predicts. The data in Table 8 indicate that this is indeed the case. How can we use this information to obtain more accurate customer forecasts for days after holidays? From Table 8, we find that the average value of Actual/Forecast for days after a holiday is 1.15. Thus, for any day after a holiday, we obtain a new forecast simply by multiplying our previous forecast by 1.15.

TABLE 8
Bank Traffic on Day after Holiday

Day after Holiday	Actual	Forecast (rounded)	Actual/Forecast
January 2	431	399	1.08
June 1	432	354	1.22
July 5	615	549	1.12
September 3	459	392	1.17
November 29	701	596	1.18
December 26	491	430	1.14

PROBLEMS

Group A

1 Suppose the bank is a college credit union and that on days when the college's professors get paid, bank traffic is much higher than usual. Assuming that college professors are paid on the first weekday of each month, how could we incorporate this fact into the forecasting procedure described in this section?

2 Suppose again that the bank is a college credit union, but now the staff gets paid every other Friday. Again, bank traffic is much higher than usual on staff paydays. How could we incorporate this fact into the forecasting procedure described in this section?

3 Suppose that the number of customers entering the bank is growing at around 20% per year. How could we incorporate this fact into the forecasting procedure described in this section?

15.6 Simple Linear Regression

Often, we try to predict the value of one variable (called the **dependent variable**) from the value of another variable (the **independent variable**). Some examples follow:

Dependent Variable	Independent Variable
Sales of product	Price of product
Automobile sales	Interest rate
Total production cost	Units produced

If the dependent variable and the independent variable are related in a linear fashion, simple linear regression can be used to estimate this relationship. In Section 15.7, we will discuss how to estimate nonlinear relationships.

To illustrate simple linear regression, let's recall the Giapetto problem (Example 1 in Chapter 3 of Volume 1). To set up this problem, we need to determine the cost of producing a soldier and the cost of producing a train. Let's suppose that we want to determine the cost of producing a train. To estimate this cost, we have observed for ten weeks the number of trains produced each week and the total cost of producing those trains. This information is given in Table 9.

The data from Table 9 are plotted in Figure 9. Observe that there appears to be a strong linear relationship between x_i (number of trains produced during week i) and y_i (cost of producing trains made during week i). The line plotted in Figure 9 appears, in a way to be made precise later, to come close to capturing the linear relationship between units produced and production cost. We will soon see how this line was chosen.

TABLE 9
Weekly Cost Data on Trains

Week	Trains Produced	Cost of Producing Trains
1	10	$257.40
2	20	$601.60
3	30	$782.00
4	40	$765.40
5	45	$895.50
6	50	$1,133.00
7	60	$1,152.80
8	55	$1,132.70
9	70	$1,459.20
10	40	$970.10

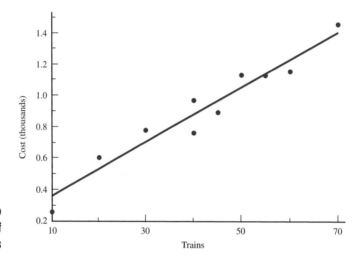

FIGURE 9
Scatterplot of Cost of Producing Trains

To begin, we model the linear relationship between x_i and y_i by the following equation:

$$y_i = \beta_0 + \beta_1 x_i + \varepsilon_i$$

where ε_i is an error term representing the fact that in a week during which x_i trains are produced, the production cost might not always equal $\beta_0 + \beta_1 x_i$. If $\varepsilon_i > 0$, the cost of producing x_i trains during week i will exceed $\beta_0 + \beta_1 x_i$, whereas if $\varepsilon_i < 0$, the cost of producing x_i trains during week i will be less than $\beta_0 + \beta_1 x_i$. However, we expect ε_i to average out to 0, so the expected cost during a week in which x_i trains are produced is $\beta_0 + \beta_1 x_i$.

The true values of β_0 and β_1 are unknown. Suppose we estimate β_0 using $\hat{\beta}_0$ and estimate β_1 using $\hat{\beta}_1$. Then our prediction for y_i (since the average value of $\varepsilon_i = 0$) is given by $\hat{y}_i = \hat{\beta}_0 + \hat{\beta}_1 x_i$.

Suppose we have data points of the form $(x_1, y_1), (x_2, y_2), \ldots, (x_n, y_n)$. How should we choose values of $\hat{\beta}_0$ and $\hat{\beta}_1$ that yield good estimates of β_0 and β_1? We select values of $\hat{\beta}_0$ and $\hat{\beta}_1$ that make our predictions $\hat{y}_i = \hat{\beta}_0 + \hat{\beta}_1 x_i$ close to the actual data points (x_i, y_i). To formalize this idea, define $e_i =$ error or residual for data point $i = $ (actual cost y_i) − (predicted cost $\hat{y}_i$) $= y_i - \hat{\beta}_0 - \hat{\beta}_1 x_i$. We now choose $\hat{\beta}_0$ and $\hat{\beta}_1$ to minimize

$$F(\hat{\beta}_0, \hat{\beta}_1) = \sum e_i^2 = \sum (y_i - \hat{\beta}_0 - \hat{\beta}_1 x_i)^2$$

The values $\hat{\beta}_0$ and $\hat{\beta}_1$ minimizing $F(\hat{\beta}_0, \hat{\beta}_1)$ are called the **least squares estimates** of β_0 and β_1. As described in Example 19 in Chapter 12 of Volume 1, we find $\hat{\beta}_0$ and $\hat{\beta}_1$ by setting

$$\frac{\partial F}{\partial \hat{\beta}_0} = \frac{\partial F}{\partial \hat{\beta}_1} = 0$$

The resulting values of $\hat{\beta}_0$ and $\hat{\beta}_1$ are given by

$$\hat{\beta}_1 = \frac{\sum (x_i - \bar{x})(y_i - \bar{y})}{\sum (x_i - \bar{x})^2} \qquad \hat{\beta}_0 = \bar{y} - \hat{\beta}_1 \bar{x} \tag{18}$$

where $\bar{x}$ = average value of all x_i's and $\bar{y}$ = average value of all y_i's.

We call $\hat{y}_i = \hat{\beta}_0 + \hat{\beta}_1 x_i$ the **least squares regression line.** Essentially, if the least squares line fits the points well (in a sense to be made more precise later), we will use $\hat{\beta}_0 + \hat{\beta}_1 x_i$ as our prediction for y_1.

Usually, the least squares line is determined by computer. Excel, Minitab, and many other popular packages will provide $\hat{\beta}_0$ and $\hat{\beta}_1$. For the sake of completeness, however, the computations needed to determine $\hat{\beta}_0$ and $\hat{\beta}_1$ for the data in Table 9 are given in Table 10, where we have used

$$\bar{x} = \frac{\sum x_i}{10} = 42 \qquad \text{and} \qquad \bar{y} = \frac{\sum y_i}{10} = 914.97$$

From Table 10 (which can easily be implemented on a spreadsheet), we find that $\sum (x_i - \bar{x})(y_i - \bar{y}) = 53,756.6$ and $\sum (x_i - \bar{x})^2 = 3,010$. From (18), we now find that

$$\hat{\beta}_1 = \frac{53,756.6}{3,010} = 17.86 \qquad \text{and} \qquad \hat{\beta}_0 = 914.97 - (17.86)42 = 164.88$$

Our least squares line is $\hat{y} = 164.88 + 17.86x$. Thus, we estimate that each extra train incurs a variable cost of $\hat{\beta}_1 = \$17.86$.

Our predictions and errors for all ten weeks are given in Table 11. To illustrate the computations, consider the first point (10,257.4). The predicted cost is $\hat{y}_1 = 164.88 + 17.86(10) = 343.5$, and the error is given by $e_1 = 257.4 - 343.5 = -86.1$.

Every least squares line has two properties:

1 It passes through the point $(\bar{x}, \bar{y})$. Thus, during a week in which Giapetto produced $\bar{x} = 42$ trains, we would predict that these trains would cost \$914.97 to produce.

TABLE 10
Computation of $\hat{\beta}_0$ and $\hat{\beta}_1$ for Train Cost Data

x_i	y_i	$x_i - \bar{x}$	$y_i - \bar{y}$	$(x_i - \bar{x})(y_i - \bar{y})$	$(x_i - \bar{x})^2$
10	257.4	−32	−657.57	21,042.24	1,024
20	601.6	−22	−313.37	6,894.14	484
30	782.0	−12	−132.97	1,595.64	144
40	765.4	−2	−149.57	299.14	4
45	895.5	3	−19.47	−58.41	9
50	1,133.0	8	218.03	1,744.24	64
60	1,152.8	18	237.83	4,280.94	324
55	1,132.7	13	217.73	2,830.49	169
70	1,459.2	28	544.23	15,238.44	784
40	970.1	−2	55.13	−110.26	4

TABLE 11
Computations of Errors

x_i	y_i	$\hat{y}_i$	e_i
10	257.4	343.5	−86.1
20	601.6	522.1	79.5
30	782.0	700.7	81.3
40	765.4	879.3	−113.9
45	895.5	968.5	−73.0
50	1,133.0	1,057.8	75.2
60	1,152.8	1,236.4	−83.6
55	1,132.7	1,147.1	−14.4
70	1,459.2	1,415	44.2
40	970.1	879.3	90.8

2 $\sum e_i = 0$. The least squares line "splits" the data points, in the sense that the sum of the vertical distances from points above the least squares line to the least squares line equals the sum of the vertical distances from points below the least squares line to the least squares line.

How Good a Fit?

How do we determine how well the least squares line fits our data points? To answer this question, we need to discuss three components of variation: **sum of squares total** (SST), **sum of squares error** (SSE), and **sum of squares regression** (SSR). Sum of squares total is given by SST $= \sum(y_i - \bar{y})^2$. SST measures the total variation of y_i about its mean $\bar{y}$. Sum of squares error is given by SSE $= \sum(y_i - \hat{y}_i)^2 = \sum e_i^2$. If the least squares line passes through all the data points, SSE $= 0$. Thus, a small SSE would indicate that the least squares line fits the data well. We define sum of squares regression to be SSR $= \sum(\hat{y}_i - \bar{y})^2$. It can be shown that

$$\text{SST} = \text{SSR} + \text{SSE} \tag{19}$$

Note that SST is a function only of the values of y. For a good fit, SSE will be small, so (19) shows that SSR will be large for a good fit. More formally, we may define the **coefficient of determination** (R^2) for y by

$$R^2 = \frac{\text{SSR}}{\text{SST}} = \text{percentage of variation in } y \text{ explained by } x$$

Equivalently, (19) allows us to write

$$1 - R^2 = \frac{\text{SSE}}{\text{SST}} = \text{percentage of variation in } y \text{ not explained by } x$$

From computer output, we find that SST $= 1,021,762$ and SSE $= 61,705$. Then (19) yields SSR $=$ SST $-$ SSE $= 960,057$. Thus, we find that $R^2 = \frac{960,057}{1,021,762} = 0.94$. This means that the number of trains produced during a week explains 94% of the variation in the weekly cost of producing trains. All other factors combined can explain at most 6% of the variation in weekly cost, so we can be quite sure that the linear relationship between x and y is strong.

A measure of the linear association between x and y is the **sample linear correlation** r_{xy}. A sample correlation near $+1$ indicates a strong positive linear relationship between

x and y; a sample correlation near -1 indicates a strong negative linear relationship between x and y; and a sample correlation near 0 indicates a weak linear relationship between x and y.

By the way, if $\hat{\beta}_1 \geq 0$, then r_{xy} equals $+\sqrt{R^2}$, whereas if $\hat{\beta}_1 \leq 0$, the sample correlation between x and y is given by $-\sqrt{R^2}$. Thus, in our cost example, $r_{xy} = \sqrt{0.94} = 0.97$, indicating a strong linear relationship between x and y.

Forecasting Accuracy

A measure of the accuracy of predictions derived from regression is given by the **standard error of the estimate** (s_e). If we let $n =$ number of observations, s_e is given by

$$s_e = \sqrt{\frac{\text{SSE}}{n-2}}$$

For our example,

$$s_e = \sqrt{\frac{61{,}705}{10-2}} = 87.8$$

It is usually true[†] that approximately 68% of the values of y will be within s_e of the predicted value $\hat{y}$, and 95% of the values of y will be within $2s_e$ of the predicted value $\hat{y}$. In the current example, we expect that 68% of our cost estimates will be within \$87.80 of the true cost, and 95% will be within \$175.60. In actuality, for 80% of our data points, actual cost is within s_e of the predicted cost, and for 100% of our data points, actual cost is within $2s_e$ of the predicted cost.

Any observation for which y is not within $2s_e$ of $\hat{y}$ is called an **outlier.** Outliers represent unusual data points and should be carefully examined. Of course, if an outlier is the result of a data entry error, it should be corrected. If an outlier is in some way uncharacteristic of the remaining data points, it may be better to omit the outlier and re-estimate the least squares line. Since all the errors are smaller than $2s_e$ in absolute value, there are no outliers in our cost example.

t-Tests in Regression

Using a *t*-**test,** we can test the significance of a linear relationship. To test H_0: $\beta_1 = 0$ (no significant linear relationship between x and y) against H_a: $\beta_1 \neq 0$ (significant linear relationship between x and y) at a level of significance α, we compute the *t*-statistic given by

$$t = \frac{\hat{\beta}_1}{\text{StdErr}(\hat{\beta}_1)}$$

[†]Actually, approximately 68% of the points should be within

$$s_e \sqrt{1 + \frac{1}{n} + \frac{(x - \bar{x})^2}{\sum(x_i - \bar{x})^2}}$$

of $\hat{y}$, and 95% of the points should be within

$$2s_e \sqrt{1 + \frac{1}{n} + \frac{(x - \bar{x})^2}{\sum(x_i - \bar{x})^2}}$$

of $\hat{y}$.

TABLE 12

Percentage Points of the *t*-Distribution†

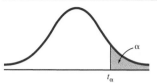

df	*a* = 0.1	*a* = 0.05	*a* = 0.025	*a* = 0.01	*a* = 0.005
1	3.078	6.314	12.706	31.821	63.657
2	1.886	2.920	4.303	6.965	9.925
3	1.638	2.353	3.182	4.541	5.841
4	1.533	2.132	2.776	3.747	4.604
5	1.476	2.015	2.571	3.365	4.032
6	1.440	1.943	2.447	3.143	3.707
7	1.415	1.895	2.365	2.998	3.499
8	1.397	1.860	2.306	2.896	3.355
9	1.383	1.833	2.262	2.821	3.250
10	1.372	1.812	2.228	2.764	3.169
11	1.363	1.796	2.201	2.718	3.106
12	1.356	1.782	2.179	2.681	3.055
13	1.350	1.771	2.160	2.650	3.012
14	1.345	1.761	2.145	2.624	2.977
15	1.341	1.753	2.131	2.602	2.947
16	1.337	1.746	2.120	2.583	2.921
17	1.333	1.740	2.110	2.567	2.898
18	1.330	1.734	2.101	2.552	2.878
19	1.328	1.729	2.093	2.539	2.861
20	1.325	1.725	2.086	2.528	2.845
21	1.323	1.721	2.080	2.518	2.831
22	1.321	1.717	2.074	2.508	2.819
23	1.319	1.714	2.069	2.500	2.807
24	1.318	1.711	2.064	2.492	2.797
25	1.316	1.708	2.060	2.485	2.787
26	1.315	1.706	2.056	2.479	2.779
27	1.314	1.703	2.052	2.473	2.771
28	1.313	1.701	2.048	2.467	2.763
29	1.311	1.699	2.045	2.462	2.756
30	1.310	1.697	2.042	2.457	2.750
40	1.303	1.684	2.021	2.423	2.704
60	1.296	1.671	2.000	2.390	2.660
120	1.289	1.658	1.980	2.358	2.617
240	1.285	1.651	1.970	2.342	2.596
inf.	1.282	1.645	1.960	2.326	2.576

†Computed by P. J. Hildebrand. Reprinted with permission of PWS-KENT Publishing Company.

StdErr($\hat{\beta}_1$) measures the uncertainty in our estimate of β_1; it can usually be found on a computer printout. We reject H_0 if $|t| \geq t_{(\alpha/2,n-2)}$, where $t_{(\alpha/2,n-2)}$ is obtained from Table 12. For our cost example, StdErr(β_1) = 1.6 (found from a computer printout), so $t = \frac{17.86}{1.6} = 11.16$. Using $\alpha = 0.05$, we find $t_{(.025,8)} = 2.306$, so we reject H_0 and again conclude that there is a strong linear relationship between x and y.

Assumptions Underlying the Simple Linear Regression Model

Statistical analysis of the simple linear regression model requires that the following assumptions hold.

Assumption 1

The variance of the error term should not depend on the value of the independent variable x. This assumption is called **homoscedasticity.** If the variance of the error term depends on x, then we say that **heteroscedasticity** is present. To see whether the homoscedasticity assumption is satisfied, we plot the errors on the y-axis and the value of x on the x-axis. Figure 10 illustrates a situation where the homoscedasticity assumption is satisfied; the figure indicates no tendency for the size of the errors to depend on x. In Figure 11, however, the magnitude of the errors tends to increase as x increases. This is an example of heteroscedasticity. Using ln y or $y^{1/2}$ as the dependent variable will often eliminate heteroscedasticity.

Assumption 2

Errors are normally distributed. This assumption is not of vital importance, so we will not discuss it further.

Assumption 3

The errors should be independent. This assumption is often violated when data are collected (as in our example) over time. Independence of the errors implies that knowing the value of one error should tell us nothing about the value of the next (or any other) error.

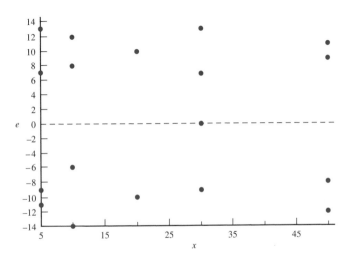

FIGURE 10
Homoscedasticity

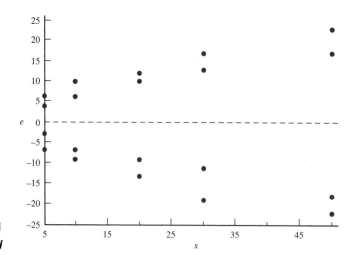

FIGURE 11
Heteroscedasticity

The validity of this assumption can be checked by plotting the errors in time-series sequence. In Figure 12, we find that the errors had the following signs: $+ + + + + + -$ $- - - - -$. This sequence of errors exhibits the following pattern: a positive error (corresponding to underprediction of the actual value of y) is usually followed by another positive error, and a negative error (corresponding to overprediction of the actual value of y) is usually followed by another negative error. This pattern indicates that successive errors are not independent; it is referred to as **positive autocorrelation.** In other words, positive autocorrelation indicates that successive errors have a positive linear relationship and are not linearly independent. If the sequence of errors in time sequence resembles Figure 13, we have **negative autocorrelation.** Here, the sequence of errors is $+ - + - + - + -$ $+ - + -$. This indicates that a positive error tends to be followed by a negative error, and vice versa. The conclusion is that successive errors have a negative linear relationship and are not independent. In Figure 14, we have the following sequence of errors: $+ + -$ $+ + - + - + + + -$. Here, no obvious pattern is present, and the independence assumption appears to be satisfied. Observe that the errors "average out" to 0, so we would expect about half our errors to be positive and half to be negative. Thus, if there is no pattern in the errors, we would expect the errors to change sign about half the time. This observation enables us to formalize the preceding discussion as follows.

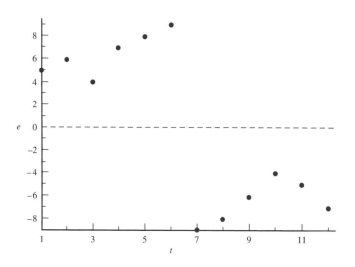

FIGURE 12
Positive Autocorrelation

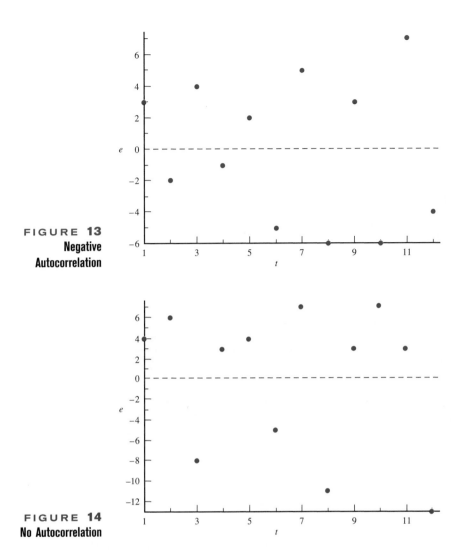

FIGURE 13
Negative
Autocorrelation

FIGURE 14
No Autocorrelation

1 If the errors change sign very rarely (much less than half the time), they probably violate the independence assumption, and positive autocorrelation is probably present.

2 If the errors change sign very often (much more than half the time), they probably violate the independence assumption, and negative autocorrelation is probably present.

3 If the errors change sign about half the time, they probably satisfy the independence assumption.

If positive or negative autocorrelation is present, correcting for the autocorrelation will often result in much more accurate forecasts. See pages 215–221 of Pindyck and Rubinfeld (1989) for details.

Running Regressions with Excel

Cost.xls

Figure 15 (file Cost.xls) illustrates how to run a regression with Excel. We have input the data from Table 9 in the cell range A2:B11 and then invoked the Tools Data Analysis

	A	B	C	D	E	F	G	H
1	trains	cost	yihatr	e				
2	10	257.4	9.32060032	248.0794	COST	REGRESSION		
3	20	601.6	18.6412006	582.958799	EXAMPLE			
4	30	782	27.961801	754.038199				
5	40	765.4	37.2824013	728.117599				
6	45	895.5	41.9427014	853.557299				
7	50	1133	46.6030016	1086.397				
8	60	1152.8	55.9236019	1096.8764				
9	55	1132.7	51.2633018	1081.4367				
10	70	1459.2	65.2442022	1393.9558				
11	40	970.1	37.2824013	932.817599				
12								
13								
14								
15		SUMMARY OUTPUT						
16								
17		*Regression Statistics*						
18		Multiple R	0.9693343					
19		R Square	0.9396089					
20		Adjusted R Squ	0.93206					
21		Standard Error	87.824643					
22		Observations	10					
23								
24		ANOVA						
25			*df*	*SS*	*MS*	*F*	*Significance F*	
26		Regression	1	960057.16	960057.16	124.4698887	3.72837E-06	
27		Residual	8	61705.344	7713.168			
28		Total	9	1021762.5				
29								
30			*Coefficients*	*Standard Err*	*t Stat*	*P-value*	*Lower 95%*	*Upper 95%*
31		Intercept	164.87791	72.743329	2.2665708	0.05317426 4	-2.8686199	332.62443
32		trains	17.859336	1.6007855	11.156607	3.72837E-06	14.1679151 4	21.550756
33								
34								
35								
36								

FIGURE 15

Regression Command.[†] Fill in the dialog box as shown in Figure 16. The Y range B1:B11 contains the name of the dependent variable and the values of the dependent variable. The X range A1:A11 contains the name of the independent variable and the values of the independent variable. Since the first row of the X and Y ranges include labels, we checked the Labels box. We checked cell B15 as the upper left-hand corner of the Output Range. We did not check the Residuals box. If we had, we would have obtained the predicted value and residual for each observation. Figure 15 shows the results of the regression.

Let's examine what the important numbers in the output mean. (We omit discussion of the portions of the output that are irrelevant to our discussion of regression.)

R Square   This is $r^2 = .939609$.

Multiple R   This is the square root of r^2, with the sign of Multiple R being the same as the slope of the regression line.

Standard Error   This is $s_e = 87.82$.

Observations   This is the number of data points (10).

[†]If the Analysis Tool Pak does not show up when you select Tools Data, go to Tools Add Ins and check the Analysis Tool Pak and Analysis Tool Pak Vba boxes.

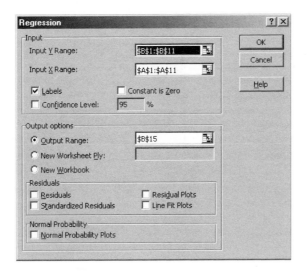

FIGURE 16

SS column The Regression entry (96,057.16) is SSR. The Residual entry (61,705.34) is SSE. The Total entry (1,021,762.5) is SST.

Coefficients column The Intercept entry (164.88) gives the value of $\hat{\beta}_0 = 164.88$, and the trains entry (17.86) gives the value of $\hat{\beta}_1 = 17.86$.

t stat This gives the observed t-statistic (coefficient/standard error) for the Intercept and the Trains variable.

Standard Error column The Intercept entry gives the standard error $\hat{\beta}_0 = 72.74$, and the Trains entry gives the standard error $\hat{\beta}_1 = 1.60$. The coefficient entry divided by the standard error entry yields the t-statistic for the intercept or slope (tabulated in the next column).

P-value For the intercept and slope, this gives Probability($|t_{n-2}| \geq$ |Observed t-statistic|). If, for example, the p-value for Trains is less than α, we reject H_0: $\beta_1 = 0$; otherwise, we accept $\beta_1 = 0$. For $\alpha = .05$, we reject $\beta_1 = 0$. For p-value $= .05$, it is borderline whether or not to accept the hypothesis that $\beta_0 = 0$.

In cell C2, we obtain $\hat{y}_1$ by inputting the formula =D$14+A2*C$20. In cell D2, we obtain e_1 by inputting the formula =B2−C2. Copying from the range C2:D2 to C2:D11 creates predictions and errors for all observations.

Obtaining a Scatterplot with Excel

To obtain a scatterplot with Excel, let the range where your independent variable is be the X range. Then let the range where your dependent variable is be the Y range. Then select X-Y Graph.

15.7 Fitting Nonlinear Relationships

Often, a plot of points of the form (x_i, y_i) indicates that y is not a linear function of x. In such cases, however, the plot may indicate that there is a nonlinear relationship between

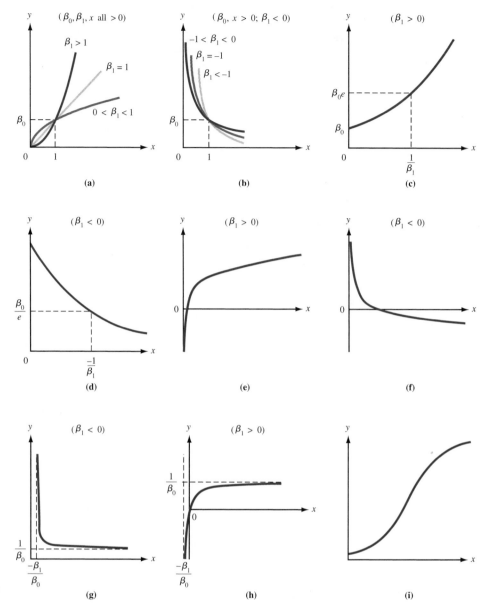

FIGURE 17
Graphs of Linearizable Functions†

†Reprinted by permission from C. Daniel and F. Wood, Fitting Functions to Data, Copyright 1980, John Wiley and Sons.

x and y. For example, if the plot of the (x_i, y_i) looks like any of parts (a)–(i) of Figure 17, a nonlinear relationship between x and y is indicated.

The following procedure may be used to estimate a nonlinear relationship.

Step 1 Plot the points and find which part of Figure 17 best fits the data. For illustrative purposes, suppose the data look like part (c).

Step 2 The second column of Table 13 gives the functional relationship between x and y. For part (c), this would be $y = \beta_0 \exp(\beta_1 x)$.

Step 3 Transform each data point according to the rules in the third column of Table 13. Thus, if part (c) of the figure is relevant, we transform each value of y into $\ln y$ and trans-

TABLE 13
How to Fit a Nonlinear Relationship

If Graph Looks Like Figure 13 Part	We Have the Functional Relationship	Transform (x_i, y_i) into	Estimate of Functional Relationship
(a) or (b)	$y = \beta_0 x^{\beta_1}$	$(\ln x_i, \ln y_i)$	$\hat{y} = \exp(\hat{\beta} + s_e^2/2)x^{\hat{\beta_1}}$
(c) or (d)	$y = \beta_0 \exp(\beta_1 x)$	$(x_i, \ln y_i)$	$\hat{y} = \exp(\hat{\beta}_0 + \hat{\beta}_1 x + s_e^2/2)$
(e) or (f)	$y = \beta_0 + \beta_1(\ln x)$	$(\ln x_i, y_i)$	$\hat{y} = \hat{\beta}_0 + \hat{\beta}_1(\ln x)$
(g) or (h)	$y = \dfrac{x}{\beta_0 x + \beta_1}$	$\left(\dfrac{1}{x_i}, \dfrac{1}{y_i}\right)$	$\hat{y} = \dfrac{x}{\hat{\beta}_0 x + \hat{\beta}_1}$
(i)	$y = \exp\left(\beta_0 + \dfrac{\beta_1}{x}\right)$	$\left(\dfrac{1}{x_i}, \ln y_i\right)$	$\hat{y} = \exp(\hat{\beta}_0 + \dfrac{\hat{\beta}_1}{x} + s_e^2/2)$

form each value of x into x. Given the relationship in the second column of Table 13, the transformed data in Table 13 should, if plotted, indicate a straight-line relationship. For part (c), for example, if $y = \beta_0 \exp(\beta_1 x)$, then taking natural logarithms of both sides yields $\ln y = \ln(\beta_0) + \beta_1 x$, so there is indeed a linear relationship between x and $\ln y$.

Step 4 Estimate the least squares regression line for the transformed data. If $\hat{\beta}_0$ is the intercept of the least squares line (for transformed data), $\hat{\beta}_1$ is the slope of the least squares line (for transformed data), and s_e is the standard error of the regression estimate, then we read the estimated relationship from the final column of Table 13. Thus, if part (c) were relevant, we would estimate that $\hat{y} = \exp(\hat{\beta}_0 + \hat{\beta}_1 x + s_e^2/2)$.

To illustrate the idea, suppose we want to predict future VCR sales for an appliance store. Sales for the last 24 months are given in Table 14 and are plotted in Figure 18 (where each dot indicates actual sales). We will use x = number of the month as the independent variable. Figure 18 indicates an S-shaped relationship between x = number of the month and y = sales during the month (like part (i) of Figure 17). So according to Table 13,

$$y = \exp\left(\beta_0 + \frac{\beta_1}{x}\right)$$

Following the third column of Table 13, we now estimate the least squares regression line for the points $(\frac{1}{1}, \ln 23)$, $(\frac{1}{2}, \ln 156)$, ..., $(\frac{1}{24}, \ln 3{,}495)$. We find $\hat{\beta}_0 = 8.387$, $s_e = .276$, and $\hat{\beta}_1 = -5.788$. From the last column of Table 13, we obtain the estimated relationship between x and y:

$$\hat{y} = \exp\left(8.387 + .5(.276)^2 - \frac{5.788}{x}\right)$$

$$= \exp\left(8.425 - \frac{5.788}{x}\right)$$

To illustrate how this formula could be used to predict future sales, suppose we want to predict VCR sales during month 26. For $x = 26$, we would forecast that

$$y = \exp\left(8.425 - \frac{5.788}{26}\right) = 3{,}649.3 \text{ VCRs}$$

would be sold.

REMARKS   **1**   For the regression on the points $(\frac{1}{1}, \ln 23)$, $(\frac{1}{2}, \ln 156)$, ..., $R^2 = 0.95$. This means that 95% of the variation in $\ln y$ is explained by variation in $\frac{1}{x}$. Unfortunately, this does not tell us anything about how accurate our predictions of actual sales (y) are likely to be. For this, we compute the

TABLE **14**
VCR Sales

Month	Sale of VCRs
1	23
2	156
3	330
4	482
5	1,209
6	1,756
7	2,000
8	2,512
9	2,366
10	2,942
11	2,872
12	2,937
13	3,136
14	3,241
15	3,149
16	3,524
17	3,542
18	3,312
19	3,547
20	3,376
21	3,375
22	3,403
23	3,697
24	3,495

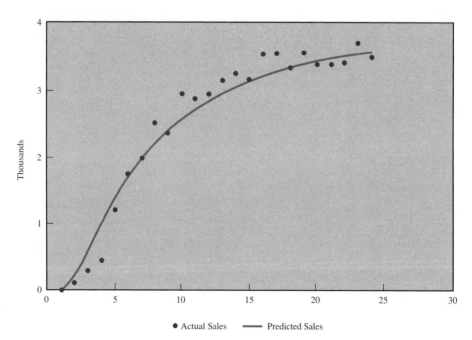

FIGURE 18
VCR Sales

predicted sales for each month, and $e_i =$ (actual month i sales) $-$ (predicted month i sales). By averaging $|e_i|$ for the 24 months, we find the MAD of our predictions to be 170.3. By applying (17), we may estimate the standard deviation of our forecasts to be 1.25(170.3) = 212.88. Thus, 95% of the time, we would expect our predictions for VCR sales to be accurate within 2(212.88) = 425.76 VCRs.

2 If we had mistakenly tried to fit a straight line to this data, we would have obtained $s_e = 546$, so fitting the S-shaped curve has greatly improved our forecasts.

Using a Spreadsheet to Fit a Nonlinear Relationship

VCR.xls

Figure 19 (file VCR.xls) shows how we can use a spreadsheet to fit a curve to the data in Table 14. We input the data in the cell range A2:B26. In columns C and D, we create the

FIGURE 19

A	A	B	C	D	E	F	G	H
1		MAD=	170.2927		VCR	EXAMPLE		
2	MONTH	SALES	1/MONTH	LNSALES	PREDICT	ERROR	ABSERR	
3	1	23	1	3.1354942	13.969974	9.0300255	9.0300255	
4	2	156	0.5	5.049856	252.37307	-96.37307	96.373067	
5	3	330	0.3333333	5.7990927	662.20451	-332.2045	332.20451	
6	4	482	0.25	6.1779441	1072.6714	-590.6714	590.6714	
7	5	1209	0.2	7.0975489	1432.6874	-223.6874	223.6874	
8	6	1756	0.1666667	7.4707938	1737.5658	18.434166	18.434166	
9	7	2000	0.1428571	7.6009025	1994.303	5.6969922	5.6969922	
10	8	2512	0.125	7.8288345	2211.4573	300.54274	300.54274	
11	9	2366	0.1111111	7.768956	2396.5747	-30.57475	30.574747	
12	10	2942	0.1	7.9868449	2555.765	386.23499	386.23499	
13	11	2872	0.0909091	7.9627639	2693.8455	178.15445	178.15445	
14	12	2937	0.0833333	7.9851439	2814.5945	122.4055	122.4055	
15	13	3136	0.0769231	8.0507034	2920.9846	215.01543	215.01543	
16	14	3241	0.0714286	8.0836372	3015.3711	225.62887	225.62887	
17	15	3149	0.0666667	8.0548402	3099.6364	49.363637	49.363637	
18	16	3524	0.0625	8.167352	3175.2979	348.70211	348.70211	
19	17	3542	0.0588235	8.1724468	3243.5904	298.40965	298.40965	
20	18	3312	0.0555556	8.1053075	3305.5268	6.47315	6.47315	
21	19	3547	0.0526316	8.1738575	3361.9455	185.05446	185.05446	
22	20	3376	0.05	8.1244469	3413.5452	-37.54522	37.545218	
23	21	3375	0.047619	8.1241506	3460.9127	-85.91273	85.912726	
24	22	3403	0.0454545	8.1324127	3504.5442	-101.5442	101.54423	
25	23	3697	0.0434783	8.215277	3544.8619	152.13809	152.13809	
26	24	3495	0.0416667	8.1590887	3582.2271	-87.22711	87.227114	
27								
28								
29						Regression Output:		
30					Constant			8.3867886
31					Std Err of Y Est			0.2761082
32					R Squared			0.9527748
33					No. of Observations			24
34					Degrees of Freedom			22
35								
36					X Coefficient(s)		-5.787996	
37					Std Err of Coef.		0.2747317	
38								
39								
40								
41								

transformed variables 1/MONTH and LNSALES. In C3, we input the formula 1/A3. In D3, we input the formula =LN(B3). Copying from the range C3:D3 to C3:D26 yields the transformed values of x_i and y_i. We now run a regression with the X range C3:C26 and the Y range D3:D26. The results of this regression are used to predict VCR sales in each month. In cell E3, we enter the formula =EXP(C\$47+C\$48/A3+.5(C\$37^2)) to generate a forecast for month 1 VCR sales. In cell F3, we determine e_1 with the formula =B3−E3. In cell G3, we determine $|e_1|$ with the formula =ABS(F3). Copying from the range E3:G3 to the range E3:G26 generates forecasts and errors for all 24 months. In cell C1, we compute the MAD for all 24 months with the formula =AVERAGE(G3:G26).

Utilizing the Excel Trend Curve

The Excel Trend Curve makes it easy to fit an equation to a set of data. After creating an X-Y scatterplot, click on the points in the graph until the points turn gold. Then select Chart Add Trendline. See Figure 20.

- Choosing Linear yields the straight line that best fits the points.
- Choose Logarithmic if the scatterplot looks like (e) or (f) in Figure 17. Then Excel yields the best-fitting equation of the form $y = \beta_0 + \beta_1(\ln(x))$.
- Choose Power if the scatterplot looks like (a) or (b) in Figure 17. Then Excel yields the best-fitting equation of the form $y = \beta_0 x^{\beta_1}$.
- Choose Exponential if the scatterplot looks like (c) or (d) in Figure 17. Then Excel yields the best-fitting equation of the form $y = \beta_0 e^{\beta_1 x}$.
- Choosing Polynomial of order n ($n = 1, 2, 3, 4, 5,$ or 6) yields the best-fitting equation of the form $y = \beta_0 + \beta_1 x + \beta_2 x^2 + \cdots + \beta_n x^n$

Before having Trend Curve fit the curve, select Options and select Display Equation on Chart and Display R^2 on Chart. The R^2 value displayed is the R^2 associated with the linear regression based on the transformed (x_i, y_i) listed in the third column of Table 13. For the Linear, Polynomial, and Logarithmic options, choosing Intercept = 0 will set $\beta_0 = 0$.

Figure 21 shows the results obtained from Trend Curve when applied to find the best-fitting straight line for the data in worksheet Cost.xls. Note that the R^2 and equation estimates match those we obtained from the Analysis Tool Pak.

Cost.xls

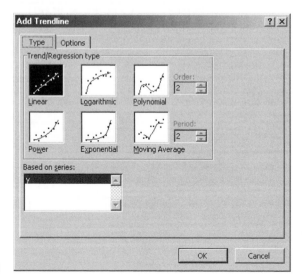

FIGURE 20

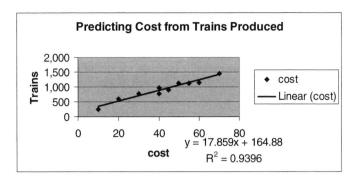

FIGURE 21

15.8 Multiple Regression

In many situations, more than one independent variable may be useful in predicting the value of a dependent variable. We then use **multiple regression.** For example, to predict the monthly sales for a national fast-food chicken chain, we might consider using the following independent variables: national income, price of chicken, dollars spent on advertising during the current month, and dollars spent on advertising during the previous month.

Suppose we are using k independent variables to predict the dependent variable y and we have n data points of the form $(y_i, x_{1i}, x_{2i}, \ldots, x_{ki})$, where x_{ji} = value of jth independent variable for ith data point and y_i = value of dependent variable for ith data point. In multiple regression, we model the relationship between y and the k independent variables by

$$y_i = \beta_0 + \beta_1 x_{1i} + \beta_2 x_{2i} + \cdots + \beta_k x_{ki} + \varepsilon_i$$

where ε_i is an error term with mean 0, representing the fact that the actual value of y_i may not equal $\beta_0 + \beta_1 x_{1i} + \beta_2 x_{2i} + \cdots + \beta_k x_{ki}$. β_j may be thought of as the increase in y if the value of the jth independent variable is increased by 1 and all other independent variables are held constant. Thus, β_j is analogous to $\frac{\partial y}{\partial x_j}$, where x_j is the jth independent variable.

Estimation of the β_i's

Suppose we estimate β_i ($i = 0, 1, 2, \ldots, k$) using $\hat{\beta}_i$. Then our prediction or estimate for y_i is given by

$$\hat{y}_i = \hat{\beta}_0 + \hat{\beta}_1 x_{1i} + \hat{\beta}_2 x_{2i} + \cdots + \hat{\beta}_k x_{ki}$$

As in Section 15.6, we define $e_i = y_i - \hat{y}_i$ and choose $\hat{\beta}_0, \hat{\beta}_1, \ldots, \hat{\beta}_k$ to minimize Σe_i^2. Usually, these least squares estimates of $\beta_0, \beta_1, \ldots, \beta_k$ will be obtained from a computer package such as Minitab or Excel. We call

$$\hat{y}_i = \hat{\beta}_0 + \hat{\beta}_1 x_{1i} + \hat{\beta}_2 x_{2i} + \cdots + \hat{\beta}_k x_{ki}$$

the **least squares regression equation.**

EXAMPLE 1   **Truck Maintenance**

We want to predict maintenance expense (y) for a truck during the current year, from the independent variables x_1 = miles driven (in thousands) during the current year and x_2 =

TABLE **15**
Truck Maintenance Data

y	x_1	x_2
$832	6	8
$733	7	7
$647	9	6
$553	11	5
$467	13	4
$373	15	3
$283	17	2
$189	18	1
$96	19	0

TABLE **16**
Computer Output for Example 1

Variable	Coefficient	Standard Error	t-Value
Constant	17.73846	31.0271	0.57171
x_1	4.061538	1.56742	2.59123
x_2	98.50769	2.756428	35.73744

Standard error of estimate = 2.106157

age of the truck (in years) at the beginning of the current year. We are given the information in Table 15.

Solution   Computer output for this example is given in Table 16. Reading down the Coefficient column and rounding to two decimal places, we obtain $\hat{\beta}_0 = 17.74$, $\hat{\beta}_1 = 4.06$, and $\hat{\beta}_2 = 98.51$. Thus, we would predict annual maintenance cost for a truck from

$$\hat{y} = 17.74 + 4.06x_1 + 98.51x_2 \tag{20}$$

For a five-year-old truck that is driven 10,000 miles during a year, we predict annual maintenance costs by $17.74 + 4.06(10) + 98.51(5) = \$550.89$.

From (20), we conclude that (holding the age of the truck constant) driving an extra thousand miles during a year increases annual maintenance costs by $\hat{\beta}_1 = \$4.06$, and that an increase of one year in the age of the truck (holding miles driven constant) increases annual maintenance costs by $\hat{\beta}_2 = \$98.51$.

Goodness of Fit Revisited

For multiple regression, we define SSR, SSE, and SST as we did in Section 15.6. We also find that $R^2 = \frac{\text{SSR}}{\text{SST}}$ = percentage of variation in y explained by the k independent variables and $1 - R^2$ = percentage of variation in y not explained by the k independent variables. If we define the standard error of the estimate as

$$s_e = \sqrt{\frac{\text{SSE}}{(n - k - 1)}}$$

then (as in Section 15.6) we expect approximately 68% of the y-values to be within s_e of $\hat{y}$ and approximately 95% of the y-values to be within $2s_e$ of $\hat{y}$. We have already seen from Table 16 that $s_e = 2.106$. Thus, 95% of the time, we expect our predictions for annual truck maintenance expenditures to be accurate within $4.21.

Hypothesis Testing

If we have included independent variables $x_1, x_2, \ldots, x_k$ in a multiple regression, we often want to test

$H_0\colon \beta_i = 0$ (x_i does not have a significant effect on y when the other independent variables are included in the regression equation)

against

$H_a\colon \beta_i \neq 0$ (x_i does have a significant effect on y when the other independent variables are included in the regression equation)

To test these hypotheses, we compute

$$t = \frac{\hat{\beta}_i}{\text{StdErr}(\hat{\beta}_i)}$$

where $\text{StdErr}(\hat{\beta}_i)$ measures the amount of uncertainty present in our estimate of β_i. $\text{StdErr}(\hat{\beta}_i)$ (and often the t-statistic) is read from computer output. At a level of significance α, we reject H_0 if $|t| > t_{(\alpha/2, n-k-1)}$. From Table 16, we find that (t for x_1) = 2.59 and (t for x_2) = 35.74. Suppose $\alpha = 0.05$. Since $t_{(.025, 9-2-1)} = 2.447$, we reject H_0 for each independent variable and conclude that both miles driven and age of the truck have a significant effect on annual maintenance cost.

Usually, variables included in a regression equation should have significant t-statistics. ($\alpha = 0.10$ or $\alpha = 0.05$ are commonly used levels of significance in regression analysis.) If an independent variable has an insignificant t-statistic, we usually remove the independent variable from the equation and obtain new least squares estimates. To illustrate this idea, suppose we have the data shown in Table 17 on sales at the Bloomington Happy Chicken Restaurant during the last 20 years (file Chicken.xls). (POP = population within 10 miles of Happy Chicken Restaurant, AD = thousands of dollars spent on advertising during the current year, LAGAD = thousands of dollars spent on advertising during the previous year, and SALES = sales in thousands of dollars.)

We attempt to estimate the model

$$\text{SALES} = \beta_0 + \beta_1\text{YEAR} + \beta_2\text{POP} + \beta_3\text{AD} + \beta_4\text{LAGAD} + \varepsilon$$

We are using YEAR as an independent variable in the hopes of picking up a possible upward trend in sales. LAGAD is used as an independent variable because we believe that last year's advertising might affect this year's sales. We obtain the following estimated regression equation (t-statistics for each independent variable are in parentheses):

$$\widehat{\text{SALES}} = 10{,}951.51 + 169.51\ \text{YEAR} - .059\ \text{POP} + 122.38\ \text{AD} + 276.93\ \text{LAGAD}$$
$$\qquad\qquad\qquad (1.91) \qquad\quad (-.70) \qquad\quad (13.84) \qquad\quad (28.92)$$

(21)

We cannot use the first year of data, since LAGAD is undefined. Since $t_{(.05, 19-4-1)} = 1.761$, we find that all independent variables except for POP are significant for $\alpha = 0.10$. Thus, YEAR, AD, and LAGAD appear to have a significant effect on sales. After

TABLE **17**
Happy Chicken Sales Data

Year	POP	AD	LAGAD	Sales
1	96,020	30	—	13,000
2	102,558	20	30	15,713
3	101,792	15	20	12,937
4	104,347	25	15	12,872
5	106,180	30	25	16,227
6	106,562	15	30	15,388
7	105,209	25	15	13,180
8	109,185	35	25	17,199
9	109,976	40	35	20,674
10	110,659	20	40	20,350
11	111,844	25	20	14,444
12	111,576	35	25	17,530
13	113,784	5	35	16,711
14	112,482	12	5	9,715
15	116,487	16	12	12,248
16	117,316	21	16	13,856
17	117,830	22	21	15,285
18	118,148	24	22	15,620
19	118,481	26	24	17,158
20	121,069	28	26	17,800

dropping the insignificant variable POP from the equation, we obtain the following estimated regression equation:

$$\widehat{\text{SALES}} = 5150.94 + 108.58 \text{ YEAR} + 121.59 \text{ AD} + 274.30 \text{ LAGAD}$$
$$\qquad\qquad\quad (8.29) \qquad\qquad (14.10) \qquad\qquad (31.72)$$

All independent variables are significant. Also, we find $R^2 = 0.99$ and $s_e = 309$. Thus, we are reasonably satisfied with this equation and expect that 95% of the time, our prediction for sales will be within $618,000 of actual sales.

Choosing the Best Regression Equation

How can we choose between several regression equations having different sets of independent variables? We usually want to choose the equation with the lowest value of s_e, since that will yield the most accurate forecasts. We also want the t-statistics for all variables in the equation to be significant. These two objectives may conflict, in which case it is difficult to determine the "best" equation. If the available computer printout contains the C_p statistic, then the regression chosen should have a C_p value close to (number of independent variables in the equation) + 1. For example, if a regression with three independent variables has $C_p = 80$, we can be sure that it is not a "good" regression. Actually, if a regression has C_p much larger than p, it means that at least one important variable has been omitted from the regression. (See Daniel and Wood (1980) for a discussion of the C_p statistic.)

Multicollinearity

If an estimated regression equation contains two or more independent variables that exhibit a strong linear relationship, we say that **multicollinearity** is present. A strong linear relationship between some of the independent variables may make the computer's estimates of the β_i's unreliable. In certain circumstances, multicollinearity can even cause a variable that should have β_i positive to have $\hat{\beta}_i$ substantially less than 0. The Happy Chicken example illustrates multicollinearity. We began with both YEAR and POP as independent variables, and as POP and YEAR both increase over time, we would expect a strong positive linear relationship to exist between them. Indeed, the correlation between YEAR and POP is 0.98. To see that the estimates of β_{YEAR} and β_{POP} are unreliable, note that in (21), $\hat{\beta}_{\text{POP}} < 0$, indicating that an increase in the number of customers near Happy Chicken decreases sales. This anomaly is due to multicollinearity. The strong linear relationship between YEAR and POP makes it difficult for the computer to estimate β_{POP} and β_{YEAR} accurately. After we drop POP from the estimated equation, the multicollinearity problem disappears, because there is no strong linear relationship between any of the remaining independent variables.

By the way, if an *exact* linear relationship exists between two or more independent variables, there are an infinite number of combinations of the $\hat{\beta}_i$'s which will minimize the sum of the squared errors, and most computer packages will print an error message. For example, if we let $x_1 = $ U.S. consumer expenditure during a year, $x_2 = $ U.S. investment during a year, $x_3 = $ U.S. government expenditure during a year, and $x_4 = $ U.S. national income during a year, it is well known that $x_4 = x_1 + x_2 + x_3$. In this case, we cannot use x_1, x_2, x_3, and x_4 as independent variables; at least one should be dropped from the equation.

Dummy Variables

Often, a nonquantitative or qualitative independent variable may influence the dependent variable. Some examples are as follows:

Dependent Variable	Categorical Independent Variable
Salary of employee	Race of employee
Consumer expenditures during year	Whether it is a wartime or a peacetime year
Customers entering bank on a given day	Day of the week
Air conditioner sales during a given month	Month of the year

In each of these situations, the independent variable does not assume a numerical value, but it may be classified into one of c categories. To illustrate, for monthly air conditioner sales, $c = 12$, and for the consumer expenditure example, $c = 2$.

Let the possible values of the categorical variable be listed as value 1, value 2, . . . , value c. To model the effect of a categorical variable on a dependent variable, we define $c - 1$ **dummy variables** as follows:

$x_1 = 1$ if observation takes on value 1 of categorical variable

$x_1 = 0$ otherwise

$x_2 = 1$ if observation takes on value 2 of categorical variable

$x_2 = 0$ otherwise

$\vdots$

$x_{c-1} = 1$ if observation takes on value $c - 1$ of categorical variable

$x_{c-1} = 0$ otherwise

TABLE **18**
Customer Traffic at University Credit Union

Day Number	Day of Week	University Payday?	Customers Entering Credit Union
1	Monday	No	515
12	Tuesday	No	360
18	Wednesday	Yes	548
23	Wednesday	No	386
24	Thursday	No	440
46	Monday	Yes	687
48	Wednesday	No	350
52	Tuesday	No	430
54	Thursday	No	370
55	Friday	No	496
70	Friday	No	506
81	Monday	No	509
89	Thursday	Yes	508
104	Thursday	No	396
106	Monday	No	600
108	Wednesday	No	266
122	Tuesday	No	360
130	Friday	Yes	521
152	Tuesday	No	398

We now include $x_1, x_2, \ldots, x_{c-1}$ (along with any other relevant independent variables) in the estimated regression equation.

To illustrate the use of dummy variables, suppose we are trying to predict the number of customers to enter the University Credit Union each day. The bank manager believes that bank traffic is influenced by the day of the week (the credit union is open Monday through Friday) and by whether or not the day is a payday for university employees. We are given the number of people to enter the bank during 18 randomly chosen days. (Day 1 is the present, day 6 is a week from now, and so on.) The relevant information is given in Table 18.

In this situation, there are two categorical variables of interest: the day of the week ($c = 5$) and whether or not a day is a payday ($c = 2$). For the day of the week, we define value 1 = Monday, value 2 = Tuesday, value 3 = Wednesday, value 4 = Thursday, and value 5 = Friday. Then we let

$$x_1 = 1 \quad \text{if day is a Monday} \qquad x_1 = 0 \quad \text{otherwise}$$
$$x_2 = 1 \quad \text{if day is a Tuesday} \qquad x_2 = 0 \quad \text{otherwise}$$
$$x_3 = 1 \quad \text{if day is a Wednesday} \qquad x_3 = 0 \quad \text{otherwise}$$
$$x_4 = 1 \quad \text{if day is a Thursday} \qquad x_4 = 0 \quad \text{otherwise}$$

For whether or not a day is a payday, we define $c - 1 = 1$ dummy variables. Letting value 1 = payday and value 2 = not a payday, we define

$$x_5 = 1 \quad \text{if day is a payday} \quad x_5 = 0 \quad \text{otherwise}$$

To account for a possible trend in the number of customers, we include T = number of the day as an independent variable. To illustrate how this information would be coded on the computer, we "code" the last two observations:

T	x_1	x_2	x_3	x_4	x_5	Customers
130	0	0	0	0	1	521
152	0	1	0	0	0	398

We obtain the following estimated regression equation (t-statistics in parentheses):

$$\hat{y} = 496.1 - 0.36T + 71.1x_1 - 78.5x_2 - 122.5x_3 - 74.8x_4 + 127.1x_5$$
$$\quad\quad\quad (-1.29) \quad (1.87) \quad (-2.04) \quad (-3.17) \quad (-1.99) \quad (4.43)$$

For $\alpha = 0.10$, we find that all independent variables except T are significant. After eliminating T from the estimated equation, we obtain the following equation:

$$\hat{y} = 466.2 + 80.4x_1 - 79.2x_2 - 109.8x_3 - 68.8x_4 + 124.3x_5$$
$$\quad\quad (2.1) \quad (-2.0) \quad -(2.86) \quad (-1.79) \quad (4.24)$$

(22)

For $\alpha = 0.10$, all independent variables are significant, so this equation appears satisfactory. We also find that $R^2 = 0.85$ and $s_e = 48.8$. Thus, on 95% of all days, our prediction for the number of customers entering the credit union should be accurate within $2(48.8) = 97.6$ customers.

Interpretation of Coefficients of Dummy Variables

How do we interpret the coefficients of dummy variables? To illustrate, let's determine how whether or not a day is a payday affects credit union traffic. On a payday, $x_5 = 1$, and we predict that $466.2 + 80.4x_1 - 79.2x_2 - 109.8x_3 - 68.8x_4 + 124.3$ customers will enter the credit union. On a day that is not a payday, $x_5 = 0$, and we predict that $466.2 + 80.4x_1 - 79.2x_2 - 109.8x_3 - 68.8x_4$ customers will enter the credit union. Subtracting, we find that on a payday, we predict (all other things being equal) that $\hat{\beta}_5 = 124.3$ more customers will enter the credit union than on a day that is not a payday.

To see how the day of the week influences credit union traffic, we note that (22) yields a different prediction for each day of the week. For Monday, $x_1 = 1$, $x_2 = x_3 = x_4 = 0$, and we obtain $\hat{y} = 466.2 + 80.4 + 124.3x_5 = 546.6 + 124.3x_5$. For Tuesday, $x_2 = 1$, $x_1 = x_3 = x_4 = 0$, and we obtain $\hat{y} = 466.2 - 79.2 + 124.3x_5 = 387 + 124.3x_5$. For Wednesday, $x_3 = 1$, $x_2 = x_1 = x_4 = 0$, and we obtain $\hat{y} = 466.2 - 109.8 + 124.3x_5 = 356.4 + 124.3x_5$. For Thursday, $x_4 = 1$, $x_1 = x_3 = x_2 = 0$, and we obtain $\hat{y} = 466.2 - 68.8 + 124.3x_5 = 397.4 + 124.3x_5$. For Friday, $x_1 = x_2 = x_3 = x_4 = 0$, and we obtain $\hat{y} = 466.2 + 124.3x_5$. Thus, we find that (all other things being equal) credit union traffic is heaviest on Mondays, next heaviest on Fridays, third heaviest on Thursdays, fourth heaviest on Tuesdays, and lightest on Wednesdays.

Multiplicative Models

Often, we believe that there is a relationship of the following form:

$$Y = \beta_0 x_1^{\beta_1} x_2^{\beta_2} \cdots x_k^{\beta_k}$$

(23)

To estimate such a relationship, simply take logarithms of both sides of (23). This yields

$$\ln Y = \ln \beta_0 + \beta_1(\ln x_1) + \beta_2(\ln x_2) + \cdots + \beta_k(\ln x_k)$$

Thus, to estimate (23), we run a multiple regression with the dependent variable being $\ln Y$ and the independent variables being $\ln x_1$, $\ln x_2$, ..., $\ln x_k$. To illustrate the idea, suppose we want to determine how the annual operating costs of an insurance company depend on the number of home insurance and car insurance policies that have been written.

TABLE 19
Insurance Company Branch Data

Branch	Annual Operating Cost	Number of Home Insurance Policies	Number of Car Insurance Policies
1	$124,000	400	1,200
2	$71,000	350	360
3	$136,000	600	800
4	$219,000	800	1,800
5	$230,000	900	1,600
6	$75,000	200	1,000
7	$56,000	120	900
8	$110,000	340	1,100
9	$120,000	490	900
10	$144,000	700	800

Branch.xls

Table 19 gives relevant information for ten branches of the insurance company (file Branch.xls).

To fit the model $Y = \beta_0 x_1^{\beta_1} x_2^{\beta_2}$, where Y = annual operating cost, x_1 = home insurance policies, and x_2 = car insurance policies, we would input into the computer points of the form (ln 124,000, ln 400, ln 1,200), and so on. The least squares estimates obtained are

$$\text{Constant term estimate} = 5.339$$
$$\text{Estimate for } \beta_1 = 0.583$$
$$\text{Estimate for } \beta_2 = 0.409$$

This regression yields an R^2 of 0.998, indicating a very good fit. The constant term estimate is an estimate of $\ln \beta_0$, so our actual estimate of β_0 is $e^{5.339} = 208.3$, and we estimate that $Y = 208.3 x_1^{0.583} x_2^{0.409}$. To illustrate the use of this equation, we predict annual operating cost for an insurance branch writing 500 home policies and 1,200 car policies. For this branch, we obtain $Y = 208.3(500)^{0.583}(1,200)^{0.409} = \$141,767$.

Heteroscedasticity and Autocorrelation in Multiple Regression

By plotting the errors in time-series sequence, we may check (as described in Section 15.6) to see whether the errors from a multiple regression are independent. If autocorrelation is present and the errors do not appear to be independent, then correcting for autocorrelation will usually yield better forecasts.

By plotting the errors (on the y-axis) against the predicted value of y (on the x-axis), we can determine whether homoscedasticity or heteroscedasticity is present. If homoscedasticity is present, the plot should show no obvious pattern (that is, the plot should resemble Figure 10), whereas if heteroscedasticity is present, the plot should show an obvious pattern indicating that the errors somehow depend on the predicted value of y (perhaps as in Figure 11). If heteroscedasticity is present, the t-tests described in this section are invalid.

Implementing Multiple Regression on a Spreadsheet

Credit.xls

In Figures 22 and 23 (file Credit.xls) we have run the regression for the data in Table 18. In the cell range A3:A21, we input the customer count for each day, and in the cell range B3:B21, the number of each day. In the cell range H3:H21, we input the day of the week for each observation (1 = Monday, ... , 5 = Friday). In the cell range G3:G21, a dummy variable indicates whether each day is a payday. We then used =IF statements to create the dummy variables for the day of the week. In cell C3, we input the formula =IF(H3=1,1,0). This places a 1 in C3, indicating the first observation is on a Monday. In cell D3, we enter the formula =IF(H3=2,1,0); in cell E3, =IF(H3=3,1,0); and in cell F3 =IF(H3=4,1,0). Copying from the range C3:F3 to the range C3:F21 generates the values of the dummy variables for all observations. To run the regression, select the Y range of A2:A21 and the X range of B2:G21. The regression output has the following interpretation:

Intercept   This is $\hat{\beta}_0 = 496.0857$.

Standard Error   This is $s_e = 48.84517$.

R Square   This is $R^2 = .845927$. This means that together, all the independent variables in the regression explain 84.6% of the variation in the number of customers arriving daily.

Observations   This is the number of data points (19).

Total df   This is the degrees of freedom ($n - k - 1 = 19 - 6 - 1$) used for the t-test of $H_0: \beta_i = 0$ against $H_1: \beta \neq 0$.

Coefficients   For each independent variable, this column yields the coefficient of the independent variable in the least squares equation. For example, $\hat{\beta}_T = -0.36222$.

Standard Error   For each independent variable, this row yields StdErr $\hat{\beta}_i$. For example, StdErr $\hat{\beta}_T = 0.279852$. The X Coefficient divided by the Std Err of Coef. yields the t-statistic for testing $H_0: \beta_i = 0$ against $H_1: \beta_i \neq 0$.

t Stat   This gives the observed t-statistic (coefficient/standard error) for the Intercept and all independent variables. For example, the t-statistic for Monday is 1.87.

Standard Error column   The Intercept entry gives the standard error $\hat{\beta}_0 = 37.66$, and the coefficient entries give the standard error for each independent variable. For example,

	A	B	C	D	E	F	G	H
1		CREDIT		UNION	EXAMPLE			
2	CUSTOMER	DAY#	MON	TUES	WED	THUR	PAYDAY?	DAYWK
3	515	1	1	0	0	0	0	1
4	360	12	0	1	0	0	0	2
5	548	18	0	0	1	0	1	3
6	386	23	0	0	1	0	0	3
7	440	24	0	0	0	1	0	4
8	687	46	1	0	0	0	1	1
9	350	48	0	0	1	0	0	3
10	430	52	0	1	0	0	0	2
11	370	54	0	0	0	1	0	4
12	496	55	0	0	0	0	0	5
13	506	70	0	0	0	0	0	5
14	509	81	1	0	0	0	0	1
15	508	89	0	0	0	1	1	4
16	396	104	0	0	0	1	0	4
17	600	106	1	0	0	0	0	1
18	266	108	0	0	1	0	0	3
19	360	122	0	1	0	0	0	2
20	521	130	0	0	0	0	1	5
21	398	152	0	1	0	0	0	2

FIGURE 22
Credit Union Example

	B	C	D	E	F	G
22	SUMMARY OUTPUT					
23						
24	Regression Statistics					
25	Multiple R	0.9197432				
26	R Square	0.8459275				
27	Adjusted R Square	0.7478813				
28	Standard Error	51.017121				
29	Observations	19				
30						
31	SUMMARY OUTPUT					
32						
33	Regression Statistics					
34	Multiple R	0.9197432				
35	R Square	0.8459275				
36	Adjusted R Square	0.7688912				
37	Standard Error	48.845174				
38	Observations	19				
39						
40	ANOVA					
41		df	SS	MS	F	iignificance
42	Regression	6	157192.73	26198.789	10.980899	0.0002823
43	Residual	12	28630.212	2385.851		
44	Total	18	185822.95			
45						
46		Coefficients	tandard Err	t Stat	P-value	Lower 95%
47	Intercept	496.08565	37.660289	13.172646	1.7E-08	414.03093
48	DAY#	-0.362218	0.2798522	-1.294318	0.2199096	-0.971963
49	MON	71.076945	38.076099	1.8667076	0.0865578	-11.88375
50	TUES	-78.47826	38.509376	-2.0379	0.0642282	-162.383
51	WED-	122.5236	38.6517	-3.16994	0.0080705	-206.7384
52	THUR	-74.82254	37.669971	-1.986265	0.070328	-156.8984
53	PAYDAY?	127.10854	28.682168	4.4316224	0.0008188	64.615462
54						

FIGURE 23

standard error $\hat{\beta}_{\text{Monday}} = 38.07$. The coefficient entry divided by the standard error entry yields the t-statistic for the intercept or slope (tabulated in the next column).

P-value For the intercept and each independent variable in a regression with k independent variables, this gives Probability($|t_{n-k-1}| \geq |$Observed t-statistic$|$). If, for example, the p-value for Wednesday is less than α, we reject H_0: $\beta_{\text{Wednesday}} = 0$; otherwise, we accept $\beta_{\text{Wednesday}} = 0$. For $\alpha = .05$, we reject $\beta_{\text{Wednesday}} = 0$.

The Data Analysis Regression Tool can handle a maximum of 15 independent variables. The data for the independent variables must be in adjacent columns.

PROBLEMS

Group A

1 For the years 1961–1970, the annual return on General Motors stock and the return on the Standard and Poor's market index were as given in Table 20 (file Beta.xls).

a Let Y = return on General Motors stock during a year and X = return on Standard and Poor's index during a year. Financial theory suggests that $Y = \beta_0 + \beta_1 X + \varepsilon$, where β_1 is called the **beta** for General Motors. Give an interpretation for the beta of a stock (in this case, General Motors), and use the data in Table 20 to estimate the beta for General Motors.

b Does the Standard and Poor's index appear to have a significant effect (for $\alpha = 0.05$) on the return on General Motors stock?

c What percentage of the variation in the return on General Motors Stock is explained by variation in the Standard and Poor's index?

d What percentage of the variation in the return on General Motors stock is unexplained by variation in Standard and Poor's index?

e During a year in which the Standard and Poor's in-

TABLE **20**

Year	Return on General Motors Stock	Return on Standard and Poor's Index
1961	12%	21%
1962	2%	−3%
1963	38%	15%
1964	26%	20%
1965	18%	12%
1966	−10%	0%
1967	0%	10%
1968	9%	10%
1969	−2%	2%
1970	−1%	−15%

dex increased by 15%, what would we predict for the return on General Motors stock?

2 We are trying to determine the number of labor hours required to produce a unit of a product. We are given the information in Table 21 (file Learn.xls). For example, the 2nd unit produced required 517 labor hours, and the 600th unit produced required 34 labor hours.

a Try to determine a relationship between the number of units already produced and the labor hours needed to produce the next unit. Why is this relationship called the **learning curve?**

b How many labor hours would be needed to produce the 800th unit?

c We are 95% sure that the prediction in part (b) is accurate within _____ hours.

3 Quarterly sales for a department store over a six-year period are given in Table 22 (file Sales.xls).

a Use multiple regression to develop a model that can

TABLE **21**

Cumulative Production	Labor Hours Needed for Last Unit
1	715
2	517
10	239
20	174
40	126
60	104
100	82
150	68
200	59
300	47
500	37
600	34

TABLE **22**

Year	Quarter	Sales (millions)
1984	1	$50,147
1984	2	$49,325
1984	3	$57,048
1984	4	$76,781
1985	1	$48,617
1985	2	$50,898
1985	3	$58,517
1985	4	$77,691
1986	1	$50,862
1986	2	$53,028
1986	3	$58,849
1986	4	$79,660
1987	1	$51,640
1987	2	$54,119
1987	3	$65,681
1987	4	$85,175
1988	1	$56,405
1988	2	$60,031
1988	3	$71,486
1988	4	$92,183
1989	1	$60,800
1989	2	$64,900
1989	3	$76,997
1989	4	$103,337

be used to predict future quarterly sales. (*Hint:* Use dummy variables and an independent variable for the number of the quarter (quarter 1, quarter 2, . . . , quarter 24).

b Letting Y_t = sales during quarter number t, discuss how to fit the following model to the data in Table 22:

$$Y_t = \beta_0 \beta_1^t \beta_2^{x_2} \beta_3^{x_3} \beta_4^{x_4}$$

where $x_2 = 1$ if t is a first quarter, $x_3 = 1$ if t is a second quarter, and $x_4 = 1$ if t is a fourth quarter. (*Hint:* Take logarithms of both sides.)

c Interpret the answer to part (b).

d Which model appears to yield better predictions for sales?

4 To determine how price influences sales, a company changed the price of a product over a 20-week period. The price charged each week and the number of units sold are given in Table 23 (file Price.xls). Develop a model to relate sales to price.

5 Confederate Express Service is attempting to determine how its shipping costs for a month depend on the number of units shipped during a month. For the last 15 months, the number of units shipped and total shipping cost are given in Table 24 (file Ship.xls).

TABLE 23

Price	Units Sold
$1	1,145
$2	788
$3	617
$4	394
$5	275
$6	319
$7	289
$8	241
$9	259
$10	176
$11	179
$12	232
$13	183
$14	181
$15	222
$16	212
$17	186
$18	110
$19	183
$20	172

TABLE 24

Month	Units Shipped	Total Shipping Cost
1	300	$1,060
2	400	$1,380
3	500	$1,640
4	200	$740
5	300	$1,060
6	350	$1,190
7	460	$1,520
8	480	$1,580
9	120	$540
10	760	$2,420
11	580	$2,200
12	340	$1,470
13	120	$790
14	100	$720
15	500	$1,960

a Determine a relationship between units shipped and monthly shipping cost.

b Plot the errors for the predictions in order of time sequence. Is there any unusual pattern?

c We have been told that there was a trucking strike during months 11–15, and we believe that this may have influenced shipping costs. How could the answer to part (a) be modified to account for the effects of the strike?

After accounting for the effects of the strike, does the unusual pattern in part (b) disappear?

Group B

6 In Example 1, we ran a regression with only x_1 (miles driven) as an independent variable. We found the coefficient of x_1 in this regression to be -51.68. This appears to indicate (contrary to what we would expect) that increasing the miles driven will lead to decreased maintenance costs. Explain this result. (*Hint:* Estimate the correlation between x_1 and x_2.)

7 Suppose we are trying to fit a curve to data, and part (i) of Figure 17 is relevant. Explain why the points of the form $(\frac{1}{x_i}, \ln y_i)$ should, when plotted, indicate a straight-line relationship.

8 Consider the regression in which we estimated cost of running an insurance company as a function of the number of home and car insurance policies. If there were a 1% increase in the number of car insurance policies, by what percentage would we predict that total costs would increase?

9 In the example in which we predicted the number of customers to enter the credit union, suppose that we had used five (instead of four) dummy variables to represent the days of the week. What problem would have arisen?

SUMMARY   Moving-Average Forecasts

$$f_{t,1} = \text{average of last } N \text{ observations}$$
$$e_t = x_t - (\text{prediction for } x_t)$$
$$\text{MAD} = \text{average value of } |e_t|$$

Choose N to minimize MAD.

Simple Exponential Smoothing

$$A_t = \text{smoothed average at end of period } t$$
$$= f_{t,k} = \text{forecast for period } t + k \text{ made at end of period } t$$
$$A_t = \alpha x_t + (1 - \alpha)A_{t-1}$$

Choose α to minimize MAD.

Holt's Method

Holt's method is used when trend is present, but there is no seasonality.

$$L_t = \text{estimate of base at end of period } t$$
$$T_t = \text{estimate of per-period trend at end of period } t$$
$$L_t = \alpha x_t + (1 - \alpha)(L_{t-1} + T_{t-1})$$
$$T_t = \beta(L_t - L_{t-1}) + (1 - \beta)T_{t-1}$$
$$f_{t,k} = L_t + kT_t$$

Winter's Method

Winter's method is used when we believe that trend and seasonality may be present.

$$s_t = \text{estimate for month } t \text{ seasonal factor at the end of month } t$$

$$L_t = \frac{\alpha x_t}{s_{t-c}} + (1 - \alpha)(L_{t-1} + T_{t-1})$$

$$T_t = \beta(L_t - L_{t-1}) + (1 - \beta)T_{t-1}$$

$$s_t = \frac{\gamma x_t}{L_t} + (1 - \gamma)s_{t-c}$$

$$f_{t,k} = (L_t + kT_t)s_{t+k-c}$$

For all extrapolation methods, we expect 68% of our predictions to be within $s_e = 1.25$ MAD of the actual value and 95% of our predictions to be within $2s_e$ of the actual value.

Simple Linear Regression

Given data points $(x_1, y_1), \ldots, (x_n, y_n)$, we estimate a linear relationship between x and y by $\hat{y} = \hat{\beta}_0 + \hat{\beta}_1 x$, where

$$\hat{\beta}_1 = \frac{\Sigma(x_i - \bar{x})(y_i - \bar{y})}{\Sigma(x_i - \bar{x})^2} \quad \text{and} \quad \hat{\beta}_0 = \bar{y} - \hat{\beta}_1 \bar{x}$$

$$R^2 = \frac{\text{SSR}}{\text{SST}} = \text{percentage of variation in } y \text{ explained by } x$$

$$r_{xy} = \text{sample linear correlation between } x \text{ and } y$$
$$(r_{xy} \text{ indicates the strength of the linear relationship between } x \text{ and } y)$$

$$s_e = \sqrt{\frac{\text{SSE}}{n - 2}}$$

We expect 68% of our predictions to be within s_e of the actual value and 95% of our predictions to be within $2s_e$ of the actual value.

A t-statistic exceeding $t_{(\alpha/2, n-2)}$ in absolute value is evidence (at level of significance α) that there is a significant linear relationship between x and y.

Fitting a Nonlinear Relationship

Step 1 Plot the points and find the part of Figure 17 that best fits the data.

Step 2 The second column of Table 13 gives the functional relationship between x and y.

Step 3 Transform each data point according to the rules in the third column of Table 13.

Step 4 Estimate the least squares regression line for the transformed data. If $\hat{\beta}_0$ is the intercept of the least squares line (for transformed data) and $\hat{\beta}_1$ is the slope of the least squares line (for transformed data), then we read the estimated relationship from the final column of Table 13.

Multiple Regression

Multiple regression is used when more than one independent variable is needed to predict y.

$$R^2 = \text{percentage of variation in } y \text{ explained by the independent variables}$$

Reject H_0: $\beta_i = 0$ at a level of significance α if $(t \text{ for } x_i) \geq t_{(\alpha/2, n-k-1)}$, where k is the number of independent variables being used to predict y.

If there is a strong linear relationship between two or more independent variables, then $\hat{\beta}_i$ may be an unreliable estimate of β_i. In such cases, we say that **multicollinearity** is present.

If a nonquantitative or qualitative independent variable (such as the day of the week or the month of the year) is believed to influence a dependent variable, **dummy variables** may be used to model the effect of the qualitative independent variable on the dependent variable. If the qualitative variable can assume c values, use only $c - 1$ dummy variables.

REVIEW PROBLEMS

Group A

1 Table 25 gives data concerning pork sales (file Pork.xls). Price is in dollars per hundred lb sold, quantity sold is in billions of pounds, per-capita income is in dollars, U.S. population is in millions, and GNP is in billions of dollars.

    a Use this data to develop a regression equation that could be used to predict the quantity of pork sold during future periods. Is autocorrelation, heteroscedasticity, or multicollinearity a problem?

    b Suppose that during each of the next two quarters, price = $45, U.S. population = 240, GNP = 2,620, and per-capita income = $10,000. Predict the quantity of pork sold during each of the next two quarters.

    c 68% of the time, we expect our prediction for pork sales to be accurate within _____.

    d Use Winter's method to develop a forecast for pork sales during the next two quarters. (Use the first two years to initialize.)

2 We are to predict sales for a motel chain based on the information in Table 26 (file Motel.xls).

    a Use this data and multiple regression to make predictions for the motel chain's sales during the next four quarters. Assume that advertising during each of the next four quarters is $50,000.

TABLE 25

Quarter	Year	Price of Pork	Quantity Sold	Per-Capita Income	U.S. Population	GNP
1	1975	39.35	30.44	8,255	212	1,549
2	1975	46.11	29.23	8,671	213	1,589
3	1975	58.83	25.12	8,583	214	1,629
4	1975	52.2	28.35	8,649	215	1,669
1	1976	47.99	28.95	8,775	216	1,718
2	1976	49.19	27.83	8,812	217	1,768
3	1976	43.88	29.53	8,884	218	1,818
4	1976	34.25	35.9	8,967	219	1,868
1	1977	39.08	32.94	9,036	220	1,918
2	1977	40.87	31.86	9,125	221	1,978
3	1977	43.85	30.74	9,280	222	2,038
4	1977	41.38	34.99	9,399	223	2,098
1	1978	47.44	32.43	9,487	224	2,148
2	1978	47.84	32.65	9,530	225	2,218
3	1978	48.52	31.58	9,622	226	2,288
4	1978	50.05	35.40	9,732	227	2,338
1	1979	51.98	33.98	9,813	228	2,398
2	1979	48.04	37.58	9,778	229	2,448
3	1979	38.52	38.59	9,809	230	2,478
4	1979	36.39	43.47	9,867	231	2,508
1	1980	36.31	41.24	9,958	232	2,539
2	1980	31.18	43.00	9,805	235	2,598
3	1980	46.23	37.57	9,882	235	2,598

TABLE 26

Quarter	Potential Customers (thousands)	Advertising (thousands of dollars)	Season	Sales (millions)
1	100	30	Winter	1,200
2	105	20	Spring	880
3	111	15	Summer	1,800
4	117	40	Fall	1,050
5	122	10	Winter	1,700
6	128	50	Spring	350
7	135	5	Summer	2,500
8	142	40	Fall	760
9	149	20	Winter	2,300
10	156	10	Spring	1,000
11	164	60	Summer	1,570
12	172	5	Fall	2,430
13	181	35	Winter	1,320
14	190	15	Spring	1,400
15	200	70	Summer	1,890
16	210	25	Fall	3,200
17	221	30	Winter	2,200
18	232	60	Spring	1,440
19	243	80	Summer	4,000
20	264	60	Fall	4,100

TABLE 27

Year	Jan.	Feb.	Mar.	Apr.	May	June	July	Aug.	Sept.	Oct.	Nov.	Dec.
1965	38	44	53	49	54	57	51	58	48	44	42	37
1966	42	43	53	49	49	40	40	36	29	31	26	23
1967	29	32	41	44	49	47	46	47	43	45	34	31
1968	35	43	46	46	43	41	44	47	41	40	32	32
1969	34	40	43	42	43	44	39	40	33	32	31	28
1970	34	29	36	42	43	44	44	48	45	44	40	37
1971	45	49	62	62	58	59	64	62	50	52	50	44
1972	51	56	60	65	64	63	63	72	61	65	51	47

b Use the Holt method to make forecasts for the motel chain's sales during the next four quarters.

c Use simple exponential smoothing to make predictions for the motel chain's sales during the next four quarters.

d Use Winter's method to determine predictions for the motel chain's sales during the next four quarters.

e Which forecasts would be expected to be the most reliable? (*Hint:* Use advertising, lagged by one period, as an independent variable.)

3 Table 27 gives the following data for monthly U.S. housing sales (in thousands of houses) for 1965–1972.

a Use the years 1965–1966 to initialize the parameters for Winter's method. Then find values of α, β, and γ that yield a MAD (for 1967–1972) of less than 3.5. (*Hint:* It may be necessary to use $\alpha > 0.5$.)

b We would expect 68% of our forecasts to be accurate within _____ and 95% of our forecasts to be accurate within _____.

c Check to see whether the data are consistent with the answer to part (b).

d Although we have not discussed autocorrelation for smoothing methods, good forecasts derived from smoothing methods should exhibit no autocorrelation. Do the forecast errors for this problem exhibit autocorrelation?

e It has been stated that if only trend and seasonality are important factors, then α should be at most 0.5. Explain why this problem required $\alpha > 0.5$.

f At the end of December 1972, what is the forecast for housing sales during the first three months of 1973?

Note: This assignment is a snap on a spreadsheet. The spreadsheet might be set up as in Table 28. In B14, enter = A\$3*A14/D2+(1-A\$3)*(B13+C13). Insert analogous formulas in C14, D14, E14. Remember that the forecast must be made before "seeing" A14. In F14, enter =A14-E14. In G14, enter =ABS(F14). Copy from B14:G14 to ??. To compute MAD, average the absolute errors for each month (rows 14–85).

4 Using x as the independent variable and y as the dependent variable, find the least squares line for the following three data points:

x	y
1	2
4	5
7	2

TABLE 28

Row	A	B	C	D	E	F	G
1	SALES	BASE	TREND	SEASON	FORE	ERR	ABSERR
2	ALPHA	BETA	GAMMA	S-11			
3	.1	.2	.3	S-10			
.							
.							
.							
13		LO	TO	SO			
14	29						
15	32						
.							
.							
85	47						

TABLE 29

x	y
0	100
1	130
2	170
3	200
4	260
5	300
6	305
7	330
8	380

TABLE 30

Humidi	Temp	Press	Hard
40	1	0	148
60	1	0	209
50	1	0	177
70	1	0	208
80	1	0	262
60	1	1	248
65	1	1	253
70	1	1	263
35	1	1	184
45	1	1	220
70	0	0	129
28	0	0	53
49	0	0	98
89	0	0	170
90	0	0	172
34	0	1	80
56	0	1	90
77	0	1	151
23	0	1	58
56	0	1	107

5 We are trying to predict the number of uses of automatic bank teller machines as a function of time. The data are given in Table 29. Here, x = number of years after 1980 and y = number of monthly uses of ATMs (in millions) during the given year. The estimated regression equation is $\hat{y} = 102.3 + 34.8x$. We are given that SST = 74,100, SSE = 1,298, and $\text{StdErr}(\hat{\beta}_1) = 1.76$.

a Test H_0: $\beta_1 = 0$ against H_a: $\beta_1 \neq 0$ for $\alpha = 0.05$. Interpret the result.

b Find the correlation between x and y.

c Is the 1987 entry an outlier?

d If present trends continue, what is the approximate probability that during 1990, more than 470 million ATM transactions per month will occur? (*Hint:* Use the fact that the errors are normally distributed.)

6 Carboco puts metal coatings on jet propeller blades. The harder the coating, the higher the quality of the coating. The coating is shot onto the blade using pressurized gas contained in an F-gun. Carboco can control the temperature and gas pressure in the F-gun and can also control the room humidity. To see how gas pressure, temperature, and humidity influence hardness, Carboco engineers have run a regression for which the dependent variable is

$$\text{HARD} = \text{hardness of a coating}$$
and the independent variables are

$$\text{HUMIDI} = \text{room humidity}$$
$$\text{TEMP} = 1 \quad \text{if temperature level is high}$$
$$= 0 \quad \text{if temperature level is low}$$
$$\text{PRESS} = 1 \quad \text{if gas pressure is high}$$
$$= 0 \quad \text{if gas pressure is low}$$
$$\text{T*P} = \text{product of TEMP and PRESS}$$

We assume that temperature and gas pressure have only two possible levels, low and high. The relevant data are given in Table 30 (file Temp.xls).

a Ignoring considerations of heteroscedasticity, multi-collinearity, and autocorrelation, which equation should be used to predict hardness? Explain.

b What combination of gas pressure and temperature setting will maximize hardness?

c Explain how changing temperature from a low level to a high level affects hardness. Be specific!

7 We have been assigned to determine how the total weekly production cost for Widgetco depends on the number of widgets produced during the week. The following model has been proposed:

$$Y = \beta_0 + \beta_1 X + \beta_2 X^2 + \beta_3 X^3 + \varepsilon$$

where X = number of widgets produced during the week and Y = total production cost for the week. For 15 weeks of data, we found that SSR = 215,475 and SST = 229,228. For this model, we obtain the following estimated regression equation (t-statistics for each coefficient are in parentheses):

$$\hat{y} = -29.7 + 19.8X - 0.39X^2 + 0.005X^3$$
$$(0.78) \quad (0.62) \quad (1.25)$$

a For $\alpha = 0.10$, test H_0: $\beta_i = 0$ against H_a: $\beta_i \neq 0$ ($i = 1, 2, 3$).

b Determine R^2 for this model. How can the high R^2 value be reconciled with the answer to part (a)?

8 Let Y_t = sales during month t (in thousands of dollars) for a photography studio (SALES in Table 31) and P_t = price charged for portraits during month t (PRICE). Use a computer to fit the following model to the data in Table 31 (file Portrait.xls):

$$Y_t = \beta_0 + \beta_1 Y_{t-1} + \beta_2 P_t + \varepsilon_t$$

Thus, last month's sales and the current month's price are independent variables.

a If the price of a portrait during month 21 is $10, what would we predict for month 21's sales?

b Does there appear to be a problem with autocorrelation, heteroscedasticity, or multicollinearity?

TABLE 31

Month	Sales	Price
1	400	5
2	1,042	4
3	1,129	8
4	1,110	6
5	1,336	6
6	1,363	10
7	1,177	9
8	603	8
9	582	12
10	697	9
11	586	8
12	673	9
13	546	10
14	334	11
15	27	8
16	76	9
17	298	10
18	746	6
19	962	7
20	907	8

TABLE 32

Year	GNP
1975	1,060
1976	1,170
1977	1,305
1978	1,455
1979	1,630
1980	1,800
1981	2,000
1982	2,220
1983	2,450
1984	2,730

b When the regression on transformed data is done, we find that $\hat{\beta}_0 = 6.86$ and $\hat{\beta}_1 = 0.105$. What is the prediction for 1985 GNP?

Group B

10 Suppose the true relationship between Y and time t is given by

$$Y = \beta_0 e^{\beta_1 t} + \varepsilon$$

where $\beta_1 > 0$. If we try to fit our usual linear model $Y = \beta_0 + \beta_1 t + \varepsilon$ to the data, are we likely to encounter autocorrelation? Heteroscedasticity? Multicollinearity?

9 The U.S. GNP during the years 1975–1984 is given in Table 32 in billions of dollars (file GNP.xls).

a Plot x = years after 1974 against GNP, and use the plot to describe how to fit a curve that could be used to predict GNP during future years.

REFERENCES

For further information on extrapolation methods, we recommend:

Hax, A., and D. Candea. *Production and Inventory Management.* Englewood Cliffs, N.J.: Prentice Hall, 1984.

Makridakis, S., and S. Wheelwright. *Forecasting: Methods and Applications.* New York: Wiley, 1986.

For further information on regression, we recommend:

Chatterjee, S., and B. Price. *Regression Analysis by Example.* New York: Wiley, 1990.

Daniel, C., and F. Wood. *Fitting Equations to Data.* New York: Wiley, 1980.

Montgomery, D., and E. Peck. *Introduction to Linear Regression Analysis.* New York: Wiley, 1991.

Pindyck, R., and D. Rubinfeld. *Econometric Models and Economic Forecasts.* New York: McGraw-Hill, 1989.

16

Brownian Motion, Stochastic Calculus, and Stochastic Control

Stock prices, inventory levels, interest rates, the size of an economy, and other quantities are often modeled as a **Brownian motion** by means of a **stochastic differential equation (SDE).** This chapter will discuss those topics, as well as Ito's Lemma, which is used to analyze SDEs. We also show how Black and Scholes used SDEs and Ito's Lemma to derive the option pricing formulas discussed in Chapter 14. We then discuss the optimal control of SDEs. Stochastic control is a very useful tool for determining optimal decisions when a random variable of interest is governed by an SDE.

16.1 What Is Brownian Motion?

Suppose $x(t)$ is a random variable whose value tells us the value of some quantity of interest (say, the price of a stock) at time t. The collection of random variables ($x(t)$: $0 \leq t \leq \infty$) is said to be a Wiener process or Brownian motion if

$$x(0) = 0 \tag{1}$$

$x(t)$ has **stationary** increments. This means that the distribution of $x(t) - x(s)$ depends only on $t - s$. $\quad$ (2)

$x(t)$ has **independent** increments. This means that for $t_1 < t_2 < \cdots < t_k$, the random variables $x(t_1), x(t_2) - x(t_1), \ldots x(t_k) - x(t_{k-1})$ are mutually independent. $\quad$ (3)

For every $t > 0$, $x(t)$ is normally distributed with mean μt. μ is called the **drift coefficient.** $\quad$ (4)

Note that $x(t + \Delta t) = \{x(t + \Delta t) - x(\Delta t)\} + x(\Delta t)$. Then (3) implies that var $x(t + \Delta t) =$ var$\{x(t + \Delta t) - x(\Delta t)\}$ + var $x (\Delta t)$. Define $u(t) =$ var $(x(t))$ and note that $u(0) = 0$. Now (2) implies that $u(t + \Delta t) =$ var $x(t + \Delta t) = u(t) + u(\Delta t)$. It can be shown that the only nonnegative solution to this functional equation satisfying $u(0) = 0$ is $u(t) = \sigma^2 t$ (where σ^2 is an arbitrary constant). Thus, from (4), the density of $x(t)$, $f_t(x)$, is given by

$$f_t(x) = \frac{1}{\sigma\sqrt{2\pi t}} \exp\left[-\frac{1}{2}\left(\frac{x - \mu t}{\sigma\sqrt{t}}\right)^2\right]$$

We can now easily answer questions about the probability of an event involving $x(t_2)$ conditional on the value of $x(t_1)$ (here, $t_2 > t_1$). Consider a Brownian motion with $\mu = 4$ and $\sigma = 3$. Assume $x(3) = 1$. What is the density of $x(7)$?

The density of $x(7)$ conditional on $x(3) = 1$ will be normally distributed, with mean $1 + 4(7 - 3) = 17$ and variance $(7 - 3)(3)^2 = 36$.

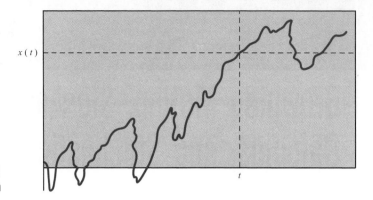

FIGURE 1
Brownian Motion

Even during a small interval of time Δt, a Brownian motion jumps around very often, and the jumps are very big, relative to Δt. A sample path of a Brownian motion might look like Figure 1. To show how fast a Brownian motion jumps around, note that for a Brownian motion with $\sigma^2 = 1$ and $\mu = 0$, $x(t + \Delta t) - x(t)$ is normal with mean 0 and variance Δt. Then the random variable

$$\frac{x(t + \Delta t) - x(t)}{\Delta t}$$

represents the "slope" of the Brownian motion at t. But

$$\frac{E(x(t + \Delta t) - x(t))}{\Delta t} = \frac{1}{\Delta t}(0) = 0$$

so

$$\text{var} \frac{(x(t + \Delta t) - x(t))}{\Delta t} = \frac{1}{(\Delta t)^2}\Delta t = \frac{1}{\Delta t}$$

Thus, as $\Delta t \to 0$, the "slope" of the Brownian motion has infinite variance. This argument can be formalized to show that a Brownian motion is nowhere differentiable.

16.2 Derivation of Brownian Motion As a Limit of Random Walks

To get a better feel for what Brownian motion means, we derive Brownian motion as a limit of random walks. Let $x(t)$ represent the price of a stock at time t. Assume that $x(0) = 0$ and that the stock price changes at times $\Delta t, 2\Delta t, \ldots, k\Delta t$ (where k is any positive integer). With probability $p = \frac{1}{2}\left(1 - \frac{\mu\sqrt{\Delta t}}{\sigma}\right)$, the stock price increases by an amount $k = \sigma\sqrt{\Delta t}$. With probability $q = 1 - p = \frac{1}{2}\left(1 - \frac{\mu\sqrt{\Delta t}}{\sigma}\right)$ the stock price decreases by an amount $k = \sigma\sqrt{\Delta t}$. Note that by the independence of successive changes in stock price, $\text{var}(x(t))$ will be approximately $\left(\frac{t}{\Delta t}\right)\sigma^2\Delta t = \sigma^2 t$. This follows, because by time t there are approximately $t/\Delta t$ steps, and each step has a variance of $\sigma^2\Delta t$. Thus, as $\Delta t \to 0$, the variance of $x(t)$ approaches $\sigma^2 t$. Thus, we can obtain a Brownian motion as a limit of random walks.

16.3 Stochastic Differential Equations

In what follows, $z(t)$ represents a standard Wiener process (i.e., $\mu = 0$, $z(0) = 0$, and $\sigma = 1$). Let $x(t)$ be a stochastic process. What does it mean to say that $x(t)$ is a solution to the following?

$$dx = f(x, t)dt + \sigma(x, t)dz \tag{5}$$

This is called a **stochastic differential equation.** Stochastic processes that satisfy (5) are called **Ito processes.** For a stochastic process to satisfy (5), it must be true that for small Δt, the random variables $x(.)$ and $z(.)$ satisfy

$$\Delta x = x(t + \Delta t) - x(t) = f(x(t), t)\Delta t + \sigma(x(t), t)(z(t + \Delta) - z(t)) + o(\Delta t)$$

$$\text{where } \lim_{\Delta t \to 0} \frac{o(\Delta t)}{\Delta t} = 0 \text{ (with probability 1)} \tag{6}$$

For example, $(\Delta t)^2$ is $o(\Delta t)$, but $(\Delta t)^{1/2}$ is not $o(\Delta t)$.

Essentially, the $f(x(t), t)\Delta t$ is the mean rate of change in x during $(t, t + \Delta t)$, while $\sigma^2(x(t), t)\Delta t$ is the variance of the change in x during $(t, t + \Delta t)$. The fact that (6) completely characterizes some stochastic processes $x(t)$ is not easy to prove. A lot of measure theory and advanced probability is required. In short, (6) is equivalent to Δx being normally distributed with mean $f(x, t)\Delta t$ and variance $\sigma^2(x, t)\Delta t$.

For example, if $dx = 2dt + 3dz$ and $x(0) = 0$, $x(t)$ will turn out to be a Wiener process with $\mu = 2$ and $\sigma = 3$. (See Section 16.4).

If we are going to model a situation via a stochastic differential equation, how do we determine what $f(x, t)$ and $\sigma(x, t)$ should be? Consider an investment in either a risky asset (such as a stock) or an asset with a deterministic and known return (such as a treasury bill). Let

W = total wealth

w = fraction of wealth invested in the risky asset

s = rate of return on the sure asset

a = expected rate of return on the risky asset $(a > s)$

σ^2 = variance per unit of time in return on risky asset

That is, in time Δt, the return on \$1 invested in the risky asset has variance $\sigma^2 \Delta t$. The risky asset earns return dr in $(t, t + \Delta t)$ given by $dr = adt + \sigma dz$.

c = consumption rate

Find a stochastic differential equation for $W(t)$, the investor's wealth at time t.

Note that at time $t + \Delta t$, we add to time t wealth the money earned by the risky and sure portions of the investments between t and $t + \Delta t$. Then

$$W(t + \Delta t) = W(t) + s(1 - wW)\Delta t + \Delta r(wW)$$

Since $\Delta r = a\Delta t + \sigma \Delta z$, we find that

$$W(t + \Delta t) - W(t) = \Delta t(s(1 - w)W) + awW\Delta t + \sigma wW\Delta z$$

or

$$dW = dt(s(1 - w)W) + awW \, dt + \sigma wW dz$$

16.4 Ito's Lemma

For most uses of stochastic differential equations, the key question is as follows. If $x(t)$ is described by (5), what does the distribution of $y(x, t)$ look like for some arbitrary function $(y(x, t))$ of t and x? The following result, **Ito's Lemma,** gives us the answer. In what follows, subscripts represent partial derivatives. For example, $y_t = \dfrac{\partial y}{\partial t}$ and $y_{xt} = \dfrac{\partial^2 y}{\partial x \partial t}$.

If $y(t) = y(x, t)$ is a continuously differentiable function of t and twice continuously differential in x and x satisfies $dx = f(x, t)dt + \sigma(x, t)dz$, then $y(t)$ satisfies the following stochastic differential equation:

$$dy = (y_t + y_x f(x, t) + \tfrac{1}{2} y_{xx} \sigma^2(x, t))dt + y_x \sigma(x, t)dz \tag{7}$$

We now give an intuitive explanation of Ito's Lemma. From Taylor's Theorem, we may write

$$\Delta y = y_x(f\Delta t + \sigma \Delta z) + y_t \Delta t + \tfrac{1}{2}(f^2 \Delta t^2 + 2f\sigma \Delta t \Delta z + \sigma^2 \Delta z^2)y_{xx} + \tfrac{1}{2}(\Delta t)^2 y_{tt} +$$
$$(f\Delta t + \sigma \Delta z)\Delta t y_{xt} + \tfrac{1}{6}(f\Delta t + \sigma \Delta z)^3 y_{xxx} + \cdots$$

To simplify Δy, we need the following two ideas.

Idea 1   Find the term in Δy with the largest variance. (It turns out to be Δz.) All random terms with variance equal to a higher order of Δt than the variance of Δz may be considered to be constants.

Idea 2   The mean of something like $\Delta t + \Delta t^2 + \Delta t^3 + \cdots$ is Δt. Observe that Δy has terms involving Δt, Δz, $\Delta z \Delta t$, Δz^2, Δz^3, $\Delta z^2 \Delta t$, Δt^2, $\Delta z \Delta x^3$, and Δt^3. Table 1 gives the means of many terms appearing in our Taylor series expansion. Terms denoted by an * in Table 1 are the terms that "stick around" in Formula (7).

By Idea 2, terms involving Δt^2 or Δt^3 may be disregarded. Now Δz is a normal random variable, with mean 0 and variance Δt^3. It can be shown that Δz^2 is a χ^2 random variable, with mean $E(\Delta z^2) = \Delta t$ and variance

$$E(\Delta z^4) - E(\Delta z^2)^2 = 3\Delta t^2 - \Delta t^2 = 2\Delta t^2$$

Since this term has higher-order variance than Δz, it may be thought of as a constant. Δz^3 has mean 0 and variance $E(\Delta z^6) = 15\Delta t^3$. Thus, Δz^3 may be considered a constant.

$(\Delta z^2)\Delta t$ has mean $(\Delta t)^2$ and variance Δt^4, so it may be ignored. Now Idea 1 implies that the random part of Δy may be thought of as involving only the Δz term; other random terms are of higher order. The mean of Δy just involves Δt, Δz^2, and terms of higher order. Thus, we may write $\Delta y = y_x f \Delta t + y_t \Delta t + \tfrac{1}{2}\sigma^2 y_{xx}\Delta z^2 + \sigma y_x \Delta z$. But Δz^2 is a "constant" with mean Δt, so we may write

$$\Delta y = \Delta t\{y_x f + y_t + .5\sigma^2 y_{xx}\} + \sigma y_x \Delta x$$

which is the desired result.

Ito's Lemma may also be proved using the well-known rules of stochastic calculus

$$(dz)(dz) = dt$$
$$dz(dt) = 0$$
$$dt(dz) = 0 \tag{8}$$
$$dt(dt) = 0$$

Note that these rules match up with Table 1. To obtain Ito's Lemma via these rules, again do a Taylor series expansion of $y(t, x)$. Then

$$dy = y_t dt + y_x dx + \frac{1}{2} y_{tt}(dt)^2 + \frac{1}{2} y_{xx}(dx)^2 + y_{xt}(dx)dt + o(dt) \qquad (9)$$

Since $dx = fdt + \sigma dz$, we may substitute this in (9) to obtain

$$dy = y_t dt + y_x(fdt + \sigma dz) + \frac{1}{2} y_{tt}(dt)^2$$

$$+ \frac{1}{2} y_{xx}(f^2 dt^2 + 2f\sigma dtdx + \sigma^2 dz^2) + y_{xt}dt(fdt + \sigma dz) + o(dt) \qquad (10)$$

TABLE 1

Term	Expected Value
Δt^*	Δt
Δz	0
$(\Delta t)^2$	$(\Delta t)^2 = o(\Delta t)$
$(\Delta t)(\Delta z)$	0
$(\Delta z)^2$	Δt
$(\Delta t)^3$	$(\Delta t)^3 = o(\Delta t)$
$(\Delta t)^2(\Delta z)$	0
$(\Delta t)(\Delta z)^2$	$(\Delta t)^2 = o(\Delta t)$
$(\Delta t)^3$	0
$(\Delta t)^4$	$(\Delta t)^4 = o(\Delta t)$
$(\Delta t)^3(\Delta z)$	0
$(\Delta t)^2(\Delta z)^2$	$(\Delta t)^3 = o\Delta t$
$(\Delta t)(\Delta z)^3$	0
$(\Delta z)^4$	$3(\Delta t)^2 = o(\Delta t)$
$(\Delta t)^5$	$(\Delta t)^5 = o(\Delta t)$
$(\Delta t)^4\Delta z$	0
$(\Delta t)^3(\Delta z)^2$	$(\Delta t)^4 = o(\Delta t)$
$(\Delta t)^2(\Delta z)^3$	0
$(\Delta t)(\Delta z)^4$	$3(\Delta t)^3 = o(\Delta t)$
$(\Delta z)^5$	0
$(\Delta t)^6$	$(\Delta t)^6 = o\Delta t$
$(\Delta t)^5(\Delta z)$	0
$(\Delta t)^4(\Delta z)^2$	$(\Delta t)^5$
$(\Delta t)^3(\Delta z)^3$	0
$(\Delta t)^2(\Delta z)^4$	$3(\Delta t)^4 = o(\Delta t)$
$(\Delta t)(\Delta z)^5$	0
$(\Delta z)^6$	$15(\Delta t)^3 = o(\Delta t)$

These are derived from the following results. For a normal distribution Δz, with $\mu = 0$ and $\sigma = \sqrt{\Delta t}$:

$$E(\Delta z) = 0$$
$$E[(\Delta z)^2] = \Delta t$$
$$E[(\Delta z)^3] = 0$$
$$E[(\Delta z)^4] = 3(\Delta t)^2$$
$$E[(\Delta z)^5] = 0$$
$$E[(\Delta z)^6] = 15(\Delta t)^3$$

Also, if corr $(\Delta z_1, \Delta z_2) = \rho$, then $E(\Delta z_1, \Delta z_2) = \rho dt$.

From (8), the only non-$o(dt)$ terms in (10) are

$$dy = y_t dt + fy_x dt + \left(\frac{\sigma^2}{2}\right) y_{xx} dt + \sigma y_x dz \qquad (11)$$

This agrees with Ito's Lemma.

A special case of Ito's Lemma occurs if we wish to find dy for $y = y(z, t)$. From (5), we plug in Ito's Lemma with $f(x, t) = 0$ and $\sigma(x, t) = 1$. Then (11) yields

$$dy = \left(y_t + \frac{1}{2} y_{xx}\right) dt + y_z dz \qquad (12)$$

A multiple-dimensional analog of Ito's Lemma may also be stated. Let $x = (x_1, \ldots, x_n)$, where

$$dx_i = g_i(t, x) dt + \sum_{k=1}^{m} \sigma_{ik}(t, x) dz_k \qquad i = 1, 2, \ldots, m \qquad (13)$$

where $z_1, \ldots, z_m$ are standard Weiner processes and the correlation between dz_i and dz_j is $p_{ij}(p_{ii} = 1)$. This is equivalent to

$$E(z_i(t) z_j(t)) = p_{ij} t \qquad \text{or} \qquad E[(\Delta z_i)(\Delta z_j)] = p_{ij} \Delta(t)$$

If y is a function of t and x, then

$$dy = \sum_{i=1}^{i=n} y_i dx_i + y_t dt + \frac{1}{2} \sum_{i=1}^{i=n} \sum_{j=1}^{j=n} y_{ij} dx_i dx_j \qquad (14)$$

where the products $dx_i dx_j$ are computed from (13) and the rules $dz_i dz_j = p_{ij} dt$ and $dz_i dt = 0$. For two dimensions, this may be simplified as follows.

Let x_1, x_2 evolve through time according to the stochastic differential equations

$$dx_1 = \mu_1(x_1, x_2, t) dt + \sigma_1(x_1, x_2, t) dz_1$$
$$dx_2 = \mu_2(x_1, x_2, t) dt + \sigma_2(x_1, x_2, t) dx_2$$

Let $y(x_1, x_2, t)$ be a twice continuously differentiable function of x_1 and x_2 and a continuously differentiable function of t. Then the evolution of y is according to

$$dy = y_1 dx_1 + y_2 dx_2 + y_t dt + \frac{1}{2} y_{11} (dx_1)^2 + y_{12}(dx_1 dx_2) + \frac{1}{2} y_{22}(dx_2)^2 + o(dt)$$

with the following formal multiplication rules:

$$(dz_1)^2 = dt \qquad (dt)^2 = 0$$
$$(dz_1)(dt) = 0 \qquad (dz_1)(dz_2) = p_{12} dt$$

where $\rho \equiv Cov[z_1(T) - z_1(t), z_2(T) - z_2(t)]/(T - t)$ is the correlation of the two Wiener processes. Then we get

$$dy = [y_1 \mu_1 + y_2 \mu_2 + \frac{1}{2}(y_{11}\sigma_1^2 + 2y_{12}\rho\sigma_1\sigma_2 + y_{22}\sigma_2^2) + y_t] dt + y_1 \sigma_1 dz_1 + y_2 \sigma_2 dz_2 \qquad (15)$$

Here are some examples of Ito's Lemma in action. If $dx = -\frac{1}{2} dt + dz$ and $y(t) = e^{x(t)}$, find dy.

$$y_t = 0$$
$$y_x = y_{xx} = e^x$$
$$f = -\frac{1}{2}$$
$$\sigma = 1$$

Therefore,

$$dy = \left(0 - \frac{1}{2} e^x + \frac{1}{2} e^x\right)dt + e^x dz$$

or

$$dy = e^x dz$$

As another example, suppose $y(0) = 0$ and let y satisfy the following stochastic differential equation:

$$dy = adt + \sigma dz \tag{16}$$

Show that $y(t, z)$ should behave like a normal distribution with mean at variance $\sigma^2 t$. That is, $y(t, z) = at + \sigma z(t)$ should satisfy (16).

From Formula (11), this is easily checked, because $y_t = a$, $y_z = \sigma$ and $y_{tt} = y_{zz} = 0$. Thus, (11) yields $dy = adt + \sigma dz$. If we didn't realize that this $y(t)$ should work, however, could we have figured it out? Via (11), we see that $y(t,z)$ can only satisfy (16) if $y_t + \frac{1}{2} y_{zz} = a$ and $y_z = \sigma$. The latter implies that $y(t, z) = \sigma z + f(t)$. Then the former yields $f'(t) = a$ or $f(t) = at + k$. Thus, $y(t, z) = \sigma z(t) + at + k$. Since $y(0, z) = 0$, $k = 0$, and we obtain the desired result.

EXAMPLE 1   Population Growth

Suppose we are developing a model of population growth. Let $y(t, z)$ denote the population of a nation at time t. Then it seems reasonable that y should satisfy

$$dy = aydt + \sigma ydz \tag{17}$$

This equation implies that the mean population growth and variance per unit time of the population growth are always a constant percentage of the current population. (Do you see why this is true?) This model is often called *geometric* Brownian motion. Use Ito's Lemma to help determine $y(t, z)$.

Solution   By (11), $y(t, z)$ will satisfy (17) only if both the following are satisfied:

$$y_z = \sigma y \tag{18}$$

$$\left(y_t + \frac{1}{2} y_{zz}\right) = ay \tag{19}$$

From (18),

$$\frac{\partial y}{\partial z} = \sigma y \quad \text{or} \quad \frac{\partial y}{y} = \sigma(\partial z)$$

$$1ny = \sigma z + f(t) \quad \text{or} \quad y = e^{\sigma z + f(t)}$$

To satisfy (19), we will need

$$f'(t)y + \frac{1}{2} \sigma^2 y = ay \quad \text{or} \quad f'(t) = a - \frac{1}{2} \sigma^2$$

$$f(t) = \left(a - \frac{1}{2} \sigma^2\right)t + k_1$$

Thus, $y(t, z) = e^{k_1 + \sigma z + (a - .5\sigma^2)t}$. Let $e^{k_1} = y(0)$. We get

$$y(t, z) = y(0)e^{\sigma z + (a - .5\sigma^2)t}$$

so

$$1ny(t, z) = 1ny(0) + \sigma z + \left(a - \frac{1}{2}\sigma^2\right)t$$

Thus, for a given t, $1ny$ is normally distributed. That is, for any t, $y(t, z)$ is lognormal. This model is also considered relevant for stock prices. Note that $y(t, z) < 0$ is impossible, because if $y(t, z) = 0$, then for all $t > T$, $y(t, z) = 0$ will hold.

EXAMPLE 2

Solve $dy = .25dt + \sqrt{y}dz$ subject to $y(0) = 1$.

Solution We must solve both of the following:

$$y_z = \sqrt{y} \tag{20}$$

$$y_t + \frac{1}{2}y_{zz} = \frac{1}{4} \tag{21}$$

From (20), $\dfrac{\partial y}{\sqrt{y}} = \partial z$. Integrating both sides of the last equation yields

$$2\sqrt{y} = z + \bar{f}(t)$$

If we define $f(t) = \left(\dfrac{1}{2}\right)\bar{f}(t)$, then the last equation may be rewritten as

$$y = \left\{\frac{z}{2} + f(t)\right\}^2$$

Plugging this into (21) yields

$$2f'(t)\left\{\frac{z}{2} + f(t)\right\} + \frac{1}{4} = \frac{1}{4}$$

This will work if $f'(t) = 0$. Thus, $y(t) = \left\{\dfrac{z(t)}{2} + k\right\}^2$. From $y(0) = 1$, we find $k = \pm 1$, so the solution to the SDE is

$$y(t) = \left\{1 + \frac{z(t)}{2}\right\}^2$$

EXAMPLE 3 Fischer's Demand for Index Bonds Model[†]

An individual's portfolio consists of real bonds, risky assets, and nominal bonds.

Q_1 = value of real (or index) bonds in portfolio. (A real bond pays a real return r_1—that is, a nominal return of r_1 plus the realized rate of inflation.

Q_2 = value of risky assets in portfolio. (A risky asset pays an expected nominal return R_2 on equity per unit time and has a variance $(s_2)^2$ per unit time.)

Q_3 = value of nominal bonds in portfolio. (Nominal bonds yield a deterministic nominal return of R_3 per unit time.)

If P is price level, we assume

$$\frac{dP}{p} = \pi dt + s_1 dz_1 \tag{22}$$

[†]Based on Fischer (1975).

Thus, we have

$$\frac{dQ_1}{Q_1} = r_1 dt + \frac{dP}{P} = R_1 dt + s_1 dz_1 \qquad \text{where } R_1 = r_1 + \pi \tag{23}$$

$$\frac{dQ_2}{Q_2} = R_2 dt + s_2 dz_2 \tag{24}$$

$$\frac{dQ_3}{Q_3} = R_3 dt \tag{25}$$

where dQ_i represents the change in value for the portion of the individual's portfolio in asset i, and z_1 and z_2 are standard Wiener processes with correlation ρ. Use Ito's Lemma to answer the following questions.

(a) What is the real return on the risky asset?

(b) What is the real return on a nominal bond?

Solution

(a) We apply the two-dimensional form of Ito's Lemma (15) to $Y(t, P, Q_2) = \dfrac{Q_2}{P}$. After noting that

$$Y_{Q_2} = \frac{1}{P}, \ Y_P = \frac{-Q_2}{P^2}, \ Y_t = 0, \ Y_{PP} = \frac{2Q_2}{P^3}, \ Y_{Q_2 Q_2} = 0, \ Y_{Q_2 P} = -\frac{1}{P^2}$$

(15) yields

$$\frac{dy}{Q_2/P} = \frac{1}{Q_2/P} \left[\left\{ \frac{R_2 Q_2}{P} - \frac{Q_2}{P^2} P\pi - \rho \left(\frac{1}{P^2} \right) P Q_2 s_1 s_2 + \frac{1}{2} \left(\frac{2Q_2}{(P)^3} \right) (s_1)^2 P^2 \right\} dt \right.$$

$$\left. + \frac{1}{P}(s_2 Q_2) dz_2 - \frac{Q_2}{P^2}(Ps_1) dz_1 \right] \tag{26}$$

$$= \{ R_2 - \pi - \rho s_1 s_2 + (s_1)^2 \} dt + s_2 dz_2 - s_1 dz_1$$

Assume that $\rho = 0$. Then the expected real return on equity is *not* the expected nominal return minus the expected inflation rate ($R_2 - \pi$). It's bigger! We will discuss this more after completing the example.

(b) We now work out the real return on a nominal bond, $\dfrac{d(Q_3/P)}{Q_3/P}$, by letting $y(t, Q_3/P) = Q_3(t)/P(t)$. We will do it by a Taylor series expansion rather than using (15).

$$dy = y_{Q_3} dQ_3 + y_P dP + y_t dt$$

$$+ \frac{1}{2} y_{PP}(dP)^2 + y_{PQ_3} dP dQ_3$$

$$+ \frac{1}{2} y_{Q_3 Q_3} (dQ_3)^2$$

We need

$$y_{Q_3} = \frac{1}{P} \qquad y_P = \frac{-Q_3}{P^2}$$

$$y_{Q_3 P} = \frac{-1}{P^2} \qquad y_{Q_3 Q_3} = 0$$

$$y_{PP} = \frac{2Q_3}{P^3} \qquad y_t = 0$$

Now (21) and (24), plus the stochastic calculus rules (8), yield

$$dy = \left(\frac{1}{P}\right)Q_3 R_3 dt - \left(\frac{Q_3}{P^2}\right)\{P\pi dt + P s_1 dz_1\}$$

$$+ 0 - \left(\frac{1}{P^2}\right)\{P\pi dt + P s_1 dz_1\}\{Q_3 R_3 dt\}$$

$$+ \frac{1}{2}\left(\frac{2Q_3}{P^3}\right)\{P\pi dt + P s_1 dz_1\}^2 + 0$$

so

$$dy = \frac{Q_3 dt}{P}\{R_3 - \pi + s_1^2\} - \frac{Q_3}{P}s_1 dz_1$$

$$\frac{d(Q_3/P)}{Q_3/P} = \{R_3 - \pi + (s_1)^2\}dt - s_1 dz_1$$

Again notice that the expected real return on a nominal bond is *not* $R_3 - \pi$, but bigger. Let's examine this anomaly by considering a nominal bond with $R_3 = 0$ (i.e., holding cash). Suppose the current price level is l and there is a $\frac{1}{2}$ probability that the price level one period hence will be P_1 and a $\frac{1}{2}$ probability it will be P_2. Then next period's expected price level is $\dfrac{P_1 + P_2}{2}$, and with probability $\frac{1}{2}$, the real value of the bond will be $\dfrac{1}{P_2}$. Thus, the expected value of the bond one period hence will be

$$\frac{1}{2}\left(\frac{1}{P_1} + \frac{1}{P_2}\right) = \frac{P_1 + P_2}{2P_1 P_2}$$

If the price level were to change in a deterministic fashion, such that next period's price were always $\left(\dfrac{P_1 + P_2}{2}\right)$, then the real value of the bond would simply be

$$\frac{1}{\dfrac{P_1 + P_2}{2}} = \frac{2}{P_1 + P_2}$$

Notice that it is always true that

$$\frac{P_1 + P_2}{2P_1 P_2} \geq \frac{2}{P_1 + P_2}$$

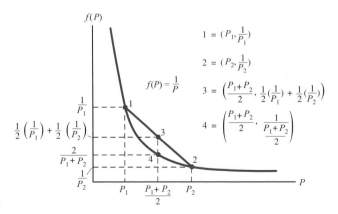

Thus, a mean preserving the increase in uncertainty will always increase the expected real value of the bond. This is why the $(s_1)^2 \, dt$ terms enter (26). Figure 2 illustrates this reasoning. Clearly, point 3 has a higher y-coordinate than point 4, so

$$\frac{1}{2}\left(\frac{1}{P_1}\right) + \frac{1}{2}\left(\frac{1}{P_2}\right) \quad > \quad \frac{1}{\dfrac{P_1 + P_2}{2}}$$

(Expected value of   (Expected value of
cash with random   cash with deterministic
price)   price)

16.5   Using Ito's Lemma to Derive the Black–Scholes Option Pricing Model

In this section, we show how Ito's Lemma was used by Black and Scholes (1973) to derive the option pricing formulas for puts and calls that we used in Chapter 14. Consider a European call option that enables the owner to purchase a single share of stock at time t for an exercise price c at time t. We assume that the price of the stock follows a geometric Brownian motion. That is,

$$\frac{dP}{P} = \alpha dt + \sigma dz \tag{27}$$

Define $w(p, t)$ to be the price of the call option at time t if the price of the stock is p at time t. Black and Scholes' brilliant insight was the following. Suppose we continuously adjust our portfolio so that we always own one call option and are short $\dfrac{\partial w}{\partial p}$ shares of stock. Then our portfolio is riskless! By arbitrage, this portfolio must always yield the risk-free rate of return r. To see why this is true, note that if the stock price changes by an amount ΔP, the option will increase in value by an amount $\dfrac{\partial w}{\partial p}(\Delta P)$. Since we lose an amount ΔP on each share of stock we have shorted, the stock portion of our portfolio will increase by an amount $-\dfrac{\partial w}{\partial p}(\Delta P)$. Thus, for any value of ΔP, our portfolio will have no risk. In any small time Δt, our portfolio should yield the risk-free rate of return. At time t, our portfolio is worth $w - \dfrac{\partial w}{\partial P} P$. Between time t and time $t + \Delta t$, the value of our portfolio will change by $dw - \dfrac{\partial w}{\partial P} dP$. For our portfolio to yield the risk-free rate, it must be true that

$$\frac{dw - \dfrac{\partial w}{\partial P} dP}{w - \dfrac{\partial w}{\partial P} P} = rdt \tag{28}$$

Now (27) and Ito's Lemma yield

$$dw = \left(\frac{\partial w}{\partial t} + \frac{\partial w}{\partial P}\, \alpha P + \frac{1}{2}\frac{\partial^2 w}{\partial P^2}\, \sigma^2 P^2\right)dt + \frac{\partial w}{\partial P}\, \sigma P dz \tag{29}$$

Together, (29) and (27) yield

$$dw - \frac{\partial w}{\partial p} dP = \left(\frac{\partial w}{\partial t} + \frac{1}{2}\frac{\partial^2 w}{\partial P^2}\, \sigma^2 P^2\right)dt \tag{30}$$

Together, (30) and (28) yield Black and Scholes' famous partial differential equation (PDE) for $w(P, t)$:

$$\frac{\partial w}{\partial t} = rw - rP\frac{\partial w}{\partial P} - \frac{1}{2}\sigma^2 P^2 \frac{\partial^2 w}{\partial P^2} \tag{31}$$

Clearly $w(P, T) = \max(P - c, 0)$ follows from the definition of a call option. Combining this **boundary condition** with (31), Black and Scholes were able to derive the Black–Scholes formula. After defining

$$d_1 = \frac{Ln\left(\dfrac{S}{X}\right) + \left(r + \dfrac{\sigma^2}{2}\right)t}{\sigma\sqrt{t}}$$

$$d_2 = d_1 - \sigma\sqrt{t}$$

we may write the call price C as

$$C = Se^{-yt}N(d_1) - Xe^{-rt}N(d_2)$$

where $N(x)$ is the cumulative normal probability for a standard normal random variable.

Note that since α does not appear in (31), the stock's growth rate α does not influence the option price. This seems counterintuitive. One would think that given two stocks with the same volatility, the stock with the higher growth rate would tend to end up at higher price levels and yield a higher price for a European call. But this does not happen because information about the stock's growth rate is already embedded in the stock's current price.

16.6 An Introduction to Stochastic Control

Consider the following problem.

$$\max_u \ E\left\{\int_O^T f(t, x, u)dt + B(x(T), T)\right\}$$

$$\text{s.t.} \quad dx = g(t, x, u)dt + \sigma(t, x, u)dz \tag{32}$$

$$x(0) = x_0$$

Here, x is the state variable, and u is our decision variable. We begin by assuming that u is one-dimensional. Define $J(t_o, x_o)$ to be the maximum value of (32) that can be obtained if we begin the problem at t_o in state $x(t_o) = x_o$. Then

$$J(t_o, x_o) = \max_u E\left\{\int_{t_o}^T f(t, x, u)dt + B(x(T), T)\right\}$$

$$\text{s.t.} \quad dx = g(t, x, u)dt + \sigma(t, x, u)dz \tag{33}$$

$$x(t_o) = x_o$$

We can rewrite (33) as

$$J(t_o, x_o) = \max_u E\left\{\int_{t_o}^{t_0+\Delta t} f(t, x, u)dt + J(t_o + \Delta t, x_o + \Delta x)\right\} \tag{34}$$

Now

$$\int_{t_o}^{t_0+\Delta t} f(t, x, u)dt = f(t_o, x_o, u)\Delta t + o(\Delta t) \tag{35}$$

Then, via the familiar Taylor series expansion, we find that

$$J(t_o + \Delta t, x_o + \Delta x) = J(t_o, x_o) + J_t(t_o, x_o)\Delta t + J_x(t_o, x_o)\Delta x$$
$$+ \frac{1}{2}J_{xx}(t_o, x_o)\Delta x^2 + J_{xt}(t_o, x_o)\Delta x\Delta t + J_{tt}(t_o, x_o)\Delta t^2 + o(\Delta t) \quad \textbf{(36)}$$

After substituting (35) and (36) in (34) and using the stochastic calculus table to take expectations, we may rewrite (34) as

$$J(t_o, x_o) = \max_u \{ f(t_o, x_0, u)\Delta t + J_t(t_o, x_o)\Delta t + J(t_o, x_o)$$
$$+ J_x(t_o, x_o)g(t_o, x_0, u)\Delta t + \frac{1}{2}J_{xx}(t_o, x_o)\sigma^2(t, x, u)\Delta t + o(\Delta t)\} \quad \textbf{(37)}$$

After subtracting $J(t_o, x_o)$ from both sides of (37), dividing both sides by Δt, and letting Δt approach 0, we obtain the fundamental result of stochastic control theory—the **Hamilton–Jacobi–Bellman equation.** We now drop the subscripts on x_o and t_o.

$$0 = \max_u\{ f(t, x, u) + J_t(t, x) + J_x(t, x)g(t, x, u) + \frac{1}{2}J_{xx}(t, x)\sigma^2(t, x, u)\} \quad \textbf{(38)}$$

Since $J_t(t, x)$ is independent of u, we may rewrite (38) as

$$- J_t(t, x) = \max_u\{ f(t, x, u) + J_x(t, x)g(t, x, u) + \frac{1}{2}J_{xx}(t, x, u)\sigma^2(t, x, u)\} \quad \textbf{(39)}$$

The boundary condition is $J(T, x(T)) = B(x(T), T)$.

To obtain a multidimensional analog of (38), the only step of the derivation that changes is the structure of (36). More specifically, consider the case of the state variables x_1 and x_2, where

$$dx_1 = g_1(t, x, u)dt + \sigma_1(t, x, u)dz_1$$
$$dx_2 = g_2(t, x, u)dt + \sigma_2(t, x, u)dz_2$$
$$\text{where } \rho = \text{correlation between } dz_1 \text{ and } dz_2$$

Then we obtain $J(t_o + \Delta t, x_o + \Delta x)$ from (15) of Ito's Lemma. The analog of (39) becomes

$$-J_t(t, x) = \max_u\{ f(t, x, u) + J_{x_1}(t, x)g(t, x, u)$$
$$+ J_{x_2}(t, x)g_2(t, x, u) + \frac{1}{2}J_{x_1x_1}(t, x, y)\sigma_1^2(t, x, u)$$
$$+ J_{x_1x_2}(t, x, u)\rho\sigma_1(t, x, u)\sigma_2(t, x, u)$$
$$+ \frac{1}{2}J_{x_2x_2}(t, x, u)\sigma_2^2(t, x, u)\} \quad \textbf{(40)}$$

The generalization to more than two state variables should be clear.

We now consider a discounted infinite-time-horizon problem in which f, g, and σ are independent of time. That is, for any t, $dx = g(x, u)dt + \sigma(x, u)dz$, and reward is earned at a rate $f(x, u)$. Let $V(x_0)$ be the expected present value (at time 0) of all rewards received on the interval $[0, \infty]$. If the state at time 0 is x_o, then for any t_o, the time independence of the parameters implies that

$$V(x_o) = \max_u E \int_{t_0}^{\infty} e^{-r(t-t_0)} f(x, u)dt$$

$$\text{s.t. } dx = g(x, u)dt + \sigma(x, u)dz, \ x(t_0) = x_0$$

Then we have that for all $t_o \geq 0$,

$$J(t_o, x_o) = e^{-rt_0}V(x_o) \quad \textbf{(41)}$$

From (41), we find that

$$J_t(t, x_o) = -re^{-rt}V(x_0)$$
$$J_x(t, x_o) = e^{-rt_0}V'(x_o)$$
$$J_{xx} = e^{-rt}V''(x_o)$$

Substituting these results into (40) yields

$$re^{-rt}V(x_o) = \max_u[f(x_o, u)e^{-rt} + e^{-rt}V'(x_o)g(x_o, u) + \frac{1}{2}e^{-rt}V''(x_o, u)\sigma^2(x_0, u)]$$

Canceling the e^{-rt} terms and replacing x_o with x yields

$$rV(x) = \max[f(x, u) + V'(x)g(x, u) + \frac{1}{2}\sigma^2(x, u)V''(x)] \qquad \text{(42)}$$

This is called the *current value form* of the Hamilton–Jacobi–Bellman equation. As a boundary condition in the infinite-horizon case, it is believed that one should use

$$\lim_{t \to \infty}[ExJ(x, t)] = 0$$

Note that (40), (41), and (42) are necessary conditions for an extremum. If $f(x, u)$ is a concave function of x and u, then (40)–(42) will yield a maximum. If $f(x, u)$ is a convex function of x and u, (40)–(42) will yield a minimum. Let's now examine stochastic control in action.

EXAMPLE 4   **Stochastic Control**

$$\min E \int_o^\infty e^{-rt}(ax^2 + bu^2)$$

$$\text{s.t. } dx = udt + \sigma xdz$$
$$a > 0, b > 0, \sigma > 0$$

Let's first try (40), and to check our answer we'll use (42). Equation (40) becomes

$$-J_t = \min_u\left\{e^{-rt}(ax^2 + bu^2) + J_xu + \frac{1}{2}\sigma^2x^2J_{xx}\right\} \qquad \text{(43)}$$

To solve for the minimizing value of u from (43), we solve

$$2ube^{-rt} + J_x = 0 \qquad \text{or} \qquad u(x) = -J_xe^{rt}/2b$$

Substituting $u(x)$ into (42) yields a partial differential equation for $J(t, x)$

$$-e^{rt}J_t = ax^2 - \frac{J_x^2e^{2rt}}{4b} + .5\sigma^2x^2J_{xx}e^{rt} \qquad \text{(44)}$$

When solving a partial differential equation for $J(t, x)$, begin by trying functional forms for $J(t, x)$, such as

$$f(t)\,g(x) \qquad \text{or} \qquad f(t) + g(x) \qquad \text{or} \qquad f(t)^{g(x)}f(x)^{g(t)} \qquad \text{(45)}$$

Plugging (45) into (44) yields

$$-e^{rt}f'(t)g(x) = ax^2 - f(t)^2g'(x)^2e^{2rt}/4b + \left(\frac{1}{2}\right)\sigma^2x^2g''(x)f(t)e^{rt}$$

For this equation to work out, each term will need an x^2 term. Thus, $g(x) = kx^2$ is necessary. Also, the exponential factors in each term must cancel. This forces $f(t) = e^{-rt}$. Thus, we try $J(t, x) = kx^2e^{-rt}$. Plugging this into (44) yields

$$\frac{k^2}{b} + (r - \sigma^2)k - a = 0$$

or

$$k = \frac{b\left\{\sigma^2 - r \pm \left[(r - \sigma^2)^2 + \frac{4a}{b}\right]^{1/2}\right\}}{2} \tag{46}$$

Since $J(t, x) \geq 0$ is required, we choose $+$ rather than $-$ in (46). Using (46) in (44), we find $u(x) = -\frac{kx}{b}$. Thus, in a stochastic control model with quadratic costs, the optimal action is a linear function of the state. Solving our problem via (42), we get

$$rV(x) = \min_u[ax^2 + bu^2 + V'(x)u + .5\sigma^2x^2V''(x)] \tag{47}$$

The minimizing u is given by

$$u(x) = \frac{-V'(x)}{2b} \tag{48}$$

Substituting (48) back into (47) yields

$$rV(x) = ax^2 + \frac{V'(x)^2}{4b} + .5\sigma^2x^2V''(x) \tag{49}$$

To solve (49), all terms must end up with an x^2, so we must try $V(x) = kx^2$. Substituting this in (49) yields $kr = a - k^2/b + k\sigma^2$ or $k^2/b + k(r - \sigma^2) - a = 0$, as before. Thus we see that (42) yields an ordinary differential equation, while (40) yields a partial differential equation. Also, note that $ax^2 + bu^2$ is a convex function of x and u, so we do get a minimum.

REVIEW PROBLEMS

Group A

1 Suppose $y(t, z)$ is a stochastic process satisfying $y(0) = 8$ and $dy = 3y^{1/3}dt + 3y^{2/3}dz$. Solve for $y(t, z)$.

2 Suppose $x_1(t)$ and $x_2(t)$ are stochastic processes governed by

$$dx_1 = \mu_1dt + \sigma_1dz$$
$$dx_2 = \mu_2dt + \sigma_2dz$$

Determine $d(x_1x_2)$.

3 For Fischer's bond model, determine the real return $\frac{d(Q_3/P)}{Q_3/P}$ on a nominal bond. Show that if $s_1 > 0$, the real return exceeds the nominal return minus the inflation rate.

4 This problem explains (to some degree) what people mean when they stop you on the street and say $dz^2 = dt$. As

usual, let $z(t)$ represent a standard Wiener process with $\mu = 0$ and $\sigma^2 = 1$,

$$S_n = \sum_{i=0}^{i=n} (z(t_1 + 1) - z(t_1))^2$$

where $t_1 = 0$ $t_n = T$

and $\Delta_n = t_{i+1} - t_i = \frac{T}{n}$. Using Table 1 from the proof of Ito's Lemma, show that

$$E(S_n) = n\Delta_n$$
$$\text{var}(S_n) = n(2\Delta_n^2)$$

Thus, as $n \to \infty$, $E(S_n) \to T$ and var $(S_n) \to 0$. Thus $\lim_n S_n$ may be considered to be a constant T. This is the real meaning of $dz^2 = dt$. (*Hint:* To find $E(X_n)^t$ and var S_n, work on a term-by-term basis and then use standard rules for determining the means and variance of sums. Make sure to explain which steps of your proof use the assumptions of stationary and independent increments.)

5 Suppose we want to model the cash balance of a bank by a stochastic differential equation. The average rate at which cash is demanded at time t is $(1.02)^t$, and the variance in the demand for cash is proportional to the average rate at which cash is demanded. Also assume that the bank receives a constant inflow of cash at a rate c per unit time. Write a stochastic differential equation that might provide an appropriate description of the bank's cash balance.

6 Suppose $x(t)$ is a Wiener process with 0 drift and a given σ. Show that the stochastic process $x(16t)/4$ will also be a Wiener process with 0 drift and an identical σ.

7 Consider a Wiener process $x(t)$ having $\mu = 0$ and $\sigma = c$. Show that for any two times s and t satisfying $s < t$,

$$E\{x(s)x(t)\} = c^2 s$$

(*Hint:* $x(t) = x(s) +$ {distance Wiener process moves between s and t}. Make sure you justify each step in your argument.)

8 Suppose $y(t,z)$ is a stochastic process satisfying $y(0) = 8$ and $dy = 3y^{1/3}dt + 3y^{2/3}dz$. Solve for $y(t, z)$.

9 Assume that $r(t)$, the interest rate at time t, may be described by the following SDE:

$$dr = b(r)dt + a(r)dz$$

Consider the following four possible specifications of $b(r)$ and $a(r)$.

	$a(r)$	$b(r)$
(1)	μ	σ
(2)	$\mu - r$	σ
(3)	$\mu - r$	$\sigma r^{1/2}$
(4)	r	$\sigma r^{1/2}$

a Which of these four specifications for $a(r)$ and $b(r)$ could lead to negative values of $r(t)$?

b Between (1) and (2), which model do you think would more often lead to very large values of $r(t)$?

10 Suppose $dx(t) = -\sigma^2 dt/2 + \sigma dz(t)$ and $y(t) = e^{x(t)}$. Find $dy(t)$.

11 In the Black–Scholes derivation, suppose that the SDE governing the stock price P is

$$\frac{dP}{P} = gdt + s_1 dz_1 + s_2 dz_2$$

where $z_1(t)$ and $z_2(t)$ are standard Wiener processes having correlation ρ. For this specification of the stock price, derive a PDE for the price of w.

12 Merton (1971) has developed a stochastic economic growth model. Let

$K(t) = $ capital level at time t

$L(t) = $ labor level at time t

$n = $ growth rate of labor

$s = $ percentage of output that economy saves

Merton assumes that $L(t)$ satisfies the following SDE:

$$dL = nLdt + \sigma LDz$$

In equilibrium, savings during the interval between t and $t + t$ must equal the economy's investment during the interval between t and $t + \Delta t$. This leads to the differential equation

$$dK = sF(k, L)dt$$

a Explain why this SDE expresses the fact that savings must equal investment.

b Determine a stochastic differential equation satisfied by the per-capita capital level $(K(t)/L(t))$. Using this SDE, determine whether an increase in the variability of the labor supply growth rate will increase or decrease the average value of the per-capita capital.

13 Consider the problem

$$\max \int_0^\infty e^{-rt}R(q)dq$$
$$\text{s.t. } dx = -qdt + \sigma xdz$$
$$x(0) = 1$$

where

$q = $ extraction rate of a resource

$R(q) = q^{1/2} = $ profit rate when extraction rate is q

$x(t) = $ stock of exhaustible resource at time t

Define $J(x) = $ maximum expected discounted profit earned over an infinite horizon if initial energy in stock is x, and $q^*(x) = $ optimal extraction rate whenever x units of resources are available. Find a closed-form solution for $J(x)$ and $q^*(x)$. Does an increase in the variability of the energy supply increase or decrease the optimal extraction rate?

REFERENCES

Arnold, C. *Stochastic Differential Equations.* New York: Wiley, 1974.

Black, F., and Scholes, M. "The Pricing of Option and Corporate Liabilities," *Journal of Political Economy* (May/June, 1973):637–654.

Fischer, S. "The Demand for Index Bonds," *Journal of Political Economy* (June, 1975):509–534.

Merton, R. "Optimal Consumption and Portfolio Rules in a Continuous Time Model," *Journal of Economic Theory* (Dec., 1971):373–413.

@Risk Crib Sheet

@Risk Icons

Once you are familiar with the function of the @Risk icons, you will find @Risk easy to learn. Here is a description of the icons.

Opening an @Risk Simulation

This icon allows you to open up a saved @Risk simulation. I do not recommend saving simulations. Instead, I paste results into a spreadsheet.

Saving an @Risk Simulation

This icon allows you to save an @Risk simulation, including data and simulation settings.

Simulation Settings

This icon allows you to control the settings for the simulation. Clicking on this icon activates the dialog box shown in Figure 1. There follows a description of what each of the tabs can do.

Iterations Tab

Various options are associated with the Iterations tab.

#Iterations　#Iterations is how many times you want @Risk to recalculate the spreadsheet. For example, choosing 100 iterations means that 100 values of your output cells will be tabulated.

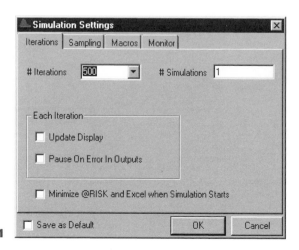

FIGURE 1

#Simulations   Leave this at 1 unless you have a =RISKSIMTABLE functon in the spreadsheet. In this case, choose #Simulations to equal the number of values in SIMTABLE. For example, if we have the formula =RISKSIMTABLE({100,150,200,250,300}) in cell A1, set #Simulations to 5. The first simulation will place 100 in A1, the second simulation will place 150 in A1, and the fifth simulation will place 300 in A1. #Iterations will be run for each simulation.

Pause on Error   Checking this box causes @Risk to pause if an error occurs in any cell during the simulation. @Risk will highlight the cells where the error occurs.

Update Display   Checking this box causes @Risk to show the results of each iteration on the screen. This is nice, but it slows things down.

See Figure 2. The Sampling tab options are as follows.

Sampling Type   While a little slower, Latin Hypercube sampling is much more accurate than Monte Carlo sampling. To illustrate, Latin Hypercube guarantees for a given cell that 5% of observations will come from the bottom 5th percentile of the actual random variable, 5% will come from the top 5th percentile of the actual random variable, etc. If we choose Monte Carlo sampling, 8% of our observations may come from the bottom 5% of the actual distribution, when in reality only 5% of observations should do so. When simulating financial derivatives, it is crucial to use Latin Hypercube.

Standard Recalc   If you choose Expected Value, you obtain the expected value of the random variable unless the random variable is discrete. Then you obtain the possible value of the random variable that is closest to the random variable's expected value. For instance, for a statement

$$=RISKDISCRETE(\{1,2,\},\{.6,.4\})$$

the expected value is $1(.6) + 2(.4) = 1.4$, so Expected Value enters a 1.

If you choose the Monte Carlo option, *when you hit F9, all the random cells will recalculate. This makes it much easier to understand and debug the spreadsheet.* Thus, with Monte Carlo selected,

$$=RISKDISCRETE(\{1,2,\},\{.6,.4\})$$

will return a 1 60% of the time and a 2 40% of the time.

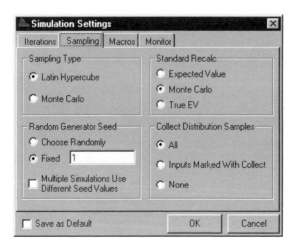

FIGURE 2

If you choose the True EV option, then the actual expected value of the random variable will be returned. Thus,

$$=RISKDISCRETE(\{1,2,\},\{.6,.4\})$$

will yield a 1.4.

Collecting Distribution Samples   Check All if you want to get Tornado Graphs, Scenario Analysis, or Extract Data. Also check this box if you want statistics on cells generated by @Risk functions. You can always check this box if you like, but if you have many @Risk functions in your spreadsheet, checking the box will slow down the simulation. Checking Inputs Marked With Collect will collect data on a subset of your risk functions marked with Riskcollect.

Random Number Generator Seed   When the seed is set to 0, each time you run a simulation, you will obtain different results. Other possible seed values are integers between 1 and 32,767. Whenever a nonzero seed is chosen, the same values for the input cells and output cells will occur. For example, if we choose a seed value of 10, each time we run the simulation, we will obtain exactly the same results.

Autoconvergence

Under #Iterations, you may select Auto. See Figure 3. You may then select a percentage such as 1%. Then @Risk keeps running until during the last 100 iterations, the mean and standard deviation change by at most 1%. This can be a lot of iterations! I prefer to choose the number of iterations myself by setting

$$\frac{2s}{\sqrt{n}}$$

equal to the desired level of accuracy for the output cell's mean. Here, s = standard deviation of output cell for a trial simulation (say, 400 iterations). For example, if a trial simulation yields $s = 100$ and I want to be 95% sure that I am estimating the population mean within 10, I need

$$\frac{2(100)}{\sqrt{n}} = 10$$

or $n = 400$.

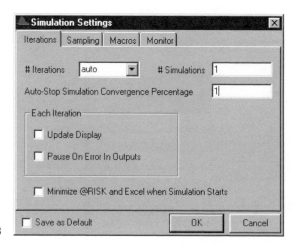

FIGURE 3

Macro Tab

See Figure 4. The Macro tab enables @Risk to run a macro before or after each iteration of a simulation. For example, checking After Each Iteration's Recalc and entering Macro1 after evaluating each ouput cell would result in the following sequence of events:

- Compute @Risk functions and calculate output cells.
- Run Macro1.
- Compute @Risk functions and calculate output cells, etc.

Select Output Cells

This icon enables you to select an output cell or cells for which @Risk will create statistics. Simply select a range of cells and click on the icon to select the range as output cells. You may select as many ranges as you desire.

List Input and Output Cells

This icon lists all output cells. Also listed are cells containing @Risk functions. These are called input cells. From this list, you can change the names of output cells or delete output cells.

Run Simulation

This icon starts the simulation. The status of the simulation is shown in the lower left-hand corner of your screen. Hitting the Escape key allows you to terminate the simulation.

Show Results

This icon allows you to see results. There are two windows:

- Summary Results, containing Minimum, Mean, and Maximum for all input and output cells.
- Simulation Statistics, containing more detailed statistics.

Clicking the Hide icon will send you back to your worksheet. To paste your statistics into your worksheet, simply select a window and Edit Copy Paste it into the worksheet.

Define Distribution Icon

This icon allows you to see the mass function or density function for any random variable. You may also use this icon directly to enter any @Risk formula into a cell.

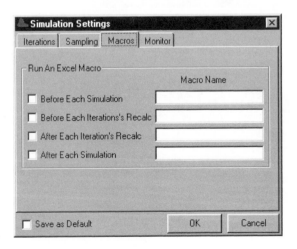

FIGURE 4

712   APPENDIX 1   @Risk Crib Sheet

Graphing

To obtain a graph, right click on the cell from the Explorer interface. Then choose the type of graph desired. To copy the graph into Excel, right click on the graph and select Copy or Graph in Excel.

A histogram gives the fraction of iterations assuming different values. The histogram in Figure 5 was generated for a cell containing the formula

$$=RISKNORMAL(100,15)$$

The histogram indicates that the input cell was bell-shaped and that the most common values of the input cell were around 100.

For a cumulative ascending graph (Figure 6), the y-axis gives the fraction of iterations yielding a value $\leq$ the value on the x-axis. Thus, about 50% of all iterations in this case yielded a value $\leq$ 100.

For a cumulative descending graph, the y-axis gives the fraction of iterations yielding a value $\geq$ the value on the x-axis. In Figure 7, this input cell exceeded 85 about 84% of the time.

An area graph replaces bars with smooth areas. A fitted curve smooths out the variation in bar heights before creating an area graph.

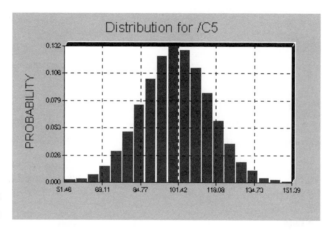

FIGURE 5
Histogram

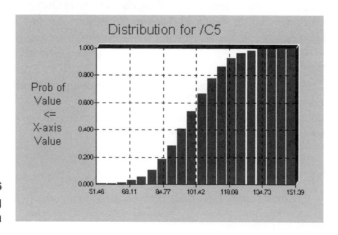

FIGURE 6
Cumulative Ascending Graph

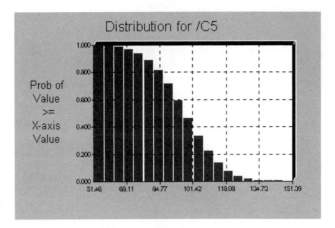

FIGURE 7
Cumulative Descending
Graph

Targets

At the bottom of the Simulation Statistics window is a Target option. You may enter a Value or Percentile, and @Risk fills in the one you left out. For

$$=RISKNORMAL(100,15)$$

we obtained the following results:

```
Target #1 (Value) =        85
Target #1 (Perc%) =        15.87%
Target #2 (Value) =        130
Target #2 (Perc%) =        97.75%
Target #3 (Value) =        114.9159
Target #3 (Perc%) =        84%
```

- We entered Target#1(Value) of 85, and @Risk reported that the cell was ≤85 15.87% of the time.

- We entered Target#2(Value) of 130, and @Risk reported that the cell was ≤130 97.75% of the time.

- We entered Target#3(Perc%) of 84%, and @Risk reported that 84% of the time, the cell was ≤114.92.

Extracting Data

Sometimes you may want to see the values of @Risk functions and output cells that @Risk created on the iterations run. If so, check Collect Distribution Samples under Simulation Settings and then click on Data in the Results window. You can then Edit Copy Paste the data to your spreadsheet and subject it to further analysis.

Sensitivity

If you want a Tornado Graph, right click on the output cell and select Tornado Graph. This also requires that you check Collect Distribution Samples. You may choose either a Correlation or a Regression graph. Tornado graphs let you know which input cells have the largest influence on your output cell(s).

@Risk Functions

We now illustrate some of the most useful @Risk functions.

The RISKDISCRETE Function

This generates a discrete random variable that takes on a finite number of values with known probabilities. See Figure 8. First, enter the possible values of the random variable

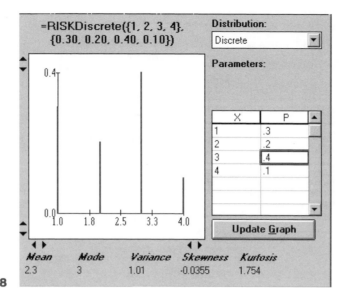

FIGURE 8

and then the probability for each value. Thus, =RISKDISCRETE({1,2,3,4},{.3,.2,.4,.1}) would generate 1 30% of the time, 2 20% of the time, 3 40% of the time, and 4 10% of the time.

If the values and probabilities were entered in A2:B5, we could have entered this random variable with formula

$$=RISKDISCRETE(A2:A5, B2:B5)$$

The RISKSIMTABLE Function

Suppose we enter

$$=RISKSIMTABLE(\{100,150,200,250,300\})$$

in cell A5, and #Iterations is 100. If we change #Simulations to 5, then on the first simulation, 100 iterations are run with 100 in cell A5. On the second simulation, 100 iterations are run with 150 in cell A5. Finally, on the fifth simulation, 100 iterations are run with 300 in cell A5. If the five arguments for the =RISKSIMTABLE function were in B1:B5, we could have also entered the =RISKSIMTABLE function as

$$=RISKSIMTABLE(B1:B5)$$

The RISKDUNIFORM Function

See Figure 9. We use the RISKDUNIFORM function when a random variable assumes several equally likely values. Thus,

$$=RISKDUNIFORM(\{1,2,3,4\})$$

is equally likely to generate 1, 2, 3, or 4. If 1, 2, 3, 4 were entered in A1:A4, then we could have entered

$$=RISKDUNIFORM(A1:A4)$$

The RISKBINOMIAL Function

See Figure 10. Use the =RISKBINOMIAL function when you have repeated independent trials, each having the same probability of success. For example, if there are 5 competi-

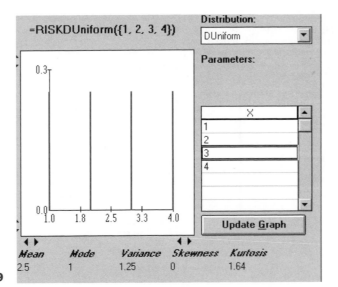

FIGURE 9

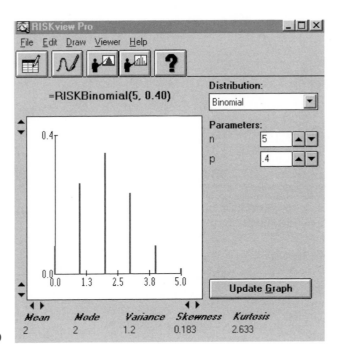

FIGURE 10

tors who might enter an industry this year, each competitor has a 40% chance of entering, and entrants are independent, then we could model this situation with the formula

$$=RISKBINOMIAL(5,.4)$$

The RISKNORMAL Function

See Figure 11. Use this function to model a continuous, symmetric (or bell-shaped) random variable. The formula

$$=RISKNORMAL(100,15)$$

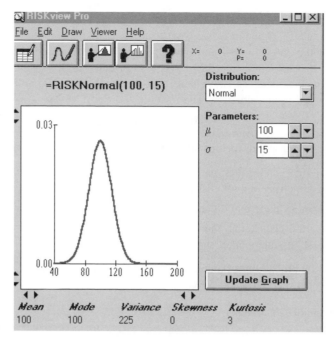

FIGURE 11

will yield

- a value between 85 and 115 68% of the time
- a value between 70 and 130 95% of the time
- a value between 55 and 145 99.7% of the time

The RISKTRIANG Function

See Figure 12. This function enables us to model a nonsymmetrical continuous random variable. It generalizes the well-known idea of best-case, worst-case, and most likely scenarios. For example,

$$=RISKTRIANG(.2,.4,.8)$$

could be used to model market share if we felt that the worst-case market share was 20%, the most likely market share was 40%, and the best-case market share was 80%. Note that the probability that the market share is between 30% and 40% would be the area under this triangle between .3 and .4. The entire triangle has an area of 1. This fact determines the height of the triangle.

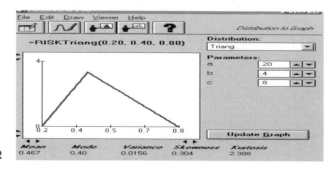

FIGURE 12

The RISKTRIGEN Function

See Figure 13. Sometimes we want to use a triangular random variable, but we are not sure of the absolute best and worst possibilities. We may believe that there is a 10% chance that market share will be less than or equal to 30%, that the most likely share is 40%, and that there is a 10% chance that share will exceed 75%. The RISKTRIGEN function is used in this situation. The formula

$$=RISKTRIGEN(.3,.4,.75,10,90)$$

would be appropriate for this situation. Then @Risk draws a triangle that yields

- A 10% chance that market share is less than or equal to 30%. This requires a worst possible market share of around 20%.
- A most likely market share of 40%.
- A 10% chance that market share is greater than or equal to 75%. This requires a best possible market share of around 95%.

Again, the probability of a market share between 20% and 50% is just the area under the triangle between 20% and 50%.

The RISKUNIFORM Function

See Figure 14. Suppose a competitor's bid is equally likely to be anywhere between 10 and 30 thousand dollars. This can be modeled by a uniform random variable with the formula

$$=RISKUNIFORM(10,30)$$

Again, this function makes any bid between 10 and 30 thousand dollars equally likely. The probability of a bid between 15 and 28 thousand would be the area of the rectangle bounded by $x = 15$ and $x = 28$. This would equal $(28 - 15)(.05) = .65$.

The RISKGENERAL Function

What if a continuous random variable does not appear to follow a normal or a triangular distribution? We can model it with the =RISKGENERAL function.

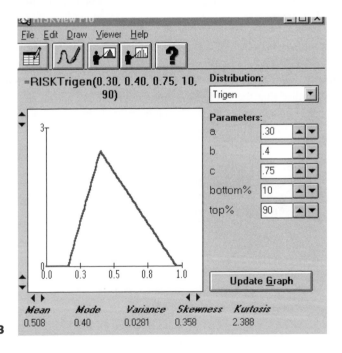

FIGURE 13

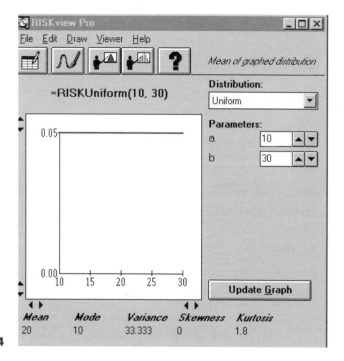

FIGURE 14

Suppose that a market share of between 0 and 60% is possible, and a 45% share is most likely. There are five market-share levels for which we feel comfortable about comparing relative likelihood. (See Table 1.) Thus, a market share of 45% is 8 times as likely as 10%; 20% and 55% are equally likely; etc. Note that this distribution cannot be triangular, because then 20% would be (20/45) as likely as peak of 45%, and 20% would be .75 as likely as 45%. To model this, enter the formula

$$=\text{RISKGENERAL}(0,60,\{10,20,45,50,55\},\{1,6,8,7,6\})$$

The syntax of RISKGENERAL is as follows:

- Begin with the smallest and largest possible values.
- Then enclose in {} the numbers for which you feel you can compare relative likelihoods.
- Finally, enclose in {} the relative likelihoods of the numbers you have previously listed.

Running this in @Risk yields the output shown in Figure 15. Note that 20 is 6/8 likely as 45; 10 is 1/8 as likely as 45; 50 is 7/8 as likely as 45; 55 is 6/8 as likely as 45; etc. In be-

TABLE 1

Market Share	Relative Likelihood
10%	1
20%	6
45%	8
50%	7
55%	6

	A	B	C	D	E	F	G
1		EXAMPLE OF					
2		RISKGENERAL					
3		DISTRIBUTION					
4							
5				Minimum	0		
6				Maximum	60		
7				Specified Points			
8				10	1		
9				20	6		
10				45	8		
11				50	7		
12				55	6		
13			45.28889	=RISKGENERAL(0,60,{10,20,45,50,55},{1,6,8,7,6})			

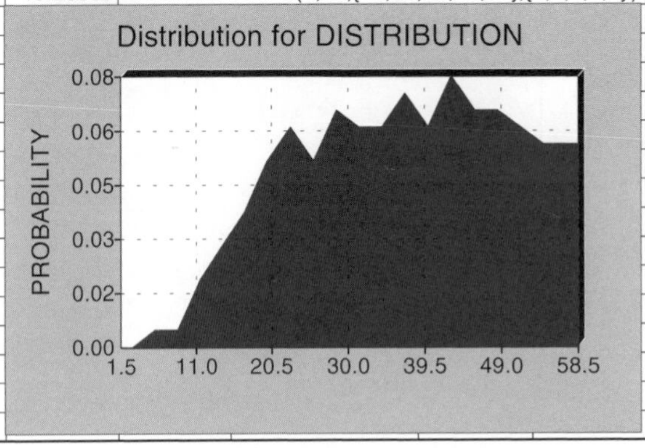

Distribution for DISTRIBUTION

FIGURE 15

tween the given points, the density function changes at a linear rate. Thus, 30 would have a likelihood of

$$6 + \frac{(30 - 20)*(8 - 6)}{(45 - 20)} = 6.8$$

Modeling Correlations

Suppose we have three normal random variables, each having mean 0 and standard deviation 1, correlated as follows:

- Variable 1 and variable 2 have .7 correlation.
- Variable 1 and variable 3 have a .8 correlation.
- Variable 2 and variable 3 have a .75 correlation.

To model this correlation structure, we use the =RISKCORRMAT command. Simply enter your correlation matrix somewhere in the worksheet. In Figure 16, we chose C27:E29.

	B	C	D	E	F	G	H
25							
26							
27		1	0.7	0.8			
28		0.7	1	0.75			
29		0.8	0.75	1			
30							
31	1	Variable 1	1.793028	risknormal(0,1,riskcorrmat(c27:e29,1))			
32	2	Variable 2	-0.449129	risknormal(0,1,riskcorrmat(c27:e29,2))			
33	3	Variable 3	-0.521328	risknormal(0,1,riskcorrmat(c27:e29,3))			

FIGURE 16

For each variable, type in front of the variable's actual distribution the syntax

=Actual Risk Function, RISKCORRMAT(Matrix, i)

Here, Matrix (C27:E29 in this case) indicates where the correlation matrix resides, and i is the column of the correlation matrix that contains the correlations for variable i. Thus, for variable 1, the correlations come from the first column of the correlation matrix.

If you run a simulation and extract the data for cells D31:D33, you will find that

- Each cell has a mean of around 0 and a standard deviation around 1.
- Each cell follows a normal distribution.
- D31 has around a .7 correlation with D32.
- D31 has around a .8 correlation with D33.
- D32 has around a .75 correlation with D33.

Truncating Random Variables

Suppose you believe that market share for a product is approximately normally distributed, with mean .6 and standard deviation .1. This random variable could exceed 1 or be negative, which would be inconsistent with the fact that market share must be between 0 and 1. To resolve this, you may enter the random variable from the Define Distribution icon as shown in Figure 17.

You could also type in formula

=RISKNORMAL(.6,.1,RISKTRUNCATE(0,1))

Then @Risk generates a normal random variable with mean .6 and standard deviation .1. If the random variable assumes a value between 0 (the lower truncation value) and 1 (the upper truncation value), that value is retained. Otherwise, another value is generated. The truncation values must be within 5 standard deviations of the mean.

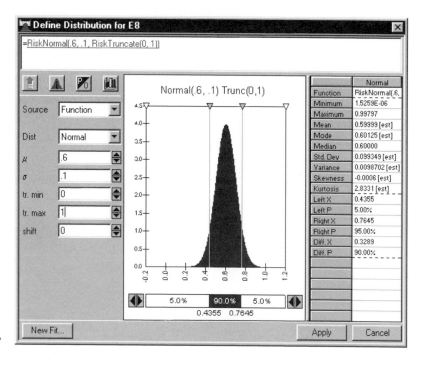

FIGURE 17

The RISKPERT Function

This function is similar to the RISKTRIANG function. The RISKPERT function is used to model the duration of projects. For example,

$$=\text{RISKPERT}(5,10,20)$$

would be used to model the duration of an activity that always takes at least 5 days, never takes more than 20 days, and is most likely to take 10 days. Whereas RISKTRIANG has a piecewise linear density function, the RISKPERT density has no linear segments. It is a special case of a Beta random variable.

Common Error Message

The error message "Invalid number of arguments" means that an incorrect syntax has been used with an @Risk function. For example, =RISKDUNIFORM({A1:A7}) may have been used instead of =RISKDUNIFORM(A1:A7).

Index